At
the
Desert's
Green
Edge

At the Desert's Green Edge

An Ethnobotany of the Gila River Pima

Amadeo M. Rea

With a Foreword by
Gary Paul Nabhan

Sumi-e Illustrations by
Takashi Ijichi

Linguistic Consultant,
Culver Cassa

The University of Arizona Press

Tucson

The University of Arizona Press
Copyright ©1997
The Arizona Board of Regents
All rights reserved
Sumi-e illustrations copyright ©1997, Takashi Ijichi
♾ This book is printed on acid-free, archival-quality paper
Manufactured in the United States of America
First printing

Library of Congress Cataloging-in-Publication Data
Rea, Amadeo M.
  At the desert's green edge : an ethnobotany of the Gila River Pima
/ Amadeo M. Rea ; with a foreword by Gary Paul Nabhan ; sumi-e
illustrations by Takashi Ijichi.
      p. cm.
  Includes bibliographical references and index.
  ISBN 0-8165-1540-9 (acid-free)
  1. Pima Indians—Ethnobotany. 2. Ethnobotany—Gila River Valley
(N.M. and Ariz.). I. Title.
E99.P6R43 1997
581.6'3'097917—dc21   97-4577
                                        CIP

British Cataloguing-in-Publication Data
A catalogue record for this book is available from the British Library.

The publication of this book was made possible by generous grants from the Labriola National
American Indian Data Center and the Wallace Research Foundation.

To the memories of Aldena and Fumika,
who taught their sons to appreciate beauty

Contents

Part 1   *The Pima and Their Country*

Part 2   *Gila Pima Plants*

Appendices

Illustrations

Figures

Sumi-e (Japanese Ink Paintings)

Sumi-e (Japanese Ink Paintings) cont.

Tables

Text Tables

Appendix Tables

Foreword

We live in a time when more American children claim to learn what they know about "the environment" from electronic media than from their schools, from the elders in their homes, or from their own explorations of their local surroundings. This book is an antidote to the despair and ignorance which that trend might otherwise bring us, for it conserves and honors a single culture's specific knowledge about a particular place. The heart of that place is the riparian corridor of the middle Gila River in the northern Sonoran Desert. Such detailed local knowledge is as endangered as the very plants, animals, and habitats in the Gila watershed that it attempts to describe. Today, we are witnessing the most rapid loss of biological diversity in human history; but we are also witnessing the most rapid loss of culturally encoded knowledge about the natural world since our human ancestors began to speak and shape symbols about the plants and ani-

mals with which we share the earth. In short, time-tested means of recognizing, valuing, and sustainably managing local resources are slipping away as we forget the vocabulary that developed over millennia to keep the need for protracted care of these resources firmly planted in our consciousness.

This monumental work by Amadeo Rea and his Piman-speaking collaborators reveals the immensity of the lexicon of O'odham plant names and associated terms that have guided them in the use, conservation, and management of the Gila River habitats. Dr. Rea conservatively estimates that the Akimel O'odham, or Pima, still retain names in their native language for more than half the flora within their reach. This proportion was undoubtedly much greater prior to the cataclysmic habitat destruction, biotic depletion, and linguistic acculturation that the Pima and their homelands have suffered during the last twelve decades.

Despite these insults, living Pima elders in the Gila River Indian Community can still call out more than 260 native names for plant species and their varieties, as well as some 70 names for birds, at least 30 for mammals, 8 for fish, a dozen for reptiles and amphibians, and another dozen for invertebrates. These names are often, but not always, in one-to-one correspondence with the Linnaean taxa recognized by Western scientists. Nonetheless, imagine this: many elderly Pima men and women over the last several centuries could probably recognize well over 350 plants and animals that have inhabited their homelands, legends, and feasts. To recognize and recall names for so many species is a feat that only a small percentage of academically trained biologists can achieve today.

Names. They are but a starting point for encoding a vast amount of information regarding the distribution, abundance, behavior, and spirit of various elements of a flora and fauna. The indigenous scientific knowledge of the Pima does not culminate in but merely begins with these folk taxonomies. The Pima also have a familiarity with patterns that one might consider in studies of biogeography, the phenology of flowers and fruits, the ecology of plant populations, the agronomy of drought-escaping crops, the horticultural tending of trees, the management of wildlife in hedgerows, the procurement of medicinal plants, and the fixing of dyes. All these other indigenous botanical sciences extend from their fundamental recognition of the diversity of plants that have resided just beyond their front doors.

The Piman language is not one of the hundred or so native languages spoken north of Mexico that is at immediate risk of becoming extinct or obsolete. It is being spoken, written, and taught—along with another hundred Native American languages—and revitalized thanks to a number of talented O'odham linguists and educators, including Albert Alvarez, Susie Enos, Ofelia Zepeda, Sally Pablo, Daniel Lopez, Tony Channa, Rosilda Manuel, Philip Salcido, Culver Cassa, the late Josiah Moore, and Dorothy Lewis. Nevertheless, Dr. Rea and his Pima consultants in farming, gathering, and etymology have recorded Pima plant names, habits, and distributions at a critical time, when we are witnessing the passing of the last generation of O'odham who grew up subsisting on wild-foraged plants and cultivated native crops on the Gila and Santa Cruz floodplains. Younger Pima, Tohono O'odham, and Maricopa individuals now residing in that area may still hold a few native foods in high esteem and may cherish certain songs and stories about desert life, but the proportion of members of the Gila River Indian Community who now depend on desert resources is but a shadow of what it was even two generations ago.

If any book can truly salvage remnants of an oral tradition, this work makes a courageous attempt to do so. It weaves together vestiges of the Pima oral tradition so that future generations interested in the natural history of their reservation will have the option to revive, renew, or build upon it as they see fit, given the changing context of their lives.

It has been far too easy for outsiders to take a quick look at the Gila River Indian Community—replete with satellite dishes, videos, and expressway access to the sterile heart of metropolitan Phoenix—and dismiss current conditions as ones in which such traditional knowledge of the desert is bound to wither and die. The irony of such categorical dismissals is that they contain exactly the same sentiments Amadeo heard when he began this project. The old saw is that the Pima were already too acculturated or "Coca-Colanized" to be able to offer much of anything fruitful about desert natural history.

The depth of this one work alone should enlighten anyone who maintains that the Pima long ago lost all their traditions, as well as their interest in desert life. And yet, as Amadeo admits, ethnobotany is but one realm of the scientific knowledge that the Pima integrate into their traditions. Moreover, this book is but one of many possible approaches to the ethnobotanical lore of the O'odham. It draws upon only a handful of Pima elders for the bulk of its information. Each of those elders might have dictated his or her own individual account of some aspect of Pima life, with plants integrated into that story in ways quite different from the present text. A number of the primary consultants for this work should be lauded as fine scholars in their own right—the late Joseph Giff, George Kyyitan, and Sylvester Matthias, Sally Pablo, and Culver Cassa, among others. To Amadeo's credit, we can begin to hear each of these distinctive voices nestled into his running text and can savor the vernacular tone with which they speak of the plant world. This should inspire others—including young Pima tribal members—to experiment with different ways of bringing their elders' voices into our ears, our hearts, and our minds.

This book should also awaken certain moribund scholars who have proclaimed that the Indians of the Southwest are now too modernized for any major theoretical advance to emerge out of the study of their ethnobiological knowledge. Amadeo has added his unique genius to this collaboration by demonstrating how the detailed study of the Pima and their knowledge of plants can provide insight into several hotly debated issues: the prevalence of adult-onset diabetes among Native Americans; the historical diffusion of crops and weeds; the Pima-Hohokam continuum (or lack of it); the timing of the desertification of desert riparian habitats; and linguistic continuity with the Tepiman language family. Thus Dr. Rea's scholarship will be appreciated beyond the orphan discipline of ethnobiology. It will contribute to topics in epidemiology, ethnonutrition, archaeology, linguistics, environmental history, natural resource management, conservation biology, and ethnology. It will inform and refresh scholars in such disciplines for many decades to come, for it has none of the superficiality of the "hit-and-run" studies that have made indigenous communities justifiably skeptical of certain kinds of academic exercises that offer native peoples no lasting benefits.

I cannot read this book without remembering the lasting friendships upon which it was built. The richness and precision of the knowledge held by certain Pima naturalists has been admirably complemented by decades of tenacious attentiveness and the profound respect that Amadeo Rea has extended to these friends. Few will ever realize the depth of personal growth and financial sacrifice Amadeo has lived through to bring this work together. I had the benefit of beginning my own career in ethnobiological research, conservation, and cross-cultural communication by watching and listening to him interview the late Joe Giff and Sylvester Matthias about the cliff-hanging plants of the Sierra Estrella and the lost fish of the

Gila River. As I listened with awe, Amadeo carefully encouraged these remarkable naturalists to recall habitats they had not seen in decades and organisms that were long gone from their reach. The details gradually emerged from the deep recesses of their memory and painted scenes that I had at first assumed Arizonans would never see again.

Yet as I became enraptured by their stories, I realized that the three of them had some special, collective alchemical talent to call up lost worlds, to make them tangible and whole once more. I can only hope that someday the scenarios presented in this masterwork will inspire the Pima and their neighbors to begin an ecological and cultural restoration that will renew the lovely, diverse world that once coursed through their lives. I hope that someday, someone will be reading this book next to a calmly flowing Gila River, replete with its attendant flowers and migratory creatures. As Culver Cassa learned when he asked one of his elderly Pima neighbors what the Gila River was once like, "We didn't think of it as desert back then. . . . It was paradise, paradise."

Gary Paul Nabhan

Acknowledgments

This work has benefited from the help of a great many people. Because it was so long in preparation, I know that I may unintentionally miss some in expressing my debt of appreciation, and I offer my apologies to anyone I have overlooked.

My first and foremost thanks go to the primary native consultants in this work: the late Rosita and David Brown, Ruth and the late Joseph Giff, the late George Kyyitan, the late Sylvester Matthias, Carmelita Raphael, and the late Francis Vavages. These were my primary teachers. Whenever I went to their homes with my notebook and tape recorder, they generously and graciously set aside whatever they were doing to spend hours explaining Pima culture to me. Later they patiently answered questions such as "What did you mean here when you said . . ." or "Explain this about the . . ." or "How again do you pronounce. . . ." Asking, verifying, reverifying—it went on for years, and they never complained.

Rosita, David, and Joe were gone before I ever thought of writing an ethnobotany. Had they lived longer, this would have been a richer, more complete work. But Ruth, Sylvester, Carmelita, George, and Francis contributed their knowledge from the beginning to the end of the writing. I truly believe that they all saw this book as a vehicle for transferring their fund of knowledge, their heritage, to future generations.

Other Pima consultants also contributed to this work: Helen Allison, Cornelius Antone, Myrtel Harvier, Irene Hendricks, Mary Narcia, Herbert Narcia, Sally Pablo, and Leonard Soke, to name a few. I worked with some of them just briefly and others occasionally over a period of years. Other Pimans who were not Gila River Pimas also contributed words, folk taxonomies, and information that helped to put the Gileño materials into a broader context. Among the Tohono O'odham, the three who helped most with ethnobotanical problems were Luciano Noriega of Quitovac, Sonora; Dolores Lewis of Ge Oidag, Arizona, and especially Frank Jim, originally from Charco 27, on the Tohono O'odham Reservation. Among the remnant Pima Bajo at Onavas, Sonora, doña María Córdoba and don Pedro Estrella, both now deceased, worked with me over a period of years. Major consultants among the Mountain Pima on several brief visits were Cruz Castellano, Isidro Cruz, Donasieno Cruz, and Lorenzo Cruz. All these people provided the materials that have allowed me to tell their story.

I have discussed a great many of the ideas and concepts in this book with dear friends and professional acquaintances; they have also had a hand in the telling.

Among them are Brent Berlin, Cecil H. Brown, Tony Burgess, Richard S. Felger, Bernard L. Fontana, Catherine S. Fowler, Wendy Hodgson, Joseph Laferrière, Eric Mellink, Gary Paul Nabhan, Sally R. Pablo, Richard Purcell, Karen Reichhardt, David Shaul, Wade C. Sherbrooke, and Jan Timbrook.

I must mention also my ornithological mentor, the late Dr. Allan R. Phillips, who first suggested so many years ago that I begin asking the Pima about the former ecological conditions along the Gila and the birds that might once have been there. That necessarily led to an investigation of folk taxonomies, the essential tool for the discussion of organisms between any two language groups. All this now seems a critical resource to tap, but biologists, whether taxonomists or ecologists, were not (and largely still are not) in the habit of asking Indians or any other "uneducated" peasants about their environment. Worldwide, folk science remains largely untapped because it does not come in a form authenticated by academic degrees.

Native speaker Culver Cassa of Sacaton spent many long hours checking for accuracy every Pima word that appears here. This task is bigger than it might seem. He had to determine which terms were legitimate variations among different consultants, which were regional variations within modern Pima, and which were just the result of my mistranscription.

Dr. Robert L. Nagell read and commented on all the chapters, offering many helpful suggestions for improving the clarity of presentation and flow of topics. Because he is not involved in either field biology or anthropology, he was a valuable critic. Dr. Gary Paul Nabhan reviewed chapters 1 and 7 and contributed many additional insights. Dr. David Shaul, a specialist in colonial period Uto-Aztecan languages, read parts of chapter 8 and provided abundant suggestions for its improvement, as did Culver Cassa.

Chapter 4, "The Pima Cultural Ecosystem," is an outgrowth of joint field studies with Gary Paul Nabhan among several Piman groups in Chihuahua, Sonora, and southern Arizona. Earlier versions appeared in *Environment Southwest* (no. 484) and *Ethnobiology: Implications and Applications: Proceedings of the First International Congress of Ethnobiology* (Belém, 1988), vol. 1. Chapter 7, "A Dietary Reconstruction," appeared in an abbreviated form in the spring-summer 1991 issue of *Arid Lands Newsletter*. An earlier version of appendix D was published in the 1988 Belém proceedings, vol. 2: it benefited from comments by Cecil Brown, Gary Nabhan, Thomas R. Van Devender, Robert Gasser, John Reeder, Charlotte Reeder, Charles H. Miksicek, Robert A. Bye, Jr., Wendy Hodgson, David Shaul, Jan Bowers, and Brent Berlin.

The chapter epigraph for chapter 2, by Eugene Linden, is © 1991 Time, Inc., and is reprinted by permission.

During this study a considerable amount of botanical material was collected as ethnobotanical vouchers and to document the vegetation from throughout Pima country in the broad sense—that is, those areas once used by the Pima in the course of their annual activities. The bulk of the botanical identifications were made by Dr. Geoffrey A. Levin, curator of plants at the San Diego Natural History Museum. He never complained of the workload (and the imposition on his own work), and he somehow managed to keep up with my fieldwork. Other identifications were by Richard S. Felger, John Reeder and Charlotte Reeder (grasses), James Henrickson (chenopods), Reid Moran, Gilbert Voss (onions), and the late Jack Reveal.

Susan Liston Breisch mounted and labeled a great backlog of botanical specimens during 1984–1985. After her, Linda Allen, James C. Dice, Judy Gibson, Melanie Howe, Annette Winner, Marcia Everton-Graham, Juda Sakrison, Alison Voss, Patricia Gordon-Reedy, and Ruth Miller prepared the specimens. Several of these were volunteering their time to the project.

Certain specialists read and commented on particular species accounts. Among these were Wayne Armstrong (duckweed), Karen Adams (tobaccos and Carrizo), Richard Felger (mesquites), Gary Nabhan (the major cultivars and Chiltepín), Andrew Weil (daturas), Paul Fryxell (cotton), Wendy Hodgson (Saya), Robert A. Bye (quelites, or eaten greens, and daturas), Kevin Dahl (Corn Smut), James Henrickson (chenopods), Laura Merrick (cultivated and wild cucurbits), Robert L. Gilbertson (fungi), and Thomas H. Nash (lichens).

The always gracious librarians Jane Bently of the San Diego Museum of Man and Carol Barsi of the San Diego Natural History Museum helped with bibliographic information throughout this study. In addition, Carol Barsi served as my computer specialist when technology tried to take matters into its own hands.

When the San Diego Natural History Museum divested itself of its senior curators, Mr. Kenneth E. Hill, a devoted bibliophile who was then on the museum's board of trustees, promised that he would see this book through to publication. He personally provided all the secretarial and computing salary necessary for its completion. I owe him a great debt of gratitude. His assistance rescued the book from a stillbirth.

The efforts of two other people put many pounds of my original drafts into the computer. Marjorie Rea (no relation; she pronounces her name with a single syllable) typed most of the species accounts (I must confess that at times she could read my handwriting even when I could not). Her compulsion for precision in the most minute details has made this a more exact work than it would have been had it been left to my broad-brush methods. (We both survived.) Marilyn Ellis, Mr. Hill's secretary, typed all the supporting chapters and the last dozen species accounts and saw the entire manuscript through its final editing stages.

During the years that I was doing most of the fieldwork and actual writing of this book, my home away from home was the Franciscan Team House in Sacaton, Arizona. Here Fr. Richard Purcell and Sister Anne Fischer provided me a quiet place in the desert among the Pima where I could work without interruption. When Richard transferred to other work, Fr. Gary Swirczynski continued the hospitality. These three had a sensitive finger on the pulse of contemporary Pima life.

Bill Evarts of San Diego traveled to the Gila River to photograph some of the major Pima consultants in their own homes. Bradley Collins, the editor of *Extension Magazine,* kindly supplied the photographs of the late David Brown and the late Joseph Giff by George Lundy and authorized their reuse here.

In addition, when I worked in the University of Arizona's library and herbarium, Dick Felger provided lodging and hardheaded advice.

Marilyn Ellis donated a much-needed dictating machine to the project. Others who have helped in various ways include Lupita Anderson, Daniel F. Austin, Deborah A. Bell, Janice E. Bowers, Alan Ferg, Takashi Ijichi, Phil Jenkins, Charles T. Mason, Jr., Donald J. Pinkava, Charles W. Polzer, Jon Rebman, Andrew C. Sanders, Ed L. Turcotte, Raymond Turner, Phil Unitt, Rebecca Van Devender, and F. Doug Wilson.

I received general support from the National Science Foundation (BNS 77-08-582, Felger and Rea) from 1977 to 1979 to study Sonoran Desert ethnobiology and in 1984 (BNS 83-17-190, Nabhan and Rea) for work among the Mountain Pima. During 1985 a grant from the National Geographic Society funded fieldwork, collection preparation, and native consultant fees. The financial assistance of these organizations is much appreciated.

Publication of this book was made possible in part by generous financial contributions by Christy R. Walton and Michiko Munda and City Beautiful of San Diego.

This manuscript was submitted in November 1991. A few changes in botanical nomenclature published subsequently and considered of lasting importance by Richard Felger have been incorporated.

Finally, I am grateful to the University of Arizona Press for undertaking the publishing of such a massive work. In particular, working with my editor, Dr. Christine R. Szuter, has been a joyous experience.

Introduction

What did you go out in the desert to see?
A reed shaken by the wind?
Then what did you go out to see?
A man dressed in soft garments?
Why then did you go out?
Matthew 11:7–9

First I want to tell you that I am not a botanist and I do not speak Pima. With that said, I suppose I should explain why I undertook the writing of an ethnobotany of a people living along the middle Gila River in southern Arizona's Sonoran Desert. I am a zoologist by training, specifically a taxonomic ornithologist. In August 1963 I went to the Pima reservation to teach high school. The Pima have been teaching me ever since. Perhaps you could call it a case of reverse osmosis. By the time I finished my study of the local birds and the changes in their habitat and distribution wrought by the various human cultures of the area (Rea 1983), I realized that there was a bigger story to tell, with even more baseline data and cultural information, about the plants of Pima country. Since childhood I have had an intense interest in plants—here was an opportunity to put this interest to good use.

As for the language, during the last three decades I have learned a lot of Pima vocabulary, if not much grammar. Folk taxonomies have been the main focus of my interest in Pima. Just as taxonomy is the initial step of any biological study, folk taxonomy is the essential beginning of any attempt to understand a people's relation to their natural environment.

Previous Studies

This is the first attempted comprehensive ethnobotany of the Gila Pima written from an emic perspective. Linguists distinguish between phonetics (the objective descriptions of the sounds of human speech) and phonemics (the sounds that a native speaker considers significant in his or her own language). Expanding on this, ethnographers have distinguished between etic studies of a culture (how it is viewed by an outsider) and emic studies (which try to bring out the people's views of their culture). L.M.S. Curtin's incomplete work (1949) was the first ethnobotany of these people; various other works have dealt with aspects of Pima naming, knowledge, and use of plants, primarily from an etic perspective.

The first of these was Frank Russell's classic ethnography (1908, reissued in 1975). Russell began his fieldwork in November 1901 and worked through June of the following year; he died of tuberculosis in November 1903 at the age of 35.

He was assisted by José Lewis Brennan, a Tohono O'odham from the Sells area. It was undoubtedly due to Brennan's efforts that an extensive body of oral literature (songs and formal speeches) was recorded, because this would have required the intermediary of a native speaker. About 77 plants are listed by their Pima names in two sections (pp. 69–78, 79–80), and several additional names are scattered throughout the text. Sixteen of Russell's plants were scientifically unidentified; all but five of these have now been determined. J. J. Thornber, the newly hired botanist of the University of Arizona, accompanied Russell on one trip and collected about 50 specimens from the Gila River area. A few of these early vouchers still exist in the University of Arizona Herbarium. Russell's list is intentionally an economic botany, not a folk taxonomy, but nonetheless it captures some valuable snapshots in time, providing information that would otherwise have been lost.

The ethnobotanist Edward Palmer was in Gila Pima country on at least three occasions: in 1867, probably in August; in August or September 1869; and about 7 February–3 March 1885 (see McVaugh 1956). He collected botanical specimens around Sacaton in 1869. These were reputed to have been lost at sea, but they may have reached their destination; because of inadequate labeling of specimens from this period, it is often difficult to determine dates and localities (McVaugh 1956). During his spring 1885 fieldwork, Palmer shipped 17 boxes of archaeological and ethnological materials to the U.S. National Museum. But as far as I know, none of his botanical collections from Pima country is presently identifiable as such. Manuscript notes Palmer compiled on the Gila Pima are at the University of Arizona Library (special collections).

A physical anthropologist, Aleš Hrdlička, worked among the Gila Pima in 1902 and again in 1905. The Pima sections of his comprehensive work of 1908 cover 47 biological species used as food or medicine. Most of these are accompanied by their Pima names, though apparently no voucher specimens were taken.

Father Antonine Willenbrink, a Franciscan friar, worked among the Gileños from 1916 until his death in 1946. He was well known to all the Pima with whom I worked, who often remarked that he knew Pima so well his speech was indistinguishable from that of a native speaker (see also Curtin 1949:23). In 1935, Fr. Antonine prepared *Notes on the Pima Indian Language,* which was mimeographed and copyrighted but apparently never published. Among various vocabulary lists of such categories as fruits, farm products, and weeds, he recorded the names of 90 Pima plants. His work lists the Pima names and their English equivalents (sometimes recorded as "roots used for dyeing" or "wild greens" or "English unknown"). There are no additional ethnographic data, nor were there voucher specimens. But I have been able to verify among living Pimas nearly every taxon he gave. Several names are known only from his lists.

The next work to deal with the plants of the Gila Pima was *Pima and Papago Indian Agriculture* (1942) by Edward F. Castetter and Willis H. Bell, botanists/agronomists with the University of New Mexico who conducted fieldwork during the three autumn seasons from 1938 to 1940. (They were absent during the actual growing season, Gary Nabhan notes.) From an etic standpoint this is a thorough work. But it is not an ethnobotany; there is scarcely a Piman word in the entire book. This makes it difficult (usually impossible) to correlate what the Pima at that time were telling the authors with contemporary Pima plant knowledge and taxonomy. Materials collected by Castetter and Bell are preserved in the Castetter Laboratory at the University of New Mexico, Albuquerque. Seed samples from the various southwestern tribes with which they worked are coded, but I have had little success deciphering the code. I searched the University of New Mexico Herbarium in the hope of locating voucher specimens of some of the plants Castetter and Bell discussed from the early 1940s that no longer occur among the Pima, but again I had little success.

Leonora Curtin's *By the Prophet of the Earth* (1949, reissued 1984) is the classic ethnobotany of the riverine Pima. Curtin worked on both the Salt and Gila River Reservations during four seasons (1939 to 1942). Her species accounts cover 75 folk taxa, with several additional taxa mentioned in passing. (Only two cultivated plants are included: bottle gourd and wheat.) Most of these are documented by vouchers that were transferred among various institutions in Arizona until being incorporated into the Phoenix Desert Botanical Garden Herbarium in 1988. Evidently Curtin intended to add to the book, but World War II terminated her fieldwork. Considering the interrupted nature of her fieldwork, there are surprisingly few errors in her ethnobotany: she confused *Trianthema* with *Portulaca,* *Malva* with *Lupinus, Cercidium floridum* with *C. microphyllum,* and evidently *Cyperus ferax* with *Scirpus americanus* and *S. maritimus;* also, she incorrectly applied the same Pima name to *Atriplex polycarpa* that she correctly gave for *A. lentiformis.* In addition, she was reluctant to believe what the Pima were trying to tell her about two species of willow, which now prove to be *Salix gooddingii* and *S. exigua* (see Group E accounts). However, one must be cautious in using her ethnographic data because not only riverine Pima but also Maricopa, Tohono O'odham, and even Yaqui and Mexican people living in the region provided information (Curtin 1949:27–45); often these are not identified by tribe in her species accounts.

In 1980, Deanna Francis, a Native American student from Maine, worked on a special botany project with Donald Pinkava, a botanist at Arizona State University. She studied spring medicinal plants with Diane Enos, a Pima from the Salt River Indian Reservation. Voucher specimens were prepared for perhaps 80 plants; a number of these, with Pima names and usages, I retrieved by manual search through the herbarium. Unfortunately, the term paper and accompanying catalog are apparently no longer available.

Interestingly, major ethnobotanical work has been conducted among the riverine Pima every four decades. And there has been some continuity of native consultants from one investigator to the next. Children of some of Russell's turn-of-the-century consultants worked with Curtin or Castetter and Bell. Several of the people with whom Curtin worked I knew well from my early years at Komatke. Descendants of people with whom Russell worked were my high school students. And the Franciscan Father Bonaventure Oblasser, who helped Castetter and Bell (as well as many other researchers), was one of my first teachers of things Piman. Nearly all of Curtin's interviewees were well known to Sylvester Matthias, one of my main consultants.

No comprehensive ethnobotany of the Tohono O'odham has yet been undertaken; however, various incomplete but very useful works have dealt with this desert branch of the northern Pimans. Hrdlička (1908) included notes on the Tohono O'odham. Castetter and Underhill produced a monograph (1935, reissued 1978) that includes about 55 folk species. This is an economic botany rather than an ethnobotany, but the ethnographic information is rather detailed. Apparently no voucher specimens were collected, hence a number of misidentifications crept in.

Lexicographers preparing word lists have included many Piman plant names in their dictionaries (Mathiot 1973 and n.d.; Saxton and Saxton 1969:esp. pp. 170–178; Saxton, Saxton, and Enos 1983). Although these works have exemplary transcriptions of the native names from one or more of the dialects, the English identifications are sometimes fanciful, and the works are not supported with vouchers.

Appendix II of Gary Nabhan's dissertation (1983b) summarizes the dictionaries and includes his own elicitations, cited with specimen numbers. To my knowledge, his are the only folk taxa from Tohono O'odham country that are scientifically verifiable. His list has been of great use to me in field elicitations with riverine Pima. Folk taxa that are seemingly unique to these desert Pimans (or at least now forgotten by Gileños) appear in this work as appendix table A.2.

Nabhan and Richard Felger have prepared an incomplete and still unpublished ethnobotany of the Quitovac Papago of northern Sonora (see Nabhan and associates 1982:132–134; Felger and associates 1992). I have drawn comparative lexemes (words) from various studies among several other Piman groups. David Shaul kindly brought to my attention an unpublished manuscript from the Pinart collection in the Bancroft Library, compiled about 1774 or earlier by some Friar Antonio (perhaps Benz) from the village of Attí in northern Sonora. Although only about half the document is preserved, it is particularly important because it represents a long-extinct Piman speech community, the nearest riverine Pima to the Gileños for whom a colonial linguistic work has come to light.

To the south is the village of Onavas, on the southern Rio Yaqui, representing a group called in colonial times the Névome and more recently the Pima Bajo (see Pennington 1979, 1980). Voucher specimens by Campbell W. Pennington for the Névome–Pima Bajo are in the University of Texas Herbarium. When I began fieldwork in this pueblo in 1978, only two native speakers survived: doña María Córdoba B. (d. March 1986) and don Pedro Estrella T. (d. December 1988). During the next eight years, I made several trips a year to work with these speakers and compiled an ethnobotany of some 350 Pima taxa from there, about two-thirds of these supported with voucher specimens.

During 1984, I made several trips into Mountain Pima country along the Sonora-Chihuahua border, accompanied by Gary Nabhan, Eric Mellink, Daniel Lopez, Renée Levin, and Geoffrey Levin. Our work centered in the Nabogame rancheria near Yepachi. We compiled a very preliminary list of folk taxa. Pennington kindly supplied us with word lists he had compiled at Maicoba, Sonora, and Yecora, Chihuahua, two major areas of Mountain Pima differentiation. Subsequently, Joseph Laferrière spent considerable periods of time at Nabogame, gathering data for his doctoral dissertation on Mountain Pima cultural ecology (1991). He made extensive ethnobotanical investigations and supplied me with many cognates for inclusion in the Gileño work. His voucher specimens are in the University of Arizona Herbarium.

Farther south among the Pimans called the Tepehuan, much work remains for a folk taxonomist. Pennington published a comprehensive volume dealing with the Northern Tepehuan (1969). Many plant names are included in this volume, as well as in his medicinal plant paper (Pennington 1963). Gary Nabhan and Laura Merrick have supplied me with some preliminary information, mostly on crop names, obtained by José Muruaga-Martínez, a Mexican researcher. Southern Tepehuan, including Tepecano, is the southern extension of the lowland Piman continuum west of the Sierra Madre Occidental. J. Alden Mason's extensive linguistic studies (1917) indicate close similarities with northern Piman, at least in word morphology. I am aware of no folk taxonomic studies from this area but have gleaned words incidentally included in texts from this area (Mason 1917; Mason and Agogino 1972).

Just as taxonomy is the starting point of any biological study, folk taxonomy is the absolutely essential tool for any ethnobiological study, whether folk medicine, dietary study, cultural ecology, or mythology. Such studies can proceed only after the Western scientist knows the names (lexemes) and how the domains of these names are mapped to the objective genetic world of plants and animals. The lexemes are the vocabulary of the culture; the biological or Linnaean species are the alphabet of that vocabulary. Without lexemes, words, there is no story. The individual species accounts that constitute part 2 of this book are attempts to map these words with precision as well as to tell the story.

Wedding the Emic with the Etic

When my friend and former student Gary Paul Nabhan heard that I was undertaking an ethnobotany of the Gileños, he advised me, "Let the people tell their story." Perhaps he had been reading Edward Castetter and Willis Bell's *Pima and Papago Agriculture* (1942) and was frustrated by the approach they had taken: their story was told almost entirely from an outsider's point of view. At any rate, I hope you will agree that Gary gave me very sound advice.

The old people with whom I was working had grown up in a period when the Pima were largely dependent on subsistence agriculture, gathering wild foods, and hunting. In their youth the wagon trip to Phoenix was still an occasional and strenuous affair. Trading posts or reservation stores were few and far between. The Pima of the time were desperately poor after decades of drought. Even though their ecological base was slipping out from under them with the loss of access to water, the Pima of the 1900s and the next several decades still depended mostly on their own resources. (In retrospect they were probably better off with a subsistence economy than with the wage economy—and its consequences—that eventually developed.)

These people had stories to tell about nearly every plant in their botanical lexicon, their assemblage of named plants. I discovered a richness of which I had never dreamed; it seemed to be limited only by my ability to frame questions and listen to answers. I hope that what you now read is more than an encyclopedia of facts assembled by an outsider: I have attempted to keep it "their story."

Historically, many so-called ethnobotanies have been little more than economic botanies: laundry lists of what the native people call a given biological species and how they use it. We are fortunate that this information has been recorded, because with every death of an old one who has spent a lifetime close to the earth, a segment of the fund of human knowledge is lost forever. But an ethnobotany, as well as an ethnozoology, should reflect the people's worldview and organization of biological phenomena. It should include the folk science of the people, not just some facts to fill the slots on an elicitation sheet of an academically trained Western investigator who comes from a culture that views the natural world basically as an exploitable resource. For this reason I have tried to produce a work that weds the emic with the etic.

Some of my experiences in Pima country have aided this perspective. I spent five years in the high school classroom at Komatke and found that most Pima students knew little of their cultural heritage and still less about the natural world that surrounds them. During my six-year residence on the reservation I attended nearly every summer social dance and seldom missed a village saint's day feast. While writing this book, I talked to old people at their lunch programs, particularly about their Piman neighbors south of the border, about whom they knew practically nothing. I also spent many evenings after village services and potlucks talking with old people about what they raised and how they used to live. With this exposure, I hope that I have captured here some of the emics of Pima ethnobotany, not just the facts.

I have tried to make this book as useful to nonspecialists as possible, keeping in mind a readership of the Pima themselves, particularly the young. One of my goals is to offer the cultural information about plants back to the Pima. In particular, I hope that young Pimas with searching minds will be able to use this book to incorporate some of the old ways into their lives.

About the Text

Throughout this work I have used the word *plant* in its broadest vernacular sense

to include algae, fungi, and lichens as well as Plantae, sensu stricto. While this practice takes some liberties with the boundaries of at least three kingdoms, it brings together all the various organisms traditionally covered in an ethnobotany. More important, it allows us to see how the Pima classify living organisms that are neither animal nor human. Even after grouping creatures into *ha'ichu doakam* 'something alive' in the sense of animate, and *ha'ichu vuushdag* 'something that grows up' or emerges from the ground as a plant, the Pima folk taxonomist is left with a residue of organisms he or she would call living things but would put into neither of these two unique beginner folk categories.

The English and Spanish common names of plants have been capitalized when used in the specific sense (for example, Spiny Aster or Velvet Mesquite) but not when used in a generic sense (aster or mesquite). By convention, the names of crop varieties but not crops are capitalized (such as Sacaton Aboriginal versus cotton). This approach may seem inconsistent, but at least it is consistently inconsistent.

The scientific names of mammals, reptiles, fish, and birds that appear in the text are given in appendix C.

The orthography employed here for Piman words is a practical one with few diacritics. For instance, because most inhabitants of North America speak English or Spanish, the simple *sh* and *ch* have been used in place of *š*, *č*, and other esoteric symbols. Accents are normally on the first syllable of a Pima word; exceptions have a written accent. More detailed information on the Pima language is given in chapter 8.

In this work the terms *Gileño, Gila Pima,* and *Akimel O'odham* (literally translated as River People) are used interchangeably to indicate the Piman people surviving along the middle Gila River. (Some Gila Pimas settled along the Salt River near Mesa in 1872 and 1873. I have not worked directly with this community, and I would presume that some dialectical change has occurred.) The term *O'odham* by itself includes both Akimel O'odham (Pima) and all the dialectical groups of Tohono O'odham (formerly called the Papago). The self-designation *O'odham* means simply the People, so when you see the term *the People* anywhere in the text, it means the O'odham, the Pimans, the true

People. The terms *Piman* and *Pimans* designate all groups from both Pimería Alta and Pimería Baja who speak the Pima language. In most cases this includes the Northern and Southern Tepehuans as well, though to avoid confusion I have generally used the combined word *Tepiman* for both portions of the inclusive group. Words that appear cognate throughout this area, at least as far as is known, are said to be pan-Piman.

Some purists may object to my using the word *Papago* in this book because that tribe officially changed its title to *Tohono O'odham* in 1986. However, I have been listening to how the Gila Pima use the terms: when speaking English, they invariably refer to their neighbors as *Papago,* and when speaking O'odham, they almost always use the designation *Tohono O'odham.* As a result, both terms can be found throughout this book, as people themselves have spoken.

That the Akimel O'odham and the Tohono O'odham are essentially the same people, speaking the same language but differing in some basic strategies of cultural ecology, has long been recognized. But by accident, through the drawing of the political border between Mexico and the United States, the historical connection between the U.S. Pimans and the Tepiman groups of northwestern Mexico has been largely obscured. A central theme of this book has been to reconnect the northernmost Pimas with their immediate relatives southward in the Mexican part of Pimería Alta, with the Pima Bajo and Mountain Pima of Pimería Baja, and with the disjunct Tepehuans still farther south.

An important resource for the Pima oral history used throughout this book is the body of calendar-stick records. These are notched sticks that narrators of oral history once used to recount significant events of a particular year. The year for northern Pimans began with the Saguaro harvest moon (late June–early July). Ethnographer Frank Russell (1908:35) discovered "no fewer than five notched calendar sticks among the Pimas." Most begin in 1833 with the meteor shower of 13 November. Russell combined the narrations of Salt River, Gila Crossing, and Blackwater. (He does not say what became of the remaining two accounts.) The Blackwater account, narrated by Juan Thomas, was independently published by C. H. Southworth (1931), chief engineer of the San Carlos Irrigation Project. This account begins in 1850 and runs through 1913. Historian Sharlot Hall (1907) recorded the Salt River annals from Chukuḍ Naak (identified by the photos included in the publication), as did Russell. Carl Lumholtz (1912:73) believed that the only Tohono O'odham stick was from San Xavier. Leslie Spier (1933:139–142) discovered the last calendar stick among the Maricopa but was able to obtain information about only a few of the principal events from its keeper. This is unfortunate; a parallel account from a tribe living adjacent to the Gileños would have been interesting.

When anthropologist friends ask what I am working on and I tell them, many make a wry face and mutter something like "Pima! Why, there's nothing left of Pima culture to study. Frank Russell wrote everything there was to write in 1902. And even then he was just picking up the pieces of a badly eroded culture. Pima—really!" I hope this work dispels that myth. For instance, presented here are several firsthand accounts of how the Gileños constructed weirs or brush dams to divert the Gila into the canal system, descriptions of reverences that must be made on gathering first crops, liturgical remnants of the wine feast ceremony, and even the censored parts of legends that earlier Euro-Americans felt compelled to leave out.

Although such fleeting glimpses into the lives of these desert Indians are interesting, they are indeed something of the past. The Pima do not live a subsistence lifeway any longer. Few farm, still fewer gather, almost none hunt, and there are no longer any fish in the now dry Gila. Nonetheless, an ethnobiology is not merely an exercise in recalling the past. There is real wisdom here, something intrinsically valuable, that has been lost to the younger generations.

Take, for instance, how the Pima dealt with their land and resources, described in chapter 4. At least in some circles, there is an increased interest in the wisdom of traditional agriculture. Pima agriculture incorporated an approach found among many small-scale farmers of the world—the maintenance of crop diversity, or polyculture. Gileños traditionally preserved various landraces (locally adapted varieties) of both post-Columbian and aboriginal crops. Elderly Pimas recalled for me more than half a dozen named varieties of maize, at least four varieties of

teparies, and an array of muskmelon and watermelon varieties. The details can be found in the individual species accounts in Group I (*e'es*, crops). Our own legacy to the world consists of hybrid strains and monoculture. This continued loss of landraces as the world turns to mechanized farming and agribusiness is called genetic erosion.

The Pima exploited the resources of the desert oasis that they themselves helped to create and sustain by their agricultural system. All the evidence suggests that the system was balanced within the harsh restraints of the local environment, yet within a century the methods of Euro-Americans brought ecological disaster to Pima country in the form of desertification, with its drying up of rivers, salinization of the soil, continuing loss of topsoil to wind and water erosion, loss of native vegetation, and continuing subsidence of groundwater levels (Sheridan 1981). (These processes are taken up in chapters 5 and 6.)

This study is also valuable because of its implications for human health. More than half of the Pimas of the Salt and Gila River reservations now suffer from adult onset diabetes (NIDDM), a phenomenon that has mushroomed in the past half-century. Other Native American groups that have lived for millennia in marginal xeric habitats may not be far behind, depending on their degree of abandonment of the native diet and their adoption of the poor eating habits of non-Indians. (This problem is taken up in chapter 7.)

The process of discovery is at the heart of all science. For me, much of this book has been just such a process, of patiently tracing leads and solving mysteries, or at least some of them. I did not just sit down with a Pima, turn on the tape recorder, and say, "Now tell me all that you know." The process is what makes much of this type of research interesting. In many cases, the story of a folk taxon as given here shares the discovery process with you.

Not all the mysteries have been solved, however. The 28 Pima-named plants that have not yet been identified biologically are listed in appendix B.

This work is not a cookbook. I have described the processing and preparation of various food plants only in the broadest outlines. For one thing, it is difficult or impossible to pin a Pima cook down to

even moderately specific cooking times and quantities. I would hope that an actual cookbook would emerge from the Pima community.

Nor was my purpose to write a sociological treatise on the status of the Gila River Pima as they are found today, in a wage economy and to a great extent living in small, centralized urban developments—with all the social and physical ills that such a life holds for a people who evolved as gatherer-agriculturalists living in dispersed rancherias. My interest, as a biologist and ethnobiologist, is in reconstructing the relationship that the People, the O'odham, had with their land and its biotic resources, a symbiosis that survives only to a limited and diluted extent in the late 20th century. To a large degree, as with much salvage ethnography, this is a reconstruction based on the memories of key older Pimas. In this respect it is oral history, not a study of contemporary Pima life.

Pimas born after 1920 know only half to a third of the folk taxa familiar to those born earlier. Likewise, the Ta'i O'odham, or upriver Pimans, know fewer taxa, both plant and animal, than the Kuiva O'odham living at the west end. The generation of Pimas born early in the 20th century was the first one exposed to comprehensive elementary education in classrooms. But they were still involved intimately with subsistence agriculture, gathering, and at least small-game hunting. By the 1920s and 1930s, the exposure to traditional ways and the associated language was largely eclipsed by elementary and secondary education, often at off-reservation boarding schools. The conspicuous difference in knowledge between upstream and downstream peoples is related to the ecological changes discussed in chapter 5. For the most part, the folk taxa unknown to upstream people are directly or indirectly associated with streamflow and the riparian community.

A Word about the Paintings

Westerners accustomed to standard botanical illustrations that attempt to capture details may be surprised at the illustrations in this book. These are in a Japanese style of painting called *sumi-e*, literally 'ink painting'. *Sumi-e* and calligraphy both have roots in China, though they have attained a degree of specialization in

Japan. Both have had a long history of coevolution with Zen.

Sumi is a black vegetable ink that comes in small bars and is ground on an inkstone with water just as the artist is about to begin painting. The painting technique depends on *nijimi,* the flow of the ink into the paper, creating the various black and gray tones. There is a spontaneity in the technique that comes with learning how to control the density and dilution of ink in the brush as it moves across the paper. Whereas a conventional pen-and-ink illustration may require hours of work, a *sumi-e* may be executed in a relatively short time. Either the desired effect is achieved or it is not; there is no reworking, only a discarded painting.

The paintings included here are by Takashi Ijichi, who was originally from the Kagoshima Prefecture of Kyushu, Japan. Takashi often began a study by sketching a plant in the field, then working in the studio from his sketches, frequently with the aid of fresh or pressed specimens and photographs of the species. He seldom painted on the reservation, because in the desert the water in the brush dries too quickly to achieve the correct ink flow.

The process of moving from the field sketch to the studio painting helps the artist to abstract the essence of the plant. A good ink painting combines the seemingly contradictory process of capturing both the generic essence of a species and the structure of a particular plant. Some species, such as Finger-leaved Gourd *(Cucurbita digitata),* have been presented in a very abstract, stylized form; others, such as some of the eaten greens and cactus, seem almost anatomical in attention to detail. (But even where it was necessary to distinguish between similar plants, the detail is more suggested than actualized.) A few of Takashi's paintings are in outline: for example, White Brittlebush *(Encelia farinosa)* and Desert-holly *(Acourtia nana).* This is useful for white flowers and pale gray plants. But for the most part he used free brush strokes.

Sumi-e painters say that the achromatic grays and blacks are capable of evoking the colors of a flower—for instance, the deep red of a rose or the sky blue of a delphinium. We'll let you be the judge.

At
the
Desert's
Green
Edge

Part 1

The Pima and Their Country

Upon their emergence I'itoi and his followers danced and sang:

Together we emerge with our rattles

Bright-hued feathers in our headdresses

We have come! We have come!

The land trembles with our dancing
and singing.

So they went on, slowly, camping at one
place, sometimes, for many days or several
weeks, making their living by hunting.
Pima Creation Epic

Chapter 1

The Gila Pima

[Tribal] peoples have lived with rhythms of the planet for tens of thousands of years, fashioning skills and procedures over the ages that must not be lost in our rush to pave the Earth asphalt. They established basic attitudes toward the primary realities of this world, and fashioned traditions that have endured difficulties that we can hardly imagine. They forged rituals and initiation rites to remind them that this Earth is sole provider and sustainer of all lifeforms. They have accumulated knowledge that we can not do without, wisdom that, should we lose it, we could never reproduce.

Brian Swimme, *The Universe Is a Green Dragon*

Following the landfall of Cristóbal Colón on the West Indian island of San Salvador in 1492 and the conquest of the great Aztec nation and its emperor Moctezuma II in 1521, the Sword and the Cross moved rapidly in all directions in the Spanish Crown's campaign to Christianize and colonize the New World. By 1580, the first Pimans were encountered, and one Franciscan friar had already learned the language (Pennington 1969:16–17). These southern Pimans were the various groups the Spaniards labeled as Tepehuanes, a dialectically diverse lot inhabiting parts of Zacatecas, Jalisco, Sinaloa, Durango, and Chihuahua. The southernmost were called Tepecano, a group that survived as a speech community until early in the 20th century (Mason 1917). The more northern people on the frontier of Nueva Vizcaya (now the states of Sinaloa, Sonora, and Arizona) the colonizers called Pima, supposedly from their frequent negation *pi, pim,* or *pima* in response to questions (Velarde 1931:113).

Of course, these people did not call themselves Tepehuan or Tepecano or Pima. These were names, for whatever reasons, that the Spaniards applied to different segments of people in different mission districts. In their own language, the people for the most part called themselves Odami, Ootoma, O'odaam, or O'odham, depending on their local dialect. They are part of the Uto-Aztecan language family, one of the largest language families in the New World (Miller 1983), a family that includes Shoshone and Hopi to the north and Nahuatl (also known as Mexicano or Aztec) to the south.

When missionaries reached the area of what is now northern Sinaloa, they found a break in the Piman (or Tepiman) continuum (fig. 1.1). In the lowlands of the Fuerte, Mayo, and Yaqui Rivers were Cahitans, now known as the Yaqui and Mayo nations. Upstream, as sort of a Taracahitan linguistic bridge, were the

Guarijío or Warihio, to the colonizers primitive and barbaric people. And in the headwaters were the Tarahumara bands. Immediately north of this break, in what the Spaniards came to call Pimería Baja, were more peoples of Piman tongue. Even though geographically disjunct, the north and south segments were recognized as an essential unit. The terms *Pimas* and *Tepehuanes* were sometimes used interchangeably (González 1977:28n.5; Sauer 1934). As early as 1616, a Spanish officer, Diego Martínez de Hurdaide, said that the Névome or central Rio Yaqui Pima were of the Tepehuan tongue (Pennington 1980:4). Likewise, a missionary in Chihuahua in 1677 called the Mountain Pima by the name Tepehuan (Pennington 1980:20). To recognize this linguistic and cultural unity, scholars now often use the hybrid cover term Tepiman to designate all the Tepehuan groups and Piman groups as a unit.

Arriving in Pimería Alta, the Spaniards were surprised at the great houses and other evidence of a former complex culture. They were familiar with the migration myths of the Aztecs. Was not the ancestral Aztlan of Nahuatl legend supposed to be in the north? The Europeans, starting with Fr. Eusebio Francisco Kino and Lieutenant Juan Mateo Manje, saw these great ruins and those near Janos, Chihuahua, as material evidence of Moctezuma's (or Montezuma's) travels southward through this desert country. To a man, the Jesuits found these ruins to be completely different from anything they had encountered so far in northwestern New Spain. They were not the work of the local folk. That two different cultural entities were involved was never questioned.

The local Gila Pima, moreover, unequivocally disassociated themselves from the builders of these great ruins and other archaeological features, both to Hispanic missionaries (for example, Manje in Bolton 1936:370; Velarde 1931:131; Sedelmayr 1939, 1955) and to the earliest Anglo itinerants who inquired (Whittemore 1893:51; Bartlett 1854:283; McClintock 1916:1:9). By the time of his studies in 1901–1902, ethnographer Frank Russell said, "[The Pima] now frankly admit that they do not know anything about the matter" of the prehistoric ruins in the vicinity.

On the contrary, Pima tradition provided rather specific calendric times for their own arrival. Reverend Isaac T. Whittemore (1893:51) was told that as a group the northern Pimans (Akimel and Tohono O'odham, Koahadk, and Chuuvĭ Kuá'adam) had arrived 350 years ago, or about A.D. 1540. J. Ross Browne (1869:107) was told that the Pima reservation land had been continuously in cultivation for 300 years, an idea evidently picked up from the Gileños in his company, including Chief Antonio Azul. This would mean the Pima had lived here since about A.D. 1565. These dates fall well after the "Classic Abandonment" of the Salt-Gila River sites (around A.D. 1450) and are a far cry from the frequently heard phrase that describes Pima occupation "from time immemorial."

In time the ones who left ruins in the Salt-Gila valley acquired a label. Russell (1908:24) formalized the word in his Pima ethnography: "The term Hohokam, That which has Perished, is used by the Pimas to designate the race that occupied the pueblos that are now rounded heaps of ruins in the Salt and Gila river valleys. As there is no satisfactory English term, the Pima name has been adopted throughout this memoir." (Because Russell wrote *u* as *o* and usually substituted *h* for *g*, the word is really *Huhugam*.)

It is important to make a distinction between what archaeologists mean by their word *Hohokam* and what Pima mean by their term *huhugam*. In anthropology, the word *Hohokam* designates a Sonoran Desert people of unknown linguistic stock who occupied a heartland centered in the Salt-Gila basin. The people were skilled irrigationists, constructing hundreds of kilometers of great canals. They produced distinctive figurines, painted ceramics, and stone sculptures. They carved shell and made turquoise mosaics. They grew cotton and other Mesoamerican crops and maintained captive macaws of a southern Mexican species. They cremated their dead. Eventually they constructed ball courts and platform mounds and acquired copper bells and slate palettes. Apparently they maintained contact with Mesoamerican neighbors in the south for well over a millennium (Doyel 1991:245-246; Haury 1976:343–348).

In contrast, the Pima word *huhugam* is devoid of any cultural or temporal connotation. *Huhugam* is the plural of *hugam*, the polite way of referring to a person who is deceased to avoid using a dead person's name in the presence of relatives or friends. Doing so would be a serious breach of etiquette. The misuse of the term was repeatedly criticized by the major Pima consultants in this book.

In speaking O'odham, elderly Pima usually referred to the earlier inhabitants of the area as the Vipishaḍ, or Little People. (George Kyyitan called them the Hekĭ Hu O'odham 'Long-time-ago People'.) They associated the Vipishaḍ with some puebloid people who went to the north, perhaps the Hopi or Zuni, a tradition the Pima maintained since the time of earliest contact with Europeans (see, for instance, Manje in Bolton 1936:370; Garcés 1900:386–387; Sedelmayr 1939:105–106, 1955:22; Russell 1908:229; McClintock 1916, 2:6). Interestingly, both the Hopi and the Zuni maintain a parallel tradition that some of their clans originated in the heart of Gileño country, or at least in the south (Cushing 1896:342, 386–388; Cushing 1988:53–58, 93–95, 149–150; Mindeleff 1891:38–39; Alfred F. Whiting field notes; Barton Wright, personal communication). These clans are usually associated with Water, Corn, or Seed People.

The Vupshkam, or Wupshkam, are the Emergenti of the Creation Story who invaded out of the east or south. "Then who are the Pima?" I asked Joseph Giff. "The Wupshkam—we are they—the ones who are left. After the battles that are recorded in Pima myths. The Vipishaḍ, the small people, builders of the ruins, inhabited the earth before the Pimas." Years later, when I asked Ruth Giff who the Vupshkam were, she replied emphatically, "The ones who came up, the modern Pima, where *we* came from." Sylvester Matthias and George Kyyitan agreed. George reiterated his disregard for the much-used term *Hohokam* as a categorization of a tribal entity: "Now they say *huhugam*. That's just a young word . . . not very old." Sylvester Matthias shook his head, saying, "It's wrong, very wrong!"

The Hohokam-Pima continuum has, however, been the traditional view of most Anglos, with a strong following among many archaeologists and even younger Pimas. Piman scholar Paul Ezell (1983:149–150) summarized the situation while putting his finger on the major problem: "Since the work of Bandelier, . . . the consensus among Southwesternists has been that such a connection must have

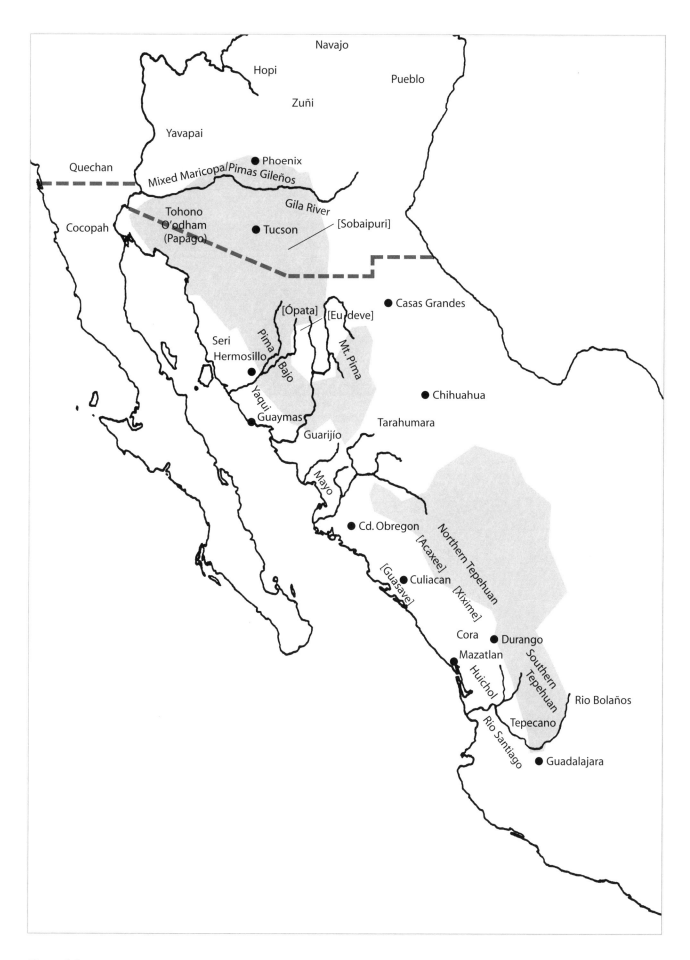

Figure 1.1 *Distribution of Tepiman languages and culture groups (bracketed groups are extinct)*

existed. . . . Yet the evidence to support an assertion that such a continuum existed was not forthcoming, so the idea of a 'gap,' a kind of Dark Ages between the end of the Hohokam Classic period at about A.D. 1450 . . . and the beginning of the written record, continued to be entertained."

Another problem with the Hohokam-Pima continuum hypothesis is that it explains Pima origin, development, and history merely as a southern Arizona phenomenon. This theory does not take into account the dispersal and differentiation of Tepiman languages throughout a 900-mile corridor to the south. About 95% of the area Pimans occupied at the time of European contact lies outside the Hohokam region. If Piman-speaking Hohokam migrated southward with the collapse in the mid 1400s, how is the rather considerable dialectical diversification of the upland groups, Mountain Pima and Northern Tepehuan, explained? Uto-Aztecan specialist Wick Miller (1983:118 and personal communication) estimates that a thousand-year separation was necessary to produce these Piman languages.

Then how do the Pima explain their affinities to the earlier people? There is, of course, no written history of this preliterate people. However, there are several sources of internal views. The first of these is the Creation Epic; the second the Pima oral-historic interpretation of the words and events of the formal tale. There exist two full-length published versions of the Gileño Creation Epic, both originating from a professional storyteller, Komalk Hogĭ (the name translates as Thin Leather or Buckskin), as well as another long though somewhat Christianized version obtained from Juan Smith, storyteller of Bapchule (Hayden 1935; Lloyd 1911; Russell 1908:206–248). Several parts relate the emergence, migration, and conquest of I'itoi's (or Se'ehe's) people (Lloyd 1911:147–161; Russell 1908:226–229). Not all the earlier puebloid people were killed, because Russell's (1908:229) version of the conquest concludes that some migrated to the north to form the Rio Grande pueblos. Russell (1908:26) credits Fr. Francisco Garcés as being the first to suggest (in 1776) a possible Gila origin of the Hopi (see Garcés 1900:386–387).

Two Pimas have set their reminiscences and tales to paper: George Webb and Anna Moore Shaw. Both were born at Gila Crossing in the 1890s. Both of these Pima authors equivocate in the origin stories (Shaw 1968:8–14, 27–28; Webb 1959:53). Their ambiguity might have been clarified by the use of the Pima categories Vipishaḍ/Vupshkam. In a sense, Webb's and Shaw's conquered and conquerors seem to be the same people—perhaps in deference to those Pimas who hold opposing views.

Evidence supporting O'odham ethnohistory comes from physical anthropology. Using multivariate analyses of 29 dental traits from human skeletal remains and 16 from living tribal persons, Christy G. Turner (1993) found that statistically the Hopi are "3 times more like the Classic Period Hohokam than are the Pima." Turner concludes that "the Hohokam may have moved in among the Hopi"— an idea, as we have already seen, that correlates with both Hopi and Pima traditions. In another study of Phoenix-area remains, Turner and Irish (1989) found, "The Pima and Hohokam [teeth] are not sufficiently similar to favor hypothesizing an ancestral-descendant relationship between them."

Rather than viewing the riverine Pima as the degenerate remnants of a pre-existing Puebloan or puebloid desert culture, then, I see them as the northernmost representatives of what we might term the northwestern Mexican rancheria cultural group, which includes the Mayo, Yaqui, Guarijío, Tarahumara, Cora, Huichol, Ópata, and Eudeve, as well as all other Tepimans. Their northward expansion into former Hohokam country I see as relatively recent, perhaps stimulated by Old World diseases that swept through northwestern New Spain every few years (Reff 1991:16, 132–274; Roberts 1989). This view is not inconsistent with archaeological evidence to date (Ravesloot and Whittlesey 1987). The protohistoric Piman period in southern Arizona thus may be temporally shallow.

The first European to have any lasting contact with the northernmost Pima at the fringes of their historic distribution was Fr. Eusebio Francisco Kino. Working from the mission of Nuestra Señora de los Dolores, established in 1687 on the Rio San Miguel, he wasted little time in beginning to explore Pimería Alta. Kino and other Jesuits used a number of names for these northern people who lived along the desert streams—Pima, Sobaipuri, Pimas Altos, Pimas Gileños, and so forth.

For instance, during his 1697 entrada, Kino referred to all the people of the middle Santa Cruz, San Pedro, and middle Gila Rivers west of Casa Grande as Pimas Sobaipuris or Soba y Jipuris (Bolton 1936:371). In this early period there were three groups of Sobaipuri: those living on the San Pedro from Fairbanks to its confluence with the Gila, those (called Koahadk) on the middle Santa Cruz between Bac (Vak) and Picacho, and those (called Pimas Gileños) on the Gila from the Casa Grande ruins westward at least to Gila Bend. In contrast, the "Pima proper" were the ones living nearest to Spanish settlements, in rancherias along the San Ignacio, Altar, Sonora, San Miguel, and Cocospera Rivers northward to the headwaters of the San Pedro and Santa Cruz Rivers—that is, south of the Sobaipuri (Bolton 1936:246–248; Ezell 1961:14). The Pimans living west of the "Pima proper" on the lower San Ignacio and Altar Rivers and along the coast of the Gulf of California were then known as the Soba or Desnudos. We would err in labeling all these subgroups as tribes: "The ethnographic reality," wrote Piman scholars Daniel Matson and Bernard Fontana, "appears to have been that there were at least two or three dozen Piman groups in northern Sonora [including what is now southern Arizona], each independent of the other and each with its own local dialect" (in Bringas 1977:23). Only by population attrition and other historical accidents have the Gileños (the Gila River Pima of this book) become known as the Pima.

The Spaniards distinguished the Gileños from the desert Pimans, groups of people who migrated between winter and summer settlements, whom the Spaniards called Papabotas, Papagos, and so forth. The name is apparently derived from a reduplicated form of the word bavĭ or bawĭ 'tepary', one of the Desert People's main crops (Velarde 1931:128).

In their own language, these groups called themselves O'odham, a word signifying the People, as the *real* people, to the exclusion of all others. The term *O'odham* has several levels of meaning, however. It is most frequently used to designate the Akimel O'odham (Pima) or Tohono

O'odham (Papago) in contrast to all others who do not speak this language. Hispanics are Jujkam, Anglos are Miligáán or Milgáán, Apaches and some others are Oob or Oobĭ, but none of these are the People, O'odham. Within the O'odham domain, the Gila Pima recognize several subdivisions. They themselves are the Akimel O'odham, River People, and are most frequently contrasted with the Tohono O'odham, Desert People. The Desert People, lacking irrigation and fertile fields, have to move around to scratch out a living. As traditionally understood by the River People, this is a term of pity; the Desert People are seen as the marginal people, while the River People are the resource-rich elite. The German Jesuit Ignaz Pfefferkorn (1949:30) noted this Pima attitude as early as 1763: the Gileños would plant extra wheat, and their "poor relatives," the Tohono O'odham, would come and help with the difficult harvest. It was almost a caste system. Not infrequently, early Anglo writers commented that the Tohono O'odham were more willing workers, more industrious, than the Pima (Hoover 1929:40). The River People's designation aof the Tohono O'odham as a separate group has been reinforced during the Anglo period by the establishment of separate reservations for the two peoples, with different political jurisdictions.

But the word *Tohono* does not really describe a single entity. It is, rather, a term that the River People applied to a collection of dialectical groups living without permanent water resources. Included were eastern groups living at higher elevations where more rain fell, especially in summer, and where a greater variety of plant and animal resources was available. At the western edge of the continuum were the people the River People pitied the most, the Sand-root Eaters (or Sand Papago, as Anglos named them). These nomadic people lived near the Gulf of California, abutting the non-Uto-Aztecan Seri of the gulf coast and the Yuman people of the Colorado River.

There has always been considerable intermarriage between the Akimel O'odham and the Tohono O'odham. Indeed, several of the native consultants contributing to this book had a parent or grandparent from Tohono O'odham country. One might conclude that by now the distinctions between River People and Desert People might be blurred or may even have disappeared altogether, but such is not the case. The emic distinction remains strong. George Kyyitan, for instance, had paternal grandparents "from up that way." (His father was Koahadk.) When I asked, "Well, what did your father speak?" he replied, "He spoke Pima, because he was raised here." In parallel manner, British, Scottish, or Irish immigrant children raised in the United States, who speak like their peers rather than their parents, are thought of as American. Ultimately, the language people speak is what identifies them.

Even though the languages of the Akimel O'odham and Tohono O'odham are mutually intelligible, the little differences of cadence, tone, word morphology, and even grammatical construction are considered important. Many years ago, on a visit to Lucy Jackson, a Pima of Komatke, I was demonstrating my counting abilities in O'odham, thinking I was doing pretty well. When I stopped, she said with disgust, "Aagh! You sound just like an old Papago." At the time, I was indeed learning O'odham from Joe Antone, my barber. He was from the village of Supi Oidag 'Cold Fields' on the Tohono O'odham Reservation. Even now, when a consultant does not recognize some folk taxon I ask about, the response often is, "That must be Papago."

However, the two categories "Akimel O'odham" and "Tohono O'odham" do not exhaust the linguistic domain of O'odham. Upstream on the Santa Cruz River (south of what is now the Gila River Reservation) lived a Piman people called the Koahadk, or Kohadk (see Dobyns 1974; Russell 1908:186, 200; Whittemore 1893:51). Their cultural ecology was intermediate between that of the River People and the Desert People. I asked several Pimas who consulted on this book whether they considered the Koahadk to be Akimel or Tohono O'odham. "Neither one," they said. "They're Koahadk. But they're O'odham." Linguistically, the Koahadk are more closely related to the Gila Pima (Saxton and Saxton 1969:183). Like Gileños and unlike the Tohono O'odham, they have both *v* and *w* sounds.

There is still another group of northern Pimans called the Chuuvĭ Kuá'adam, the Jackrabbit Eaters. I have never discovered who these people are or where exactly they live. They are mentioned as early as 1893 as being one of the four segments of the Pima (Whittemore 1893; see also Lloyd 1911:15). Were they people from northern Sonora living away from river valleys? Apparently they are not the Pima Bajo, at least not as represented by the Névome. Could they be remnant Sobaipuri from either the Santa Cruz or the San Pedro River? I have found no Pima who recognized the colonial term *Sobaipuri*. Occasionally I hear secondhand accounts of Pimas who have encountered Chuuvĭ Koa'adam from Sonora, and Sylvester Matthias can mimic their speech, but he does not know where they live.

A few older Pimas have a vague concept that there are O'odham in Sonora, in addition to the handful of Tohono O'odham villages on the Sonoran side of the international border. During the Mexican Revolution, a family of Pima Bajo lived as refugees in the area of Laveen, Arizona. In 1978 we made contact with the last two speakers of Névome or Pima Bajo, who were living on the middle Rio Yaqui. Sally Pablo, Ruth Giff, and Sylvester Matthias made a number of visits with me to their village. Pedro Estrella of that settlement called himself Pima when speaking Spanish and O'odham (without any further designation) when speaking his own language. In his mind, all Pimans were O'odham without distinction. At one time the Gila Pima had a firmer concept of their connection with the southern Pimans, judging from some historic Anglo accounts, but time and an international border have erased this awareness.

At another level of contrast, the term *O'odham* is used more broadly for Indian (Native American) as distinct from all non-Indians. It can even be used in the sense of "man": when I was a Franciscan novice, Fr. Regis Rhoder, O.F.M., said I would be called U'uhig O'odham 'Bird Man' if I ever worked among Pimans. And the term can be used in a much narrower sense to designate a Piman person who embodies the ideals that the Akimel O'odham uphold. Here is how George Kyyitan explained it: "Most of them, they really don't know what O'odham is, what that *means*. But the way I noticed a long time [ago] was that it means that somebody is really a person who cares for things, not to steal. . . . And that's when the People will say, 'We'll choose this guy, . . . [elect him] for this or that, because he, he really is *heg o wuḍ si o'odham*.' See, this is the meaning."

The Gileños were recognized as rancheria people by the colonizing Europeans. This settlement pattern differs from what the Spanish-speaking Europeans recognized as pueblos to the north—clusters of usually contiguous dwellings much like the small towns and hamlets of their European homeland. Pima rancherias were named entities of loosely dispersed settlement groups on the floodplains. At the time of Kino's entradas to the Gila, the Pima villages may have been in the process of becoming more compact because of an increasing Apache threat; on the basis of historical documents, Joseph Winter (1973:69) says that at contact there were "scattered individual family houses fifty or so yards apart."

Gileño rancherias were located on the upper terraces of the Gila, above their irrigated fields on the lower terraces. This put their homes out of reach of the floods. The areas surrounding the villages were kept clean of mesquite and other brush so that raiding parties would not have cover. Still, the fields, with their mazes of tall brush fences and living fence rows (see chapter 6), provided ample opportunities for surprise attacks on workers, and the calendar sticks—their records of oral history—attest that the Apaches attacked isolated workers.

Within a rancheria were smaller ranchos occupied by a man, his wife or wives, and his sons and their wives. Their houses, storage sheds, and arbors, along with their adjacent house gardens and nearby fields, formed the nucleus of these ranchos. Joseph Giff provided a description of the relationship between the family ranchos and the rancheria:

> The houses were just far enough away for privacy, but they could call to the next house if help was needed. A man could get on top of his house and call if the Apaches were seen, and the word could be spread that way. There was one man with a good voice, a strong voice, who was the town crier. Early in the morning, when it was still almost dark, he would get on top his house and announce the work for the day. If they were going to have a rabbit drive,

he would announce that—where they were going to go, what they were going to hunt. Or he would announce that they were going to clean ditches or repair the dam on the river. If anyone was sick or needed help, the town crier would announce that. Everyone would help each other. And if there was an attack on a villager, a runner would go to the next village and announce that, and all the Pimas would come.

Sylvester Matthias also described the relation of the smaller groupings to the larger settlement:

> Relations live together in a bunch. [They] had a common house, and every night the male folk had conversation, talk about the news [in the] *jeeñ kii* [men's communal meeting house]. Men might spend [the] whole night there [and] tell stories. [Settlements were] called *kavuḍ kiikam*. These were less than a mile apart. Like Komatke, [a village] had at least four *kavuḍ kiikam*. There were elders at each place—someone responsible. Anyone needs help—anyone sick in the family—[he] could make rounds from 3 a.m. to daylight.
>
> Everyone had two homes—[a] summer home and [a] winter home. In fields were summer homes; after [the] crop is in, then [they] move up to [the] winter home [on higher ground]. The *huhulga kii* [menstruation hut] was used by families in the settlement; that is, women of various households shared a hut.

One documentary account reports the effectiveness of the intervillage communication system that Joe Giff described. In July 1852 John Bartlett and several of his boundary survey crew, guided by two Maricopas, traveled up the Salt River to visit other walled ruins then still standing northeast of the South Mountains. On their return, they cut directly southward over the plain between the Salt River and the Pima settlements. Camping overnight, they resumed their journey at 4:30 the next morning and reached the Gila in two miles:

> As we entered the first fields of the Pimos [*sic*], the sentinels in the outskirts, seeing us approach in long single

file, mistook us for Apaches and gave the alarm accordingly; a very natural mistake, as no party of emigrants or travellers had ever entered their country from the north. We heard the alarm given, and echoed in all voices, from one tree or house-top to the other, until it reached their villages. "Apaches! Apaches!" was the cry from every mouth; and when it reached the first village, it was borne onward to every part of the community, even to their allies the Maricopas. The two Indian guides who were with us, discovered the stampede we had so unintentionally caused among their Pimo brethren, and seemed to enjoy the joke much. In a few minutes we saw the Pimos mounted, bounding towards us in every direction, armed and ready for the contest; others, on foot with their bows and arrows, came streaming after them; and in a short time, the foremost horseman, who was doubtless striving to take the first Apache scalp and bear it as a trophy to his people, reined [in] his steed before us. As he and those about him perceived their mistake, they all burst into a hearty laugh, which was joined in by the rest as they came up. (Bartlett 1854, 2:249–250)

At the time of Hispanic contact, the primary Pima villages of the middle Gila were clustered between the communities now called Casa Blanca and Sacate. Later, in the 19th century, the Gileños extended their settlements farther up- and downstream. The impetus for this expansion was the declining quantity of water in the Gila River (see chapters 5 and 6), coupled, perhaps, with a reduced Apache threat. At least from the last decade of the 1600s through the 1870s, the Apaches influenced Gileño activities. In November 1774, Fr. Pedro Font noted: "The [Gileño] Indians were asked why they lived so far from the river, since formerly they had their pueblo on the banks, whereas now they had moved it to a place apart. They replied that they had changed the site because near the river, with its trees and brush, they fared badly from the Apaches, but now being far away they had open country through which to follow and kill the Apaches when they came to their pueblo" (Bolton 1930, 4:46).

When I arrived in 1963, the rancheria pattern was still the norm for a good number of Pima families, though the dispersal was no longer defensive after 90 years of Apache containment. Clusters of mud houses were set in fields, well separated from neighbors by fence rows and mesquite bosques. The rancho usually consisted of at least one small rectangular mud "sandwich" house and a *vatto* 'arbor' with an Arrow-weed *(Pluchea sericea)* roof, under which most of the daily activity took place. Brilliant red and green pomegranate *(Punica granatum)* bushes were common in yards, and houses were often shaded by large Athel Tamarisks *(Tamarix aphylla)*—both introduced species. An outhouse, a corral or two, and a *vachkĭ,* or pond of water for domestic and animal use, usually completed the rancho. But round houses, the circular Arrow-weed cooking enclosure, the menstruation hut, and the storage shed for crops were already things of the past. A grid of dusty roads in the dry season— muddy when wet—connected these family clusters, as did small irrigation ditches. Everything else was a patchwork of small cultivated fields, recently abandoned fields, and old abandoned fields becoming bosques, or mesquite woodlands.

During the past quarter-century, the individual ranchos have, for the most part, been cleared for mechanized farming. Sometimes a cluster of Athel Tamarisks and a crumbling sandwich house, if spared by the bulldozer, mark the old oasislike dwelling place. The tribe has initiated its own type of urban sprawl in the form of housing projects. Cinder-block houses, no match for the long hot summers, are crowded together with paved streets and scant yard space—sterile places for children as well as plant life. Now there is indoor plumbing that sometimes even works. The transformation of rancheria peoples into urbanites in a generation or two is resulting in various societal and medical problems.

A group of rancherias apparently had a spokesman who exercised authority by mutual consent. Consensus was reached by adult males meeting in the large community round house at night. The Spaniards—accustomed to rigorous hierarchies in civil, military, and ecclesiastical matters—gave "rods of authority" to local headmen of each settlement, thereby establishing formal avenues of communication between local indigenous governments and the colonial government. The

Pima labeled each of these designated authorities the *uusgakam* 'owner [or possessor] of the stick'.

It would seem that this Spanish practice reinforced the authority of existing village headmen, investing it with new symbols, but the Pimas and the Europeans probably looked upon this authority quite differently. Piman society was egalitarian. Only a few individuals in addition to the *tobdam,* or "village crier," mentioned earlier were in positions of authority. The ditch boss, called in Pima *huis* or *shuudagĭ ñuukuddam* 'water caretaker', was charged with organizing the village crew to build the weirs, or brush dams, across the river and to maintain the ditches to the fields. Apparently he was also responsible for the distribution of water to the fields. In times of offensive war, there was a specific war leader, the *chukuḍ namkam* 'Great Horned Owl meeter', who directed the warrior activities primarily through communication with these owls *(Bubo virginianus),* who were said to be reincarnated deceased Pimas who knew the locations of Apache camps.

Some confusion exists in the minds of younger Pimas today between the concepts of clans and moieties. There are five Gileño clans, or sibs, each identified by what a child calls his or her father. Each clan has a mascot or totem animal (table 1.1), a particular color of *uihimal,* or Velvet-ant (actually the wingless females of a type of wasp in the family Mutillidae). By contrast, the *ñui* (Vulture) and *ban* (Coyote) moieties divided the villages into two approximately equal groups for sports and other social activities but did not influence marriage. Clans may once have had other functions, but historically they affected only the terms children applied to their father. Descent in both clan and moiety is patrilineal. (For the Tohono O'odham, the bear may be the totem of the Ogolgam clan [Saxton and Saxton 1973:371; Saxton, Saxton, and Enos 1983:122], but this is not the case with

the Akimel O'odham.) Two or possibly three clans were present farther south among the colonial Névome, as indicated by the terms *apapa, mama,* and *ogga* in referring to the father (Pennington 1979:88). These Névome words do not sort out conveniently to one or the other moiety as would be expected if one moiety represented Vupshkam invaders and the other Hohokam residents.

How did the outsiders who first encountered the Gileños perceive them? Jesuit and later Franciscan missionaries pushing northward in Pimería Alta found riparian rancheria people quite like the ones with whom they had already been working intimately for 70 years (Spicer 1962:87–88). Often they had with them on the Gila some Piman servants who spoke the language and could communicate readily with these northern cousins. (The major sound shifts that today distinguish Akimel and Tohono O'odham from Pima Bajo apparently had not yet occurred.) The new arrivals left us their impressions of the Gila Pima.

On his first visit in 1694, Kino (1919:128) stated his impressions plainly: "All were affable and docile people." In 1697, Kino noted that the leader of San Andrés del Coatóydag, Juan de Palacios, "welcomed us with all affection, and with so many arches and crosses that they reached for more than two leagues" (Kino 1919:173). The place-name is revealing: it is derived from the terms *Koahadk,* the name for the Pimans living on the Santa Cruz wash between the Picacho Mountains and the Gila, and *oidag* 'field'. Apparently these dry-farming cousins of the Gileños also had fields on the Gila.

The Swiss priest Fr. Philipp Segesser, working among Pimans farther south on the Santa Cruz (at Guevavi and Bac), bemoaned that his charges painted themselves with pigment or blood from a slaughtered cow and danced nightly, making them "lazy and inactive." But he noted their virtues as well: "The Pimas are very

Table 1.1   Northern Piman Clans and Moieties

Clan	Term for Father	Color	Moiety
Apapagam	*apap, apapa*	white or yellow	*ban*
Apk(i)gam	*apk(i)*	white or yellow	*ban*
Maamgam	*maam*	red	*ñui*
Vaavgam	*vaav*	red	*ñui*
Ogolgam	*ogol*	red	*ñui*

generous, share everything, and own everything in common, even their clothing. When one is having a meal and has not enough even for himself and another comes along, the visitor nevertheless gets [a] half share" (Segesser 1945:146). Also, he observed that

> there are those among the Pimas who are very skillful and who fabricate things of fibers, straw, cotton, and other material in a way that moves Europeans to wonderment. So, for example, they make black and white fiber baskets covered with all kinds of figures, and woven tightly enough that they are water-tight and water can be stored in them as in a cask. They know how to make wool and cotton sashes which are as pretty as costly ones. These they bind about their bodies and heads. Also they make cotton tablecloths and other things which are only a little inferior to those made in my fatherland. What they see they imitate, insofar as they are ordered to do so. If one asks them to make something one receives the invariable reply that they do not know how, but if one orders them to make something they make it, because what the Pima has once seen made, he can imitate. (Segesser 1945:147)

We can only guess whether there were any substantial differences between these Sobaipuri and the adjacent Gileños. My hunch is that the differences were minor and were comparable to those between Akimel and Tohono O'odham.

Another Jesuit, Fr. Luis Velarde, had long experience among the various Piman groups in the northern parts of Pimería Alta. In 1716 he observed, "Their customs are not as irrational as is promised by their barbarity, and although polity does not exist, they all salute each other and give each other the hand, even upon first sight. They are generous and liberal, as far as they can be in their poverty, and no one who visits their *rancherías* or homes, be he one of themselves or a stranger, will lack necessities. They live in a community together in the winter, and in the summer each one in his hut." Velarde, not a very astute ethnographer, missed some subtleties. "They have no government, nor laws, traditions or customs [with] which to govern themselves; and so each one

lives in liberty," he noted. Although Velarde did recognize the war chief and the hunt coordinator (Rea 1979; Underhill 1946:96, 264), as well as an apparently emerging local leader whose office was gained by popular consent and personal charisma, he nevertheless added, "they have no authority to speak of" (Velarde 1931:134–135).

In January 1774, Captain Juan Bautista de Anza made his preliminary trip from the Tubac presidio, south of Tucson, to San Gabriel, California. This was a small expedition of 34 men, including 20 soldiers and 2 Franciscan friars (Juan Díaz and Francisco Garcés) as well as 140 mounts and 65 beeves. On the return trip, Anza reached the Pima villages on 22 May. At Uturituc (Anza wrote it *Juturitukan*), they encountered "about three thousand persons, all of whom live closely united, [who] have been harried by the last attack of the Apaches. The fields of wheat which they now possess are so large that, standing in the middle of them, one cannot see the ends, because of their length. They are very wide, too, embracing the whole width of the valley on both sides, and the maize fields are of similar proportions" (Bolton 1930, 2:127). Some think Anza was engaging in hyperbole, but other explorers of the time also made reference to the unfenced fields away from the river, and the Pima by this date may have had more than three-quarters of a century of experience with this new miracle grain from the Old World. In spite of "the great drought and the still greater famine" that Anza mentioned a few days earlier, downriver, the Gileños were doing quite well (Bolton 1930, 2:127).

Returning by the longer route upstream, Garcés reached the Pima settlement of Uturituc a bit later than Anza; there he rested a week at the villagers' request. "You see that there is plenty to eat, and we will bring you fish," the governor told the friar, who needed little persuasion (Bolton 1930, 2:389). Garcés no doubt enjoyed the full week of recuperation among his old friends, whose language he spoke, and we can only wish he had left us with more than a sentence or two covering his sojourn.

The following year, 1775, a much larger colonizing expedition left Tubac, arriving at the Pima villages in November. Anza wrote, "The affability and friendly

treatment which I experienced from these people in my last expedition I have found repeated on this occasion. They all had the good manners to come to salute me and to prepare a bower or arcade of five naves in which to lodge us, and where they voluntarily supplied us with an abundance of water, wood, and some provisions of the kinds which they use" (Bolton 1930, 3:17–18). Even Fr. Font was impressed: "These Pima Indians of the Gila are gentle and of good heart, and to show their appreciation for our coming they begged permission from the commander to dance, and then they went from tent to tent of the soldiers dancing, the women linked together in their fashion. In short, these people manifested great pleasure at seeing us in their country" (Bolton 1930, 4:43). These were not easy words for the feverish and suffering Font to spare on the Gileños. "In the afternoon I went with Father Garcés, accompanied by the Pápago [sic] governor of Cojat, to visit the pueblo and see the fields. The latter are fenced in with poles and laid off in divisions, with very good irrigating ditches, and are very clean. They are close to the pueblo and on the banks of the river" (Bolton 1930, 4:43).

In the next century, itinerants from the United States also recorded their impressions of the Pima villages. Usually they had crossed the continent, and not without perils. From the last strongholds of "civilization" in New Mexico, they traversed Apache country, where the dangers were quite real. Whether they traveled down the Gila directly or took the somewhat easier route across to the pueblo of Tucson and then north to the Gila, they were exposed to ambush and raids until they arrived at the Pima villages. The Pima and their adjacent western neighbors, the Maricopa, shocked the travelers out of their prejudices about Native Americans.

In November 1846, Brigadier General Stephen Watts Kearny arrived in command of U.S. troops headed for California. Lieutenant Colonel William H. Emory (who was with Kearny) recorded his first encounters with these people:

> Where we encamped, eight or nine miles from the Pimos village, we met a Maricopo Indian, looking for his cattle. The frank, confident manner in which he approached us was in strange contrast with that of the suspicious

Apache. Soon six or eight of the Pimos came in at full speed. Their object was, to ascertain who we were, and what we wanted. They told us the fresh trail we saw up the river was that of their people, sent to watch the movements of their enemies, the Apaches. Being young, they became much alarmed on seeing us, and returned to the town, giving the alarm that a large body of Apaches were approaching.

Their joy was unaffected at seeing we were Americans, and not Apaches. The chief of the guard at once despatched news to his chief, of the result of his reconnoissance. The town was nine miles distant, yet in three hours, our camp was filled with Pimos loaded with corn, beans, honey, and zandias (water melons). A brisk trade was at once opened. (Emory 1848:82)

Emory continued, still astonished by the frank, confident manner of the people: "The camp of my party was pitched on the side nearest the town, and we saw the first of these people and their mode of approach. It was perfectly frank and unsuspicious. Many would leave their packs in our camp and be absent for hours, theft seeming to be unknown among them" (Emory 1848:82).

Although this was during the war with Mexico, there was still time for science, and Emory collected not only plants but also bird specimens. "We secured to-day our long sought bird, the inhabitant of the mezquite, indigo blue plumage, with top knot and long tail. Its wings, when spread, showing a white ellipse" (Emory 1848:83). Emory characterized the bird well. Its Pima name is *kuigam* 'owner of the mesquite'; we know it as the Phainopepla (*Phainopepla nitens*).

The second day, after visiting the local ruins, Emory and some associates hurried on to the eastern village to observe the people:

We came in at the back of the settlement of Pimos Indians, and found our troops encamped in a corn field, from which the grain had been gathered. We were at once impressed with the beauty, order, and disposition of the arrangements for irrigating and draining the land. Corn, wheat, and cotton are the crops of this peaceful and intelligent race of people. All the crops have been gathered in, and the stubbles show they have been luxuriant. The cotton has been picked, and stacked for drying on the tops of sheds. The fields are subdivided, by ridges of earth, into rectangles of about 200 x 100 feet for the convenience of irrigating. The fences are of sticks, wattled with willow and mezquite, and, in this particular, set an example of economy in agriculture worthy to be followed by the Mexicans, who never use fences at all. The houses of the people are mere sheds, thatched with willow and corn stalks. (Emory 1848:83)

Captain Abraham R. Johnston, another member of Kearny's troops, also visited the ruins. When he returned to camp, he found not only good grass but also Pimas bringing supplies. Johnston noted, "We were all struck with their unassumed ease and confidence in approaching our camp—not like the Apaches, who bayed at us like their kindred wolves, until the smell of tobacco and other (to them) agreeable things, gave them assurance enough to approach us. The Pimos and the Coco Maricopas live alongside of each other, but are distinct people, speaking different tongues. The latter once lived near the mouth of the Gila. The Pimos have long lived at their present abode." The next day Johnston wrote, "The Indians are still in camp, with their melons, corn, beans, and petiza [petaya] molasses [Saguaro syrup]; they spent the night in our camp, by the camp fires, without sleeping—talking and laughing incessantly" (Johnston 1848:598–600).

Following more slowly down the Gila trail were Kearny's wagon train and the Mormon Battalion under the command of Colonel Philip St. George Cooke. The Mormon Battalion arrived at the Pima villages in late December. "I halted one day near the villages of this friendly, guileless, and singularly innocent and cheerful people, the Pimos," wrote Cooke (1848:254). He painted a favorable picture of the people: "The Pimos, large and fine looking, seem well fed, ride good horses, and are variously clothed, though many have only the centre cloth; the men and women have extraordinary luxuriance and length of hair. With clean white blankets and streaming hair, they present mounted quite a fine figure. But innocence and cheerfulness are their most distinctive characteristics" (Cooke 1878:161–162).

The Mormon Battalion was abundantly supplied with diarists, and many made observations on the Pima and Maricopa. For example, Robert Whitworth, a 16-year-old Englishman, enlisted with his partner when they happened onto the battalion at Fort Leavenworth, Kansas. After four months, the threadbare and starving troops dragged themselves into Pima country. Young Whitworth noted in his diary, "Watermelons grow here at Christmas." He did not realize that the Pima knew how to keep these melons throughout the winter. The youth found the people different from any others they had encountered in their cross-country march:

This Tribe of Indians are called Pimos. Their villages occur along this River for 25 miles. Their Huts are shaped something like beehives. They are built of Wood and covered with mud. These People appear to be very civil in comparison to their neighbours the Apaches. They cultivate the Soil and produce very fine Cotton, which they make into Blankets. They are rich in Horses & Mules and have a fine Agricultural Country. They are a powerful Tribe and number about 30000 [*sic*; 3000]. More children amongst them than I ever saw before—These People wear no other clothing than a Breech Cloth and Blanket. (Gracy and Rugeley 1965:152)

Others in the battalion were impressed as well. Sergeant Daniel Tyler wrote in his diary,

On the 22nd, we marched ten miles and arrived at the Pima village, supposed to contain about 4,000 inhabitants. They were quite a large-sized, fine-looking race of people and very industrious and peaceable. They engaged in agriculture and manufactured blankets and other fabrics by hand. The poison of the civilized asp is unknown among them, and our American and European cities would do well to take lessons in virtue and morality from these native tribes.

Long before we reached the village we were met by the Indian women and children, many of whom were quite pretty and graceful, and walked generally by twos, with arms lovingly entwined around one another, presenting a picture of contentment and happiness that was very pleasing to look upon. (Tyler 1881:234)

Henry Standage likewise observed how open the Gileños were: "About 2 p.m., when near the river, the Pemose [sic] Indians, some 200 in number, came to trade with us, bringing meal, corn, beans, dried pumpkins, and watermelons, which they readily exchanged for old shirts, etc. They really seemed glad to see us, many of them running and taking us by the hand" (in Hayden 1965:23).

As usual, events far from the Gileños affected their contacts with non-Indians. Such an event was James W. Marshall's discovery of gold at Coloma in northern California. In the ensuing gold rush, an estimated 60,000 Anglo-Americans passed through the Southwest via the Gila trail (Spicer 1962:147). Not all made the journey successfully, but a great many more would have perished in this unfamiliar desert were it not for the supplies provided by the Maricopa and Pima. One of the hoard of 49ers left this account: "These Indians are darker than our northern races and generally exhibit stout limbs and strong muscle, and are more given to laughing, and there is almost a total absence of that rigidity of countenance" (Evans 1945:154).

A few years later, after spending more than two weeks in July 1852 visiting the Pima and Maricopa, Bartlett wrote, "There are no tribes of Indians on the continent of North America more deserving of the attention of philanthropists than those of which I am speaking. None have ever been found further advanced in the arts and habits of civilized life. None exhibit a more peaceful disposition, or greater simplicity of character, and certainly none excel them in virtue and honesty" (Bartlett 1854, 2:264). Similarly, in a February 1855 report for the Southern Pacific Rail Road Survey, Colonel A. B. Gray wrote, "Quiet and peaceful, they have no fears except from their enemies, the Apaches, and are very industrious, much more so than the lower order of Mexicans and live far more comfortably" (Gray 1856:85).

The Bartlett-Conde line of the original boundary survey proved unsatisfactory to the U.S. government and to railroad

interests. James Gadsden renegotiated the treaty in 1853. As part of the ongoing surveys, Lieutenant Nathaniel Michler passed through the Pima villages in May 1855. Michler found that the Pima

[are] further advanced in the art of agriculture and are surrounded with more comforts than any uncivilized Indian tribe I have ever seen. Besides being great warriors, they are good husbandmen and farmers, and work laboriously in the field. The women are very industrious, not only attending to their household duties, but they also work superior baskets, cotton blankets, belts, balls, &c. Their huts are very comfortable, being of an oval shape, not very high, built of reeds and mud, and thatched with tule or wheat straw. They are the owners of fine horses and mules, fat oxen and milch cows, pigs and poultry, and are a wealthy class of Indians.

As we journeyed along this portion of the valley of the Gila we found lands fenced in, and irrigated by many miles of acequias, and our eyes were gladdened with the sight of rich fields of wheat ripening for the harvest—a view differing from anything we had seen since leaving the Atlantic States. They grow cotton, sugar, peas, wheat, and corn; from the last two, parched and ground, they make a meal, which, mixed with water, forms a cooling and palatable drink. (Michler 1857:117)

By September 1858 the Butterfield Overland Mail had opened its stage line for transcontinental travel between St. Louis and San Francisco. Waterman L. Ormsby was the only passenger to make the full maiden passage. Traveling north from Tucson, the stage intercepted the Gila at the Sacaton station, where Ormsby noted,

Forty miles beyond the [Picacho] pass the company have a station where I saw the first Indians in their wild native costume—much resembling that of our New York model artists. They were a band of fifteen Pimos, engaged in dressing a beeve which they had just sold to the station keeper. The dexterity with which they separated the various parts and sliced up the animal into strings of meat to be dried was quite

remarkable. The men were generally in the costume of Adam, with a dirty cloth in the place of the fig leaf. The women, of which there were three, had cloths slightly larger, and a little cleaner, but down to the middle of the body wore beads on their necks and arms, and "didn't wear anything else." They all of them had fine muscular developments and were the very picture of health. (Ormsby 1942:97)

In spite of admiring the Pimas' rich land, irrigation, and crops, Ormsby entertained no romantic notions about these Native Americans. While breakfasting and changing horses for the next stage station, Maricopa Wells, he made more notes on the Pima: "The Indians are hideous looking objects in their filthy scraps of clothing and naked brown bodies, and frequently frightened our mules as they passed by. The men are lazy and take good care to make the women do all the work. We saw numbers of sovereign lords walking along, or riding, and making their squaws carry the loads—a spectacle which would give one of our women's rights women fits instanter. The women are, however, lazy too, and the men have about as much work to drive them as they would to do the work themselves" (Ormsby 1942:98).

On a military tour of inspection, Brigadier General James F. Rusling visited the Pima villages, which he said then numbered ten, with two additional Maricopa villages. He wrote,

Both tribes are a healthy, athletic, vigorous looking people, and they were decidedly the most well-to-do aborigines we had yet seen. Unlike most Indians elsewhere, these two tribes are steadily on the increase; and this is not to be wondered at, when one sees how they have abandoned a vagabond condition and settled down to regular farming and grazing. They have constructed great acequias up and down the Gila, and by means of these take out and carry water for irrigating purposes over thousands of acres of as fine land as anybody owns. Their fields were well fenced with willows, they had been scratched a little with rude plows, and already (March 9) they were green with the fast-springing wheat and barley. In addition, they raise corn, beans, melons, etc., and have horses and cattle in considerable numbers. One drove of their livestock, over 2,000 head, passed down the road just ahead of us, subsequently when en route to Tucson, and we were told they had many more. (Rusling 1874:369–370)

About this time, though, the Pimas' riverine oasis—marveled at by numerous travelers—began to dry up, and their ecosystem started to collapse. Water was unscrupulously usurped upstream from their villages, and the Gila watershed, the source of the river, was irrevocably damaged for the sake of short-term profit (see chapter 5).

The foregoing view of Pima life during the past three centuries is a view from the outside—a view by Europeans and Euro-Americans, based on visits ranging from a half-hour stage stop to a three-day recuperative stint at the villages. A few stayed a week or two. Although Jesuits and Franciscans had taken up long-term residence among Pimans somewhat farther south and some had left extensive relaciones of their observations and experiences, no missions were established among the Gileños during the Hispanic period.

It is perhaps presumptuous to venture an emic or insiders' view of the Pima world. But at least an attempt may be warranted, to balance the descriptions by outsiders that were based on casual contact. I will try to synthesize briefly the teachings I have heard from Pimas born early in the 20th century, principally information supplied by Joseph Giff, supplemented by details from Sylvester Matthias, George Kyyitan, and Francis Vavages. As with all teachings, these are the cultural ideals, handed down from one generation to the next. The description of such societal norms is the realm of ethnographic science; the manner in which people fall short of such values is the realm of literature.

Pimans once shared a common cosmology, as evidenced by their Creation Story. This story relates how the earth and sky came into being through the agency of Jeweḍ Maakai (Earth Maker or Earth Shaman), who then made Ñui (Turkey Vulture) to be his helper. The Sun and Moon begot Ban (Coyote). Last, out of the north came the fourth powerful being, who had two names: Se'ehe (Elder

Brother) and I'itoi. The sacred number four was then complete.

Some younger Pimas, unfamiliar with the finer points of the story, believe that the name Se'ehe is Pima and I'itoi is Tohono O'odham. Yet Font gave the name of this being as El Bebedor, the Drinker, from a Gileño narrator in 1775 (Font 1931:38–41; see also Bolton 1930, 1:256–259; Russell 1908:212–213); ii'i is the verb 'to drink'. And the Pima narrator Thin Leather gave both names (Lloyd 1911:31).

The Pima understood that everything was created for a purpose. We may not know what that purpose is, and it may not be useful to us, but still all things were put here for a purpose. "Just to beautify the desert" was an oft-repeated phrase I heard when I asked about the use of some plant.

O'odham are part of the natural world, not something separate from it. The People must be in some kind of a balance with their surroundings. There are sacred things around us—shrines, animals, even some plants—that can cause illness when we insult them in some way: physically molesting them, walking on their tracks, or even talking harshly about them. Thus, the tobdam who announces the shaada, communal hunt, must speak kindly and indirectly about the animals to be hunted that day. Butchering itself can be either abusive or respectful. Likewise, sacred animals must be handled respectfully; even their bones must be kept away from dogs and menstruating women.

Insults may cause a variety of problems, physical or psychological. They may produce certain illnesses; they may cause a nursing woman's milk to dry. Indeed, while his wife is pregnant, a man must take special precautions about what he hunts and how he kills it to prevent birth deformities. (The "sacred things" and the violations that Gila Pima talk about are quite similar to those Don Bahr and associates [1974] studied among the Santa Rosa Tohono O'odham.)

A maakai who can diagnose illnesses must be called to determine what had been offended, how the individual disrupted his or her relationship with the rest of the world. Once the cause is determined, usually someone who knows the proper curing song is brought in to effect the actual curing, restoring the harmony the individual had lost. Still other healers take care of ordinary illnesses such as pains, snakebites, wounds, and broken

bones; these are the people who use herbal remedies.

In the old days, the grandfathers and the grandmothers would teach the young people of the family. Children were important: "We don't harm our children, just talk to them. We don't spank them the way white people do, because then the one who gave them to us, our Maker, would think we don't like them and take them away from us." Early in the morning, before it was even light, they would begin teaching, talking to the grandchildren before they got up. A boy was taught to be hardworking and industrious, to be unafraid of the Apaches, to be brave. Young men were expected to practice all the time with their arrows, to be good marksmen, and with their shields, dodging arrows. They were taught to always be ready and to not draw attention to themselves. A girl was taught that she must grind corn and wheat, work hard for her family, and provide them with food; that she must be careful in all that she does and make good baskets; that she must get up early in the morning and not be afraid of the cold, even if there is ice on the water. These lessons were repeated over and over, day after day.

The first basket that a girl made would be given to her mentor, an older woman who instructed her. The first rabbit or other game that a boy shot would be given to someone older who helped train him; he could eat none of it himself.

In the communal meeting house, a man was expected to wait his turn to talk if he was old enough to participate; younger men were to listen. If a warrior was shot, he was to drop back behind the others and pull out the arrow if he could, or die quietly without making a fuss. If a warrior killed an Apache, he was to stay away from the others and from the village for 16 days. During this time he would fast and would have to bathe in the river even if it was winter and there was ice on it, and not complain.

Still other teachings had to do with everyday generosity. Pimas were taught not to store food just for themselves. If anyone needed food, it was to be given freely. Likewise, in a communal hunt, game was shared equally among all the participants. But a Pima could not be in a hurry or be too direct in asking for something. A Pima who needed to borrow something, perhaps a tool, from a neighbor could not just go and ask for it but instead would sit down and eat if offered food (even if he or she had just eaten). Then a man would maybe have a smoke. Then after an hour, maybe, he could mention what he needed to borrow. Pimas also were expected to be especially generous to their relatives, the Tohono O'odham: "They don't have very much in their country. They don't have rivers. So when they come here to work, we send them home with perhaps some extra sacks of wheat."

Other cultural teachings also had to do with relationships among the Pima. For example, Pimas were expected to be able to laugh at themselves. If someone fell down, observers and the fallen alike were to laugh. That would be the proper thing. Perhaps a more important example of interpersonal relationships is that when a young man married, he was to bring his wife to live with his family and build his house close by. If he went to her place, the Pima would say, "*Ha-at eḍa vaak*" 'he entered his in-laws' rear-end'.

What is my own perception of the Gileños after working more than a quarter of a century among them? I can give no simple answer, but I can add a few details to the rough image I have sketched out in this chapter.

I have found them generally open and informative, especially the older people. The amount of cultural information varies considerably even among the oldest generation; merely being old is no assurance of tribal wisdom. Moreover, a few older people were "stingy with their knowledge," as one Pima woman put it, and were this way even among their own people. By contrast, the ones who made this book possible spent enormous amounts of time explaining things and going over them. They were models of generosity.

The Pima treat material things casually. In the past, almost anything could be made from materials available in the local environment. Except for coiled baskets and some stone tools, everything could be readily fashioned in a short time. Even a house could be constructed in a few days. Pfefferkorn, farther south in the Pimería, related that the Pima exhibited a detachment that extended even to their houses: "Much more often [than fires] these huts are destroyed by the mighty windstorms which range terribly during the rainy season. More than once I saw a Sonoran hut,

in which an entire family was gathered, torn loose, blown to shreds, and scattered far and wide by a gust of wind. Sonorans are not only undismayed by such accidents but laugh with all their might" (Pfefferkorn 1949:193).

J. William Lloyd spent all of August and September 1903 with the Pima, recording the Pima Creation Story as told by Thin Leather. Between each narrative episode, he left some intimate observations of the Gileños. This was his overall impression: "The Pimas are fond of conversation and often come together in the evenings and have long talks. Their voices are low, rapid, soft and very pleasant and they laugh, smile and joke a great deal. They are remarkable for calmness and evenness of temper and the expression of the face is nearly always intelligent, frank, and good-natured. They are noticeably devoid of hurry, worry, irritability or nervousness" (Lloyd 1911:57).

Sixty years later I arrived at another Pima village one hot August night. Though I stayed a great deal longer, my impression has been much the same as Lloyd's, and I could not improve on his description of these unobtrusive people. Still, when they sit around outside at night, I wonder what they are laughing about.

Chapter 2

Native Consultants

Stored in the memories of elders, healers, midwives, farmers, fishermen and hunters in the estimated 15,000 cultures remaining on earth is an enormous trove of wisdom. This largely undocumented knowledge base is humanity's lifeline to a time when people accepted nature's authority and learned through trial, error and observation.

Eugene Linden, *Time Magazine,* 1991

This ethnobotany was possible only because of the careful, painstaking, patient efforts of the people who shared with me their experiences of living along the Gila and Santa Cruz Rivers. These were Pimas who were immersed in the ways of their ancestors, who were among the last to remember the old ways, even if sometimes only in an attenuated form. These were people who cherished their own traditions in subtle and unobtrusive ways. For more than a dozen years they were my teachers.

All the consultants in this book spoke O'odham as their first language. When they conversed with their spouses and other native speakers, O'odham was their language of choice. With one exception, they spoke O'odham to their children as well. When I asked them if they had known any English before they started school, the invariable response was, "Not a word!" Yet they had mastered the English language, and many spoke it more correctly and more precisely than did Anglos of similar socioeconomic background. In addition, most had at least a rudimentary skill in conversational Spanish.

Most native consultants for this book were long-time friends of mine, in some cases for 10 or 20 years, before I thought of putting together an ethnobotany. Francis Vavages was the only exception; I met him while I was writing at Sacaton. They all had known me long enough and well enough to be uninhibited about disagreeing with me when necessary, correcting and elaborating on texts I brought back to them for verification.

The Principal Consultants

Rosita Brown
1923–1977 (Santa Cruz)

When I first met David and Rosita Brown, shortly after I arrived on the reservation in August 1963, they lived in a mud house among fields and mesquite bosques in the rancheria of Santa Cruz. There was only an unpaved road to the village then, passable if the mud or fine dust was not too deep. Although she was a young woman, Rosita was deeply rooted in Pima traditions. She was a basket maker at a time when the craft had waned and only a few women in each village made an occasional basket. Perhaps even more intriguing, on some of my visits she would be parching wheat, making *haakĭ chu'i*, pinole. Although I have since watched several other women parching wheat, her skill was unexcelled. The dexterous flips of her parching basket always seemed to send the wheat grains and the bits of glowing mesquite embers and coals right to where she wanted them.

Several years after I met them, the family moved to St. Catherine's Church at the foot of the Sierra Estrella bajada. Here they occupied the building that was once the school.

Rosita was a good teacher of things Piman, always with a subtle humor. She had a way of cutting through one's defenses.

Rosita Brown

One hot afternoon I was returning from a climb in the Estrellas when I stopped to have a chat with the Browns, who were resting under their *vatto* 'arbor', sharing a six-pack of beer. Rosita offered me a can. I refused it, saying, "I haven't eaten all day, and I'll get drunk." But that was just a polite excuse; it was inappropriate for a schoolteacher to be seen drinking. "Ah, Mr. Rea," chided Rosita, "I thought you were our friend!" I had the beer, and it could not have tasted better.

In summer, different Pima villages sponsor dances lasting from nightfall to midnight. If there is a basketball court, it becomes the dance floor. If not, any flat circular area is cleared of rubbish and large pebbles, and the dancers' shoes accomplish whatever further smoothing is necessary. Either local bands or bands brought in from the Salt River or Tohono O'odham Reservation provide the music. There are two basic types—a conventional music called chicken scratch and a two-step polka called *choodi*, derived from *schottische* both linguistically and musically. Many Pimas may be heavy, but they are light on their feet and could dance all night if the law did not require that dancing stop at midnight. (So fond are they of dancing that some couples in the summer travel to Tohono O'odham country, where dancing is allowed until the sun comes up!) Once the band warms up, the "floor" is crowded with couples of every age, from ones so little they can scarcely walk, to grandparents and great-grandparents willing to forget their arthritis for the night. And of course, every dog in the village, not wanting to miss the excitement, is out there too, yelping when kicked or stepped on.

At one of these dances I spied Rosita sitting on a bench with some other women. "*Tto tvailla?*" I asked. "Do you want to dance?" "Sure," she said. Off we went. After a few great circles around the floor, I noticed Rosita reaching with her free arm around the back of her waist, grabbing at something. "I'm tired," she said. She danced me over to her sister and said a quick word to her, and I danced off with a new partner. When the tune ended, I said to Rosita, "You didn't get tired!"

"Well, next time I'll bring a safety pin!" Then she confessed, "The elastic on my slip came loose."

David Brown
1917–1981 (Santa Cruz)

Like his wife, Rosita, David was strongly steeped in Pima tradition, although he was somewhat younger than most of the other major consultants on this work. Over the years, he was the source of considerable information about both ethnozoology and ethnobotany, as well as diseases and folk medicine.

For the 16 years that I knew him, he was the sexton at St. Catherine's Church in Santa Cruz. Three times each Sunday morning he sent the bells pealing out on the quiet desert air.

One day I spent hours talking to David about plants. His eyes were red, I thought perhaps from a dry winter dust storm. Marcella Giff drove into the yard, bringing blankets for a funeral. She spoke in Pima.

*David Brown
(courtesy* Extension Magazine; *photograph
by George Lundy)*

When she left, I asked David who had died.

"My son—died of pneumonia yesterday."

Not long after, David was killed while walking down the road at night. Someone else rang the bells at the base of the Estrellas.

Joseph Giff
1907–1982 (Komatke)

In his career, Joe was a farmer, a cowboy, a carpenter, a tribal council member, a lieutenant governor, and a policeman. Had he grown up in another time and place, he might have been a college professor. The name Giff comes from the Pima word *gevĭ*, meaning 'snow', 'frost', or 'ice'.

He was a natural teacher, explaining details with a comprehensive and philosophical mind. But it was perhaps his closeness to the earth and his culture that contributed the most to his wisdom. Joe knew plants, crops, and forage, and he added considerably to both my ethnobotanical and my ethnozoological studies. And he was a singer, a person who not only knew the words to songs and had a good voice but could also explain the details of this oral literary form. He was a natural etymologist, dissecting words, analyzing syllables, contrasting meanings, explaining nuances. He took particular exception to the Hohokam roots that archaeologists had invented for the Pima. His knowledge of history was likewise comprehensive, and it is unfortunate that his fund of ethnohistory was never really tapped.

Joe's humor was both subtle and dry. When his daughter-in-law admired a new car passing by, he asked her, "Would you like to have a car like that? Well, you will—someday." (That is, she might eventually have one from a used-car lot.)

Once when I was hunting birds in Joe's cornfields, I noticed a pile of coarse hair that I thought was from a butchered deer. Deer are rare in Pima country, but I knew that Joe occasionally hunted them.

"Did you get a deer, Joe?" I asked, showing him the hair.

"No, that's *tasi'ikol* 'Javelina'. Sally got them, two young ones."

I was puzzled. I had hunted with his grandsons for years, but his daughter always claimed she could not hunt. Joe saw my confusion but made no effort to clarify his statement.

"But she tells me she can't shoot a gun!"

"Well, she didn't exactly shoot them."

"No?"

"No. She was on her way to a funeral yesterday morning, going up Williams Field Road [now Pecos Road], when this mother ran out with two young ones. By the time Sally stopped the car, she had hit the young, one on each side. So she just put them in the trunk and went on to the funeral."

In his later years, Joe went on dialysis, as do many diabetic Pimas. Then he lost a lower leg, which restricted his activities, and he finally lost his eyesight. Still, when he had a window of vision, perhaps just a spot, he would examine a bird or mammal or plant I brought to him to identify.

My last visit to Joe was in late December 1981. I asked him about the song he knew about Coral-bean *(Erythrina flabelliformis)*. Joe answered in his typical panoramic manner, bringing together linguistic, poetic, cultural, and geographic factors. He explained how this song fit with other Blue Swallow Songs and how these as a series paralleled the Psalms in praise of Creation.

While Joe spoke, Takashi Ijichi borrowed the keys to my truck to drive into Phoenix to buy a jacket. When he returned a few hours later, I was still sitting there listening.

Joseph and Ruth Giff
(courtesy Extension Magazine; *photograph*
by George Lundy)

Takashi admonished me: "You shouldn't be tiring him."

"I'm not. He's still answering the question that I asked before you left."

Joe died of a heart attack a month later. At that time I had not yet begun writing an ethnobotany, although I had been collecting information on Pima ethnobiology in general, and more specifically on hunting, fishing, and ornithology. Had he lived, this would have been a more thorough work than it is, particularly in its agricultural aspects.

Joe's wake lasted all night, but I lasted only until 4:30 a.m., when I left for a bit of sleep. As a traditional singer, he was known among singers even of different tribes. Years earlier, some Mohave friends from the Colorado River had promised that they would sing for him if he died before they did. There was some expectation that they might come to his wake, even though they had to drive halfway

across Arizona. Sometime after midnight they arrived, and the Mohave Blue Bird Singers gave him a proper send-off, even though these two groups were not only of different language families but once were traditional enemies (see Kroeber and Fontana 1986, esp. p. viii).

Joe was the most philosophical of the Gileño consultants I worked with. He was thoroughly grounded in the genesis aspects of the Pima Creation Epic, from which flowed the Piman perspective, the expected practices, the proper mode of behavior. Joe's cosmological vision paralleled a thoroughly Franciscan mysticism. This is aptly demonstrated in a long Pima discourse Joe published on the Blue Swallow Songs (which Don Bahr translated with him into English; Giff 1980). A sample of his commentary (Giff 1980:128, 131) follows:

See, there is the Gila Monster which is spotted and nice to see, many things are like that. See, they are just to be seen and known as Jeweḍ Maakai's [Earth Shaman's] products, as gifts to people. . . .

[A]s I have already explained many things are not eaten; they have a hidden nature which is Jeweḍ Maakai's knowledge of a mysterious something which we people will never find.

Ruth Giff
1910– (Komatke)

Ruth Giff is an active woman, given to much work and little talk. There is a certain sense of urgency about her. She came from the little village of Santa Cruz, nestled between the Estrellas and the Gila. The people here have an admixture of "Papago," and many remember it. Actually, the rancheria was settled by Huhu'ula peoples from what is now the northern part of the Tohono O'odham Reservation. Ruth was of the *ñui* moiety and the Vaavgam clan.

While her husband, Joseph Giff, lived, Ruth seemed to be eclipsed by his erudition. She usually said little during my interviews with him. But at times, when I sensed she had something to contribute, I asked her about it when just the two of us were together. Joe and Ruth raised four children and celebrated their 50th wedding anniversary in 1981. After Joe passed on, I spent more time with Ruth and found she had much to contribute.

In her mid 70s she began farming several acres of land at their old ranch. Perhaps this was stimulated, in part, by our interest in traditional crops, for once after both Gary Nabhan and I had been talking with her, she said she wanted to farm again. She asked a neighbor with a tractor to plow a field for her. Then, to beat the stifling heat of the southern Arizona desert in summer, she would rise at four o'clock in the morning, walk a mile to her fields, and work about four hours, returning when the morning coolness vanished.

Sometimes I would catch her in her fields, near the end of her work, to learn about Pima farming, weeds, and so forth. Sometimes we picked melons and young squash, *haal maamaḍ*, among the rambling vines. "See, that one is going to develop," she would say, pointing a crooked index finger. "We were taught never to point like that [with a straight finger] or the fruit would fall off." Even when visiting my garden in San Diego, she would point indirectly to some fruit on a tree. This was another example of Piman indirectness. Not even a developing squash or melon should be indicated directly lest it be embarrassed.

Many times when I arrived at her house, Ruth was working on her coiled baskets or preparing material for them. Countless hours go into preparations. In her garden, Ruth was constantly selecting the best qualities from her devil's claws (*Proboscidea parviflora*), large clammy plants grown for the black fibers that can be stripped from the pods and woven into the famed Pima baskets. One year she would talk about pliability or the ease in stripping the black "skin" from the "claw." Another year she would be saving seed from plants with extra long pods, giving her longer fiber strips. She exercised a constant, deliberate process of selection and improvement of her crop, because she knew the qualities she desired.

When not gardening or making baskets, she would sometimes be off with the village women making quilts to be sold at the annual bazaar, a money raiser for St. John's Indian School.

One time I asked Ruth to make me an all-black tray basket, an item exceedingly rare in collections. The black devil's claw is the most difficult of the fibers to work with, but it is the most durable in the finished product. From time to time I would see the start of it among the things she was working on. It grew to about six inches in diameter, then disappeared. She later told me that she had been asked on rather short notice to make a tray basket for the Pope to use at Mass in Phoenix, so she continued the design from the black base she had started for me. I never got an all-black basket.

Ruth was never stingy with her information. When I brought her sections that I had written, she patiently checked through them with me. Our afternoon conversations would sometimes last far into the night. She contributed more than any other O'odham woman to this book.

Once I asked Ruth if I could make a film of her preparing each type of weaving material and coiling a basket. When the time came, she took me outside her kitchen to where there was good light. There were great red ants everywhere in her yard, coming and going from a nearby hill. Their stings are intensely painful. She calmly went over with a hoe, piled a bit of dirt over their hole, and said, "There. That should keep them busy until we're through." No more eloquent lesson in the Piman way could have been provided.

Sylvester Matthias
1913–1997 (Komatke)

There is a yarn about a traveler passing through the West who, when he stepped off a stagecoach at some dusty station, noticed an old Indian resting in the shade nearby. Having heard of the phenomenal

Ruth Giff
(© 1987 Bill Evarts)

memory of Indians, the traveler introduced himself to the one in the shade and asked, "What did you have for breakfast on this morning a year ago?"

"Eggs," came the instant and unelaborated response. The Anglo went off satisfied.

Several years later this traveler was passing through the stage station again and spied the same Indian resting in the shade. Wondering if the Indian might remember him, the traveler approached and by way of greeting said, "How!"

"Fried," shot back the Indian.

While this story is doubtless apocryphal, it might very well have applied to Sylvester Matthias. I have been interviewing and tape-recording Sylvester since

June 1968. His command of details and dates is incredible. Sessions recorded half a decade or even a decade apart have been nearly verbatim. At times he would pick up on a conversation of many years earlier to elaborate on a point. Sometimes I would let the discussion of a plant lapse for two or three years or more before quizzing him again about it in the field. Even with a plant he did not know the name of in Pima, such as Desert-straw *(Stephanomeria pauciflora)* and *Tidestromia,* he would often remind me that I had asked about it at such-and-such a place before.

When I was teaching high school at St. John's Indian School, Sylvester was the school's gardener. He hobbled around, bowlegged and a bit lame from some childhood injury, keeping the palms, oleanders, and bird-of-paradise bushes alive and the Bermuda Grass lawns trimmed.

I knew that the Franciscans often consulted him when they were unable to locate some baptismal or marriage record in the parish archives. "Oh, that would probably be about in 1923," he would say. "She probably would have been using the name Baptisto, not Lewis, at that time." When my ornithological mentor, Dr. Allan R. Phillips, suggested that I ask local Pimas about the early conditions of the Gila and the birds it may once have supported, Sylvester was the first candidate. During his free time we sat down with a tape recorder, bird books, and piles of bird skins from my collection. The modest little manuscript that resulted that summer eventually grew into my doctoral dissertation and ultimately into a book, *Once a River* (Rea 1983).

Sylvester's knowledge was different from Joe Giff's. Whereas Joe, who was conscious of the interplay of the dominant cultures around him, often expressed things in terms of intertribal dynamics, Sylvester focused more on the details of what was happening within the Gila communities. Where Joe could discourse on the etymology of words, Sylvester was often unable to break a word into its components. Still, Sylvester could mimic any Piman dialect that he had ever heard and would do so with great humor. On our trips to Pima Bajo country in southern Sonora, he and Pedro Estrella would spend days talking. Sylvester quickly picked up the sound differences between Pima Bajo and Pima Alto and would modify his own speech to Pedro's way of speaking.

After his retirement from St. John's, Sylvester became the most accessible consultant for either short or extended trips. He had no fields to look after, and his wife, an invalid, lived in her own house a stone's throw away from Sylvester's little mud hovel. After he had chopped her a supply of mesquite wood, he was ready to be off. Sylvester usually made up his mind about something quickly, and there was little use trying to persuade him to change it. One time I was going up the Gila to collect birds and plants where the river was still running. "I'm going up above the Buttes for a few days, Sylvester," I told him. "It's pretty rough getting in, and we have to camp. Do you want to come along?"

"Yep," came the answer before the words were out of my mouth.

Thinking he had either not heard me correctly or not understood my question, I repeated it.

"I said I'll go." And so he did, though he had to crawl into the camper shell to sleep at night because he was too arthritic to get up and down from a camp cot.

Traveling cross-country with Sylvester was always a pleasure. We could drive across the desert for half an hour just enjoying the beauty, neither of us feeling a need to break the silence. Then Sylvester might say, "That hawk that was back there. Did it have a white tail, or a white back?"

"The dark one? Yeah."

"That's what we call *vakav.* It used to be here [at Komatke] when the Gila was running." And so we went, for many years.

I sometimes wondered at Sylvester's great knowledge of things Piman. I found that even peers of similar age admired his knowledge. Ruth Giff, who was three years older, explained that Sylvester's mother had abandoned the family when he was little, so his father and grandmother raised him. Consequently, his grandmother was the direct source of much of the 19th-century information related here. Sylvester was of the *ban* moiety and the Apapagam clan.

I suspect also that his family was part of the faction of traditionalists mentioned in the Gila Crossing calendar-stick accounts as "trouble makers" in the late 1800s. These families maintained Catholic traditions from the Hispanic era and continued the annual October pilgrimage to Magdalena, Sonora, at a time when the reservation was becoming properly Presbyterian. (This was fortunate for us, because with his experience in the Baboquivaris and northern Sonora, Sylvester was acquainted with a number of plants and birds that no one else on the Gila knew.) And also, they danced! They carried out rain ceremonies and wine feasts after such things had been forbidden on the reservation.

One particular ceremony, the Navĭchu celebration, bears all the earmarks of a Hopi kachina cult, including masked dancers that look and behave very much like kachina beings. On one outing, Sylvester casually mentioned something about how a mask was made. I learned that Sylvester was a Navĭchu by inheritance on the male line. Although he had never participated in the ceremony in the capacity of this lead kachina, he was somewhat familiar with the ceremony, called Wiigida in Tohono O'odham, and he knew how to make the gourd masks used by the various dancers. Over the years he made some three dozen for me.

Sylvester made many other artifacts as well. He was always extremely traditional in his methods, unlike Francis Vavages, who was innovative. In fletching arrows, he used sinews. In painting them red, he used powdered ochre or wet crepe paper. (Crepe paper is probably what someone early in the century had used as a source of paint, and he continued to use it.) In making Arrow-weed *(Pluchea sericea)* quail traps, he would bind them with willow bark, if it was available, and otherwise with strips of rags or baling wire.

Probably as a result of their strong missionization, these normally earthy Indians became quintessentially prudish. Over the years, though, Sylvester became less guarded with me. As a result I have been able to include in this book some of the symbolism and imagery of Pima culture recorded nowhere else, and even some of the more bawdy passages purged from earlier transcriptions of legends.

Pimas always enjoy a good laugh, but I once learned to be cautious in my own joking. One winter morning as Sylvester and I were taking our leave, I offhandedly said to his wife, "Well, Irene, why don't we just tie your wheelchair on top of my pickup camper with a rope, and you can come along."

To my astonishment, she immediately started giving orders. "Go get that thing back there and fold it up." I went and found a collapsible wheelchair. "Put it in the back of your truck. Now get that block and put it next to the door." It was a sawed-off piece of railroad tie that she could climb up on with a couple of people helping her. "Now help me up." The next thing I knew, everyone was somehow crammed into the old Chevy, wheelchair, crutches, and all.

We were off to collect plants on the floodplain, and along the way we discovered a pond full of winter ducks, which the Matthiases had fun locating with binoculars. We traveled as high as a vehicle could go on the Sierra Estrella bajada, where I made a hot lunch. People who noticed us out later remarked that they had not seen Irene outside her house in years.

As work progressed on this book, I asked Sylvester if he wanted to come to San Diego during the summer. He said he had not been there since such-and-such a day 40 years earlier and decided to come for a week's visit. The drive gave us a chance to look for plants along the way. When it was time for the return trip, I told him, "The bus takes 13 hours to Phoenix, the plane 55 minutes. The price is almost the same. How do you want to go?"

"I'll fly."

"Have you ever been on an airplane before?"

"No."

He flew a number of summers after that also. I do not know what attracted him, whether it was the flight or all the attention he got from the crew because they had to board him in a special chair— he was too arthritic to climb the stairs.

Each day during his stay, he would work mornings by himself, engrossed in his crafts or mask making. In the afternoon and evening we would generally spend four hours working on taped sessions on a wide range of topics. With him, as well as the other consultants, this seemed to be a good interview period, adequate to cover the subject properly.

Often Sylvester would open the next day's session with, "Now, something I forgot to mention yesterday when we were talking about . . ." Sometimes he would even pick up on something from a year before.

He particularly enjoyed eating fish during his visits. "This is *real* Pima food," he would always say. This was a sad commentary. Sylvester was from the part of the reservation where the river had endured the longest. "It was running full swing till about 1927," he once said. Then it died slowly during the next two decades. Today, not even a guppy could survive on the reservation part of the Gila.

In his home village, Sylvester is noted for a certain simplicity rather than for his knowledge. He used to accompany the Franciscan priests on their rounds as a sort of sexton or adult acolyte. One frequently told story is that at the end of a requiem Mass, many years ago, Sylvester got up, opened the casket, then announced in his distinctively inflected, rather tenor voice, "And now . . . you can in-ter-view . . . the bo-dy."

Sylvester was the person primarily responsible for developing the Pima classification system of plants that appears in this book (see fig. 8.7). He understood the concept of hierarchy, of nesting categories within categories. And he was quite comfortable using covert categories—groups not formally named in Pima. We began with little cards, each bearing the name of a terminal taxon (what an organism or group of organisms is called) in Pima and English. Sylvester worked these into 10 or 12 "families," as he called them. When he was not personally familiar with a species, he asked me the plant's characters, then assigned it. His questions about these unknowns gave me insights into the criteria he was using in making classifications. One day we got down to such difficult things as lichens and mosses, and he puzzled over the card marked "*kumul*/mushroom." His forehead was knit for some time until he finally turned to me and said, "How would *you* put it if you were writing it?" I had to laugh. We both did. But it was clear from his question that the Pimas were writing *their* book.

Carmelita Raphael

Carmelita Raphael
1921– (Santa Cruz)

In my early days on the reservation, I used to attend services periodically at St. Catherine's Church in the isolated village of Santa Cruz. Though the church was not far by horseback or as the crow flies from Komatke and Gila Crossing, the road to it was in those days a challenge until it snaked up onto the sandy bajada. Here perched one of the most beautiful buildings in Gila Pima country. Inside, the cream-colored walls were decorated with complicated mission-style friezes. A foot-powered organ provided the music. The organist was Carmelita Raphael. In later years a friar had the walls painted over with sterile white and sold the organ (this is called progress), but a delightful group of people remained, quiet and unobtrusive.

Carmelita had a husband and a series of sons, any one of whom could have modeled for Handsome Man of the Creation Story. Most fell to one tragic accident after another. But she was a woman of strength who continued on without self-pity.

Long after most other Pima families had abandoned their ranchos for the new tribal housing projects, Carmelita and her surviving children still had horses, cattle, corrals, and gardens surrounding a cluster of Athel Tamarisks (*Tamarix aphylla*) and a few buildings. Her place served as the model for the rancho, the subunit of the rancheria, described in chapter 1. When I first came to the reservation, these family nuclei were everywhere along the middle Gila, but by the time I began to write about them in the late 1980s, they were almost gone.

When I would drive into Carmelita's yard, she would come out of her adobe house in a fresh, bright cotton dress, and I would ask if she was getting ready to go to town. (I did not want to hold her up, and a Pima would be too polite to tell me she had other plans.) She would say she was not going anywhere. This scene was repeated many times, so I knew she was telling the truth.

Carmelita claimed no extensive knowledge of wild plants, but she still farmed and knew crops. In the spring we would go out on the bajadas to pick cholla buds for pit roasting, and there I learned more about the wild flora. Like many others in her village, she had family connections in

Tohono O'odham country. Her mother was from Menager's Dam near the border, and as a child Carmelita spent some time with relatives there. Carmelita is of the *ñui* moiety and the Vaavgam clan.

George Kyyitan
1912–1996 (Bapchule)

I drive into the Bapchule village compound, a collection of older mud houses and newer plastered ones, and park next to one that is so bright pink that it would seem more at ease on some Mediterranean island. Promptly a short, white-haired man hurries over from the *vatto* at the feast house where he was sitting and talking with several older men.

"When did you get in?" he asks. I never know how I am supposed to answer that, so I say, "Night before last," and that seems to serve. George Kyyitan, trim in his late 70s, wastes no time in setting up. He goes into the house and retrieves a long orange extension cord, and we plug in my stereo recorder. (I prefer transcribing from stereo tapes, especially when Pima is involved.) We sit down and begin a session that will probably last for the next three or four hours. I do most of the listening, keeping us on course when necessary, giving the session some structure.

George's father was Koahadk, from the village of Vaiwa Vo'o (Cocklebur) on the Tohono O'odham Reservation; his mother was a local Pima. George said he was of the *ñui* moiety and the Apkïgam clan. He explained the origin of his surname: "My dad [was] from Kohadk. They usually come over here and work at [the] wheat harvest. My dad had lots of horses. When they do that stationary thrashing, he brings horses to help them thrash. And he plays accordion for the dances. Miguel Cajetan. They don't know how to spell it on [the] census, so they spell it *Kyyitan*." And so it remains.

I knew George and his wife, Dorothy, for more than 25 years and ate at their house dozens of times a year, but only recently did I begin asking him questions about plants for the book. Most of my information had come from west-end villages, especially Komatke. The west end, far from the agency, was relatively conservative. It was also the last area where the Gila went dry. But Bapchule, too—at least when Frank Russell was observing there (1901–1902)—was a hotbed of traditionalism, a thorn in the side of both missionaries and government bureaucrats. The *mamakai* 'shamans' still held sway there, Russell said, and the ancient rituals to bring rain were still conducted. Julian Hayden recorded the last version of the Creation Story near there in 1935. Furthermore, Sylvester Matthias kept reminding me of linguistic differences between east-end and west-end people, many of these involving plant names. So I began taking George out with me to look for plants.

George's close-knit family all spoke Pima. Only the television was in English, and during warm weather we escaped from that outside. Even George's English reflected Pima, more so than any other consultant I worked with. The conversations with George I have included throughout the folk generic accounts will give you a good insight into the structure and thinking of the language, particularly with pronouns. Take, for instance, this verbatim transcription of George's response when I brought him some Mormon Tea (*Ephedra):* "That one is something like a bush—got leaves on there, but *this* one here is *uus tii*. That's what they call *uus*— 'sticks', 'limbs', or whatever. I think this is

what *they* call. Because *she* said, that time, 'Have you seen that *uus tii?*' 'No,' I said. 'Well, it's over there. You go over there and get some, drink.' And I tried to go. Of course, I don't know it. I only know *that* one, the other kind. And I couldn't find it. But this one here, I saw this one there, right in the wash there. And I think this is what she meant, 'cause this is like an *uus* there, 'stick'." (This transcription is close to hearing a Pima discourse—but using English words!)

Who was this fellow with the leprechaun smile? He was raised in an *olas kii*, or traditional round house, at a time when these no longer existed in most villages. In many other ways, as well, he was a step back into an earlier generation. A widower, he remarried, raised eight children, and a few years ago celebrated his 50th anniversary with Dorothy, a kind, gentle woman.

George worked at various jobs as a laborer in Phoenix and the surrounding agricultural districts. Then he spent 22 or 23 years working with archaeological surveys. He was part of Emil Haury's crew that excavated Snaketown, the Hohokam village (see figures in Haury 1976:x, 131). He was nearly 80 when he retired.

One day while we were outside taping, a young mother and her child walked into the yard. "If you're busy, I can come back later," the girl said. They spoke a few words quietly in Pima, and when I indicated to George to go ahead, he said, "Excuse me. The doctor has to go to work."

This was the first of many times that I was able to watch George practice the healing art of *koa i geesig i vamigida*, or fixing "fallen fontanelles" (see Group E, *Prosopis velutina).* If the child is not properly attended to, he or she can die. Having

George Kyyitan (©1987 Bill Evarts)

lost one of his own children, George decided to learn the technique himself. George assumed this community responsibility without ever demanding "green frog skins," as he called money. People pay of their own volition. "I won't say how much," he told me. "I never ask them unless they want to give. I can't say how much. It feels some way. I'll never say that—'You owe me so much.'" George never accepts anything for follow-up treatments. One time a couple visiting Sacaton from Tucson came to have their baby repaired and handed George a whole roll of bills. George refused it all: "You don't know if something might happen to your car, that you have to pay for that." George explained to them, "I'm not paying for this [ability]." Then George told me, "When I learn it, I just learn it some way. I don't know how I learn it. I'm just watching [them] doing it. Some people do that. I know there are some people, they charge 50 dollars or 15 dollars. At *once!*"

George's is the Pima way. Too much is not a good thing, as explained in the *jeweḍ hiósig* account (see Group L).

Francis Vavages
1906–1995 (Sacaton)

I met Francis while I was writing folk generic accounts at Sacaton. He was the only person I had not known for years before our first interviews. I sometimes hesitate to approach someone completely unknown to me.

"You should go talk to Francis," urged Fr. Richard Purcell. "He's a flute maker. He'd love to talk to you and might have something to contribute to your book." I did need more information from upriver Pimas, because most of my studies over the years had been among the west-end

Francis Vavages

people. So one morning I drove the one long block over to Francis's red-brick house. He was at work outside, with half-finished Carrizo *(Phragmites australis)* flutes and mesquite spoons in various stages. I had taught his now-married daughter when she was still in high school, so we were not exactly strangers. He was a lanky man at least six feet tall, and I liked him at once. He promptly began explaining the production of the Pima flute, played a funeral song, and imitated doves and Great Horned Owls *(Bubo virginianus)* with a new instrument, all obligingly on tape. "I've been trying to teach some young Pima [this keening song]," he said. "I'm not going to be around forever. But they're not interested." Francis had assumed the role of a village elder, which he played with both forthrightness and humility. He was frequently

called upon to address Pima children in classrooms, so he had a repertoire of subjects he covered. I told him what I was working on. "Well, what do you want to know about plants?" he asked. "I don't know very much, but I know some things. Maybe we can go out on the desert and I'll show you some."

Although Francis's overall knowledge of plants was not great, at least as far as folk taxa go, he was a valuable consultant because he would readily admit the limits of his knowledge. As a youngster he had had close contacts with his grandparents, so at times it was if I were listening to some of the people Frank Russell had interviewed 85 years earlier in the same settlement.

Francis grew up at the nearby settlement of Haashañ Keek (Sacaton Flats). He started boarding school at Sacaton not knowing a word of English—as did all the others in this book. He graduated from the Phoenix Indian School in 1928. For 30 years he was with the San Carlos Irrigation Project, delivering water to the upper parts of the reservation. It was perhaps in this capacity, wearing rubber boots, that he acquired his well-known Pima name, Kakawáádi Shuushkam 'peanut shoes he has'.

Unlike Sylvester Matthias, Francis was an innovator in his crafts. Whereas Sylvester bound his flutes with soft, white cotton string, as has been done for ages, and burned the holes with a heated wire, Francis bound his with bright red yarn and added small figures with a wood burner. "I like to jazz them up a bit," he told me once, as he inserted some dyed red feathers at the lower end. His mesquite spoons, finely carved, bore little resemblance to the traditional pattern found throughout the rancheria tribes of northwestern Mexico (see Fontana 1979:29, 36; Russell 1908:100–101).

Francis, a widower, was active even in his mid 80s. I would have to catch him at work in the morning because at about 10 o'clock he would be off to the community building to have lunch with the old folks. On other days he had appointments at schools. In retirement he was a busy person. I would sometimes catch sight of him walking briskly across town with the stride of a man decades younger.

Thanks to him we have a fine account of how weirs were built on the river, how reverences were made to a plant before its

fruit was picked, and details of tobacco cultivation, as well as several songs and stories (the origin of Foothill Paloverde [*Cercidium microphyllum*] and Saguaro [*Carnegiea gigantea*], for example), among other things. Francis was of the *ñui* moiety and the Vaavgam clan.

Other Gila Pima Consultants

Several other Gila Pimas are mentioned in the folk generic accounts as having supplied information on plants or Pima life. These people either were younger, had more limited information, or were less accessible for interviews.

Irene Hendricks
1924– (Casa Blanca)

While I was writing at Sacaton, Irene would occasionally come by to visit. She took an interest in the plant specimens I was preparing and offered to go into the field. Her knowledge of both Pima taxonomy and plant uses, particularly medicinal uses, was extensive. She was the first to identify several folk taxa for me. With an appetite for the latest village news, she often strayed from folk taxonomy to other subjects, but she carried her eccentricity with aplomb.

She liked to gather wild greens. One day she asked me, "Why are you staying in California if you study *iivagĭ* ['greens']? There's no *iivagĭ* in California; it's just cement! I don't think there's any *iivagĭ*, any soil." For Irene, California meant Los Angeles, where she occasionally visited a daughter.

Sally Giff Pablo
1934– (Komatke)

Like Dante's guide Beatrice, it was Sally who opened a window for me from the white man's sphere into the world of the Pima. A well-trained professional, she was committed to the Pima way. For instance, she was devoted to the collection and preparation of traditional foods, and much of my exposure to these things was thanks to her. (Also, she was a great dance partner.)

Sally was an articulate person in both Pima and English who always spoke her mind. Active for decades in tribal politics, serving many years on the tribal council, she attempted to infuse into it some sense of an environmental ethic. This did not

always win her friends either among tribal politicians or among white entrepreneurs accustomed to taking advantage of Indians.

Sally earned a master's degree in health education at the University of California at Los Angeles and worked to raise the awareness of her people in the areas of diet and maternal health care.

Sally's parents, Joe and Ruth Giff, were major consultants for this book.

Myrtel Mails Harvier
1902–1993 (Sacaton)

I had heard that the Senita (*Lophocereus schottii*) that botanists had collected in the 1930s and 1940s still grew in Myrtel's yard. One day I paid her a visit. She was just a bit of a woman, nearly blind, but with a clear mind and a powerful voice. Her late husband, Frank, had been a Baptist minister. Myrtel at 85 was a model hostess, concerned that we had something to eat and were comfortable. She knew the colony of *cheemĭ* (Senita) growing in the yard and its history and gave a grand account of how her grandmother had used the little puffballs on her injured finger when she was a child. Although she was at home with Pima, her English diction was a model of precision. "I will pray for you," she said as we left, and I am sure she did.

Herbert Narcia
1921– (Komatke)

Holy Family Church in Blackwater is an artless box of a building. An older Pima with a crew cut plays a portable electric organ and hurries through some Gregorian chant for the service. Herb Narcia, the organist, compensates for being the sole choir member by projecting a loud, clear voice. Later he proudly tells me that he can still belt out a good *Missa de Angelis* when his services are called for.

Herbert was born and raised in Komatke, where he graduated from St. John's Indian School. After attending Lamsdon Business College in Phoenix, he became a specialist in pumps and moved to Blackwater, within the San Carlos Irrigation Project area. Both he and his sister, Mary Narcia Juan, married into local families at the far eastern edge of the reservation.

Their grandfather was the locally famous Keli Manol, Old Man Manuel,

a Tohono O'odham *maakai* 'shaman'. Apparently he knew techniques of both the "first aiders" and the shamans working with *kaachim mumkidag* 'staying sickness', and Herb had some peripheral knowledge of his grandfather's treatments, although he never learned the techniques himself.

Herb is talented as well in Pima songs, which he sings with clarity and vigor. He is highly conscious of the obligations that go with being a traditional singer. "The old men would say, 'When you sing, you never change the words of an Indian song—sing it just like you learn it.' But I guess whatever they're afraid of has lost its power." He sings for a local Pima dance group and has a number of young apprentices learning his songs. Some of his songs about plants appear in this book.

Other Piman Consultants

A number of Pimans who are not Gila River Pima contributed information on plants, helping to put this work into broader perspective. Some of these I worked with for long periods of time (for eight years in the case of Pedro Estrella), others for only a few days.

Frank Jim
Tohono O'odham; 1916– (Charco 27)

In high school I taught a number of Frank's children. For about the last 10 years he served as my "outgroup comparison." When the family lived off-reservation at Roll, in Yuma County, I would often stop to visit on the long trips between San Diego and Pima country. Frank always made me welcome and treated me almost as if I were one of his boys. "If we are gone when you come back through, here are the keys to the house. Up here is the key to the freezer." That was for any specimens I might have collected. "Make yourself at home. There's food in the refrigerator." He would always tell me this before I left the house.

Frank had spent a good part of his younger life as a cowboy in the Ajo Mountain region, where he learned much about the local flora and fauna. And his knowledge of ceremonialism was extensive as well. Although he was not a *maakai*, he knew many songs and could sing with others in certain curing sessions. (Piman songs may be sung by a group of healers who know them.) One night he casually brought out his box of sacred healing objects and spread them out on the kitchen table, explaining what type of "staying sickness" each was used for. Then he divided them into two groups. "These I know the songs for, and I can sing with them. These I don't know." When his own family has problems with *kaachim mumkidag,* he takes them to a *maakai.* (These curings, I might add, do not involve any herbal remedies.)

Some Gila Pima folk taxa were not well known in the north, either because they were more widespread in Tohono O'odham country or because they were trade items from there. Frank was helpful in the eventual botanical identification of *siv u'us,* Bladder Sage *(Salazaria mexicana).*

Pedro Estrella Tánori
Névome, Pima Bajo; d. 1988 (Onavas)

After some years of searching fruitlessly for native speakers in historical Pima settlements in Sinaloa, Sally Pablo and I finally located two in the village of Onavas on the Rio Yaqui, in southern Sonora. It was in this colonial settlement, probably in the 1660s, that the *Vocabulario en la lengua Névome* was composed (Pennington 1979).

When we arrived, we inquired about Pedro, whose name had been given us in Tónichi as a native speaker. Someone pointed out his house. A little man with a shock of white hair responded, and we explained our intention of wanting to speak Pima with him. "*Pues, vamos a practicar,*" was his instant response, as he pulled up some stools for us on his dirt floor. I did not speak Pima, and Sally did not know Spanish, but the three of us soon had a lively conversation going. The soft-spoken Pedro found almost anything amusing, including the differences between his dialect and Sally's. After several centuries of separation, the languages had drifted apart. It became evident to me after I listened to the two for some time that, with three major sound shifts, Pima Alto could quite serviceably be converted into Pima Bajo.

Over the years, I managed to make two or three annual trips to visit Pedro and collect plants throughout the neighboring country, which is Sinaloan Thornscrub vegetation. We compiled a list of about 350 plant folk taxa, but some of these are back-translations from Spanish to Pima. (Pedro was not hesitant about doing this.)

This gave me the most complete corpus of plant names from any other Piman group to compare with those from the Gila Pima. The cognates and contrasts are incorporated into the individual folk generic accounts of this work.

Pedro was probably in his early 80s when we first began work. So he must have been in his early 90s when he died a month after visiting the Gila River Pima.

María Córdoba
Névome, Pima Bajo; d. 1986 (Onavas)

María was the only other speaker of Pima Bajo living in Onavas by the time we arrived. She and Pedro had both helped John R. Cornell with language studies that were eventually compiled by Ken Hale into the "Breve vocabulario del idioma Pima de Ónavas" (unpublished).

María was a tiny woman who lived with her family in the older, more colonial part of the little pueblo. She was always dressed in black and was working whenever we called. In contrast to the more impish (but always proper) Pedro, María was a grave woman. She would always remind us that she had little opportunity to speak Pima and so had forgotten much. Nevertheless, she would reflect, then answer in a clear and surprisingly powerful voice. She never went into the field with me, as Pedro frequently did, but I would occasionally bring her plants to identify. In a number of instances, she was the sole source of the names for flora and fauna.

As a title of respect, the people in this mixed blanco and Piman pueblo always referred to her as doña María and to Pedro as don Pedro. María was the *maestra* who chanted the ancient Latin hymns during the local Semana Santa ceremonies.

Chapter 3

Historic Pima Habitats

Un muy ameno paiz con sus esteros,
cienegas, carizales mucha arboleda de
sauces y alamos.

Fr. Jacobo Sedelmayr, 1746

The land of the Gila River Pima has been
so radically altered during the three
centuries of European contact that the
aboriginal or precontact conditions can
scarcely be envisioned today. To get a
picture of what Pima country was like, we
must glean from the documents left by
European missionaries, explorers, trappers, gold seekers, military men, boundary
surveyors, itinerants, and early settlers;
the Pimas left no written records of earlier conditions (although some changes
are reflected in Pima oral tradition).
Fortunately, the documents are fairly
numerous. They fall into two time periods,
each reflecting Euro-American political
and environmental changes: the Hispanic
period and the Anglo period.

Hispanic Period (1694–1853)

The Hispanic period further subdivides
into three sections: the initial Jesuit
period, 1694–1767; the Franciscan period,
1768–1821; and the relatively brief period
of secularization following the Mexican
Revolution (which ended with the
Gadsden Purchase), 1821–1853. Effective
Anglo-American influence on the
Pima began in 1846 with the Mexican-
American War.

In general, the Jesuit period provided
the most scholarly relaciones, accounts
containing ethnographic and linguistic
information from contact and early missionization times. These were written
from the biases of 17th- and 18th-century
European clerics whose sympathies with
the Native Americans ranged broadly from
sheer tolerance to scientific fascination,
even if they were unable to make the final
step to outright admiration.

No missions were founded in Gileño
country during the Hispanic period, so
the people there remained on the fringes,
occasionally visiting their southern relatives, who lived within the more pervasive
European sphere. These northern Pimas
freely selected which crops and technolo-
gies interested them; they borrowed words
for some and invented names for others
(see appendix D).

The Europeans made various entradas,
or expeditions, northward from the established outposts in the more southern parts
of Pimería Alta. A major objective, in
addition to cartography, was to fortify the

frontier against enemies of the Cross and the Crown. What attracted the attention of these explorers were locations suitable for the founding of missions: dependable water sources, good land for agriculture, and pasture for cattle, horses, and perhaps sheep. This perspective must be kept in mind when reading the descriptions of Pima country written by any of these men, Jesuit or Franciscan.

The Jesuit Fr. Eusebio Francisco Kino made at least four entradas to Gileño country, beginning in November 1694. On Kino's second entrada, in November 1697, he traveled down the San Pedro River valley, which was then occupied by 11 villages of Sobaipuri, to the Gila River, and then followed the Gila and its "very large cottonwood groves" westward for three days to the Casa Grande ruins (Kino 1919:171–172). On his last entrada Kino traveled up the Gila from the Colorado in the spring of 1699. On the lower Gila, near Mohawk Mountain, Yuma County, lived both the Pima and the Coco-Maricopa. Here Kino remarked, "All its inhabitants are fishermen, and have many nets and other tackle with which they fish all the year, sustaining themselves with the abundant fish and with their maize, beans, and calabashes" (Kino 1919:195). And when he reached the present Pima villages, Kino added, "In some places they gave us so much and so very good fish that we gave it as a ration to the men, just as beef is given where it is plentiful" (Kino 1919:197).

Lieutenant Juan Mateo Manje accompanied Kino on a number of the explorations as a military escort. At the Pima villages Manje wrote in his diary, "On the fourth [of March 1699] we continued toward the east and passing through the settlement of Encarnación [Tussonimo], nine leagues [about 22 miles] after this [area of Maricopa Wells], we slept in a pasture and broad fields, abundant in grass, where our horses ate well. We were informed that the other side of the river is much better for a cattle ranch and for horses; but I pass judgement on what I saw" (Hayden 1965:8). We must not imagine that these entradas were small expeditions. On this entrada, for instance, Kino recorded 90 pack animals, three Europeans, and some servants. When a document says there was ample pasturage it means grass for a large contingent.

A number of relaciones, such as those of Fr. Luis Velarde written in 1716 and Fr. Philipp Segesser in 1737, provide rich ethnographic information on the Upper Pima but little ecological information. The next relación that gives some insight on the environmental conditions of the Gila is that of Fr. Jacobo Sedelmayr, written in 1746:

> Today there dwell in the basin of the Gila, not very distant from the Casa Grande, a branch of the Pima tribe divided into three rancherías. That farthest east is called Fuguissan [Tuquisan]; four leagues down stream in [sic, is] Tussonimo. Still farther on the river runs entirely underground in hot weather, and where it emerges there is situated the great rancherías called Sudac-sson [Sudacson]. All of these rancherías on either bank of the river and on the islands enjoy broad acres for the cultivation of crops. These Indians raise corn, beans, squash, and cotton which they use for clothing. Those of Sudac-sson raise wheat by irrigation.

> Leaving behind these Pima settlements and trekking down stream we came upon broad savannas of reed grass and clumps of willow and a beautiful spring with good land for pasture. We named the place Santa Teresa [Maricopa Wells]. Passing on down river another five or six leagues and keeping it always in view with its willows and cottonwoods, we came to its confluence with the Río de la Asunción [Salt River], which in its turn is formed by the Salado and the Verde. A very pleasant country surrounds this fork of the rivers. Here the eye is regaled with creeks, marshes, fields of reed grass and an abundant growth of alders [sic, willows] and cottonwood. (Sedelmayr 1955:23–24)

Sedelmayr appears to have been one of the few to have explored the area below the Salt-Gila confluence, including the great bend in the river, which he said he discovered in 1744. Moving downstream he noted, "From the confluence to the first settlement the distance is about 12 leagues. The rancheria is one thickly populated . . . with Pimas and Cocomaricopas, who for the most part know both languages."

Thus, at this time there was an uninhabited section of the Gila running for some 24 leagues (55–66 miles) starting at the westernmost Pima settlement of Sudacson.

In 1767 came the decree from King Charles III of Spain suppressing the Jesuits within Spanish holdings in the New World; the Jesuits were abruptly rounded up and incarcerated until they could be returned to their native countries. Some laboring in the Pimería did not obtain their freedom until a decade had passed; some used this time in preparing lengthy accounts of their mission labors.

On his return to Germany, Fr. Ignaz Pfefferkorn, S.J., who had been stationed at Guevavi on the upper Santa Cruz, published a description of the Sonoran province. Below the Casa Grande he said,

> Following down the Gila beyond this, spread along both sides of the river, are the still unconverted Pimas. This tribe is separated into three populous communities, of which the largest inhabits a pleasant, abundantly tree-covered country fourteen miles long and irrigated by aqueducts, which are built from the river to the surrounding country with little difficulty because the land is so level.

> From the habitation of the Pimas it is approximately twelve miles to the above-mentioned Río de la Asunción. The region where this river flows into the Gila is very beautiful, is entirely level, and is exceptionally good for raising all kinds of grain and plants. Both sides of these two rivers are inhabited by the Cocomaricopas. (Pfefferkorn 1949:28–29)

Within a year, Franciscan friars took over the Jesuit missions of Pimería Alta. Two of the most memorable (and unalike) friars figure importantly in this period of Gila Pima history: Fr. Pedro Font and Fr. Francisco Garcés. Font, a Catalán, was musical and artistic and not in good health. He could also be bitingly sarcastic (see Font 1975); if there was anything bad to be said about the Pimas or their country, he was the one to say it.

The Aragonese peasant Garcés, by contrast, enjoyed not only his work, but the people and the country as well. Perhaps our best picture of him comes from observations by Font:

Father Garcés is so well fitted to get along with the Indians and go among them that he appears to be but an Indian himself. Like the Indians he is phlegmatic in everything. He sits with them in the circle, or at night around the fire, with his legs crossed, and there he will sit musing two or three hours or more, oblivious to everything else, talking with much serenity and deliberation. And although the foods of the Indians are as nasty and dirty as those outlandish people themselves, the father eats them with great gusto, and says that they are good for the stomach, and very fine. In short, God has created him, as I see it, solely for the purpose of seeking out these unhappy, ignorant, and rustic people. (Bolton 1930, 1:291)

While stationed at San Xavier del Bac, Garcés made five entradas as far as Gileño country between 1768 and 1775. In January 1774, Garcés made his fourth entrada, accompanying Captain Juan Bautista de Anza on the first overland expedition from the northern Piman missions near Tucson to San Gabriel, California.

On Anza's return trip in late May 1774, Fr. Juan Díaz (Bolton 1930, 2:302–305) wrote of the six villages along the Gila between Sutaquison (Sudacson) on the west and Uturituc on the east: "These are the Pimas whom we call the Gileños, all of whom are confined to the space of three leagues on one bank of the river and the other, because thus united they are better able to withstand the continuous assaults of the Apache." Then traveling three leagues (6.6–7.8 miles) more upstream from the last village they "passed the siesta at a place of plentiful pasturage close by the river," which Anza said was two leagues (4.4–5.2 miles) below the famed ruins.

On his return trip in June 1774, Garcés took the river route northeast of Gila Bend rather than the almost universally traveled jornada cross-country directly to the Pima villages with Anza and his crew. Apparently by this date the inhabitants of the great bend were all Opas, as the Gila River Yumans were then called. Other than remarking on the great abundance of water "almost entirely from the Rio Azul

[Verde] and the Salado," Garcés did not comment on the habitats in the vicinity of the Salt-Gila confluence. On 22 June he encountered the people from Sudacson, who were five leagues (12–13 miles) downstream from their village gathering Saguaro fruit. "I passed through many pastures and reached the village of La Encarnación del Subtaquisson [sic]. There are three pueblos with many people. Doubtless with its pastures, which are scarce on the Gila, this is the best site [for a mission and presidio]." He then visited the large village of Nacub on the north side of the Gila and Tuburs Cabors on the south side, near the river. Garcés delighted in the Gileños and accepted their offer to remain there and rest for a week. They fed him fish and game (Bolton 1930, 2:388–390).

The most detailed information of this period comes from the 1775 trip, which was Anza's expedition to Monterey with a group of colonizers. Because of its large size, the group had to move slowly, giving Anza and the two friars ample time for documenting their observations.

In addition to having general poor health made worse during this arduous trek, Font was suffering from a fever. He did not let compliments about the Pima and their land slip from his pen readily. One evening he noted in his journal, "They are rather corpulent Indians, and are very ugly and black, especially the women. And perhaps because they eat much péchita, which is the mesquite pod ground and made into atole, the tornillo [Screwbean, *Prosopis pubescens*], grass seeds, and other coarse things, when they are assembled together one perceives in them a very evil odor." Still, Font had to admit, "These Pima Indians of the Gila are gentle and of good heart" (Bolton 1930, 4:43). The next day he added, "Since the soil here is thin, a very sticky dust is raised, as a result of which, and of their coarse foods, these Indians are very ugly, dirty, and evil smelling. The Gila River at this place was dry, and water was taken from it by making wells in the sand. Only in the time of floods is it useful for the grain fields and corn fields of the Indians. On its banks it has a continuous cottonwood grove, but the trees are not very large." Another day the feverish Font complained, "The climate appears to me to be very cold in winter and very hot in summer, and from what I saw the region

does not offer the best of advantages. Only on the banks of the river and by the use of much water can harvests such as the Indians reap be obtained; for building there is no timber except that of the grove along the banks of the river, which is not very large; and for cattle and horses the land is very short of pasturage" (Font 1931:34). Evidently by this time the Pima villages were more consolidated because of Apache threats, as Díaz had noted the previous year, for Font reiterated that the Pima were gentle and lived in established pueblos or towns (1931:49–50). Font thought this would be a strategic place for founding missions if a good presidio could be provided, but "aside from the expense of maintaining it, here is the difficulty, that all the country is so lacking in pasturage, as I have said above, not only for the horses but even for the cattle."

The same night, perhaps just a tent away, Garcés wrote, "In gentleness, pleasantness of manner, and aptitude for living together in their villages they surpass all others of their nation. They took care of our wants and feasted us extravagantly, for they have sheep very like those of the Hopi Indians or perhaps the same. . . . They have chickens; horses, also, some of which they bartered with the soldiers for [red] baize. They brought us the drinking-water needed for all of us and they served us in everything just as might be expected from old Christians and very faithful subjects of the King. Tobacco and beads were distributed to them" (Garcés 1965:7).

The views of the two Franciscans are somewhat different. Was the phlegmatic, "Indianlike" Garcés painting too rosy a picture?

In the largest of the tents sheltering this unusual crew, Anza wrote that same night,

The affability and friendly treatment which I experienced from these people in my last expedition I have found repeated on this occasion. They all had the good manners to come to salute me and to prepare a bower or arcade of five naves in which to lodge us, and where they voluntarily supplied us with an abundance of water, wood, and some provisions of the kinds which they use. This good treatment I reciprocated with an abundance of glass beads and tobacco, distributing them amongst all those who assembled, who were more than a thousand. In this pueblo there is a good piece of pasturage, a circumstance to be appreciated because of the usual lack of it. (Bolton 1930, 2:17–18)

Anza thus substantiated Garcés's impression. Most important in our present consideration is Anza's remark on "a good piece of pasturage" at Uturituc, west of the modern village of Sacaton near Casa Blanca. Two days earlier, Anza had written, "We arrived at the Gila River at a site with abundant pasturage and water which by its inhabitants is called Comari." In Pima the word *komalĭ* means 'flat', and the area was somewhere between Sacaton and the Casa Grande ruins, probably the flats southwest of Blackwater village. When Anza spoke of "a good piece of pasturage" and "abundant pasturage" we must keep in mind that he was leading an expedition of 240 people accompanied by 695 horses and mules and 355 head of cattle: he needed enough pasturage each day for 1,050 head of livestock (Bolton 1930, 4:23).

Garcés noted the general lack of pasturage on the middle Gila near the Pima villages, except near Sudacson where "there is enough to keep a presidio supplied" (Garcés 1900:108–109, 1965:8). Apparently these *pastos* (pastures) were on the north side of the river.

Twenty years later, another Franciscan, Fr. Diego Bringas, in the capacity of apostolic visitor, explored Gileño country. He had hopes for establishing missions in these northern fringes of Pimería Alta. In 1796–1797 he wrote a massive report to his potential sponsor, King Charles IV of Spain. Of the Pima villages he wrote,

The natural products of the country as far as to the banks of the Gila are kinds of bushes: chamizo [Four-wing Saltbush, *Atriplex canescens*], ocotillo *[Fouquieria splendens]*, hediondilla [Creosote Bush, *Larrea divaricata*], and others which are unknown. Pasturage is scarce except for some points right beside the river, and in a cienega which is to the west of the pueblos. As for trees, the banks of the river are covered with cottonwoods and willows which are the only timber for construction, although at a distance of 25 leagues almost directly north there is an abundance of pine. Mesquites, creosote bushes, and saguaros are found in the open country, as well as quail, rabbits,

hares, and deer. The river abounds in fish of various species. As for harmful animals, none are seen but the coyote. The fruits which are actually grown and cultivated by the inhabitants are various: beans, maize, wheat, watermelons, melons, squash, and cotton. If you wish to consider others which can be grown there, every species of grain, tree, and legume would do well because of the mild climate and even temperature. The river can fertilize these beautiful tracts of land with its waters. These can easily be conducted anywhere for farming. Even the gentiles steal a portion of its water by means of a poorly built dam which feeds a main ditch along their fields and distributes the water to small fields cultivated by each family. (Bringas 1977:90–91)

Bringas's elegant report was destined never to be delivered to the king. With Spain and England at war beginning in 1790, interest in the frontera dimmed. Then the Mexican War of Independence began in 1810. The missions were secularized in 1813, and the Franciscans abandoned Pimería Alta by 1824, when Spanish-born missionaries were expelled.

There are few descriptions of Pima country dating from the Mexican period (1821–1853). The most notable exception is the journal of James Ohio Pattie (1930). Beaver trappers such as Pattie descended the Gila from New Mexico, working all the way to the lower Colorado River. It was a difficult time for itinerants, however, because the Apaches, peaceful for about three decades, resumed their raiding in 1831.

Anglo-American Period (1853–present)

The Mexican-American War, initiated in 1846, ended with the signing of the Treaty of Guadalupe Hidalgo in February 1848. By this treaty, Mexico ceded to the United States all of Arizona north of the Gila as well as the modern states of California, Nevada, Utah, most of New Mexico, and parts of Colorado and Wyoming.

With the U.S. declaration of war on Mexico, General Stephen Watts Kearny led the Army of the West to California. They descended the Gila, reaching the Pima villages in November 1846. Lieutenant Colonel William H. Emory

left an informative account of their brief sojourn among the Pima and Maricopa (1848:80–85). Somewhere between the Buttes, where Emory noted, "The Gila at this point, released from its mountain barrier, flows off quietly at the rate of three miles an hour into a wide plain," and the Casa Grande ruins, the troops camped near "a long meadow, reaching for many miles south, in which the Pimos graze their cattle." As they moved downstream Emory continued to paint a picture of what he found:

We travelled 15½ miles and encamped on the dividing ground between the Pimos and Maricopas. For the whole distance, we passed through cultivated grounds, over a luxuriantly rich soil. The plain appeared to extend in every direction 15 or 20 miles, except in one place about five miles before reaching camp, where a low chain of hills comes in from the southeast, and terminates some miles from the river. The bed of the Gila, opposite the village, is said to be dry; the whole water being drawn off by the zequias [acequias] of the Pimos for irrigation; but their ditches are larger than is necessary for this purpose, and the water which is not used returns to the bed of the river with little apparent diminution in its volume.

Most of the Maricopa were reported to be mounted by this time, and probably the Pima as well. The local people, at least, knew where to find sufficient pasture for their own livestock.

Henry Smith Turner was another member of the Kearny battalion who kept a diary. After emerging from the Buttes on 9 November, he described the discovery of grass: "Encamped on a good spot of grass, something we had no expectation of finding, having been told by our guide [Kit Carson] we should find no grass before reaching the Pima Village, above which we now suppose ourselves to be 30 miles. We also found good grass at noon today" (Turner 1966:104–105). West of the Pima settlements, apparently among the main branches that converge to form the Santa Cruz Wash near what was later named Maricopa Wells, Turner wrote, "We are encamped on a spot where the grass is excellent, wood sufficiently abundant, and water a short distance off. We have left the

Gila several miles to the north of us, and will not see it again for 2 days. We are now encamped in a slough that empties into the Gila."

John S. Griffin, M.D., another diarist with Kearny and the Army of the West, noted, upon emerging from the Buttes, "No man can imagine our relief, who has not toiled through them for the last forty days on tired and damn contrary mules— the bottom almost destitute of grass and covered with musquite [sic] wood. After marching some 20 miles we found a patch of coarse grass—where we encamped." Again, before reaching Casa Grande ruins, Griffin wrote, "We were agreeably disappointed in finding an occasional patch of grass. . . . About 6 o'clock we encamped in a grassy bottom some six miles above the Pimas village" (Griffin 1943:32).

Captain Abraham R. Johnston, accompanying the Army of the West, kept a detailed diary that parallels the accounts of Emory, Turner, and Griffin (1848:595–602). After emerging from the spectacular cliffs, the Buttes, he noted, "We came upon a fine spot of grass, one mile from the river, where we nooned; all the rest of the plain was naked, except the mesquite, creosote, and other bushes, which covered perhaps one-third of the land." About another 12 miles downstream, Johnston said, "We encamp in soda grass, quite abundant, running a mile or more along the direction of the road."

The area was rich in archaeological remains. Six miles farther the next morning brought them parallel to the Casa Grande ruins, which Johnston sketched, together with its ballcourt and "pyramid," or platform mound. "When I got to camp, I found them on good grass, and in communication with the Pimos, who came out with a frank welcome." According to Johnston's account, there were three grasslands between the Buttes (near modern Price) and the easternmost Pima village somewhere west of modern Sacaton. These grasslands may have been composed of Desert Salt Grass (Distichlis spicata), Big Sacaton (Sporobolus wrightii), or Alkali Sacaton (S. airoides).

The next day Johnston visited other ruins north of the Gila that had a ballcourt and mound. He noted, "We returned towards the village, and found the camp in some of their corn-fields, which are separated by fences, and are all cultivated by irrigation, apparently with care." After

passing through the villages, the army must have stopped at Maricopa Wells before undertaking the jornada to Gila Bend on 13 November. Johnston wrote, "A string of cotton-woods border the river, and throughout the country there are no other trees. The road was dusty and dry; our camp in an extensive pasture, reaching for miles, under the mountains." Although treeless and waterless, "for the first two miles we had a grass plain of salt grass, the ground in places crusted with salt and occasional pools of water."

Five weeks later the Mormon Battalion straggled in after their 40-day passage through the upper Gila. On 23 December, Colonel Philip St. George Cooke wrote, "The march [through the Pima villages] was fifteen miles. The whole distance was through cultivated grounds and a luxuriantly rich soil; there is a very large zequia well out from the river; the plain appeared to extend in every direction fifteen or twenty miles." The troops camped that night at the Maricopa village to the west. Cooke wrote, "The hospitality and generosity of these allied tribes is noted; they feed and assist in every way travelers who are in need" (Cooke 1878:165).

During the California Gold Rush, thousands of 49ers passed through the Pima villages. With each contingent and wagon train were horses, mules, and oxen; finding water and forage for the livestock was essential. With these great numbers it is not surprising that the "hoofed locusts" ate every blade of grass available along the Gila from Pima country to Yuma.

On the last day of May 1849, John E. Durivage neared the river from Tucson, the usual route at that time. He wrote,

After a wearisome ride I saw the wagons and the tall cottonwoods of the Gila, & when within half a mile of it, my tired mule smelt the running water. She picked up her ears, gave one long bray, and struck a bee line for the Gila directly through the thick chaparral. I hung on to her back like death to a deceased African, and away we went like the wind to the banks of the Gila, into which she plunged her head and never raised it till her sides were distended like a hogshead. . . . The Gila at this point is narrow, not more than one hundred yards, and flows at the rate of six miles an hour. The grass is coarse and not abundant. (Durivage 1937:217)

A few days later the company moved west before undertaking the jornada to Gila Bend. "The meadow in which we camped is 3 miles from the Gila and has a number of springs in it. The water is nearly all saline, and one spring is strongly impregnated with sulphur," noted Durivage.

Benjamin B. Harris and company arrived on the Gila in early July 1849. "The Pima Indians meet the men ten or fifteen miles above the [Pima] villages with gourds of water, roasted pumpkin, and green corn. Serving these, they hurried forward for relieving the others. Broken-down or abandoned stock they took in hand for bringing along" (Harris 1960:80). Harris left several brief but important verbal snapshots of the local environment. "Next morning, we descended the rich Gila bottom through a dense forest of mesquite to the Pima Village, meeting relief parties along the way." Soon they were welcomed by head chief Juan Antonio Lunas—"a fine, handsome, statesmanlike, elderman"—who assured the Anglo-Americans they could turn over the horses to his people to be attended to. "So, our animals were surrendered to be grazed and herded at good pasture at a distance of two or three miles and, for the first time since starting, were removed from our sight and supervision."

On 22 August 1849, George Evans camped with a contingent of 49ers about eight miles above the Pima villages. "Here we found a little grass, and our poor animals are once more feeding with a good appetite. . . . The Gila River opposite our present camp is a deep, narrow, and rapid stream of warm, muddy water, with the banks covered with a dense growth of wild willow and weeds, tall cottonwoods, and the low willow tree, known as the water willow [Seep-willow, Baccharis salicifolia]. A dam has been constructed and by small canals the water is conveyed over the bottoms and thrown into the fields. Here they raise nearly every kind of grain that is usually found in a warm climate" (Evans 1945:152–154). The scattered Pima huts extended for 12 miles along the river. "The grass is extremely short and grows upon salt ground, spots of which are perfectly white and covered with salt." About a mile out of camp the next morning, apparently near Maricopa Wells, Evans's group "found a well of water and better grass, and, having forty miles to pass without water or grass, are encamped and remain here until this evening."

A month later, John Woodhouse Audubon, son of the famous artist-ornithologist, arrived on the Gila with 46 companions and what was left of their weary horses, mules, and oxen. Audubon described their country briefly in his journal: "The river bottom here forms a great flat, which was, I think, once irrigated; at all events, it is cut up by a great many lagoons, nearly all muddy. . . . The country is nearly flat, and on the light sandy soil there is found grass, in some places very sparse and thin, and in others pretty good" (Audubon 1906:157).

In October, Lorenzo D. Aldrich and company passed west to the goldfields. By this time the hordes of gold seekers with their animals had cleared out what remained of the season's pasturage: "Grass being unobtainable, we fed the animals on corn stalks which we purchased of the Indians" (Aldrich 1950:53). By late October the Pima and Maricopa harvests were well under way. "They came into our camp with green corn, pumpkins, melons, kiln-dried wheat, peas, beans, dried corn, etc., sold in baskets or bags at such prices as you could bargain for."

Robert Eccleston (1950:207) arrived at the Pima villages on 18 November 1849. "It was not long before the road came close to the long-looked-for Gila. I rode in to see it, as the cottonwood, willow, etc., obstruct the view, and found a swift stream about 40 feet [12 m] wide, not as clear as I expected to find it, but perhaps this may have been caused by the late rain."

The Quaker 49er Charles E. Pancoast left an account generously spliced with imaginative mixings of details from tribes both east and west of the Pima (1930:243–247). Perhaps his ecological descriptions are sound. Approaching the villages from Tucson, he wrote, "We came to a long stretch of Marsh Land, now dry, but grown up with Reeds and Flags, and in the evening came out upon the shores of the Gila River. Here we made Camp, burning dry Reeds for fuel, while our Cattle fed on Flags." Apparently they were then along the banks of the Little Gila, overgrown with cattails *(Typha domingensis),* bulrushes *(Scirpus* spp.), and Carrizo *(Phragmites australis).* "This River, although much less in volume, bears some resemblance to the Missouri, in that it spreads out over much land,

with numerous Channels and sandy Islands, and has a rapid Current of water mingled with mud and sand, which give it a yellow tint. We did not rest on The Sunday, but travelled on through the Wilderness of Reeds, and reached the Pima Villages on Monday at Sundown."

At the time of the California Gold Rush, Mexico still owned southern Arizona, including the Pima country through which the 49ers were traveling. Although the United States acquired lands north of the Gila through the treaty of Guadalupe Hidalgo, which was ratified in July 1848, a satisfactory boundary was yet to be worked out by the two countries. Late in 1850 the boundary commission, headed by General Pedro García Conde of Mexico and John Russell Bartlett of the United States, began the joint survey in El Paso.

Bartlett and his expedition forged one of the most colorful chapters in Southwestern history (Hine 1968). They had sufficient pack mules and horses to haul 160 tons of freight for the survey. Bartlett had the impossible task of managing more than 100 men, from artists and surveyors to carpenters and mule skinners. The literate and artistic Bartlett had to preside over drunken brawls, knifings, murders, and summary courts and hangings. One French artist, who found the sex ratio of the crew greatly to his liking, was castigated by a puritanical assistant surveyor and another crewman, who subsequently had to be pensioned by Congress as "a confirmed lunatic." The Frenchman detached himself from the U.S. camp in a dispute over wages, took up residence with Conde's brother, and then vanished south of the disputed border, taking his valuable sheath of paintings along with him. Somehow Bartlett kept together the rest of his centrifugal crew and in the summer of 1852 spent more than two weeks among the Pima and Maricopa, while he and his remaining artists sketched, painted, and wrote. These were the first graphic portrayals of Gileño country. Traveling at night to avoid the excessive heat of day, Bartlett and part of his crew completed the jornada from Gila Bend to the villages on 30 June 1852:

At daylight we passed the southern end of a range of mountains [Sierra Estrella] which extend to the Gila, terminating near the mouth of the Salinas

[Salt] River; and at half past six we reached some water holes [Maricopa Wells], about a mile from the first Coco-Maricopa Village, thus making the journey of forty-five miles in thirteen hours. . . . It was indeed a pleasant sight to find ourselves once more surrounded by luxuriant grass. Although we had met with a little salt grass *[Distichlis spicata]* in one or two places on the march, which no animal would eat if he could get anything else, we had not seen a patch of *good* grass since leaving our camp at San Isabel *[sic],* fifty-six miles from San Diego [California]. . . . As it would yet be several hours before we could look for the wagons and the remainder of the party, we turned the mules out to luxuriate on the rich pasture before them, and creeping under some mezquit *[sic]* bushes, soon fell asleep. . . . The water here is found in several holes from four to six feet below the surface, which were dug by Colonel Cooke on his march to California. In some of these holes the water is brackish, in others very pure. The Gila passes about two miles to the north; for one half of which distance the grass extends, the other half being loose sand. (Bartlett 1854, 2:210–211)

The remainder of the crew took the long way around, upriver to the mouth of the Salt, then southeast to the villages, to survey the entire boundary. They reported this route to be thickly wooded and unsuitable for wagons.

There was grass but no shade at this camp, so the crew, accustomed to the luxury of a daily swim in the Gila, persuaded the commissioner to strike camp and move the several miles to the river, even though the local people warned that no grass would be found there: "After crossing a deep arroyo of sand [Vekol or Greene's Wash?], which is filled by the river at its floods, and pushing our way through a thick underbrush of willows, we at length reached the bank of the [Gila] river, when I found the statement of the Indians too true. There were many fine large cottonwood trees, beneath which we stopped, and which afforded us a good shade from the scorching rays of the sun, but there was not a blade of grass to be seen, and, what was worse, *the Gila was dry!*" (Bartlett 1854, 2:214–215).

During the sojourn Bartlett discovered more about the local condition of the Gila:

The dryness of the river was produced by the water having been turned off by the Indians to irrigate their lands, for which the whole stream seemed barely sufficient. It is probable, however, that, with more economical management, it might be made to go much further. The valley or bottom-land occupied by the Pimos and Coco-Maricopas extends about fifteen miles along the south side of the Gila, and is from two to four miles in width, nearly the whole being occupied by their villages and cultivated fields. The Pimos occupy the eastern portion. There is no dividing line between them, nor anything to distinguish the villages of one from the other. The whole of this plain is intersected by irrigating canals from the Gila, by which they are enabled to control the waters, and raise the most luxuriant crops. At the western end of the valley is a rich tract of grass, where we had our encampment. This is a mile or more from the nearest village of the Coco-Maricopas. On the north side of the river there is less bottom land, and the irrigation is more difficult. There are a few cultivated spots here; but it is too much exposed to the attacks of their enemies for either tribe to reside upon it. (Bartlett 1854, 2:215, 232–233)

Twelve miles above the easternmost Pima village and cultivated fields, Bartlett made the last camp on the Gila before the 90-mile jornada to Tucson. He noted, "The river was here much contracted, with steep banks fifteen feet high, and completely overhung with willows and cotton-woods, the latter from the opposite banks, meeting at the top. Its width was less than fifty feet, and its greatest depth did not exceed nine inches. It moved sluggishly along, was well charged with mud, and uncomfortably warm to bathe in. The bottom lands were three quarters of a mile in width on the south side, where we encamped, with a rich soil, and filled with mezquit trees" (Bartlett 1854, 2:260).

George Thurber was botanist for the boundary commission. He wrote of the Saguaro, "The fruit of this Cereus is an important article of food among the Indians of this region, who collect it in large quantities and roll it into balls, which keep well without preparation" (Asa Gray 1855:305). In Pima country Thurber collected "a curious thorny shrub" which Asa Gray described as the new genus and species *Holacantha* (now *Castela*) *emoryi*, Crucifixion-thorn, as well as a new genus and species of the leguminous tree *Olneya tesota*, Desert Ironwood.

The Gila was not destined to be the dividing line between Mexico and the United States for long. After being recalled to Washington, Bartlett was dismissed in a cloud of acrimony. He had dreamed of a lavish multivolume governmental publication of his survey but instead had to publish his two-volume *Personal Narratives* commercially.

The United States chafed at the proposed Bartlett-Conde line, by which it would lose to Mexico the proposed southern route for a transcontinental railroad. James Gadsden was sent to Mexico to negotiate with the Mexican president, Antonio López de Santa Anna. At length, the Gadsden Purchase was ratified in June 1853, the boundary through the Southwest was established at its present position, and the Pima found themselves, along with most of the Tohono O'odham, under a new foreign government.

Emory, promoted to major, was the U.S. boundary commissioner of the new survey. Groundwork was finished in 1855, and a three-part work soon appeared.

Dr. C. C. Parry of the new boundary commission surveyed the geology of the Gila in March 1855 for a proposed transcontinental railroad. Where the Gila opened into Pima country he noted the alluvial floodplain:

This latter consists of an upper level, supporting a shrubby growth of mezquite, and a lower bottom subject to overflows. On these upper portions the Indians usually construct their dwellings, thus overlooking the lower cultivated fields. The amount of land here capable of cultivation is quite extensive, forming a belt on each side of the river often several miles in width, and extending east and west for 20 miles or more.

The stream of water, then at its average height, (in early March, [1852]) measured about 40 yards in width with an average depth of 2 feet, the volume, however, being considerably diminished by the extensive irrigating ditches drawn from above.

The line of the river bank is at this season set off with lagoons and marshes, and everywhere bordered with a dense willow growth, rendering it difficult of approach. (Parry 1857:20)

The organic content of desert soils is usually slight, from 0.1% to 1.0% under natural local conditions (Fuller 1975). The geologist for the survey, Dr. Thomas Antisell, found that the soil at Maricopa Wells had an upper half-inch-thick layer of humus. The wells were then 7 feet deep, with the water rising at that time to within 2½ feet of the surface. "The influence of this body of subterranean water is marked by differences of vegetation, which, as soon as the hill slope commences, ceases to bear grass, rushes, canchalagua [a gentian, probably *Centaurium calycosum*], and mesquite" (Antisell 1856:137).

In late April 1858 a group of nine Anglo-Americans crossed the waterless 90 miles north of Tucson. One of the itinerants, John C. Reid, left this account: "The Gila river and its course, is told a long way off by the green cottonwoods which fringe its banks. It is a bold, clear little river of uniform volume, and fordable in most places. Its temperature was at mid-day, but little below that of the atmosphere. I have before noticed the fact that this stream occasionally disappeared from the surface channel" (Reid 1858:228). Before undertaking the jornada west to Gila Bend, he wrote, "we had fed our animals to repletion on corn and *alkalied* grass, and induced them to drink as freely as possible of alkalied water." But as with Anza and his expedition, there were unfortunate consequences. "Soon after leaving the Maricopa Wells, several of our animals gave evidence of having been injured by the grass and water obtained there. None suffered so much as my own, a mustang mare purchased at Calabazas. The alkali depleted and enfeebled her so suddenly that I was forced to walk and lead her much of the first hundred miles" (Reid 1858:230–231).

When Pimería Alta changed hands in 1824, the Mexican government had recognized the land rights even of indigenous peoples as they had existed under the crown. With the Gadsden Purchase, the Gileños wondered about these newer, even paler foreigners: "A few of the old braves and head men, were importunate to learn our opinion as to whether Uncle Sam would confirm their titles, made by the Mexican Government, to the land occupied by them" (Reid 1858:229).

In February 1859 an act of Congress appropriated money for a survey of Pima and Maricopa lands under Colonel A. B. Gray. The initial reservation (confirmed by executive order the same year) comprised 64,000 acres, or 100 square leagues, of actually farmed land. The Pima and their neighbors the Maricopa had possession of a strip of land 4 miles wide extending 25 miles along the Gila. Pima leaders were puzzled. This, they noted in wonderment to the Anglo commissioners, was hardly what they needed for their grazing, fishing, hunting, and gathering. They *used* a much greater area than what they had under actual cultivation. The present reservation was subsequently built by increments. For various reasons (see chapter 5) the water supply began to fail, and the Pima established new villages in areas not included in the original reservation survey. The survey of 1869 added 81,140 acres, and that of 1876, after the Pima extended their villages upstream, added 9,000 more acres (Russell 1908). By 1990 the reservation totaled 371,933 acres (fig. 3.1).

Works soon appeared promoting the newly acquired territories in the West. For instance, Captain William F. Colton praised the grazing potential of the Gila Valley (1869:316–317), saying that "east of a line drawn from Baboquivari Peak to the Gila above Sacaton" the land "[is] covered at all times of the year with a magnificent growth of grama grass [*Bouteloua* spp.]" while west of this line "bunch or gieta grass [Big Galleta, *Hilaria rigida*] is abundant, and furnishes, in addition to the valley grasses, excellent grazing."

Colton was one of the few to give details on the seasonal change in water level in the Gila. "During the months of July and August a few showers cool the heated traveller, and give a temporary freshness to the vegetation; and during the month of December one or two heavy rains may be expected, which raise the streams, and sometimes flood portions of the valley. At such times the Gila River, at the Pima villages, is from 50 to 75 yards wide and about 10 feet deep, while near its mouth it attains a width of 150 yards, with a depth of about 12 feet" (Colton 1869:316–317).

Lands adjacent to the taller riparian timber, dominated by grand cottonwoods, were either cleared fields or mesquite woodlands. One bosque in particular merited attention by early travelers because it lay between Casa Grande ruins and what later became Sacaton. Harris (1960:80) noted this "dense forest of mesquite" upstream from the Pima villages in 1849. In July 1852, Bartlett (1854, 2:271–272) gave this picture of the bosque: "This bottom is a continuation of that occupied by the Pimos, although much narrower, it being only a mile in width. . . . It is thickly covered with mezquit-trees from twelve to twenty feet in height; among those on the plateau, the tall and graceful petahaya [Saguaro] occasionally thrusts forth its thorny arms, like so many solitary columns or giant candelabra." He found the ruins "rising above a forest of mezquit."

In 1859, Samuel W. Cozzens (1874:189) left the Pima villages with two local guides to visit the ruins. He wrote, "Keeping along the bank of the river, travelling through dense groves of mesquite and cottonwood, we made during the day about eighteen miles, and camped at night in a beautiful grove on the banks of the Gila." They had several miles yet to go before they reached the mother acequia feeding the ruins, which they "traced for a long distance, through a plain . . . overgrown by mesquite." Cozzens recognized that the mesquite that had infiltrated the long-abandoned Hohokam fields was different from the dense bosques along the river.

Likewise, in 1863, author J. Ross Browne visited the Pima villages in the company of various local notables of the period, both Indian and white. Some of the group made a side trip to visit Casa Grande ruins. Brown noted, "Following the banks of the river next morning through dense groves of mesquit, keeping in view, a little to the left, a peculiar conical peak, which forms a prominent landmark, we travelled some eight or ten miles"; before reaching the ruins they "struck the remains of an ancient acequia, very large and clearly defined. Mesquit trees, apparently falling into decay from age, now stand in the bed of the main acequia" (Browne 1869:114).

Still later in the century, around 1875 or 1880, Herbert Brown (1906) reported attending a Pima-Maricopa ceremony in this "mesquite forest" between Casa Grande ruins and Blackwater village.

Pima fields accounted for a great deal of the anthropogenic habitats along the middle Gila. There are no reliable estimates of the amount of land aboriginally under cultivation on the middle Gila or of how many Pimas once lived there: counts early in the Jesuit period appear to have followed disastrous waves of Old World diseases. About 3,000 acres were under cultivation in 1854, but this was after the Pima and Maricopa were stimulated by the outside economy, supplying troops and 49ers with food and forage. It has been estimated that in historic times subsistence farming required between 2 and 5 acres per Pima family (Castetter and Bell 1942:54, 132), and Pima families at the turn of the century were estimated to cultivate between 1 and 5 acres each (Russell 1908:87). A few years later, C. H. Southworth (1914–1915) plotted nearly every field on the reservation; family fields then ranged from 5 to 15 acres, and plots were 100–200 "steps" in width, a step being equivalent to about five feet (Russell 1908:88,93). Euro-Americans marveled at the richness of these fields, which were continuously under cultivation without crop rotation or manuring (Cozzens 1874:186; Russell 1908:86–87).

Fences probably date from the introduction of European livestock. Pima folk taxonomy recognizes several types of fences: *sha'i koli* (brush fences), *wiis koli* (branches piled up), and *vainam koli* (barbed wire). Stick fences, *uus koli*, are mentioned in Hispanic documents dating at least to the Franciscan period (for example, Font in 1775). One reference from the mid 19th century states that there were living fences, as well. Gray wrote, "It is astonishing with what precision they construct their *acequias*—irrigating canals—some of them the *acequias madre* of very large size and without the use of leveling apparatus, but simply by the eye. Their gardens and farms too are regularly ditched and fenced off into rectangles and circles with hedges and trees planted as if done by more enlightened people" (Gray 1856:85). Such fence rows extended the riparian oasis effect far out onto the terraces of the Gila. Only a few years later, Cozzens also noted, "They have nearly a thousand separate enclosures, which are divided by very excellent fences, made of crooked sticks and mesquit" (Cozzens 1874:187). Cozzens does not say whether the mesquites were living trees or the

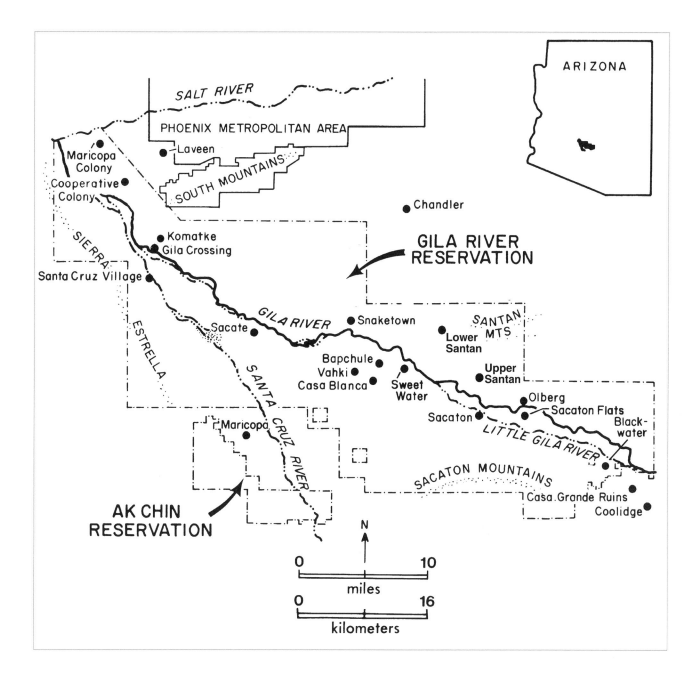

Figure 3.1 *Gila River Indian Reservation*

trimmed thorny branches; young mesquite saplings are stoutly armed and well
suited for fencing.

Joseph Giff (see chapter 4) described
these fields as completely walled areas
with small entrances through which the
worker had to crawl on hands and knees.
In the humorous episode he recounted, an
intruder was unable to turn around within
the passage to make a retreat. Less humorous were the 1838 and 1840 calendar stick
accounts of a Pima worker within a field
being unable to see Apache raiders outside
his fence trying to get in (Hall 1907).

In 1864 Judge Joseph Pratt Allyn visited
Ammi White at Casa Blanca, leaving us
this account: "From the village to the river,
a half mile to a mile, are the cultivated

fields. This tract is marked off by *acquias
[sic]* and perfectly impenetrable brush
fences, presenting a striking contrast to
the desert that stretches interminably in
every direction: outside, sand and bushes;
inside, the wheat several inches high
and the ground moist and rich" (Allyn
1974:108–109). The young judge encountered the *wiis koli* that Giff described in
his story of the watermelon thief. He
wrote, "These brush thickets around the
fields are pierced by apertures through
which you have to crawl on your knees,
just the reverse of the English style or
stile. It's not either a dignified or comfortable mode as I learned one evening,
when after taking a walk, I started back
in a direct light line and found myself

entangled in a perfect network of alternating ditches and fences, which compelled me to go about twice as far, before
I got home, as I would have had to if I
had gone back as I came."

Summary of Historic Habitats

From this album of verbal snapshots
covering two centuries, a composite
picture of what Gila Pima country was
like before abuses of the land radically
altered the Pima environment comes into
focus. Four habitats in particular stand
out for characterization: (1) the river
and aquatic habitats; (2) the riparian

woodlands; (3) grasslands; and (4) the mesquite bosques. Loss of water resources pushed these habitats to the point of no return (chapter 5).

The River and Its Aquatic Habitats

At one time the Gila ran for much of its length, disappearing below the surface during times of low water flow downstream from the protohistoric villages of Tussonimo (in modern Gila Pima, Cheshoñĭ Mo'o) and Uturituc (Hejel Jeg) but arising again as surface flow at the rancheria of Sudacson (*shuudagĭ shon* 'water base' or 'water origin', signifying the place where the river re-emerges). The middle Gila was well supplied with islands. Even as late as the first decade of the 20th century, U.S. Geological Survey maps show eight islands in Pima country upstream from Pima Butte and Sacate; several were one to two miles long. The Gila was confined to a channel with defined banks (Bryan 1922:72). During low-water flow, fords had to be sought for crossing even on horseback. Spaniards seldom crossed to the north bank even though there were villages and pastures on that side. The river rose twice annually, flooding the islands and the lower terraces, after the spring snowmelt in the headwaters and again during the midsummer monsoons. Flood irrigation of these lowlands and islands was the major, if not exclusive, agricultural method of the Gileños. Before its watershed was damaged, the Gila above its confluence with the Salt was described as being between 40 and 300 feet wide and from 1 to 3 feet deep. Even as late as the 1870s, after watershed damage was well initiated, Hiram C. Hodge (1877:38) could write, "The total length of the Gila, including its many windings, is fully 650 miles. For 400 miles, at low water, the Gila has an average width of about one hundred feet and a depth of one to two feet." (Hodge's figures for the Salt River were twice these amounts.)

Fish were traditionally an important resource for the Gileños. Joseph Giff and Sylvester Matthias from Komatke remembered eight folk taxa of fish and described several other species whose Pima names they had forgotten. Water insects and aquatic birds were significant images in Pima sung poetry. The names of a dozen aquatic birds survive in the Pima lexicon

(Rea 1983). There remains a significant folk taxonomy of aquatic plants and emergent vegetation (10 folk taxa; Groups A and B), and we have no way of knowing the attrition of such folk taxa that may have occurred during the past century of desiccation. By the time of my arrival in 1963 most of the aquatic biota were known only to Pimas at the western villages, where the river lasted longest. In the early 20th century, botanists at the Sacaton field station collected many species of aquatic plants in the nearby Gila bottoms. Their specimens are now deposited in the University of Arizona Herbarium (see chapter 6).

Sloughs (called *cienegas* or *esteros* in Hispanic accounts of Pima country) occurred in various places along the rivers. One, more than three miles long, was on the lower Santa Cruz River, between Maricopa Wells and Santa Cruz village (Spier 1933:20–21).

Some surface flow and riparian habitat remained during the first half of the 20th century. Sylvester Matthias knew the lower Santa Cruz, and Joseph Giff, when young, ran cattle in New York Thicket. According to them there were long sloughs standing all the time. Many small "creeks" emerged along the base of the Sierra Estrella and fed the river. There were great canebrakes of Carrizo *(Phragmites australis)* so thick, Joe complained, that the only way to get through was by following cow trails. Interspersed among mesquite groves were open "meadows" covered with dock *(Rumex* sp.), salt grass *(Distichlis spicata),* and great patches of Yerba Mansa *(Anemopsis californica).* According to Ruth Giff, a major source of crude salt was found here.

Fr. Celestine Chin, O.F.M., told me that they called the Gila bottoms below Bapchule and S-totoñig "Louisiana" because the vegetation was similar to that of the bayou country, with thick vegetation and vines hanging between the trees. Apparently Milkweed Vine *(Sarcostemma cynanchoides)* was abundant here among old-growth mesquite.

Riparian Woodlands

Dense timber, with a canopy predominantly of Frémont's Cottonwood *(Populus fremontii)* and Goodding's Willow *(Salix gooddingii),* fringed much of the Gila. This was mixed with occasional Velvet Ash

(Fraxinus velutina). In places itinerants found this bottom timber dense enough to obscure the view of the river. Above the Pima villages the channel was so narrow that the cottonwood-willow canopy was described as meeting at the top. (This was probably where the river was split into a main channel and the "Little Gila.") In places mesquite bosques must have abutted the riparian woodlands, as they still do above the Buttes, where the Gila is yet a living river (Rea 1983:65–75).

Grasslands

Because there are no local analogs, it is difficult to imagine today grasslands, meadows, and pastures in Pima country. Yet these are attested to frequently in the documents from the Hispanic and Anglo periods. There is a consensus that these were spotty swales, that the entire country was not covered with grasses, as were areas in southeastern Arizona that some itinerants had passed through. These localized grasslands were consistently mentioned in two areas south of the Gila: near Maricopa Wells, and southeast of the Sacaton Mountains. According to Pima oral history and inferences from Hispanic accounts, there was another grassland on the Queen Creek drainage (fig. 3.2). We must not underestimate the sizes of these pastures, which were described as having abundant pasturage for several hundred to more than a thousand head of livestock for a period of days, at least. The eastern grassland was described as reaching many miles southward.

The documentary accounts agree with ethnohistoric information from older Gila Pimas. For instance, Joseph Giff recalled when he was young, working upstream from Santa Cruz village: "I rode down the road or trail—must have been [the Butterfield] stage coach road—looking for cattle to round up. Up at the other end was a village called Nakshel Vavhia ['scorpion well'], where a few families lived. *Vavhia* [means] 'well', you know. Grass was high all the time—lots of feed [for cattle]. [The place was also called] Yaanos. People from there moved to Ak Chiñ."

In 1991 I asked Sylvester Matthias about grasslands. He knew of them formerly above Santa Cruz village and New York Thicket. "Yaanos is right close to Hidden Valley Tavern. Real name is Nakshel

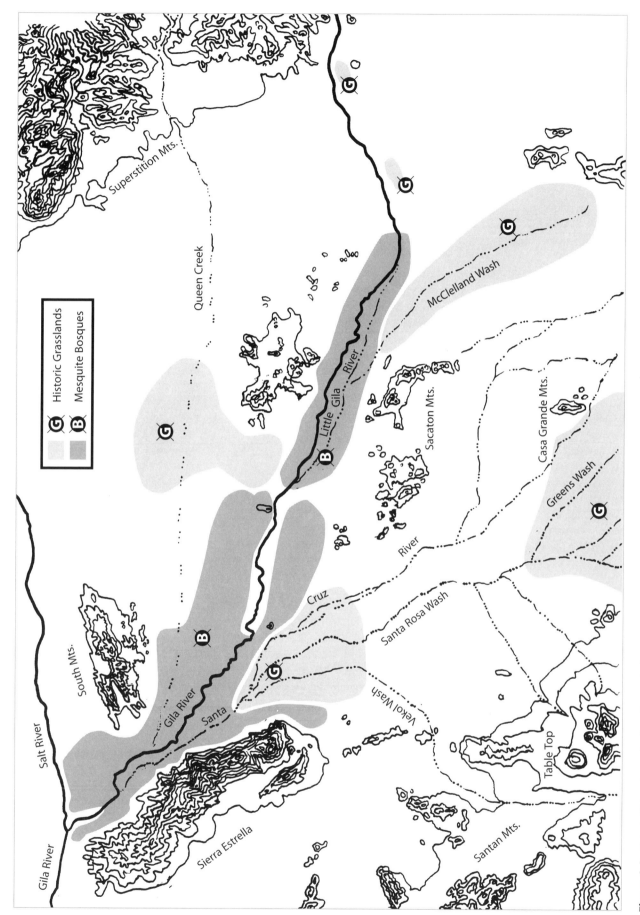

Figure 3.2 *Extent of historic grasslands and mesquite bosques*

Vavhia 'Scorpion Well'. Kui Kuup ['closed mesquite']—that's where the mesquite thickets [are] now. Used to be just [an] open place with those streams. . . . It's a long stretch. Creeks, seepage water; the water's just even with the ground. At Hannamkam Wecho ['beneath a place that has chollas'], a little butte there; west of Casa Grande [city] right under there somewhere is good grazing land."

These were the grasslands southeast of the southern end of the Sierra Estrella that were mentioned so frequently by both Hispanic and Anglo travelers. I asked Sylvester if he knew of any north of the Gila.

"Up here where Chandler is—they call it Toota Muḍadkam. It's the name for a grass *and* the name for place, now [called] Chandler. [It's] something like a *vashai* ['grass']. Likes moisture, lots of moisture. Where Chandler is now—Queen Creek floods all that area. I think the stock like it." Sylvester's Toota Muḍadkam appears to be the place north of the historic Pima rancheria Sudacson where Fr. Garcés had noted sufficient grass to support a presidio.

In the quarter-century that I have been working on the Gila River Indian Reservation, I have seen even-aged stands of mesquite replace agricultural fields that have been abandoned by the Indians. Grasslands probably suffered a similar fate. The drawing by John Bartlett sketched in 1852 and John Pratt's watercolor based on it (see Rea 1983:26; Hine 1968:pl. 35) show the area between the Estrellas and Pima Butte as open, with small trees fringing only the immediate bank of the Santa Cruz and larger riparian timber along the Gila in the background. An engraving of Maricopa Wells made several years later by A. H. Campbell also shows an open plain between the wells and the distant Estrellas (Parke 1857:pl. V). These figures confirm Sylvester Matthias's offhand remark that the open areas, presumably grasslands, around Maricopa Wells later closed in with mesquite.

Two factors may account for these grasslands at the low elevations at which they were found (1,000 to 1,200 feet). The first factor is physical. They occurred in fine, deep soils where the water table was high as a result of washes reaching the Gila. In the case of the Maricopa Wells area, various branches of the re-emerging Santa Cruz River converged with the Santa Rosa, Vekol, and Greene's Washes. The

more eastern meadows were perhaps associated with McClellam Wash and a playa off the southeast tip of the Sacaton Mountains (Ezell 1958). Before the advent of grazing, the relatively low watersheds of these washes were in much better condition to supply underground water, if not surface flow, most of the year.

The other factor is anthropogenic. The Pima practiced fire drives, called *kuunam*, as a hunting method (Rea 1979). I believe these were carried out in these moist, grassy areas—away from the cultivated fields and rancherias, which had brush fences and houses that would have easily ignited. Choice rodents, especially the cotton rat, and other small game would have proliferated here. Fire drives would have suppressed woody growth, such as mesquites, acacias, and saltbushes, that compete with grasses. There are no eye-witness documentary accounts of Gileño fire drives, but there are for Pimans living farther south, and these agree with Gila Pima oral history.

The first account is from the *relación* of the Swiss Jesuit Fr. Philipp Segesser, penned in 1737:

> The grass here is never mowed, one reason being that it need not be stored for the winter. Real, lasting winters are unknown. High grass is sun-dried down to the roots and when the Pimas skirmish with enemies in the mountains and light the grass, wind carries the fire along unchecked until it reaches any large brook, when it is extinguished. Such conflagrations cause great damage to the country, destroying cattle pastures, and so on. The Pimas also customarily start such fires to corner game, although this practice is prohibited and is heavily punished.
>
> It is another matter when dry grass is intentionally set afire immediately before the rainy season. Then, because of the rain, large quantities of young tender grass sprouts forth from the ashes. Then cattle find their paradise and grow fat. (Segesser 1945:157)

Segesser had been stationed at a number of northern Piman missions—San Ignacio on the Rio Magdalena, in northern Sonora, and Guevavi and Bac on the Santa Cruz River, in southern Arizona.

The German Jesuit Fr. Ignaz Pfefferkorn, who was stationed at several northern

Piman missions, left an even more detailed account of these fire drives:

> In various places in Sonora there are large areas covered with zacatón. This thick brush is infested with large numbers of rats and mice which the Sonorans sometimes hunt. Twenty or thirty and sometimes more Sonorans assemble and surround a given circle of brush. They start fires, setting the dry brush ablaze in a circle, and the animals hidden therein are forced to take flight. As the fire advances, the animals retreat more and more to the center and the Indians in turn close the circle on them. In this manner the hunt is continued, until finally a large number of rats and mice is driven together into the center. Of these, the heat has already killed some and burned others; the rest are killed by the Indians with clubs. Then the distribution is made. Each Indian fastens to a string by their tails the mice which have fallen to his share. He hangs this string over his shoulder like a bandoleer. Thus attired, shouting and leaping joyously, the entire company returns home, where the game they have brought affords them a splendid feast. (Pfefferkorn 1949:198–199)

The Jesuit then added, "An Indian wishing to make me a present once offered me some pieces of this booty. When I begged to be excused because I was not used to such delicacies, he wondered that I was so ignorant of dainties."

Several points merit emphasis here. The sanction on the fire hunts that Segesser noted reflects the strong European bias against burning the landscape. Fire suppression, which has continued to be a strong element in Anglo-American land management, probably helped to eliminate the Gila Pima fire drives in their historic grasslands.

Also, Pfefferkorn specified that the drives were conducted in areas with *zacatón*, which he described carefully: "Zacatón in shape and in color is like straw, though it is noticeably thicker. Also it does not grow in single shafts but rather in shrub-like bunches. It grows to a height of more than three ells [about six feet]" (Pfefferkorn 1949:192). What he is describing is Big Sacaton (*Sporobolus wrightii*), a plant Ruth Giff and Sylvester Matthias recalled as still growing along the lower Salt River in their youth.

Mesquite Bosques

Bosques or almost pure stands of Velvet Mesquite *(Prosopis velutina)* occurred across Pima country on the lower alluvial plains of the Gila, Salt, and Santa Cruz Rivers as well as in spots along the major washes. Anglo-Americans usually described these bosques as being dense. The Pima resorted to these woodlands to collect one of their major wild foods, mesquite pods. Old bosques were the major source of the heavy construction posts the Gileños used in their round houses, ramadas, storage sheds, and community meeting houses. Pimas who consulted on this book recalled that there were bosques in the flats between Blackwater and the Sacaton Mountains, along the lower Queen Creek drainage at Interstate Highway 10 west to Gila Crossing village, and along the north side of the Gila from Maricopa Road to the Gila Crossing Presbyterian Church (fig. 3.2). Mesquite stumps still standing in these areas are mute confirmation of the Pima oral history (see photos in Rea 1983:35, 39). Another thicket starting near Maricopa Wells and extending between the Santa Cruz and Gila Rivers nearly to Gila Crossing was so large that the Pima, with their characteristic dry humor, called it New York Thicket. Ornithologist Johnson A. Neff of the U.S. Biological Survey spent the summers of 1938 and 1939 here studying the breeding of White-winged Doves. He described his study site:

> Near the mouth of the Santa Cruz river southwest of Laveen is a great alluvial flat with intermittent heavy growths of mesquite that is called by the local Pima Indians the New York thicket. In some places it is said to be fully six miles in diameter. Nowhere is the growth continuous, however, mesquite thickets covering several acres being interspersed with openings of similar size. The thickets are connected by strands of mesquite growing along the banks of the numerous washes that meander in from the south. At the margins of the big thicket the growth of mesquite is young and short, but in older sections the mesquite and screwbean reach heights up to about 40 feet.

> The entire thicket is on the Gila River Indian Reservation, a few miles east of the Sierra Estrella mountains, along the base of which is a narrow belt of giant cactus that furnishes food for the white-wings. (Neff 1940:281)

The land occupied by the Gileños, and before them the Hohokam, is a composite of mountain ranges, bajadas, and floodplains. The flatlands were once a rich and varied mosaic of bosques, saline flats, sand dunes, grasslands, marshes, and riparian timber. Superimposed on the natural vegetation were anthropogenic habitats, the result of centuries of land manipulation by native people. By a network of canals, ditches, and living fences enclosing acres of cultivated fields, the oasislike riverine country of the middle Gila, lower Salt, and lower Santa Cruz Rivers was extended across the lower terraces of the floodplains.

Chapter 4

The Pima Cultural Ecosystem

Even the most primitive tribes have a
larger vision of the universe, of our place
and functioning within it, a vision that
extends to celestial regions of space and to
interior depths of the human in a manner
far exceeding the parameters of our own
world of technological confinement.
Thomas Berry, *The Dream of the Earth*

An ecosystem is normally thought of as
the complex formed by the interaction of
environment (the nonliving component:
earth, air, water, sunshine) and the biotic
community (the living component, or
biological organisms). It is a system
because the plants, animals, and abiotic
processes are interdependent and mutually
sustaining. Ecologists generally leave out of
this formula people and human cultures.

Are people part of the ecosystem?
For instance, can we speak of a city such
as the Pimas' neighbor, Phoenix, as an

ecosystem? Cut off its supply of imported
petroleum, piped-in water, and trucked-in
food, and this city of nearly two million
would wither in a very short time. Allow
Phoenix to grow, and it does so at the
expense of the naturally occurring assemblage of plants and animals. Phoenix is
neither mutually sustaining nor interdependent with the local biota. It is not a
community in any ecological sense but
a body kept alive in the desert by high-
powered electrodes and intravenous
transfusions. But among societies there
are many different levels of cultural
complexity. Some human communities
can and should be considered as cultural
ecosystems, not just as populations.
This concept has both metaphorical and
biological reality.

From myth to medicine, the Pimas'
metaphor of themselves was as part of
the desert, a component of it rather than
something separate—and especially not
as superior to it. Personal and community
health were viewed as part and parcel of
the larger natural community. Instances
of this cultural self-perception are found
throughout the individual accounts of
plant species, the folk genera.

In a biological sense, as well, we can
view the riverine Pimans who lived in
dispersed rancherias as a cultural ecosystem, an interdependent complex forming
a functioning whole in the Sonoran
Desert. This cultural system increased the
density and biological diversity of plants,
fishes, reptiles, mammals, birds (Rea 1983;
see also Nabhan and associates 1982).

The Pima rancheria ecosystem no
longer exists in any viable form today, but
this sketch is based on conversations with
Pimas born early in the 20th century who
remember it. The Pima ecosystem was

based on three subsistence strategies: (1) the harvest of wild crops, (2) double- and triple-cropping agriculture, and (3) fishing and hunting.

Two wild crops were the backbone of the gathering strategy: Saguaro *(Carnegiea gigantea),* a tall, columnar cactus; and Velvet Mesquite *(Prosopis velutina).* Saguaro cactus fruits ripen in late June or early July. The fruit usually was produced in greater quantities than the O'odham could ever use. The harvest was the time for the annual wine feast, a religious celebration intended to bring the clouds and to continue the rains necessary for planting and growth. Syrup, jam, and seed meal were additional Saguaro products that could be stored for use throughout the year. Saguaro productivity was occasionally poor in spite of good flowering, as in the summer of 1991 on the reservation. Early summer rains could also ruin the oncoming crop; the O'odham hoped for the rains to come after the crop was harvested.

Velvet Mesquite produced two annual crops, one early in the summer, the other late in the summer. When the dry pods fell, they were gathered in enormous quantities and stored in rooftop granaries. The flour produced from the pod mesocarp, rich in proteins and carbohydrates, was used to make bread, pudding, and a whole series of other dishes and drinks. Although Frank Russell (1908:66) was told that the mesquite crop in certain years might fail, some of my consultants contested this notion. Various factors might adversely affect pollination, producing a negligible crop, but in a few months the trees would reflower to produce a late summer or fall crop. Nor would the lack of rain guarantee crop failure; the water table was high, even if surface flow in the Gila or Santa Cruz should cease, and the mesquite is a deep-rooted tree.

Agave was a third important crop, though its harvest usually took the Gileños far from the safety of their rancherias into Apache country. Cutting the heads or hearts for pit roasting was labor intensive, but large quantities of very sweet, storable food resulted. Dried mescal was also available through trade with the Tohono O'odham, who had more ready access to several species of agaves. In historic times, the only place in Pima country proper where agaves are found is in the Sierra Estrella.

An annual cycle of harvesting wild greens (primarily agrestals, or disturbance plants) was significant and persists among more traditional families (see chapter 7 and Group I). Besides greens, the riverine Pima harvested many other wild crops, some still popular in the late 20th century, including the flower buds of Buckhorn Cholla *(Opuntia acanthocarpa)* and the fruits of wolfberry *(Lycium* spp.). In most years the cholla blooms prolifically, and in late spring the wolfberry bushes are festooned with small tomatolike berries. I have known only one year when one species or another of wolfberry did not produce well—even with a drastically altered water table.

Perhaps it is in agriculture that the riverine Pima showed their finest tuning to their desert environment (Castetter and Bell 1942). Aboriginally they double-cropped, planting first in the spring as soon as the danger of frosts was over and while the river ran high from snowmelt in the watersheds, and again during the late summer rainstorms. With scarcely any manipulation of the terrain, floodwater farming was possible on the lower terraces and on the several large islands that were present in the Gila. At the time of contact, the Sobaipuri on the San Pedro were irrigating their fields with ditches, as were other Pimans on the Altar and Santa Cruz Rivers (Bolton 1936:246). However, there is no documentary indication of canal irrigation among Gileños until 1744 (Winter 1973): 50 years after Kino's initial entrada, Sedelmayr (1939:106, 1955:23) mentioned irrigation at a single village, Sudacson, where the underground Gila re-emerged and where wheat was being grown. Although this difference between Pimans on adjacent streams seems anomalous, one version of the Creation Story says specifically that the Pima originally did not use ditches (Hayden 1935).

Once irrigation was established, water was diverted from the river by means of brush weirs, which was possible when the Gila flowed through a confined bed. Three eyewitness Pima accounts of how these weirs were constructed are narrated in part 2 (see Group D, *Atriplex polycarpa* and *Pluchea sericea;* Group I, *Triticum aestivum*). One earlier Anglo account, the only one written before serious environmental degradation, provides additional insight: "The dams, which serve the purpose of drawing off the irrigating water,

are constructed of old willow trunks and snags; these, in the course of time entangling the loose soil and sediment borne down by the river, furnish a bed for the willow growth, thus becoming more permanent with age" (Parry 1857:20).

Headgates for the main canals were usually several miles above the primary cultivated areas they served. The ditches drew off a considerable volume, sometimes all the surface flow of the Gila (Bartlett 1854, 2:215; Emory 1848:38; Parry 1857:20). The best water, Pima farmers said, was somewhat silty, to enrich the fields. If it was too muddy, the soil surface would crack on drying. Enough water had to be turned in to leach accumulated salts from the soil.

Fields were well fenced, with either a wattle of spiny branches or living plants, such as mesquites, wolfberries, and Graythorn *(Ziziphus obtusifolia).* Often a mesquite or two was left growing in the field; the Gileños appreciated their value as fertilizing agents (see Group E, *Prosopis velutina*). Apparently no other dressing was used.

Also, the riverine Pima practiced some *akĭ chiñ* farming. (This was the only farming strategy available to the Tohono O'odham, who had no permanent rivers in their country.) In the *akĭ chiñ* method, local summer rains collected in the arroyos (the *a'akĭ*) dissecting the bajadas and ran through them onto the floodplains. This water carried along rabbit and rodent manure and the nitrogen-rich leaflets of leguminous trees such as mesquites, Catclaw Acacia *(Acacia greggii),* Desert Ironwood *(Olneya tesota),* and paloverdes that grew in and around these arroyos. Small check dams at the lower end slowed the flow, spreading the water out over the *chiñ,* the 'mouth' of the wash, and letting it soak in. Here on these deltas the crop seeds were planted (Nabhan 1979a, 1983b).

The Pima farmer used two wooden tools. The first was a simple digging stick, called the *eskuḍ,* flattened and sharpened at one end. It reached about as high as the worker's elbow and was made from a straight piece of Graythorn or Desert Ironwood. "It was used in planting maize and other crops, as a lever to pry out bushes when clearing the ground, as a pick when digging irrigating ditches, and in case of surprise it made an effective weapon of defense" (Russell 1908:97).

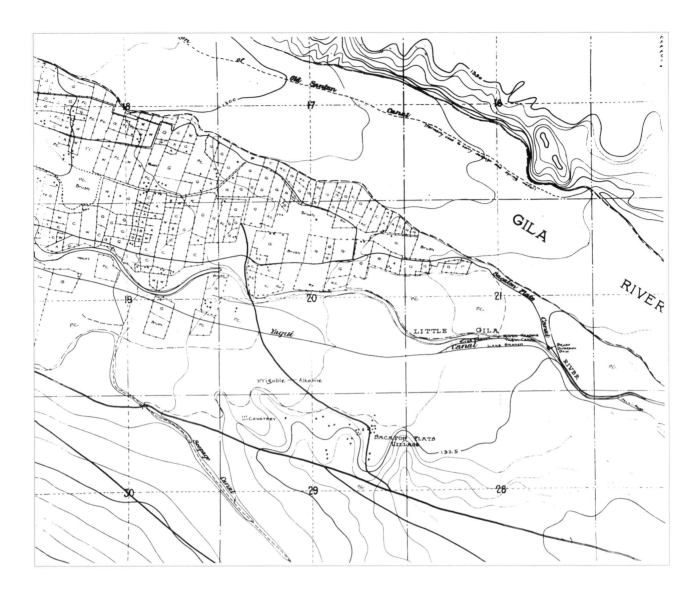

Figure 4.1 *Southworth's map of Pima fields (courtesy Arizona Historical Society)*

The second implement, the *giikĭ*, has been variously termed a 'hoe' or 'weeding stick'. It was about as long as a man's arm, usually made of ironwood (see Castetter and Bell 1942:135–137; Russell 1908:97). It might be compared to a modern machete in both size and shape. To use it, the worker had to be kneeling (see Underhill 1938b:45).

With the introduction of oxen by the Spaniards, a primitive wooden plow was used for scratching the soil surface: it took the same name as the handheld hoe, *giikĭ*. The construction of the wooden plow, *uus giikĭ*, was amazingly similar throughout the rancheria tribes of northwestern Mexico. In the 1980s I found it still in use among the Mountain Pima and the Tarahumara of the Sierra Madre.

In the winter, freezing temperatures can occur from November through at least February or early March. Nevertheless, the Gileño farmer could anticipate an average growing season of 263 continuous frost-free days (King and Loomis 1932). During the hot months of summer (May through September), high temperatures of 105°F (40.5°C) can be expected daily, and during the hottest months, the temperature usually reaches around 110–112°F (43–44.5°C).

The annual rainfall in Pima country is bimodal, with about half of the precipitation originating from gentle Pacific coastal storms in winter, the other half during more violent convection storms (commonly called monsoons) from July through September. The normal annual range of precipitation is about 4–10 inches (90–225 mm), but as little as 3 inches (67 mm) may occur during drought years, followed by perhaps 12 inches (270 mm) the next year.

The Pima had selected for fast-maturing, drought-resistant crop varieties that produced a minimum of foliage and a maximum of seed. The first crop was planted after the danger of frost had passed, in late March or early April. When the mesquite put forth new leaves—*kui i'ivagĭdag mashad* 'mesquite leafing-out moon'—the People knew it was safe to plant. After the first harvest, the second crop was planted, timed to the rise in the river from the summer rains falling in the headwaters. According to widespread tradition in northwestern Mexico (and in Gileño country as well), the summer rains should begin around San Juan Day, 24 June. More realistically they would be expected to come four or five weeks later. The Pima had to make their second planting early enough that crops would have time to mature before the first killing frost in fall. The Feast of the Assumption, 15

August, was considered the latest time to plant safely. The Pima watched the position of the sun setting on the Sierra Estrella. When it reached a certain named peak on its return south (the particular peak differed from village to village), they knew that the second crops must be in.

The double-cropping included a number of major crops. Maize was grown in at least the seven varieties that are still remembered. All of these were 60-day corns, in contrast to two named hybrid varieties the Pima obtained commercially, which are 90-day corns. The Pima maize plant is small, almost sorghumlike, growing to about shoulder height if there is enough water but smaller where less water is available. It produces less foliage than modern varieties and as a result loses less water through transpiration.

At least three species of bean were grown. In addition to the common bean, there was the much more important tepary (Nabhan 1979b; Nabhan and Felger 1978). The tepary is a superbly desert adapted, locally domesticated little bean that produces four times more seeds on just one or two good waterings than the common bean that Euro-Americans are raising on much more water. Furthermore, this bean has superior amino acid content compared to the common bean. The Pima also have their own local landrace of lima bean that has high salinity and drought tolerance. In addition to these, the Pima cultivated as food crops one species of grain chenopod (now lost), at least one species of grain amaranth (also now lost), and two or three species of squash.

When Kino first arrived in Pimería Alta in the late 17th century, he introduced wheat and several other winter crops (legumes) that made triple-cropping possible. Wheat became a significant Gileño crop by 1744. So in postcontact times, the Pima farmed on a year-round basis. The Gileños adapted what they liked, selecting over time for the heat, aridity, and salinity of their local desert growing conditions.

Frank Crosswhite (1981:64) noted what he called a "second garden" where tailwater from Pima fields was allowed to spread out to irrigate an area with wild and semiwild greens. "The degree to which it was a separate entity varied. Sometimes it was merely a weed patch at the end of the

field. Often it was rather extensive. This second garden didn't require the intensive cultivation of the planted field and generally was not leached of salts like the primary field. As a result most plants were relatively salt-tolerant. Nevertheless, some agricultural management seems to have occurred." Crosswhite added, "The importance of the second garden was that it provided vitamins and minerals which were not so abundant in the starchy crops of the main field. Also, a well-managed second garden provided food to take home almost every day of the year, whereas the crops of the main field matured and had to be harvested at very specific periods of short duration."

Pima hunting and fishing also showed every indication of being fine-tuned to their desert ecosystem. Fishing was most important. Quite a few species lived in the Gila and its tributaries. Indeed, two species (Colorado Squawfish and Razorback Sucker) attained lengths of three to six feet (1–2 m), and one weighed nearly 100 pounds (6–20 kg). These huge fish sometimes got into the Pima irrigation ditches. Fish were captured in communal drives, in traps, or by hand, but not with the Old World technique of hook and line.

Game animals, both birds and mammals, were taken by a number of techniques that were formally named in Pima (Rea 1979). A man might leave the rancheria quietly and go to the mountains or even distant mesas in search of Desert Bighorn, Pronghorn, Mule Deer, White-tailed Deer, or Javelina. Only certain individuals engaged in this activity, perhaps one in each settlement. Private hunting was almost a religious vocation, like shamanism. It required family support so that a man's fields were attended to in his absence. His support groups at home shared in the kill.

Most Gileño hunting was a social activity, coordinated by the *tobdam*, the town crier or hunt leader. There were several types of social hunting: the *shaada*, the *kuunam*, and the *kuushada*. These were described to me by Joseph Giff and Sylvester Matthias.

The *shaada* was the communal rabbit hunt or surround conducted on foot. It was carried out on flatlands dominated by brush and coarse grass, a vegetation type called generically in Pima *sha'i* (see Group D). Here the dominant plants were

Desert Saltbush *(Atriplex polycarpa)*, wolfberries, Triangle-leaved Bursage *(Ambrosia deltoidea)*, occasional mesquites, and perhaps Alkali Sacaton *(Sporobolus airoides)* and another unidentified grass. This hunt produced primarily jackrabbits and cottontails along with some rodents. Younger men ran the animals toward the center of the circle or to some predetermined spot, killing them as they went, while older, less agile men prepared cooking pits and cleaned and roasted the game. All shared equally in the catch.

The *kuunam* was the fire drive, described by Pfefferkorn in the 1760s. This was held away from the village, perhaps in grassy swales or grasslands where rodents proliferated and the vegetation could be ignited. The main rodent sought was the cotton rat but packrats or wood rats, as well as jackrabbits and cottontails, were also taken.

After the Hispanic introduction of horses, mounted Pima engaged in another social hunt called the *kuushada*, a form of surround that covered a greater terrain than the *kuunam*. A *kuushada* is occasionally still conducted to gather jackrabbits for a feast day. (The word is etymologically related to *kuush* 'to scorch', and it is tempting to suspect that the term was originally applied to the fire drive but became displaced in postcontact times when the hunting methods diversified.)

There were also several types of hunts conducted in fields and along fence rows. Torches of bound Arrow-weed *(Pluchea sericea)* were used at night, particularly in winter, to flush Abert's Towhees, Gambel's Quail, White-crowned Sparrows, and other animals that congregated there. Later, other types of homemade lanterns and even flashlights were used in field hunting.

Two types of traps, a cage trap and a flat lattice-type trap, were in operation at nearly every home. Lark Buntings, White-crowned Sparrows, quails, and doves—attracted by spilled grain and other seed—were the game sought.

Younger boys often passed their time traveling along the riparian timber with slingshot or bow and arrow, hunting doves or other game to supplement their diet. Both boys and girls collected cicadas to eat.

The oasis farming communities maintained by the Gila Pima promoted dietary supplements such as semiwild greens that were gathered in an annual cycle. The same was true for hunting. After fish, the next most esteemed food for the Gileño was the cotton rat, abundant in old fields and weedy areas. Most of the rodent and bird species that the Pima preferred were the ones that proliferated as a result of Pima farming methods, which produced weedy canal banks, living fence rows, dense thorny brush fences, and fallow fields (Rea 1979, 1983). This phenomenon is what Olga F. Linares (1976) has termed "garden hunting." Even the communal hunts and fire drives concentrated on the floodplains (Rea 1979).

As noted in the preceding chapter, Pima fire drives for small game probably maintained and even promoted the local grasslands that historic visitors, both Hispanic and Anglo, reported. Because of fire suppression and a falling water table, the open grasslands and cienegas have been converted to brush steppes now dominated by Burro-weed or Jimmyweed (*Isocoma* spp.), Desert Saltbush, and stunted mesquites.

With the social, economic, and environmental changes that have occurred in the 20th century, the Pima ecosystem is a thing of a past. Only a few relics of the system remain.

Chapter 5

A People without a River

But at first the whole slope of the world was westward, and though there were peaks rising from this slope there were no true valleys, and all the water that fell ran away and there was no water for the people to drink. So Jeweḍ Maakai, Earth Doctor, sent Ñui, Buzzard, to fly around among the mountains, and over the earth, to cut valleys with his wings, so that the water would be caught and distributed and there might be enough for the people to drink.

Pima Creation Story, told by Thin Leather

Even otherwise well informed people often take a stream or river for granted as something that is just "there." They do not see this surface flow as a *process* initiated by the complex interaction of precipitation, vegetation, microorganisms, soil, and geology. Many, including not a few Pimas, think that a river such as the Gila stopped because someone built a dam upstream but do not see this damming of a river as a last-ditch effort to save what remnants of flow might survive from a stream already in its death throes. Rivers are not magical things that just happen. They are the product of a watershed—all the upland area that collects and releases precipitation, both snow and rain, slowly throughout an annual cycle. The health of a river is no better than the health of its watershed. Because the watershed is sometimes far away from the river that most people see and use in the lowlands, this essential component is easily forgotten.

Permanent streams depend, of course, on precipitation that comes more or less on schedule in fluctuating annual amounts. There are short-term cycles in the amount of precipitation, producing wet years and drought years; and there are long-term alterations in precipitation and temperature, climatic fluctuations that are seen in geological time scales, or over the span of centuries. If rainfall were to fall on sheer rock in the desert, the waters—propelled by gravity—would run off to the ocean in a matter of days or weeks; there would be no permanent streams. In effect, this is indeed what happens (usually in a matter of hours) in the arroyos (see chapter 6) when the short and violent summer convection rainstorms strike.

In a healthy watershed such as the historic Gila uplands of eastern Arizona and adjacent western New Mexico, precipitation may first hit the canopy—the leaves, needles, and small branchlets—of oaks, junipers, and piñons or at higher elevations pines, firs, and spruce. Then there are the smaller trees and shrubs of the midstory and understory, and finally the grasses and forbs that protect the soil with a network of fine leaves and roots. Snowfall may pack in shady areas, melting slowly. The vegetation acts as the first "sponge," absorbing water for slower release.

As the slowed precipitation penetrates past this vegetative sponge, it is absorbed by the next sponge, the soil. Soil (like the stream) is often perceived as just there—just dirt underfoot—rather than as something evolving from centuries of vital activity. The soil is the result of the weathering of rock and the decay of plant materials through the activity of micro-organisms and insects and other arthro-pods in the ground. Desert soils are notoriously poor in organic matter, but in the upper-elevation watersheds, the soil is rich in accumulated humus.

Moisture, still propelled by gravity, slowly percolates deeper into porous inorganic layers, the next sponge, to flow over impermeable strata. Dendritic under-ground seepages converge until moisture re-emerges at springs and seeps as surface flow. An embryonic stream is born, to be joined by others of its kind, eventually to become tiny creeks in the higher eleva-tions, then larger streams and finally rivers in the lowlands.

But there is some magic in all this. Watersheds in the arid and semiarid uplands are fragile entities whose very existence is something of a miracle. The woodlands and forest communities as well as the soils that serve as sponges are themselves relicts of cooler and wetter Wisconsin (Late Pleistocene) climatic regimes. In the Holocene—the last 10,000 years—these communities have retreated up the mountains to elevations of about 3,300 feet, or more than a thousand meters (Van Devender and associates 1987). Although they can maintain themselves under the more xeric modern conditions, once removed, they may be unable to replace themselves: ecological succession may be stopped. For example, anthro-pogenic changes in arid vegetational communities may be clearly visible even centuries later, and archaeologists sensitive to the clues from vegetation changes can still locate ancient fields and other human modifications in these semiarid uplands.

Watershed Damage

The watershed of the upper Gila—that is, the area above the San Carlos River (2,500 feet)—embraces 12,020 square miles (Olmstead 1919), of which about half is in New Mexico. After the European colonization of the Southwest, various modifications of the headwaters spelled death to the middle and lower Gila and the Pima way of life (Dobyns 1978, 1981; Hoover 1929; Rea 1983). The disturbances were many but interrelated, and the story of the death of the Gila is complex.

Trapping Out the Beaver

Although North American Beaver are not usually thought of as desert animals, they occurred throughout the Gila drainage from the cool mountain mead-ows and canyons of New Mexico down through the hot country of Arizona to the confluence of the Gila and Colorado Rivers and beyond. Beaver build dams, thus impounding water, slowing runoff, increasing percolation, settling silt, halt-ing stream degradation (Stabler 1985). They were an integral part of the Gila watershed system and had been so throughout the wetter Pleistocene, the last 1.8 million years.

With the Louisiana Purchase of 1803, the United States acquired a large parcel of land west of the Mississippi. Spain still held California, Arizona, and New Mexico, and access to those lands was via passport obtained from the Hispanic governor of Santa Fe. Beginning in 1821 with the Mexican War of Independence, local officials, short on money, were amenable to Anglo-American entrepreneurs with a nose for quick riches from beaver pelts. During the next 10 years, enormous numbers of beaver were trapped from the Gila and all its tributaries. There was no thought whatsoever of a sustained-yield harvest; by 1826 there was a veritable stampede of fur trappers into the Gila watershed (Weber 1971:112), some teams consisting of two to three dozen men. In several instances, James Ohio Pattie (whose account of fur trapping along the Gila is perhaps the most famous, if not always the most accurate) emphasized that they trapped the rivers clean (Pattie 1930:213, 226). In both the uplands and the desert, trappers were often able to take 30 or more animals in a single night. By 1834

the beaver population had collapsed. Although the species was not entirely extirpated, decades passed before local populations could reach anywhere near their original numbers.

Meanwhile, the water control system supplied by these animals was eliminated. As Henry Dobyns (1981:115–116) wrote in his historical analysis of the demise of the Gila, "As beaver dams broke and beaver ponds drained or silted in, those streams accelerated. Then they carried larger volumes of flood water because original storage capacity had been reduced. They became muddier and muddier as floods descended from deforested, hoof-trampled steep slopes."

Mining and Deforestation

Arizona has long been known for its mining industry, yet few are aware of the enormous impact this activity had on the semiarid uplands of the Gila drainage. Because of a lack of manpower during the Hispanic period, mining was probably not a sufficiently large enterprise for wholesale destruction in the watershed. But with the Apaches effectively contained by 1873, Anglo-Americans turned their attention to the rich resources of silver, gold, and copper in Arizona Territory.

Fuelwood was in enormous demand—especially oak and mesquite. The amounts of wood extracted for mining are aston-ishing (Dobyns 1981:96–102). Hardwood stoked the boilers that ran the stamp mills to crush ore; it fueled the smelters as well. As mines were excavated below the earth's surface, timbers were needed to shore up mine shafts. Wood cooked the foods in the mess halls and heated the barracks of the armies of workers who flooded the territory. In addition to stripping the woody vegetation from the steep arid hillsides, mining support crews needed roads and burros to haul it to camp—two more sources of erosion. It was an onslaught with no thought to the short- or long-term consequences to soil or the watershed.

Livestock and Overgrazing

A goal of the Jesuit system had been to establish permanent pueblos of Indians concentrated at mission sites where the *indígenes* could be indoctrinated into the Christian faith, a process called *reducción*

(Matson and Fontana in Bringas 1977:13–14). This required the establishment of agricultural centers, which was no problem among tribes that had been farming the desert for centuries. New crops were added to the native repertory, including some arid-adapted Mediterranean and African cultigens, but the basic technology of crop agriculture was already in place. The major Spanish innovation was the introduction of Old World livestock—horses, mules, donkeys, sheep, and cattle. Fr. Kino, a leading figure in opening Pimería Alta, was an especially skilled cattle rancher. Northern Pimans as well as the neighboring Ópata took well to this new enterprise.

But so did Apaches, who were attracted by the new source of food. With the final Sobaipuri abandonment of the San Pedro valley in 1762 and the Hispanic abandonment of the great ranches in the southeastern section of the state, cattle became feral, free-ranging herds. In 1846 the Mormon Battalion was attacked by a herd of wild bulls on the San Pedro southeast of Tucson, the only "battle" the Saints engaged in on the entire trip from Council Bluffs, Iowa, to California. Men were wounded, but no one was killed; at the end of the "great bull fight" between 10 and 50 or more feral cattle (according to which account you read) lay dead on the banks of the river. There, three years later, the Texas Argonauts (early 49ers from that state) encountered a wild herd estimated at between 5,000 and 15,000 head (Harris 1960). At about the same time, Henry William Bigler (1962:31) reported 4,000 wild cattle and, several days later, bands of wild horses on the San Pedro. Geologist C. C. Parry noted in 1852, "The numerous bodies of wild cattle now running at large over this section of the country [San Bernardino Ranch in southeastern Arizona] are the remains and offspring of domestic herds, now widely scattered and hunted by Indians" (Parry 1857:17).

However, native desert plants and Eurasian herbivores did not co-evolve. As Apache fighter Captain John G. Bourke (1891:40) observed, "The wild grasses of Arizona always seemed to me to have but a slight root in the soil, and . . . the presence of herds of cattle soon tears them up and leaves the land bare." The local flora apparently had little defense against the new predators. Native American horses of

other species, presumably with niches and eating habits similar to the newly arrived species, had disappeared from the continent at the end of the Pleistocene or in the early Holocene, 8,000–10,000 years ago. But cattle were entirely new to these arid lands. They concentrated in the stream bottoms, especially during the hotter parts of the day, devouring succulent aquatics and emergent vegetation as well as the saplings of riparian trees such as cottonwoods and willows. Their hooves, supporting heavy bodies, trampled more vegetation, churning the stream banks into mud holes. Elsewhere they cut trails in dry, newly exposed soils, setting up more erosional processes (Behnke and Raleigh 1978).

The grasslands of the upper and middle Gila were a source of wonder to the Anglo-Americans, who began claiming these watersheds for ranching as soon as the local Apaches were confined in 1873. In his elegant study *A Legacy of Change: Historic Human Impact on Vegetation of the Arizona Borderlands,* Conrad Bahre noted the extent to which the windmill and railroad had contributed to the cattle ranching boom: "In 1881 the Southern Pacific advertised for settlers; soon after, ranchers from overgrazed areas in Texas, New Mexico, and the Mexican states of Durango, Chihuahua, and Sonora began moving their herds into southern Arizona. By 1890 the entire region must have looked like one huge cattle ranch" (Bahre 1991:116). Sheep were reported to have outnumbered cattle before 1892. Bahre (1991:123) concludes, "Probably no single land use has had a greater effect on the vegetation of southeastern Arizona or has led to more changes in the landscape than livestock grazing and range management programs."

Like many other entrepreneurs in the Southwest, cattlemen saw a resource to be exploited. They did not realize that the native grass species had evolved in response to a different suite of herbivores; that plant communities in place in the 19th century had been established during more moist geological times; and that native fire drives might have played a role in maintaining these grasslands by suppressing woody shrubs. Only much later, through the study of the tree-ring record, did scientists show that the Gila basin and its headwaters were sustaining a period of

drought that began approximately in the 1860s and continued for the next half-century until reversed by a wet trend beginning in 1906 (see Fritts 1965). These were all esoteric factors that have become evident in hindsight. But most important, and most devastating, the ranges were grossly overstocked.

The stark realities of grazing the arid and semiarid country have been minutely detailed by historian Jay J. Wagoner, who observed, "The fame of Arizona's unequaled grazing facilities had been widely spread. Hundreds of emigrants were coming into the new cow country to begin an experimental exploitation of the luxuriant grasslands, which would culminate in the deterioration of the range" (Wagoner 1952:38–39). "The primary objective of cattlemen up to 1885 was numbers; overstocking was the inevitable result of unrestricted use of the federal range for grazing."

Even more severe a critic was journalist and resident Arizona ornithologist Herbert Brown, who minced no words in fingering one of the major culprits in the general drama of decline in southern Arizona:

The stock business at one time promised enormous profits and because of this the country was literally grazed to death. During the years 1892 and 1893 Arizona suffered an almost continuous drouth, and cattle died by the tens of thousands. From 50 to 90 per cent of every herd lay dead on the ranges. The hot sun, dry winds and famishing brutes were fatal as fire to nearly all forms of vegetable life. Even the cactus, although girdled by its millions of spines, was broken down and eaten by cattle in their mad frenzy for food. This destruction of desert herbage drove out or killed off many forms of animal life hitherto common to the great plains and mesa lands of the Territory. Cattle climbed to the tops of the highest mountains and denuded them of every living thing within reach. The ranges were foolishly overstocked, and thus many owners of big herds were financially ruined by their covetousness. (Brown 1900)

Nor were the Euro-Americans the only ones to overstock the Old World grazing animals. The Tohono O'odham also adopted the "cowboy culture," and their herds increased, especially in the 20th century; parts of the Tohono O'odham Reservation have been drastically overgrazed (Bauer 1971; Fontana 1976, 1983b; Hackenberg 1983).

Pima Livestock Raising

Throughout the three centuries of European contact, Pima lands probably withstood relatively little grazing, unlike the higher country to the south and east. While the Gileños vigorously took up winter wheat agriculture, they seem to have been less enthusiastic about livestock. Although no missions were established on the Gila during the colonial period, the Pima did have some animals. Horses were more prized by the Gileños than were cattle, undoubtedly because of the military advantage they provided. (I am told by older Pima that they ate horses as well, and horse fat was esteemed for its special lubricating properties; see also Russell 1908:81). Franciscan missionary Fr. Pedro Font encountered 18 mounted Pimas in 1775 (Russell 1908:84), and Fr. Diego Bringas (1977:89), another Franciscan, reported in 1795 that the Gila Pima "raise some horses and some kine, both large and small, although in small numbers. Under better guidance, however, these may increase so far as land will permit."

In 1846, Captain William Emory noted that the Pima had "but few cattle, which are used in tillage and apparently all steers, procured from the Mexicans" (Emory 1848:84). That same year, Henry Bigler (1962:37) observed that these agricultural people had "some poultry, . . . a few cattle and a good many fine ponies, some mules and jackasses." B. B. Harris (1960:81), on his July 1848 trek through Pima country, observed, "They raised horses—but not cows—and worked the soil with plow-oxen obtained from the Mexicans. . . . Each able-bodied man was required to keep a horse for war purposes against the ever devilish Apache." A dozen years later, Samuel Cozzens (1874:187) likewise said, "They have but few animals." Other mid-century Anglo-American itinerants noted that the Pima, while generous with wheat and other commodities, parted with their beloved animals with great reluctance—and at considerable price.

The calendar-stick oral history transcribed by Professor A. J. McClatchie and published by historian Sharlot M. Hall (1907) provides some important snapshots of the development of animal husbandry among the Pima and neighboring tribes. Pima oral histories recorded the presence of mounted Pimas during an Apache skirmish in 1838. The horse also served to attract the Apache, as the 1838 entry also notes: "Another Pima had a field near and some horses in it. The Apaches tried to steal all the horses. . . . The Pima men fought and two Apaches were killed. They did not get any horses. They came again to get the horses—this time they tried to ride. The horses jumped the river-bank and threw them off. A Pima shot an Apache in the arm. The river-bank was so high, the Pimas could not get down to kill the Apaches, but they got all the horses back." Apache horse raiding is recorded in the calendar-stick record until 1863.

At least by 1851, Gileños were close-herding their livestock, as recorded on the calendar stick: "One night four Apaches came to a Pima village to steal horses, but the Pimas had tied their horses close to their kees [kiik] (huts of brush and grass.) So the Apaches came to the corral and turned out the cattle and drove them toward Tempe Butte." Boundary Commissioner John Bartlett (1854, 2:237) likewise noted, "There seemed to be numbers of horses among them, which with the cattle are left to graze near the villages during the day, and at night are brought into the corrals, or yards, for safety." Until the Apaches were subjugated and confined to reservations, a major reason for the close-herding of cattle was protection from theft. Even into the 20th century, although cattle ranged freely during the day, they were rounded up and penned during the night, as Joe Giff, who had been a cowboy in his youth, explained. Corrals were massive affairs of stacked mesquite logs.

Sala Hina (Sally Fish) told Russell (1908:85) that her father and uncle, both Koahadk, brought the first cattle to the Pima about 1820. Russell noted, "The custom of killing and eating the cattle at the death of their owners contributed materially toward preventing increase in Pima herds." This practice likewise would have contributed to the preservation of local habitats. Russell added, "Oxen were very scarce for half a century after their

introduction, and the old men and women speak sadly of the weary waiting for their turn to use the single ox that dragged the wooden plow for perhaps a whole village."

Indian agent Silas St. John in 1859 recorded 850 horses and 799 cattle among the Pima (Hackenberg and Fontana 1974; Kroeber and Fontana 1986:67). That translates to 8 cows per square mile, given the original reservation size. The grasslands used for pastures (see fig. 3.2), however, were almost certainly outside this designated area, which was primarily the farmed terraces along the river. Pima fields were carefully enclosed by brush fences to keep livestock out.

Although their numbers were not high in Pima country, livestock nevertheless ultimately contributed to the depauperization of the vegetation of the floodplains of Pima country. Grazing animals fed on Carrizo *(Phragmites australis),* for example, helping to eradicate it. In addition, horses dispersed mesquite seeds nearly everywhere, which helped to crowd out grasses as the water table dropped and the nonwoody plants struggled for survival. Although grazing by the Gileños' livestock may have affected the composition of the local flora, it probably had no significant effect on the river's surface flow or on the water table of the region.

Floods, Droughts, and Ecological Damage

Biannual flooding of the Gila, caused by spring snowmelt and late summer storms (commonly called monsoons), was a normal cycle of river flow on which the Pima depended. Desert precipitation can be highly variable; a dry year may be followed by another with three or more times the average annual amount of rainfall. More important, a single storm episode may drop as much rain as would normally fall during several months. In a healthy watershed, these extremes are usually checked. But we have seen that the Gila watershed was under siege and far from healthy. The removal of beaver and the consequent collapse of their water storage system, the loss of woody vegetation from steep hillsides, the incessant attack on grass and other palatable vegetation by hordes of livestock, the impacting of the bare soils left the watershed subject to both sheet and gully erosion.

The alternating dry and wet years then wreaked havoc on the Gila hydraulic system. A century of devastating floods continued the ecological damage into the middle Gila country where the Pima lived and the lower Gila reaches that had for the most part already been abandoned. The new Anglo-American communities on the Gila and lower Salt likewise came to feel the brunt of shortsighted practices in the headwaters.

Henry Dobyns (1981:167–201) has amassed details on 15 years of serious floods in the Gila drainage between 1833 and 1941. A number of these floods were of such magnitude that they entered the oral history kept by the calendar-stick annalists (Russell 1908:38–66), such as in the 1833 account:

In the early winter the meteoric shower took place. This event was followed by heavy rains that caused floods in the Salt and Gila rivers. The spectacle of falling stars was to the Pimas an augury of disaster, and the succeeding floods were regarded as a punishment for sins which they had committed. What the sins might be they did not know, but concluded that they must have offended some medicine-man who possessed great magic power. Many thought it must be the medicine-man Kake who brought this calamity upon them because they had not shown him the respect that he thought was due him. It is said that when the flood was at its height he climbed a cottonwood tree and thence proclaimed in a loud voice that he would perform certain miracles that would prove disastrous to them if they did not listen to him and show him respect.

Others declared that the floods were caused by the two sons of an old goddess, Takwa-artam [Toa Kuá'adam]. When she saw the flood threatening to overwhelm the Pimas and Maricops [sic] she said to her sons: "Give me back my milk and then you can drown my people. The land is yet what it was when it was new." This puzzled the two brothers. They knew that they could not return the milk that had nourished them in infancy, so they did not allow the flood to rise any higher, but caused it to go down.

What ecological damage might have occurred in the 1833 flood is unknown. The Gila was reported to run two miles wide at Casa Blanca (see Olmstead 1919:9). Given the buildup of feral cattle in southeastern Arizona coupled perhaps with Apache animal husbandry in the headwaters of the Gila, erosion and arroyo gullying must have begun. The recently decimated beaver population would have taken years to recover: these animals produce only two to four young a year after a three-month gestation period, and the kits stay with the family group nearly two years. This flood no doubt began silting in the water control systems of these animals.

The immediate effect of these rains in Pima country was beneficial. The following year "was long remembered because of bountiful crops of wheat, corn, squashes, pumpkins, and watermelons that were raised. The desert mesas [bajadas] were carpeted with flowers and the bloom of cacti further transformed them into gardens" (Russell 1908:38–39). The Pima annalist Chukuḍ Naak 'Great Horned Owl's ear' added, "Our people worshiped the gods in grateful recognition for their protection; we danced unmolested by the murderous Apaches; we looked after the welfare of our households" (Russell 1908:38–39).

The next flood episode to merit comment by a calendarist was that of 1868 (in Russell 1908:52–53). "A heavy rain caused a flood which destroyed the store at Casa Blanca. This was known as the Vamati Tcoki [vamaḍ juukĭ], Snake rain." Ethnographer Frank Russell added, in a footnote, "The store was more than 2 miles south of the channel of the river, but it had been built at the foot of a little rise upon which the present village is located and was within the reach of the flood. This is but one of many instances where the white settlers of Arizona have not profited by the experience of the natives, ancient and modern, who have located their homes beyond the reach of the freshets that transform the shallow beds of blistering sand into irresistible torrents that overrun the bottom lands which may have been untouched by flood for many years." He appended this comment by historian Hubert Howe Bancroft: "The flood of September, 1868, was perhaps the most destructive ever known,

destroying three of the Pima villages and a large amount of property on the lower Gila."

The Pima are almost obstinately deliberate in decision making. They had had at least 200 years of experience watching the rise and fall of the Gila. And yet three of their villages were within reach of the floodwaters. It is probable that this event exceeded anything they had previously witnessed during their sojourn on the Gila. What the Pima of the time could not realize was that the Gila was entering its death throes. By this time the natural system of water catchment and absorption was in disarray, and its condition worsened during the next three decades.

For the year 1872–1873 the calendar-stick annalist narrated, "For several years the Pimas had had little water to irrigate their fields and were beginning to suffer from actual want" (Russell 1908:54). With the declining streamflow the Pima began searching for new, better-watered places to establish rancherias and fields. In 1873, some Pimas from upriver relocated at Komaḍk Wecho 'under the Estrellas'. (This village later became Catholic Komatke and Protestant Gila Crossing.) Frank Russell (1908:54) remarked, "There is an unfailing supply of water at this place; the Gila, after flowing 75 miles beneath the surface, rises to form a stream large enough to irrigate several hundred acres." There had once been a protohistoric rancheria in this vicinity; Kino mapped the village in 1696 as Comacson and recorded the place-name Comac in 1698, and Manje noted a population of 200 the following year. In the early 20th century, J. W. Hoover (1929:44) wrote, "That portion of the Gila below the railway bridge at Pima Butte is perennial, and the waters except when high are clear. The perennial flow of the river here is primarily due to underground drainage deflected from the Salt River. There is also some extra seepage from the Roosevelt irrigation project. [Thus] . . . at the lower end of the reservation there is always water in the river, and there is no real want; but the floods wash out the dams and make diversion more difficult." In discussing relocation along the drying Gila, Joseph Giff explained, "Look at Komaḍk—that's where there will be water for a long time. Because the South Mountains and Komaḍk [Sierra Estrella] joined [underground], rock bottom, like a lake, so there's water to use for a long time." It was on this portion of the reservation where Joe and Ruth Giff, Sylvester Matthias, David and Rosita Brown, and Carmelita Raphael grew up.

At the same time, Mormons began colonizing along the Salt River upstream of Phoenix. Desiring a buffer against the Apache, they invited Pimas from the Gila River to settle near them. About 300 from Shuuḍk (Wetcamp) at the middle of the reservation moved "and cleared fields along the river bottom south of their present location. . . . Water was plentiful in the Salt and the first year's crop was the best that they had ever known," noted the Pima annalist (Stout 1873).

In January 1874 and again in February, record storms swept across the Gila-Salt watersheds (Dobyns 1981:175–178). The event entered the Pima oral historical account obliquely, with wry Pima humor: "The Pimas went on a campaign against the Salt River Apaches soon after a heavy rain. When they reached the Salt river it was too high to be safely forded, so they built a raft and tried to take their saddles and blankets across upon it. The raft sank and they lost all their effects" (Russell 1908:55).

In February 1891 came the third major flood of the century. At Florence, just upstream from the Pima, the southern edge of the Gila channel was widened by another 200 feet. The San Pedro was severely gullied (Dobyns 1981:189–195). The calendar-stick annalist recorded, "In the spring of 1891 occurred the last and most disastrous of the Gila floods. The Maricopa and Phoenix Railroad bridge was swept away and the channels of both the Gila and Salt rivers were changed in many places. The destruction of cultivated lands led to the change of the Salt River Pimas from the low bottoms to the mesas" (Russell 1908:62).

Still other descriptions of the Gila also record the changes in the river and the surrounding vegetation at the end of the 19th century. After interviewing elderly Indians on the Gila, J. W. Hoover wrote that as late as the 1880s, "the river occupied a narrow channel and its bed was covered with large stones. There were no such stretches of bare and sandy waste as are found everywhere along the river now. Tall grass sometimes several feet

high, covered the whole countryside yearly. But the once grassy swales are now for the most part salt bush steppe or desert. Lakes or ponds were once common over the river flats where there is no trace of them today" (Hoover 1929:44). Clyde P. Ross, reporting on the Gila water supply for the U.S. Geological Survey, interviewed John Montgomery, a rancher from Arlington, Arizona, not far downstream from the reservation. When 12 years old, Montgomery was in camp at nearby Powers Butte in the summer of 1889. Ross (1923) summarized the conditions Montgomery had seen on the Gila below the Salt-Gila confluence:

At that time the river had a well-defined channel with hard, sloping banks lined with cottonwoods and bushes. The water was clear, was 5 or 6 feet deep, and contained many fish. The grazing lands near the river were in much better condition then than now. Several varieties of grass then abundant have since died out. Mr. Montgomery attributes the change in the character of the river largely to the practice of cattlemen of burning the heavy brush that once covered its banks in order to drive out wild cattle which had sought shelter there. This destroyed the natural protection and left the soft silty soil exposed to rapid erosion. The disastrous floods of 1890 and 1891 did much to break down the river's confining banks, partly filled the channel with sediment, and in general interfered with the equilibrium that had been established.

The flood of 1891 was the death toll for the Gila, for the Pima way of life, and for much of the local flora and fauna as well. Despite remnants, the fabric of the oasis ecosystem was rent. Hoover (1929:45) summarized the situation well:

During the eighties there was a series of wet years with abundance of natural forage. The ranges built up rapidly, and overgrazing resulted. During the same period the mountains of the Upper Basin in southeastern Arizona were being rapidly stripped of their timber for use in the mines. The hills were barer than now, because with the advent of the railways better mine

timber was brought in from the outside; but the cutting for fuel continued. At the end of this series of wet years came the disastrous flood of 1891. Before this, flood waters of the Gila merely spread out over the flats and irrigated them. Now with the banks of the river unprotected by brush and grass, the channel suddenly widened, and many good ranches along the river were cut out. Smaller branches of the river in the upper basin cut channels as much as twelve feet deep. On the desert deep channels appeared in what had been grassy swales.

Originally, much of the Gila bottom-lands had been covered with dense riparian growth. Frémont's Cottonwood (*Populus fremontii*) was the major canopy species, but there were several species of willow (*Salix* spp.) as well, and even an occasional Velvet Ash (*Fraxinus velutina*). Seep-willow (*Baccharis salicifolia*), Mexican Elder (*Sambucus mexicana*), Graythorn (*Ziziphus obtusifolia*), Screw-bean (*Prosopis pubescens*), and vines such as Virgin's Bower (*Clematis drummondii*) and Climbing Milkweed (*Sarcostemma cynanchoides*) diversified this plant community. Floods scoured much of the riparian growth, and drought weakened the rest. Riparian timber retreated from the unstable sandy meander of the river to the edge of the lower terrace. In time even the edge proved unsuitable for these hydrophilic species. The final retreat for the last great cottonwoods and occasional willows was along Pima ditch banks on the terraces (Rea 1983:28–60, figures).

Cultural and Ecological Effects of the Loss of Water

Along with the gradually diminishing river flow, the Gila Pima had to contend with another problem. Anglo-Americans and Mexican-Americans began settling upstream from the Pima rancherias beginning in the 1860s. Resentment ran high against them; indeed, 400 Pimas moved into Mexican fields upstream from the reservation and harvested the Mexicans' crop. But the invaders persisted; by 1869, a year after the founding of Florence, 400–500 settlers were taking Indian water. In 1873, Reservation Agent J. H. Stout wrote, "The settlements above the reservation are still increasing, and in a few years the farmers there will need and appropriate all the water that the river affords during the warm season preceding harvest. The reservation does not afford a sufficient quantity of water for the support of all the Indians belonging to it, and some of them in consequence have left it in order to get a living. About 1300 members of these tribes are thus living outside the lines of their reserve—about one thousand just above on the Gila, and some 300 have moved to the Salt River Valley" (Stout 1873:281).

In 1882 the U.S. Army was attempting to confine Chiricahua Apache groups that were escaping from their reservation. Lieutenant Thomas Cruse was surprised when his detachment was ordered to Maricopa "to police the Pima Indians" (Cruse 1941). En route the puzzled troops conferred: "There was no record of a Pima warring against a white man. Yet now their new Agent was fearing that the Pimas were on the verge of an outbreak." The details of this seldom-mentioned footnote on Arizona history from an officer involved in the incident are worth repeating:

When we reached the Agency at Sacaton we discovered that the trouble had its roots in that ancient cause of warfare—water rights. White settlers had been flocking into the country of the Pimas, locating homesteads, beginning the wonderful farms and orchards that now surround Phoenix. They had ignored the Indians' rights to water, and the Pimas had been deprived of irrigation until they looked out on barren fields and faced starvation.

No rations had ever been provided for the Pimas. Washington seemed to move on the theory that only he who refused to work would be fed. The Pimas—unlike the Apaches—had supported themselves and given no trouble, so they had been virtually ignored. Now, robbed of the water they had always owned, they were ready to fight anybody and everybody.

For three weeks we worked to settle the Indian grievance. The officers of our force, the Agent, old settlers and newcomers, leaders of the Pimas,

held conferences. At last we arranged adequate and fair water supply for the Indians, and they went cheerfully back to their fields.

In actuality, more was said than done. Reservation agents complained, but the federal government seemed paralyzed in dealing justly with the problem. Legal action was never taken against the water usurpers. By 1886 the Florence Canal Company had completed its waterworks, and nearly all surface flow was then diverted from the Gila upstream of the Maricopa and Phoenix Railroad; within a year the bulk of the reservation was left high and dry (Spicer 1962:149).

In the same year, Chief Antonio Azul and a delegation of Pimas went to Washington, D.C., to plead their case. A proposal was made to remove the tribe to Indian Territory (now Oklahoma), and the delegation on its return trip visited the proposed relocation site. Back home the ever-conservative Pima, still clinging to a life of hunting, gathering, and traditional desert agriculture as much as the deteriorating environment would allow, resisted being wrenched from their homeland in the Sonoran Desert.

By the time Frank Russell arrived in November 1901 to write an ethnography of the Gila Pima, he found the people in most rancherias destitute. The calendar-stick annalist for the previous year noted laconically, "A woman at Blackwater was fatally bitten by a rattlesnake." In a footnote Russell told the story that the oral historian had glossed over: "This woman had gone far out on the desert to search for mesquite beans, as she was without food; indeed the whole community was starving because of the failure of the crops owing to the lack of water in the river for their ditches" (Russell 1908:65). And the preceding year when the oral historian at Blackwater said simply, "There was no crop this year," Russell added,

The water of the Gila had been so far utilized by white settlers above the reservation, for the most part more than a hundred miles above, that there was none left for the Pimas. It is difficult to obtain accurate information at this time of the number who perished either directly or indirectly by starvation. During this and the following year

five persons are known to have died from this cause, and it is probable that there were others. Most of the Pimas will not beg, however desperate their need may be, so that not all cases were reported.

In one case a wood chopper tried during the hot season to cut mesquite for sale, but he was too weak to withstand the heat and the exertion and was found dead in the chaparral. An old couple were found dead in their house with no food of any kind in their storehouse, and it is supposed that they preferred to starve rather than beg. A man riding to Salt River was too weak from hunger to keep his saddle and fell and perished. (Russell 1908:65)

By Russell's day the grasslands noted by Hispanic and Anglo itinerants had vanished, and livestock suffered as a result.

As the fields now yield an insufficient supply of food for their owners, it follows that there is little grain for the horses, which grow poor and thin in winter; indeed, many die of starvation. Their principal food during that season is saltbushes.

The once famous grassy plains that made the Pima villages a haven of rest for cavalry and wagon-train stock are now barren, and it is not until the mesquite leaves appear in April that the horses can browse upon food sufficiently nourishing to put them in good condition. As the mesquite beans ripen, in June and July, live stock fattens rapidly. (Russell 1908:84–85)

George Webb summed up the plight of his people in the early 20th century: "Where everything used to be green, there were acres of desert, miles of dust, and the Pima Indians were suddenly desperately poor" (Webb 1959:122).

Dense Velvet Mesquite (*Prosopis velutina*) bosques once covered much of the unfarmed sections of the lower terrace of the Gila from Blackwater to the Salt-Gila confluence, a distance of more than 65 miles as the river runs (fig. 3.2). Many Pima farmers, left without water, turned to woodcutting as a source of livelihood. For the year 1896–97 one Pima annalist stated, "The River practically dry, the

Blackwater Indians were forced to leave homes to sell wood" (Southworth 1931:50). The importance of woodcutting continued: "In 1895, 462 cords of mesquite wood were cut and sold for firewood by Indians whose crops had failed," and "by 1905, nearly 12,000 cords a year were being cut and sold in Phoenix" (Hackenberg 1983:173). The local two steam-powered flour mills also helped consume the mesquite from the central part of the reservation. The *Phoenix Gazette* of 16 April 1900, perhaps with some journalistic hyperbole, reported, "There are 30,000 cords of mesquite wood cut and piled between Maricopa Junction and Phoenix." Hoover likewise noted this important contribution to the local economy in the early quarter of the 20th century: "During the several past years the Indians in the middle of the reservation have had to depend largely upon cutting and selling mesquite wood. A man required two or three days to assemble a load which he may sell for about $6.00. Then he must haul it 15 to 20 miles and peddle it, which takes two more days unless there is a prompt sale. With more plentiful water the Indians leave off wood cutting and go back to their fields" (Hoover 1929:49).

Normally this hardy tree would have been able to resprout from stumps and to reseed itself from horse-dispersed seed. However, beginning in 1902, wells drilled on the reservation and around its periphery began lowering the water table drastically. Not only did mesquite regeneration become impossible, but some of the standing bosques, deprived of water, became huge mesquite cemeteries. Miles of saltbush steppe with stumps and snags are all that is left to mark these former woodlands.

Webb made an important ecological observation: "Today there are no mesquite trees left on the Reservation that are not second growth. If you look at the base of any mesquite tree you will find a dry stump. There is where a much bigger tree once grew" (Webb 1959:122). He was correct: I have traveled much of the reservation floodplain on foot and never encountered a native old-growth bosque. Indeed, I have seen but few relatively large mesquites, although ancient ones still survive upstream in the narrows east of Florence between the Buttes and Cochran (Rea 1983:67–75).

Live bosques do still occur along the Gila from Santa Cruz village to the Gila-Salt confluence at Monument Peak and 115th Avenue, but the bosques on the upper two-thirds of the reservation are all gone. It is difficult to imagine they could ever have been there, but a careful search of the saltbush-wolfberry flats today reveals the stumps. Remnants of these dead bosques lie along Blackwater School Road, along the dry, dirt channel once known as the Little Gila, in the drainage east of Snaketown, and from above the site of Maricopa Wells through the famous New York Thicket nearly to Santa Cruz village.

Casa Grande Ruins National Monument at the southeastern edge of the reservation is surrounded today by a dead bosque. This is the forest described by B. B. Harris (1960:80) in 1849 and J. Ross Browne (1864) in 1863, among others (see chapter 3). Karen Reichhardt (1992 and personal communication) studied the vegetation here in 1987 and found an average of 0.04 living mesquites per acre, whereas in a 1931 study there were 18.2 trees to an acre. She attributed the loss to grazing and groundwater pumping.

Whatever the causes of the death of the river, the effects were wide reaching. The flood of 1891 and flooding during the wet years of 1906–1917 evidently scoured most of the submergent and emergent vegetation from the degraded Gila River channel. Indeed, resident ornithologist M. French Gilman failed to find the most common marsh birds breeding there during his stay (1907–1915). Native consultants from the upper villages recognized but few of the folk taxa from the aquatic and emergent plants (see part 2, Groups A and B).

On 31 July 1903, J. William Lloyd arrived at the middle of the reservation to begin translating and recording the Pima Creation Story. He noted, "The famous Gila is not a very imposing stream at any time, and now was no stream at all, but a shallow dry channel, choked with desert dust, or paved with curling flakes of baked mud which cracked like bits of broken pottery under our ponies' feet. But I afterwards many times saw it a turbid torrent of yellow mud, rushing and foaming from the mountain rains; perilous with quicksands and snag, the roaring of its voice heard over the chaparral for miles to windward" (Lloyd 1911:3–4).

Attempts by Indian commissioners and local agents and others to rectify the abuse of Pima water rights were completely ineffective. Perhaps a reason legal action was never pursued was that the Gila began to fail to supply a sufficient flow for even the upstream settlers. After years of negotiation by Arizona's Senator Carl Hayden, Congress finally approved the San Carlos Irrigation Project in 1924, authorizing the construction of Coolidge Dam on the Gila on the San Carlos Apache Reservation. The water stored in this upstream reservoir was to supply the irrigation needs of both Indians and whites (Hayden 1965). The dam, a technological answer to an environmental abuse, was duly built and dedicated by President Calvin Coolidge in 1930. Webb summarized the results: "When the dam was completed there would be plenty of water. And there was. For about five years. Then the water began to run short again. After another five years it stopped altogether" (Webb 1959:123). The San Carlos Irrigation Project was without water; engineers had underestimated the dryness of the Southwest.

Water for project lands has since been pumped from deep wells beneath Pima country. Much of the underground water in this region was deposited in the wetter, cooler Pleistocene. The capacity of the land to recharge underground water storage has been greatly diminished with the local changes in vegetation, such as the loss of the grasslands and mesquite bosques, and the inability of the watersheds to provide water sufficient to produce aboveground flow in the desert streams and washes. As a consequence, water pumping exceeds recharge several times over, and water tables continue to fall year after year: underground water is considered an exploitable resource to be mined rather than a renewable resource to be harvested (Bowden 1977; Powledge 1982:147–159).

In 1922, hydrologist Kirk Bryan summarized the changes that had occurred on the middle Gila during the preceding four decades: "Early reports indicate that up to about 1880 the Gila flowed in a relatively deep channel through its flood plain, overflowing it only in times of flood. There was also a considerable low-water flow. At present the channel is a sandy waste with many tortuous subsidiary channels, constantly shifting in position, and there is no low water flow except in favored places. There seems to be a greater proportionate load of sediment, which under present conditions is silting up the channel" (Bryan 1922:72–73).

In time the water flow to even the lower villages vanished, although earlier writers had referred to the "unfailing supply" of water there (Hoover 1929:44; Russell 1908:54n). By 1950 even these "favored places" were dry, and the people once living at the desert's green edge had become a people without a river.

George Webb wrote a passage that serves as an apt eulogy for the Gila River:

> In the old days, on hot summer nights, a low mist would spread over the river and the sloughs. Then the sun would come up and the mist would disappear. . . .
>
> The red-wing blackbirds would sing in the trees and fly down to look for bugs along the ditches. Their song always means that there is water close by as they will not sing if there is not water splashing somewhere.
>
> The green of those Pima fields spread along the river for many miles in the old days when there was plenty of water.
>
> Now the river is an empty bed full of sand.
>
> Now you can stand in that same place and see the wind tearing pieces of bark off the cottonwood trees along the dry ditches.
>
> The dead trees stand there like white bones. The red-wing blackbirds have gone somewhere else. (Webb 1959:124–125)

Chapter 6

Modern Gila Pima Habitats

The corn, the beans and the pumpkins sent forth such a pleasant freshness that it is not to be wondered at the tired, wing-sore birds after a night's flight, should be attracted by such a scene of peace and plenty.

Ornithologist George F. Breninger, 1901

If you drive across Gileño country today, your eyes are not likely to be regaled by a pleasant country with great cienegas, creeks, and canebrakes such as Sedelmayr reported in 1744. In the Gila you will not find any of the fish that Kino said were used as rations for the troops in 1699. Nor will you normally find water in that stream for a swim such as Bartlett's boundary survey crew were accustomed to take during those hot July days of 1852. Do not look for grasslands where Anza pastured more than a thousand head of livestock in 1775; you would be lucky to

find patches of Desert Salt Grass (*Distichlis spicata*) a few feet across. Do not expect to shade yourself under the large cottonwoods and willows the California-bound 49ers discovered fringing the river. And if you were to make the trip from Sacaton to the Casa Grande ruins today, you would not need a local guide to get you through the dense Velvet Mesquite *(Prosopis velutina)* bosques.

From an ecological perspective, much of Pima country today is disturbed wasteland. The floodplains have suffered severe desertification (xerification). Nearly all surviving habitats are anthropogenic, that is, they are the result of human modifications of the plant cover, soil, salinity, and soil moisture. Except during rare wet years any surface flow in the rivers today is sewage effluent or irrigation wastewater, often water deposited in wetter geologic times and then pumped from deep wells before being used and discarded.

The vegetation on the bajadas of the four local mountain ranges has survived much better. Hardwoods, heavily harvested at the turn of the 20th century, have regenerated. Shrubs and forbs seem not to be too severely grazed, and native vegetation, except for grasses, has survived the influx of exotic species. The desert mountain ranges have survived the Columbian Conquest with their vegetational associations largely unaltered, thanks to their steepness and inaccessibility.

So what are the habitats of Pima country that are to be found at the end of the 20th century? We can examine the plant communities and floral associations starting from the higher elevations. Here we will concentrate on the dominant, usually perennial, plants. (The following section focuses on the individual histories of certain critical species in these communities.)

Pima country is in the Basin and Range Province—a country dominated by bare ranges of rock, usually basaltic, arranged in a generally northwest-to-southeast direction. These rise like rugged islands above a sea of alluvium. Here the classic demarcation of desert mountain, bajada, and floodplain is particularly well defined (see fig. 8.1).

Desert Mountains

Four rugged ranges frame Gila Pima country. At the eastern end of the reservation, on opposite sides of the river, are the Sacaton Mountains on the south and the larger and geologically more diverse Santan Mountains on the north. The Sacaton Mountains are a loose assemblage of ridges arising from the alluvial flatlands. The higher peaks exceed 2,000 feet in elevation, with the highest at nearly 3,000 feet. The peak of Santan Mountain is slightly higher, just over 3,000 feet in elevation. To the northwest are the South Mountains (earlier called the Salt River Mountains), which block off most of the sprawling metropolis of Phoenix from Pima lands; the elevation of this ridge is about 2,000 feet. At the southwest is the Sierra Estrella, the longest, highest range in Pima country, with elevations exceeding 4,000 feet. The dominant vegetation on these four desert ranges includes

Catclaw Acacia, *Acacia greggii*

Triangle-leaved Bursage, *Ambrosia deltoidea*

Saguaro, *Carnegiea gigantea*

Foothill Paloverde, *Cercidium microphyllum*

White Brittlebush, *Encelia farinosa*

Mormon Tea, *Ephedra aspera*

Bush Buckwheat, *Eriogonum fasciculatum*

Ridge Barrel Cactus, *Ferocactus cylindraceus (F. acanthodes)*

Ocotillo, *Fouquieria splendens*

Desert-lavender, *Hyptis emoryi*

Pima Ratany, *Krameria parvifolia*

Creosote Bush, *Larrea divaricata*

Teddy-bear Cholla, *Opuntia bigelovii*

Engelmann's Prickly-pear, *Opuntia engelmannii*

globemallow, *Sphaeralcea* spp.

The major range is the Sierra Estrella, which extends some 20 miles across the desert and is biologically the most diverse range in the Gileño area. The Gila and Salt Rivers join at its northernmost corner, where the combined river swings westward once again. Some north-facing geologic amphitheaters of the sierra are particularly mesic. Certain cliffs and deep canyons may receive but a few hours of sunlight each day during most of the year. This is the only range in Gila Pima country that has vegetation of the Upper Sonoran Zone.

Many species in the Estrellas are found nowhere else in Gileño country. At lower to intermediate levels of the north-facing slopes are rather dense and brushy stands of plants such as

Desert Agave, *Agave deserti*

Desert-oregano, *Aloysia wrightii*

Elephant-tree, *Bursera microphylla* (in the southeast)

Desert Hackberry, *Celtis pallida*

Crossosoma, *Crossosoma bigelovii*

Bush Penstemon, *Keckiella antirrhinoides microphylla*

Arizona Rosewood, *Vauquelinia californica*

Although these plants were readily within the reach of the Pima on hunting and gathering excursions, few were named—or perhaps their names had been forgotten as the Pimas' daily regimen concentrated in the floodplains, especially during the past century.

Higher in the sierra are islands of Upper Sonoran Zone vegetation, particularly on the more mesic north-facing slopes. North of Butterfly Mountain, southeast of Montezuma Sleeping, is a forest of Crucifixion-thorn (*Canotia holacantha*). Above this, in areas shaded almost year round, are species such as

Silver Wormwood, *Artemisia ludoviciana*

Virgin's Bower, *Clematis drummondii*

Red-seeded Juniper, *Juniperus erythrocarpa*

Desert Scrub Oak, *Quercus turbinella*

Oak-belt Gooseberry, *Ribes quercetorum*

These communities are relicts of cooler, more moist Pleistocene regimes. During the glacial periods they probably dominated much of this desert range but have since retreated upslope to their present tenuous footholds. Some of these species have viable populations, while others, such as the junipers, are only ancient individuals, evidently unable to reproduce successfully in the hotter, drier modern climate.

The ruggedness of these desert ranges has protected them from grazing by livestock. Other than rodents and lagomorphs, the only herbivores are Mule Deer, Desert Bighorn, and perhaps White-tailed Deer. As a consequence, the grassy flora remains relatively rich and includes

Purple Threeawn, *Aristida purpurea*

Cane Beardgrass, *Bothriochloa barbinodis*

Tanglehead, *Heteropogon contortus*

Littleseed Muhly, *Muhlenbergia microsperma*

Bush Muhly, *Muhlenbergia porteri*

Alkali Sacaton, *Sporobolus airoides*

Sand Dropseed, *Sporobolus cryptandrus*

Desert Needlegrass, *Stipa speciosa*

Near the south end of the Sierra Estrella is a large depression about a quarter-mile in diameter called the Upper Basin, which is drained by an arroyo to the east. Its floor, at 3,000 feet, supports a Saguaro–Creosote Bush–paloverde community with Ocotillo and Desert Agave. Thickets of Desert Hackberry and patches of Alkali Sacaton grow at springs, as do several other smaller species unique to the reservation.

In the early 1990s, a threat to the reservation sierras proved even more pernicious than grazing. I began noticing over several years the steady advance from the lowlands of annual Old World grasses, particularly Red Brome (*Bromus rubens*). During the winter and spring, this fuel accumulated in the more moist areas, particularly in the shade of trees. Hot summer fires then swept up desert slopes, sometimes burning for many days. While fires are commonplace on the disturbed floodplain below, Pima native consultants said they had never seen the mountains burn in their lifetime. The ultimate effect to non-fire-adapted sierran vegetation remains to be determined, but the floral composition of burned habitats has been altered.

Bajadas

Bajadas are the well-drained piedmonts of gravelly soil and loose rocks. These degrading slopes skirt the desert peaks and sierras. Several plants of economic significance grow principally here, such as Saguaro, Buckhorn Cholla (*Opuntia acanthocarpa*), and Creosote Bush. Other plants include

> Triangle-leaved Bursage, *Ambrosia deltoidea*
>
> White Bursage, *Ambrosia dumosa*
>
> Foothill Paloverde, *Cercidium microphyllum*
>
> White Brittlebush, *Encelia farinosa*
>
> Janusia, *Janusia gracilis*
>
> White Ratany, *Krameria grayi*
>
> Pencil Cholla, *Opuntia arbuscula*

Arroyos originate high in the rocky mountain outcrops and cut, sometimes deeply, through the bajadas before opening out into the alluvial floodplain in small deltas called *akĭ chiñ* 'wash mouth'. Desert Ironwood (*Olneya tesota*) is a characteristic tree of these drainages. On the upper bajadas the arroyos support Catclaw Acacia, Desert-lavender, and Chuparosa (*Justicia californica*). Lower on the bajadas, the more mesophytic Blue Paloverde (*Cercidium floridum*) may be common, giving these washes a more riparianlike appearance. (Some desert writers use the term *riparian* to label the vegetation of these arroyos, which carry water only during local storms and for some hours after. I prefer to restrict this term to the vegetation growing along permanent streams, as the Latin word designates— even if the flow now is only below the surface much of the time.) Other important plants of the arroyo bottoms include the gracile, often sprawling Christmas Cholla (*Opuntia leptocaulis*), Black-stemmed Wolfberry (*Lycium macrodon*), Canyon Bursage (*Ambrosia ambrosioides*), and Burro-brush (*Hymenoclea salsola*).

In the spring the parasitic Desert Broomrape (*Orobanche cooperi*) is often common on the sandy arroyo floors. The shady bank of an arroyo is often festooned with Desert Bedstraw (*Galium stellatum*), particularly where there is a canopy of trees.

Some plants in Pima country grow primarily in the clean, coarse, deeper sands of arroyos, especially in their lower reaches at the bases of bajadas or where they open out into the floodplains, the *akĭ chiñ*. Many of these plants are root perennials with major growth during the mid to late summer thunderstorms:

> Indian Root, *Aristolochia watsoni*
>
> Finger-leaved Gourd, *Cucurbita digitata*
>
> Desert Datura, *Datura discolor*
>
> Desert Tobacco, *Nicotiana trigonophylla*
>
> Chinchweed, *Pectis papposa*
>
> Coyote's Devil's Claw, *Proboscidea altheaefolia*

The Floodplains

The bulk of Pima country is flatland with fine pale gray alluvial soil (the Laveen series). When wet the soil is slippery and soft, and a vehicle can be engulfed like in quicksand. When dry it rises in clouds of fine dust when a horse, wagon, or auto passes. The sickly Fr. Font in 1775 was not exaggerating when he complained of the clouds of very sticky dust (Bolton 1930, 4:33). The amounts of sand, salinity, and moisture in the soils vary from place to place, affecting plant growth. The agricultural activities of both Indians and whites (on leased land) in Pima country are wholly within the floodplains. About half the flatlands are farmed.

Most of the flatland still in native vegetation is an open steppe with Desert Saltbush–wolfberry–Velvet Mesquite association. Desert Saltbush (*Atriplex polycarpa*) dominates as low, dense grayish shrubs. Interspersed are low, sprawling *Lycium californicum*, as well as some taller species of wolfberry. Single mesquite trees may occur here, but they are widely spaced and stunted. There is insufficient water in these saltbush flats to produce either large or dense growths of mesquite.

Individual saltbushes are usually perched on small, spaced mounds or hummocks about a foot above the surrounding flatlands that are formed by

windblown sands (Shantz and Piemeisel 1924). The interstitial soil is somewhat saline and often fine textured so that rainwater or surface floodwater may accumulate, producing growths of annuals such as

> Desert Blue Aster, *Machaeranthera arida*
>
> Desert Cambess, *Oligomeris linifolia*
>
> Chinchweed, *Pectis papposa*
>
> Woolly Plantain, *Plantago insularis*
>
> Camel Grass, *Schismus arabicus*
>
> Mediterranean Grass, *Schismus barbatus*

Several species of cactus are found principally in this habitat. Saltbush Barrel Cactus (*Ferocactus wislizeni*) may be locally common; where fires have not occurred for long periods, it may reach waist or even shoulder height. Engelmann's Hedgehog Cactus (*Echinocereus engelmannii*) is generally well distributed throughout these flats. However, with the accumulation of the nonnative Mediterranean and Camel Grasses during wet winters, and subsequent spontaneous fires during hot summers, this low cactus suffers. The all but invisible Desert Night-blooming Cereus (*Peniocereus greggii*) is restricted to these flats, where it usually grows in the protection of the short-statured Velvet Mesquite. Occasionally found here as well is the even rarer Thornber's Fishhook Cactus (*Mammillaria thornberi*), a sprawling colonial cactus that also tends to favor some shrubby protective covering. In places, Chain-fruit Cholla (*Opuntia fulgida*) and its hybrid, Gila Cholla (*Opuntia x kelvinensis*), may be abundant. These cholla "forests" are mostly on the eastern two-thirds of the reservation. One of the reservation's rarest birds, the LeConte's Thrasher, reaches the northeastern corner of its breeding range in the saltbush flats of Pima country. Gila Monsters live there as well.

Much of this *Atriplex-Lycium* association was formerly mesquite bosque, as the large stumps attest. In places, Hohokam fields and ruins are now just saltbush flats. The areas most often cleared today for mechanized farming are the saltbush-wolfberry flats; indeed, perhaps half of the flats on the reservation have now been converted to fields irrigated by pumped groundwater. Aboriginally, the

Pima probably farmed only the lowest terraces adjacent to the river, as well as its islands.

Pima fields that have been farmed in the traditional manner—that is, lightly plowed and intercropped—within a few years have a suite of weeds. While in active cultivation such fields may host native plants such as

> Palmer's Carelessweed, *Amaranthus palmeri* (summer)
>
> Bursage Ragweed, *Ambrosia confertiflora*
>
> spiderling, *Boerhavia* spp.
>
> Yellow Nut-grass, *Cyperus esculentus*
>
> Golondrina, *Euphorbia* spp.
>
> Camote-de-raton, *Hoffmanseggia glauca*
>
> Patota, *Monolepis nuttalliana* (winter)
>
> Black-seeded Devil's Claw, *Proboscidea parviflora*
>
> White Horse-nettle, *Solanum elaeagnifolium*
>
> globemallow, *Sphaeralcea* spp.
>
> Horse-purslane, *Trianthema portulacastrum*

The fields may also have nonnative plants such as

> Nettleleaved Goosefoot, *Chenopodium murale*
>
> Field Bindweed, *Convolvulus arvensis*
>
> Bermuda Grass, *Cynodon dactylon*
>
> Cheeseweed, *Malva parviflora*
>
> Sweet Clover, *Melilotus indicus*
>
> Russian-thistle, *Salsola australis*
>
> Mediterranean Grass, *Schismus barbatus*
>
> London Rocket, *Sisymbrium irio*
>
> Spiny Sowthistle, *Sonchus asper*
>
> Annual Sowthistle, *Sonchus oleraceus*

Generally, coarser plants are found growing in field borders, on ditch banks, and even in earth ditches (especially laterals) that run only occasionally. These are primarily native plants, including

> Spiny Aster, *Aster spinosus*
>
> Wright's Saltbush, *Atriplex wrightii*
>
> Desert-broom, *Baccharis sarothroides*
>
> Horseweed, *Conyza canadensis*
>
> Wild Sunflower, *Helianthus annuus*

> Quail-plant, *Heliotropium curassavicum*
>
> Jimmyweed, *Isocoma (Haplopappus) acradenia*
>
> White Horse-nettle, *Solanum elaeagnifolium*
>
> Seepweed, *Suaeda moquinii*
>
> Common Cocklebur, *Xanthium strumarium*

Nonnative plants in these disturbed areas include Johnson Grass (*Sorghum halepense*), Salt-cedar (*Tamarix ramosissima*), and Tree Tobacco (*Nicotiana glauca*).

If fallow fields are not reclaimed, a rank growth eventually covers the entire area. At this stage the dominants usually include plants such as the following native species:

> Spiny Aster, *Aster spinosus*
>
> Four-wing Saltbush, *Atriplex canescens*
>
> Quail Brush, *Atriplex lentiformis*
>
> Desert-broom, *Baccharis sarothroides*
>
> Jimmyweed, *Isocoma (Haplopappus) acradenia*
>
> Velvet Mesquite, *Prosopis velutina*
>
> Seepweed, *Suaeda moquinii*

The nonnative Bermuda and Johnson Grasses are also likely to be dominant species here.

Where the water table is sufficiently high, the fields finally revert to nearly impenetrable mesquite bosques with Graythorn (*Ziziphus obtusifolia*), especially along the former fence rows, and Jimmyweed and globemallow in small openings and along road edges or trails that may be maintained. More poorly watered areas may have only clumps of Graythorn, occasional well-spaced mesquites, and some of the woody or suffrutescent shrubs from the previous list.

Abandoned fields without sufficient groundwater to support bosque growth may remain bare much of the year, with hard-packed exposed soil. Yet they are not always barren. Good winter rains, in particular, may produce a dense growth of low annuals, several of economic importance to the Pima, such as Patota and Woolly Plantain.

Perhaps the most fragile and vulnerable habitats on the floodplains, now that riparian and marsh habitats have been lost, are the sand dunes, found principally

along the Gila between Maricopa Road and Gila Crossing village. These aeolian, or wind-deposited, soils are stabilized in dunes 10–15 feet tall. According to U.S. Geological Survey maps based on ground surveys of 1903–1904, these dunes once extended along the Gila throughout most of the reservation. There are ethnohistorical accounts of dunes in the area between Laveen and the reservation, areas today all under agricultural development (see Group G, *Hesperocallis undulata*). Significant perennial vegetation on the remaining dunes includes

> White Bursage, *Ambrosia dumosa*
>
> Desert Lily, *Hesperocallis undulata*
>
> Big Galleta, *Hilaria rigida*
>
> Desert Night-blooming Cereus, *Peniocereus greggii*
>
> Velvet Mesquite, *Prosopis velutina*

Living fence rows have been a distinctive feature of Pima fields, since at least the 19th century. At one time, fields were practically fortified with thick walls of thorny branches piled to head height. In time even a barbed-wire fence becomes a living fence because birds perch on the fence and drop seeds from fruits they have eaten, establishing new plants along the fence row. These fence rows were important to wildlife as well as to the Pima, who hunted birds and small mammals there (Rea 1979, 1983:46–49). Plants making up these living fences include

> Spiny Aster, *Aster spinosus*
>
> Quail Brush, *Atriplex lentiformis*
>
> wolfberry, *Lycium* spp.
>
> Velvet Mesquite, *Prosopis velutina*
>
> Mexican Elder, *Sambucus mexicana*
>
> Climbing Milkweed, *Sarcostemma cynanchoides*
>
> Seepweed, *Suaeda moquinii*
>
> Graythorn, *Ziziphus obtusifolia*

Small drainages through the alluvial flatlands that are not seasonally flushed with a volume of water may accumulate salts. Some plants commonly found in these situations are

> Pickleweed, *Allenrolfea occidentalis*
>
> Narrow-leaved Saltbush, *Atriplex linearis*

Desert Salt Grass, *Distichlis spicata*

Greasewood, *Sarcobatus vermiculatus*

Seepweed, *Suaeda moquinii*

Fine alluvial, often saline soils with good subsurface water—as are found around seeps and sumps, along the lower edges of cultivated fields, and along ditches—may support dense stands of Arrow-weed (*Pluchea sericea*) and patches of Quail Brush. The Arrow-weed thickets are usually without admixtures of other species. Once a component of almost every Pima construction, the plants are now sought primarily to cover *vaptto*, the ramadas or wall-less shade structures found in every traditional Pima yard.

The bed of the Gila River today is either bare or choked with thickets of Salt-cedars (*Tamarix ramosissima*), which sometimes extend for miles. A few young Frémont's Cottonwoods (*Populus fremontii*) grow west of Interstate 10, opposite Gila Butte. Several Goodding's Willows (*Salix gooddingii*) survive at the Olberg Diversion Dam, where irrigation water is returned to the Gila channel. Plants found on the bare, sandy stretches of the riverbed not monopolized by Salt-cedars include (along with the nonnative Tree Tobacco, *Nicotiana glauca*) native species such as

Seep-willow, *Baccharis salicifolia*

Desert-broom, *Baccharis sarothroides*

Sweetbush, *Bebbia juncea*

Blue Paloverde, *Cercidium floridum*

Sacred Datura, *Datura wrightii*

Burro-brush, *Hymenoclea monogyra*

Jimmyweed, *Isocoma acradenia (Haplopappus)*

Thurber's Sandpaper Plant, *Pentalonyx thurberi*

Desert-straw, *Stephanomeria pauciflora*

If there are good winter rains, various annuals may carpet the sands of the river channel. Some of the conspicuous ones are

Sand-verbena, *Abronia angustifolia*

Pepper-grass, *Lepidium* spp.

Blazing Star, *Mentzelia* spp.

Arizona Evening-primrose, *Oenothera avita* var. *arizonica*

Dune Evening-primrose, *Oenothera deltoides*

sorrel, dock, *Rumex* spp.

The natural marshes have all vanished from Pima country. Wastewater, particularly irrigation tailwater that runs with some frequency over a period of years, may produce marshes with emergent vegetation where the soil becomes boggy, not drying out between irrigations. Such areas on the reservation are rare. I reported on one between Co-op and Maricopa colonies (Rea 1983:53–56). Smaller areas occur along Casa Blanca Road near Sacate and west of Sacaton. Aquatic species that may occur in these areas include

Junglerice, *Echinochloa colonum*

spike-rush, *Eleocharis* spp.

Water Pennywort, *Hydrocotyle verticillata*

Yellow Waterweed, *Ludwigia peploides*

Knot Grass, *Paspalum paspalodes*

Salt-marsh Fleabane, *Pluchea odorata*

Willow Smartweed, *Polygonum lapathifolium*

dock, *Rumex* spp.

Hard-stem Bulrush, *Scirpus acutus*

Salt-marsh Bulrush, *Scirpus maritimus*

Southern Cattail, *Typha domingensis*

Water Speedwell, *Veronica anagallis-aquatica*

Nonnative aquatic species include Barnyard Grass (*Echinochloa crusgalli*) and Rabbitfoot Grass (*Polypogon monspeliensis*).

Riparian timber no longer occurs along the reservation portion of the middle Gila but still exists (as a reestablished plant community) along the lower Salt River where it forms the northern boundary of the reservation, that is, from about 83rd Avenue to 115th Avenue west of Phoenix. Since the late 1950s this woodland has been fed by permanent effluent from Phoenix sewage plants. The dominant trees of this community are Frémont's Cottonwood and Goodding's Willow, with an intermediate story of Salt-cedar, Seep-willow, and Desert-broom.

The edges of the river may have extensive stands of such emergent vegetation as

Southern Cattail, Tule *(Scirpus validus, S. acutus)*, smaller bulrushes (such as *S. americanus, S. pungens,* and *S. maritimus*), spike-rushes, and Yellow Waterweed *(Ludwigia peploides)*. Formerly, dense stands of Carrizo *(Phragmites australis)* also grew at the edges of the river. With loss of surface flow, most of the Gila channel now supports extensive stands of Salt-cedar to the exclusion of almost all other species. In places there is not even enough subsurface water for this exotic species.

Modern Pima habitats are a far cry from what they were at the time of the Hispanic *entradas* or the Anglo-American boundary surveys or, for that matter, early in the 20th century when the consultants for this book were growing up. Most severely altered are the habitats in the floodplains where the Pima and Maricopa live. The loss of the rivers and their hydrophilic plant communities (see chapter 5) is the most obvious change. Half the flatlands have been converted to modern mechanized farming (Rea 1983:98–110). The environment into which new generations of Gila River Pima are born has lost much of its resemblance to that recorded in tradition. The game, the fish, the wild plants once at the core of Pima daily life, song, myth, and ritual are no longer part of a young Pima's consciousness.

The Native Vegetation and Its Desolation

Baseline data at the plant species level are confined almost exclusively to the 20th century, a turbulent period both culturally and ecologically. In Gila Pima country, documented botanical specimens—that is, permanent vouchers deposited in various U.S. herbaria—begin in 1901 when Professor J. J. Thornber collected in the Pima villages for Frank Russell (1908:18). (Dr. Edward Palmer, briefly in 1867 and 1869 and again in 1885, was the first ethnobotanist to make collections there, but his botanical specimens were either lost or were distributed in eastern herbaria without accurate locality data; see McVaugh 1956.)

Many botanical collectors have worked in Gila Pima country (table 6.1). My own collections began soon after my arrival in 1963, and important specimens were deposited at Arizona State University in

Tempe. Material collected for this book is vouchered at the San Diego Natural History Museum, with replicates at the University of Arizona in Tucson, Arizona State University, or the Desert Botanical Garden in Phoenix.

The part of the reservation in which collections have been most thorough is the Sacaton area, especially the local "Gila bottoms." The U.S. Field Station (operated under cooperative agreement by the Office of Indian Affairs and the U.S. Department of Agriculture's Bureau of Plant Industry) was near the river, and R. H. Peebles, T. H. Kearney, and other botanists stationed there collected continuously during the 1920s and 1930s. It is unfortunate that there were no notations with the rarer specimens from these bottomlands indicating whether species represented permanent populations or merely waifs that did not become successfully established.

The mountains above these bottomlands have not been neglected. Most of the botanical treasures of the mesic island peaks of the Sierra Estrella were discovered in 1935 by Peebles and in 1936 by Leslie N. Goodding and his associates. Eric G. Sundell's master's thesis (1974) on the flora of the Estrella Regional Park added considerably to the knowledge of the northern part of this botanically rich range. Botanical notes and accompanying specimens (at ASU) made by E. Linwood Smith during his extended surveys of Desert Bighorn also added to the flora of the sierra. Thomas Daniel and Mary Butterwick (1992) provided a complete flora of the South Mountains. Field Station botanists also collected in the Santans.

Altogether, a flora of more than 525 species of vascular plants has been recorded for the Gila River Reservation and its immediate vicinity. This number includes both native and naturalized plants but not species occurring only in cultivation such as crops and ornamentals. All the Sierra Estrella is included, although more than two-thirds of this range falls outside the reservation boundaries. Most of the South Mountains lies outside the reservation to the north.

Distributional Significance of Species

The Gila Pima flora is unique for several reasons. A number of plants are found here at the very northern or northeastern edges of their natural distributions in Arizona (and in some cases in the entire United States). These include

Desert Bush Snapdragon, *Antirrhinum cyanthiferum*

White-stemmed Milkweed, *Asclepias albicans*

Butterfly Bush, *Buddleia scordioides*

Elephant-tree, *Bursera microphylla*

Pringle's Buckwheat, *Eriogonum pringlei*

Emory's Barrel Cactus, *Ferocactus emoryi (F. covillei)*

Desert-olive, *Forestiera phillyreoides (F. shrevei)*

Yellow Felt Plant, *Horsfordia newberryi* (nearly)

Thornber's Fishhook Cactus, *Mammillaria thornberi*

Screwbean Mesquite, *Prosopis pubescens*

Yerba-de-la-flecha, *Sapium biloculare*

Table 6.1 Botanical Collections from the Gila River Indian Reservation and Vicinity

Abbreviation	Botanist	Date	Location
Historic			
PHK	Peebles, Harrison, Kearney & associates		
DDP	D. D. Porter	1926	Sacaton
GJH	G. J. Harrison		
THK	Thomas H. Kearney		
RHP	Robert H. Peebles		
JAH	J. A. Harris	1928	Sacaton
HFL	H. F. Loomis & Edith Loomis	1926	
RRH	Robert R. Humphrey		
ARL	A. R. Leding		
ES	E. G. Smith	1935	
IT	I. Tidestrom	1927	Sacaton
JJT	J. J. Thornber	1901–1918	
FS	Forrest Shreve	1934, 1941	
G&H	L. N. Goodding & Humphrey	1936	Sierra Estrella
G&G	L. N. Goodding & Gunning	1936	Sierra Estrella
MFG	M. French Gilman	1907–1915	Sacaton, Santan, & Blackwater
GWP	George W. Plumb	1936	Gila Crossing
RAD	Robert A. Darrow	1940	
LSMC	Leonora S. M. Curtin	1940–1942	
SHH	S. H. Hastings	1916	vic. Sacaton
WLP	W. L. Pinney	1912	Gila River
KGS	K. G. Smith	1936	Sacaton
FAT	F. A. Thackery	1928	Sacaton
C&B	Castetter & Bell	1939–1940	Sacaton
Recent			
VLB	Vorsila L. Bohrer	1965–1966	Sacaton, Snaketown
H&T	Rodney Hastings & Ray Turner		
GPN	Gary Paul Nabhan	1974–1978	
ELS	E. Linwood Smith	1975–1976	Sierra Estrella
AMR	Amadeo M. Rea	1963–1993	
PF	Paul A. Fryxell		
TI	Takashi Ijichi	1985	Sierra Estrella
EGS	Eric G. Sundell	1973–1974	
EL	Eleanor Lehto	ca. 1960s	lower Salt River

Some plants with a more westerly distribution reach only as far east as Gila Pima country. These include

Desert Agave, *Agave deserti* (±)

Desert Fleabane, *Erigeron oxphyllus*

Desert Lily, *Hesperocallis undulata*

Bush Penstemon, *Keckiella antirrhinoides microphylla*

Colorado River Fishhook Cactus, *Mammillaria tetrancistra*

Diamond Cholla, *Opuntia* X *ramosissima*

Spanish Needles, *Palafoxia linearis*

Mohave Sage, *Salvia mohavensis*

Still other species range westward only as far as the reservation. These include

Bunched Ayenia, *Ayenia compacta* (*A. filiformis*)

Little-leaf Ayenia, *Ayenia microphylla*

Gila Cholla, *Opuntia* X *kelvinensis*

Staghorn Cholla, *Opuntia versicolor*

Rock Sage, *Salvia pinguifolia*

The first Arizona record of Alkali Weed (*Nitrophila occidentalis*) was from the vicinity of Gila Crossing village, apparently in the saline flats of the lower Queen Creek drainage (*George W. Plumb, 75; RHP, 13232, ARIZ*). Agricultural developments in the Lone Butte area have altered the local ecology, and this plant has not been found there since, although both Pickleweed (*Allenrolfea occidentalis*) and Greasewood (*Sarcobatus vermiculatus*)—two other halophytes—still occur. In 1964 I collected the rare *Cressa truxillensis* at Komatke (*AMR, s.n., ASU*) but have not rediscovered it since the Indian school sewage system was improved. This was also on the Queen Creek drainage. There were two earlier collections from the reservation. Near the eastern edge of St. John's Cemetery is the northernmost known colony of Thornber's Fishhook Cactus.

The vegetation of the Sierra Estrella is particularly interesting. Many species reach their northern, northeastern, or western distributions in this range. In addition, a northern species of wild-heliotrope, *Phacelia rotundifolia,* was recorded (Sundell 1974) at its southern limits in the Estrellas. The upper slopes of the sierra preserve relictual communities

of plants that in the Pleistocene would have been the dominants over much of the lower area that is now desertscrub (Brown 1978; Van Devender and associates 1987). Relicts in this community include

Crucifixion-thorn, *Canotia holacantha*

Narrow-leaved Goldenweed, *Ericameria laricifolia (Haplopappus)*

Red-seeded Juniper, *Juniperus erythrocarpa*

Bush Penstemon, *Keckiella antirrhinoides microphylla*

Shrub Live Oak, *Quercus turbinella*

Oak-belt Gooseberry, *Ribes quercetorum*

Mohave Sage, *Salvia mohavensis* (?)

Rock Sage, *Salvia pinguifolia*

Yerba-de-la-flecha, *Sapium biloculare*

Arizona Rosewood, *Vauquelinia californica*

With the warming and drying climate during the Holocene, these species have increasingly been found only on the more mesic northern- and eastern-facing slopes of the higher peaks. In addition, the only colonies of Alkali Sacaton (*Sporobolus airoides*) I found anywhere on the reservation were in the Upper Basin, at 2,950 feet. The grasses (24 species) and ferns (8 species) are notably rich in this range.

Desolation of Native Species

The flora of the Gila River reservation, most especially that in the lowlands, has undergone historic changes, particularly from 1875 to 1975, that can be described only as catastrophic (see table 6.2). Ten of the locally extinct plants are aquatics—emergents, submergents, or floating plants. Another seven or eight are typically from saturated soils along streams or in cienegas. Not only have individual species been driven to extirpation (local extinction), but entire plant communities have been eradicated. At work is a process called desertification, which is a phenomenon of both arid and semiarid country.

The earmarks of desertification (Sheridan 1981) are especially well marked in Gila Pima country: (1) reduction of surface streamflow, (2) declining water table, (3) salinization of topsoil and water, (4) severe soil erosion, and (5) desolation

of native vegetation. According to one study, "approximately 10% of the United States land mass is in a state of severe or very severe desertification. The actual acres *threatened* by severe desertification, however, are almost twice that amount" (Sheridan 1981:4). What has happened on the middle Gila is thus a microcosm of what is happening or what may happen in other arid and semiarid regions.

As a consequence of the four earmarks of desertification, the fifth—the desolation of native vegetation—has occurred to an alarming degree along the Gila (table 6.2). In Gila Pima country we can see the replacement of native species by exotic species, the loss of emergent and submergent vegetation, the loss of cienega-type vegetation, the loss of riparian timber, the loss of most mesquite bosques, the loss of grasslands, and the loss of traditional farming with its manmade oases.

The replacement of native plants by aggressively colonizing exotics is less evident in the actual number of species than in the enormous area, principally on the floodplains, that has been invaded. Some exotic plants that have taken advantage of the disturbed ecological conditions in the lowlands include

Giant Reed, *Arundo donax*

Five Hook Bassia, *Bassia hyssopifolia*

Wild Mustard, *Brassica tournefortii*

Malta Star Thistle, *Centaurea melitensis*

Field Bindweed, *Convolvulus arvensis*

Filaree, *Erodium cicutarium*

Cheeseweed, *Malva parviflora*

Bur Clover, *Medicago polymorpha*

Sweet Clover, *Melilotus indicus*

Tree Tobacco, *Nicotiana glauca*

sorrel, *Rumex dentatus*

Russian-thistle, *Salsola australis*

Milk-thistle, *Silybum marianum*

London Rocket, *Sisymbrium irio*

Salt-cedar, *Tamarix ramosissima*

In addition, most of the common grasses of the lowlands are Eurasian weeds:

Wild Oat, *Avena fatua*

Red Brome, *Bromus rubens*

Bermuda Grass, *Cynodon dactylon*

Table 6.2 Plants Collected in the Early 20th Century on the Reservation, Subsequently Extirpated

Scientific Name	Collector[1]	Herbarium[2]	Scientific Name	Collector[1]	Herbarium[2]
Aizoaceae			Plantaginaceae		
Sesuvium verrucosum	MFG	ARIZ	*Plantago lanceolata*	PHK	ARIZ
Alismataceae			*Plantago major*	PHK	ARIZ
Sagittaria cuneata	PHK	ARIZ	*Plantago rhodosperma*	PHK	ARIZ
Asteraceae (Compositae)			Poaceae (Gramineae)		
Eclipta prostrata (alba)	PHK	ARIZ	*Phragmites australis*		
Brassicaceae (Cruciferae)			*Sporobolus wrightii*	FR, AMR	US[3], SD
Nasturtium officinale	PHK	ARIZ	Polygonaceae		
Rorippa curvisiliqua	PHK	ARIZ	*Rumex violascens*	SHH	ARIZ
Rorippa hispida	PHK	ARIZ	Portulacaceae		
Rorippa islandica	PHK	ARIZ	*Calandrinia ciliata*	JJT	ARIZ
Rorippa obtusa	PHK	ARIZ	Potamogetonaceae		
Chenopodiaceae			*Potamogeton pectinatus*	IT, SHH, PHK	ARIZ
Nitrophila occidentalis	GWP, PHK	ARIZ	Primulaceae		ARIZ
Cucurbitaceae			*Androsace occidentalis*	PHK	
Brandegea bigelovii	MFG, EGS	ARIZ, ASU	*Samolus parviflorus*		
Cyperaceae			(*S. floribundus*)	PHK	ARIZ
Eleocharis macrostachya	PHK	ARIZ, US	Ranunculaceae		
Scirpus acutus	EL	ASU	*Myosurus cupulatus*	PHK	ARIZ
Scirpus americanus	PHK	ARIZ	*Myosurus minimus*	JJT, PHK	ARIZ
Scirpus maritimus	PHK, RRH	ARIZ	*Ranunculus scleratus*	PHK	ARIZ
Scirpus validus	PHK	ARIZ	Salicaceae		
Gentianaceae			*Salix exigua*	PHK	ARIZ
Centaurium calycosum	PHK	US	Saururaceae		
Eustoma exaltatus	JWT	ARIZ	*Anemopsis californica*	LSMC	DES
Haloragaceae			Scrophulariaceae		
Myriophyllum aquaticum			*Bacopa rotundifolia*	WLP	ARIZ
(*brasiliense*)	PHK	ARIZ	*Mimulus glabratus*	PHK	ARIZ
Juncaceae			*Veronica anagallis-*		
Juncus interior			*aquatica*	PHK	ARIZ
var. *arizonicus*	PHK	ARIZ	Solanaceae		
Juncus macrophyllus	PHK	ARIZ	*Nicotiana attenuata*	PHK, C&B	ARIZ, NM?
Juncus tenuis var. *dudleyi*	PHK	ARIZ	*Solanum americanum*	PHK	ARIZ
Lemnaceae			Zygophyllaceae		
Lemna gibba	WLP, PHK	ARIZ	*Kallstroemia californica*	PHK	ARIZ
Oleaceae			*Kallstroemia grandiflora*	PHK	ARIZ
Fraxinus velutina			*Kallstroemia parviflora*	PHK	ARIZ
var. *glabra*	JJT, PHK	ARIZ			

[1] See table 6.1.
[2] Herbaria containing middle Gila River collections: ARIZ = University of Arizona, Tucson; ASU = Arizona State Museum, Tempe; DES = Desert Botanical Garden, Phoenix; NM = University of New Mexico, Albuquerque; SD = San Diego Natural History Museum; US = U.S. National Museum, Smithsonian Institution.
[3] Artifact: hairbrush made from roots, collected by Frank Russell.

Wild Barley, *Hordeum leporinum*

Foxtail, *Hordeum murinum*

Littleseed Canary Grass, *Phalaris minor*

Rabbitfoot Grass, *Polypogon mon-sp;eliensis*

Camel Grass, *Schismus arabicus*

Mediterranean Grass, *Schismus barbatus*

Johnson Grass, *Sorghum halepense*

Most of the noxious grasses and other weeds with which the Pima must contend fall into this roster of exotics. Not all are invaders; some weedy species such as Desert-broom (*Baccharis sarothroides*) and Spiny Aster (*Aster spinosus*), for instance, are natives. But these Old World species are the dominants in the contemporary lowlands.

Emergent and submergent vegetation is gone from the Gila River channel between the Salt-Gila confluence and the Ashurst-Hayden Dam east of Florence. In other words, there is no permanent habitat left for water plants on any part of the Gila now enclosed by the reservation.

The cienega-type (marsh) vegetation also has all been lost. The major cienega in Gila Pima country was an area called Santa Teresa by Hispanic itinerants and Maricopa Wells during early Anglo settlement. There the water table was high; the soils were permanently saturated and probably mostly alkaline. The plants expected to have been there (they disappeared before being collected) include

Yerba Mansa, Yerba-del-manso, *Anemopsis californica*

Slim Aster, *Aster subulatus (A. exilis)*

Canchalagua, *Centaurium calycosum*

flat-sedge, *Cyperus* spp.

False Daisy, *Eclipta prostrata (E. alba)*

spike-rush, *Eleocharis* spp.

Catchfly Gentian, *Eustoma exaltatum*

Camphor-weed, *Heterotheca subaxillaris*

Water Pennywort, *Hydrocotyle verticillata*

rush, *Juncus* spp.

Yellow Waterweed, *Ludwigia peploides*

monkeyflower, *Mimulus* spp.

mousetail, *Myosurus* spp.

Alkali Weed, *Nitrophila occidentalis*

paspalum, *Paspalum* spp.

Salt-marsh Fleabane, *Pluchea odorata*

Willow Smartweed, *Polygonum lapathifolium*

Celery-leaved Buttercup, *Ranunculus sceratus*

cress, *Rorippa* spp.

dock, *Rumex* spp.

bulrush, *Scirpus* spp.

Salt Marsh Sand Spurry, *Spergularia marina*

Big Sacaton, *Sporobolus wrightii*

Blue Streamwort, *Stemodia durantifolia*

Southern Cattail, *Typha domingensis*

Water Speedwell, *Veronica anagallis-aquatica*

This list of probable component species is based on analogy with similar areas with saturated soils on or adjacent to the reservation (Rea 1983:53–56; see also 84–85).

Gone also from the reservation is the riparian timber mentioned in historical documents and oral historical accounts (chapter 3). Two trees dominated this woodland: Frémont's Cottonwood (*Populus fremontii*) and Goodding's Willow (*Salix gooddingii*). The understory probably included

Seep-willow, *Baccharis salicifolia*

Velvet Ash, *Fraxinus velutina*

Velvet Mesquite, *Prosopis velutina*

Coyote Willow, *Salix exigua* (possibly)

Mexican Elder, *Sambucus mexicana*

Many of the species from the cienega probably occurred at the edges of the riparian timber in the small creeks and marshes or on the riverbanks.

Mesquite bosques have disappeared from all but the lower end of the reservation, where the water table is somewhat higher. Over most of the remainder of Gila Pima land, mesquites occur only as scattered individual trees. These bosques are predominantly single-species stands with admixtures of Quail Brush (*Atriplex lentiformis*), Four-wing Saltbush (*A. canescens*), and Graythorn (*Ziziphus obtusifolia*). Mesquite was one of the major economic species for the Pima, and dead trees are still a significant fuel source.

Pima grasslands, as noted in earlier chapters, have disappeared without a trace. Two of the species that would have occurred in them are Desert Salt Grass (*Distichlis spicata*) and Alkali Sacaton (*Sporobolus airoides*). Pima fire drives in areas distant from rancherias and adjacent fields may have contributed to the

maintenance of these desert grasslands by suppressing woody plants such as Jimmyweed (*Isocoma acradenia*) and mesquites.

Cultural Impact of Species Losses

Species losses affect not only the biotic diversity and stability of the local environment but also the cultural stability of the people who have come to rely on that environment. Many species lost or significantly reduced in numbers were of cultural significance to the Akimel O'odham (table 6.3). Some were merely named, that is, were linguistically noticed; and their purpose was said to be "just to beautify the desert." But others had economic importance to the riverine Pima, such as cattails and willows. Such plants are now confined to irrigation ditches and wastewater runoff areas and to the lower Salt River, which forms the northern boundary of the reservation. One major medicinal plant, Yerba Mansa, can no longer be found in Gila Pima country or even the surrounding vicinity. The cultural details for each of these species are given in the individual folk generic accounts.

Finally, with the demise of traditional Pima farming, the rancheria oases disappeared. Each individual rancho or family settlement was an oasis with a house cluster, lateral canals, *charcos* (small ponds

Table 6.3 Species of Cultural Significance, Extirpated or Nearly Lost from the Reservation

Group	Scientific Name	Significance	Extirpated	Nearly Lost
A	*Azolla filiculoides*	named	X	
A	*Lemna* spp.	named	X	
A	*Zannichellia palustris*	named		X
B	*Phragmites australis*	multiple uses		X*
B	*Sporobolus wrightii*	brushes	X	
E	*Castela emoryi*	medicine		X
E	*Chilopsis linearis*	named		X*
E	*Fraxinus velutina*	named	X	
E	*Maclura pomifera*	bows, living fences		X*
E	*Populus fremontii*	multiple uses		X
E	*Salix exigua*	storage basket base	X	
E	*Salix gooddingii*	multiple uses		X
E	*Sambucus mexicana*	medicine, food		X
F	*Eremalche exilis*	greens	X	
F	*Rumex violascens*	greens	X	
G	*Acourtia nana*	named		X*
G	*Anemopsis californica*	medicine	X	
G	*Nicotiana attenuata*	ceremonial smoking	X	
G	*Rumex hymenosepalus*	medicine		X

*Known from a single plant or clone.

of water for domestic use), fields, and fence rows (see chapter 4). Traditional ranchos that were still functioning when I arrived in the early 1960s sometimes included large cottonwoods and mesquites, pomegranates, Mexican Elder, Graythorn, and wolfberries. These have almost entirely been replaced by what one reservation agricultural agent proudly called "clean farming": large fields with clean borders maintained by herbicides, monoculture cropping, and concrete ditches from which phreatophytes (water-loving plants such as cottonwoods) have been carefully eradicated. There, sterility and monotony abound. The Gileños, meanwhile, have been resettled from their discreetly spaced ranchos into new cinder-block housing projects. For various reasons, few Pimas can now farm tribal lands. Health problems increase, especially diabetes and its consequences, because traditional foods are now seldom eaten (see chapter 7).

What's Left?

The flora in the three major ranges in Gila Pima country is largely unmodified. The floristically richest range, the Sierra Estrella, is protected by its ruggedness; all but the most vigorous visitors (human and animal alike) are dissuaded from ascending its ridges. To my knowledge, the only grazing in this range (as well as the Santan and Sacaton Mountains) is by a few deer and Desert Bighorn and smaller native herbivores. The north end of the sierra is a park, as is nearly all of the South Mountains; this status gives the desert mountains some protection from human intrusion.

The more gently sloping bajadas are more vulnerable; they are sometimes the scenes of drinking parties and the vandalizing of native plants and petroglyphs. In addition, they are accessible to grazing, but I suspect that wild horses may be the only large herbivores there regularly. (While writing this book I counted more than 140 wild horses in a large herd crossing Maricopa Road at the Gila River.) Plants on the floodplains, by contrast, are severely impacted by human activities, especially the wholesale clearing of saltbush flats for mechanized farming.

Regenerative Capacities

Most plant species lost through desertification in Gila Pima country are associated with riparian and marsh habitats. To a certain extent, regeneration is possible (Rea 1983:52–60, 1988). One example is Barehand Lane marsh between Maricopa and Co-op colonies, where a biologically diverse aquatic habitat redeveloped slowly from irrigation tailwater. Another well-documented case is the lower Salt River at the north boundary of the reservation. There, treated sewage effluent that was turned into the riverbed eventually produced a cottonwood-willow gallery forest with some understory. Shallow waters in places harbored marshes with cattails, bulrushes, flat-sedges, and other emergent vegetation (table 6.4). The current for the most part was too swift and the water turbidity too great for submergent or floating species. Many riparian birds, both breeding and wintering, colonized these areas.

Several other species have made spontaneous comebacks elsewhere in Pima country. For instance, Big Sacaton (*Sporobolus wrightii*) in Upper Santan,

Table 6.4   Plants That Have Colonized or Recolonized the Reservation since 1960

Scientific Name	Barehand Lane Marsh	Lower Salt River
Apiaceae (Umbelliferae)		
Hydrocotyle verticillata	X	X
Asteraceae (Compositae)		
Eclipta prostrata (alba)		X
Pluchea odorata (purpurascens)	X	X
Brassicaceae (Cruciferae)		
Brassica tournefortii		X
Cyperaceae		
Cyperus erythrorhizos	X (sp.?)	X
Cyperus odoratus		X
Eleocharis caribaea		X
Eleocharis parvula		X
Scirpus acutus		X
Scirpus maritimus		X
Fabaceae		
Sesbania exaltata		X
Lemnaceae		
Lemna gibba		X
Spirodela punctata		X
Onagraceae		
Ludwigia peploides	X	X
Poaceae (Gramineae)		
Arundo donax		X
Echinochloa crusgalli		X
Leptochloa uninervia		X
Polygonaceae		
Polygonum lapathifolium	X (sp.?)	X
Polygonum persicaria		X
Rumex crispus	X	
Rumex dentatus	X	X
Pontederiaceae		
Heteranthera dubia	X	
Ranunculaceae		
Ranunculus scleratus	X	
Salicaceae		
Salix gooddingii		X
Scrophulariaceae		
Stemodia durantifolia		X
Veronica anagallis-aquatica	X	
Solanaceae		
Solanum americanum	X	X

Note: All AMR collections, deposited in ARIZ or SD.

Coyote Willow *(Salix exigua)* in Sacaton, and Carrizo *(Phragmites australis)* in Sacaton Flats have appeared. (But in each case the colonizing plants were eradicated.)

Sensitive Habitats, Threatened Species

Our concept of endangered species is an unfortunate one: it concentrates attention on species that are so rare by the time that they are recognized that they may be beyond hope, diverting attention from the larger problem of sensitive habitats that are being devastated by human impact. (This is another case in our technological society where we find comfort in treating symptoms rather than diseases and their underlying causes.) Biologically devastated habitats on the reservation include pond, marsh, cienega, riparian, grassland, and bosque communities, as noted above. Many of these no longer exist there in any recognizable form.

This kind of misplaced attention is well illustrated by an experience I had one late spring day while studying breeding birds on the Gila River in Yuma County near Tacna, Arizona (Rea 1983:108). The quiet morning was shattered by the sound of great tractors coming up the riverbed dragging rigs that plowed the dense marsh vegetation into the sand. Soon there was nothing left of the marsh except one small patch of cattails. I inquired of the governmental personnel what was the purpose of the destruction and was told that this was a "flood control" project.

"But why are the tractors avoiding that one little spot out there?" I asked.

"Oh, that. There's a Clapper Rail breeding in that patch. It's an endangered species."

Overlooked was the fact that an endangered species is part of a complex community of organisms all dependent on an intact marsh. A single patch of cattails with one Clapper Rail nest (which I am sure was quickly abandoned) is biologically useless.

Because of habitat destruction on the reservation, many species of plants and animals are threatened with local extinction:

Acleisanthes longiflora

Desert-holly, *Acourtia nana*

Crucifixion-thorn, *Castela emoryi*

Virgin's Bower, *Clematis drummondii*

Commicarpus scandens

Finger-leaved Gourd, *Cucurbita digitata*

Saltbush Barrel Cactus, *Ferocactus wislizeni*

Desert Lily, *Hesperocallis undulata*

Colorado River Fishhook Cactus, *Mammillaria tetrancistra*

Thornber's Fishhook Cactus, *Mammillaria thornberi*

Cane Cholla, *Opuntia spinosior*

Staghorn Cholla, *Opuntia versicolor*

Desert Night-blooming Cereus, *Peniocereus greggii*

Yellow Mistletoe, *Phoradendron tomentosum*

Canaigre, *Rumex hymenosepalus*

Mexican Elder, *Sambucus mexicana*

Also threatened are most native grass species in the bajadas and floodplains and any surviving members of the families Cyperaceae and Juncaceae (sedges and rushes). This is a highly conservative list of only the more outstanding examples. The list includes some plants that are widespread but in extremely low numbers (such as the Colorado River Fishhook Cactus; I have found only three individuals on the reservation in more than a quarter-century of fieldwork). Other plants have very narrow habitat requirements. The Desert Lily, for example, clings for survival on a few sand dunes in the Lone Butte area. "Development" of this area could eradicate the easternmost populations of this species. Other species, such as Crucifixion-thorn, Desert Night-Blooming Cereus, and Saltbush Barrel Cactus, grow in habitats that are prime targets for agricultural development.

Ecological destruction surrounds Pima country in the urbanization and agricultural development of the Phoenix metropolitan area on one side and the Casa Grande and Coolidge area on the other. It remains to be seen whether the reservation can preserve an interesting and unique local biota as well as a people and their way of life.

Chapter 7

A Dietary Reconstruction

Their regular fields, well-made irrigating ditches, and beautiful crops of cotton, wheat, corn, pumpkins, melons, and beans have not only gladdened the eye but also given timely assistance to the thousands of emigrants who have traversed Arizona on their way to the Pacific.

Sylvester Mowry, 1850

It is not known for certain when the Gileños arrived at their riverine oases, but all versions of the Pima Creation Epic, as well as their oral history, tell of a Piman conquest of a puebloid people who were living in these desert valleys. The invaders were coming from somewhere within the Tepiman heartland, most likely the northern half, and they were culturally preadapted to utilize the flora and fauna of their new home at the northern fringes of the Sonoran Desert.

Even their linguistic baggage, the simplest monosyllabic and bisyllabic plant names, labeled species found in the low, hot deserts of northwestern Mexico and the adjacent sierran foothills as well as in their new home along the Gila River (see chapter 8). No adjustment time was necessary.

The O'odham living at the desert's green edge made a good living. In spite of what appears to the outsider to be a rigorous environment, the Gila River Pima maintained a stable culture in their desert riparian ecosystem for at least three centuries. Numerous documentary accounts indicate that the population was well within the local carrying capacity; Jesuit and Franciscan travelers in the Hispanic period and Anglo-Americans in the 19th century marveled at the industrious and provident people they encountered on the Gila. Early in the Anglo period, the Gileños responded to floods of military and civilian Anglo immigrants, producing prodigious amounts of grain and other food supplies for commerce far in excess of their own needs.

But in the 20th century it is not for their highly successful adaptation to desert living that the Gila River Pima and their kinfolk the Salt River Pima are most famous. Rather, it is for having the highest known incidence of adult-onset diabetes of any ethnic group in the world. Non–insulin dependent diabetes mellitus, known by the acronym NIDDM, is part of a syndrome of lipid and glucose metabolic dysfunctions that includes gallbladder disease, obesity, and hypertension. Although the Pima now have the highest rates, they appear to be merely the tip of the iceberg of ethnic populations predisposed to this condition. As world diets become "modernized," probably all indigenous peoples will be at high risk except Europeans, north Africans, and western Asians.

For more than 99 percent of human history, cultures have lived as hunter-gatherers (Lee and DeVore 1968:3). Genetic insulin resistance was apparently advantageous under most aboriginal conditions. There seems to have been selection for insulin sensitivity only relatively recently among specific populations with a long association with animal husbandry and intensified cereal agriculture.

Diabetes was nearly unknown among the Pima early in the 20th century. But by World War II, significant changes had occurred in the Pima environment, culture, and eating habits. Through a series of land-use changes, primarily in the watersheds, the Gila suffered irreversible damage (see chapter 5). The River People became a people without a river, and the economic backbone of the tribe was destroyed. During the first half of the 20th century, the O'odham switched from what was essentially a subsistence economy to a cash economy, and their diet changed radically. Physiologies that had evolved through millennia for diets obtained in arid and semiarid environments were suddenly (in a genetic and evolutionary sense) confronted with new metabolic challenges. By the late 20th century, more than 75 percent of Pima Indians over 35 years old were overweight or obese, and more than half suffered from diabetes (Ravussin and Knowler, 1993).

The nutritionally poor modern diet (Smith and associates, n.d.; Smith and Pablo, 1993) is a dim reflection of what the aboriginal and early colonial Pima once ate. One might wonder if it is even possible this late in the acculturation process to reconstruct the aboriginal diet of the River People. Qualitatively, and to a certain extent quantitatively as well, I think the answer is yes. The historical subsistence diet is known in considerable detail. Older people living today grew up when a cash economy still hardly affected the diet. Roads at the time were scarcely passable during parts of the year, and flooding rivers sometimes cut off travel altogether. Most store commodities required a rough wagon trip into Phoenix or one of the other off-reservation Anglo settlements. Farming and gathering were everyday activities for the Gileño men and women

who tell their stories in the folk generic accounts that follow. And all the Pima men who were major consultants for this book said that in their youth they hunted with bows and arrows, keeping the family pot filled with game.

Protohistoric Subsistence Diet

Permanent Jesuit contacts in the colonial period were limited to southern and central Pimería Alta, and documentary accounts of Gila country are not sufficiently detailed to offer insights into the subsistence economy of the Gileños. However, using ethnographic analogy and Pima oral history, we can reconstruct the Pima diet as it was before the introduction of Eurasian crops and animals and before extensive ecological changes occurred in the riverine habitat. As a baseline, the protohistoric Gileño dietary is assumed to include all indigenous crops and all native plants and animals known to have been harvested during historic times.

At least 99 biological species of plants entered the dietary of the Gila River Pima: 30 are postcontact species, including 16 introduced annual crops and 14 fruit trees (the list excludes crops and products that were known but were not named, such as olives, pepper, and okra, occasional dietary items). This leaves 69 native plants that are known to have been eaten at some time during the Gileños' annual cycle; at least 58 native wild species of plants were harvested to a greater or lesser extent at some time in history.

In an attempt to evaluate the relative importance of foods in the Pima diet, I have used the following terminology. I consider a plant to be a significant food (of either major or minor importance) if it was gathered and taken home for preparation and serving. A staple means a food that was stored and available for use throughout the annual cycle; some staples could be stored for years. Other dietary items were snack foods, eaten as they were encountered during the course of daily activities. Certain foods, both plant and animal, were considered specifically children's foods, sought by growing youngsters. Many plant and animal resources reached their lowest ebb in the dry Sonoran Desert spring, an anticipated annual event: stress-period foods were eaten during this late spring period when most stored resources were low.

Finally, a few were famine foods used only during times of starvation, a relatively infrequent event.

Wild Foods

Despite being an agricultural people whose skills were praised by both Hispanic and Anglo itinerants, the Gileños were until the 20th century dependent on wild harvest for more than half their annual diet. The major wild crops (table 7.1) included Velvet Mesquite (*Prosopis velutina*), Saguaro (*Carnegiea gigantea*), agaves or century plants (*Agave deserti* and perhaps other species), and chollas (*Opuntia* spp.).

Mesquite bosques were once found throughout Pima country. Excessive woodcutting during the late 1800s and early 1900s—the drought years—cleared most. Subsequent downpumping of groundwater prevented even these deep-rooted trees from reaching moisture in most parts of the reservation. Many bosques never regenerated. Before these ecological changes, the Pima stored great quantities of mesquite pods in large Arrow-weed (*Pluchea sericea*) baskets or bins. Usually two crops appeared each season, the first about June, the other later, about September. Mesquite pods that were crushed and then soaked in water produced a drink called *vau*. It was so frequently used at meals that native consultants compared it to modern coffee. Pods might be cooked with other foods such as the pulp from barrel cactus (*Ferocactus* spp.). The flour from crushed mesquite pods was made into an uncooked cake or loaf that was frequently served at meals. Various other dishes were made from Velvet Mesquite or Screwbean (*Prosopis pubescens*) pod mesocarp. Both ethnohistorical and documentary evidence indicate that mesquite must have been the staff of life for the Gila Pima.

The Saguaro provided another abundant and seldom failing crop, available in midsummer. So important was this plant in the annual cycle that the Piman new year began with the Saguaro harvest moon. The sweet pulp is filled with seeds rich in protein and fat. The fruit was dried whole as well as prepared as jam and syrup. The ground seeds were mixed with grains to make a porridge or made into a paste resembling peanut butter. Some of the syrup was fermented for the annual wine feast, an elaborate liturgical celebration intended to bring rain and continue it through the monsoonal growing season.

In addition to mesquite *vau,* the Pima made several other beverages. Wolfberry (*Lycium* spp.) crushed and boiled with water and sometimes whole wheat or white flour was drunk cold. Ripe Saguaro fruit made a similar cold drink.

A third major wild harvest was cholla buds, gathered in spring. The principal species, Buckhorn Cholla (*Opuntia acanthocarpa*), was ready in March or April, the secondary species, Pencil Cholla (*O. arbuscula*), about a month later. The green calix with the entire unopened flower was picked, pit-roasted with Seepweed (*Suaeda moquinii*), then eaten fresh or dried for future use. When dried, the roasted buds occupied little storage space and lasted indefinitely.

Pit-roasted agave hearts may have been the fourth major wild resource in aboriginal times. Although their procurement was labor intensive, the yield was great. The dried product was very sweet and could be stored indefinitely. The drawback was that the agave harvest took the Gileños away from the relative security of their villages into the outlying desert ranges, where they were vulnerable to attacks by the Apache. (Athabascan raiding probably became a serious threat about the time of the arrival of the Jesuits in northern Pimería Alta. Earlier the Pima probably exploited with impunity all the desert ranges lying in their territory.)

Wild greens were an important part of the Pima diet (see Group F). Most of the species were eaten either boiled or else boiled, strained, and fried. Many of the species gathered grew in profusion as a result of the Gileños' agricultural practices. The annual cycle of greens covered most months of the year. If winter rains came from the Pacific, Patota (*Monolepis nuttalliana*), a small chenopod, grew on bare, disturbed soils of the floodplains. In late winter and early spring, Star Mallow (*Eremalche exilis*), wild-heliotrope (*Phacelia* spp.), and fiddleneck (*Amsinckia* spp.) grew on the bajadas, and wild sorrel or dock (*Rumex* spp.) in the marshes. (All four of these vanished from the Pima dietary; they are likely to have been spring stress-season foods.) In the hot spring, several goosefoots (*Chenopodium* spp.) and two annual saltbushes (*Atriplex* spp.) were (and by some Pimas still are)

harvested in abundance. During summer, Pima fields produced two other important greens: Verdolaga (*Portulaca oleracea*) and Palmer's Carelessweed (*Amaranthus palmeri*). The leafier species, especially the saltbushes and amaranths, were dried and stored for year-round use. Between the fresh and dried products, the Pima were never without their greens throughout the year.

Many kinds of seeds were prepared by being parched with live coals (*haak*), then ground into flour (*chu'i*). The resulting pinole was easily transportable and could be eaten simply by adding water. In addition to *huuñ haakĭ* (maize pinole) and later *haakĭ chu'i* (wheat pinole), roasted flour could be made from the seeds of both cultivated and wild amaranths, canary grass (*Phalaris* spp.), Chia (*Salvia columbariae*), Desert Ironwood (*Olneya tesota*), Patota, and Tansy-mustard (*Descurainia pinnata*). During famines, even Pickleweed (*Allenrolfea occidentalis*) and Quail Brush (*Atriplex lentiformis*) seeds could be parched.

Several seed plants were used to make mucilaginous drinks that we know were highly esteemed, even though they have completely fallen from the Pima dietary during the past half-century: these include Chia and Indian-wheat or Woolly Plantain (*Plantago* spp.). Certain conditions, particularly winter and spring rains, produced great crops of wild plants. The Pima fire drives (Rea 1979) also probably stimulated the growth of wild plants in specific areas, making harvesting easier. A drink was made from Tansy-mustard, a native plant, and later an introduced mustard, London Rocket (*Sisymbrium irio*), which is now abundant in Pima country. The seeds of both of these were mixed with water or were parched before being mixed with water to make a kind of pinole. The seeds of all four of these species absorb great quantities of water when immersed, making a mucilaginous mass. With changing taste prejudices, these fell out of favor. One old Pima described the Chia drink to me as "something nasty, like when you have a cold."

Certain foods were snack foods eaten primarily by children. These included the tubers of several bulrushes (*Scirpus* spp.) and nut-grasses (*Cyperus* spp.), the bulbs of Bluedicks or Covena (*Dichelostemma pulchellum*), the pulp of a *Mammillaria* cactus, the fruits of Engelmann's Hedgehog

Cactus (*Echinocereus engelmannii*) and Wright's Ground-cherry (*Physalis acutifolia*), and adult cicadas (Cicadidae). Later the fruit of Northern Fan Palm (*Washingtonia filifera*) was added.

Other foods were snacks indulged in by all age groups. Both the white and the amber sap from Velvet Mesquite and Screwbean were relished. About February the catkins or male flowers of Frémont's Cottonwood (*Populus fremontii*) were eaten as is, as were willow catkins and mesquite flowers later in spring. Squash blossoms were cooked with crushed wheat as a conventional soup. The oily seeds of both the wild and domesticated devil's claws (*Proboscidea* spp.) were a welcome addition to the diet, as were cotton (*Gossypium hirsutum*) and various cucurbit seeds. The female pods of the cottonwood, before they matured, provided a chewing gum.

Besides mesquite, the Gila Pima made use of several other arboreal legumes. The peas from Little-leaved or Foothill Paloverde (*Cercidium microphyllum*) were harvested and eaten raw while still tender or were cooked when halfway mature. Dry Desert Ironwood seeds provided another crop in midsummer; leached and roasted, they tasted like peanuts. Flour processed from Screwbean pods was used somewhat like that from the more abundant Velvet Mesquite.

Pima women dug the small tubers of Camote-de-raton, or Hog-potato (*Hoffmanseggia glauca*), even though these grew at considerable depths. Occasionally even the flesh of barrel cactus was prepared as a special dish with mesquite pods.

Both Graythorn (*Ziziphus obtusifolia*) and Mexican Elder or Desert Elderberry (*Sambucus mexicana*) grew abundantly in Pima fence rows; both produced fruits the Pima gathered. Graythorn was used until recently. With the drying of Pima country, the water-loving elderberry was almost forgotten, but earlier Hispanic missionaries had complained that the Upper Pimas in northern Sonora brewed a wine from this tree that kept the Indians drunk for days.

Two parasitic plants and one fungus also contributed to the aboriginal Pima diet. In late winter, Desert Broom-rape (*Orobanche cooperi*) appeared in washes and other sandy places. Just as this parasite began to emerge, while most of the succulent stalk was below ground, it was

Table 7.1 Wild Plants in the Gila River Pima Dietary

Scientific Name	Pima Name	Seasonality	Value/Importance	Source	Part(s) Eaten	Preparation
Acacia greggii	*uupaḍ*	3	Starv	L	S	?
Agave deserti	*a'uḍ, a'oḍ*	2	M-st	Ex	Heart	Pit
Allenrolfea occidentalis	*iitañ*	3,4	Starv	L	S	Par,B
Amaranthus palmeri	*chuuhuggia*	5,st	Major	L	L	B
Amoreuxia palmatifida	*shaaḍ*	3,4?	Minor	T	U	V
Amsinckia spp.	*cheḍkoadag*	2	Stress	L	L	B
Atriplex elegans	*toota onk*	2	Minor	L	L	B
Atriplex lentiformis	*eḍam*	3	Starv	L	S	Pit,Par,Wa
Atriplex wrightii	*onk iivagĭ*	5,st	Major	L	L	B
Capsicum annuum	*olas kokol*	5,st	M-st	T	F	R,B
Carnegiea gigantea	*haashañ*	5,st	M-st	L	F	V
Cercidium microphyllum	*kuk chehedagĭ*	2	Minor	L	S	R,B
Chenopodium berlandieri	*huáhi*	2	Minor	L	L	B
Chenopodium pratericola	*chuális*	2	Minor	L	L	B
Cucurbita digitata	*aḍavĭ*	5,st?	Snack	L	S	Ro
Cyperus spp.	*vashai s-uuv*	5	Snack	L	U	B
Descurainia pinnata/ Sisymbrium irio	*shuu'uvaḍ*	2	M-oc	L	S	Wa
Dichelostemma pulchellum	*haad*	2	Ch	L±	U	R
Echinocereus engelmannii	*iisvikĭ*	2	Snack	L	F	R
Eremalche (Malvastrum) exilis	*ñiádam*	2	Stress	L	L	B
Ferocactus spp.	*chiávul*	5	Minor	L	Pulp	B
fungi (Willow Mushroom?)	*kumul*	—	Minor	L	All	B
Hoffmanseggia glauca	*iikovĭ*	5	M-oc	L	U	B,Ro
Lycium spp.	*kuávul, kuáwul*	2	M-se	L	F	R,B
Mammillaria microcarpa	*ban mauppa*	5	Ch	L	Pulp	R
Monolepis nuttalliana	*opon*	1	M-Se	L	L	B
Olneya tesota	*ho'idkam*	3	Minor	L	S	Par
Opuntia acanthocarpa	*hannam*	2	M-st	L	Fls	Pit
Opuntia arbuscula	*vipnoi, vipĭnoi*	3	Minor	L	Fls	Pit
Opuntia engelmannii	*iibhai*	4	Minor	L	F	R
Opuntia leptocaulis	*navĭ, a'ajĭ navĭ*	4,1	Snack	L	F	R
Opuntia spinosior	*kokavĭ hannam*	2	Minor	L	Fls	Pit
Orobanche cooperi	*mo'otaḍk*	2	M-se	L	U	R,Ro
Phacelia spp.	*havañ taataḍ*	2	Stress	L	L	B
Phalaris minor, P. caroliniana	*baabkam, (pl.) baahapkam*	2	Minor	L	S	Par
Phoradendron californicum	*haakvoḍ*	2,3,4	Snack	L	F	R,B
Physalis acutifolia	*kekel viipiḍ*	4	Ch	L	F	R
Plantago insularis, P. purshii	*muumsh*	2	Major	L	S	Wa
Populus fremontii	*auppa*	1	Snack	L	Fl	V
Portulaca oleracea, P. retusa	*ku'ukpalk, ku'ukpaḍ*	3	M-Se	L	L	B
Proboscidea spp.	*ban ihugga*	3,4	Snack	L	S	R
Prosopis pubescens	*kuujul, kuujuls*	3,4	M-St	L	F	V
Prosopis velutina	*kui*	3,4	M-St	L	F	V
Quercus turbinella	*viyóōdi*	4	Snack	T	F	R
Rumex violascens, R. crispus	*vakoandam*	2	Minor	L	L	B
Salvia columbariae	*daapk*	2	M-oc	L	S	R,Par
Sambucus mexicana	*dahapdam*	3,4	Snack	L	F	R
Sarcostemma cynanchoides	*viibam*	2,3,4	Ch	L	Sap	Ro
Scirpus maritimus var. *paludosus*	*vak*	5	Snack	L	U	R

(continued on next page)

Table 7.1 Wild Plants in the Gila River Pima Dietary   *(continued from previous page)*

Scientific Name	Pima Name	Seasonality	Value/Importance	Source	Part(s) Eaten	Preparation
Sonchus asper	*s-ho'idkam iivagĭ*	2	Minor	L	L	R
Sonchus oleraceus	*huai hehevo*	2	Minor	L	L	R,B
Suaeda moquinii	*chuchk onk*	2	Minor	L	L	Pit
Typha domingensis	*uḍvak*	5	Snack	L	U	R
Washingtonia filifera	*hevhodakuḍ*	3,4	Ch	L	F	R
Yucca baccata	*hovij, hovich*	5,st	Minor	T	F	R,Pit
Ziziphus obtusifolia var. *canescens*	*u'us chevaḍbaḍ*	3	Minor	L	F	R,B

Abbreviations:
Seasonality: 1 = winter; 2 = spring; 3 = summer; 4 = fall; 5 = year-round (stored).
Value/Importance: Ch = children's food (primarily); Major = major food; Minor = minor food; M-se = major food, seasonal; M-oc = major food, occasional; M-st = major food, staple; Snack = snack food; Starv = starvation food; Stress = stress-season food (spring).
Source: L = local gathering; Ex = long-distance expeditions; T = trade (Tohono O'odham).
Part(s) Eaten: F = fruits; Fls = flowers; L = leaves; Pulp = stems of succulent plants; S = seeds; U = underground parts (roots, tubers, bulbs).
Preparation: B = boiled; Par = parched; Pit = pit roasted; R = raw; Ro = roasted (baked on coals); V = various methods; Wa = mixed with water.

gathered and eaten cooked or raw. Desert Mistletoe (*Phoradendron californicum*) often infested arboreal legumes. The pinkish berries were eaten cooked or raw. One mushroom (probably *Lepiota rachotes*) that grew on the trunks and fallen logs or willows along the river was made into a gravy.

Through trade with the Tohono O'odham or desert Pimans, the Gila Pimas acquired Saya (*Amoreuxia palmatifida*), acorns, Chiltepines (*Capsicum annuum*), cakes of Banana Yucca (*Yucca baccata*), and several other plants that did not grow in Pima country.

The Pima were not without sweets, although some sources supplied only very small amounts. In addition to Saguaro fruit products and the dried yucca cakes and agave heads and hearts already mentioned, the provisions of several insects were sources of sweets. A carpenter bee that bored into dried agave stalks stored pollen and nectar for its larvae; another bee put regurgitated nectar into chambers in the ground. Both were robbed. In addition, a scale insect secreted a kind of honeydew on the leaves of certain plants that could be licked or scraped off. A Pima described this to me as "like manna in the Bible."

Cultivated Crops

At least six food crops formed the backbone of Gila Pima aboriginal agriculture (see Group I): maize or corn, squash and pumpkins, teparies, limas, and grain amaranth and chenopod. Because of the sparse archaeological record, it is not known how long these were among the protohistoric Pima, but the first four (corn, squash, and the two species of beans) are well attested to in Piman song, symbolism, ceremony, myth, and oral tradition. The two pseudo-cereals, amaranth and chenopod, persisted into the late 19th century, then vanished. Also, it is not known whether the Pima cultivated the common bean or kidney bean in this early period. There is internal evidence that they did not; unlike the tepary and lima, the common or kidney bean is not well adapted to the local environmental conditions. There were additional nonfood crops such as cotton and devil's claw, whose seeds were secondarily snack food. Domesticated or semidomesticated grasses may also have been among the crops, but if they were, they fell out with the introduction of wheat and Eurasian herbivores.

All these crops were readily stored for use throughout the year. Even the flesh of the two cucurbits was dried for later use, as were their seeds. At least one pumpkin stored well whole.

Maize was undoubtedly the major grain food. Its symbolism pervaded myth, song, and ceremony. In legend, it is a fickle deity. Presumably this reflects its variable productivity in reality.

The spring crops were planted whenever the mesquites leafed out, when they would be safe from freezing weather. At that time, the snows melting in the Gila headwaters raised the level of the river so that fields on the lower terraces and islands could be readily irrigated.

A second planting came at the beginning of the summer monsoon season when the river rose once again. During this "short planting" the Pima hoped to gather a second crop before the late autumn frosts. With their 60-day corn varieties and other quick-growing localized landraces, they were usually successful.

There are hints that some native grasses were cultivated. These might have been winter crops, a third planting. A recent suggestion has been made that the Upper Pima, like their Hohokam predecessors, were cultivating agaves on terraced bajadas or piedmonts of the Southwest (Dobyns 1988). There is no ethnohistorical or ethnographic evidence among the Gileños for such a hypothesis, even though this was a widespread practice among many (if not most) Mesoamerican groups.

Animal Foods

Of the major forms of animal protein procurement—garden hunting, communal drives, fishing, and big-game hunting—the last was the least significant for the Gila Pima. On a day-to-day basis, fishing and garden hunting of small rodents and lagomorphs (rabbits and hares) kept animal meat in the pot, as did the continuous trapping of small granivorous birds such as migratory sparrows and buntings in winter and quail and doves year-round. In other words, the Pima exploited those animals that exploited the lush local conditions that the Pima themselves created or helped to maintain by their water management practices.

Fishing was once a mainstay of animal protein for the Gileños, as attested by the earliest explorers. Older Pimas described to me 9 named species of fish, 7 of them native. Of these 9, 7 were important food resources for a fish-loving tribe. (Not one of these survives on the middle Gila today!)

Stress and Famine Foods

In spite of wild crops, cultivated crops, and both hunting and fishing, the Pima complained of a stress period or hard time in the spring. During April and May (and, aboriginally, perhaps June), stored goods were at their lowest ebb and game animals least abundant. This may have been the time of the highest consumption of wild greens. But it was also the time of the seasonal availability of wolfberries and cholla buds, two major wild crops that have maintained their cultural importance even today.

There were actual starvation times in addition to the annual spring stress period. Drought could affect the local soil moisture and the production of wild crops as well as the volume of water available in the river for cultivated crops. Floods also might interrupt the farming system, as Captain Martín Bernal noted in November 1697 (Smith and associates 1966:42). Apache raids might deplete stores of staple foods (Basso 1983:466; Goodwin 1971:19). At such times the Gileños were forced to rely on starvation foods such as the seeds of Pickleweed and Quail Brush. Seeds of Catclaw Acacia (*Acacia greggii*) may likewise belong in this category. Some of the wild greens that have now fallen into disuse may have been starvation foods. Examples are Wheelscale Saltbush (*Atriplex elegans*), wild-heliotrope, and fiddleneck. The few old Pima who recall these being eaten have commented that they either were tasteless or had objectionable textures.

Researchers have established two minimal requirements of famine foods: "First, they have to be edible, and second, they must be available even when more frequently consumed rations cannot be acquired. Famine foods can often barely claim the first and may require substantial processing to make them edible or reduce their toxic constituents" (Minnis 1991:232).

Myth, ritual, and story were important in perpetuating knowledge of famine food use (Minnis 1991:250); indeed, famines may be so infrequent that the knowledge about famine foods may not be passed from one generation to the next without such oral tradition. The story of *toota hannam* is such an example: the oral Creation Epic was lost in the first half of the 20th century, and predictably, no one with whom I worked had any inkling that Chain-fruit Cholla (*Opuntia fulgida*) might be edible.

Various plants known to have been used as food or seasoning by other tribes in the Southwest apparently did not enter the Gila Pima dietary although they were known to (and most named by) the Gileños. These plants and the parts used from them include

spiderling, *Boerhavia* spp., greens

Watergrass, *Echinochloa colonum*, seeds

Barnyard Grass, *Echinochloa crusgalli*, seeds

Mexican Gold Poppy, *Eschscholzia mexicana*, greens, seeds (?)

Wild Sunflower, *Helianthus annuus* var. *lenticularis*, seeds

Desert Lily, *Hesperocallis undulata*, bulbs

Pepper-grass, *Lepidium* spp., greens, fruits (condiment)

Blazing Star, *Mentzelia* spp., seeds

evening-primrose, *Oenothera* spp., greens, seeds

Desert Cambess, *Oligomeris linifolia*, seeds

Chinchweed, *Pectis papposa*, greens (condiment)

Desert Night-blooming Cereus, *Peniocereus greggii*, tuber

Marsh Nightshade, *Solanum americanum*, greens, fruits

Horse-purslane, *Trianthema portulacastrum*, seeds

puffballs, *Tulostoma* spp., fruiting body

Some of these may have been used at one time but have since been lost from the oral historical record. Others may have been famine foods that were unknown to contemporary consultants.

Postcontact Dietary Changes

The Hispanic Period (1694–1853)

With the arrival of Hispanic colonizers in Pimería Alta in the 1680s, profound changes affected the lives of the native peoples on the frontier. Even before 1694, when Fr. Kino encountered the Gileños, changes were already under way that modified the Pimas' physical and cultural environments. The Gila Pima were far enough out on the fringes of the Spanish colonial frontier that they selectively absorbed into their culture what they wished while maintaining the rest of their culture intact. Nonnative crops were among the things that they quickly adopted (see Groups I and J).

Some of these became major crops. Wheat, in particular, filled an apparently vacant niche through the addition of a third (winter) planting. Wheat proved so successful that it eventually eclipsed all other crops in importance. With the 1849 Gold Rush and a continuing flow of Anglos to the new state of California, the Pima had a ready market for surplus wheat. Various pulses or legumes such as garbanzos, black-eyed peas, lentils, garden peas, and perhaps common beans also entered the Pima fields and diet during the Hispanic period, though their relative importance varied.

From a suite of Old World seasonings the Pima selected several (onion, garlic) but declined most (anise, cumin, cilantro, mustard, mints, and others; see Morrisey 1949; Nentvig 1971). Even onion and garlic may have been adapted much later, in the Anglo period. Salt and chile have remained the ubiquitous northern Piman seasonings.

The watermelon and muskmelon, both introduced novelties, proved so popular that they dispersed northward in advance of the Spaniards. Sweet sorghum was an esteemed addition as well; the crushed stalks produced a juice for making syrup. The tomato may also have been adopted at this time.

Many of the new crops were additive, but wheat largely eclipsed maize as the basic crop. Wheat tortillas became the staple at most meals; even the oldest people today cannot remember anyone making corn tortillas. Wheat replaced corn and mesquite flour in many other preparations as well. Wheat pinole (parched and finely ground wheat) became a staple for both those traveling and those at home. It needed only to be mixed with water to make a hearty cereal drink.

One of Kino's main goals during the late 17th century was to establish cattle ranching as a firm economic footing for the mission endeavors in Pimería Alta. Although never missionized during the colonial period, the Gileños acquired some horses and cattle, as well as chickens. At this time, both horses and oxen seem to have been used principally as transport and draft animals (Ezell 1961).

Grapevines and various fruit trees such as figs, apricots, peaches, and date palms were introduced at Pima missions to the south; some of these new crops may have found their way north at this time. Pomegranates likewise may have come in early because they were preadapted to desert oasis conditions. Most fruits, however, seem to have been adopted only during the Anglo period. After all, the desert supplied an abundant crop of Saguaro and wolfberry fruit, and these did not have to be pampered and irrigated, just harvested.

The Anglo Period (1853 to present)

The most pervasive contact with Europeans came when enormous throngs of Anglo-Americans began passing through Pima lands. Early on, beaver trapping along the desert streams drew some. The goldfields drew others; indeed, the southern route to California brought thousands. In about two decades, whites started putting down roots in the desert surrounding Pima country. Florence (founded in 1868), Phoenix (1869), Maricopa Station (1879), Tempe (1879), Casa Grande (1880), and other communities sprang up. Soon the rivers began drying up. The compound abuses of over-grazing, timber cutting in the headwaters, and depletion of native beaver dealt the mortal wounds to the Gila (see chapter 5), though it was to writhe in its death throes for another three-quarters of a century. Between 1875 and World War II, the Pima way of life underwent many changes that affected the diet.

Use of the staples Velvet Mesquite, Screwbean, Indian-wheat, Chia, Tansy-mustard, Saguaro, and agaves nearly disappeared. Many wild greens fell out of the dietary, perhaps partly as a result of drying and abandoned fields. Table 7.2 gives the approximate time when various foods are believed to have dropped from significance in the Pima dietary.

Steam flour mills at Sacaton and Casa Blanca ensured a ready supply of wheat flour and an abandonment of corn flour, which was produced by difficult manual labor. In the early 1900s the Pima ceased processing corn dishes with lime or ash, which are essential for the release of niacin, thus further impoverishing their diet. Trading posts appeared on the reservation and stores in the new Anglo settlements nearby. Sugar and lard became regular commodities as the Pima discovered new ways of preparing foods. Flour, sugar, and lard became the new trinity in a diet impoverished both varietally and nutritionally.

The progressive drying of the Gila meant that the Pima were dependent less and less on their own resources and more on government rations and a wage economy. By the turn of the century, many Pima eked out a living cutting and selling mesquite wood, almost denuding the reservation (see chapter 5.). Water pumping far in excess of recharge helped kill many of the surviving mesquite trees. Many of the wild food items from the floodplains became rare or unavailable. Government farm advisors taught the Indians, who had been farming the desert for centuries, "clean agriculture" along corporate models. There was little if any appreciation for the ancient desert-adapted crop varieties. Many vegetables and fruits that entered the Pima diet—including sweet-potatoes, lettuce, lemons, grapefruits, oranges, and plums—were grown at the Sacaton Field Station but only occasionally elsewhere on the reservation.

The advent of Eurasian herbivores meant a more dependable source of animal protein for many Gileños. The Pima ate burro, mule, and horse meat as well as beef. But with the good came the bad. Wild game is decidedly lean throughout much of the year, but beef, especially, is fat. Beef lard became available from their own animals as well as from trading posts and Anglo towns. And the Pima relished it. A new kind of cooking appeared: grease frying, unknown in aboriginal cuisine. Deep-fried dough became the famous "Indian" fry bread. Boiled desert greens were drained, then fried in lard. Even cholla buds were mashed and cooked in lard as a final stage of preparation.

Joseph Giff was a strong advocate of traditional Pima foods. In explaining Pima cooking methods to me one day, he developed this linguistic argument on the types of cooking:

> *Chuáma* means 'to pit roast something'; *chuámai* is 'something that is roasted in a pit', like *chuukug chuámai,* the way they pit roast the beef at the feast. *Chuukug ga'i* means 'meat cooked on open coals', but you can *ga'a* anything, like *huuñ ga'i* 'roast corn'. Some call *chemit* ['tortillas'] *ga'i. Hidoḍ* [is to] stew or boil anything—beans, corn; [it means] 'cook by boiling'. *Gakijid*— 'jerk'—[means] anything you dry, spread it out to dry—[as] *chuukug gakïchu. Hikchuvulid* 'to cut into strips', like *hikchulda,* that squash. Or *simin* 'to cut an object', like *simna,* the squash. Frying is a newer thing—they never had that before—*manjugĭ ch-eḍ bahijida.* I don't think they had lard at all. That word is from Spanish— *manteca.* I guess if they had it [animal fat or lard] they'd call it *giig,* that's 'fat'.

Thus by running through all the native terms for cooking or preparing food for cooking, Joe contrasted them with the term for frying or cooking in grease, which is based on a loan word from Spanish, not the Pima word for fat.

The most significant changes in Pima diet came in the decades before and after World War II, when over half of the traditional foods fell from common use. Abundant calories from the new simple-carbohydrate and high-fat diet soon changed the tall, lean Pima into an obese people. The switch from native to Euro-American foods also had effects other than obesity. Older Pima I worked with perceived the deleterious effects of the new diet. One day while discussing what he called "natural foods," Sylvester Matthias told this story:

> When the older people started to eat white man's foods, starchy food, that's what happened to every case. And they called the doctor where the [Gila Crossing] clinic is now. It was a little old frame house. That was the doctor's house. So. Anybody had trouble—was

Table 7.2 Losses from the Gila River Pima Dietary

Scientific Name	English Name	Part(s) Used	Period of Loss			
			Contact (1694–1865)	Drought (1866–1906)	Transition (1907–1945)	Modern (1946–1990)
Acacia greggii	Catclaw Acacia	seeds	X			
Agave spp.	agaves	pulp			X	
Allenrolfea occidentalis	Pickleweed	seeds		X		
Amaranthus hybridus	grain amaranth	seeds		X		
Amoreuxia palmatifida	Saya	roots			X	
Amsinckia spp.	fiddlenecks	greens		X		
Atriplex lentiformis	Quail Brush	seeds		X		
Carnegiea gigantea	Saguaro	fruits & seeds				X
Cercidium microphyllum	Foothill Paloverde	seeds			X	
Chenopodium berlandieri	Huauzontle	seeds		X		
Chenopodium pratericola	Narrow-leaved Goosefoot	greens				X
*Cicer arietinum**	chick-pea	seed				X
Cucurbita digitata	Finger-leaved Gourd	seeds		X		
Cyperus spp.	flat-sedges, nut-grasses	tubers			X	
*Descurainia/Sisymbrium**	wild mustards	seeds			X	
Dichelostemma pulchellum	Bluedicks	bulbs				X
Echinocereus engelmannii	Engelmann's Hedgehog	fruits				X
Eremalche exilis	Star Mallow	greens			X	
Ferocactus spp.	barrel cactus	pulp		X		
Gossypium hirsutum	cotton	seeds			X	
Hoffmanseggia glauca	Hog-potato	tubers				X
*Lens culinaris**	lentil	seeds			X	
Monolepis nuttalliana	Patota	seeds		X		
mushroom (*Lepiota rachotes?*)	mushroom	aboveground parts			X	
Olneya tesota	Desert Ironwood	seeds			X	
Opuntia x *kelvinensis*	Gila Cholla	flower buds			X	
Opuntia arbuscula	Pencil Cholla	flower buds				X
Opuntia leptocaulis	Desert Christmas Cholla	fruits			X	
Orobanche cooperi	Desert Broom-rape	whole plant			X	
Phacelia spp.	wild-heliotropes	greens			X	
Phalaris spp.	Canary Grass	seeds		X		
Phaseolus acutifolius (some varieties)	tepary	seeds			X	
Phoradendron californicum	Desert Mistletoe	fruits			X	
Physalis acutifolia	Wright's Ground-cherry	fruits			X	
Plantago spp.	Indian-wheat	seeds		X		
Populus fremontii	Frémont's Cottonwood	male flowers		X		
Portulaca spp.	Verdolaga	seeds	X			
Proboscidea spp.	devil's claws	seeds				X
Prosopis velutina	Velvet Mesquite	pods			X	
		seeds	X(?)			
Prosopis pubescens	Screwbean	pods		X	?	
Rumex crispus, R. violacens	Curly Dock, sorrel	greens			X	
Salix gooddingii	Goodding's Willow	male flowers		X		
Salvia columbariae	Chia	seeds			X	

(continued on next page)

Table 7.2 Losses from the Gila River Pima Dietary *(continued from previous page)*

Scientific Name	English Name	Part(s) Used	Period of Loss			
			Contact (1694–1865)	Drought (1866–1906)	Transition (1907–1945)	Modern (1946–1990)
Sarcostemma cynanchoides	Climbing Milkweed	sap				X
Sambucus mexicanus	Desert Elderberry	fruit		X		
Scirpus maritimus	Salt-marsh Bulrush	tubers			X	
Sonchus spp.*	sowthistles	greens				X
Typha domingensis	Southern Cattail	shoots, rhizomes		X		
*Vigna unguiculata**	Black-eyed Pea	seeds				X
Yucca baccata	Banana Yucca	fruit			X	
Zea mays						
(most native cultivars)	corn, maize	ears				X
Ziziphus obtusifolia	Graythorn	fruits			X	?

Sources: Reports of native consultants (in part 2); also Castetter and Bell 1942; Curtin 1949; Hrdlička 1906, 1908; Russell 1908.
*Postcontact acquisition.

sick—he'd go and tell the doctor. So. "What were the symptoms?" The first thing, he'd give them castor oil. [Sylvester began laughing.] Then he'd go check. "How is it?" "Alright! Alright!" And then, next patient would come, and then—now he knows, they'd been eating starchy food. They were constipated. They [formerly] lived on natural food. So, after so many cases, every time he's called, he'd have to give castor oil. And finally, the people around all over—because he's been all over giving them castor oil—called him Siipuḍ Maakai ['Anus Doctor']. And all that stuff that you mentioned, *ka'akwulkada, hohi-wuichda* [mesquite dishes], and that *vau,* they're all natural foods. But then they started eating starchy food. That's why it make them constipated.

"How long ago was that story told about Siipuḍ Maakai?"

"That was right after World War I, in the twenties."

"Do you think that's when the diet started changing away from mesquite and natural foods?"

"Right. Mostly, they were using that *vau,* that beverage, with their meals, in the early twenties."

The 20th century has seen a continuing decline in the use of native foods, a decline in both soluble and insoluble fiber, and an increase in fat and simple carbohydrates in the Pima diet. In addition, an active, hardworking people have become a sedentary people as farming, gathering, and hunting have become activities of the past and wage earning comes now more often from an office job. The daily regimen includes many hours of television. A young Pima today is likely to be raised on deep-fried chicken, hamburgers, greasy popovers, white bread, and canned sodas or syrupy punch. Few now know the wild greens, cholla buds, roasted corn, whole-wheat dishes, fish, or wild game of years past.

Chapter 8

Words for Mapping the Natural World

And the English that is spoken by [Piman] Indians sounds like Indian, anyway, resting in the hollows of the throat, each word coming out soft as a baby rabbit. [It] has the right sound for people of the desert. It holds the quiet of the mesas; it rolls words into the natural shape of breathing so they come from the mouth still soft around the edges, small winds that might blow across a hill of summer weeds. That kind of sound.

Byrd Baylor, *Yes Is Better Than No*

The cognitive features of ethnobiology have become the central focus of ethnobiology in the late 20th century, particularly through the works of such prodigious writers and thinkers as Scott Altran, Brent Berlin, Cecil Brown, Ralph Bulmer, R. F. Ellen, Terence Hays, and Eugene Hunn, to name but a few. Much of this discussion has been synthesized by Brown (1984) and Berlin (1992). The primary question of these folk taxonomists is, How do people conceptualize their environment and how do they linguistically map this conceptualization to the biological world? Put more simply, how does any group of people with a common language *see* and *talk about* nature?

This chapter focuses on some cognitive and lexical aspects of how the Pima deal with their desert home. To talk, one must have a set of sounds (phonemes) making up words (lexemes). Pimans speak a subset of the Sonoran branch of the Uto-Aztecan language family. Piman (or Tepiman) was once spoken for nearly 900 miles, from the Rio Santiago in northern Jalisco to the Gila River in southern Arizona (Kroeber 1934; Sauer 1934). The single interruption, the Taracahitan bridge (fig. 1.1) may be a protohistoric or even historic intrusion into the Tepiman continuum. To appreciate the mapping of cognitive elements, some understanding of the Piman language and its sound system is in order.

There are many attributes of the world to think and talk about. The characteristics of most concern to us here are those relating directly or indirectly to the world of plant-human interactions, or what is called ethnobotany. The mapping of biological species is qualitatively different from the naming of parts of plants and animals (partonomy) as well as the mapping of the nongenetic environment; yet these interrelate. The accounts of this book have been arranged by folk generics, cognitively and lexically recognized discontinuities of the natural world. How these folk genera are arranged into a hier-archy of "folk families" or life forms is also part of the mapping process. One such Pima arrangement (others may be possible) has been used for the sequence of the accounts in this book. (The Gila Pimas' task of naming the new plants arriving in their country as a result of the Columbian encounter is dealt with sepa-rately in appendix D.)

The Pima Language

It would be unfortunate to read this book and not savor the words that make a folk taxonomy a living reality; it would be like eating a nice stew with lots of herbs but having a cold and being unable to taste anything. To begin with, Pima words for the most part are very easy for English speakers to pronounce.

Most of the sounds are found in English or Spanish; only a few will give difficulty. For example, the *i*, written according to local convention *e*, is not the *e* vowel of Romance languages. If you say the English word *pull* or *open* with the corners of your lips spread rather than rounded, you will be close. (The language lacks the conventional *e* sounds.) The vowel *u* is pronounced like the *u* in *brute* (or the *oo* in *food*). The Pima vowel *o* can be pronounced like the *a* in *all* (or the *o* in *dog*); but there is a more rounded *o* pronounced like the *o* in *gold* that I have written in this book as *ō*. The vowel *i* also has two variants; the shorter *i*, as in *stick*, occurs especially in last syllables preceding an ending consonant (*s-totoñik*), but else-where in the word it is pronounced like the *i* in *police*. The vowel *a* is pronounced like the *a* in *father*. The *l* is a sound located on the sound map somewhere between *l*, *d*, and *r*; it may be nearest the single flap *r* of Spanish, as in *entrada*. The *ñ* sound, common in Spanish and borderlands English, is pronounced like the *ny* in *canyon*. There are three consonants that require attention: *d* is a soft dental stop with the tip of the tongue, *ḍ* is the retroflex *d* made with the tip of the tongue on the alveolar palatal area, and *t* is a stronger voiceless dental stop made with the tip of the tongue. The terminal *g* in Gila Pima has a sound somewhere between *g* and *k*, as in *hiósig* 'flower', and it is often difficult to determine which is which. Gila Pima also has a distinct termi-nal *k*, as in *s-onk* 'salty'.

It has been said that the Akimel O'odham use *v* where the Tohono O'odham use *w*, but this generalization is not correct. Whereas most dialects grouped as Tohono O'odham have only the *w* sound, both sounds occur in Pima. It is true that Gileños say *chuuvĭ, bavĭ,* and *viv,* while their southern cousins trail off in breathy *chuuwĭ, bawĭ,* and *wiw.* There are words in Gila Pima that can be pro-nounced either way, such as Vupshkam or Wupshkam 'the Emerging People'. However, there are also Pima words that are always pronounced with a *w,* such as *wecho* 'beneath' and *dahiwua* 'to sit down'.

Terminal vowels, and even occasionally word medial vowels, may be unvoiced. They are still there, but they are whis-pered. This is similar in English to the difference between voiced and unvoiced consonants, as in *gut* (voiced) and *cut* (unvoiced) or *then* (voiced) and *thin* (unvoiced). In this work, devoiced vowels are written with a breve (˘): *bavĭ* 'tepary'. I have noticed considerable variation in pronunciation of devoiced vowels among individual speakers on the Gila; some speakers pronounce a very distinct and full blown vowel at the end, as if saying *bavĭ,* while others use a very short termi-nal vowel that you must listen carefully to hear, and still others seem to drop the terminal vowel altogether, as in *bav.* Some speakers vary. Perhaps these differences once had dialectical importance. Dean Saxton, Lucille Saxton, and Susie Enos (1983:114–115) discuss devoicing in Tohono O'odham, where this phenomenon is more pronounced. The presence or absence of the voiced terminal vowel may modify the meaning of a lexeme, as in *al* 'small' versus *ali* 'child'. Also, voicing or nonvoicing may change meaning, as in *iivagĭ* 'eaten greens' versus *iivagi* 'to bud, leaf out'.

The accent or major stress in Pima words is usually on the first syllable, such as in Ákimel Ó'odham, Tóhono Ó'odham. It makes little sense to produce the sounds of a word correctly if the accent is mis-placed: a native Piman speaker will pro-bably not recognize a word erroneously accented. The stress marks will be indi-cated in this work only when they fall somewhere other than in the first syllable, as in *vilgóōdi* 'apricot'.

A hyphen (-) is used to show the addition of a particle or enclitic such as in *s-haḍam* 'to be sticky'. The hyphen has no phonetic value but lets the reader know that the sounds in the two words keep their separate values.

All Pima words beginning with a vowel are preceded by a glottal stop (') or closure of the throat. Because this is obligatory in the language, the glottal stop in this position is omitted to avoid having excess diacritics. There are, however, glottal stops within words that must be made for cor-rect meaning: these are similar to the stop made in the English exclamation *'uh-'oh.* Examples of some changes in meaning due to the absence or presence of the glottal stop are the words *shai* 'to go on a communal hunt [or] surround' versus *sha'i* 'brush', or *tai* 'fire' versus *ta'i* 'uphill, upward'. Additionally, there are both single and double vowels without a glottal stop. For instance, *aan* is the Desert-willow, while *a'an* are the wings or wing feathers of a bird. Putting this all together we have such distinctions as *o* 'or, is, are', *oo* 'back [sing.]', *o'o* 'back [pl.]', and *oo'o* 'bones'.

Pima is spoken much more slowly than standard American English. Even the single vowel, therefore, seems like a lengthened vowel to most English speak-ers. Therefore, to be understood by a native speaker, Pima words with double length vowels (written sometimes as *a:, i:,* and so forth) need to be strongly emphasized.

The speech pattern of Gila Pima is also greatly inflected. Listening to old people talk to one another in O'odham is almost like listening to song. Both the pace and the inflection carry over into English. To reflect this, I have italicized the very strongly inflected words of native consul-tants in the conversations recorded here. Pima is music, the leisurely lilting music of desert dwellers, not the hurried clipped speech of urbanites. If you expect a Piman speaker to understand you, remember to give each vowel breathing room.

Some Pima Perceptions of the Natural World

But words and sounds are not the only components of folk taxonomy, which is a cultural phenomenon in itself, a way of viewing the world. The physical and biological world that the Pima perceive is both similar to and dissimilar from what English-speaking people perceive. Perception is strongly modified by linguistic categories in any language. But environment modifies language as well.

Biological species, whether they be birds, mammals, lizards, or insects, represent discrete entities in nature. It should come as no surprise that folk taxonomists and Western biologists tend to recognize the same boundaries in the noncontinuum of living organisms and that there is a high correlation between folk categories and biological species. The degree of correspondence is modified by size, commonness, utility, and number of closely related species, as noted in a great body of literature (for example, Berlin 1992; Berlin and associates 1973). Differences arise because Piman folk determinations are influenced by smell, growth form, seasonality, and habitat, whereas the Linnaean taxonomist constructs a classification of plants based largely on reproductive parts that are often microscopic.

Outside the realm of genetic systems, there will be less correspondence in linguistic categories and perceptions between different cultural groups. The boundaries of many natural phenomena—such as geomorphology, colors, tastes, plant associations, and plant anatomy—are either fuzzy or arbitrary. The words mapping some of these domains are relevant to ethnobotany.

The partitioning of many physical phenomena, such as the light spectrum, is purely arbitrary. The linguistic division of the landscape, likewise, with a bit of imagination, can be seen as basically arbitrary. Other areas, such as the division of plant parts and body parts, have a more objective basis in physical reality. All cultures seem to recognize flowers, leaves, and stems and to label these with words. But what about petioles, stamens, and pistils, or fruits, seeds, nuts, and berries? Ultimately, a speech community makes decisions, and the ethnobiologist must be sensitive to avoid eliciting distinctions where none exist.

Geomorphology

In Pima country the three primary desert landforms—mountains, bajadas (piedmonts), and floodplains—are distinctly characterized (fig. 8.1). Mountains such as the Santans are *do'ag,* and a mountain range is *do'ag an vavañ.* A smaller hill or peak separated from a range is called *kavulk* or *kawulk.* (These usually had specific names, which often provided the name for a nearby rancheria: Caborca, Sonora, is a surviving example.) An isolated lava or cinder cone is a *kavichk.* The bajada is *to'otonk;* the junction of the mountain and the bajada is the *do'ag shon.* The arroyo or dry wash that cuts through

the bajada is called the *akĭ* (pl. *a'akĭ*). The place where the *akĭ* opens out onto the floodplain is called *akĭ chiñ* 'arroyo mouth'. Desert farmers, particularly the Tohono O'odham, developed a special technology of dry farming utilizing the summer rainfall that collected in the *akĭ* and spread out in the *akĭ chiñ.* This geomorphological term supplied the name for a number of Piman settlements, including the colonial period Koahadk village Aquituni, between the Picacho Mountains and the Gila.

Flat lands are called *s-sheliñ jeveḍ.* The fields here are *oidag.* Piman summer settlements often took the name of specific fields: some surviving names in Tohono O'odham country are Ge Oidag 'Big Fields', Supi Oidag 'Cold Fields', and Al Oidag 'Little Fields'. The entire landscape is said to be *jeveḍ kaachim* 'earth remaining there'.

For a desert people, the Pima have a broad vocabulary relating to water and moisture. A river such as the Gila is called *meldadam,* from the verb *meḍ* 'running' or more frequently *akimel* 'wash running'. A smaller stream, such as the Santa Cruz River, is a *vo'oshañĭ.* This word implies a large wash with tributaries. A spring or creek is *shongam* (pl. *shoshongam*), derived from the word *shon* 'beginning, base, starting point, trunk'.

Figure 8.1 *The Pima Landscape*

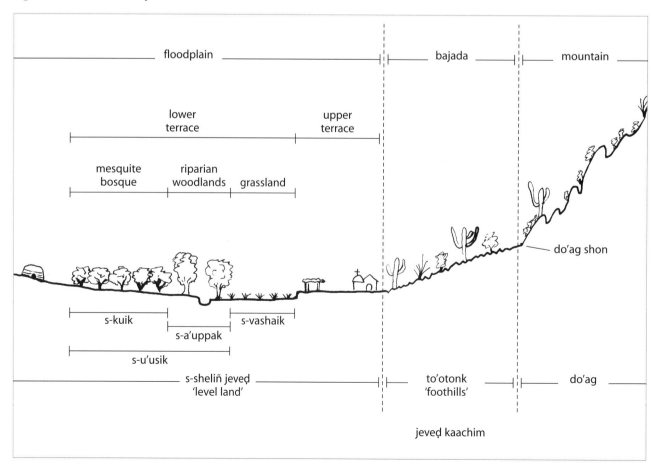

The ocean or a large inland sea such as the Salton Sea is *kaachkĭ;* usually in song or other formal literature, the ocean is called descriptively *kaachim shuudagĭ* 'remaining there water'. A large lake is called *hejelko shuudagĭ. Shuudk* signifies a pond, lake, or muddy place where moisture collects. A pond constructed to hold water at a ranch is *vachkĭ* (in Tohono O'odham usually called *vo'o*). A seep might be termed *s-va'uandam jeveḍ,* from the verb stem *va'u* 'to moisten, dampen'. This was compared to a meadow and described as "a place where it's always damp and there's grass and cows can graze in there." An area with saline water and high evaporation, which may appear white, is an *ongam,* or salt flat.

A mesquite bosque is *s-kuik.* A cottonwood grove is *s-a'uppak.* Any woodlands, regardless of species composition, is *s-u'usik.* A grassland is *s-vashaik* or *s-vashik.*

Colors

The human eye sees colors as a continuous spectrum of visible light, as in the rainbow. Each culture makes decisions on how to divide this continuum into conceptual units, labeled by words. Although we may feel that our own way (English, for instance) is somehow objective, the divisions are purely arbitrary and culturally determined. When asked for the colors, the riverine Pima generally give the following:

s-toha, (pl.) *s-toota,* white

s-chuk, (pl.) *s-chuchk,* black

s-koomagĭ, (pl.) *s-kokomagĭ,* gray

s-vegĭ, (pl.) *s-vepegĭ,* red, maroon

s-oám, (pl.) *s-o'am,* yellow, orange, brown

s-cheedagĭ, s-chehedagĭ, blue, green

Thus the achromatic colors are divided into three, and the chromatic colors also into three. Some people will offer several additional colors: *s-vegium,* (pl.) *s-vepegium,* pink; *s-iibhaimagĭ,* which may be lavender, bright pink, or purple; *s-mo'oialmagĭ,* violet. Two of these are derived from plant names: *s-iibhaimagĭ* comes from the fruit of the prickly-pear (*iibhai*), while *s-mo'oialmagĭ* is from the purple-black fruit of nightshade, *mo'oial.* The *s-* is usually deleted before older proper names in Pima.

Understanding the range of colors included in any language is important for folk taxonomy. For instance, the tepary, *s-oám bavĭ,* is a bean with a yellowish brown seed coat. But the still unidentified Pima plant *s-oám hiósigam* might be either orange or yellow (or any shade that is not obviously red or pink). And so with the blue-green segment of the Pima color terms. The name *cheedagĭ vishag* is sometimes rendered as 'green hawk', leading one to believe the bird is mythical. But as Joseph Giff explained to me, this is not a hawk at all, but rather a falcon; nor is it green, but bluish. Apparently the Prairie Falcon, common in Pima country, is *vishag;* the Peregrine Falcon, a rarer bird that once occurred here along the river, is *cheedagĭ vishag* (see Rea 1983:135–136).

Tastes

There are five labeled categories of taste: *siv,* bitter; *s-he'ek,* sour; *s-i'ovĭ,* sweet; *s-onk,* salty; and *s-ko'ok,* "hot" or *picante.* These words appear in the Gileño ethnobotany either as plant name modifiers (for example, *s-i'ovĭ che'ul* 'sweet willow'; *siv u'us* 'bitter sticks') or as descriptions of plants, foods, and medicines.

The last of these, *s-ko'ok,* describes the sensation of substances such as chiles, called *picante* in Spanish. Curiously, English has not evolved a word to describe this taste, substituting instead the word *hot* (as in "hot sauce"), even though the sensation is not contrasted with cold. Some menus now substitute the word *spicy,* another inadequate comparison because spices include a range of aromatic seeds and herbs, very few of which are *picante.* In Pima there is no ambiguity: the word *s-toñ* 'hot' is used to describe temperature and *s-ko'ok* to describe the taste of *ko'okol* 'chiles', for . . . well, *picante.*

Plant Associations

Language elements can also be used to describe where or how plants grow. In Pima the suffixes *-ik, -ag, -ig* can be used to mean 'place of', such as in the Pima village of S-totoñik 'very many ants place of'. At the west end there were once settlements or segments of rancherias called S-navag (from *navĭ,* Christmas Cholla, *Opuntia leptocaulis*) and S-kuávulig (from *kuávul,* wolfberry, *Lycium* spp.). The same form is used to designate plant associations. For example, a mesquite bosque is

s-kuik, from *kui* 'mesquite'. The term *s-u'usig,* from *uus* 'tree', designates a woods or thicket. A grassland or patch is *s-vashaig;* it was described to me as a marsh, a natural grassy low place, and derives from the word *vashai* 'grass'. The cottonwood gallery forest along the river is called *s-a'uppak,* from *a'uppa* 'cottonwoods'.

Kinds of fields are likewise distinguished. Small fields near a house, with chiles, onions, tomatoes, and pepo squash, for instance, are called *al oidag.* Larger fields with regular crops such as corn, wheat, teparies, and larger squash and pumpkins are termed *oidag.* (This is similar to the difference in English between *garden* and *field.*) Container planting, prominent among Hispanics of the Southwest, was apparently unknown and is infrequent even today.

Plant Morphology

The Pima have developed numerous ways of talking about plants generally and about parts of plants specifically (fig. 8.2). A plant is called *ha'ichu vuushdag* (variant, *ha'ichu wuushdag*) 'something that grows up'. Sprouts of a plant coming up are *vuvhaidag.* A seedling, whether monocot or dicot, is *vuushdag.* A volunteer, a plant that comes up by itself, is *hejel vuushkam.* The shoots of a plant are *i'ipoñig.* On a woody plant such as a *sha'i* 'shrub' or *uus keekam* 'tree', the trunk is called *shon* (this word can also mean the stump after a tree is cut). Branches and stems are *mamhaḍag* (as is the petiole of a palm leaf). The bud at a branch tip is *kug* (from *kuugĭ* 'to have an end'). Leaves are *haahag,* the word being both singular and plural. The roots of any kind of plant are *tatk. Eldag* can mean the bark of a plant, the husk of corn (*huuñ eldag*), or the skin of any animal (such as *huai eldag* 'Mule Deer hide'). A knot in a tree or wood is *wuhi,* the same word as for 'eye'. A tendril is the *biiwualig.* A forked branch is a *sha'alk.*

There is only one word to indicate a flower: *hiósig.* This includes the calyx, sepals, petals, stamens, and pistil, which are not separately distinguished. (No one could give me a name for pollen, called *chu'i* in Tohono O'odham; I suspect *chu'i* was the Gileño name too, now forgotten.) The unopened flower bud and large calyx of the cholla, for instance, are considered together as the *hiósig.* This more inclusive terminology has probably developed in response to the harvesting of the

Figure 8.2 *Parts of Plants*

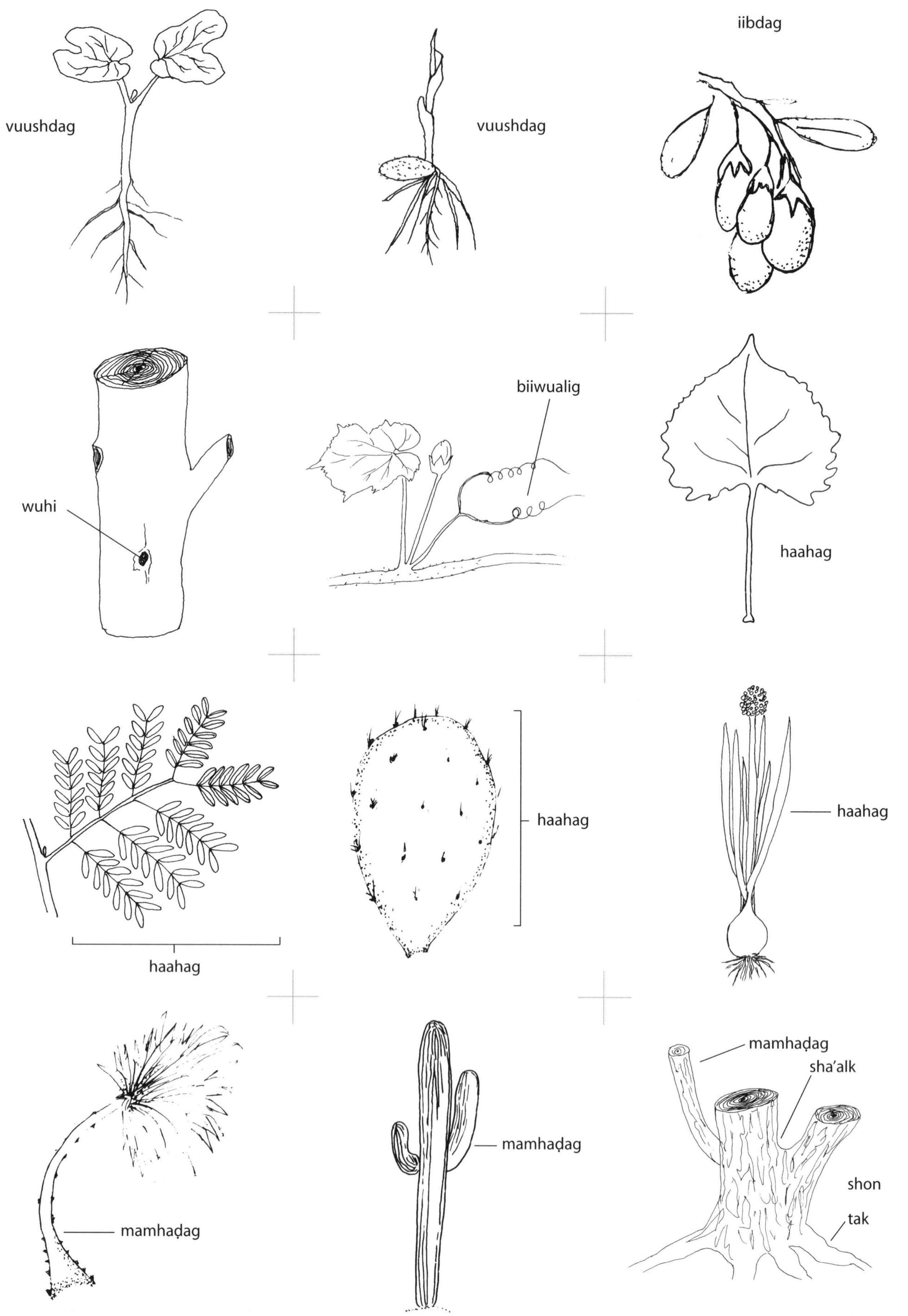

vuushdag

vuushdag

iibdag

wuhi

biiwualig

haahag

haahag

haahag

haahag

mamhaḏag

mamhaḏag

mamhaḏag
sha'alk

shon

tak

Figure 8.2 *Parts of Plants (cont.)*

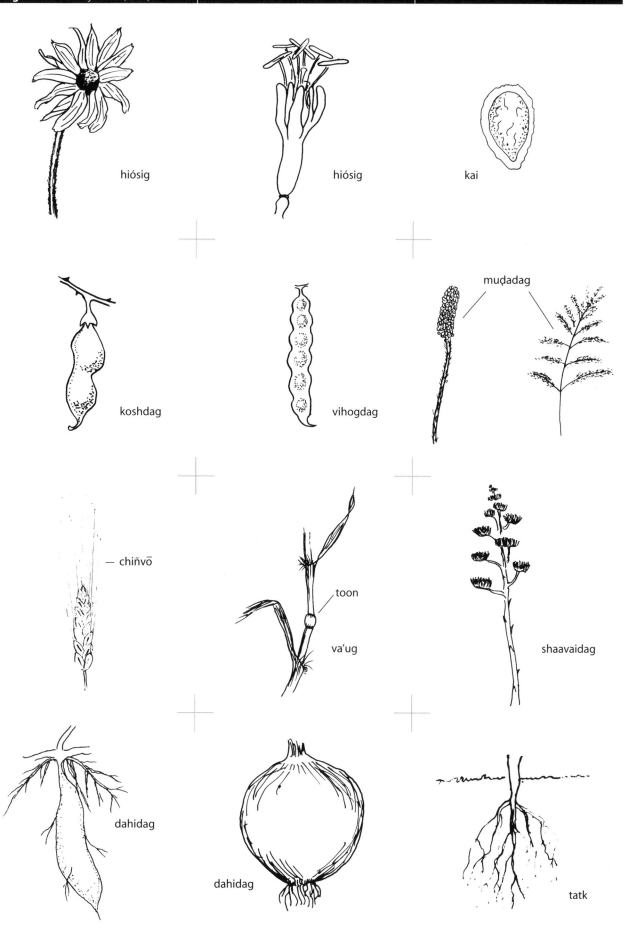

hiósig

hiósig

kai

koshdag

vihogdag

muḍadag

— chiñvō

toon

va'ug

shaavaidag

dahidag

dahidag

tatk

unopened cholla bud with its calyx. The *vaḍag* is the nectar in flowers. (The word can also mean the juice from fruit or something cooked in water, such as soup.) This term was explained in the following way: "*Hiósig vaḍagaj* is what the *vipismal* [hummingbird] comes to drink." The Pima, especially youngsters, sucked the *vaḍag* from several species of flowers. The phrase *eḍa shuudagĭ* 'water inside' might be used, particularly in sung poetry, in place of *vaḍag*. Madeleine Mathiot (n.d.:99) gave *nakvaḍag* as the term for the pistil of a flower as well as for ear wax and walnut, but no one among the Akimel O'odham whom I asked about this term recognized the botanical usage.

The emerging flower or fruit, not yet full blown, is the *do'idag* (from the word *do'i* 'raw, unripe'); this is roughly equivalent to 'bud' but somewhat more inclusive. The unripe fruit of anything is the *iibdag*. (The same word means the heart of an animal.) This term is restricted to fleshy or somewhat fleshy fruits. The seeds of any plant, domesticated or wild, are called *kai*. Seeds saved for planting are *kaijka*. The hard seedpod or hull of such plants as the paloverde or ironwood are *koshaj* (the same word as *kosh* 'nest'), or *koshdag*, but the soft pod of such legumes as peas or beans is *eldag* 'its skin', as in *vihol eldag*, *muuñ eldag*. Leguminous pods may also be called *vihogdag*, except those of mesquite, which are specifically *vihog*. The spores inside a puffball are called *kuubdag* 'dust' or 'smoke'.

In plants classified as *vashai* 'grasses', the flower is generally called *muḍadag*, which is often rendered as 'tassels'. Thus the flowers as well as the seed heads of such plants as true grasses, sedges, *Oligomeris*, chenopods, amaranths, and plantains all have a *muḍadag*. Plants that have *muḍadag* do not have *hiósig*, according to Ruth Giff. Grasses also have a *toon* 'joint', the same word used for the knee of a person or animal. The stalk (straw) is called *va'ug*. The long awns of grasses such as wheat and barley are *chiñvō*. The larger flowering stalk of certain plants (yuccas, agaves, and at least Canaigre, *Rumex hymenosepalus*, among the sorrels and docks) is the *shaavaidag*.

Some special terms are used for the underground parts of the plant. An ordinary root is *tak* (pl. *tatk*), no doubt etymologically related to *taḍ* 'foot'. The

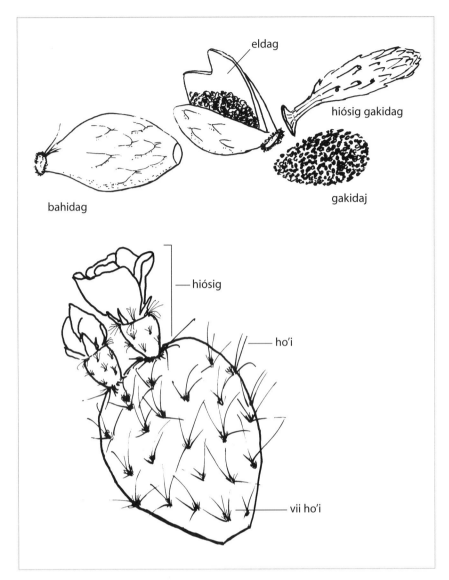

Figure 8.3 *The morphology of cactus*

dahidag is a bulb or tuber. This term includes only enlargements of the central part, as in a lily or carrot, but not other underground enlargements such as a potato or sweet-potato.

When referring to something still on a plant, the form is somewhat different, taking the suffixes *-aj* or *-ij* to show possession. Consequently one says *tatkaj* 'its roots', *shonaj* 'its trunk', *mamhagaj* 'its branches', *hiósigaj* 'its flower(s)', *iibdaj* 'its fruit', *haahagaj* 'its leaves', *kaij* 'its seeds', and so forth. After harvesting maize, wheat, or sorghum, all that remains, 'its straw', is *va'ugaj* or *moogaj*. (Similarly marked forms are used to designate parts or organs of an animal.)

There is some specialized vocabulary associated with cactus (fig. 8.3). While the spine or spines of any plant are *ho'i*, the glochids are *vii ho'i* 'little spines'. (The adjective *vii* 'small', while commonly used in Pima Bajo, is rarely used in Upper Pima speech today and appears here as an apparent archaic usage.) When ripe, the entire fruit of the Saguaro is *bahidag*, while the red pulp and seeds that come out are the *gakidag*. The empty skin is *eldag*. The large dry flower that falls away is the *hiósig gakidag*. The woody ribs, the dried vascular bundles of a Saguaro or other columnar cactus, are *vaapai*.

The Pima characterization of the morphology of the corn (maize) plant (fig. 8.4) is at least as complicated as in American English. The word *huuñĭ* means polysemously corn as a species, a crop, a plant, an ear off the plant, and individual kernels. The stem of the plant is *va'ug*, a term that can be used for the stalk or straw of other grassy plants. The fresh leaves are *huuñ haahag* (a fixed form in permanent plural), but when dry they are called *huuñ gakidag*. The tassels are called *huuñ muḍadag*. On the tassels grow the *huuñ hiósgga* 'corn's flowers', the pollen-producing anthers, according to Sylvester. The joint of the cornstalk is the *mamadaj* ("because that's where the ear comes out," said Ruth Giff; this word is etymologically related to *mamaḍho* 'to give birth'). The ear of corn growing on the plant is *huuñ maḍ* (pl. *huuñ maamaḍ*) 'corn baby'. (For some speakers this term does not apply after the ears are picked; then they are just *huuñ*.) The husks are *huuñ eldag* 'corn skin' or 'corn bark'. The silk is *huuñ chiñvō* 'corn mouth hair' or 'corn mustache'. The kernels, called just *huuñ*, grow on the *huuñ oág* 'corn marrow' or 'corn brain'. After the ear is eaten, however, the empty cob is called *huuñ kumkuḍ*. (*Kum* means 'chewed', and *kumkuḍ* is used whether the corn has been chewed or shucked manually.) A row of kernels is *huuñ vavañim*.

Cucurbits also have some specialized terminology (fig. 8.5). The central cavity is the *edaj* 'its insides'. Here the seeds, *kai*, are supported by the *hihi* 'intestines'. The squash or pumpkin as a whole, as well as its flesh, is *haal*. The groove found in certain forms is *s-vopoks*. The peduncle is the *vaksig*. A squash sits on its *at* 'bottom' or 'buttocks'.

On the cotton plant (fig. 8.6), the prominent green calyx enclosing the flower or boll is the *koshdag*, a term also applied to a nutshell or legume hull. (In most other cases, the sepals or calyx are not designated as separate from the flower.) The open boll, ready to be picked, is the *bahidag*, the same word as used for a ripe fruit.

To pollinate something, such as dates, is *ha-doomgid* (from *doom* 'to copulate [with something]', indicating that the Pima understood the genetic implications of the process). The term includes both the natural process, such as by bees, and artificial pollination.

As might be expected from the pivotal role mesquite played in the lives of these desert riverine people, the Pima have a complex vocabulary describing its stages of development (fig. 8.7). When the tender, feathery green leaves sprout in the spring, around early April, the term

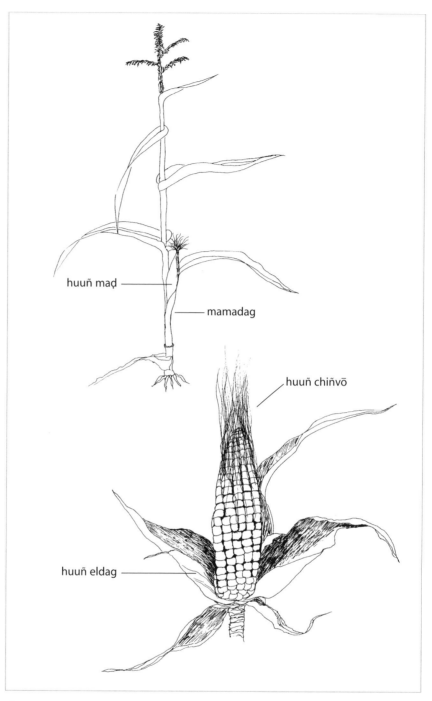

Figure 8.4 *The morphology of maize*

kui i'ivagĭdag is used to describe this stage. The verb *iivagi* means 'to put forth new leaves'. The still-green flower buds of the mesquite are *kui hulkadag* (often in the form of *kui hulkadaj* 'mesquite its flower buds'). The term *hulkadk* is used specifically for certain trees such as mesquite, acacia, and cottonwood and means 'to be in bud'. The stage with opened yellow flowers is called *kui hiósig*. Certain individual flowers become pollinated, producing the embryonic green

fruits; this stage was apparently named also, but no one could remember what it was called. When the pods elongate and are pendant but still green and sufficiently transparent that the immature seeds are visible within, they are called *vihog moikdag* (from *moik* 'soft'). The dried mature pods are *kui vihog*, whether on the tree or fallen or harvested. One refers to the mesquite season or harvest time as *kui baakam*. *Ha'ichu baakam* (from *baha, baak* 'to ripen') means the time when

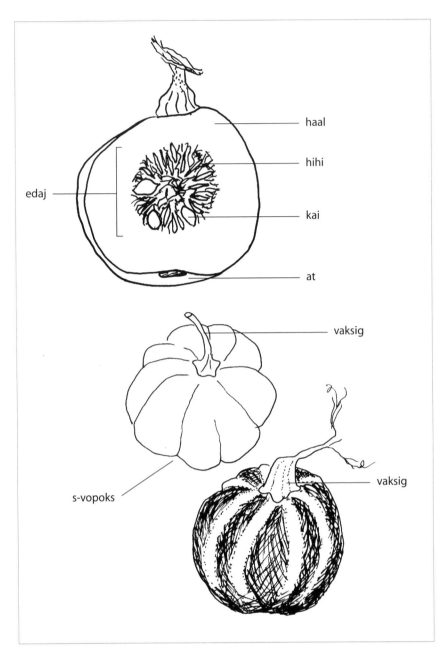

Figure 8.5   *The morphology of cucurbits*

something is ripe, ready for harvest. The Seri on the Sonoran coast recalled eight named stages of mesquite development (Felger and Moser 1985:338).

Perceiving and Naming Plants

In contrast to colors and anatomical parts, the folk taxonomy of biological organisms deals with objective discontinuities in nature; the plants and animals that a tribal person names are the same ones that Linnaean taxonomists classify and name. A people may construct order out of marshes, springs, arroyos, creeks, and rivers as they do with social phenomena, but the naming of organisms is fundamentally different. "When human beings function as ethnobiologists," Brent Berlin (1992:8) emphasizes, "they do not construct order, they discern it."

The fundamental unit in folk taxonomies is the folk genus. Folk genera are the most salient units naming the discontinuities. Often they are terminal taxa, that is, they are not further subdivided in the classification system. Frequently they correspond to biological genera. There are around 240 identified Pima folk generics and another two dozen unidentified ones.

Several of the taxa that I have treated in separate accounts may be only folk species of a genus, that is, *kui chepelk* (Honey Mesquite, *Prosopis glandulosa*) may be merely a kind of *kui* (Velvet Mesquite, *Prosopis velutina*); the same might be true of the two willows (*Salix* spp.), the poplar *(Populus nigra)* and cottonwood *(Populus fremontii)*, and three of the chollas (*Opuntia* spp.). Lumping these would reduce the folk generics by five. While by linguistic criteria alone the relegation to folk species would seem to be appropriate, I am not sure this would be psychologically true. I doubt that a Pima would consider *toota hannam* (Chain-fruit Cholla, *Opuntia fulgida*) really a "kind of" *hannam* (Buckhorn Cholla, *Opuntia acanthocarpa*) in the same sense that someone would consider the folk specifics *toota ko'okol* and *s-vepegĭ ko'okol* to be kinds of cultivated chiles (*Capsicum* spp.). (For an excellent example of how psychological reality differs from linguistic data, see sugarcane, *Saccharum officinarum*, Group G.)

Most of the folk generics subdivided into folk specifics are domesticated plants, as would be expected (table 8.1). But one wild bush, the wolfberry (*Lycium* spp.), is unequivocally divided into four folk specifics. This was an important food, ripening during the spring stress period when many other resources were low. There are many local biological species of wolfberry, perhaps seven in Pima country, several being excellent eating and several others inferior in either fruit quality or productivity or both. As Pima familiarity with this food has declined, predictably the generic name *kuávul* is generally remembered, while the folk specifics are almost forgotten. I am not aware of any folk varietals in Pima, that is, specifics that are further divided. Among the Seri, a nonagricultural Sonoran Desert group, Felger and Moser found varietals with only two plants: *Opuntia arbuscula* and *Pectis*.

How much of their local flora did the Gileños formally (that is, linguistically) recognize? We can make some rough calculations. There are 234 folk taxa, mostly folk generics, identified and treated in this book and an additional 27 unidentified generics (appendix B) for a total of 261 folk taxa. However, 78 of these generics label cultivated, ornamental, or extra-limital biological species. Thus, 183 native or naturalized plants are currently named in the local flora.

I have recorded at least 525 wild biological species from the reservation or immediately adjacent areas. But for various reasons this base number is too large an estimate of what the Pima would have both encountered and perceived or remembered. First, some are restricted to the local sierras, where I was unable to take elderly Pima to ask about them. Furthermore, some of these sierran species are quite rare, and one could hike for years without stumbling across them. For instance, it took me 29 years to discover *Forestiera*, a small tree in the Estrellas. Also, I encountered a number of species that no botanist doing formal surveys had previously collected, even when we worked in the same wash and basin. Second, because of the ecological changes in Pima country, the people have not seen many plants, particularly hydrophils, in half a century or more. In this time some plants once familiar have been dropped from awareness. Third, many biological species are cryptic and could not be distinguished without an optical aid. (I must sometimes remind botanist friends that Indians do not carry around hand lenses or dissecting microscopes when constructing folk taxonomies.) The many species of *Euphorbia, Cuscuta,* and *Juncus,* to name but a few, are not readily distinguished in the field. Finally, there are many species of grasses, for instance, that I never asked Pima consultants about. These categories account for at least 170 species from the local flora. This leaves about 355 potentially named biological species from Pima country. This number is still probably high because it includes waifs from the Sacaton river bottoms that may never have become established and others, especially weedy species, that have invaded Gileño country only in recent decades. (The naming of species new to the local biota stopped around the middle of the 20th century.)

Be that as it may, the Gila Pima linguistically recognize at least half of their local flora: this is a conservative amount that does not account for cultural erosion. For example, 32 historic folk genera were no longer recognized by any Pima I worked with. An additional 20 taxa were known to but a single consultant. (Most of these were verified outside the language community—among Pima Bajo or Tohono O'odham, or from historic sources.) I suspect that at one time the Pima linguistically recognized three-quarters of the local species, exclusive of microscopically differentiated ones.

Undifferentiated taxonomic residuals occur when a category is not completely partitioned: for example, a life-form category that is not completely partitioned by all the included folk generics (Berlin 1992:114–116). In Gila Pima folk taxonomy, residuals are most often seen among invertebrates. For instance, there are several named generic beetles, then all the rest; among *mumuval* there are several folk generics covering conspicuous bees, wasps, and flies, then all the rest of the named category. In each case, the unassigned species are by far the largest component of the domain. I am aware of no plant examples of undifferentiated residuals at the folk generic level. But there are semantic ways of treating new or unassigned biological species. When asked about certain plants that are not covered by folk generics, a Pima may say, *ha'ichu sha'i* '[just] some kind of bush'.

Historical Aspects of the Pima Folk Taxonomy

Languages evolve, generally from simpler to more complex. Folk taxonomies are no exception. The vigorous cross-language studies of Brown and associates have added greatly to our understanding of how sequences of complexity are encoded into taxonomies. In searching for patterns, Brown has demonstrated that these stages are quite predictable in the world's languages (1984); these sequences apply to the encoding of ethnobiological life forms as well.

The earliest plant names in a language tend to be primary unanalyzable lexemes, such as *oak, elm,* and *pine* in English. Polymorphemic and linguistically analyzable lexemes such as *Emory Oak* and *Ponderosa Pine* are later additions to a people's lexicon. Gila Pima, like English, has a large inventory of simple primary lexemes. Primary plant lexemes, with their corresponding Linnaean species, are listed below:

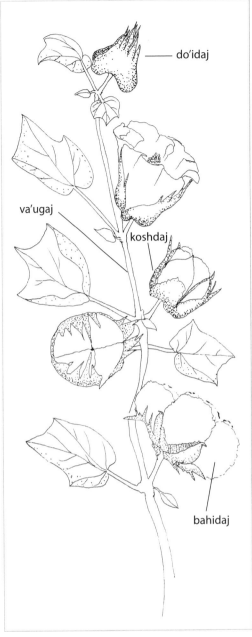

Figure 8.6 *The morphology of cotton*

a'agovĭ, Castela emoryi, Crucifixion-thorn

aan, Chilopsis linearis, Desert-willow

aḏavĭ, Cucurbita digitata, Finger-leaved Gourd

a'uḏ, Agave deserti, Desert Agave

auppa, Populus fremontii, Frémont's Cottonwood

avhaḏ, Lepidium spp., Pepper-grass

bavĭ, Phaseolus acutifolius, tepary

bawui, Erythrina flabelliformis, Coral-bean

bitoi, Fraxinus velutina, Velvet Ash

Table 8.1 Polytypic Pima Folk Genera

Scientific Name	Folk Generic	Number of Folk Species
Noncultivated Plants		
Lycium spp.	wolfberry	4
Prosopis spp.	mesquite	2
Salix spp.	willow	2
Domesticated Plants		
Capsicum spp.	chile	6
Citrullus lanatus	watermelon	4
Cucumis melo	muskmelon	6
Cucurbita spp.	squash	5
Gossypium hirsutum	cotton	2
Lagenaria siceraria	bottle gourd	2
Phaseolus acutifolius	tepary	3 (4?)
Phaseolus lunatus	lima bean	3
Phaseolus vulgaris	common or kidney bean	4
Sorghum bicolor	sweet sorghum	2
Sorghum bicolor	milo maize	2
Triticum aestivum	wheat	6
Zea mays	maize[1]	10

[1] Not all may have been deliberately maintained; several may be named mixings.

cheemĭ, Lophocereus schottii, Senita Cactus

che'ul, Salix gooddingii, Goodding's Willow

chuchuis, Stenocereus thurberi, Organ-pipe Cactus

eḍam, Atriplex lentiformis, Quail Brush

giishul, Mammillaria thornberi, Thornber's Fishhook Cactus

gōhi, Morus microphylla, Desert Mulberry

haad, Dichelostemma pulchellum, Bluedicks or Covena

haal, Cucurbita spp., squash, pumpkin

haashañ, Carnegiea gigantea, Saguaro

hannam, Opuntia acanthocarpa, Buckhorn Cholla

havul, Phaseolus lunatus, lima bean

hovij, Yucca baccata, Banana Yucca

huk, Pinus spp., pine

huuñ, Zea mays, maize, corn

ihug, Proboscidea parviflora var. *hohokamiana*, White-seeded Devil's Claw

iibhai (?), *Opuntia engelmannii*, Engelmann's Prickly-pear

iikovĭ (?), *Hoffmanseggia glauca*, Camote-de-raton

iisvikĭ (?), *Echinocereus engelmannii*, Engelmann's Hedgehog Cactus

iitañ, Allenrolfea occidentalis, Pickleweed

ki'akĭ, giád, Amaranthus hybridus, grain amaranth

koom, Celtis pallida, Desert Hackberry

kotḍopĭ (?), *Datura* spp., datura

kovĭ, Chenopodium berlandieri, cultivated chenopod

kui, Prosopis velutina, Velvet Mesquite

kumul, mushroom

kuupag, Ephedra, Mormon Tea

mamtoḍ, Algae

melok, melhog, Fouquieria splendens, Ocotillo

mōhō, Nolina microcarpa, Bear-grass

mo'oial (?), *Solanum americanum*, Marsh Nightshade

muumsh, Plantago spp., Indian-wheat

muuñ, Phaseolus vulgaris, common or kidney bean

navĭ, Opuntia leptocaulis, Christmas Cholla

noḍ, Sporobolus wrightii, Big Sacaton

oidpa, Pectis papposa, Chinchweed

opon, Monolepis nuttalliana, Patota

pihul, Lophophora williamsii, Peyote

shaaḍ, Amoreuxia palmatifida, Saya

shegoi, Larrea divaricata, Creosote Bush

takui, Yucca elata, Soaptree Yucca

toa, Quercus spp., oaks

toki, Gossypium hirsutum, cotton

umug, Dasylirion wheeleri, Sotol

uupaḍ, Acacia greggii, Catclaw Acacia

uupio, Juglans major, Arizona Black Walnut

vaapk, Phragmites australis, Carrizo

vaas, Jatropha cardiophylla, Limber Bush

vaiwa, Xanthium strumarium, Common Cocklebur

vak, Scirpus spp., bulrush

vi'al, Gaura parviflora, Velvetweed

viopal, Hyptis emoryi, Desert-lavender

viv, Nicotiana spp., tobacco

These are monomorphic, that is, they are apparently unsegmentable. They would appear to be the core Piman plant taxa. An additional smaller number are partly analyzable but would appear to be ancient as well:

chuuhuggia (*chuhug*, night, dark), *Amaranthus palmeri*, Palmer's Carelessweed

dahapdam (*-dam* [agent]), *Sambucus mexicana*, Mexican Elder

hivai (*hiv*, to rub against something), *Helianthus annuus*, Wild Sunflower

ko'okol (*ko'ok*, to be painful), *Capsicum annuum*, Chiltepín

ko'otpidam (*-dam* [agent]), *Bursera microphylla*, Elephant-tree

no'oshkal (*nonha*, eggs [?]), *Aster spinosus*, Spiny Aster

sivijuls (*siv*, bitter), *Rumex hymenosepalus*, Canaigre

tohavs (*toha*, to become white), *Encelia farinosa*, White Brittlebush

uḍvak (*vak*, bulrush), *Typha domingensis*, Southern Cattail

vakoa (*va*, water-related), *Lagenaria siceraria*, bottle gourd

vavish (*va*, water-related), *Anemopsis californica*, Yerba Mansa

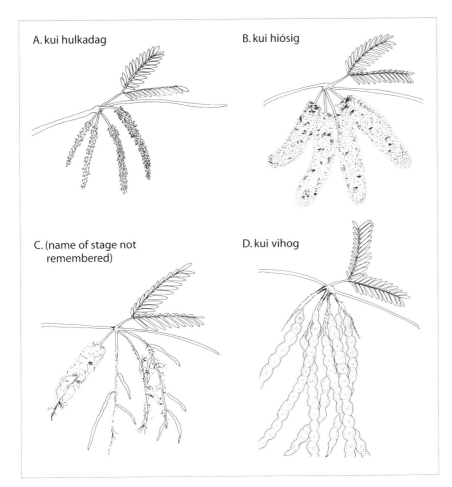

A. kui hulkadag

B. kui hiósig

C. (name of stage not remembered)

D. kui vihog

Figure 8.7 *Stages of mesquite pod development*

Even some of the analyzable segments, such as *va*, a root word referring to water, are ancient in Uto-Aztecan languages. These 72 folk generics may offer some insight into Gileño origins.

This assemblage may represent a proto-Piman plant lexicon. A few basic water plants are represented *(Algae, Phragmites, Scirpus)*, in addition to a large grass *(Sporobolus)*. There is one mushroom. Several small annuals with abundant edible seeds *(Lepidium, Monolepis, Plantago,* and probably *Descurainia)* are included. (Perhaps Chia, *Salvia columbariae*, should be as well. Its Pima name, *daapk*, means smooth, slippery, wet, or naked. In water, Chia seeds become mucilaginous. But no one knows which came first, the adjective or the plant name.) There is also a basic vocabulary for small bushes *(Ephedra, Allenrolfea, Atriplex, Jatropha, Larrea)*. The life-form "bush" is marked by the unanalyzable lexeme *sha'i*. Large bushes and small trees *(Erythrina, Celtis, Castela, Chilopsis, Hyptis, Morus)* are also well represented. And there is a basic repertory of large trees *(Acacia, Fraxinus, Juglans, Pinus, Populus, Salix, Prosopis, Quercus)*. Cactus and cactuslike plants are well represented (15 names, the largest number in any grouping). A basic lexicon of cultivars (maize, lima bean, tepary, common bean, squash, cotton, amaranth, chenopod, and tobacco) is labeled with monolexemic terms. These are all *e'es* 'crops'. The lexeme *viv* 'tobacco' may indicate either wild or domesticated forms of *Nicotiana*, so we cannot argue that it is necessarily an ancient crop.

The list includes a number of plants found essentially in the uplands surrounding the deserts *(Celtis, Fraxinus, Juglans, Morus, Pinus, Dasylirion, Nolina, Quercus, Yucca)*. Some of these plants are southern in distribution, not even reaching Gila Pima country *(Amoreuxia, Erythrina, Lophophora, Stenocereus, Lophocereus, Jatropha)*; all of these can be found, however, in Tohono O'odham country except *Lophophora*.

Among the Gila Pima, 26% of folk generics are unanalyzable (61 of 235).

Among the Seri of coastal Sonora, the only other group with comparable data, 36% of the folk generics are unanalyzable monolexemic names (110 of 310).

Some Taxonomic Loose Ends

Some plant names are in use among the Tohono O'odham but not among the Gileños (appendix table A.2). In some cases, the two groups have different names or labels for the same biological referent. A number of plant species do not reach as far north as Pima country. In a few instances the Gila Pima recognize the Tohono O'odham name but use their own name instead. These situations are treated in the text of the individual folk generic accounts.

A number of plants, for one reason or another, should appear in the Gila Pima plant lexicon but are absent (appendix table A.3). Some are plants that occur conspicuously near and in villages, such as Narrow-leaved Saltbush *(Atriplex linearis)*, Salton Saltbush *(Atriplex fasciculata)*, Woolly Tidestromia *(Tidestromia lanuginosa)*, and various spiderlings *(Boerhavia* spp.). The Pima are quite aware of these plants but are unable to provide names for them. "They just have no name" is the only explanation I heard. Furthermore, they are often closely related to other plants in the local environment that are named. Narrow-leaved Saltbush, for instance, grows with *kokomagĭ sha'i*, Desert Saltbush *(Atriplex polycarpa)*, and the casual observer would not likely notice that the two are different. Similarly, the spiderlings *Boerhavia erecta* and *B. intermedia* are called *makkom jeej*, while the even more widespread and conspicuous *B. coccinea* and *B. coulteri* are nameless.

Perhaps this is folk taxonomic erosion: the loss of names over time with culture change. Circumstantial reasoning would point to this conclusion. The Pima have indeed moved from a purely subsistence lifestyle as farmers and gatherers to a wage economy. A people who formerly spent nearly all their time out of doors have now become indoor people.

But not all the evidence is negative. No one with whom I worked remembered the name for several plants that had been recorded at the turn of the century, such as Limber Bush *(Jatropha cardiophylla)*, Peyote *(Lophophora williamsii)*, and the cultivated grain chenopod. A few plants

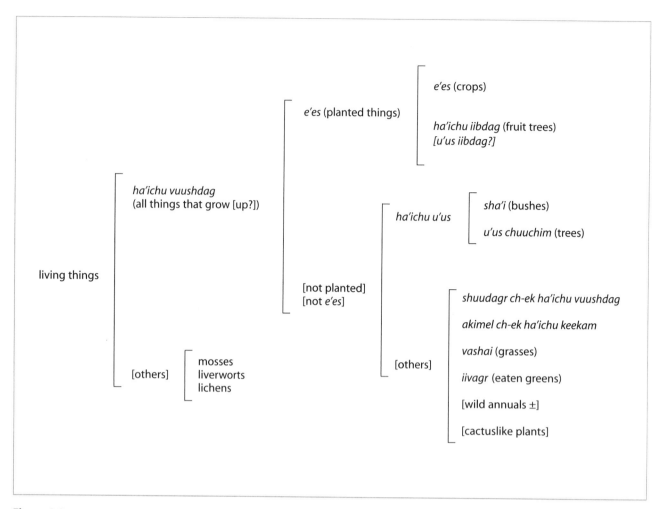

Figure 8.8 *Sylvester Matthias's arrangement of plant categories*

we encountered in the field, such as Water Pennywort *(Hydrocotyle verticillata)* and Salt-marsh Fleabane *(Pluchea odorata)*, Sylvester Matthias thought had a Pima name, but either he could not remember it or perhaps he had not learned it or even heard it. As far as I have been able to tell, a number of current folk taxa are now remembered by but a single speaker. Moreover, some plants in the Pima lexicon remain unidentified (appendix B). A real puzzle is the number of plants (appendix table A.4) that are abundant where riparian or marsh conditions have been recreated but that are not recognized by any of the Pima consultants with whom I worked. These include Yellow Waterweed *(Ludwigia peploides)*, Willow Smartweed *(Polygonum lapathifolium)*, and monkeyflower *(Mimulus* spp.), as well as a few smaller plants such as spike-rush *(Eleocharis* spp.), Blue Streamwort *(Stemodia durantifolia)*, and Water Speedwell *(Veronica anagallis-aquatica)*. Originally these must have been part of the local aquatic communities.

Cognates of certain plant names are pan-Piman. Others appear to be, at least insofar as the evidence goes. Major agricultural crops such as *muuñ, huuñ,* and *baví* are in this category. But there are many examples among wild plants as well, such as *naví, ho'idkam, kui,* and *vashai.* Considerably more field work must be done to map the geographic spread of various lexemes among Tepiman languages. When a biological referent drops out of the local environment, such as the columnar cactus in the Sierras, does the word for it also drop out, or is it applied to a different but somehow related referent? At least in some cases, the lexeme is conserved and the referent is changed. Examples are given in the individual accounts.

Further folk taxonomic comparisons between Tepiman groups are tantalizing, but at this time necessarily far from possible. Ethnobotanies are needed for various Tepehuan groups. Some colonial texts on northern Piman remain to be transcribed and analyzed. Even

comprehensive elicitations among Tohono O'odham dialects are needed. Then the entire Tepiman folk taxonomy can be synthesized. It will make a fascinating story.

Plant Categories above the Generic Level

What Linnaean taxonomists classify as "plants" in the broad sense the Pima put into 12 categories, including two covert categories and one group of miscellaneous unaffiliated plants and one group of unassigned organisms (Group L). These are generally called life-form classes by ethnotaxonomists. Sylvester Matthias, who provided the hierarchy (fig. 8.8), consistently called them families.

Groups A and B may well have been segregated protohistorically from Group C, possibly as a result of a more desertic people encountering a species-rich riparian habitat during their northward migration. Group A includes submergent plants as well as floating ones, such as duckweed *(Lemna* spp.). Group B maps directly to what botanists call "emergents." The Pima word for Group C, *vashai,* etymologically is derived from *va-,* a root word referring to water, and *sha'i,* something brushy. But its proto-Piman meaning was probably more in the sense of 'water grass', and even today it is used by the desert Pima Bajo to include any grassy plant growing in wet places such as streamsides and springs. In Gila Pima today it includes not only grasses and sedges, but such pasture-type plants as Indian-wheat (*Plantago* spp.), Pepper-grass (*Lepidium* spp.) and Sweet Clover (*Melilotus indica).*

Unlike the Tohono O'odham, who have elevated the name for their most prominent tree, mesquite, to the life form 'tree', the Gila Pima use the phrase *uus keekam* (literally *uus* 'stick [or] wood' + *keek* 'standing' + *-kam,* attributive) for this life form. The word *uus* alone may have once meant 'tree', as it still does in certain archaic songs, but it has come to have a much more inclusive meaning, similar to Spanish *palo* 'stick'.

Two categories unequivocally are utilitarian rather than morphological: *iivagĭ,* eaten greens, and *e'es,* crops. These are normative in the sense of how the Pima usually deal with a species. All plants used as greens, either cooked or raw, invariably are classified as *iivagĭ.* Even plants that have not been prepared as greens within anyone's lifetime are still placed in this category if people remember that they were once (80 or 90 years ago) eaten. The other utilitarian category is *e'es,* crops or planted things. Species normally planted and raised by people are included here, even though the Pima know that sometimes or in some places the same species may be naturalized. An example is Castor-bean (*Ricinus communis).* A Pima, explaining to me that this brushy euphorb was an *e'es,* remarked that he had seen it growing "wild" around Palm Springs in southern California. There, perhaps, it would be considered *sha'i,* a bush, but not in Pima country, where it is always somebody's plant.

Group J has a domain that is entirely postcontact: planted fruit trees. It includes all the citrus and rosaceous trees. Before the Jesuit period, the Pima raised no trees for their fruits, although some native bushes and the Saguaro cactus provided wild gathered fruits. Groups I and J are nested, with the category name *e'es*

indicating both planted things generally and, more specifically, all crops that are not fruit trees.

Groups K and L are interesting in how they contrast anomalous organisms. Group K includes mushrooms and puffballs (*Tulostoma, Podaxis,* and *Battarrea* spp.), Corn Smut (*Ustilago maydis),* Sand-root (*Pholisma sonorae),* and perhaps Peyote (*Lophophora williamsii).* This is a category of unaffiliated folk generics considered to be *ha'ichu vuushdag* 'things that grow up' that do not fit any of the previous 10 life-form or utilitarian categories. There are some surprises. How can Corn Smut be considered something that grows up? The association seems very perceptive: mushrooms and puffballs belong here, and Corn Smut is considered related to these, so it goes here too, even though it does not behave properly in growth form.

What seems counterintuitive to me is Group L, biological organisms not considered in Pima to be *ha'ichu vuushdag* yet still considered living things. In other words, the vernacular English dichotomy between plants and animals does not exhaust what the Pima would include as living things. This unnamed assemblage includes mosses, liverworts, and lichens. They are clearly leftovers, not united by any morphological similarities. While Sylvester had no hesitation about including atypical plants such as duckweed (*Lemna* spp.) and Mosquito Fern (*Azolla* spp.), as well as algae, in his Group A, he excluded the Bryophytes and mold from the unique beginner category plants.

Part 2

Gila Pima Plants

Hiósig Ñe'i

mumuwui va hiosinge
id damai jevenai wushañime
wecho keekam donovange
heg amjen amai si chemoho
wesiko s-ap tahadkam
(in song language)

mu'i hiósig id daam jeveḍ o wushañim
wecho keekam do'ag
heg amjeḍ am si cheemo'o
wesko, s-ap taahadkam
(in spoken Pima)

Many flowers are coming up on this land.
[Morning] Below standing mountains,
From there they fill the land.
Everywhere there is happiness.
(free translation)

Sung by Culver Cassa, 1993

Plan of the Folk Generics (Biological Species Accounts)

Folk generics are usually terminal taxa, that is, what any given plant is called. Approximately 250 of these are known to the Pima presently (or were, in the recent past). These "are the basic building blocks of any folk taxonomy, are the most salient psychologically, and are likely to be among the first taxa learned by the child" (Berlin 1992:15–17). To a large extent, at least with Pima, these correspond to Linnaean species, so to the biologist what follows are the individual species accounts. These form the core of the ethnobotany.

Methods

I elicited folk taxonomies primarily in the field. Whenever possible, plants were identified when I and the primary consultants with whom I worked found them growing throughout Pima country. Some species required trips to the higher country of central and northern Arizona or southward toward Tucson and the more elevated Tohono O'odham country: the Pima

formerly traveled widely in hunting, foraging, and military activities and so were familiar with many plants of these higher elevations. In addition, because of the near-complete loss of their riparian vegetation, I found it necessary to go far afield to find analogs to the historic plant communities along the central Gila. This included taking individuals to the Cochran area east of Florence where the Gila still runs, to the lower Colorado River near Yuma, and even to desert streams in eastern San Diego County, California. Additionally, we made trips to the magnificent oases in the desert of northern Sonora between Nogales and Imuris, formerly Pima country. Several Gila Pimas made trips with me to the middle Rio Yaqui in southern Sonora to visit the remaining two speakers of Pima Bajo. Sylvester Matthias and I spent weeks there roaming about the country with Pedro Estrella collecting plants and Pima names and folk science.

However, in some cases it was necessary to bring plant specimens to people for identification. Characteristically, the first step a Pima makes in identifying a plant is to crush and smell its foliage. For this reason, I tried to use nothing but freshly collected specimens in elicitations. Only as a last resort did I rely on dried pressed plants, photographs, or drawings.

Whenever possible I tried to elicit names in the context of biological contrast sets. These are sets of two or more related or morphologically similar plants that might be included in or excluded from the domain of a Pima folk taxon. For instance, when talking about paloverdes it is necessary to establish which of the two local species a person is talking about. Whereas an outsider might readily lump the two, for the Pima they are distinct entities with unrelated names (that is, one is not a "kind" of the other). But I was not able to use biological contrast sets for all folk generics. Of the three species of *Mammillaria* (fishhook cactus) on the middle Gila, for example, *M. tetrancistra* is exceedingly rare, while the other two are more common. *M. microcarpa* and *M. thornberi* have quite different growth forms. I never found anyone who knew both the common species as a linguistic contrast set, so I am not sure whether their Pima names contrast two species or are folk synonyms for small species of fishhook cactus.

Another good example of a biological contrast set is the two weedy species of annual saltbush, *Atriplex wrightii* and *A. elegans.* Only by asking someone about both could one be sure of the domain. As it turns out, each has its own Pima name. But the common and somewhat similar *A. fasciculata* is excluded from both of these folk taxa. Failure to consider biological contrast sets led Curtin (1949:108) to miss the distinction Gileños made between two local willow species.

The ubiquitous brushy saltbushes of Pima country illustrate another lesson in contrast sets. Three species (*Atriplex canescens, A. lentiformis,* and *A. polycarpa*) are distinguished by Gila Pima, but another (*A. linearis*) is unequivocally excluded as unnamed. However, I was given four different names by different people for the most abundant saltbush (*A. polycarpa*). Are there really four folk synonyms, alternate names used currently in the speech community? This is rare among the Pima for native plants (see appendix D), but it may have resulted from a collapsing taxonomy where the original referent of a name became rare or unfamiliar and the name was then applied to the most common species. I suspect that true folk synonymies may be involved, however, because people who used one name for *A. polycarpa* did not recognize the alternative names.

Nearly every conversation appearing in this book was tape recorded and is transcribed here verbatim, with bracketed words [such as these] inserted for clarity. Each Pima speaks in his or her own style, and this is preserved in both Pima and English. Frequently I took sections of text back to someone for clarification or elaboration. Words or phrases that were unclear

or were missing from the original tape have been inserted from subsequent interviews. At times entire parts of a taped session were unclear to me, so I had to go back and repeat the question or ask more details about obscure sections. In these cases, I spliced two and sometimes three conversations, even though these may have been recorded over a period of years. If the subsequent conversation took a different turn, with new or different information, I have usually included both old and new information. In some instances where my tape recorder was unavailable or where stopping an important conversation to go unpack it may have interrupted the train of thought, I scribbled the conversation longhand, attempting to preserve the words and word order. Later the same night, I would either type or rewrite the session. In all cases, the aim has been to preserve the information just as presented, in each consultant's own manner.

I tried to be largely nondirective in interviews, though the success of this method varied with the individual. Some stayed within the subject matter more readily than others. But tangents often indicate associated concepts in Piman thinking, and if one of the goals is to write an emic ethnobotany, then tangents have a definite purpose: these tangents have added immeasurably to the richness of this work. These aspects of Pima folk science would have been lost with a rigid format for data elicitation. I also tried not to be leading in my questioning in such a way as would unintentionally coerce people into agreeing to something they did not agree with but which presented the path of least resistance.

Sometimes, where I thought I had been given wrong information about something or the wrong name for an organism, I waited two or three years before bringing up the subject again. "Now, tell me again about these little yellow things growing here in the wash. What do you call them? Are they good for anything?" (I may have known very well, for instance, what the Tohono O'odham call the plant and what they use it for.) If this "failed" after a number of years, I might say, "Well, the Papago call this such-and-such." And everyone I worked with knew me well enough to voice disagreement readily when necessary: "No, I never heard of

that," or "No, we never used that here; maybe they did over there."

Whenever possible, I collected voucher specimens to document what we were talking about. I took these repeatedly when there might be confusion of similar botanical taxa, such as for grasses, sedges, saltbushes, chenopods, spiderlings, and dock and sorrel. I have prepared both positive and negative vouchers—plants that document a folk taxon and those that represent plants that are unnamed in Gila Pima or are excluded from a folk taxon.

Arrangement of Folk Generics

The grouping of individual species accounts, the folk generics, is by emic categories—that is, those categories the Gila River Pima perceived—rather than by the categories of Western botanists, which are based on evolutionary relationships (see chapter 8). The classification begins with an unlabeled category, living things, which includes all animals and humans as well as all plants. The next level is a dichotomy between 'things that grow up' (*ha'ichu vuushdag*) and other living things such as mosses and lichens (Group L). All 'things that grow up' are subdivided into *e'es* 'planted things' and an unlabeled category of things that are not planted, that do not belong to anyone, that just grow by themselves. Planted things are divided into *e'es* in the sense of 'crops' (Group I) and *ha'ichu iibdag,* or fruit trees (Group J). Naturally growing things, things that are not planted, can be divided into two intermediate categories: *u'us,* or woody plants, and a covert or unlabeled category that includes all other plants (Group K). There are two kinds of *u'us: sha'i,* or bushes (Group D), and *u'us chuuchim,* or trees (Group E). Plants that are not woody and that do not fall into the "bush-tree" category constitute six groups: *shuudagĭ ch-eḑ ha'ichu vuushdag,* plants growing in or on the water (Group A); *akimel ch-eḑ ha'ichu chuuchim,* plants standing in the water (Group B); *vashai,* grasses and grassy plants (Group C); *iivagĭ,* eaten greens (Group F); and two covert categories of wild annuals (Group G) and cactuslike plants (Group H). The category of leftover organisms (Group K) that do not fit any known category but are still considered plants (or at least living things) include fungi and Sand-food, a root parasite. These are unaffiliated folk generics. A final group contains organisms considered to be living things but not plants, by Pima definition (Group L).

Headings

Within each Pima group, species are arranged alphabetically by scientific name. The scientific name is followed by the name of the describer (the "authority"). Scientific names undergo changes for a number of reasons: a botanist working on a group may reclassify it; a genus may be split into several genera, or several genera may be lumped into one; new studies may show that a species has been misplaced altogether; someone may find that the plant had already been classified but the older name was overlooked (by the rules of priority, the oldest name clearly applying to a species must be used); a botanist may discover that populations once considered a single species are in reality several different biological species; a segregate may require a new name to distinguish it. To enable the reader to follow the scientific nomenclature that has been used from the beginning of ethnobotanical studies among the O'odham, other names by which the plant was identified in writings on either Akimel or Tohono O'odham follow the current scientific name and are bracketed and preceded by =. In some cases, the scientific name used in Piman literature was a misidentification of the folk taxon, rather than a change in classification or nomenclature: early workers did not have the aid of comprehensive floras with the classification keys that we take for granted in North America today. Where this occurs, the synonymous entry is bracketed and preceded by ≠.

A Pima folk taxon may map to several different biological species. For instance, *muumsh* includes *Plantago ovata* and *P. patagonica*. In these cases, the Piman domain is broader and more inclusive than the botanical. Where a Pima name includes only two botanical species, each of these is given in the heading. Where more than two species are included in a Pima name, as with the globemallows, the scientific name is given as the genus, such as *Sphaeralcea* spp. Usually, the actual biological species that have been identified as belonging to the Pima taxon are given under the heading "Vouchers" at the end of the species account.

In some cases, a folk taxon has never been verified by a field identification, but the botanical identification is reasonably certain from the biological evidence, from historical works, or from the ethnobotanical evidence of other Piman groups such as the Tohono O'odham or the Pima Bajo using the same name. When the biological referent of a folk taxon is not absolutely certain, the scientific and English common names are followed by a question mark. And finally, some folk taxa have not yet been identified to biological species, nor is there even a reasonably solid guess what the biological referents might be. These are listed in appendix B.

The vernacular or common name in English or Spanish accompanies each terminal folk taxon; these common names are followed by the family name. All proper names in English and Spanish of wild plant species have been capitalized. A species name may be unitary (Saguaro) or compound (Organ-pipe Cactus). Lowercase indicates a plant name in the generic rather than specific sense (palo-verde, as opposed to Blue Paloverde or Foothill Paloverde). This practice is in keeping with a growing trend among botanists and a universal convention among ornithologists and some other zoologists. The immediate, practical benefit of capitalization is to distinguish a specific epithet from a word combination that is merely descriptive (a blue bird perches on a great gray oak—neither the bird nor the oak species is indicated). But there is also a philosophical reason for capitalizing the names of organisms: the designation and capitalization of proper nouns bestows some kind of dignity. We capitalize our own names, artificial areas we draw on maps (Arizona), and even random units of time such as days of the week. Yet long before our states were thought of, the biological species were here, and here they will remain (we hope) long after our own names and other "proper" nouns are forgotten.

I have followed horticultural convention in capitalizing the names of crop varieties, such as Sacaton Aboriginal cotton, but not the names of cultivated species, such as apple, tomato, carrot. Some may find this convention inconsistent, but at least it is consistently inconsistent.

One last convention. Names that are attributed to some similar or allegorical form with a secondary linguistic marker— that is, there is no biological relationship to the base word—are hyphenated. The prickly-pear, for example, is not a kind of a pear any more than a Russian-thistle is a kind of a thistle or an Osage-orange a kind of orange.

In each folk generic heading, the primary Pima name appears first. Where there is a plural form that is used as an alternate name, this follows. For instance, some Pimas call *Polypogon monspeliensis* (Rabbitfoot Grass) *shelik bahi,* while others call it *sheshelik baabhai;* the first is singular, the second plural. In addition, sometimes more than one name is used for the same plant by different speakers. Some call *Sorghum halepense* (Johnson Grass) *kaañu chu'igam,* for example, while others call it *vaapk chu'igam.* Usually these folk synonyms occur with plants that are postconquest introductions (see appendix D). Often these synonyms break down geographically, with speakers in one part of the reservation using one name and those in another using the alternative name. Occasionally there are variants of the same name. For instance, speakers from the west end, Kuiva O'odham, say *miiliñ* for muskmelon while those from the east end, Ta'i O'odham, say *miiloñ.* These differences have usually been pointed out by speakers themselves.

Text

Information included in the text for each Pima folk genus includes a number of different topics. The sequence has been loosely arranged according to the narrative. Usually each account opens with a description of the plant and where it grows, unless the plant is a well-known cultivar. Plants have been described in nontechnical terms as far as possible. (Technical descriptions can be found in various floras, for example, the classic Shreve and Wiggins 1964.) As much as possible, measurements in this section are given in terms that anyone likely to be wandering through the desert might readily evaluate without a measuring instrument: a plant may grow to ankle-height and have flowers as wide as your thumbnail, or a bush may grow to between waist-high and chest-high and have fruits as long as your thumb.

Measurements of plants are given in a system I devised that requires no special instruments (listed below with English and metric equivalents):

fingernail, ½ inch, 12 x 12 mm

thumbnail, ⅝ inch, 15 x 15 mm

finger joint, ½ inch, 35 mm

thumb-length, 2 ½ inches, 65 mm

finger-length, 4 1/4 inches, 110 mm

palm, 3½ x 4½ inches, 90 x 110 mm

palm + fingers, 8 x 8 inches,
120 x 120 mm

ankle-height, 5 inches, 130 mm

knee-height, 2 feet, 60 cm

hip-height, 3½ feet, 1 m

chest-height, 4 feet, 1.3 m

shoulder-height, 5 feet, 1.5 m

head-height, 6 feet, 1.85 m

Where these "autometric" units are impractical, as in long distances, the English system is generally used because this is still the vernacular on the reservation. Plants are described as they grow in Pima country. They may behave differently elsewhere; for example, in other parts of their ranges, they may grow taller or shorter or even have different habitat requirements.

The remaining parts of each folk generic account generally begin with the name (and etymology) of a folk taxon in Gileño and related languages. Often the tracking of a plant's name, a folk taxon, was like a detective story. Clues came from people still living, earlier writers, colonial documents, legends and creation stories, and related words among Piman groups in Mexico.

The balance of each folk generic account focuses on historical and ethnographic information. The folk knowledge of a half-dozen Gila Pimas who were born early in the 20th century usually forms the bulk of the account, as discussed earlier. This is Pima folk science. Interwoven with this folk knowledge is historical information gleaned from Hispanic and Anglo documents and more recent writings on the Gila Pima. Often the historic data validate what a Pima was telling me in the 1980s. Sometimes the new information shows a great elaboration over what was recorded earlier. At other times we see instead the process of cultural erosion: some of the early writings record items that have completely slipped from the oral history of living consultants. Where only

parts of a cultural complex survive, as with the Saguaro wine feast and the Peyote cult, historical records help interpret the whole. The student of cultural change will be interested in which types of information, which practices, have survived at least among some individuals and families into the 1980s and which have been completely forgotten.

The symbolic role a plant played among the Pima may also be learned from sources other than contemporary native consultants. Piman oral literature preserves some symbolism. Foremost among these genres is the Creation Story (Hayden 1935; Lloyd 1911; Russell 1908:202–236), in which certain plants such as maize, tobacco, Catclaw, and Ocotillo played a role. Other noncanonical narrations tell of the origin of mesquite, Foothill Paloverde, and Saguaro, for instance. There must have been many of these stories at one time.

In addition, the symbolic roles of plants may be learned from another form of oral literature: song. Only a small fraction of the vast resource of sung poetry has been recorded. Pima songs are precise, tightly structured utterances, often in a language style quite different from ordinary speech. Portions of some published songs relating to plants have been included here, with the orthography standardized as far as possible. Others were sung for me by Joseph Giff, Francis Vavages, Herbert Narcia, and George Kyyitan. These have been transcribed and translated by Culver Cassa, a native speaker.

Following the folk generic account may be some additional information. The heading "Technical Notes" segregates matters of interest primarily to taxonomic botanists. Herbarium specimens that are positive vouchers for a folk taxon are listed at the end of each species account after the heading "Vouchers": ARIZ is the University of Arizona, Tucson; ASU is the Arizona State Museum, Tempe; DES is the Desert Botanical Garden, Phoenix; SD is the San Diego Natural History Museum; US is the National Museum of Natural History, the Smithsonian Institution. Collector initials (see table 6.1 for their names) and collection numbers are italicized. (AMR specimens are given with collection number only: the first of each set has been deposited in SD, with replicates in ARIZ, ASU, or DES.) Following in parentheses are the initials or name of the

Piman consultant who identified the specimen and supplied its folk name (see chapter 2). If multiple biological species are included in a folk taxon, the species names are also given.

Some voucher specimens are from the reservation proper but many, as the text indicates, are from "Pima country," a broader area that tribal ethnohistory indicates was formerly used in hunting, fishing, and gathering activities. This includes all the Sierra Estrella and the South Mountains, largely outside the present reservation boundaries. Because of the degraded condition of the modern lowland flora, it was sometimes necessary to gather vouchers from elsewhere in the Sonoran Desert, including Sonora and southern California.

Group A, Shuudagĭ ch-eḍ Ha'ichu Vuushdag

plants growing in or on the water

Sylvester Matthias segregated a few Pima folk genera as a special category of plants that he called *shuudagĭ ch-eḍ ha'ichu vuushdag* (literally, 'water in/on something sprouts/shoots'): two types of algae, a water fern, a pondweed growing under the water, and duckweed, which floats on the surface of still water. Holly-leaved Water-nymph (*Najas marina* L.) belongs here also, but Sylvester could not recall its Pima name. Undoubtedly, there were once additional species included in this group, but the long years of desiccation have wiped these from Piman memory. In fact, most Pimas now recognize only one water plant: *mamtoḍ*. The remainder were remembered only at the western villages where surface flow persisted the longest.

This group of plants contrasts with the next set of water plants that we would designate in English *emergents*.

Algae
algae, hairlike
Green Algae
kuup cheveldakuḍ

Although the Gila once ran swiftly year-round, now it scarcely runs at all. Occasionally there are still good rains in the headwaters of the Gila and Salt Rivers, and the dams upstream must release water for months or even years. Once in Yuma County when the Gila was running for a long period, I had to cross it in the vicinity of Tacna. I noticed great strands of hairlike green algae in the Gila and stopped to pull some out. To my surprise these formed slender ropes, some 50 or more feet long, undulating in the swiftly moving clear water.

Sylvester Matthias told me about this type of algae: "*Kuup cheveldakuḍ* is green, hairlike algae in the running rivers. I don't know what *kuup* means. Maybe it's a water plant, too. *Cheveldakuḍ* means 'stretched out, stretchers' or 'growers or growing'. It's two words, but I don't know the first one."

Not surprisingly, I have never discovered a Tohono O'odham who recognized this word, nor is it given in any of the dictionaries. But the Tohono O'odham had no permanent rivers in their country. They had to be content with the still-water algae, *mamtoḍ.*

kuup cheveldakuḍ algae

The Pima name for damselfly is iókos.

Algae
algae, mushy spongy type
Green Algae
mamtoḍ

After the summer rains rush down the *a'akĭ* (arroyos), water stays for weeks in tinajas (natural stone basins) in the Sierra Estrella and other desert ranges. These water holes, called *chechepo* in Pima, become a focus of life for birds, mammals, and insects. For some, it is just an opportunity for free water. For others, it is an essential part of a life cycle. Often there are sandy "beaches" at the edges of these tinajas. In these unlikely locations, water-loving annuals such as Yellow Monkeyflower *(Mimulus guttatus)* and Blue Streamwort *(Stemodia durantifolia)* spring up. But the users of the tinajas—Bighorn Sheep and Rock Wrens, diving beetles and mosquito wigglers, and even the showy annuals—are likely to distract your attention from a green mass growing in the water, a plant the Pima have named and even sung about, the spongy algae, *mamtoḍ.*

Sylvester Matthias told me, "*Mamtoḍ* is different [from *kuup cheveldakuḍ*]. It grows in slow water or pool[s]. We used to make balls, to make snowballs from it. It has no form, just there under the water like [slime]." Ruth Giff was standing nearby and said she knew *mamtoḍ* but never heard of *kuup cheweldakuḍ.*

It is surprising how often this seemingly insignificant plant comes up in Pima oral literature (unfortunately often glossed 'moss'!). For example, in the oration "Elder Brother as he restored himself to life" (Russell 1908:341), there is reference to spring waters covered with "*mamathat.*"

Of the five water plants in this group, *mamtoḍ* was the only one recognized by George Kyyitan, Francis Vavages, and most other people from the eastern portions of the reservation. Perhaps this should not be surprising, for the eastern area is where the decline of the Gila first began to affect the Pima. This algae grows relatively quickly, even in temporary ponds and water holes in the desert, so it remained widely known.

There appears to be strong dialectical variation in this plant's name among northern Pimans. Frank Jim, a Tohono O'odham, described *mamkmaḍ* as "green stuff just floating when water stands." In the dictionaries it has been written as *mamtoḍ* (Saxton and Saxton 1969:170)—glossed 'mildew', 'mold', 'algae', 'scum' (Saxton, Saxton, and Enos 1983:40)—and *mamdhod* or *mamadhod* (same glosses), with *mamadhodmagĭ* as the color of green algae (Mathiot n.d.:44). I have found no one among the Gila Pima who recognized this color term. In Gileño, mildew and mold are considered separate from algae, although the words are closely related, sharing a common origin (see Group L).

Vouchers: *1467, 1717* (SM).

Azolla filiculoides Lam.
Duckweed Fern, Mosquito Fern
Azollaceae
shuudagĭ eldag

When discussing *hodai eldag* and *jeveḍ eldag*—the mosses, liverworts, and lichens of rocks and soil—Sylvester Matthias once mentioned the plant *shuudagĭ eldag* 'water skin'. He noted that livestock, especially horses, might eat it. On another occasion, he explained, "Where the water drops, water falls, and a pool [forms] there, water moss [algae] grows there. That's where the *shuudagĭ eldag* comes in, on the pond."

"What does it look like?"

"It's green, something like. . . . I can't explain what it looks like."

"Is it different from *mamtoḍ?*"

"Different! It's like a 'chile seed', like *ko'okol kai* [duckweed], but it's *together.* On *top* of water. Like *ko'okol kai* but together. Sometimes you could see it floating."

Another time Sylvester said, "It's green, on top of water, floating like *ko'okol kai.* Not used [for anything]. The stock can have a feast on that *shuudagĭ eldag*—the horses; I don't think the cows bother it." He said the plants lacked flowers and were not rooted to the bottom. "It's something like a moss. It's only found in a pool of water where it's not running."

What he described suggested Duckweed Fern (*Azolla* spp.). This plant floats like duckweed (*Lemna* spp.) but has minute overlapping leaves and often forms a blanket on the water's surface. Nitrogen-fixing blue-green algae, *Anabaena azollae,* symbiotically inhabit this Pteridophyte. But it had never been collected in Gileño country nor to my knowledge found any closer than the Salt-Verde confluence, some 45 miles north.

On a mid September day in 1995, Culver Cassa and I decided to take a jaunt east of Florence up the dry sandy stretch of what was once the Gila, to the Ashurst-Hayden Dam; Culver had never seen the Gila as a live stream. The riparian community extends this far west, to where the entire flow of the river is diverted into a huge canal to feed irrigation ditches in local communities and on the upper end of the reservation. Excess flow ends up in Picacho Reservoir. At the base of the diversion dam some water seeped into shallow pools in the gravel. There, through my binoculars, I spotted some bright green floating plants that looked for all the world like water fern. Some at the edges had turned reddish and even bronzed-purple. We scrambled down the bank and found miniature fernlike fronds with dangling roots floating in water an inch or so deep. In places they formed carpets of contiguous plants. We photographed them and gathered samples in plastic containers.

The next day we took our specimens to the rest home in Maricopa Colony where Sylvester was staying. By that time he was confined to a wheelchair, could see very little, heard poorly, and talked slowly. But his memory was still excellent. I showed him the plants. "I can't see them," he said; there was not enough light in his room. It was time for him to go eat, so we left.

A week or two later we were in the area again, and this time I wheeled Sylvester out into the sunshine that he so enjoyed. I handed him the plastic container with the floating water ferns. When he finally found a spot of vision he said, "Oooh. Where did you find this?"

"At the Ashurst-Hayden Diversion Dam east of Florence."

"Was the water very still, no current, shallow?"

"Yeah."

"This is *shuudagĭ eldag,*" he pronounced definitively.

It was if someone had just handed me a thousand dollars; I had waited nearly two decades to verify this Pima name.

Vouchers: *1972* (SM).

Lemna spp.
duckweed
Lemnaceae
ko'okol kai

Duckweed is a plant that has reduced life to the barest of essentials: a simple floating body with chlorophyll, a single rootlet that dangles like a white thread into the water, and, very rarely, a stamen and a pistil, as sort of a reminder to the world that they are indeed seed-bearing plants. The fronds or individual bodies are so small that half a dozen could rest on your fingernail. These float on the surface of quiet ponds and streams, where they make up for their simplicity by their productivity. Most of the time, they reproduce asexually, each little fat, flat-topped glob of green cells producing at its rounded edge miniatures in its own likeness that ultimately break free. In a short time they will cover a very still pond, but where there is wind, they get bunched up along shorelines.

I knew that as late as 1935 some duckweed had been collected on the Gila, so I kept searching for it on charcos, on pump ponds, and along the lower Salt where the river always ran, sustained by sewage effluent. Years went by, and I never found duckweed. A friend, Wayne Armstrong of Palomar College, specializes in this taxonomically difficult group of plants. He gave me some samples in plastic bags to show to Sylvester Matthias and others. Sylvester recognized them, saying, "There *is* a name in Pima," but he could not remember it.

ko'okol kai   *Lemna* spp.

The insects are mosquitos, vaamog *in Pima*

Then one fall I brought back from Picacho Reservoir a prickly water plant, purplish and green, that I had found floating in the still water. We were at that time trying to identify *kuukvul huch,* a troublesome water weed in Pima ditches. Sylvester looked over the purplish plant, then said, "This *does* have a Pima name. It grows in still water, in ponds, like the places where you find *ko'okol kai.*" And he stopped, started laughing. Of course! The name for duckweed is *ko'okol kai* 'chile seeds', which is just what the little green plants look like.

After we had a Pima name for the tiny plant, we could talk about it. Ruth Giff said, "We used to pop those little things." Wayne said that in some parts of the world duckweeds were harvested for food, but Sylvester said, "*Ko'okol kai* [is] not used for anything. I don't think the stock touch it." No one else had heard of any use for it either.

Eventually the Salt-Gila confluence at 115th Avenue was colonized by duckweed. At first there were just a few individual plants here and there, and I collected them gingerly one by one from the water's edge, but six months later they formed thick mats in many different places, with other relatives among them.

Technical Notes: Wayne Armstrong determined that the sample taken at the Salt-Gila confluence consisted mainly of Inflated Duckweed (*Lemna gibba* L.), some Duck-mat (*Spirodela punctata* (G. Mey.) Thomp.)—an exceedingly rare occurrence in Arizona—and apparently Little Duckweed or Water-lentil (*Lemna minor* L.).

Vouchers: *954, 955* (SM).

Zannichellia palustris L.
Horned Pondweed, Common Poolmat
Zannichelliaceae
kuukvul huch, kuukul huch

The Vupshkam, the people of the Creation Epic who marched in from the east, are said to have settled among the ruins of the earlier people that they had conquered; here they came to call themselves the Akimel O'odham, the River People. Alongside them in their new home were hundreds of miles of canals that had fallen into disuse. The story continues, "And

those who lived where the mound now is between Phoenix and Tempe were the first to use a canal to irrigate their land. And these raised all kinds of vegetables and had fine crops. And the people of the Gila country and the people of the Salt River country at first did not raise many vegetables, because they did not irrigate, and they used to visit the people who did irrigate and eat with them; but after a while the people who lived on the south side of the Salt River also made a canal and you can see it to this day" (Lloyd 1911:120). But these inexperienced people had trouble with their survey, and the water would not run upslope. At last, through the agency of a woman *maakai* 'shaman' who walked through the canal seeking the high spots, the canal began to run.

Ditch irrigation agriculture has been the principal livelihood of the Gila Pima ever since. And in these ditches was a water plant, *kuukvul huch,* that caused the People problems. Carmelita Raphael said the plant rooted and had to be cleaned from the ditches every few months.

Sylvester Matthias said, "*Kuukvul huch* will just block the water. It's something like a weed. Have to reach in there [in the ditch], grab a bunch and throw it out. Oh, boy! That's a job. Sometimes *kuup cheweldakuḍ* gets bad, too. Have to pull them by hand. Get in there. *Kuukvul huch* has roots—it'll come right back again. Has roots somewhere, that's why it spreads. What you got to do is grab a bunch, throw it out."

"Was there any particular time of the year when you used to clean those ditches?"

"I guess it's year-round, but it's only the summertime we can get into the water. Several times go there and get the *kuup cheveldakuḍ* and *kuukvul huch.* Pull it out. Santa Cruz [village] had lots of problems with their ditch, that *kuukvul huch.* They had lots of it. It would slow down their water. They do just like we do with heavy rakes. Just pull the *kuukvul huch* out."

Another time he told me, "*Kuukvul huch* is in slow water; caused problems in the ditches. Has something sharp at the tips, like claws [of the little owl]." This plant's Pima name is derived from *kuukvul* or *kuukul* 'Western Screech-owl' (which is common in this country) and *huch* 'claw' or 'nail'.

I was long puzzled by the correct botanical identification of this folk taxon. Could Sylvester have been describing to

me Horned Pondweed *(Zannichellia palustris),* a plant of ponds, ditches, and slow streams, which had been collected at Gila Crossing and Sacaton early in the century? Or perhaps Sago Pondweed *(Potamogeton pectinatus),* another water plant once found in the area?

A specimen *(AMR 410)* turned up that we had collected many years earlier in a stream above Sunflower, Arizona, and that I had quite forgotten. Sylvester had called it *kuukvul huch.* Our museum botanist, Geoffrey Levin, identified it as *Chara vulgaris* L., Common Stonewort. He gave me some references on this unusual Charophyte, an algae. In one I read, "*Chara* may also be identified by its musky, skunklike, or garlicky odor" (Pritchard and Bradt 1984). I asked Sylvester if his plant had a smell. "*Kuup cheweldakuḍ* has no smell, but the *kuukvul huch,* you can smell it! It's a peculiar odor. I can't tell you what it's like. But it's funny, we never get the smell on our hands, on our bodies, as we pull it. It doesn't smell like mud or anything. It's a funny smell. Sometimes some other things will give you itch. But it doesn't bother. No itch."

The funny smell proved to be a red herring in my search. For the most part, only west-end Pima knew the O'odham name for this plant. But Culver Cassa discovered Horned Pondweed on the tribal farms near Bapchule. Checking it with older coworkers, he verified that this abundant and weedy plant growing in the muddy ditch bottoms was *kuukvul huch.* When he showed me some plants in a drained ditch, it was immediately evident why the Pima had named the plant 'screech-owl's claws'. In the axils between the stem and slender filamentous leaves were clusters of three or four tiny seeds, each two or three millimeters long, with a curved apex about a millimeter long. These looked precisely like the feathered toes and claws of the screech-owls that I had once raised in my reservation classroom. Some Pima with excellent observational skills had named this water plant appropriately. And the plant's smell confirmed Sylvester's comments. (This event has cautioned me about the dangers of relying on a single field identification with a consultant.)

Vouchers: *1969* (CC and Jonathan Jose).

Group B, Akimel ch-eḍ Ha'ichu Chuuchim
plants standing in the river, emergents

Looking today at the middle Gila and its tributaries, with their usually dry beds choked with nonnative Salt-cedars *(Tamarix ramosissima),* it is difficult to imagine that the Gila Pima once lived in a land of lush riparian habitats with well-developed emergent aquatic vegetation. But this is what early Spanish chroniclers such as Sedelmayr saw and older Pimas remember well.

Carrizo (a bamboolike reed), cattails, a small bulrush, and a giant bulrush (Tule) are included in this category, which Sylvester Matthias called *akimel ch-eḍ ha'ichu keekam* (literally, 'river in some-thing standing'). (Perhaps it is not by coincidence that these same four plants are mentioned together under *tule* in the 1774 vocabulary from Attí, northern Sonora.) They grew in great thickets along desert streams, in the water. In this group Big Sacaton *(Sporobolus wrightii)* is anomalous. It grows in highly saline marshes, not necessarily in the water, but Sylvester Matthias consistently included it here, excluding it from Group B, *vashai* (grassy plants). He debated about Seep-willow *(Baccharis salicifolia),* which might grow in streams, but he finally decided it fit better in Group D, *sha'i* (bushes or shrubs). It is not an obligate water plant as are the others, and it may be found (today usually is found) in the sandy beds of streams. Arrow-weed *(Pluchea sericea)* is likewise somewhat ambiguous but is better considered *sha'i.*

Phragmites australis (Cav.) Trin.
    [= *P. communis* Trin.]
Common Reed, Carrizo
Arundo donax L.
Giant Reed
Poaceae (Gramineae)
vaapk

Today you will find growing in many dry yards of the Phoenix and Tucson region, as well as in the Pima villages, thickets of a tall coarse plant resembling bamboo. But the plant has none of the grace of bamboo, and its broad heavy leaves are dull green. This coarse plant is called Giant Reed, and no one seems certain how long it has been around. Sometimes it escapes and is found in wild places such as along the lower Salt River bordering the reservation on the north.

Few people realize that aboriginally a graceful, more grassy plant—Common Reed or Carrizo—once grew all along the banks of the Gila and lower Santa Cruz Rivers, and probably along some of the canal banks as well. As so often happens, though, something native and delicate is replaced by something exotic

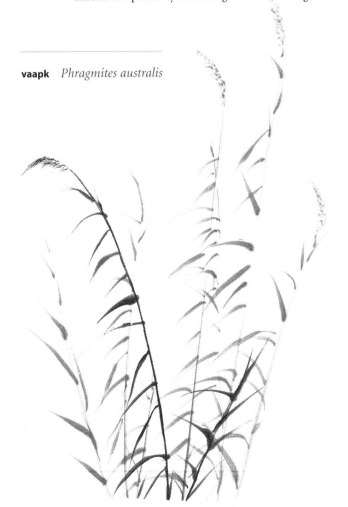

vaapk *Phragmites australis*

and gross. We seem to be making that an ecological rule.

The two reeds apparently are not linguistically distinguished in modern Pima. *Arundo donax* is planted for windbreaks and a bit of shade around homes. "Well, it looks the same," Ruth Giff said when I asked her about the "new" species she has growing on one side of her house.

Several traditional uses for the mature culms or stems persist.

Sylvester Matthias made several Pima flutes for me. The flute is called *vaapk kuikuḍ* ('cane' + 'musical instrument'). These are made from a piece of mature stem with a node at the middle perforated on top. There are three finger holes burned through the top of the lower section and an additional thumb hole on the bottom. With a heated wire Sylvester burned a few line designs on the flute and completed the decorations with some windings of soft cotton string, which help modulate the whistle at the middle node opening. He gave me this flute, explaining that in the old days when a Pima boy became interested in a girl and wanted to woo her, he would take his flute and play songs at a distance for her to listen to. He might sit on the riverbank while she worked there or serenade near her house during the night. I asked Sylvester if he could play such a song so I could record it, which he obligingly did. Triumphantly that afternoon I returned to Sally Pablo's house with my newest acquisitions, telling her of the song. "Let me hear it," she said, "I've never heard a Pima love song." I turned on the tape recorder and played it. "Oh, that's nothing but John Philip Sousa. Sylvester used to play in the school marching band." And so it was.

At the other end of the reservation, Francis Vavages spent a few hours every morning, in good weather, at the east side of his house making flutes or carving wooden spoons. When I first visited him, he had just finished a flute. With natural showmanship, he used his flute to imitate the father,

mother, and baby *chuchkuḍ* 'Great Horned Owls'. "That's how they communicate with each other so they stay together." His rendering of a *hoohi* 'Mourning Dove' brought a response from one in a nearby mesquite. Then he played a song for funerals, an elemental example of keening that would be recognized anywhere in the world.

Several years later, I asked Francis if any of the young Pimas were learning how to make flutes. He said they are, that he goes to the schools and teaches them. Then he took a cut section of Carrizo from his workbench to show me how they were made. Both ends were open, and in the middle was a joint (a node with two internodes). With a large knife, he quickly hacked away at what was left of the leaf base at the node, smoothing it down to the diameter of the rest of the stem. He also smoothed down the lateral bud emerging from the joint. The bud always points toward what becomes the mouth end of the flute. On the side, he marked two spots, about a finger-width above and below the node, drilled them open, then connected them with a deep groove reaching all the way to the inner chamber, leaving the septum in place. With masking tape (which would later be wrapped with yarn) he covered most of the upper end and some of the lower part of this channel, allowing air to pass from the upper section across to the lower section, producing the whistle sound. He next placed his three middle fingers on the lower half of the flute and marked the placement for the three finger holes. These he drilled open, then filed crossways, so his fingers fit smoothly across the holes. Last, with a piece of Arrow-weed *(Pluchea sericea)* he reamed out any remaining debris from the upper and lower ends of the flute, smoothed off the sawed edges of the tube openings with sandpaper, polished a few off-colored spots, then tried the tone. The flute was then essentially complete, except for wrapping and decorating.

"There, now you can make one yourself," Francis said, handing me two flutes in different stages of completion. From beginning to playing was probably less than ten minutes; the decorating took longer.

A guessing game called *vopodai* uses four game sticks, *vopodaikuḍ*, each a single segment of reed. This game, Sylvester said, was played for a long time at Bapchule after Sunday Masses. George Kyyitan of

Bapchule made me a set. He gave the following names for the pieces: *keli* 'old man', all marked; *oks* 'old woman', unmarked; *mo'o chuul* 'head corner', marked on top only; *huḏa s-chuk* 'black side', middle marked. George called the gambling sticks *vopodaikuḏ.* "*Voi [vood]* means 'to lay down', to lay that bean in there," he explained. "Every Sunday, right after Mass, they played until sundown." If the guesser named the correct reed containing the hidden bean on the first guess, he earned ten points; if he called it on the second, six points; and on the third, four points. But if the bean still remained hidden in the sand of the fourth reed to be emptied, he received no points (see Russell 1908:176–177).

According to Sylvester, the Pima version of the game is called *vaapk vopodai.* He explained it was "a guessing game, using tubes of bamboo. Harvey Enos [was] the last Pima to sing the guessing game song, in the Mission Indian language [of southern California]. His [Pima] Indian name was Yakachk. He's the only one that knows [the song]. I forgot what year he died—a hundred and some odd years [old]. Nobody knows what Yakachk means. He knows the rules. They hide something in the tube, *muuñ* or something else. There was a Pima version of the game and a Mission version just called *vopodai.*" According to Sylvester the third gaming piece is called *mootoḏo.* He said it was an old word whose meaning he did not know, "but it's something about the head." His other names correspond to those George gave.

A few months after Sylvester told me about the guessing game, I was reading some songs recorded by José Lewis and Frank Russell. One of these, "Vopatai Ñe'i" (Russell 1908:299), a game song recorded from Prairie Falcon Flying, is not in the Pima language. Russell's footnote says, "Derived from another tribe, but from which is not known. The meaning of the words is not known to the Pimas."

Phragmites australis—the original, more grasslike, water-loving species— had been declining since the river started failing. On the upper part of the reservation, Russell (1908:134) noted, "Reeds, Phragmitis *[sic]* communis, were formerly common along the Gila, but continuous seasons of drought caused them to disappear. Sleeping mats were made from them, but such mats are now rarely seen, agave *[sic]* leaf being used instead." Later

(Russell 1908:147) he reiterated, "Mats were formerly made by the Pimas of the cane, that grew in abundance along the Gila until the water supply became too scant for the maintenance of this plant." Russell was not alone in these observations. Mary Lois Kissell carried out field-work at nine Gila Pima villages during the winter of 1910–1911. She wrote, "A great many years ago, plaiting was done by the Pima, but owing to the shutting off of the headwaters of their two rivers by the white men, these streams are dry during most of the year, and the one suitable plaiting material, the river plant, *Phragmetis [sic] communis,* which formerly grew along their banks is no longer found. This river cane was a stiffer and less durable material and much more difficult to manipulate than palmea *[Dasylirion],* so that its use was limited to mattings as it was unsuit-able for baskets, or articles not flat. Hence, because of this change which cut off the material, the Pima do not plait as of old, and altered conditions have restricted the tecnic *[sic]* to the Papago" (Kissell 1916:138–139). A few pages later, she added details about preparing the stems for plait-ing: "These plants were cut down with large knives from land near the streams, carried home, and the stems dried and stored away for future use, but before plaiting the hollow stems were split lengthwise with the thumb nail, and then spread flat" (Kissell 1916:153–154). She then described the weaving method (see Kissell 1916:fig. 12).

Because of peculiarities of geology, the river survived half a century longer at the west end of the reservation (Districts 6 and 7). The people I worked with there remembered the original, wild-growing *vaapk.* Sylvester said, "*Vaapk,* bamboo [cane], grew in New York Thicket, but not lower. The Santa Cruz River came up again on the north side of Maricopa Highway. New York Thicket is [was] on the Santa Cruz. Similar word is *vaapai* 'Saguaro rib.'" Ruth Giff added, "*Vaapk* used to grow in great patches in New York Thicket [called] *vaapk ch-eḏa* ['*Phragmites* there' or] the place with a lot of Johnson *[sic]* Grass."

Joe Giff may have witnessed the coup de grace of this fine plant: "*Vaapk*—[they] used the thin ones, mashed them [for weaving]. Tall ones grew in New York Thicket. Used to weave door coverings, table mats [to eat on the ground], and

sleeping mats. They used to grow out, but [were] all eaten down by cattle." Curtin wrote of *vaapk,* "A tall, coarse perennial, with flat leaf-blades, which grows in wet ground but is disappearing from the reservation owing to subsoil pumping." She never collected a specimen of it.

Formerly, one of the most important articles of household "furniture" was the sleeping mat or *main.* I bought one in Onavas from a Pima Bajo woman, Carolina Humar; hers was made from fronds of Palmetto *(Sabal uresana).* In the Mountain Pima rancheria of Nabogame, the houses for the most part still lack European-style beds. Each evening the sleeping mats (made of Bear-grass, *Nolina microcarpa*) that have been rolled up in the corner are unrolled on the floor, the blankets spread on top; the whole family retires at once. This bed occupies nearly the whole room. And so it was, once, with the Gila Pima.

In addition to the many woven house-hold articles (from sleeping mats to door-hole covers [called *kiijeg kukpadag*]), along with the flute and guessing game sticks, ceremonial items were also made from *vaapk.* A reed "cloud blower" is shown with a set of medicine man's prayer sticks *(omina)* in Russell 1908 (p. 107, fig. 24). One of the twirling wands Sylvester made me to accompany a Navĭchu singer's mask is reed decorated with red leather and feathers from a *haupal* (Red-tailed Hawk). Some ancient songs mention *vaapk;* one of these Russell (1908:285) translates as "ceremonial reeds." In the Pima Creation Story, Jeweḏ Maakai (Earth Doctor) escaped the flood by climbing into his *vaapk* walking stick, and Ban (Coyote) by getting into a *vaapk* tube (see, for example, Lloyd 1911:38).

Like many other riparian and aquatic plants and animals, the reed has the root word associated with water, *va-,* as part of its name. The word *vaapk* is the same in Tohono O'odham, lowland Pima Bajo, and colonial Névome; the Mountain Pima say *va'apaka,* and the Northern Tepehuan, *vapakai* (Pennington n.d., 1969:121). Joseph Giff suggested a possible relation-ship to the idea of "hollow," as in *vaapagga* 'to have a hole' (see Mathiot n.d.:211), but the word for hollow (or burrow) may have derived from the word for reed. Possibly *vaapk* anciently originated from the pluralized form of *vak,* another water plant (bulrushes).

Regardless of the derivation of the word, *vaapk* played a special role in Pima mythological roots. In the Creation Story is an account of a fickle goddess, Tobacco Woman, who gambles with an important deity, Corn Man, and loses. Because she begrudges him his legitimate winnings, she abandons the People and leaves for the west. Without tobacco smoke to make rain clouds, Corn is not able to germinate, to feed the People, so he and his constant companion Squash leave for the east, where there is moisture. Some years later, Corn contemplates a return and sends a messenger to Tobacco Woman's father, a powerful *maakai,* inquiring about a possible return. When the answer is affirmative, Corn tells the messenger to instruct the People to build a special structure, a *main kii* 'mat house' made of *vaapk.* Also, the People are told to put outside all their containers, even broken ollas. At dusk Corn Man enters the village, where his bride awaits him in the ceremonial house. All night long the two sacred beings make love, and all night long there is the thump and bump of corn and squash raining from the sky, filling all the containers.

The Pima had no ceremonial *kiva* or underground chamber like the Hopi had for kachina dancers and sacred ceremonies. However, they did have a kachina ceremony, with a primary masked dancer, the Navĭchu, accompanied by three types of assistants whose masks strongly resembled some still worn by Hopi kachina dancers. The Navĭchu and his assistants emerged not from a kiva but from a *main kii* especially constructed as the temporary home for the sacred beings. To there they retired between dances.

The first recorded European penetrations of Gileño country were by the Jesuit Kino, beginning in 1694. One wonders what the Pima thought of this strange black-robed man, riding atop an animal taller than any they had ever encountered before in their environment. They must have been especially enthralled when he unpacked his leather saddlebag, took out some white clothing to cover every bit of the black, and begin a slow, stately dance before a plank draped with fine-cloth, with a shiny yellow cup and a little dried *chemat,* or tortilla. The hushed attention of even the normally loud soldiers as they kneeled must have made quite an impression on the soft-spoken Pima. Once a day the man dressed in black transformed himself to white like this. Sometimes the soldiers even burned some *ushabĭ* (dried sap, resin, pitch) to make a sweet-smelling smoke for Kino. And one day the man even went to one of their own shrines, *chuulk* (Casa Grande ruins), to do his dance. Lieutenant Cristóbal Martín Bernal in his diary one evening tells us exactly how the Gileños at the settlement of Sudacson interpreted all this: "On arriving I found all the people in two lines to render me obedience, men, children and women, for all together there were 400 souls. They had made a house of mats for our Father Kino, having pinole made of mesquite and squashes all ready for us" (Smith and associates 1966).

Technical Notes: *Phragmites* and *Arundo* may be easily confused; the flower heads are essential for positive determination. No reservation specimen of *P. australis* is known to have been preserved; the nearest to the reservation where I have found it was a colony on the north bank of the Gila where the river still runs, at the Buttes (22 miles east of the reservation). Probably some were still growing along a San Carlos Irrigation Project canal between Hashañ Keek and Olberg, but these were sprayed and did not flower.

Vouchers: *165* (sm, *A. donax*); *1633* (*P. australis,* sterile); lsmc *83* (*A. donax*).

Scirpus acutus Muhl.
Hard-stem Bulrush, Tule
Scirpus validus Vahl
Soft-stem Bulrush, Tule
Cyperaceae
ko'okpoḍk

The desert, with 110°f days for months on end, may seem like an incongruent place for marsh plants, yet they once thrived in the marshes and swamps and cienegas along the Gila from western New Mexico to the Colorado River. Among these hydrophiles were the great bulrushes called Tule (*Scirpus acutus* and *S. validus*). Plump smooth green tubes tapering to points, about head-height or more, formed thickets almost as dense as the paler green cattail colonies. At the former Barehand Lane marsh, I watched a Tule colony in a saline pond of irrigation tailwater develop from just a small patch in 1972 until it colonized nearly half an acre.

Coots and Common Gallinules broke Tule stems to make their floating nests here. I have not found this plant on the reservation, not even along the lower Salt River, since about 1979.

Joe Giff sat telling me about *ko'okpoḍk.* "They're thick, the leaves, or the stems, I guess, about like a wiener, about as thick as my thumb. That word, *ko'okpoḍk,* means 'filled up', something filled up, or inflated. You know, those stems are filled up." Mathiot (1973:496) gives an additional meaning for the verb 'to be full of air', most appropriate for the Tule stems, which are filled with white, spongy cells.

The formal Pima name for this plant is permanently in the plural—not surprising for a plant that is colonial. The Piman names for a number of organisms are fixed plurals: *totoñ* 'ants', for instance.

Ruth Giff told me, "*Ko'okpoḍk* grows in *shuudagĭ* ['water'], [the name] means 'something full'; someone eats too much, he is *kopoḍk.* They're big tall ones that grow like *uḍvak* [Southern Cattail, *Typha domingensis*]. They have little flowers near the tops."

"Odd that they wouldn't use it for weaving or anything."

"I think it's soft; I don't think it has a cane in it like *uḍvak.*" (She was referring to the cattail's woody flowering stalks, not its leaves.)

Sylvester Matthias recalled the plant. "Round one, about four feet, spreads from the base, something like pampas grass. Grows in the seepage water [near Komatke]—along the creeks, too."

"Did they use it for anything?"

"Just decoration for the creeks, no use."

Sylvester once identified *ko'okpoḍk* on the lower Colorado River as the much smaller Spiny Rush (*Juncus acutus* L.) rather than a Giant Bulrush, which we found there also. Spiny Rush likewise is cylindrical and spongy, but its leaves are only half as thick as a pencil. It grows in more discrete clumps than the spreading bulrush. However, several years later in San Diego, I drove Sylvester to a thick stand of Giant Bulrush (*Scirpus californicus*) and asked him if he knew the plant. He looked for a moment, then said "*Ko'okpoḍk.*" Regardless of the biological identification of this folk taxon, both rushes are now gone from the middle Gila.

After we had been to the luxuriant cienegas north of the one-time Pima settlement of Imuris in Sonora, Ruth told

Scirpus americanus Persoon
[= *S. olneyi* A. Gray]
Olney's Bulrush
Scirpus pungens Vahl var. *longispicatus*
(Britton) Cronquist
[= *S. americanus* auctorum, non
Persoon]
Three-square Bulrush
Scirpus maritimus L. var. *paludosus*
(A. Nelson) Kük.
[= *S. paludosus* A. Nelson]
[= *S. robustus* auctorum, non
Pursh]
[≠ *Cyperus ferax* Rich.]
Salt-marsh Bulrush
Cyperaceae
vak

Along the highway between Nogales and
Imuris, Sonora, are several cienegas, great
swampy basins with the water table so
high that an acre or more may be flooded
year-round. These are filled with emergent
water plants of various sorts and rimmed
with willows and cottonwoods. These
improbable desert oases are biotic paradis-
es, and I wanted to take some Pimas there

vak *Scirpus maritimus*

me more about three kinds of water
plants: *ko'okpoḍk,* Tule; *vak,* Three-square
Bulrush (*Scirpus pungens* var. *longispicatus);* and *vashai s-uuv,* flat-sedges (*Cyperus*
spp.). "I haven't seen *vak* [here] for a long
time. It's all dried up now."

"There's something, too, that grows this
tall," I said, putting my hand over my
head. "Would that be it?"

"There's something like that—
ko'okpoḍk . . . full of air. It seems like
ko'okpoḍk has the same kind of leaves as
the *vak,* but they were *so* high. Did the
ko'okpoḍk have that brown top?"

"Yeah, way up, near the top on the
side."

"Well, my memory is still good! I
remember those things as a child. They
grow. And those things—we used to call
them airplanes—those bugs—"

"Dragonflies?"

"Yeah. *Vashai s-uuv* is more on the
ground, in the fields."

"The *vak* is in the water, and the
ko'okpoḍk is in the water?"

"Yeah, like the *uḍvak.* Green, dark
green. Just like fat leaves—they're full of
air. They call it *ko'okpoḍk*—the stem and
all, I guess. *Vak* [grows] by the *vo'oshañ*
[stream]."

Vouchers: *1763* (SM, *S. californicus* (C. A.
Mey.) Steud., San Diego, Calif.).

ko'okpoḍk *Scirpus acutus*

to identify plants that once might have grown at similar cienega-like spots between the Gila and lower Santa Cruz Rivers. An opportunity came in 1986. On our way to Pima Bajo country, Ruth Giff, Sally Pablo, Sylvester Matthias, and I stopped at one of these Sonoran cienegas. We confirmed Pima names for Yerba Mansa *(Anemopsis californica),* Arizona Grape *(Vitis arizonica),* Inflated Duckweed *(Lemna gibba),* various algae, and Curly Dock *(Rumex crispus);* we also found two species of pennywort *(Hydrocotyle verticillata* Thunb. and *H. ranunculoides* L.). Sylvester recognized *H. verticillata,* saying, "There *is* a name in Pima," but he did not know the name. In reference to a number of the grassy emergents I collected, Sylvester said, "In general, just call them *shuudagĭ hugĭdag-an vashai* 'water edge/ along-the grass.'" These included a sedge *(Carex chihuahensis* Mack), Wire Rush *(Juncus balticus* Willd.), and a spike rush *(Eleocharis macrostachya* Britt.)

But the most abundant plant in the cienega was Olney's Bulrush. Ruth thought these were *vak.* Several years earlier she had told me, "*Vak* is something like *vashai s-uuv* [nut-grass, *Cyperus* spp.], but it is big and grows right in the water. We see them when we go swimming. Has little nuts, and little brown something on top. But *vashai s-uuv* is something that grows on the ground, but very much like *vak.* This one *[vashai s-uuv]* is small and very hard to get rid of when they get in the garden. Has some kind of nuts [too]."

Later, when we were back at Komatke, I pressed Ruth for more details on the bulrush tubers. "That plant you called *vak.* You said there were nuts underneath. Did you used to eat those nuts?"

"Uh-huh. When we were swimming in the *vo'oshañ* [lower Santa Cruz River]. That's where they grew around. It had no taste, but we used to chew it. You know, just like those *haad* [Covena, *Dichelostemma pulchellum*] up in [the] hills. It had no taste, but we go after it. But— we'd see the blossoms come out of a bush and we'd dig them out and we'd eat them. Just like that *vak.*"

"When you dug that thing underneath the *vak,* how big was that?"

"Small, very small."

"As big as your fingernail, or bigger?"

"No, a little bigger, like a little tiny onion."

"How did you get those, dig in the water?"

"No, just pull up the plants."

"The whole plant?"

"It's wet, so they just come up. They were just always small when we dig them out."

"How many on each plant?"

"I don't remember."

Curtin (1949:99) had identified *vak* as *Cyperus ferax,* a flat-sedge. But Ruth's suggestion agreed with what we had learned elsewhere. For instance, at the desert oasis village of Quitovac in northern Sonora, when Gary Paul Nabhan and I asked Luciano Noriega how the village got its name, he just took us over to the edge of the pond where Olney's Bulrush was growing, pulled up a stem, and said, "*Vak.*"

When Spanish missionaries and soldiers arrived at a native settlement, they usually Christianized the place-name by preceding the native name with a saint's name, often the one being celebrated in the liturgical calendar for that day. When Kino arrived at the Sobaipuri village Vac below what is now Tucson, he hybridized the place name to San Xavier del Bac, adding a saint's name. (The *v* and *b* are interchangeable in Spanish.) The mission once sat near a marshy bend in the stream, no doubt filled with these bulrushes. Today the famous "white dove of the desert" perches high above the dry bed of the river, whose channel has been downcut by cultures who failed to learn the hard lessons of the desert. (Saxton and Saxton 1969:45 gives the derivation of the name Bac from the verb *waak* 'to enter', yielding "where the water soaks in," implying that this is where the northward-flowing Santa Cruz River went underground. But the river ran far beyond there in the days before overgrazing.) The Akimel O'odham interpretation of the mission's name is clear from this comment by Sylvester Matthias: "San Xavier del Bac, we just call Vak, [after] that plant. That's all we knew it by. But the young generation don't know what *vak* is, and they just call it San Xavier."

This erect bulrush is relatively short (less than shoulder-height), with triangular stems slimmer than a pencil. In all these characters it is quite the opposite from *ko'okpoḍk,* the other type of *Scirpus* in the Pima lexicon.

Elliott Coues (in Garcés 1900, 1:69n.3) somehow convinced himself that *bac* was the Piman word for "house, adobe house, also ruined house, ruins, etc." and cited examples of place-names containing this word. The word for house is *kii.* Coues's mix-up may have come through the word *va'akĭ* 'ruins' or 'rain house', a place-name that still survives for the Pima village Casa Blanca. Given that Spaniards most frequently wrote Piman *va*—the root word for 'water'—and *vak,* the water plant, with the interchangeable *b,* I suspect that all the place names Coues cited refer either to water or to this bulrush. How appropriate for desert peoples to name places for stands of bulrushes.

Technical Notes: Quite a merry-go-round of ethnobotanical as well as botanical confusion surrounds this intermediate-sized bulrush. In the drying riverine and marsh conditions of this century, it has seldom been possible to obtain adequate contrast sets of aquatic plants on the Gila. Also, botanical distinction of species often requires microscopic examination of seeds, whereas folk taxa are usually based on more gross morphology and habit.

Curtin (1949:99) associated *vak* with the flat-sedge *Cyperus ferax* Rich. (= *C. odoratus* L.). Her specimen *(#36)* is sterile, hence unidentifiable to species. When Ruth Giff was presented with taller, less leafy specimens and more stocky and leafier specimens of what ultimately (and to my surprise) turned out to be the same botanical species, *C. erythrorhizos* Muhl., she identified the slender ones as "probably *vak*" (AMR 854) and the leafy ones as *vashai s-uuv* (AMR 855). These were the only forms she had to contrast.

At Quitovac oasis in Sonora, Tohono O'odham Luciano Noriega identified the abundant *S. americanus* (= *S. olneyi* A. Gray) (GPN and AMR 1873) as vak. At a cienega in northern Sonora, Ruth Giff also identified *S. americanus* Persoon (AMR 1047) as most likely *vak.* Thus, this is at least one of the botanical species included in the folk taxon.

However, Ruth noted that *vak* has underground tubers. This tuber-bearing plant is *S. maritimus* var. *paludosus,* a species I have collected at various places in Pima country including the lower Salt River and at the Salt-Gila confluence (AMR 31, 33, 131, 1043). Ruth also identified AMR 1564 and 1571 from the Gila River near Roll, Arizona, as *vak,* commenting on the black tubers.

I have taken a specimen of *S. pungens* var. *longispicatus* on the reservation (Olberg spillway, Gila River, 18 April 1988, AMR 1516).

Although the specimen has no ethnographic annotation, I doubt this species would be excluded from *vak*.

Although we can now never know what full biological contrast sets may have included from the intact Gila flora, at least the above species of bulrushes are contrasted with *ko'okpoḍk*, the densely growing terete, the taller Tule species (*Scirpus validus* and *S. acutus),* and with *vashai s-uuv,* the more terrestrial flat-sedges or nut-grasses (*Cyperus* spp.).

Vouchers: *1047* (RG, SM, *S. americanus* Persoon); *1564, 1571* (RG, *S. maritimus*); *1873* (Luciano Noriega, Quitovac, Son., *S. americanus* Persoon).

Sporobolus wrightii Munro ex Scribn.
Big Sacaton, Sacaton Grass
Poaceae (Gramineae)
noḍ

Along the banks of the Gila River and in the cienegas and great alkali grasslands that were mentioned by the first Europeans to explore Gileño country was an abundant coarse bunch grass called Sacaton or, in Spanish, *zacatón*. The flowering heads of this grass are open and airy, a delicate contrast to its harsh environment. You need not look for Sacaton Grass on the reservation today: it was gone from the Gila by the time botanists first started collecting specimens here in 1901.

Sylvester Matthias told me of fire drives, called *kuunam,* that the Pima once conducted out in the grasslands and brushy areas (see chapters 3 and 4). Apparently, the grass persisted a bit longer at the western villages where the river lingered; there I was able to learn at least the plant's name and use. Sylvester Matthias told me, "*Noḍ,* the Sacaton Grass, was used for making hairbrushes; [but is] no longer found [here]." Ruth Giff also knew the plant. She said, "*Noḍ,* the clump grass; Noḍ Ch-eḍ, that's the place where a lot of Sacaton Grass grows; horses like the young shoots—*muḍadaj,* the stalks. A brush for hair was made from the Sacaton Grass from New York Thicket, called *gasvikuḍ* '[something] to comb hair with'; *gaswua* means 'to comb hair'. Noḍ grows mostly in the *ongam* ['place that has salt']."

Russell (1908:116, 118) said that the Pima "frequently smooth the hair with a brush which was formerly made from the roots of the 'Sacaton grass', *Sporobolus wrightii,* but as this no longer grows along the river, where the majority of the villages are situated, they now make use of maguey fibers or Yucca." I had assumed that even these western Pima knew no more than this, but while checking manuscript pages with Sylvester during his 1987 visit to San Diego, I read him what I had written above and asked him if he knew this grass.

"There was a place there, on Baseline Road [in the Laveen area, north of the reservation]—there was a *lot* of *noḍ* there. Where Tumbleweed Inn is—used to be something like a swamp—and there's lot of *noḍ* there."

"How tall did it grow?"

"About three feet. That's how come that ditch; it's a drainage ditch, from way up there; there were lots of cattails, [and] *noḍ.*"

"How did it grow—in bunches or just all spread out?"

"Just here and there—bunches."

"How would you describe it?"

"I guess you know what Pampas Grass is?"

"Yeah," I said, mentally picturing the great grassy mounds of *Cortaderia* sometimes grown ornamentally in Phoenix.

"It's similar to that, only they're smaller—Pampas Grass is big. But it's similar to that, only smaller, the *noḍ.*"

"So it's a fairly coarse grass?"

"Coarse grass, yeah."

"What did the *muḍadag* ['tassels'] look like?"

"No, there's no *muḍadag;* it just—" he spread his hands apart to show how it was open on top, without obvious seed heads. "It's similar to that—Papagos use it for baskets—*mōhō* [Bear-grass, *Nolina microcarpa*]".

"How long ago did it disappear?"

"Ah, let's see. . . . After they put that drainage—and everything was drained and the pumps was put up to pump that—it dried up, except what's running now—in that drainage [ditch]."

"There's a stockyard in there now," I said.

"It runs right through across there—51st [Avenue] and Baseline on down. And that place that used to be that swamp

turned into alkali—and they called it Ongam Ch-eḍ 'It Has Salt *in*', or a place with alkali. The first service station was put up there—on clear up to—I forgot what avenue it was—there's a store and a beer place, and a bar where Indians used to drink. They used to call it Noḍ Ch-eḍ. Then they called it Ongam Ch-eḍ, because it's always a salt flat in there after it dried up."

"How far west did it extend?"

"Almost to Maricopa [Colony]. East almost to 35th Av[enue]."

"Were there cottonwoods on the Salt River then?"

"Yeah, uh-huh. Cottonwoods, but mostly willows, and cattails too—on that Noḍ Ch-eḍ and Salt River."

"As they drained it, did the *noḍ* become more abundant? Or was it always there?"

"It was always there."

"When it dried, did it stay?"

"Until that drainage [was] put in and that pump put in—and so it went dry."

"Then what happened to the *noḍ*?"

"They just dried out. And now [then] it's just alkali—salt flats. Now it's put into cultivation. Now they're growing cotton and alfalfa. Now it's no longer *ongam.* They're growing, they're farming there, growing everything."

But I always like to have voucher specimens to verify a folk taxon, and when at all possible, I like to collect them in the field, with the native consultant along. Unfortunately, the likelihood of ever getting Sylvester or Ruth together with the plant in Arizona was remote. Then came a chance. In September 1989, Sylvester came to visit us in San Diego and to make some artifacts for an exhibit in New York. Takashi had come across some interesting desert acreage west of Jacumba in eastern San Diego County, California. We drove through Jacumba to a sluggish stream meandering under the freeway. The soil at the edge of the road was so salty it looked like frost and icy mud clinging to plant stems. There were lots of old plant friends there—Screwbean (*Prosopis pubescens),* Catclaw (*Acacia greggii),* Yerba Mansa (*Anemopsis californica),* Four-wing Saltbush (*Atriplex canescens),* Quail-plant (*Heliotropium curassavicum),* Desert Salt Grass (*Distichlis spicata),* cattail (*Typhus* sp.), bulrushes (*Scirpus* spp.)—all plants that Sylvester knew well in Pima. But what I wanted Sylvester to see were the scattered thick clumps of grass, looking quite like a fine-leaved Pampas Grass, just as he once described. The dense green tussocks grew about waist-high, well spaced. We found a few still with their wistrous fruiting heads that the livestock had missed or could not reach.

"What is this?" I asked.

"*Noḍ,*" pronounced Sylvester. "This is it. *Noḍ.*"

I felt privileged. Someone had just described to me a plant community that I thought had completely disappeared a century earlier. No biologist had ever seen it, no botanist had ever collected there. But two people could describe the area in detail, including its distinctive grass.

Technical Note: Many authors lump Sacaton Grass or Big Sacaton *(S. wrightii)* with Alkali Sacaton *(S. airoides)* or treat it as only a variant *(S. airoides* var. *wrightii).* Charlotte Reeder, however, considers them distinct species, having seen no intergrades, although the two may grow together. I have found the

smaller Alkali Sacaton at springs in the Upper Basin, at the south end of the Sierra Estrella. Possibly this species was included in the Pima folk taxon as well as Big Sacaton. Reeder has identified as *S. wrightii* a roadside plant Culver Cassa discovered near Santan while this book was in press (CC and AMR *1830*).

Vouchers: *1624* (SM, Jacumba, Calif.).

noḍ *Sporobolus wrightii*

Typha domingensis Persoon
[≠ *T. angustifolia* L.]
Southern Cattail
Typhaceae
uḍvak

When we look at the Gila today, it is difficult to imagine what must have greeted the early Hispanic explorers' eyes as they traversed the cienegas and oxbow marshes and finally the great Salt-Gila confluence. Acres of cattails must have been evident. Even today, given permanent water and saturated soils, cattails are one of the first and most successful emergent water plants to colonize. At sewage ponds, along the effluent channel of the lower Salt River, and at a few marshes created by irrigation runoff, Southern Cattail forms great colonies of grass-green flat leaves, usually taller than a person. The flowers begin developing in early summer from a central woody stalk. The male and female flowers are segregated at the top. Masses of yellow pollen are produced by the thinner male section. Then the rich light-brown female section below swells like a skewered sausage. This part gives the plant its English name, cattail.

The Gila Pima name for this culturally important plant is *uḍvak:* I presume that this is some archaic prefix *uḍ-,* plus *vak,* the widespread name for another emergent water plant, Tule. *Vak* is further derived from the root word for water, *va,* which appears often in the names of plants and animals associated with the riparian community. Both Fr. Antonine (1935) and Curtin (1949:64) recorded the term *udvak* (making no distinction between *d* and *ḍ*). But for the Tohono O'odham, the word was given as *uduvhag* (Mathiot n.d.:483) or *uḍawhag* (Saxton and Saxton 1969:44, 172; Saxton, Saxton, and Enos 1983:59), seemingly quite beyond the *vak* derivation.

Etymology aside, the Gila Pima make one indispensable use of *uḍvak:* the solid stem of the blooming stalk, when split and dried, forms the foundation for their excellent coiled baskets. Some basket makers dry the stem first.

One hot summer morning I took Ruth Giff and several of her granddaughters down to Barehand Lane marsh to gather *uḍvak.* "It must be in July—from beginning to end to be just right," she told me. And what a picture they made: wearing great colorful broad-brimmed hats with flowers, bright blouses, skirts tucked up in their belts, with wet brown legs and arms, cutting away with big kitchen knives, throwing the slender, arrowlike green stalks up on the ditch bank for me to load into the truck.

These get a special name in Pima, she told me. "*Wupdaj,* the flower of cattail. Not *muḍadaj*—that's what it would be called on any other plant, the 'tassels' or 'flowers' [of grassy plants]. Can be used for making pillows—*mo'ochkuḍ* is a 'pillow'. After you use it, it just turns to powder; it doesn't even last one season! Didn't ever use feathers in those days. Sometimes when we were children, our mother would give us just a sack with a bit of wheat seeds for a pillow. That word *wupdaj* is used only for the 'cattail' of the *uḍvak*."

I asked Sylvester Matthias about the word *wup[a]daj*.

"*Wupadaj*—the basket makers go by that *wupadaj*; they tell when it's ready to be picked."

"How do they know when it's ready?"

He laughs. "I don't know. My grandmother told me to go and pick a *wupadaj* and she says, 'It's too green!' Ah-ha. And we just happened to be out there [later] with my older cousins, and we're going fishing, and then we stop and [say], 'Let's go check with grandma.' And then we pick it up and take it to her. 'It's ready! It's ready! Come on—let's go get them.' So they give us knives and she grabbed a sickle. Then we went down there, down to the Gila. 'I'll show you how to cut them.' Then she went down there to a certain [patch]. You can walk on them—they're pretty strong. You can just bend them and walk on them."

"You bend the whole plant?"

"Uh-huh—bend it and cut it, and just step aside, and cut the next one—and step, because it's muddy in there—but they're pretty strong—*wupadaj*—cut them. Cut just enough to leave a stump—but as I said—you have to go to the next and step on that and then cut the *wupadaj*. Men can help split the *wupadaj*, but can't weave with them." Sylvester's reference here is to a very strong taboo, in effect even today, against a man weaving baskets. To do so would turn him into a *uikwuaḍ*, a transvestite.

"That's safe?" I asked, surprised that this would be allowed.

"It's safe."

"Have you ever heard of the Pima using the pollen, for paint or in ceremony or anything?"

"No. But I saw the Apaches do that—during the Sunrise Dance [puberty ceremony for girls]—they'll spray—spray everybody with that yellow [pollen]—just like holy water."

I inquired about the use of cattails for food, but Ruth knew of none. Sylvester Matthias also said they were not eaten, "but there *is* a tradition that when Father Francisco Garcés went through here, he ate the roots of *uḍvak*, so they knew about that, way back then. He must have learned that from some tribe, because they say that's what he ate here." This was from the friar's many trips made to the Gila Pima country between 1768 and 1775, two centuries earlier. Oral history sometimes preserves the most insignificant details.

The information given to Curtin (1949:64–65) on the collection, preparation, and cooking of cattail pollen biscuits was by a 98-year-old Maricopa woman, not by a Pima. Unfortunately this has been interpreted as "a food prized by the Pimas" (Fish 1993:246). But I would be surprised if Gileños did not once pull up and eat the stalks, as did probably all other Indians in the Southwest. In fact, old man Tashquinth of Komatke, born in the 1860s, told Curtin that "the tender white stalks and the roots were gathered all the year round and eaten raw."

Curtin was also told by some Salt River women, not identified to tribe, that cattail leaves were used to weave mats and roofs. Similarly, years ago in rural Sinaloa, when Sally Pablo and I were searching unsuccessfully for Pima Bajo settlements, she bought *petates* or sleeping mats woven from cattails. Ruth, her mother, an accomplished basket maker, assured me that *uḍvak*, cattail, was never used this way by the Gila Pima. Others agreed. Kissell (1916), who visited nine Pima villages gathering information on weaving, mentioned only *Dasylirion* and *Phragmites* as materials for mat weaving. (The material used in the plaited square baskets she collected should be checked; this technique has long since been abandoned by both Akimel and Tohono O'odham.)

An 80-year-old Pima woman, originally from the Gila but at the time living on the Salt River reservation, told Curtin in 1940 (1949:64) that "the yellow pollen was used dry to decorate face, chest, and back." This sounds like painting for some dance or ceremony, and it is unfortunate that we do not know anything of the context.

Vouchers: *419* (SM); LSMC *18* (not extant).

uḍvak   *Typha domingensis*
The dragonfly in Pima is vak chechetopĭ.

Group C, Vashai
grasses, grassy plants, forage plants, hay

This is a well-marked, widely known Piman plant category or life form that corresponds somewhat loosely with the botanical concept of grasses but is not limited to them. Etymologically, the word is composed of *va-*, a common root word associated with water in Sonoran languages, and *sha'i* 'brush'. I suspect that the compound *vashai* is linguistically ancient, preceding the divergence of the modern Pima word *sha'i*, a plant life form. Pedro Estrella used *vashai* for sacate in general and *vasho* as the base word in the names of such water plants as spike-rushes (*Eleocharis* spp.) and Yellow Monkeyflower *(Mimulus guttatus)*. In colonial times, Névome *vaso* (pl. *vapso*) indicated 'hierba, grama, o zacate' (Pennington 1979:57, 61). The word would probably appear today as *vasho* in northern Piman, and it seems to be preserved in the name for the woven straw storage basket, the *vashom*. In the colonial Attí vocabulary from northern Sonora, *zacate* appears as both *vashoi* and *vasoi*.

It is not surprising in a language that probably evolved in xeric portions of the continent (Sonora and Sinaloa) that grasses should be associated with water and damp riparian situations. In its extended meaning among Gileños today, the word can signify hay or pasture.

The plants that are now subsumed in the life form *vashai* include sedges and most of the true grasses, alfalfa (even though it is planted), clover, and several other nonwoody plants that may serve as pasturage (*Lepidium* spp., *Oligomeris linifolia,* and *Plantago* spp.). Most are associated with wet or damp areas, at least seasonally, but this does not seem to be a prerequisite. Unnamed forms of *vashai* grow on sand dunes, on dry bajadas, and even in the desert mountains. It is a reasonably safe guess that some of the more xerophilous grasses were named when men spent more time in the higher country hunting and when women spent more time there gathering.

Sylvester Matthias's inclusion of Dodder (*Cuscuta* spp.) in this category is curious. Dodder is not green, and it is usually found on woody or semi-woody shrubs such as Seepweed *(Suaeda moquinii)*. George Kyyitan would not put it here, because it has recognizable flowers and is not feed for animals.

It would be interesting to know how such semiaquatics as Water Pennywort *(Hydrocotyle verticillata)* and spike-rushes would be classified, but they have been extirpated and do not appear now in the Gileño lexicon. At a cienega in northern Sonora, we found these along with Wire Rush *(Juncus balticus),* a small sedge *(Carex chihuahensis),* and Blue-eyed-grass *(Sisyrinchium* sp.). Sylvester Matthias said these plants would be called *shuudagĭ hugĭdag-an vashai* (literally, 'water edge/along-the grass'), an intermediate category below the life form *vashai*.

Everyone agreed that Carrizo or Common Reed *(Phragmites australis),* included in the emergent water plants (Group B), is not a *vashai*, even though botanically it is a grass, as are maize and several other crops (Group I).

Mathiot (n.d.:243) gives *vashai* as 'grass, underbrush, hay (same as *sha'i*)'. If this is true for the Tohono O'odham, their concept is quite different from that of the Akimel O'odham.

Avena fatua L.
Wild Oat
Avena sativa L.
Common Oat
Poaceae (Gramineae)
koksham, aatoks muḍadkam

In the late winter and spring, one of the more conspicuous large grasses is Wild Oat, with its coarse drooping heads. Field edges and roadsides are favored places.

No one seems to know how long Wild Oat has been in this region, but I suspect that it was one of the many weeds that accompanied early Hispanic invasions. Contaminated wheat and seeds hitchhiking in the wool of sheep, for instance, probably led to a rapid dispersal of this Eurasian grass. Disturbance of the soils by the newly introduced Old World herbivores helped give these weeds a good foothold.

Joseph Giff, an experienced cattleman, knew a lot about grasses. The word *koksham,* he told me, means 'it has a coat or shell'. Ruth Giff and Sylvester Matthias also consistently identified this grass as *koksham,* giving the same gloss. Culver Cassa notes that the word is best considered coating or casing for something, from *koksh,* the plural of *kosh,* originally 'nest'. Fr. Antonine (1935) gave *mudakam* as 'wild oat', but no one at the west end recognized this name.

George Kyyitan of Bapchule walked out on the flats with me near Sweetwater. "*Aatoks muḍadkam,*" he explained, pointing to a heavy grass, "that's 'Wild Oats', means 'it has tassels hanging down': *aatoks*

'hanging down,' *muḍadaj* 'its tassels'— the 'heads'—can be of wheat or barley, or anything like this," he said, holding the robust stem of oats.

On a trip to the Sierra Estrella Sylvester Matthias told me of a grass called *toota muḍadkam.* "It means 'white tassels'. I don't know what it is—but some grass my grandmother told me about. The Chandler area was just wide open with *toota muḍadkam.*" According to Russell (1908:91), "Oats are seldom raised in that region. They are called 'white tassels' by the Pimas." This would translate to *toota muḍadkam.* Perhaps the name originally applied to some native grass and was later used for *Avena* species by people Russell worked with. I found no one any longer using *toota muḍadkam* to mean 'oats' (see appendix B).

I also have not found anyone growing oats today. Some thought that the cultivated species would be called by the same name as the wild one. But most, like Russell, said they doubted that the Pima ever raised the crop. Castetter and Bell (1942:70) did not mention oats as a Pima crop, though some was being raised then by the Tohono O'odham. I have found no specimens of cultivated *Avena sativa* from the reservation in herbaria, but wild *A. fatua* has been collected since 1902.

Sylvester recalled, "Wild Oats, *koksham,* used to grow around the [St. John's] septic tank. Makes good feed—only when it's green. When it's dry, stock won't eat it." I asked Sylvester if the Pima children used to get rolled oats to eat for breakfast. "Yes, they called it *wuulō ki'ivi,* 'the donkey chewed up'. Some Pima call the burro *wuulō,* while others pronounce it *wuulu.* But for 'mush' they can just say *mōsh.*" All the older people I asked were familiar with *wuulō ki'ivi* as the term for cooking oats, commercially obtained. (George Kyyitan called rolled oats *wuulō ki'avi* instead of *ki'ivi.*)

I asked Ruth Giff about *wuulō ki'ivi.* "Oh, yeah! That's what they call it—the cooking kind. Maybe because the way they are—smashed. That means 'the burro chewed it'. Or maybe 'the burro's chewing gum,'" she added, laughing at her pun. (*Ki'ivi* can also mean 'chewing gum'.)

Vouchers: *373, 676* (sm); *986* (gk); *1332* (ih). (All *A. fatua.*)

Cuscuta spp.
Dodder
Convolvulaceae
vamaḍ giikoa, vamaḍ givuḍ, vamaḍ vijina

From a distance, Dodder may look like a yellowish or pale orange mist that has settled on top of a bush. Or like bright, dense cobwebs. But up close it is seen to be a mass of succulent, leafless threads— stems of a parasitic plant. The host, often Seepweed *(Suaeda moquinii),* may be almost covered. Where the orange strands cross the host plant's leaves or branches, they circle and thrust tiny teeth into the green tissue to extract nutrients. Nestled in the chlorophyll-less mass are the minute flowers, like waxy white or cream stars. But a hand lens would be needed to discern them and to distinguish the different species found in Gila Pima country. The hosts are not always herbaceous. Sometimes woody shrubs are the victims, and once in southern California I saw a

koksham, aatoks muḍadkam   *Avena* sp.

tall Athel Tamarisk (*Tamarix aphylla*) crowned with orange Dodder.

The Pima name for this parasite is *vamaḍ giikoa* 'snake's headdress [or] crown' or, according to Joseph Giff, *vamaḍ givuḍ* 'snake's belt'. *Vamaḍ* means any nonpoisonous snake, as distinct from *ko'i, ko'oi*, any poisonous one. (One is not "a kind" of the other.)

Although there seems to be some aversion to this unusual plant and its unorthodox growth form, no one ever expressed this directly, so perhaps it was not a conscious apprehension. When I asked Sylvester Matthias about the plant, he said, "Gets on a tree. It's separate. Will choke it, any kind of plant. Always yellow. Called *vamaḍ giikoa* 'headdress [or] crown of the snake'. *Giikoa* is old traditional headband made out of yucca—or green corn husks to put on head. *Hakkoa* is the round thing to carry olla (*ha'a*) or basket on head. My grandmother made them."

Ten years later I pressed him for more specifics. He had none. "But what did the Pima do if they found *vamaḍ giikoa* in their fields?"

"They'd get a hoe and whack it off—because it gets into plants—and get all over it." He mentioned it was most troublesome in alfalfa. Joe noted its fondness for Seepweed, its most frequent host plant on the reservation.

Sylvester's unhesitating assignment of Dodder to the *vashai* category makes sense in light of the plant's name among some Tohono O'odham: *wepegĭ washai* 'reddish grass' (see Nabhan 1983b; Nabhan in Felger and associates 1992).

Curtin (1949:66) four decades earlier recorded information on Dodder that helps explain the unexpressed aversion that I sensed: "At Salt River Women's Club it was explained that the Pima name for dodder means 'snake crown' and Indians run away from it because they believe 'snake is underneath.' On the other hand, Lewis Manuel claimed that 'the Pima fear dodder because if a snake sees them take the plant, the snake will get after them.' At Sacaton Flats, cattle feed on strangle weed [Dodder], although Dean McArthur said they are known to have died from its effects; therefore the Indians fear the plant may be poisonous and do not touch it."

There is another name, *vamaḍ vijina*, that Fr. Antonine recorded in the 1930s. While *vijina* is generally used for 'rope', it can also mean 'thread' or 'web', such as a spider's web. This latter sense seems most appropriate for Dodder. This name is apparently nearly gone from the speech community; the only one who recognized it is a person who remembers Fr. Antonine compiling his word list when she was a girl. "Isn't it something orange?" she asked me.

> Technical Notes: Species of *Cuscuta* that have been collected in Pima country include *C. campestris* Yunck., *C. indecora* Choisy, *C. salina* Engelm., *C. tuberculata* Brandegee, and *C. umbellata* H.B.K. All species are included in the folk taxon without distinction.
>
> Vouchers: 1 (?; *C. salina*); 1459 (IH, *C. salina*); 1477 (SM, *C. campestris*); LSMC 26 (*C. indecora*).

Cynodon dactylon (L.) Persoon
Bermuda Grass
Poaceae (Gramineae)
komal himdam vashai, a'ai himdam vashai, kii wecho vashai

One of the more persistent lawn-type grasses throughout the Southwest is Bermuda Grass, which probably arrived in the Phoenix area soon after 1869. By the turn of the century, Bermuda Grass was recorded as very abundant and troublesome in the Salt River Valley (Davy 1898).

Its underground rhizomes make it particularly difficult to eradicate. The plant has good dispersal abilities. I once found it growing at a tinaja (a natural water tank) high up in the Sierra Estrella, far from any regular human activities. Herbarium specimens from the reservation go back to 1925–1926.

As the Pima encountered this introduced grass, they applied to it different names. "When the Pima [first] went into Phoenix," explained Sylvester Matthias, "they saw this grass around people's yards, so they called it *kii wecho vashai*. That's what the name means: *kii* 'house', *wecho* 'under' or 'around' [*sic*]. Later they found out that it's growing anywhere. Then they started calling it *komal himdam vashai*; *komal* means 'flatly', *himdam* means 'it spreads, spreads out.'" There is a third name, *a'ai himdam vashai*, that I found in Fr. Antonine's 1935 plant list. "It's OK, it's correct also," Sylvester said. "*A'ai* means 'spreading out in all directions'. *Komal* sounds like spreading just one way. They would use it over in that side [indicating Sacaton], where Fr. Antonine got his information." George Kyyitan knew both names. But it seems that the name *komal himdam vashai* has won ground, because it is the form that everyone I talked with on the reservation (among those still speaking the language) uses today.

Once when Sylvester and I were visiting Pima Bajo Pedro Estrella, I showed the two men Trailing Four-o'clock (*Allionia*

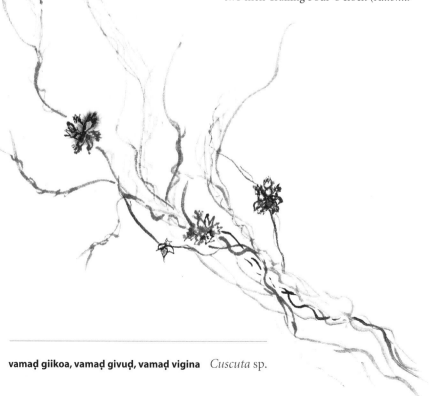

vamaḍ giikoa, vamaḍ givuḍ, vamaḍ vigina *Cuscuta* sp.

incarnata L.), hoping to get a name that might be cognate with Upper Pima. Pedro looked at the supine, spreading, purple-flowered plants and said, "*Komaḍ himdam.*" Sylvester laughed understandingly; Pedro's name means 'something spreading out'. But still I have not found a Gileño name for the four-o'clock plant that grows conspicuously in sandy places along the Gila and lower Salt Rivers.

Vouchers: *369* (SM); *1755* (RG).

Cyperus spp.
flat-sedges, nut-sedges
Cyperaceae
vashai s-uuv

The flat-sedges as a group are rather easy to identify among the marsh plants because the basal leaves are creased down the middle (plicate) and the tops of the naked triangular stalks usually have a few to many radiating leaves that give the plant the appearance of an umbrella without its covering. The flower heads may be loose or compact, depending on the species, but are usually a golden brown. They grow at the edges of marshes or in very wet soils, such as irrigated fields and lawns. The most troublesome field species are Old World introductions.

The Gila Pima name *vashai s-uuv* means 'smelly [or] scented grass'. Ruth Giff explained that the term *vashai* by itself covers all kinds of sedges and grasses in general, but *vashai s-uuv* is a particular plant. Then she added, "Vashai S-uuvak [is] old Pima name for Scottsdale. Used to eat the 'nuts' when we were kids— just chew them; not much flavor." In contrasting it with *vak (Scirpus maritimus, S. americanus,* and perhaps aquatic species of *Cyperus),* Ruth said, "*Vashai s-uuv* is something that grows on the ground, but very much like *vak,* right in the water. *Vashai s-uuv* is small and very hard to get rid of when they get in the garden. Has some kind of nuts."

I wondered if *vashai s-uuv* might have been something that just children ate, and I posed the question to George Kyyitan. "I think everybody eats them. Just once in a while, not all the time. Kind of a little

sweet." Apparently these nutlets were a snack food for anyone.

Sylvester Matthias said, "*Vashai s-uuv* is a nut-grass. [We found them] at Phoenix Indian School, reddish, pale, with black head [the 'nuts' or rhizomes]; dig them out of the ground; they get in the lawns. It's smaller than salt grass *[Distichlis spicata]* and tender." Another time he told me, "*Vashai s-uuv,* that's nut-grass. Scottsdale's name is Vashai S-uuvak, from the nut-grass—gets into yards—[they] have to dig it out."

Curtin (1949:98) was told additional uses for these plants that were not mentioned to me in the 1980s. Walter Rhodes said that the fresh or dry tubers were chewed for coughs or colds. Dean McArthur, a Pima, "asserted that a certain man at Co-op grew *vashai soof* for snakebite. A tuber of yellow nut-grass was chewed and the quid immediately applied to the wound, followed by another. When this remedy was used the wound was never lanced, as is done in most cases. In about three days the patient was walking around." Lewis Manuel, a Blackwater Pima, related that during the *kuushada* or mounted communal hunt for jackrabbits, his father would chew nut-grass tubers and spit them up his horse's nostrils to "pep it up."

The above information applies to Yellow Nut-sedge or Chufa *(C. esculentus),* a native species. This has been the species most often collected in Pima country. Purple Nut-sedge *(C. rotundus)* is probably also included in this folk taxon.

Technical Notes: Species of *Cyperus* collected in Pima country include *C. erythrorhizos* Mühlenb., *C. esculentus* L., *C. odoratus* L. (= *C. ferax* Rich.), and *C. rotundus* L. (US, Glendale, Ariz.).

Vouchers: *518* (SM, *C. esculentus*); *855* (RG, *C. erythrorhizos*); LSMC *11, 36* (sterile).

Distichlis spicata (L.) E. Greene
Desert Salt Grass
Poaceae (Gramineae)
mu'umk vashai

Desert Salt Grass grows in thickets or low swales in moist and usually alkaline places such as cienegas and streambanks. But

today on the reservation, this wiry perennial is restricted almost exclusively to canal banks because of the lack of moisture elsewhere. The yellowish rhizomes form a dense mat on the ground. The grass may be quite green, but on dryer ground it has a more bluish cast. The leaves come off the stems in two opposite rows, alternating up the stem until the whole plant reaches perhaps calf-height. The leaves are quite narrow, pointed, and rigid. When stressed, they roll inward. At certain times of the year, a few compact heads rise on wiry stems from the tops of the plants. These are either male or female, and an entire colony may be a clone of a single sex. But during much of the year the patch is barren.

The Gila Pima name *mu'umk vashai* 'sharp grass' refers to the rather rigid, pointed leaf blades. It is well known in this country, which has lost most other perennial grasses. Sylvester Matthias told me, "Grows in marshy places with a lot of alkali. Sharp points, that's *mu'umk;* no seeds *[sic].* You don't want to wrestle around it. Just spreads when cattle drop it in manure. Horses don't eat it, but cows do. Grows in patch [colony] with very tough ground."

Joseph Giff told me that once there were whole meadows of *mu'umk vashai,* as well as great thickets of *vavish* (Yerba Mansa, *Anemopsis californica*), in New York Thicket, where he used to run cattle.

In a Tohono O'odham dictionary (Saxton, Saxton, and Enos 1983:48), I came across the name *onk vashai* (literally, 'salt grass') for *Distichlis.* I asked Ruth Giff about this. "I never heard *onk vashai* used for *mu'umk vashai;* maybe [in] Papago." She guessed correctly. Sylvester and others I asked likewise had never heard this designation on the Gila.

Desert Salt Grass was mentioned frequently by Europeans traveling through Pima country to or from California (see chapter 3). It was contrasted with more luxuriant grass (particularly around the cienega of Maricopa Wells) that the livestock preferred to eat. One might well wonder what kind of grass was growing near Maricopa Wells: today there is no trace of native grasses in that area.

Desert Salt Grass is closely related to (some think possibly conspecific with) the famous Palmer's Salt Grass *(Distichlis palmeri)* found in the brackish Colorado

River delta and formerly a significant crop plant for the Cocopa Indians. There is no indication that the Pima used the less productive *mu'umk vashai* as a human food source.

Vouchers: *384* (SM); *1461* (IH).

Echinochloa colonum (L.) Link
Junglerice, Watergrass
Poaceae (Gramineae)
s-o'oi vashai

In the fall I was walking through Ruth Giff's field and noticed a coarse, sprawling grass with heavy seed heads striped green and reddish, the leaves marked with purple chevrons. It looked like the type of seed one might harvest. Certainly it was distinctive enough to be named, as indeed it was: *s-o'oi vashai*. Ruth explained the derivation of *s-o'oi* as something sort of striped with two colors, like candy.

I took the plant over to Sylvester's house. He recognized it at once. "*S-o'oi vashai*. It's spotted. Good for feed. Food for stock. In the way of Johnson Grass family. Same leaf as Johnson Grass when real young, but only spotted red and green, light green. Stock will get it. It

spreads out." He said it was not eaten by people.

At Santa Cruz village while I was going through Carmelita Raphael's small kitchen garden of chiles, melons, onions, and summer squash, I noticed in the furrow this spreading grass with purple-reddish blotches. Carmelita knew no name for it.

Apparently this Old World grass found its way into southwestern fields in the early 1900s. In irrigated lands with sorghum, cotton, alfalfa, lettuce, and melons it is a common weed (Parker 1972:42). On the reservation I have mostly found it in people's gardens.

Vouchers: *888* (RG); *1501* (SM); *1552* (SM, RG). Barnyard Grass, *E. crusgalli* (L.) Beauv., is normally excluded from this folk taxon (*847*, RG, long and dense awns). However, Sylvester Matthias included one specimen (*1027*), apparently because of its coloration.

Heteropogon contortus (L.) P. Beauv.
Tanglehead
Poaceae (Gramineae)
biibhinol vashai

On 21 February 1983, I drove Sylvester Matthias south from Santa Cruz village along the east flanks of the Sierra Estrella to look for plants. Along the lower edge of the bajada we found several grasses that he did not recognize (*Aristida adscensionis, Bromus rubens, B. carinatus*). At the base

of the sierra, growing among the huge black boulders, was a tall, stiff, perennial grass.

"I think this is what we call *biibhinol vashai*," he said. The grass was *Heteropogon contortus*, known in English as Tanglehead. The Pima word *biibhinol* 'to wrap around', the Latin *contortus*, and the English *Tanglehead* all allude to the long, twisted awns of the spikelets that become entwined with one another when they fall from the plant.

Vouchers: *262* (SM).

Hordeum leporinum Link
Common Foxtail, Wild Barley
Hordeum murinum L. ssp. *glaucum* (Steudel) Tzvelev
Glaucous Barley
Poaceae (Gramineae)
ba'imuḍkam, koson bahi

Like many of the other grasses common in the West, this foxtail is an aggressive Old World species. On the reservation it is a common annual along roadsides and field edges and in other disturbed places. The seeds of this European weed are arranged in compact opposite rows that suggest the English name, foxtail. Although the seed heads are attractive, they are annoying, especially when dry, because of the long stiff bristles with minute teeth.

The Pima name *ba'imuḍkam* comes from *ba'itk* 'throat' and *muḍadkam* 'tasseled' or 'to have tassels'. According to

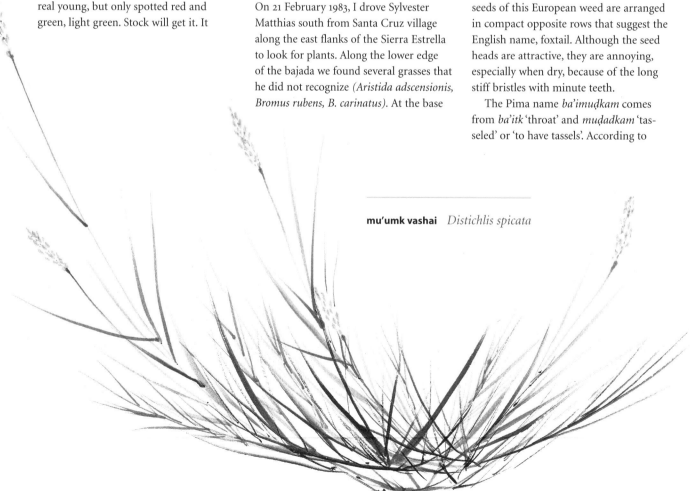

mu'umk vashai *Distichlis spicata*

biibhinol vashai   *Heteropogon contortus*

Joseph Giff, the implication of this name is that the plant has tassels that can get in an animal's or person's throat. "It can get in wheat fields and really irritate people who are trying to harvest the wheat by hand. Or those tassels can get under your jeans and just crawl along your skin. They're really itchy." He glossed the name as 'something that will stick in the throat'.

Sylvester Matthias telescoped the folk name *sheshelik baabhai* (Rabbitfoot Grass, *Polypogon monspeliensis*) into *ba'imuḍkam*, saying both names applied to the same species, *Hordeum leporinum*. "If swallowed, it's going to choke you." But Francis Vavages called Wild Barley *koson bahi* 'packrat's tail' and contrasted it with *sheshelik baabhai*, Rabbitfoot Grass. (He did not know a name for canary grass (*Phalaris minor*), which others called *baabkam*.) Leonard Soke, a Pima rancher, also said that *koson bahi* was a grass found out in the hills, whereas *sheshelik baabhai* and *baabkam* were both grasses of wet places.

Vouchers: *317* (RG, SM); *682* (SM).

Lepidium spp.
Pepper-grass
Brassicaceae (Cruciferae)
avhaḍ

Difficult to identify to species, the Pepper-grasses as a group are distinctive mustards, at least when fruiting. They thrive in disturbed places on the floodplains, particularly fallow fields, yards, and waysides. One species or another may grow up on the bajadas between the black pebbles of desert pavement. When these dry, baked places are watered by winter rains, small, globular Pepper-grass plants grow. The flowers are scarcely noticeable to the naked eye, and the basal leaves wither and drop as the pods mature. But the many stalks of the light grass-green pods will attract attention if one is watching the ground closely. Each fruit is a flat circular or oval disk supported on a slender petiole, with a distinct notch at the opposite end. The disk is really two capsules, with one seed in each.

We were on the sandy terrace of the lower Salt River below Maricopa Colony when I stopped the pickup to go check on several large spring mustards that looked unfamiliar to me. On my way back to the truck where Sylvester Matthias was waiting patiently, I chanced to scoop up off the sand a Pepper-grass plant.

"What's this?" I asked, not really expecting an answer.

"*Avhaḍ,*" came the instant reply.

"*Avhaḍ?*" I asked, surprised that he had spontaneously dredged up this name, which he probably had not said in half a century or more.

"Yes, *avhaḍ,*" he affirmed.

"Is it used for anything?"

"No, it's just a little plant that grows in the sand."

And so it was. Over the years, I have rechecked this, always with the same response. The structure of the word *avhaḍ* suggests that it is one of the primitive unanalyzable words that are basic in folk taxonomies. The various species of Pepper-grass, all lovers of dry places, are natives with which the O'odham must have had long association. The word should be pan-Piman. But so far, I have found no one else who could give me a Piman name for Pepper-grass, nor anyone else who recognizes the word *avhaḍ*.

The green seed heads have a piquant, cresslike flavor, and it would not surprise me if they were once used in Gileño cooking, though no ethnographic record of this persists. The Seri do use them in cooking (Felger and Moser 1985:288).

Technical Notes: Specimens (AMR, SD) of both *L. oblongum* Small (a South American introduction) and *L. lasiocarpum* Nutt. ex Torr. and A. Gray were identified by Sylvester Matthias as *avhaḍ*. Specimens of *L. thurberi* Wooton, as well as one reputed *L. densiflorum* Schrad. (PHK 8386 ARIZ), have been taken on the reservation. This latter appears to me to be a hybrid between *L. oblongum* and *L. lasiocarpum,* the most common reservation species. Characters distinguishing the species require a lens for determination; doubtless all *Lepidium* species are included in the folk taxon *avhaḍ*.

Vouchers: *558* (SM, *L. oblongum*); *647* (SM, *L. lasiocarpum*).

ba'imuḍkam, koson bahi   *Hordeum leporinum*

Medicago sativa L.
alfalfa
Fabaceae (Leguminosae)
s-puluvĭ chu'igam vashai, s-puluvam, s-puluvĭ vashai

Alfalfa is a perennial legume that is grown in considerable acreage on the reservation and elsewhere in the Southwest. It is an erect clover with showy purple flowers. Sometimes it escapes and is found as a roadside weed.

Its original Gila Pima name, I am told, is *s-puluvĭ* (or *s-puluvam*) *chu'igam vashai:* literally, 'Sweet Clover it looks like grass', or, in freer English, a grassy plant that resembles Sweet Clover. That was a good name when Sweet Clover was the well-known plant, perhaps naturalized during the Hispanic era. But during the 20th century alfalfa became a significant Piman crop. Its original name was unwieldy. Some shortened it to *s-puluvĭ* (downriver Pimas) or *s-puluvam* (upriver Pimas), while others have opted for merely *vashai.* The word *vashai,* then, can be used at several different levels: it can mean all grasses and grasslike plants; it can be used loosely for any kind of hay; and it can mean, with the dropping of the markers, specifically 'alfalfa'. Ruth Giff's response to my question "What do you call alfalfa?" is interesting: "I think they would just call it *vashai,* that's anything that's green like that; or maybe they would call it *s-puluv,* a clover."

I wondered when alfalfa was introduced to the Pima on the Gila. There appears to be no evidence of it from the Hispanic period, nor was it reported during the Anglo period when the Pima planted oats and barley for livestock: the difference may be that alfalfa requires large amounts of water (Paul H. Ezell, personal communication).

How early in the 20th century Pima cultivation of alfalfa began is not known, but at least by 1919 it was a significant crop, with a thousand acres planted in alfalfa, mostly Hairy Peruvian (Castetter and Bell 1942:76). In 1923 the Pima had 1,600 acres planted in alfalfa, yielding 4,790 tons of alfalfa. In 1935 there were 2,500 acres of alfalfa, and in 1940, 4,000 acres, mostly Chilean for pasturage and hay (Castetter and Bell 1942:78). Considerable acreage in the 1980s was still devoted to alfalfa. It is irrigated by flood-ing the entire field about every two weeks. Sheep may be pastured on the cut fields in winter. The Pima may get four or more crops per year. Francis Vavages explained, "*S-puluv* can have three or four mowings [a season]. As soon as you get through, you water, then it comes up again until it freezes."

Alfalfa fields indirectly contributed to the table the Pimas' favorite animal food, the Arizona Cotton Rat (*vosho,* pl. *vopsho*). Sylvester Matthias related, "You could find *vopsho* in the fields. They're good. We used to catch them when we had alfalfa. We find a *lot* of them. They can just stew them up, skin them up, then eat them, *vopsho.* Alfalfa fields or wheat fields."

As for medical uses, David Brown told me, "Alfalfa roots are boiled [to make a tea] for diabetes, they say." Sylvester said he had never heard of this. He thought the alfalfa root was a Mexican remedy rather than one used by the Pima.

Vouchers: *1357* (HN).

Melilotus indica (L.) Allinoi
Alfalfilla, Sweet Clover, Sour Clover
Fabaceae (Leguminosae)
s-puluvĭ, s-puluvam

Sweet Clover, also called Sour Clover, is a bushy, green, annual clover that enjoys deep, moist soils such as yards, roadsides passing through irrigated fields, and ditch banks. The plants are erect and many branched. Three oblong leaflets make up each leaf, and the plant has a pleasing smell. The minute yellow flowers are borne thickly on numerous spikes throughout the winter and spring. The pods are like pinheads clinging to the spikes. The only other plants on the reservation that might possibly be confused with this are Bur Clover (*Medicago polymorpha*), which is prostrate with yellow flowers, and alfalfa (*Medicago sativa*), which is erect with purple flowers. All three are from Eurasia. *Melilotus* was a very early Hispanic introduction, probably arriving with contaminated grain seed. Details of timing are lacking so far for Pimería Alta, but seeds of this plant were recovered from adobes made during the California Mission period (1769–1824; Hendry and Kelly 1925).

There is variation in the form of the Gila Pima name for Sweet Clover. At the west-end villages it is called *s-puluvĭ.* But in the eastern villages one usually hears the name *s-puluvam;* Curtin (1949:131) recorded this name from a Sacaton man.

As alfalfa has assumed significance as a crop during this century, the names for it and Sweet Clover appear to be undergoing a transition called a marking reversal (see appendix D). While the original name for alfalfa is *s-puluvĭ chu'igam vashai,* some Pima speakers have shortened the name to *s-puluvĭ,* indicating that alfalfa is the more culturally significant plant. The complete reversal comes when the much older Sweet Clover is secondarily marked as *mischiñ s-puluvam* 'wild clover' or *kiikam s-puluvam* 'home clover'. Although the word *kiikam* is derived from *kii* 'house', the sense is more of 'native'. And *mischiñ* is used to modify some organism that is normatively domestic that has gone feral or wild.

Culver Cassa tells me that this name is an analyzable lexeme. It breaks down to *s-,* an intensifier marker; *pul,* the name of some plant; and *uuv* 'smell' or *uuvam* 'smells of'. But what *pul* is remains a mystery. All that is known is that it is some native plant with which the Pima compared the incoming alien species.

S-puluvam was one of the plants used as the target in a game called *wulivga,* intended to train boys in marksmanship (see Group D, *Pluchea sericea;* also Curtin 1949:131). I asked Sylvester Matthias if *s-puluvam* had any use.

"They can mow it down. Stock won't eat it when it's green. But they can stack it up. Animals will eat it dry, in winter."

Vouchers: *648, 865* (RG); *983* (GK).

Oligomeris linifolia (Vahl) Macbride
Oligomeris, Desert Cambess
Resedaceae
vepegĭ vashai

Winter rains bring up—usually in sandy places, desert pavement, and sometimes fallow fields—a deep green, grassy plant that might easily be overlooked. The leaves are mere succulent filaments. Out on the desert pavement, where moisture is limited, the plant may grow to scarcely finger-high, with just one or several erect stems topped with knobby globular seed heads. But where there is more moisture, each plant may grow dozens of erect heads and be two or three or even more times that tall. About middle to late spring, as these dry areas begin to lose soil moisture, the whole plant becomes paler green, and the maturing, fruiting heads turn pink. Then the entire plant, leaves and all, turns coral pink. The little pods are filled with minute black seeds, perhaps the smallest of any of the abundant annuals on the reservation.

At first none of the consultants with whom I worked recognized this plant. Finally when I showed it to Sylvester Matthias during the pink stage, he remembered it as *vepegĭ vashai* 'reddish grass'.

"It's in one of these bedtime stories. I don't remember it very well. . . . Someone tempted Coyote not to eat anything. It was some animal, maybe Jevhō, the Gopher. . . . Coyote bet his arrows. They're painted red. There were four temptations. The first was *chukukui*, the stink bug [darkling beetle]. But he didn't eat it. The next was *vepegĭ vashai*. He looks at it, and looks at it. 'That must be tender.' But he didn't eat it. The third time, I don't remember what it was. Finally he was tempted with the red mesquite *vepegĭ vihog* [pods]. And he started eating them. And his painted arrows turned into *s-vepegĭ vahammaḍ*, red racers [a whipsnake], that went up into the mesquite. That's why you find the red racer climbing in a mesquite. And that's the end."

"And what happened to Coyote?" I asked Sylvester.

"He would be [was] a smart animal at first, but when he was tempted to eat *vepegĭ vashai* his life kind of changed a bit. He's not as smart as he should have been. He used to be a smart animal. [After that] he wasn't a successful hunter. He had no luck hunting. He used to be a skilled hunter. Coyote lost that."

"Did the People eat *vepegĭ vashai?*"

"No. People didn't use. . . . They don't grow in the fields. In desert soil—where sandy, not in fields."

"How about those little black seeds?"

"No."

I asked Ruth Giff if she knew *vepegĭ vashai*. "I think it's something that just comes up in the winter," she replied. No one else had heard of this plant.

Vouchers: *665, 723* (SM).

Phalaris minor Retz.
Littleseed Canary Grass
Poaceae (Gramineae)
baabkam, (pl.) baahapkam

This bluish grass should catch the attention of anyone except the truly unobservant. An annual, it comes up well spaced in areas of fine saline silt that have had standing water, such as the edges of fields or where sheet flooding has carried fine alluvium into flats, even in mesquite bosques. If the crackling mud dries quickly, this grass produces a single head on one erect culm, but where water stands for a long time, the plant will branch at the base, producing many culms about knee-height. The tight inflorescence, several inches long, is made up of many small pointed scales (glumes) enclosing large seeds. It is beardless.

This introduced Mediterranean grass is called *baabkam* or, in the plural, *baahapkam*. (Different people use different forms.) The name *baabkam* (literally, 'one's mother's father' + 'has') means 'it has a grandfather [from the mother's side]'. No one could give me any reason for its being named this, though it once must have had some poetic allusion. This grass arrived during the Anglo period. Undoubtedly the Pima name originally applied to one of the native species such as Carolina Canary Grass (*Phalaris caroliniana* Walter) or Timothy Canary Grass (*Phalaris angusta* Nees), but it was transferred when the very similar looking Old World species invaded. The native species are so rare that I have discovered them only once on the reservation, in 1991.

Sylvester Matthias said *baabkam* was used for pasture but not for human food. Apparently this was not always the case, though. Reading through Russell's (1908:76) list of plants used for food, one day I ran across an unidentified plant he called *papkam:* "The heads are tied in bunches and dried in the sun. They are then shelled, screened, the seeds parched, ground on the metate, and eaten as pinole. They are 'not sweet.'" I recall Joe Giff saying *baabkam* seeds were once eaten but did not taste good.

Another species in the genus *Phalaris* produces the familiar canary seeds of commerce. The seeds of *baabkam* are similar, though smaller.

Vouchers: *162, 163, 248* (SM); *990, 1560, 1715* (GK).

Plantago ovata Forsskal
   [= *P. insularis* Eastwood]
   [= *P. fastigiata* E. Morris]
Plantago patagonica Jacquin
   [= *P. purshii* Roemer & Schultes
      var. *oblonga* (E. Morris)]
Indian-wheat, Woolly Plantain
Plantaginaceae
muumsh

When winter rains come at the right time, a small grasslike plant with narrow woolly leaves will grow out on the barren flats and even up on the bajadas between the black rocks of desert pavement. At times these grayish plants are truly abundant in these parched places. These are not true grasses; they produce many heads of small, translucent whitish flowers with brown centers. The more moisture, the taller the plant and the more heads each

baabkam, (pl.) baahapkam   *Phalaris minor*
The young cricket or katydid is called chukugshuaḍ; *a grasshopper is* shoo'o

will produce. In a very dry place the plant may not reach three inches tall, producing only half a dozen flowering heads, but in more moist places they may be at least twice that, covered with the seed heads that give the plant its English name Indian-wheat.

Although I have heard several variations of this plant's name from the Gila Pima *(muumshō, muumsho, muumshum)*, the most frequent and I think most correct form seems to be *muumsh*. This is how Fr. Antonine (1935:12) wrote it, calling *muumsh* a "small grayish weed, good for feed."

The plants mature during what was formerly the spring famine season for the Pima. At one time they were harvested, though even elderly Pimas today are uncertain of the method of preparation. When the heads dry, they can be threshed and winnowed, producing many small, elliptical, lavender to rufous brown seeds. *Muumsh* seeds as well as those of *daapk* (Chia, *Salvia columbariae*) and *shuu'uvaḍ* (London Rocket, *Sisymbrium irio*, or Tansy-mustard, *Descurainia pinnata*) were placed in water, where they swelled to produce a mucilaginous drink. I think all of these dropped from popularity because people began to object to their texture.

Apparently *muumsh* was lost from the dietary before either *daapk* or *shuu'uvaḍ*. George Kyyitan and Sylvester Matthias (who both used the name *muumshum*) said they had eaten the other two in their youth, but not *Plantago*. This is curious because *Plantago* is by far the most abundant, proliferating in areas cleared by fire or in abandoned fields not taken over by shrubby vegetation.

Perhaps *muumsh* was not used exclusively as a food. A Tohono O'odham who had been raised on the Salt River Reservation told of its use for diarrhea: "Early in the morning, before eating, half a cup of ripe seeds, which have been gathered and stored, and half a cup of water are mixed and allowed to stand a short time. Before this mixture jells too hard, it is administered. The dose for a baby is one tablespoon of seeds to half a glass of water" (Curtin 1949:97). Unfortunately, this account does not specify whether this was a practice of the Pima or their desert cousins. A related species from the Old World, *Plantago psyllium*, is regularly sold over the counter as a bowel stimulant and lubricant. Undoubtedly there were seasons

when the aboriginal Pima diet likewise was low in fiber and the native *Plantago* species served as a useful antidote.

"Have you ever heard of *muumsh* being used as a medicine?" I asked Sylvester.

"Yeah. Mainly the Papagos. It's mainly the knowledge of the Papago. When something drop in your eye. Put a few grains in there. With that eye butter, it'll come out. And they make some kind of a tea, drink from it. That's Pima knowledge."

Ruth Giff was talking about *shuu'uvaḍ* when she remembered a little plant she thought was called *muumsh* or *muumsho* that Joe's grandmother used to prepare. "Tiny plants, white, with kind of fluff on the plants."

Joe said these were *muumshum* or *muumsho*. "We used to pull them off, then rub them when they get dry. And get the seeds out. Then boil it."

"Did she grind it?" Ruth asked her husband.

"No, just boil it. We used to go out, my grandmother and I used to come right back here anywhere, you know, and gather that, in the spring, that little plant, that *muumsho*. All that stuff that we're talking about that people don't use any more, see. They say that's what Se'ehe give them for food. That's their food—that was given them by Se'ehe. And they say when they abuse that, why, Se'ehe isn't going to give us any more rain to grow that stuff, that we eat, outside. . . . And the Indians are supposed to appreciate that and go pick it and fix it the way Se'ehe told them how for food—all the things you get out in the desert, for food. . . .

"And a lot of the things that we eat are given to us by the Maker. [He] made that for us for food, and give to us for food— the game and all the plants and all that. . . . See, nobody's paying any attention. There's plenty of it back there. . . . And somebody goes after it, somebody gathers it to make money off it. They sell it to the People now, People who cannot afford to go after it, who don't want to do it, because it takes quite a bit of time and work. They don't have that kind of time, so they'd rather buy it. Well, that's what Se'ehe said, when they start abusing things like that, the natural food that is given us, why, he's going to hold the rain back on us. Then we're going to starve. . . . Now what have we got to eat? We don't know how to eat our natural food that our Maker gave to us. If we use that, we could

still be on it—there wouldn't be no diabetics."

Technical Notes: *P. ovata*, with its array of synonyms, is the most abundant species in Pima country, both on the floodplains and elevated lands. *P. patagonica*, with its conspicuous linear bracts and longer, heavier inflorescence, is much less common. The two may be found growing side by side. Several greener, mesophytic species have been collected on the reservation, but the folk taxon and uses apparently apply only to these two grayish xerophytic species.

Vouchers: *P. ovata*: *6, 7* (JG, RG, ?DB); *1741, 1894* (GK). *P. patagonica*: *591* (RG); *1895* (GK); LSMC *59*.

Polypogon monspeliensis (L.) Desf.
Rabbitfoot Grass
Polygonaceae
shelik bahi, (pl.) sheshelik baabhai, ban bahi

Like canary grass, this European grass grows in damp alluvial places, but it prefers even more water: edges of rivers, pastures, drainages, temporary ponds. Also about knee-height, this more greenish plant has distinctive flower heads— finger-length racemes of tiny individual flowers with fine, soft awns or bristles. These give the flower heads of the plant both the appearance and the texture of a rabbit's foot, hence its English name.

At least in the west-end villages it is called *shelik bahi* or, more often in the plural form, *sheshelik baabhai*. A number of grasses throughout Piman country are named by comparison with some animal's tail, *bahi*. In this case it is the *shelik*, or Round-tailed Ground Squirrel, sometimes erroneously glossed 'prairie-dog'. The *shelik* has a thin and sparsely haired tail, but when alarmed it will fluff this up so that the tail looks more like the grass's seed head.

Leonard Soke, a Pima cattleman from Santa Cruz village, characterized *sheshelik baabhai* as a soft grass found in wet places with *baabkam* (Littleseed Canary Grass, *Phalaris minor*). He contrasted these with *koson bahi* (probably Wild Barley, *Hordeum leporinum*), which is found out in the hills.

Everyone I asked called Rabbitfoot Grass *sheshelik baabhai*, except George Kyyitan of Bapchule, who called it *ban bahi* 'Coyote tail'. George called Wild Barley *(Hordeum leporinum)* instead of Rabbitfoot Grass *sheshelik baabhai*.

Vouchers: *245* (RG, SM); *1559* (GK, *ban bahi*).

Sorghum halepense (L.) Persoon
Johnson Grass
Poaceae (Gramineae)
kaañu chu'igam, vaapk chu'igam

Johnson Grass is the bane of irrigation ditches and wet field edges in Pima country. It is a dense-rooted perennial grass that looks much like the congeneric cultivated Sudan grass *(S. sudanense)*, an annual. The coarse Johnson Grass grows in clumps or colonies from waist- to about head-high. The bright green leaves would seem to make good forage for livestock, but they are dangerous because when injured they cause prussic acid poisoning. In summer and fall the open panicles or seed heads are heavy with reddish seeds.

Johnson Grass probably arrived in the Pima area early in the 20th century: there is a 1925 specimen from Sacaton (ARIZ). It quickly became an established nuisance (Castetter and Bell 1942:178). The conflict between what a farmer wants to grow and what else wants to grow in the disturbed plant communities we call gardens and fields was escalated by the introduction of noxious Old World weeds. Perhaps it is significant that the Pima have no word for "weed." Indeed, in the 1940s, Castetter and Bell (1942:174) noted, "All informants were very definite in their opinion that anciently weeds were much less abundant than at present; that the Pimans kept their fields more free of them than now. This may well be true when we take into consideration the numerous weeds which have been introduced and the rank growths of Bermuda and Johnson grass in the Gila valley."

Joe Giff had farmed for years at the west end. Johnson Grass was one of the more troublesome weeds he had to battle. "We call it Johnson *sha'i*. Wasn't here a long time ago. Don't know when it came in, but has spread all over. It looked like

good fodder, and some Papagos took it back to their village to plant. But when it took over, people weren't so happy with them [for bringing it in], because it was hard to plow." Joe laughed because this was a practical joke on the Tohono O'odham, though done with all good intentions. But his daughter Sally Giff Pablo laughed, saying that Johnson *sha'i* was not really the name for the grass. It was just a play on words, something said jokingly; the Pima, having no word for 'weed', may call a weed *sha'i*, as did her father. And the plant is not a joke to anyone except possibly the Pima, who can laugh at anything, even what others call a disaster.

There are two names for Johnson Grass, explained Sylvester Matthias. *Kaañu chu'igam* (literally, 'cane sorghum it looks like') is usually used in the east end. The comparison here is with the so-called sugarcane, a sweet sorghum that is raised for its juice (see *e'es*, Group I). The other name is *vaapk chu'igam* ('Carrizo [or] Giant Reed it looks like'). This is also an appropriate comparison. Although the rank growth of Johnson Grass does not closely resemble the primly alternating leaves of the tall Carrizo or Giant Reed, both are perennial plants that grow back from an extensive mass of roots deep in the muddy banks of canals and streams. Even their rhizomes look similar.

The Pima may have formerly eaten the seeds (Kearney and Peebles 1960). Sylvester had never heard of this, nor had anyone else I asked.

At Sacaton, Francis Vavages said, "*Kaañu chu'igam* and *vaapk chu'igam* is the same thing. They're mean plants—they'll grow anyplace. They'll ruin your field."

This grass has invaded the Pima Bajo fields of southern Sonora also. One September day after the *temporales* (monsoonal rains), I walked through the densely overgrown *milpas* with Pedro Estrella. Noticing the tall, heavy grass almost smothering his squash, I asked what it was. "Zacate de Yonsón," he replied. The name in his Piman dialect was *Yonson vashaai*.

Frank Jim, a Tohono O'odham, said that the seeds of this grass are bigger, like those of cane sorghum. He explained that someone took the seeds to the Tohono O'odham village of Hikiwan thinking that was what they were. The weed took over

shelik bahi, (pl.) sheshelik baabhai, ban bahi
Polypogon monspeliensis

about 40 acres of pasture. "Can't get rid of it! Animals can eat it when it first comes up, but not after its growing." Perhaps this was the same incident Joe had recalled.

"What do you call that grass, Frank?"

"Johnson *vashai*."

Maybe Joe Giff's independently evolved name was not such a joke after all.

Vouchers: *374* (SM); *1932* (GK).

Sorghum sudanense (Piper) Stapf
Sudan grass
Poaceae (Gramineae)
kaañu chu'igam vashai

Sudan grass is grown for pasturage; often a whole Pima field is seeded to it. This Old World grass looks like something between the congeneric pestiferous Johnson Grass and cane or sweet sorghums. Its Pima name, *kaañu chu'igam vashai,* indicates exactly that: *kaañu* is the cane sorghum; and by itself, *kaañu chu'igam* means Johnson Grass. Only with the *vashai* 'grassy plant' tacked on does it mean Sudan grass.

Sudan grass grows about head-high and, unlike Johnson Grass, is annual. Its panicle or fruiting head is more open than that of the stocky milo maize or other grain sorghums. And it is a grassier, more open plant than either the black or the red cane sorghum.

Introduced into the United States in 1909, *kaañu chu'igam vashai* has been grown extensively by the Gila Pima at least since the 1930s. When I went to the reservation in the early 1960s, it was still a rather common forage crop in most villages, but decades later it is seldom planted; there is less livestock to feed. I asked Carmelita Raphael of Santa Cruz village about her seed source. She buys it by the pound in a Phoenix feed store.

I asked George Kyyitan what he called Sudan grass.

"*Kaañu chu'igam vashai* or [just] *kaañu vashai*—feed—cattle, horses. Boy! That one is pretty hard to load up on a wagon. It's slick! Never make a load out of it— slips out."

"When did they start raising that? When you were a kid?"

"No. I think it is . . . but after that, when they started pasturing cattle around here, they plant it here."

Sylvester Matthias related, "*Kaañu chu'igam vashai,* for feed, cows and horses, is very good because, they say, every 15 days you can pasture it again. Water it; feed [graze] it up; or mow it down; water it [again], and in 15 days another crop. It's more cheaper than alfalfa."

Sylvester classified this plant as *vashai,* emphasizing its growth form, while George considered it *e'es,* according to its status as a planted crop.

Although Kearney and Peebles (1960:143) found it growing spontaneously at the Sacaton Field Station, I have never found feral plants persisting.

Vouchers: *1485* (SM).

Group D, Sha'i
bushes

This life form, like *vashai* (Group C), is well defined morphologically; it corresponds closely with the English concepts 'shrub' or 'bush' and (collectively) 'chaparral'. Plants put into the category *sha'i* are woody to a greater or lesser extent, are usually perennial, and are smaller than desert trees. The only somewhat shrubby plants that are excluded are the yuccas, which Sylvester Matthias included in the covert category of spiny plants (Group H), along with cactus, agaves, and other armed rosette plants.

A few plants such as Bursage Ragweed *(Ambrosia confertiflora)*, Horseweed *(Conyza canadensis)*, Small-flowered Gaura *(Gaura parviflora)*, Russian-thistle *(Salsola australis)*, White Horse-nettle *(Solanum elaeagnifolium)*, and globemallows (*Sphaeralcea* spp.) may be borderline cases between *sha'i* and the covert category of wild annuals (Group G), which seems to be defined more on its negative attributes. These are all "weedy" plants, but there is no word or phrase in the Gila Pima language that corresponds to the English concept 'weed'—a usually prolific, rank plant that appears unwanted in cultivated situations. When pressed, a Pima may call a weed *sha'i* by default.

Sylvester Matthias excluded *shegoi* (Creosote Bush, *Larrea divaricata*) from the category *sha'i*. Apparently it is too special to the Pima to lump with other woody plants, and hence he placed it in its own category (although I have retained it in this group). Mason (1917), with the Tepecano (the southernmost community of Southern Tepehuan), glossed *sai* as 'hay, zacate'; this domain is labeled *vashai* (Group C) in Gila Pima.

Allenrolfea occidentalis (S. Watson)
    Kuntze
    [≠ *Atriplex* sp.]
Pickleweed, Iodine Bush
Chenopodiaceae
iitañ

In very saline areas, usually where the salts and carbonates show like powdery frost over the surface of the soil when it dries, grows a shrub with grass-green succulent stems that look like tiny sausages strung together. These dense Pickleweed shrubs reach to about shoulder- or even head-height, usually with Greasewood *(Sarcobatus vermiculatus).*

Joseph Giff was the first to tell me about Pickleweed: "In addition to *kauk u'us* [Greasewood], there's another salt bush called *iitañ,* . . . grows with it. The leaves are like [a] string of beads—guess that's supposed to be the leaves. They used to eat the seeds of *iitañ,* in the legend. It's just like powder. The *chuchk onk* [Seepweed, *Suaeda moquinii*] grows here, too."

Ruth Giff said, "*Iitañ* grows in *ongam* [salt-impregnated soil]. You know, they did [eat it], way back, when they were starving; they used all kinds of seeds to keep alive."

Sylvester Matthias told me, "When hard up for food, people ground this *iitañ* for food. They shake it into a basket to harvest; make pinole, because [it's] salty. Grows in saline areas." A few years later Sylvester added, "*Iitañ,* they say, you can grind up and use it as pinole; it's also feed for the cattle, especially the cows. Horses do eat it, but not as well as the cows, because this is salty, and cows like something salty."

Curtin's information (1949:69) both confirms and elaborates on the preparation techniques: "When seeds ripen in the summer, they were gathered in a basket and winnowed, then were roasted in a special pot [the *muta*], 'with ears on the sides,' said Sarah Jones and Juana Innes. After this, the seeds were ground on a metate, water was added, and the whole cooked like *atole.*"

Russell (1908:73), who thought the plant was one of the *Atriplex* species, gives this account of its preparation from information gathered at the turn of the century: "The heads of this saltbush are pounded up in the mortar and screened to separate the hulls. The seeds are washed, spread to dry, parched in a piece of olla, and ground on the metate. They are then ready to be eaten as pinole, or dry, as in the latter case a pinch of the meal being taken alternately with a sip of water."

It is clear from the comments of both Ruth and Sylvester that *iitañ* was essentially a starvation food. Paul Minnis (1991) noted that knowledge of famine foods is often transmitted by means of legend, as Joe pointed out in this case.

There are two existing references to this plant in the Creation Epic, both times as proper names. The Original Being, called Jeweḍ Maakai (Earth Doctor or Earth Maker), first made Ñui (Vulture or Buzzard) as a helper (Lloyd 1911:28). The second helper, Ban (Coyote) is associated with a common plant, White Brittlebush (see *Encelia farinosa* account). Still, there were only three co-creators, and the Piman sacred number calls for four for completeness. Thin Leather's version of the story (Lloyd 1911:31) continues: "But now out of the North came another powerful personage, who has two names, See-ur-huh [Se'ehe] and Ee-ee-toy [I'itoi]." In Russell's (1908:209) account there is a solitary reference to a third name for this being: "After a time the earth gave birth to one who was afterwards known as Itany [Iitañ] and later as Sieehu, Elder Brother." With that the traditional Piman complement of four was fulfilled: Earth Doctor, Buzzard, Coyote, and Se'ehe or Elder Brother, more commonly known to the Tohono O'odham as I'itoi. His original plant name, Iitañ, is never mentioned again in the myth, and one may wonder if he was borrowed from some northern Puebloan peoples and added to the Piman pantheon. (Pickleweed is a plant of the Great Basin desert, found growing along the Little Colorado River and Rio Grande where the Puebloan tribes live as well as throughout the Gila drainage.) Perhaps the Akimel and Tohono O'odham took this culture hero, each rebaptizing him with their own name. Elder Brother and Coyote assume the greater significance as the Creation Story unfolds than do either of the earlier two.

iitañ   *Allenrolfea occidentalis*

Later in the Creation Story (Russell 1908:232) there is another mention of this plant as part of a proper name. Coyote ultimately created two others like himself: Sandy Coyote and Yellow Coyote. They lived together near the mouth of the Grand Canyon. Sandy Coyote went to the Va'aki O'ob (literally, 'old house' or 'ruins' + 'enemies') to get a wife named Iitañ Uvĭ (Pickleweed Woman). Perhaps her people were Hualapai or some other Yuman group.

Sylvester told of one additional use. "*Iitañ* wood makes pretty good light fuel. Don't smoke, too. Gives more light than kerosene lamp or lantern in those days. If anybody's out of kerosene, just burn it, open the stove, and just have all the light you need."

Today, the plant has a very limited distribution across the floodplain of the western villages, but it must have been more extensive before so much of the reservation was converted to mechanized farming.

Vouchers: *160, 1659* (SM); *LSMC 80.*

Ambrosia ambrosioides (Cav.) Payne
[= *Franseria ambrosioides* Cav.]
Canyon Bursage
Asteraceae (Compositae)
ñuñui jeej

Most trees and woody plants of the Sonoran Desert, at least those growing away from the rare perennial streams and rivers, have leaves that are small, often compound, reducing water loss to a minimum. But along the sandy edges of arroyo bottoms cutting through the bajadas of the desert ranges, Canyon Bursage flaunts stiff wands covered with simple green leaves, like woolly tongues about 6–7 inches long. (These get even longer as one goes south into Sonora.) For a perennial, the plant seems almost out of place in this desert because it is so green that it makes the arroyos look like dried-up creeks. But its appearance is not its only singular characteristic. As you walk along the sandy bottoms of these arroyos in a summer dawn, you will notice a

distinctive smell, suggesting a stream in the mountains. That is the smell of Canyon Bursage.

The plant is rather widely known among the Tohono O'odham as *ñuñui jeej* 'vultures' mothers'. People from villages on both sides of the border knew this folk taxon. But for some reason, few Gila Pima could give me a name for this conspicuous plant.

I brought Carmelita Raphael some Canyon Bursage. She called it *ñuñuvĭ jeej*. I asked her whether she knew the name from the Pima or Tohono O'odham side of her parentage. She was unsure. Sylvester did not know the plant. A few years later, Ruth Giff mentioned, "Don Thomas [a Komatke singer who died in 1992] has seen that *ñuñui jeej*. He knows it. His father is Papago, his mother Pima." As far as I have been able to learn, the only Pimas who have known the name for this plant are Tohono O'odham on one side of the family and lived part of their youth in Tohono O'odham country. But the Gila River Pima, busy with their fields, seem to have forgotten many plants of the mountains and bajadas that are still well known to their desert cousins.

Two small perennial bushes on the bajadas and sand dunes and better-drained flats are in this genus: Triangle-leaved Bursage (*Ambrosia deltoidea*) and White Bursage (*Ambrosia dumosa*). They are compact, smaller-leaved, grayish to silvery plants. Though they are common, I have found only one Pima who knew a name for either one: Sylvester Matthias called Triangle-leaved Bursage *kokomagĭ shegoi* (see *A. deltoidea* account). We are left to guess whether White Bursage may once have had a name that is now forgotten or whether it was simply not named by the Gileños.

Canyon Bursage grows taller and more luxuriantly farther south. In Pima Bajo country I was unable to discover any name from Pedro Estrella except *chícura*, which is an adaptation of its widespread Spanish name *chicúra*. There it is used medicinally.

The Tohono O'odham Frank Jim associated the *jeej* plants with *a'akĭ*, arroyos. This is an interesting connection considering that *ñuñui jeej, vipismal jeej* (*Justicia californica*), and the still unidentified *ko'oi jeej* are characteristically dry-wash

plants; *makkom jeej* (*Boerhavia* spp.) and *mo'otaḍk jeej* (*Ambrosia confertiflora)* are commonly found in washes.

Vouchers: *112* (Delores Lewis, Tohono O'odham); *598* (CR); *LSMC 15.*

Ambrosia confertiflora DC.
[= *Franseria confertiflora* (DC.) Rydberg]
Slim-leaved Bursage, Bursage Ragweed
Asteraceae (Compositae)
mo'otaḍk jeej

This root perennial of yards, ditch banks, roadsides, and abandoned fields of Pima country sometimes spreads so thickly by its root system as to become a serious weed. The leaves of the young plants are divided and redivided, suggesting carrot leaves. But they are a pale blue-green and felty, and they have a pleasant mintlike odor when crushed. The pollen-bearing flowers rise on main stalks; the female flowers that develop the burs are below, among the leaves.

Its Pima name is *mo'otaḍk jeej*. *Mo'otaḍk* 'head-emerging' is Broom-rape, *Orobanche cooperi; jeej* means 'its mother'. So Slim-leaved Ragweed is '*Orobanche's* mother'. This name indicates the Pimans' intimate knowledge of botany. Broom-rape is a parasitic plant lacking chlorophyll. Deep in the sand or alluvial soil it attaches its roots to some other plant, often one of several *Ambrosia* species. So the various ragweeds and bursages are literally the mother or nurse to the thick purple *mo'otaḍk* plants.

I have never found a Broom-rape parasitic on its "mother" in Gileño country. But one day near the Pima Bajo village of Onavas, Sonora, don Pedro Estrella and I were walking across his milpa before planting time. We found a Broom-rape, here named *duit heosig* 'earth flower' in Pima Bajo. I dug it out and found its roots connected to those of a Slim-leaved Bursage. Pedro knew at once what I was looking for. He called the bursage mother plant *mo'otarĭ*, clearly an abbreviated form of its northern name. He said that a very

mo'otaḍk jeej   *Ambrosia confertiflora*

bitter tea is made from *mo'otarĭ* roots for stomachaches. Sylvester Matthias and Ruth Giff had never heard of such a use among the Gila Pima.

Vouchers: *520* (sm); *886* (?).

Ambrosia deltoidea (Torr.) Payne
[= Franseria deltoidea Torr.]
Triangle-leaved Bursage
Asteraceae (Compositae)
kokomagĭ shegoi

Out on the floodplains and the gently sloping bajadas of Pima country grow a bewildering array of small grayish bushes. To the novice they may all look alike— just "sagebrush." But not a one of them is actually sagebrush (*Artemisia tridentata*), a plant of the cooler Great Basin desert

to the north. The most common such bushes on the fine soils of the flatlands are *kokomagĭ sha'i*, Desert Saltbush *(Atriplex poly-carpa)*. But higher on the gravelly and sandy soils of the bajadas you will find two shorter types of bushes, the bursages. They may be quite common, interspersed among Creosote Bushes *(Larrea divaricata)*, paloverdes (*Cercidium* spp.), and White Ratany *(Krameria grayi)*.

Triangle-leaved Bursage is the more common of the two. It is an erect little bush with many straight stems, reaching about midway up your calf. As both the scientific and the English names imply, the pale greenish leaves are acute triangles. Their margins are toothed. Bursages are in the aster or sunflower family, but this may not be obvious without a closer look. The male flowers are borne on the tips of new branches. Being wind-pollinated, they need no ray flowers (the "petals" in this family) to attract insects. In late spring and early summer these little disks of staminate (male) flowers produce clouds of pollen, like their near cousins the ragweeds. Immediately below, on the same flowering stems, are the clusters of female flowers. These mature into small spiny burs that surround the stem in a clump.

It took me more than ten years to verify the Pima name for this plant in the field. In April 1978, David Brown, Sylvester Matthias, and I went out from Santa Cruz village looking specifically for *koko-magĭ sha'i* and *kokomagĭ shegoi*. The road skirted the lower bajada, occasionally bounding up over the piedmont and down through a wash. We found plenty of *kokomagĭ sha'i*, Desert Saltbush. Though we drove and walked for miles, we never found (or at least they never agreed that we found) *kokomagĭ shegoi*. They concurred that the medicinal plant, Creosote Bush, was *shegoi*, and that if there was any confusion with the gray bush we sought (*kokomagĭ shegoi* means 'gray Creosote Bush'), Creosote Bush could be specified as *muhaḏagĭ shegoi* ('greasy [or] grease-colored *shegoi*'). But usually conversational context distinguished the two.

Over the years the name would come up again, and I often wondered whether *kokomagĭ shegoi* was just an alternate name for *kokomagĭ sha'i*. For instance, in January 1985 we found Triangle-leaved Bursage growing next to Desert Saltbush in a wash coming down from the Estrellas, but Sylvester was not sure of the first plant. "It *may* be *kokomagĭ shegoi*" was all he would concede. Later that fall, he assured me that our mystery bush was not the same as *kokomagĭ sha'i* because each had different leaves. How they differed I could never discover. I took to his house branches of the saltbushes *Atriplex poly-carpa, A. canescens, A. lentiformis,* and *A. linearis* to check leaf shape. None of these! He had names for each of these except *A. linearis,* which he said had no Pima name, just as everyone else told me.

I tried other leads. I took branches of Triangle-leaved Bursage to Francis Vavages, Cornelius Antone, and Frank Jim (a Tohono O'odham), but none knew a name for it. Nor did any of the women. Out in the Sacaton Mountains George Kyyitan recognized neither the plant nor the Pima name.

During his visit to San Diego in 1987, Sylvester again confirmed that the two gray plants in question were indeed distinct species. "We'll find it; it grows up along the boundary line in the South Mountains where I go to get clay." In May 1988 while I was working at Sacaton, I stopped by his place one evening when I was at the west end. The topic of *kokomagĭ shegoi* again came up. "We can get it—at the water storage tank on South Mountains." It was already dark by then, but I was pretty firmly convinced that our mystery plant had to be a bursage. In early June I tried again. He directed me up a sparsely traveled dirt road toward South Mountain through Desert Ironwoods *(Olneya tesota)* and Creosote Bushes as tall as my truck. At the very base of the mountain, where the vibratingly hot desert-varnished rocks begin, he motioned me to stop.

"They're all dead!" he said. We were surrounded by Triangle-leaved Bursage bushes, most of them dried almost to a crisp in the blistering midday sun. Searching, I found one along a dry gully that still had greenish leaves. The stems of the new growth had a dark brown to blackish varnished look. We smelled the aromatic twigs and leaves. Surely these, along with Creosote Bush and the freshly moistened earth, must be responsible for the smell of a desert rain.

"Is it used for medicine or anything?" I asked.

"No, nothing. Just this regular *shegoi* here," he said, breaking off a piece of Creosote Bush. "They use this *muhaḍag shegoi* for medicine, but I never heard of anyone using *kokomagĭ shegoi* for anything."

When I opened my plant press later, I realized how appropriate the Pima name was. I had to peel the paper carefully away from the specimen. The abundant terpines of the younger branches, like those in the leaves of its namesake Creosote Bush, stuck to the paper.

But why had it taken me so long to get this folk taxon identified? True, fewer and fewer people get out into the bajadas and mountains anymore, but I had asked men who had driven cattle and hunted, and women who gathered cholla buds—all in places with abundant bursage. One reason is that plants with uses and plants growing in closer proximity to where people live are more likely to be remembered in a collapsing taxonomy than are plants that are farther away and serve no practical purpose. And plants are best identified in their natural setting; Sylvester knew the plant when he took me to a place where he expected to find it, even though he had not recognized detached branches I showed to him years earlier.

Supposedly the Tohono O'odham name for this bush is *tatshshagĭ* or *taḍshshagĭ* (see Mathiot n.d.:181), but I have found no Pima who recognizes this word. I say supposedly, because any identification of these notoriously confusing plants should be based on a preserved voucher specimen and information as to what other similar shrubs it was contrasted with. Saxton, Saxton, and Enos (1983:57) identified *tatshagĭ* as Wheelscale Saltbush (*Atriplex elegans*), a plant widely known to the Gila Pima as *toota onk* 'white salt'. Either the Akimel and Tohono O'odham names are not concordant, or the lexicographers have misidentified their referents.

Vouchers: *1561* (SM).

Aster spinosus Benth.
Spiny Aster
Asteraceae (Compositae)
no'oshkal

In the moist and usually saline silty soils of Pima field edges and ditch banks often is found a thick bushy plant, seemingly innocuous. It grows to between knee- and shoulder-height. Its few small leaves drop early, leaving masses of somewhat grayish green stems fortified with stout spines. (Were the spines lacking and the stems more rigidly erect, you might think these were *Ephedra* bushes that had taken up residence in the lowlands instead of on mountainsides.) The small flowers are suitably asterlike, with white rays and bulging yellow disks. When the flower heads mature, the fluffy white tufts atop each seed or achene look like those of the more ornamental and orthodox asters you may know.

But this wiry and aggressive bush is not much appreciated by Pima farmers, who call it *no'oshkal*. (I suspect that the prefix *no-* reflects some concept of its prolific tendencies: the Pima word for 'egg' is *nonha;* 'to become pregnant', *nonhat;* 'one who is pregnant', *nonham.*) Ruth Giff told me, "It's *very* hard to get rid of—spreads out from roots; it got into last summer's field gardens." Later she added, "I guess it's worse than *kaañu chu'igam* [Johnson Grass, *Sorghum halepense*]." Eradication is a tough job because of Spiny Aster's thick rootstocks, especially when the plants form hedges or thickets. Sylvester Matthias said, "It grows in ditch banks and gets into wheat fields; gets tangled with wheat and has to be removed. We have no use for it."

Others likewise knew no use for it, at least directly. Francis Vavages, however, noted an indirect use. "*No'oshkal* grow along the ditches. They get rid of them—trash to the fields. Mostly good for *vopsho*—[cotton] rats—people eat those. They live in that *no'oshkal*, breed in there." While this seems like a simple statement, it reveals something more complex about Piman culture and their approach to the environment. Not even a pernicious weed is an unmitigated disaster; the undesirable bush proliferating as a result of Pima land use provided cover for their most prized game, the Arizona Cotton Rat.

Wherever the Pima have transformed the river terraces into fields bordered by irrigation ditches, Spiny Aster takes over the moderately saline seepage areas. On the modern middle Gila it is found almost exclusively in anthropogenic conditions. And it would seem this was the case even centuries ago. In colonial times, the Jesuits found the Pima living in three great rancherias. Starting about 12 leagues (approximately 30 miles) downstream from the Salt-Gila confluence, various travelers listed and mapped between 30 and 40 smaller mixed villages of Cocomaricopa and Pima peoples. Although the villages have been considered primarily as Cocomaricopa (Yuman), the village names are almost all readily translatable into Pima, with allowance for a few historic sound shifts, such as *s* to *sh*. The fourth village that Fr. Sedelmayr recorded in 1744 he wrote as Noscaric ('place of many Spiny Asters'). Apparently Pima farmers were having trouble with this weedy shrub even then!

Vouchers: *37* (SM, JG); *523* (SM).

Atriplex canescens (Pursh) Nutt.
Four-wing Saltbush
Chenopodiaceae
sha'ashkaḍk iibadkam, koksvul sha'i

There are two large saltbushes in Gila Pima country—both grayish and very shrubby, both found in moist saline areas; but here their similarity ends. Quail Brush or Lens-scale (*Atriplex lentiformis*) has broad leaves and nondescript flowers and fruits and usually grows to well over head-height. The less common Four-wing Saltbush is a woody shrub of hip- to about head-height. Its very pale, almost silvery, leaves are quite linear. But what attracts one's attention are the dried fruits, which are as big as your fingernail. Each bears four large wings or bracts with somewhat toothed margins. In the fall these turn from green to a pale greenish yellow and sometimes to a reddish color. They may almost cover the female plant so that it looks like it is in bloom. In the late fall, when the mesquites have lost their leaves, Four-wing Saltbushes and Rayless Goldenrods (*Isocoma acradenia*) provide splashes of color to the otherwise dreary bosque edges.

There are two names among the Gila Pima for this large saltbush. In the west end it is called *sha'ashkaḍk iibadkam* 'rough fruit it has'. Curtin (1949:67) recorded this name also from the west end but was unable to obtain a specimen for species identification. At the east end, Irene Hendricks of Casa Blanca called the plant *koksvul sha'i*. When I asked what *koksvul* meant, she said, "The ones that the *papko'ola* [Pascola dancers] use on their legs." These are large cocoons, so the name here means 'cocoon brush'.

Sylvester Matthias told me that *sha'ashkaḍk iibadkam* apparently has no use, not even for kindling, but some of the Pima consultants with whom I worked said that if *eḍam* (Quail Brush) was not available for washing baskets, Four-wing Saltbush could be used. Curtin and Russell recorded this use also. The leaves have large amounts of saponin (Hall and Clements 1923).

sha'ashkaḍk iibadkam, koksvul sha'i

Atriplex canescens

There is another species of saltbush, *A. linearis* (Narrow-leaved Saltbush), that some botanists lump into *A. canescens* (Four-wing Saltbush) as a variety. This bush is smaller, usually less than a meter tall, has linear leaves, and is often found growing abundantly with Desert Saltbush (*A. polycarpa*) on the saline flats. It ranges narrowly on the Gila drainage from just east of the reservation to the Lower Colorado and in the Salton Sea basin or Colorado Desert—everywhere sympatric with Four-wing Saltbush. Its less conspicuous fruits have four "wings," but the bracts are smaller and differently shaped. I was told by every Pima familiar with Four-wing Saltbush that Narrow-leaved Saltbush was not included in the folk taxon *sha'ashkaḍk iibadkam*. But no one could give me a name for this smaller shrub.

Technical Notes: *A. canescens,* the species that bears a Pima name, is taller than *A. linearis.* The two may be found growing on the same flats, even with branches intertwined. *A. canescens* has yellowish green leaves that are completely flat or only slightly curved toward the margins, about 4–6 mm broad. *A. linearis* has bluish gray leaves, very scruffy above and below, narrow (2–3 mm) from base to tip, that are strongly plicate, the margins often folded until they almost meet. In growth form and leaf color, *A. linearis* resembles *A. polycarpa*, with which it grows. The four "wings" on *A. canescens* are greatly extended, papery, and with venations, but those of *A. linearis* are fleshy fingers that more or less coalesce to produce the four short "wings" on a relatively larger fruit. Stem galls resembling a small green apple with reddish marks are characteristic of *A. linearis*, while *A. polycarpa* frequently has cottony stem galls. I have found neither of these on *A. canescens.*

Vouchers: *413, 458, 1625* (SM); *1346* (IH).

Atriplex lentiformis (Torr.) S. Watson
Quail Brush, Lens-scale
Chenopodiaceae
eḍam

The largest saltbush of the Gila is Quail Brush, which is dense, intricately branched, and usually well over head-height. The bush forms a silvery amorphous mass reaching to the saline ground. The leaves are broad, larger than those of any of the other shrubby saltbushes of this region. The female heads bear dense clusters of tan fruits that fall readily. These are small, compressed, and enclosed in two bracts. Their appearance is mealy.

George Kyyitan and I were crossing a salt flat west of Sacaton that had once been a mesquite bosque but was now marked only by stumps. There were lots of canary grass (*Phalaris* spp.), a newly arrived mustard, Seepweed (*Suaeda moquinii),* and mostly the small Desert Saltbush (*Atriplex polycarpa*). George walked over to a thick, almost white bush. "This is *eḍam*—soapy things in those. They get [it] to wash their baskets—to make them very white. They do this before they take them in to market. They say they use it, too, for their clothes."

At the west end, Quail Brush is especially common along ditch banks and field edges. Here Sylvester Matthias told me, "*Eḍam* leaves, especially when tender, [are] used to wash up a freshly made basket, to polish a basket—even a used one—use cloth, especially a piece of old Levi's; will polish them up. Stalks used for summer fuel—when they don't want long coals—just quick heating of a meal—and don't want to use heavy woods."

Joe Giff also told me about it. "Used to wash our baskets with *eḍam;* a gray bush; they have seeds there, too, they once ate; pointed leaves, not long. *Big* bush, bigger than a *kuávul* [wolfberry, *Lycium* spp.]."

I asked Ruth Giff if she had ever used *eḍam* for her baskets. "Sure. It gets dirty because it takes a long time to finish. Just put basket in water and rub the leaves on it. Then put it out in the sun to dry."

"Is it a bleach as well, or just a soap?"

"A soap. That's all we use it for."

All this is confirmed by earlier accounts. In the 1940s Curtin (1949:66–67) wrote, "The leaves are rubbed in water to produce a lather with which clothing and baskets are washed, although it is too strong for the hands." According to an old man from Sacate, "in the past, the tiny seeds were roasted and eaten in times of famine."

These two suggestions of food use are verified by turn-of-the-century accounts giving preparation methods now forgotten. Hrdlička (1908:263) wrote, "*U-u-tam* (Atriplex lentiformis) is a bush growing near the Gila. The seeds are gathered and pounded up in a wooden mortar, the bran being blown away. The mass is then placed on the inside bark of the cottonwood, laid in a heated hole in the ground and covered with more cottonwood bark, all being overlaid by grass or brush. It is allowed to remain thus for two days, when the meal is taken out, mixed with water, and eaten as mush with the occasional addition of salt."

The account by Russell (1908:78) is similar: "*Urtam*, Atriplex lentiformis. The seed of this saltbush is cooked in pits which are lined with Suaeda arborescens [Seepweed], and the papery inner bark of the cottonwood moistened and mixed together. The roasting requires but one night, then the seeds are taken out, dried, parched, and laid away for future use. When eaten, it is placed in a cup and water added until a thick gruel is produced."

Russell (1908:80) mentioned an additional medicinal use: "The root is powdered and applied to sores." Neither Curtin nor I were able to verify this among subsequent generations.

Vouchers: *249* (SM).

Atriplex polycarpa (Torr.) S. Watson
Desert Saltbush, All-scale, Cattle Spinach
Chenopodiaceae
kokomagĭ sha'i, kokomagĭ shegoi, sheshgoi, toota sha'i

Some plants are so ubiquitous across the Gila floodplains that they dissolve into oblivion: to the casual observer they hardly seem to be there. Desert Saltbush is one of these. With the dropping of water tables and the century of continuous clearing of mesquite trees for firewood, Desert Saltbush has become one of the most common shrubs across the flats. These spaced, rather rounded shrubs, usually about knee- to hip-height, are sometimes called sagebrush by those careless with their vernacular.

The scruffy leaves have the same silvery suggestion of green that is seen on other woody saltbushes, but they seem to squat right down on the stems. Some are larger, three times longer than wide, while others are mere scales forming miniature "roses"

on slender twigs. The intricate branches are grayish with a weathered look. To appreciate the plant, you must get down on your hands and knees and search for the twisted, gnarled, short trunk that supports the bush. The beauty of Desert Saltbush lies in its silent, unobtrusive, gnarled witness to the dry, cold winters and the torrid summers.

The Gila Pima usually call this saltbush *kokomagĭ sha'i* 'gray brush', although some speakers say *s-kokomagĭ sha'i*. There appears to be no current cultural use for this saltbush specifically, although Ruth Giff said she used it for kindling. Sylvester said, "Yeah, for kindling. You can still get those big . . . , the bottom part. That's for fuel, too. When you just want to heat something, and you don't want to, especially in summer, you can just burn that *kokomagĭ sha'i* wood. And heat something, and there's no coals left to heat the house."

Everyone said you can identify it by the cottony "fruits," which are actually white galls caused by the Cecidomyiid fly. Sylvester said there is a name in Pima for these cottony balls, but he could not remember it. (The actual fruits are minute, maturing in stiff clusters on separate plants from the pollen-bearing ones.)

One day Francis Vavages pointed out a *kokomagĭ sha'i*, pulled off some of the galls from the top of the bush, and explained, "These are used to cure pain, relieve pain. They light it, moisten the skin," he demonstrated with saliva, "so it will stay there, then burn it on the skin."

Ruth Giff called Desert Saltbush *sheshgoi*, which is the plural form of *shegoi*, the Pima name for Creosote Bush (*Larrea divaricata*). I have heard this used several times by others and have been told that in the singular it always means Creosote Bush. This appears to be a unique case where the plural indicates a different folk taxon from the singular.

Others call Desert Saltbush *kokomagĭ shegoi* 'gray Creosote Bush'. Sylvester once said that *kokomagĭ sha'i, kokomagĭ shegoi,* and *toota sha'i* 'white bush' are all different names for the same shrub. Of course, it is entirely possible that this is a collapsing taxonomy and only two of these are really folk synonyms, and the third, perhaps *toota sha'i*, marks one of the other grayish bushes of the *Ambrosia-Atriplex* complex. (Subsequently Sylvester assigned the name *kokomagĭ shegoi* to *Ambrosia deltoidea*, Triangle-leaved Bursage.) Although

Narrow-leaved Saltbush (*Atriplex linearis*) has a growth form similar to that of Desert Saltbush and both grow together in yards and deserted fields at the west end, everyone from Santa Cruz and Komatke told me that this species is not included in *kokomagĭ sha'i*. White Bursage, *Ambrosia dumosa,* is also relatively common in Pima country. Perhaps one of these two bushes was originally named *toota sha'i*. (Curtin erred in ascribing the name *eḍam* to Desert Saltbush as well as Quail Brush [*Atriplex lentiformis*], and I suspect that the uses she ascribed to Desert Saltbush probably apply only to Quail Brush.)

This seemingly useless shrub once served an important purpose in the days of river irrigation. Here is the story Sylvester tells of his remembrance of building the brush weirs: "The *sha'i kuupa* ['brush closure (or) dam'], *kui kuupa* ['mesquite closure'], was a lot of work because they have to force it down [into the riverbed]." Sylvester demonstrated the actions more than he described them. "Takes muscles to build it. To the width of the river. Go out on the desert, cut mesquite limbs, trim it up, get the fine parts. It takes [a] lot of work. It takes—oh! I don't know how many wagons. Because they have to have those mesquite limbs and the *kokomagĭ sha'i*, sagebrush [*sic*], haul it down to the bank of the river, and then move it down to where they're going to dam. They force all those mesquite poles—posts, I'd call it." He indicated a distance of about four feet apart for the posts, then he gestured to mimic the weaving back and forth of the smaller mesquite branches between these uprights. The brush weirs were fairly permanent installations, lasting several seasons. (For another version of how weirs were built, see *Pluchea sericea* account.)

"When did you build the dams [weirs]?"

"Anytime when the river gets high and wash[es] the dam away. Can still get into the water in January. It's a cold month, but the water is warm. We had to bring that wheat straw, that, that's been threshed. That's everybody saved that straw— *moog* we call it. It's the fine stuff [chaff] after threshing. It takes a lot. Then when they build that mesquite deal, then the *kokomagĭ sha'i* goes next to it and they force it against, push it, push it, and stick [it] into the mesquite brush that's been already put up. And then comes the straw. Boy, that takes lots of work.

"Then pile sand against the *moog*. Then maybe more *sha'i*. Then pile more sand, up to the height of the water. That *moog* can really hold the sand. Sometimes when the flood washed away [the dam], they have to build a new ditch [extension] to where they put that [new] dam. Sometime we're lucky and we don't have any flood, and the water is running. It'll hold till the next flood comes in. Unless, if it's a *big* flood, then it widen the river. Then it's more work. If there's no flood, the river is narrow."

"About what year did they stop using the *sha'i kuupa* and go to the cement?"

"I wasn't around then. I was in White River 1934 to 1935. And I heard they were working on a [cement] dam. [Before], everybody uses that brush dam. When they have floodwater, they got their dams there. And can build their dam during dry season. They depend on the rain just like the Papagos—they're waiting for the floodwater."

As might be expected from a people whose livelihood depended on a desert stream, the Akimel O'odham distinguished between the normal biannual rise or flood, *ge'edag shuudagĭ* 'big water' and the occasional destructive flood, *cheshadag akimel* 'to reach the top river' or 'overflowed river'.

Vouchers: *102, 122, 261* (SM); *388, 537* (RG).

kokomagĭ sha'i, kokomagĭ shegoi, sheshgoi, toota sha'i

Atriplex polycarpa

Baccharis salicifolia (Ruiz & Pavón) Persoon
[= *B. glutinosa* Persoon]
Seep-willow, Batamote
Asteraceae (Compositae)
oágam

Seep-willow may fool you. Watch out. The straight, slender stalks with their willowlike leaves growing in sandy streambeds do, indeed, look like willow saplings. But the resinous smell of the crushed leaves distinctively suggests the Compositae or sunflower family, not the aspirinlike odor of true willows (*Salix* spp.). And the slender, tan Seep-willow canes are only shoulder- to a bit over head-height, never growing into trees. But if you are still confused, look for the flower heads. These are tightly bound purple discs without rays. The cream-colored pappus, the cluster of fine bristles, is more conspicuous. If the seeds have already blown away, there should be a few dried receptacles at the tips of some branches for certain identification.

Seep-willow is one of the pioneer colonizers of riverbeds after floods. It thrives, sometimes forming thickets, where erosion has washed away most of the accumulated alluvium.

The pan-Piman name for Seep-willow is *oágam*. Joseph Giff explained the derivation of its name: "*Oágam* means something inside, as [in] *oág* 'brain', or what's inside the skull, but can also be 'marrow', the soft stuff that's inside a bone; it refers to the pithy stems of this plant." Sylvester Matthias likewise said, "*Oágam*, from *oág*, means 'brains' or 'marrow'. Makes good arrows, but they are slow and light—they float, unlike Arrow-weed *[Pluchea sericea]*; they're just for children's play. Used to make darts, *dodak ab ha to'a*. Arrow-weeds are solid, they don't have the soft part; that's why they call this *oágam*. [It's] not medicinal."

Ruth Giff told me, "*Oágam* [is] also used for sides of houses—*kosin* or *kii*—like Arrow-weed, and plastered with mud; may be used for *vatto* ['arbor'], too. They're gone now, just like the Screwbean *[Prosopis pubescens]*. But they're not.... I don't think they last as long as the *u'us kokomagĭ* [Arrow-weed]. I think they're kind of soft in the stems. They're sticky." The *vatto* is a wall-less shade structure often found in Upper Piman yards. Traditionally it is the center of most daily activities.

Likewise, Sylvester said, "Where they [Pima] couldn't find any Arrow-weeds they used this on their *vatto*." Another time he told me, "*Oágam* used to be common. Used like Arrow-weed for thatching the *vatto*, [but] not the house. On house and siding, just use *u'us kokomagĭ* because they are stouter." At Piliñ Keek 'Bridle Standing' near Sacate, his mother's home, *oágam* was used for the *vatto*.

George Kyyitan walked over the deep fissures fracturing the dried mud on the bed of the Gila River to the willowlike wands covered at their tips with small greenish cream-colored flower buds. "This is *oágam*," he pronounced. "They use it for shade, too. They use it for housing, around [outside], put it around it. But this is not so good . . . [as] Arrow-weeds. Arrow-weeds is better than this one. This one get rot easy. See," he broke and peeled back a piece of stem, "lots inside, the 'brains', or whatever they call it. And too much in there and thin outside. That's why it gets rot so easy. And Arrow-weeds is thicker and got little bitty 'brain' inside—whatever they call it. *Oágam*. That's what they used a long time before [when] there wasn't much Arrow-weeds."

Curtin does not mention this plant. But Russell (1908:156) recorded that in the construction of the rectangular storehouse (*kosin*), "the large bush, Baccharis glutinosa, is often used." He noted (1908:127) an additional use by potters: "The final step, if the vessel is to be decorated, is to apply the black mesquite pigment with a sharpened stick, made from Baccharis glutenosa [*sic*], which has a large pithy center. The vessel is again subjected to heat for a few minutes until the decoration has assumed a deep black color, when it is finished." During my work, no one mentioned this use—but then there are no potters left among Gileños. Even early in the century, the art was in ebb.

"*Oágam* is the Pima name," explained Sylvester. "But the Papago call it *ñehol*. There's a village there, somewhere near Pisin Mo'o, they call Ñehol Keek, that's the village; but the Pimas call it Oágam Chuuchk, 'the *oágam* there, growing there'."

Vouchers: *70, 557, 885* (SM).

Baccharis sarothroides A. Gray
Desert-broom
Asteraceae (Compositae)
shuushk vakchk

Desert-broom is not likely to be confused with anything else in the desert. Found in washes or riverbeds, it is dense and bright green, hip-height to well over head-height, and completely spineless. The stems are strongly angled rather than round. The bush seems to outdo itself when blooming. The male and female flowers are on different plants. The pollen-bearing flower heads resemble little cones of yellowish flowers; the seed-bearing flowers become so thick with white pappus or "down" that the tops of the branches droop.

The Pima name is *shuushk vakchk* 'wet sandals/shoes'. The reasons for this curious name are lost in antiquity. The Tohono O'odham use the same name, most saying *wakchk*. Far to the south, at the Tohono O'odham oasis village of Quitovac, Sonora, Luciano Noriega gave us *shuushk uwakita*, a variant I have not heard elsewhere. Nabhan (in Felger and associates 1992) recorded it as *shushk kuagĭ* and *shushk kuagsig* from Philip Salcido and Delores Lewis, both Tohono O'odham.

The most obvious derivation of a word is not necessarily the correct one; occasionally it is pure coincidence. Superficially it might appear that the Pima word *shuushk* is derived from the English word *shoes*. But this is not the case. The *Vocabulario en la lengua Névome* (Pennington 1979:121), compiled during the 1660s, gives the Pima Bajo word for *zapatos* as *susca*, certainly cognate with Upper Pima *shuushk*. (The *s* to *sh* transformation is just what is expected today between these two Piman language groups.) And a dictionary defines *cacle* as a kind of sandals worn by friars, Indians, and soldiers. Likewise, the Attí northern Sonora vocabulary compiled before 1774 gives *zapato* or *cacles* as *susc, susca*. In a classic case of marking reversal (see Witkowski and Brown 1983), the sandal decreased in importance and became known historically as *kaikia shuushk* 'stringed shoe'. The Mountain Pima along the Sonora-Chihuahua border in the Sierra Madre still wear their sandals, which they call *shuushk*. Throughout all of Piman country, except for the northern

tip transferred to the United States in the Gadsden Purchase, the name for the European shoe is *zapato*. *Shuushk* thus is derived from neither English nor Spanish but is an indigenous Piman word.

Sylvester Matthias said, "Cut branch of *shuushk vakchk* and make a yard rake or broom because they're strong. Papago use them [also] when they go out to camp— when they make their *uksha* 'outside kitchen' or windbreak, the way Pimas use Arrow-weeds. When *shuushk vakchk* is dry, can be used [by Pima] for quick fuel."

While identifying Desert-broom in the field, George Kyyitan said, "Some of them [Pima] use *shuushk vakchk* for [thatching] shed, *vatto*, but not many of them."

At Blackwater one day, Francis Vavages broke off a piece of the bush and identified it in Pima, saying, "This grows at the edge of fields. They use it for cleaning seeds, after thrashing. For sweeping, cleaning their house, yard, ground, because they don't raise too much dust."

Saxton, Saxton, and Enos (1983:79) incorrectly called Desert-broom *aan*. This is instead Desert-willow (*Chilopsis linearis*), as they had correctly glossed it earlier (Saxton and Saxton 1969:170, 171). Not to be outdone, Mathiot (n.d.:359) identified *shuushk-vakchĭk* correctly as 'broomweed (a bush)', but in the same volume (Mathiot n.d.:215) she incorrectly gave *vaas* as 'broomweed', that is, *Baccharis sarothroides*, and *vaasig* as 'to be full of broomweed in one location'. *Vaas* is one of the widespread, unanalyzable, monolexemic Piman generics; it refers to the botanical genus *Jatropha*, Limber Bush. No species of *Jatropha* reaches the Gila, and the word *vaas*, as far as I can tell, is now forgotten there (see *J. cardiophylla* account).

Vouchers: *548* (?); *550* (RG); *1729* (FV); LSMC *37*.

Conyza canadensis (L.) Cronquist
[= *Erigeron canadensis* L.]
Horseweed
Asteraceae (Compositae)
vopōgsha, vopōgshakam

Although the little fleabanes or "wild-asters" often have attractive flower heads, their very close relative, Horseweed, seems to have been left out. This rank plant, said to be a native of eastern North America, is common in disturbed soils, particularly in gardens and fields. Horseweed grows straight as an arrow; the stalk is densely circled with soft, green, lance-shaped leaves. In wet soils, such as the bottoms of resting canals, these annuals may grow as tall as a person, forming thickets. Only when blooming does the tip of this leafy arrow branch out, bearing multitudes of insignificant-looking flower heads. Almost no one will give them a second look until the inflorescences are covered with small cottony or silky tufts that bear the seeds away in the wind. These disturbance plants such as Horseweed are masters at seed production and dispersal.

I was talking to Ruth Giff by phone one day when she asked me if I had ever found *vopōgsha*. She had just discovered one in the ditch above her garden. It was new to me. "*Vopōgsha* means 'stepchildren', or maybe it means 'quiver' also. It's just one stem with leaves, all around it. Stinks awful! Don't know what the flowers look like. Grows any time of year." I suspected it was Horseweed and asked her to watch the plant and press some for me when it bloomed. Later in the summer there was flooding from the rains and her canals raged, but the *vopōgsha* survived, and she made a specimen. The next year they were everywhere in her ditches.

Sylvester Matthias explained the curious name. "*Vogsha* is singular, *vopōgsha* plural. Means 'quiver' or 'stepchild'. It's the same word. This plant must have a lot of stepchildren." I pressed him for the association. "Well, someone's stepchild is an obligation, like a quiver that is strapped to your back. You can't get away from it. You just have to carry it along."

George Kyyitan of Bapchule agreed. "*Vopōgshakam* has no uses. Feed their horses with it—cattle. That's all it's good for. *Vogsha* means 'quiver' or 'stepchild'." He did not know which meaning applied to the plant.

Whether the Gila Pima name is derived from the plant's proliferating capabilities or is a reference to the straight, arrowlike stems (as found together in a quiver) is not known. One time, several years after our initial discussion, Sylvester also called the plant *vopōgshakam* 'stepchildren it has' (literally, plural marker + 'stepchild' + attributive marker), so the first meaning would seem to fit his concept. (Thus, there

are three meanings to this word.) In a later conversation George thought just *vopōgsha* was the better name for this plant.

Vouchers: *130, 867, 1022* (RG); *521* (SM); *1749* (GK).

vopōgsha, vopōgshakam

Conyza canadensis

Dyssodia porophylloides A. Gray
San Felipe Dyssodia
[?] *Porophyllum gracile* Benth.
Yerba-del-venado
Asteraceae (Compositae)
do'i-uv u'us

I was long puzzled by a plant Sylvester Matthias described to me. "*Do'i-uv u'us* [is] like Mormon Tea *[Ephedra]*, only gray; a little bush in the foothills. [The] word *do'i-uv* means something that smells 'raw', like [the] odor of blood." I asked Felix Enos about this a few days later. He knew of such a plant but did not know what it looked like. Sylvester said that once he was traveling with his father on Tamtol Do'ag 'White-crowned Sparrow Mountain', in the Sierra Estrella, when he stepped on the plant by accident and was so overwhelmed by the odor that he threw up.

I brought him some Yerba-del-venado (*Porophyllum gracile*), which is pungently aromatic and grows at the bases of the sierras, but he did not think this was the same plant. A few years later he again described the plant, saying, "Greenish plant; very tender; smells worse than blood; when I stepped on it, everything came out [he vomited]. It's not used for medicine." Later he added, "I don't think it's used for anything . . . give you a headache."

I thought perhaps he could have been referring to Turpentine Bush (*Haplopappus laricifolius* A. Gray), a distinctively resinous plant, somewhat stiff; but it is far too leafy to be compared to Mormon Tea. And it is found only on the highest peaks of the Estrellas, in relictual Upper Sonoran communities.

In mid April 1987, Takashi and I were working in the south end of the Sierra Estrella, he making sketches of plants he was going to paint, I taking notes on plants I was writing about. We were on the arroyo that plummets from the Upper Basin, a drop of nearly 1,650 feet in the space of less than a mile. I was content to work the lower reaches, but Takashi wanted to climb to the top. The upper part of the arroyo becomes so steep, the room-sized boulders so difficult to bypass, that the only practical manner of ascent is to scale the loose gravelly slopes south of the wash. I stopped occasionally on these steep slopes to press a plant in a small plant press that fits in my knapsack: *Hibiscus denudatus,* the delicate poinsettia-like *Euphorbia eriantha,* and others.

Suddenly, on the way up the dry slope, I was aware of an overpowering odor rising to my nostrils in the hot dry air. I looked down at my feet and the zigzag pattern I had taken but saw nothing unfamiliar. Sylvester's mystery plant came to mind. I continued up. Soon the odor came again. This time, on retracing my steps downslope, I saw that I had stepped on a rather tender, dense, nearly leafless knee-high bush that was just starting to show a few opening flower heads. It was a composite I had never seen before, but one I could not forget. The flower heads were tightly encased in green bracts or phyllaries, which were conspicuously dotted with scent glands. As I picked more of the bright green plant for pressing, I seemed engulfed by the odor. *Dyssodia porophylloides:* aptly named, but even more pungent than its namesake, *Porophyllum.* I stepped back and compared the recently injured plant with some nearby *Ephedra.* Indeed, Sylvester had made an apt comparison. And somehow the plant had escaped my notice until I accidently stepped on it that hot afternoon.

Tohono O'odham Frank Jim knew *do'i-uv u'us* also. "If you walk through it, the smell is going to stick to your shoes. If [a] skunk pees [sprays], get a whole bunch and burn it, stand in the smoke; or [use] that greasewood [*sic,* Creosote Bush, *Larrea divaricata*], *shegi.* We never tried it. We never bothered a skunk." He made these comments about fresh *Porophyllum gracile,* from the Sacaton Mountains; we had no *Dyssodia porophylloides* for comparison, so a contrast set was lacking. Sylvester specifically excluded specimens of *Porophyllum* as belonging to this folk taxon, but his comments about the plant being grayish and tender apply better to *Porophyllum* than to *Dyssodia.*

The origins of this interesting word should not be left dangling. Sylvester had told me on several occasions that *do'i* meant 'raw', and Culver Cassa had analyzed this as *do'i-uv* 'raw smelling'. Sylvester associated the raw smell with blood and nausea. *Do'i* may also mean 'raw, unripe, unbaked', as in *do'i huuñ* 'raw [or] unripe corn' or *do'i ha'a* 'unbaked clay olla'. The word *do'iv* is defined as 'to smell raw, wet and unpleasant (such as fish, the ocean, the wet fur of a dog, geranium leaves, etc.' (Mathiot 1973:266): for example, *s-do'iv o g kaachkĭ* 'the ocean smells "raw"'. While there is no good equivalent to this Piman word in English, the smell of crushed Yerba-del-venado or San Felipe Dyssodia will give you a good idea of the concept.

Vouchers: *1700* (SM); *1911* (Frank Jim, Tohono O'odham, *P. gracile*).

Encelia farinosa A. Gray
White Brittlebush
Asteraceae (Compositae)
tohavs

Across the bajadas and especially on the steep rocky slopes of the mountains throughout Pima country is a very common shrub with leaves so pale they look white, except when actively growing. The bush is usually a bit over knee-height. In spring these rounded bushes may be so covered with masses of deep yellow flower heads on slender stalks that from a distance entire mountain slopes appear yellow. The pleasant-smelling, scarcely woody shrubs have neither thorns nor spines nor sticking seeds to annoy the visitor. When the summer sun heats the black desert-varnished rocks to temperatures that send any human searching for shade, White Brittlebush may respond by losing most of its white leaves.

To the Pima (as well as the Tohono O'odham) this plant is *tohavs;* the word probably derived from *toha* 'to become white'. It finds a place rather early in the Creation Story, where Jeweḍ Maakai or Earth Doctor is providing for two additional helpers. The first of these is Ñui, Buzzard or Turkey Vulture, "made out of the shadow of his eyes." The second is Ban, Coyote. Thin Leather gave Lloyd (1911:31) the following account of Coyote's origin:

Now the sun was male and the moon was female and they met once a month. And the moon became a mother and went to a mountain called Tahs-my-et-tahn Toe-ahk [Tash Maihitañ Do'ag] (Sun Striking Mountain) and there was born her baby. But she had duties to attend to, to turn around and give light, so she made a place for the child by trampling down the weedy bushes and there left it. And the child, having no milk, was nourished on the earth.

And this child was the coyote, and as he grew he went out to walk and in his walk came to the house of Juhwertamahkai and Nooee, where they lived.

And when he came there Juhwertamahkai knew him and called him Tow-havs, because he was laid on the weedy bushes of that name.

Thin Leather's narratives to Lloyd (1911:31) and Russell (1908:208–209) continue with the origins of Se'ehe, also called Itany (Iitañ; see *Allenrolfea occidentalis* account). The traditional Piman complement of four original beings was thus completed: Earth Doctor, Buzzard, Coyote (often called Tohavs), and Se'ehe or Elder Brother. And so throughout Piman legend, the word *tohavs* may mean the plant or may be the alternate name for Coyote.

---

tohavs *Encelia farinosa*

This little bajada bush plays a more mundane role among the Gileños and other northern Pimans. Where the stems are damaged by insects or browsing, for instance, they exude clear drops of amber gum. These pellets may be the size of a hat-pin head or smaller, and many must be gathered to make a bit of chewing gum. The odor is distinctive, like the smell of Wild Sunflowers (*Helianthus annuus* var. *lenticularis*). The gum can be chewed indefinitely, and the flavor is delightful, making the search worthwhile.

For some, chewing gum became a more elaborate production. Ruth Giff told me, "When we get home [from school], we'd mix *tohavs* with lard, you know, not the *koji manjugĭ* [pork fat]; we never used *koji manjugĭ* way back, just the other kind [beef tallow]; we'd mix it in there so it wouldn't crumble. Same way with the *auppa* [Frémont's Cottonwood, *Populus fremontii*] berries; see, I don't know where they get the idea to mix that."

Vouchers: *1900* (GK); LSMC *70*.

Ephedra aspera Engelm.
[= *E. nevadensis* S. Watson var.
  aspera (Engelm.) Benson]
[≠ *E. antisyphilitica* Berland. ex
  C. A. Mey.]
[≠ *E. fasciculata* A. Nels.]
Mormon Tea
Ephedraceae
kuupag

In the Santans, the Sacaton Mountains, and the Sierra Estrella, you may come across a bush between knee- and hip-height that looks like a dense mass of dry, dark green spaghetti stretching skyward. There are no leaves, just an occasional minute scale at the joints. But if you climb these desert ranges in the spring, the dense masses of bare stems may be in bloom. The male plants have green flowerlike cones bearing abundant pollen sacs; the female plants have plumper cones with a single seed. These curious plants belong to neither the group called the flowering plants nor the group called the conifers; they combine some elements of each and have some characters unique to their own relictual class and order. There is only a single genus, *Ephedra*, found throughout many arid regions of the world, but there are many species.

Some older Pima are familiar with the name *kuupag* for Mormon Tea, and a few still use it. David Brown told me, "*Kuupag*, Mormon Tea, is also good for diabetes [as well as Crucifixion-thorn, *Castela emoryi*]. Also, alfalfa roots were boiled for diabetes, they say."

Sylvester Matthias asked me to bring some branches for his wife, a diabetic. He added, "*Kuupag* is for diabetes or for VD."

"Did people ever drink it just like tea or coffee?"

"Yeah, some people *do* use it just as a beverage, when they have nothing to drink. I remember my father saying they had been using it as tea. They [had] been trying to find a substitute coffee, but didn't find nothing. They try willow leaves. It does turn into like coffee, but it's the *taste!* They couldn't find no other that could be a substitute coffee."

Years later I brought him again a specimen of *Ephedra*. He said, "That's for diabetes. . . . *Kuupag!* Clean your system out, have a good urine. Some use it for VD."

Many older Pimas do not know the plant as *kuupag* but instead are familiar with its English name. I asked Ruth Giff about Mormon Tea. "I've *seen* them growing up in the mountains," she replied.

"Did you ever drink it as a tea?"

"Yeah. Just a drink. Not as a medicine. Is it a medicine too?"

I showed the plant to Carmelita Raphael of Santa Cruz village. "*Kuubak*," she said. "We used to get it and make tea out of it." I do not know whether her version of the plant's name was a dialectical variation, idiosyncratic, or perhaps a mispronunciation from not having heard the word for a long time. She is the only one I heard say the name this way.

Deanna Francis and Diane Enos recorded its use at Salt River to raise blood pressure. I found no one on the Gila who knew about this use.

Preparation for either the beverage or medicine is simple. A large handful of the green twigs is stuffed into a teapot and boiled for a while. The first pot may be weak, but the wet plants can sit for several days and more water can be added for subsequent boilings. The second and third boilings produce the best tea, and a fourth is still good. The yellow tea is good with a spoonful of honey and serves as a quick pick-me-up. The stimulant ephedrine is one of the plant's compounds.

Curtin (1949:76) 40 years ago recorded an additional use from the Maricopa Indian village of Lehi: "The roots are dried in the sun, powdered on a flat stone, and sprinkled on all kinds of sores, including those caused by 'bad disease.' George Webb [a Pima] stated that the powdered roots are applied in this way for syphilis." Sylvester heard only of the branches being used as a tea for this purpose.

Later at the Desert Botanical Garden in Phoenix, George Kyyitan and Sylvester Matthias toured the grounds with Gary Nabhan and me. George did not know the name *kuupag* when Sylvester pointed out the plant to us. Sylvester said, "Some people use it for domestic tea, but this is for diabetes *and* for the VD. Good medicine for the VD. It's mainly used for the syphilis—it can knock syphilis out! But people do use it for domestic tea."

Russell (1908:80) did not give a Pima name for the plant but recorded the Pima

kuupag   *Ephedra aspera*

use of *Ephedra* both for a beverage and as a remedy for syphilis at the turn of the century. In fact, one of the species was known botanically as *E. antisyphilitica*.

Curtin recorded two additional names for *Ephedra*. *U'us ti* 'sticks tea' is someone's descriptive name that was used in place of *kuupag*. I heard three people from the Bapchule–Casa Blanca area use the improvised name *u'us ti*. But George Kyyitan said *u'us ti* is not the *Ephedra* I showed him; it is a taller plant whose leaves make a red tea. He was unfamiliar with *kuupag*. Curtin's other name, *kuuvid nonovĭ* 'Pronghorn's forelegs', was believed by Gila people with whom I worked to be some slender cholla (*Opuntia* sp.) rather than *Ephedra*.

Vouchers: *748* (CR).

Erythrina flabelliformis Kearney
Coral-bean, Chilicote
Fabaceae (Leguminosae)
bawui

If you travel in the low desert mountains of far southern Arizona or Sonora from late May to early July, you might see a rather nondescript, thick-branched bush sporting brilliant red tubular flowers in clusters. A closer look shows that these are modified bean flowers, restructured so as to attract hummingbirds, their chief pollinators. Come back to these hills in fall and you will see the long woody pods filled with bright orange-red plump beans or seeds. (An occasional bush has dull yellowish beans.) These remain in the pods long after they have opened. In the rainy season the tree will wear leaves, each composed of several large, bright green leaflets shaped like aspen leaves. To the English-speaking, this is Coral-bean, in Mexico called Chilicote. As you go south along the foothills of the Sierra Madre, the bushes become taller so that in Pima Bajo country, Coral-bean is a good-sized tree, here and there thrusting its improbably red flowers up into the canopy of the leafless thornscrub forest. The wood is as light as balsa, convenient for making camp stools. The rock-hard seeds are toxic, causing paralysis of the throat and ultimately death.

One would not expect a plant that does not even grow in Gila Pima country to have much cultural significance for the Pimas. But it does. Coral-bean is *bawui* among all the Tepiman groups I have encountered, and so it is among the Gila Pima. These Gila people are derived from the south, from *bawui* country, and probably not that many centuries ago. And *bawui* maintains its earlier significance among them.

Joseph Giff explained about *bawui* in Tohono O'odham country: "A man had it blooming in his yard. That village, down by the Vav Giwulk [Baboquivari Mountains]—you know, called Topawa—well, it gets its name from that bush. But someone didn't know how to write it. It's supposed to be called *ḍo bawui,* this means 'this is' and 'Coral-bean'; that's what the man answered when someone asked what was growing in his yard. So they called the place Topawa, but it means 'this is *bawui, ḍo bawuĭ*. That's the real meaning of that term."

The last time I visited Joseph Giff, late in December 1981, he told me again about Coral-bean. Many Pima songs were in series and were supposed to be sung together. The songs Joe sang were the Blue Swallow Songs (Gidval Ñe'ñei, the swallow or Purple Martin Songs). Joe explained that there was a *bawui* song, "Bawui ch-eḍ Do'ag," sung by Paul Manuel, who died in July 1983. The place mentioned in the song, Bawuigam Do'ag 'Coral-bean Mountain', was somewhere south of Pima country. Joe sang the song for us, then translated it:

A blue bird was traveling.
He saw ahead of him the beautiful
Bawuigam Do'ag covered with
   blossoms;
He remained there and sang.

Unfortunately, I did not tape this song, and Joe died a month later.

But when I went to Sacaton to write on Pima plants, Fr. Richard Purcell remembered he had a tape of Joe and Sylvester, recorded about 1976 or 1977. On it Joe, a vigorous singer with a clear, powerful voice, sang the "Bawui Ñe'i":

ga maiñ hi man da
ge ve se bawui ga ma hi tovan
ganu va ge e he ka
tamano na hi wa

i ñuñ ve ka chi to van
ge ve si yo si ka ha
enamo na hi wa
avañ si ne e da

On the tape Joe provided this expanded explanation of the Coral-bean song: "This one is about the Blue Swallow [Purple Martin], that he was going along, traveling along—in the air, of course—and he saw this mountain, way ahead of him. And on this mountain was some plants—some bushes—what we call *bawui*—I don't know what they're good for, but they say pretty flowers, and they've red berries [*sic,* seeds] on them. And this mountain was covered with those bushes. And so the bird alighted on there and he liked what he saw around there. And there around him the mountain was all in blossom. And he sat there, lighted there and sang. And when he said '*si ñe'e,*' that means that he is singing joyously; he was interested, he was happy—singing happily there on the mountain.

"I used to go up there in the spring, late spring. And you see flowers in the hidden places—under the boulders, and in the washes, in places anywhere that nobody sees. Anyway, why those flowers are there nobody knows but him who put them there. And why we're here in this flat dry desert, we don't know. But him who puts us here knows." And then Joe sang.

Sylvester Matthias mentioned, "Kitt Peak is the northernmost place for *bawui*, [called] Bawuigam Do'ag. Chiltepines here also" (see Group G, *Capsicum annuum*). Once he told me, "[There is a] tale, that anyone who bothers *bawui*, it'll start raining, [it's a] rain producer. Fresnal Canyon, in [the] Baboquivaris, [is where you find it]; a swallow song about Eagle mentions *bawui*." A few years later he elaborated on *bawui* as a rainmaker: "Apparently this tradition [is] from both Pima and Papago. Don't bother the bush, the pods, or the seeds. It would be—I would call it—a sacred thing. If they needed the rain, they have to go ask somebody for rain. They put [on] the rain making dance, or the wine feast. The rain ceremony."

Sylvester said that he was familiar with *bawui*, that he knew the plant "from east of Topawa, on the west slope of Baboquivaris—way up, along the arroyos." In San Diego I had some spiny species of *Erythrina* growing in a patio pot. On a visit Sylvester commented on my *bawui*.

"How do you know that?" I asked, because mine was somewhat different from the Arizona species.

"Because it has leaves like an *auppa* [Frémont's Cottonwood, *Populus fremontii*]."

Russell (1908:292–295) recorded another song that mentions this bush: in the last stanza of the "Kikitavar Ñei" ("Gigidval Ñe'i," Swallows Song) the Coral-bean flower (*bawui yosi'ime* in song language) is mentioned in the original song line and in the word-for-word translation but is

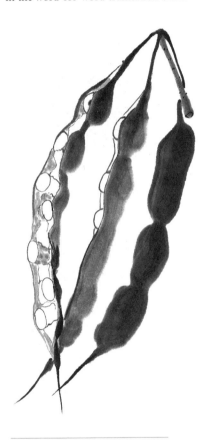

bawui   *Erythrina flabelliformis*

not incorporated into the text of the free translation. Russell (1908:176) makes an additional reference to *bawui*: "A guessing game [*vopodai*] in which a number of players act as assistants to two leaders. A small bean [*bawui*] is used by the Papagos and a ball of black mesquite gum by the Pimas. It is placed in one of four joints of reed. The reeds are then filled with sand, all being concealed under a blanket, and the opponents guess which reed contains the ball." Russell's footnote adds: "[Small bean] obtained from Sonora from the tree called paowi [*bawui*] by the Pimas and chilicoti by the Mexicans."

One of the last places on the reservation where gambling games persisted was at the village of Bapchule. I asked George Kyyitan of this village if he knew *bawui*. "Yeah. Little, red . . . like beans. It's *sure* hard! Used for that game called *vopodaikuḍ*, cane tube game." A few days later he presented me with a set of the four marked tubes, explaining that one of the red beans was hidden in one of the tubes. Each tube has different markings and different names (see Group B, *Phragmites australis*).

In the resurrection speech of I'itoi, Jeweḍ Maakai (Earth Doctor) blows smoke from a cigarette made from two different kinds of roots, *kokoi tatk* 'ghost root' and *pawois [bawui] tatk* 'Coral-bean root' (Russell 1908:345). The first plant remains unidentified, but both Akimel and Tohono O'odham I talked with who knew about this plant associated it with rattlesnakes (*ko'oi*) rather than ghosts (*kokoi*) (see appendix B).

Given the modern mystical associations that have survived around this plant, it would seem that proto-Pimans had some idea of its unusual chemical properties. It is a plant that falls into the category of sacred beings, along with datura and Peyote (*Lophophora williamsii*).

Vouchers: *1967* (GK, SM); *42, 43* (Pedro Estrella, Pima Bajo).

Gaura parviflora Hook.
Velvetweed, Small-flowered Gaura
Onagraceae
vi'al

About late April you may begin noticing here and there around the central section of the reservation some slender plants sticking their heads above the Wild Oats (*Avena fatua*), globemallows (*Sphaeralcea* spp.), and annual saltbushes (*Atriplex* spp.) growing in the moist edges between damp fields and roadsides. Sometimes they reach to between chest- and shoulder-high. If you stop for a closer look, you will not find much of a blossom. But with a hand lens, you may appreciate the pinkish flower with slender petals, which looks much like a miniature evening-primrose (*Oenothera* spp.). This

resemblance is not surprising, because the two are in the same family.

I was driving along Casa Blanca Road one spring, when Sylvester Matthias noticed the wispy spikes in a spot where irrigation water sometimes collected. We stopped to have a better look. "*Vi'al*," he said. That was all he knew: it had no use, it was just there. "There's an old man named Vial, after that plant, *vi'al*. That must be long ago before I ever come to existence. But I heard about it."

Although it is an annual, Sylvester always included it as a *sha'i*, perhaps because it is large and somewhat woody. Several times I pointed out the plant to various other Pimas and mentioned its name, but no one else seemed to know Small-flowered Gaura.

Vouchers: *386* (SM).

Hymenoclea monogyra Torr. & A. Gray
Hymenoclea salsola Torr. & A. Gray
   var. *pentalepis* (Rydberg) L. Benson
Burro-brush, Jécota
Asteraceae (Compositae)
iivdhat

Most of the year Burro-brush is a rather indistinct bush with bright grassy-green linear leaves and stems. It is somewhat resinous and shiny. There are two species in Gila Pima country. *Hymenoclea salsola*, growing to about knee-height, is the more compact shrub, found growing in sandy places such as bajadas and arroyos. True to its specific name, it resembles *Salsola* (Russian-thistle) in shape if not in color. After blooming in the spring it is covered densely with conspicuous whitish fruiting bracts that nearly conceal the plant. The other species, *Hymenoclea monogyra*, is taller and more loose and wispy. It flowers in the fall and has only very small fruits. This one grows here and there with Seep-willow (*Baccharis salicifolia*) and Desert-broom (*B. sarothroides*) in the otherwise barren sandy stretches of the major rivers. It is even more inconspicuous.

Apparently both species are included in the Gila Pima folk taxon *iivdhat*. The plant is now all but forgotten. My first contact with the name was through the Pima Bajo Pedro Estrella of Onavas, Sonora. He called the plant *i'ivdag* or *i'ivdad* in his Piman dialect and Jécota in Spanish. The plant, he said, is used commonly for the cross-thatching of roofs.

One day as Sylvester Matthias and I were traveling up toward Saguaro Lake in Arizona, he pointed out the bright green bushes in the sandy washes, calling them *iivdhat*. But on other trips he seemed not to recognize the plant, usually saying "It should be greener than this" when I called his attention to some Burro-brush. (Perhaps the difference between the two local species confused him.)

Later, on one of our outings together in southern Sonora, Sylvester and Pedro compared notes on this plant, so there is no doubt that the names each uses apply to a Burro-brush. The Onavas bushes grew to over head-height. Sylvester noted, "[Many] foot tall here; these [on Gila] way smaller, about half the height on these. This is *iivdhat*." Another time Sylvester made this contrast: "*Iivdhat* [is] similar to the *shuushk vakchk*, but smaller. No flowers. Same color or a little bit darker." He was comparing Burro-brush with Desert-broom, which blooms profusely.

Sylvester related, "*Iivdhat* [is] on the desert. Pimas used [it] for broom, like *shuushk vakchk*. Papago use [it] for outdoor cooking—to make that outdoor kitchen, the *uksha*." The Gileños had other plants, particularly Arrow-weed (*Pluchea sericea*), that were more suitable for thatching. The plant is not mentioned in any of the Tohono O'odham dictionaries. Nabhan (in Felger and associates 1992) gave *iivadhoḍ* for both species from southern Tohono O'odham living in the vicinity of Quitobaquito.

The Piman name *iivdhat* is related to a root word that signifies a concept of something leafy and green, such as newly opened buds. From this root come a number of common words: *iivagĭ* are edible greens; the month (*mashad*) when *auppa*, the cottonwood, burst forth in new green leaves was called *auppa i'ivagĭdag mashad*; the next month was the critical *kui i'ivagĭdag mashad*, the time when bare mesquite branches began bursting out in their delicate greens; something that is light green like newly opened buds is said to be *iivagim s-cheedagĭ*. And Burro-brush? It is just a nondescript bright green cloud out on the desert.

Vouchers: *687, 1838* (SM, *S. salsola*); *1054, 1279* (SM, Pedro Estrella, Pima Bajo, *S. monogyra*).

Isocoma acradenia (E. Greene) E. Greene
[≠ Bigelovia spp.]
[= Haplopappus acradenius (E. Greene) Blake]
[≠ Aplopappus heterophyllus (A. Gray) Blake]
[≠ H. heterophyllus (A. Gray) Blake]
[≠ H. pluriflorus (A. Gray) H. M. Hall]

Jimmyweed, Alkali Goldenbush, Burroweed, Rayless Goldenrod

Asteraceae (Compositae)

sai u'us

Along roads and field edges and other disturbed and somewhat alkaline areas, mostly on the floodplains, grows a rather erect, dark green, small-leaved shrub scarcely more than knee-high. Often it forms a row, like a little hedge, following a trail or ditch. The narrow leaves are resinous. One could easily overlook it. But in the fall it becomes more noticeable, its tops covered with deep yellow or pale orange flowers indicating its membership in the sunflower or aster family and explaining one of its alternate names: Rayless Goldenrod.

Its Pima name, *sai u'us*, means 'bad-smelling sticks [or] bush'. All the older consultants knew it. "*Sai* means the odor itself, or the taste," explained Sylvester Matthias. "Awful smell, or smells very bad. I don't know how we got that [name] *sai u'us*. Really, it's [a] Papago word, *sai;* 'bad smell', *sai uuv*. Pimas would say, *s-uam uuv*. Used *exteriorly* [externally] on open sores or on a rash; boil it [the branches] and use the tea [infusion] as a *wash*. Disinfects whatever is keeping [the] sore open, but it's painful, they say, because it's real bitter." Then Sylvester added, "And I heard they can use that in your shoes. When your shoes get smelly, just put it in there. Or your feet get sweaty. It'll dry [them] up. Or cut the smell."

This usage goes back in time. Hrdlička (1908:246) recorded, "Scarification is used by the Pima in localized inflammations. They make several cuts in the skin with a piece of glass, allow a little blood to flow out, and then apply the heated leaves of a plant known as *sai-u-us*." This procedure is called *goi* in Pima and appears to have been a rather common medical practice

employed for a number of different remedies as well as for "cleansing the blood" of runners before races. Curtin makes no mention of this plant. Russell (1908:79) lists an unidentified plant, *rsios* [= *shius*] (which must be a faulty transcription of *sai u'us*), saying, "Two unidentified species of Bigelovia [*sic*] are used as a dressing for scarified wounds. The bruised leaves are applied to bleeding surfaces that have been cut with broken glass."

Irene Hendricks told me, "When you cut those *sai u'us*, you can use them to keep away the *vaamug* ['mosquitos']. Cut a bunch of those, a pile of them, about a day ahead, and burn them. They get rid of the *vaamug*. But have to make lots of smoke. If too dry, won't do any good. Have to add some more green ones to the bunch."

George Kyyitan knew the plant also. "*Sai u'us* smells awful—like tequila!" He knew no medicinal uses for this plant but added that dry *sai u'us* twigs are good to start fires. However, George did know a use others had not mentioned. "They use that one, *sai u'us*. Leaves of it. I think some, that's what I heard, use it for odor. Right here." He motioned to an armpit.

"Body odor?"

"Yeah. Use it for that. That's what Chico told me one time. I did the same thing. I got the odor from the football shoulder pads. Somebody, a white man, used mine. And he got that. And it had it on there. So I wear it. So I had it. Boy! That's one thing I sure hate to smell. I came back home . . . take a bath . . . put a powder right here. Every time. Every time have to put powder. And I quit the shoulder pads. It was mine, but I didn't use it no more. So I got away from that odor there. Then Chico, over here, he told me that he did the same thing. So I went over there and got some of that. [He meant *sai u'us* leaves.] Then I put it right there. And it's gone."

"And did you use them dried or wet?"

"Wet. Green. Just sticks in there. It's kind of sticky."

Technical Notes: Central Arizona specimens, including those from the reservation, have long been known as *H. heterophyllus* (A. Gray) Blake (= *I. wrightii* (A. Gray) Rydb.) (see Kearney and Peebles 1960:862; Lehr 1978). A recent revision relegates this name to the synonymy of *I. pluriflora* (A. Gray) E. Greene of southeastern Arizona and the Chihuahuan Desert (Nesom 1991).

Specimens with entire leaf margins (except for an occasional small tooth near the base) and with a distinct resin gland near the apex of the phyllaries are *I. acradenia* (E. Greene) E. Greene (= *H. acradenius* (E. Greene) Blake). These predominate in central and northwestern Arizona, including most of the reservation. However, in more southern and southeastern Arizona is a plant with conspicuously lobed (pinnatifid) leaves and glandless phyllaries thickened at the tips. This is *I. tenuisecta* E. Greene. It appears to hybridize with *I. acradenia* in more southern and eastern areas of Pima country.

Pima consultants associate their name *sai u'us* with the species having linear, smooth-margined leaves. The conspicuously lobed form was puzzling. In the Santan Mountains, Sylvester Matthias identified *AMR 739* (with smooth leaf margins) as *sai u'us* but rejected *AMR 740* (with lobed leaf margins) from the folk taxon. The two were equally common on the bajada there, and we found no intermediates. In the Gila River bed at Sacaton, George Kyyitan picked out smooth-margined examples but said of an adjacent bush with lobed leaves (*AMR 1742*), "I don't know that one; I don't know if it's *sai u'us* too." Sylvester Matthias identified two composite samples (*AMR 1768, 1769*) from the Gila River bottoms at Santa Cruz village as *sai u'us*. These were mostly unlobed. We showed them to Carmelita Raphael, who concurred. We made several more collections beyond the southern end of the Sierra Estrella, where intermediate lobing becomes more common. Sylvester identified one composite sample (*AMR 1772*) as *sai u'us* but was more cautious about another sample (*AMR 1773*): "Would be *sai u'us* also, or in that family." These were subsequently identified as *I. acradenia* (accepted) and *I. acradenia* x *tenuisecta*, respectively. Near Blackwater, Sylvester commented on a pinnatifid specimen (*AMR 1573*), "This is in the family with *sai u'us*," while accepting the linear-leaved bushes nearby as legitimate. About a quarter of the bushes there are lobed. When shown Jimmyweed at Jacumba, San Diego County, California, Sylvester commented, "Looks like a California form of *sai u'us*." The specimen (*AMR 1626*) was *I. acradenia* (E. Greene) E. Greene var. *eremophila* (E. Greene) Nesom.

Vouchers: *250* (RG, SM); *501, 1498, 1768, 1769, 1772* (SM); *LSMC 24*.

Jatropha cardiophylla (Torr.) Muell. Arg.
Limber Bush
Euphorbiaceae
vaas

Although Se'ehe is a major deity in the Piman Creation Story, he is anything but a paradigm of perfection. After the flood, when the four major beings emerged, Jewed Maakai (Earth Maker), Ñui (Buzzard), and Ban (Coyote) acquiesced to Se'ehe's insistence upon the title of Elder Brother, although he was the youngest. The Gila Pima associate themselves most strongly with this culture hero, who in all fully recorded versions of the Creation Epic led them (the Vupshkam) up from the underworld and out of the east to their present location. Later, when the People were unable to extricate themselves from some difficulty, Elder Brother's assistance was sometimes sought. But he played tricks on people. He said one thing when he really meant another. His was a perverse sense of humor. And although the People ultimately drove him away because of his indiscretions, particularly with young girls, he is still around Pima country occasionally, and he is still playing jokes.

A slim Pima lady and I walked out across the bajada toward an abrupt desert-varnished outcrop of the Sacaton Mountains. We stooped over to look at some black ants scurrying along the sandy trail. "You know, sometimes you see Se'ehe out here," she confided. "He just comes running along, real quick. He's just *little!*" she said in her musically cadenced Pima tones.

"Ah!" I said in mock disbelief. "Have you seen him recently?"

"No. Not recently. But I *have* seen him, a *long* time ago. He's just a *little* old man. But he runs *quick*. And you better be careful—he plays tricks on you. Haven't you ever seen him?"

"No," I told her. But I lied. He had played a trick on me. On all of us. And it has taken us 80 years to catch on. It started at the beginning of the century when ethnographers Frank Russell (1908:219–221) and J. William Lloyd (1911) recorded the Pima Creation Story from

Thin Leather. In one episode, a handsome young gambler called Vandai (or Vantre) is turned into an eagle who makes his home on a cliff. After exhausting all the local game, he turns to hunting men. Their numbers being depleted, the People seek the assistance of Se'ehe, who says he will come to their assistance in four days. Entreaties continue, and at last he arrives in four years, reluctant to take action against this good-looking but evil person, who is his own relative. To reach the home of the Eagle-man, Se'ehe needs several items, but he does not tell the people directly what he wants. First he asks for a *muta*, or seed-parching pan, when he really wants obsidian. "Then," Lloyd's account (1911:82) continues, "he told them to bring him four springy sticks. And they ran and brought him all the kinds of springy sticks they could find, but he told them he did not mean any of these. And for many days they kept on trying to get him the sticks which he wanted. And after they had completely failed Ee-ee-toy told them what he wanted. It was a kind of stick called *vahs-iff*, which did not grow there, therefore they had not been able to find it. And besides vahsiff sticks were not springy sticks at all, but the strongest kind of sticks, very stiff. So they sent a person to get these." Se'ehe whittled four of these to sharp points and used them as pitons to climb up the cliff, "pulling out the sticks behind him and putting them in above."

Thin Leather gave Russell (1908:277) songs to accompany the Creation Epic. In the song that I'itoi sang while climbing Eagle-man's cliff, the name *vaasif chuchua* is translated as 'leafless stick'. In recording I'itoi's song, Danny Lopez (in Evers 1980: 124) gives the word as *waasiw*, translated simply as 'plants'.

The plant for years had me puzzled. No one recognized a tree or bush called *vahsiff*. Nor had I encountered it among the Tohono O'odham or the Pima Bajo. Se'ehe was still playing tricks.

One day I was reading through the typescript of the Creation Myth Julian Hayden had painstakingly recorded at Bapchule in 1935—a version told by Juan Smith. Included was a version of the Eagle-man story, complete with translations of the songs that go with the telling. At first Se'ehe implores the kachina Navĭchu's help. This Puebloan personage plants a gourd seed at the base of Eagleman's cliff. He sings and dances, and the gourd plant climbs the cliff, but a strong wind blows it back down and Navĭchu is unable to get it up again. Hayden noted, "Then Se'ehe looked around and he got some kind of wood which narrator Juan doesn't know but it is called 'vaas.'" Se'ehe sings,

> I am I'itoi [Se'ehe].
> *Vaas* I have stuck
> The *vaas,*
> And I walked on them
> And killed the eagle.

In that one simple syllable, *vaas,* was the answer to Elder Brother's long and successful obfuscation. I knew the mystery plant well. It belongs to a subtropical genus of bushes and trees, *Jatropha.* Various species grow in the Pima heartland far to the south, and don Pedro Estrella at the Pima Bajo village in southern Sonora had shown me several. One of these, the papery barked *Jatropha cordata* (Ortega) Muell. Arg., or Papelillo—a conspicuous tree—he called simply *vaas.* Another, a small bush with mahogany-brown bark, *Jatropha cardiophylla,* he called *s-tut vaas* 'black *vaas*'. This species extends farther north, well into Tohono O'odham country. Gary Paul Nabhan (1983b) found the Tohono O'odham calling the bush *waas* and calling another, rarer species (*Jatropha cinerea* (Ortega) Muell. Arg. in DC.), which barely crosses the international boundary, *koomagĭ waas.* But neither species reaches as far north as the Gila.

In a Tohono O'odham migration legend (Mason 1921), I'itoi (here called Montezuma!) tarries four years, then responds. "He asked the people to gather him some reeds known as *vaas,* and with these and his great machete he started off. He came to the foot of the steep mountain and sang a song. Then he stuck one of the reeds in the side of the mountain, where it remained. Pulling himself up by means of this, he stuck in another; and after repeating the process ten times, he reached the top of the rock."

So this is what Se'ehe or I'itoi or Elder Brother sent the poor Pima out to search for in vain: a plant that does not even grow in Gileño country! But Elder Brother gets the last laugh. The limbs of the *vaas* bush are not hard at all, as the storyteller is led to believe. Instead, they are among the most supple of any bush growing in the desert. The Tohono O'odham and Seri split the twigs for basket weaving. And the English name for *vaas* is Limber Bush.

Se'ehe is still playing tricks in Pima country. He does not always say what he really means. But he may. Only it may take us years to discover which.

Justicia californica (Benth.) D. Gibson
[= *Beloperone californica* Benth.]
Beloperone, Chuparosa
Acanthaceae
vipismal jeej

Beloperone is a common shrub in some of the major arroyos of the Sierra Estrella, near the northeastern edge of its range. It is at home only in high-energy streambeds that may be flooded briefly after an intense desert storm. Where the arroyos start to fan out into the floodplains, the bush stops. In the clean sands and water-sculpted boulders of the higher places it keeps company with Desert-lavender (*Hyptis emoryi*), Desert Ironwood (*Olneya tesota*), Catclaw Acacia (*Acacia greggii*), and often the beautiful vine *Janusia gracilis.* We have not found it in the Sacaton Mountains, and it is apparently absent from the Santans as well, though there seem to be some suitable washes for it there. It is also missing from some of the big arroyos in the Estrellas that appear to be perfectly suitable. It has been found in a single wash in the South Mountains (Daniel and Butterwick 1992).

The bushes are round and usually between hip- and shoulder-high. The branches are dense, velvety, and brittle. In wet seasons they are leafy, but in drought they may be almost bare. From midwinter through late spring the tips of the branches are often adorned with small clusters of scarlet tubular flowers with a lower lip turned downward at an angle for about half the length of the tube. In Hummingbird Canyon (see Rea 1983:12, map) above Santa Cruz village, the flowers are deep red, but in other arroyos the flowers range from dull orange-red through brilliant scarlet.

vaas   *Jatropha cardiophylla*

The O'odham name for this bush is *vipismal jeej* 'hummingbird's mother', and it is an apt name, indeed. This bush and Desert-lavender are the two primary food sources for the Costa's Hummingbirds that breed in the Estrellas late in the winter. These birds and the bush are co-evolved: Beloperone holds its flowers up high, a few in a bunch, where the hummingbirds can easily and efficiently take the nectar reward the plant provides; and the plant's anthers are arranged so that the feeding bird dusts its head thoroughly with pollen before moving on to the next bush. Sometimes the crowns of these hummingbirds are so loaded with pale golden pollen that they appear to be a new species.

Though I had long known the O'odham name for this bush, I began to wonder if it was really in the Gileños' lexicon; had I learned the name from Tohono O'odham students? My inquiries to Pimas such as Sylvester Matthias, Ruth Giff, and Carmelita Raphael—who lived close to the Estrellas—drew blanks: they were familiar with the bush but knew no name for it. The few Pimas who did know the name were half Tohono O'odham and had spent some of their youth on that reservation. Joe Giff originally may have been my source of the name, but he was gone, and I had no voucher to jog my memory.

One night at Holy Family Church in Blackwater, when the community gathered after Mass for a potluck, I asked Herbert Narcia if he know what *vipismal jeej* was. He had been raised in Komatke, under the Estrellas, and moved to the eastern village in 1940. "I think it's a plant, something like a rose." This was hardly definitive. Three years later at another potluck, I brought up the subject with him again. "Yeah, it's some kind of a bush with red flowers that grows over under the Estrellas." Beloperone is the only plant there that fits that description.

One day in southern Sonora, I took a branch of *Justicia californica* to don Pedro Estrella, the last speaker of Pima Bajo. He smiled in immediate recognition. He called it *vipsumal heosig* 'hummingbird flower' in his Piman dialect and Chuparosa in Spanish. Later we found in the field a similar red-flowered bush in the same family, *Jacobinia ovata* A. Gray. Pedro called this plant *vipsumal heosig* as well. He included in the taxon *vipsumal* one other small red-flowered plant, *Ipomoea coccinea* L., calling it Chuparosa-del-monte. The name is probably pan-Piman. Among the Mountain Pima, Joseph Laferrière (1991) recorded *vipishi hioshgama* for two other red-flowered hummingbird-pollinated plants (*Lobelia cardenalis* L. and *Stachys coccinea* Jacq.).

Vouchers: *1200* (Pedro Estrella, Pimo Bajo, *Justicia californica*); *1262* (PE, *Jacobinia ovata*); *1224, 1232* (PE, *Ipomoea coccinea*); *1890* (Frank Jim, Tohono O'odham, *Justicia californica*).

Krameria grayi Rose & Painter
White Ratany, Chacate, Cósahui
Krameriaceae
eeḍhō

For those who have not yet learned the secrets of the desert, there are an awful lot of plants that look alike: there are spiny, green-barked trees and bushes; there are blackish-barked trees and bushes; and there are squat, rounded, gray bushes that sometimes look almost white. They are not really all the same. For instance, the last group separates into various culturally important plants.

On the powdery tan soils of the floodplains, the common gray rounded bushes are mostly *kokomagĭ sha'i*, Desert Saltbush (*Atriplex polycarpa*). But as you move up onto the well-drained soils of the bajadas and the rocky mountains, several other small grayish bushes predominate: White Brittlebush (*Encelia farinosa*), Triangle-leaved Bursage (*Ambrosia deltoidea*), and White Bursage (*Ambrosia dumosa*). Among these you will find one of the Pimas' most important wild plants, *eeḍhō*.

White Ratany is a dense shrub reaching knee-height at most. Some of the intricate branches have sharp points. The grayish leaves are narrow, hugging the stems. In spring, the magenta flowers will positively identify this shrub. These are tricky: what appear to be petals are really sepals, the usually greenish structures that encase the developing bud. On the ratanys these five sepals of irregular length make the showy star, and the true petals are scarcely developed at the base of the stamens. The seedpods, too, are unusual. They are inflated and covered with small spines barbed at the apex. There is a single seed in each thumbnail-sized pod.

All older Pimas still know this plant. The roots had two uses: to make a red dye and to make a medicinal tea. Everyone I worked with remembered the dye, though it has fallen into disuse. They all mentioned that basket weavers once boiled willow splints with *eeḍhō* to make a pink or purple element in their designs. Even clothing, such as a white T-shirt, could be boiled in *eeḍhō* to make it pink. Russell (1908:118) mentioned its use for dyeing leather. Walter Rhodes gave Curtin (1949:92) a more detailed description of how the dye was once prepared: "Dry *oeto*

roots are ground on a metate and about a pint of the powder is added to a gallon of cold water, which is boiled for an hour. The willows are placed in this brew and boiled for half an hour, after which the shoots are removed, rinsed in cold water, and dried in the sun." This use must have died out early, for I have never seen a basket or anything else in the past quarter-century with this pink.

What has not died out, though, is the use of *eeḍhō* as medicine.

Sylvester Matthias told me on several occasions, "*Eeḍhō* roots are boiled and used as a tea for diabetes. It looks like strawberry soda, but has no taste." Another time he said, "Used for sores—ulcers of any kind—skin or stomach. Drink it. If you have a bad cold or a regular sore throat—you can drink that up for sore throat. When it gets *real* sore, they make it *real* hot—as hot as they can stand it. Take maybe a couple sips."

"How much root do you need to make a pot of tea?" I asked.

"Uhh—*that's* the question! Just a little bit. But for dye, it takes a lot. Lucy Jackson once used it for a basket. And she didn't like the color on the basket. It's too maroon. And she wanted real light, more red [pink]. So she adds commercial dye to suit the color she wants."

George Kyyitan showed me some *eeḍhō* bushes in the Sacaton Mountains. "Use this for medicine—[for] anything that's inside your stomach; [use] roots—boil them, [for] kidney. Kind of purple-reddish." Later he told me, "*Eeḍhō* is for diabetes and lots of things. Sick men or women wash their sores."

One morning when I stopped to visit Francis Vavages at Sacaton, he had among a pile of half-finished flutes and partly carved wooden spoons on his worktable a bundle of *eeḍhō* roots. Someone (not a Pima) who did not know the plant had asked him to get some. "Use same roots of *eeḍhō* three or four times till there's no color. It has a bitter taste [*sic*]. It purifies the blood. Or for arthritic joints—it's like there's maggots in there—not actually maggots, but something in the joints that cause that pain—and *eeḍhō* clears it out, purifies it. Also can use *shegoi* [Creosote Bush] for this."

In early April when we were out picking *hannam* (cholla buds) on the Sierra Estrella bajada above Santa Cruz, Carmelita Raphael pointed out White

Ratany bushes she called *eeḍhoi,* a dialectical variant of the name. She told me an amazing story about her oldest sister, who was diagnosed as having tuberculosis. After a year or so of unsuccessful medical treatment, the family took the girl home, where she started drinking *eeḍhoi* tea. Later she went back to her doctor, and he could find no trace of the tuberculosis, but her insides were pinkish. He wanted to know what she had taken, but apparently she was too embarrassed to tell her doctor about the "Indian medicine" she had been using.

I asked Ruth Giff about *eeḍhō.*

"Yeah, that's a medicine. Tony [Sabori] used that for his ulcer and it worked. Yet it's *very* mild. Just drink it for water. Not like the *shegoi. Shegoi*'s really *siv* ['bitter']."

Joe Giff also mentioned ulcers as one of the things this plant is used for. I did not ask him how it was administered, but 80 years earlier Russell (1908:80) had noted, "The root is powdered and applied to sores." About the same time, Hrdlička (1908:245) recorded a similar use: "As an antidote for the irritating effect of the *hávan tátat* plant [wild-heliotrope, *Phacelia*] on the skin (see p. 183), the Pima use another plant known as *uh-to* [*eeḍhō*], which grows in the mountains. They grind fine the root of this plant and apply it to the seat of the inflammation."

Culver Cassa and I went out to the Sacaton Mountains to see what these fabled *eeḍhō* roots were like. With one swing of a pick he caught a long root parallel to the surface. We pried it up through the loose gravelly-sandy soil. The corky, ochre-brown root bark encased a pink inner woody section. In just minutes we had a good supply of roots. *Krameria* is partly parasitic on the bursages (*Ambrosia* spp.) and Creosote Bush (*Larrea divaricata),* both growing nearby. Perhaps these corky roots were the ones running out to tap nutrients and moisture from their neighbors.

One usually assumes that plants used in native medicines are rich in aromatic oils and terpines or astringent chemicals. With the Gileños, major medicines have included *shegoi* (Creosote Bush) and, when the river was still running, *vavish* (Yerba Mansa, *Anemopsis californica)* and *sivijuls* (Canaigre, *Rumex hymenosepalus).* But *eeḍhō* has neither strong smell nor bad taste, only the bright red color "like strawberry pop."

eeḍhō   *Krameria grayi*

Technical Notes: White Ratany (*Krameria grayi)* is the species commonly found in Gila country. Pima Ratany (*K. erecta* Wild. ex Schultes, long known as *K. parvifolia* Benth), with needlelike leaves, has been collected high in the Estrellas (GPN & AMR *462)* as well as upriver from the reservation, east of Florence (AMR *1610).* The name *eeḍhō* probably applies to both species, though this needs specimen verification in the field. Frank Jim tells me that Tohono O'odham of his community distinguish two folk species: *eeḍho* and *ge'egeḍ eeḍho* 'large *Krameria',* presumably differentiating between *K. erecta*

and *K. grayi.* He said that both are used interchangeably for medicine. The rank of folk species, rare among wild plants, indicates the high cultural salience of ratany.

Vouchers: *34, 35, 720* (all SM?); *760* (CR); *979* (GK); LSMC *44.*

Larrea divaricata Cav. ssp. *tridentata* (DC.) Felger & Lowe
[= *L. tridentata* (DC.) Coville]
[= *L. mexicana* Moric.]
[= *Covillea tridentata* (DC.) Vail]
Creosote Bush
Zygophyllaceae
shegoi

"Here, let me open a new watermelon," Rosita Brown said, slicing a turgid oval melon with a big butcher knife.

"That's okay," I insisted. "I'll eat this piece. I don't want very much."

"Ah, that piece's old. Take a fresh one."

Fall was still very hot. It was the mid 1960s, and there was no electricity in Santa Cruz village then. Rosita and David kept the melon in an old-fashioned icebox. There was no ice, either.

I should have heeded her advice to eat the new one. In a few days, I was flat on my back, weak and soaked in sweat, with severe diarrhea. In Phoenix I was diagnosed with paratyphoid—and not just one but two species of *Salmonella.* My prescription was Aureomycin. When I finally got back to the classroom, I showed the pills to my Indian students: "These are fifty cents a tablet! I have to take one every six hours." I expected them to be impressed. Instead, they were universally incredulous. Why take them? Why not just drink *shegoi?* It doesn't cost anything. Surely it would work better than those little gold pills from the white doctor.

What is this plant that is universally esteemed as the most important medicine among both Akimel and Tohono O'odham? One never has to look very far on the reservation to find it: just go to any well-drained area where the soil is gravelly. On the bajadas and rocky slopes of the desert mountains, even in places so dry that other perennial bushes cannot survive, you will find an erect, slender-branched shrub growing to between waist- and head-height. The bark is gray, and the stems seem to grow in tortured segments, vegetable clocks measuring increments of time. Each little olive-colored leaf is split into two sections joined at the base. As a sure identification, crush some of the densely resin-covered leaves in the palm of your hand. They should have a pungent, almost overpowering smell, the source of the plant's name. If it is spring, you may find the branches bearing abundant small bright yellow flowers. The flowers resemble those found on the prostrate Puncture-vine *(Tribulus terrestris)* in the fields and villages—not surprising, because they are both in the same family. Or if you come late in spring, you will find the bush with a hoary look caused by abundant fruits bearing dense white fur.

You might sit and contemplate the *shegoi* bush. It will provide you scarcely any shade. But on the hottest, stillest day, the erect branches will sense the smallest motion of the dry air, and wave. Your skin may not even feel the breath, but the plant will respond.

You may notice how *shegoi* creates a microhabitat. Among its roots are holes and burrows of such desert specialists as pocket mice, Round-tailed Ground Squirrels, various lizards and snakes, and

shegoi   *Larrea divaricata*

terrestrial insects. Even desert plants may find shelter here—nursling Saguaros, for instance, and many winter-spring annuals whose ancestry is from the Pacific region eons ago.

Some day in summer while you are out on these Creosote Bush flats, the desert may treat you to one of its more theatrical displays—the building up of moist, warm monsoonal air into dramatic thunderheads. After wind and thunder and lightning, wispy spiderwebs of rain may connect cloud to earth. Then you will smell the most characteristic odor of the desert—freshly dampened soil and wet Creosote Bush leaves. My poet-scientist friend Gary Nabhan was once teaching creative writing on the Tohono O'odham reservation. He asked his young students to describe the desert. One wrote simply, "The desert smells like rain." This may seem like a contradiction to anyone who conceives of the desert as a bone-dry wasteland. But to someone growing up in the Southwestern deserts, the invigoratingly fresh smell of rain on Creosote Bush says it all.

Shegoi was the first plant created in the various Pima creation stories:
In the beginning there was no earth, no water—nothing. There was only a Person, *Juh-wert-a-Mah-kai* [Jeweḍ Maakai] (The Doctor of the Earth).

He just floated, for there was no place for him to stand upon. There was no sun, no light, and he just floated about in the darkness, which was Darkness itself.

He wandered around in the nowhere till he thought he had wandered enough. Then he rubbed his breast and rubbed out *moah-haht-tack [muhaḍagĭ]* that is perspiration, or greasy earth. This he rubbed out on the palm of his hand and held out. It tipped over three times, but the fourth time it staid straight in the middle of the air and there it remains now as the world.

The first bush he created was the greasewood [*sic*; Creosote] bush. And he made ants, little tiny ants, to live on that bush, on its gum which comes out of its stem. (Lloyd 1911:27–28)

It is perhaps not surprising that the Pimas' most important medicinal plant should hold primacy in their origin myth, but

references to the plant do not stop with this initial genesis episode.

There is a scale insect, *Tachardiella larrea,* that lives on *shegoi* branches, secreting a hard covering of dark brown lac over its body. Apparently the Pima were unaware of what caused these droplets, for they are called *shegoi ushabĭ* 'Creosote Bush sap'. But this lac had important uses in their Creation Epic. In the flood episode, the culture hero I'itoi (Se'ehe) escapes by sealing himself in an olla he has made from lac for that purpose. After the flood he must build a fire, melting the *ushabĭ* to get out of his vessel. (You can still see his blackish vessel standing upright in the Pinacates near the mouth of the Colorado River.)

The plural form *sheshgoi* is sometimes used by Gileños to mean *kokomagĭ sha'i,* Desert Saltbush *(Atriplex polycarpa);* unlike the Pima, the Tohono O'odham use the plural simply to indicate more than one Creosote Bush. *Kokomagĭ shegoi* 'gray Creosote Bush', according to Sylvester, is Triangle-leaved Bursage *(Ambrosia deltoidea).* If there would be any confusion in conversation indicating Creosote Bush, it might be clarified as *muhaḍagĭ shegoi* 'greasy *shegoi*'.

Mathiot (n.d.:335) writes the word *shegai,* a variation I have heard occasionally on the Gila. Saxton, Saxton, and Enos (1983:52) give *shegii* as a variant of *shegoi.* Russell (1908:79) wrote *shekaikukuk,* which is really two lexemes. The second word, *kukug,* indicates 'the tips thereof', according to Culver Cassa.

Lac collected from *shegoi* played very important roles among the Gileños. Sylvester called it a "solder" that was melted and used to repair things such as broken or cracked pottery. "There's a pitch on a greasewood [*sic,* Creosote Bush], too, that you can use, too, for a sealer. If your pot, your olla, leaks, you can melt it up and just rub it around. And it holds. It was also used to seal ollas of food and *haashañ sitol* 'Saguaro syrup'. Especially those Aangam [Tohono O'odham] traded to the Pima ollas of *haashañ sitol,* sealed with *shegoi ushabĭ.* The ones I saw came from Gila Bend [Reservation]."

"How large were they?"

"Let's see. About half a gallon, I think."

Whittemore (1893:55) said the Gila Pima stored Saguaro jam "in small earthen jars hermetically sealed, a foot or two underground." (For a comprehensive survey of this technique among South-western Indians, see Euler and Jones 1956.)

Joe Giff elaborated on uses of lac: "*Shegoi ushabĭdag*—melt it, to mend pottery—or to cover the kickballs; heat it and coat the ball with it, then scrape it off on a rock where it's too much, to make it perfectly round, and test it and make it right weight. They don't want it too heavy. I asked around about them, some saying they used mesquite or ironwood because it's heavy and they can kick it far; but these others said, no, they never used it. They used either palo verde or cotton-wood; it's light. They don't want it too light or too heavy. It depends on the person's power or strength, but those heavy balls—they have to run 15, 20, 25 miles and back—and that would tire a man if it's too heavy."

Russell (1908:106) collected an awl used in basket making whose handle was a one-inch ball of lac. Kissell (1916) illus-trated several others. I asked Sylvester, who made *oipij* 'awls' for basket weavers, if he knew of this. "Yes. Hilda Manuel had one. Just a small one. Maybe she got it in Papago country."

One Labor Day, George Kyyitan and I sat outside talking. There had been a torrential downpour that morning, and much of Bapchule had turned into a shal-low lake. After dark the spadefoot toads wasted no time, emerging with a mating chorus that sounded from every wet part of the village. Unfortunately—for us—mosquitos *(vaamog)* also had perpetua-tion of their species on their mind, and a blood meal from us was an urgent means to that end.

"What can you do about *vaamog,* George?"

"Can burn a rag, manure, or *shegoi* to get rid of them. Burn it green. Burns smoky."

Instead of a pebble in the mouth to allay thirst, a Pima might use a twig of *shegoi,* Sylvester said. "Scrape the bark off, enough that fits in your mouth. And it'll keep your mouth wet. Anytime when you get dry, or out in the mountains walking and couldn't find any water. You can do that."

Another use for *shegoi* has definitely dropped out of Pima culture. Hrdlička (1908:243) wrote of this: "If a Pima killed several Apache, although the act was lauded, it was believed that some of the

progeny of that man would become insane or otherwise injuriously affected. This result could be obviated by use of Apache hair, a tuft of which, tied with a chicken-hawk feather and an owl feather and burned in a certain way with greasewood [*sic,* Creosote Bush], would cure any sickness induced by the contact with the Apache." Hrdlička had identified this bush a few paragraphs earlier as *Covillea triden-tata,* but what raptor is intended is any-one's guess: *chicken-hawk* is an imprecise vernacular term.

Everyone with whom I worked had a battery of ills they treated with *shegoi.* One day in the Sacaton Mountains, I asked Irene Hendricks how she used it.

"Open sores. If fever, drink it and throw everything out."

"You have to vomit it?"

"Uh-huh. And they good for diabetes. That's why I got six months on my dia-betes [was cleared at one biannual checkup from needing to take insulin]—I drink it."

"How much do you use?"

"A pitcherful to drink and throw it out."

"How much?"

"I just wash it and put a little handful and put it in warm water until it looks like tea."

"Do you cook it or just leave it in the warm water?"

"Just leave it in the warm water. When we cook it and it's just boiling, boiling, and it's *sure* strong!"

One December day, Ruth Giff told me, "I have a sore throat and cold. I've been taking *shegoi*—drinking it during the day. I don't like it too strong, but my father used to make it *real* dark. The sore throat is going away. Now it's just in the head."

"How much do you drink?"

"Just a little bit. Maybe about ten times a day."

"How strong do you make it?"

"I just boil the water first. Then I put some of that in there and let it stand. I don't boil the *shegoi.* But after a while it gets strong."

Sylvester Matthias told of many uses. "Dry it up and pound it up and use it on open sores."

"How do you pound it?"

"Any flat rock—make powder. Somebody else told me that you get wet [damp] feet—you can put *fine shegoi* leaves in your shoes and it'll dry your feet up."

Joe Giff said for diarrhea, use *shegoi ushabĭdag.* Sylvester agreed, "You just boil it and drink, just like soup."

There is a rich historical record for uses of this plant as well. Two of the most brief are from the turn of the century. Russell (1908:79) noted, "The leaves of the creosote bush are boiled and the liquor is allowed to cool a little, when it is drunk as an emetic. The boiled leaves are also used as a poul-tice." His description parallels Irene's con-temporary one. In his medicinal treatise, Hrdlička (1908:244–245) mentioned two uses: "In stomach troubles the branches are boiled and the decoction is drunk hot. In cases of pains in the chest, and even in other parts of the body, the leaves and smaller branches are heated and applied as a dry poultice over the seat of the pain."

Curtin (1949:62–63) in the 1940s gath-ered the most complete information on the use of *shegoi.* The Pima used various parts of the plants, ranging from leaves to bark. With boiled leaves they treated dandruff, impetigo, toothache, rheuma-tism, gas, and headache caused by upset stomach, and they used the preparation as a hot drink for colds and an emetic for high fever. Powdered leaves they used to treat sores and as a deodorant for armpits; small twigs in leaf they used as antiper-spirant for the feet and as a deodorant in shoes; heated green branches they used on bruises and bound onto the seat of body pain; creosote "gum" (lac) they prepared as a decoction to treat tuberculo-sis or chewed and swallowed to relieve dysentery; and the bark they used as an intestinal antispasmodic. The last two uses Curtin gives on the authority of a secondary source (Stone 1932).

There is one curious omission from this list. The most frequent use of *shegoi* among the Pima in the 1980s is to lower the level of blood sugar, which becomes dangerously high in diabetes. At the time of Curtin's fieldwork in the early 1940s, this disease had not reached the pervasive level it now holds (see chapter 7). Russell (1908:267–268) did not even mention dia-betes from medical case histories supplied for 1892–1902. Indeed, only one case of diabetes was reported by a reservation physician treating 37 patients one month during 1902 (Hrdlička 1908). But the dis-ease probably appeared as soon as the environmental degradation in the last quarter of the 19th century caused the life-supporting Gila to fail. The Pima were forced to rely more and more on the typical white person's diet, high in fats and simple carbohydrates. The replacement of the corn and mesquite of the original diet with wheat was probably another factor initiating the epidemic.

One afternoon as we were coming back from Phoenix, Sylvester asked me to stop along a wash just inside the reservation boundary. He wanted to pick some *shegoi* for his wife, a diabetic, from a certain plant growing among hundreds of other Creosote Bushes. "This is the best," he explained. "It's darker green and more shiny." Indeed, the others about it had a duller, more yellowish olive look.

Another day, in late spring, I sat outside with Carmelita Raphael in the welcomed cool of evening. She told me that some people used *eeḏhō*, White Ratany (*Krameria grayi*), for diabetes.

"What do you use?" I asked.

"Just *shegoi*. It works for me. Sometimes I quit, but when I start using it again, everything is okay. I don't have to go to the bathroom so often. If I stop, I start again three or four days before I go in for my monthly test. Then my blood sugar is okay."

"How often do you take it?"

"Sometimes just some in the morning and again before I go to bed. Sometimes I drink it three times a day."

"How much?"

"About half a cup or a cup."

"You can stand that much?"

"It's okay. Maybe I don't make it as strong as David." I had told her David Brown's was as brown as dark tea and I couldn't drink it!

"Do you have some made now?"

"Yeah." We went inside her crowded kitchen, and she reached under a white cloth and pulled out a sealed quart jar, full to the top with a honey-colored liquid. "Here, you want to try some?" She poured about an ounce into a clean cup for me, then filled her own. I managed to get it down in about two drafts.

"Aaugh!" I said. "I don't know how you can stand it."

"When you get used to it, it's okay."

"My insides have been loose today. Will this help that too?"

"Sure. But you better have more." She took the cap off again and poured my cup half full.

I struggled with it this time; my throat burned, as if swallowing retch. "How long does this keep like this?"

"Oh, a couple days. Or either five or four."

I wrote this in April 1989. Physicians and researchers are struggling to understand the complex interaction between diet and the peculiar physiology of the human population with the highest known diabetes incidence in the world. And I was talking to an Indian woman who believes she knows a solution, a way to 'beat the system' as it were and control her own blood sugar levels, keeping off the white man's insulin by the use of nothing more than a branch of an ubiquitous shrub that has always served as the prime Pima medicine.

Vouchers: *982* (GK); *1016* (?SM); LSMC *5*.

Lycium spp.
wolfberry, squawberry
Solanaceae
kuávul, kuáwul

During the dry seasons, which span most of the year, the wolfberries are ash-colored, leafless, finely branched bushes growing commonly out on the flats among the bluish gray Desert Saltbush (*Atriplex polycarpa*) and scattered Velvet Mesquites (*Prosopis velutina*). They are ubiquitous and most of the year so nondescript that many passersby probably think they are just some dead bush. They are drought deciduous and lack chlorophyll in the bark.

A good wet period at any season will quickly draw them from dormancy, although they are dependent primarily on winter moisture. When good rains fall in the cool season, the seemingly dead bushes turn green with usually succulent, simple leaves about the size of your fingernail. In a short time many wolfberry bushes are so covered with tiny tubular greenish white or lavender flowers that they literally hum with bees. About mid April many bushes are covered with little orange berries the shape of eggplants—the resemblance is not surprising, because both are in the same family. Most fruits are nearly fingernail sized, though in

exceptional years, they may swell to almost twice that. The taste is ambiguous but not unpleasant. Squawberries might be compared to slightly bittersweet, very juicy cherry tomatoes with firm seeds.

Various species of wolfberry do well in the fine alluvial, often saline soils of the Gila floodplain. Some favor better-watered canal banks and field edges. Birds eating the juicy berries perch on barbed wire, spreading the seeds in their droppings so that fence rows on traditional Pima farms become hedges of wolfberry and Graythorn (*Ziziphus obtusifolia*). Other species are found in the bajadas and mountains, mostly along washes.

Gila Pima country has perhaps the highest species diversity of wolfberries of anywhere in the world; at least five (more likely seven) species grow there. The plant has even provided one place-name: the area between Komatke and Co-op Colony is called in Pima S-kuávulig 'place of many wolfberry bushes'. Almost every Pima who knows some native plant names knows *kuávul*. These are big bushes growing at least to waist-high, but more often are chest- to well over head-high. Yet most Pima seem to forget about the squat, often spreading little *Lycium californicum*, which grows in saline areas, in ancient fields, and on Hohokam ruins.

David Brown and Sylvester Matthias were showing me bushes at Santa Cruz village. They came to a spiny plant, scarcely knee-high, with succulent cylindrical leaves and miniature orange to red fruits, each almost filled by two seeds. This, they said, is *sisipuḏ kuávul* 'anus [pl.] *kuávul*'. Both old men said, "Don't know why it's called that." Later I asked Ruth Giff in Pima if she knew it. "Oh, yeah. That little one you can't eat." Invariably over the years when I have asked other people about the kinds of *kuávul*, they have said just one; then, when I ask about *sisipuḏ kuávul*, they always say, "Oh, yeah." They know it, but the culturally insignificant (though common) species tends to be the first to fall out of a folk taxonomy.

But was that the only species the Gileños distinguished among the welter of biological *Lycium* species found here? When I asked, Sylvester described four different kinds of *kuávul* in the Pima folk taxonomy: "just" *kuávul* (grows along fields, fences, irrigation ditches; this one is the main species harvested); *toota kuávul* 'white *Lycium*' (grows in fields, ripens about June); *chuchk kuávul* 'black *Lycium*' (grows in the desert; the one we found at the top of the bajada he said was new branch growth, so the branches were not yet very dark); *sisipuḍ kuávul* 'anus [pl.] *Lycium*' (not good to eat; dwarf; round, succulent leaves). I must plead guilty to not having resolved all the correlations between the four folk species and the six or seven biological species during the years that I have worked with Sylvester in the field. But I have always received the same information from him. In 1988, a decade after he gave me the first breakdown (above), I asked him again about the various kinds. His comments were nearly the same: "*Kuávul,* the 'regular' one, that the local People use, ripens early to the last of April. *Toota kuávul* ripens with the wheat—the tail end of May, early June. As soon as the wheat gets ripe, the *toota kuávul* ripens. *Chuchk kuávul*—I don't know about that; it grows in the mountains and the People don't use it for food, but they know it; somehow they never go after it the way they go after Saguaro and cholla buds. *Sisipuḍ kuávul* ripens just about the time the 'regular' *kuávul* here does. Can't eat *sisipuḍ kuávul;* they *can,* but they say they don't have the patience to pick them up. This woman that you saw the picture of over here, that old Papago lady, when it's April, she goes and picks it—I don't know how many days to make a serving! There used to be a lot around the other side of the cemetery. Nobody would believe it because they're too tiny. The regular *kuávul* wouldn't have the taste till it's *real* ripe."

One year, he found one that ripened in January that he did not know before. He was unsure which folk species it was. "I never heard anybody talking of a *kuávul* that ripens in January. Stems like *toota kuávul,* but they don't ripen at this time. Stems sort of halfway toward *toota kuávul.*"

Ruth Giff also had a story of the opportunistic fruiting of the "regular" *kuávul:* "Oh, some came on in September . . . *ripe!* We had a real heavy rain—and it came from it. Just full. Not as many as when they ripen in spring—just here and there—and *big!* And I thought I'd go back with my container, and I didn't go. . . . About four days after, I went—the birds had cleaned it up. And I noticed the one behind Sally's house has *lots* of green ones on it already in January."

Ruth said she did not know *chuchk kuávul,* but she was familiar with the other three. "Just plain *kuávul* grows in fields; also at Sally's [her daughter's house nearby]. *Toota kuávul* is ripening now [mid May] on 51st Avenue." By this time most of the "regular" (and most abundant) species had finished fruiting.

A year later, in January, I discussed *kuávul* and *toota kuávul* again with Ruth. "When is *toota kuávul* ripe?" I asked.

"Late! Almost summer."

"How do you tell them apart?"

"Well, it's different, the shape. I think *toota kuávul* are *round,* just round. These other, regular, are more [elongate]."

"Are the branches a different color?"

"I don't think so, . . . seems like it's the same," she said, thinking about the differences. "They used to grow along the ditch. Not like these that grow up on the desert [flats]. I don't know if there's any left." She paused, then added, "Yeah, I think some are along the freeway."

kuávul, kuáwul   *Lycium* sp.

One of the problems in identification is that botanists and Pimas use different characters in determining species. To the botanist, stamen length, corolla lobes, calyx shapes, and pubescence visible with a hand lens are all important. The botanist deals with a pressed plant in the herbarium (at best, wolfberries do not make good specimens). The Pima, by contrast, encounter the whole plant growing in their immediate environment. It is an important food source, coming at the season when other crops, at least formerly, were running low. When the bush turns orange with fruit, it is time to pick berries. In talking with Ruth Giff, Joe Giff, Sally Pablo, David Brown, and Sylvester, I found that the following diagnostic criteria seem to be used in distinguishing the folk species of wolfberry (listed approximately in decreasing value of importance): color of bark, growth form and size, shape of fruit, color of fruit, taste of fruit, time of fruiting, habitat, and size and hardness of seeds in fruit.

There is no problem identifying two of the folk species. *Chuchk kuávul* embraces several biological species: *L. macrodon, L. berlandieri,* and apparently *L. andersonii.* These are tall, very rigid and tough-wooded species found in the desert mountains or the upper parts of desert washes. The springy limbs are spiny and mostly shiny and mahogany colored, except for the new growth. The fruit is usually quite small and either bright red or red and green, unlike that of the others. The leaves may be noticeably nonsucculent and narrow for a *Lycium.* The other obvious folk species, as already noted, is *sisipuḍ kuávul,* which is *Lycium californicum.*

But there remain *L. torreyi, L. fremontii,* and perhaps *L. exsertum* in the lowlands. The abundant and widespread *L. fremontii* is surely "regular" *kuávul.* It usually has purple flowers, large juicy orange fruit, and succulent leaves. Rosita Brown, Sally Pablo, and others identified specimens of this species as the one the Pima traditionally ate. How the remaining two biological species map to folk species, I am not sure. Whichever has noticeably whitish branches (even the older ones) and fruits late in the season would be *toota kuávul.*

Large quantities of "regular" *kuávul* berries are picked, at least by a few Pima women, in the spring (generally early April through early May). They are prepared in various ways. Sally Pablo would can them. Her pancake breakfast might be smothered in *haashañ sitol* 'Saguaro syrup' or a sauce made from the crushed berries cooked in water until a bit thickened, called *kuávul hidoḍ*. Or she would cook this sauce with crushed wheat as a side dish to accompany the evening meal, much as one eats cranberries. At Sacaton where I was writing, half-gallon plastic pitchers of *kuávul* would appear regularly in the refrigerator during the fruiting season: some plain, some cooked with sugar for drinking, others cooked with wheat. Some years the bushes are more heavily loaded than others, though I have almost never seen a year, even in drought, when there were not far more berries produced along fields and fence rows than could possibly be picked even if the whole community turned out.

I had heard that *kuávul* berries were once dried. Castetter and Underhill (1935:19) considered this the usual preparation among the Tohono O'odham. Apparently this practice has been long abandoned in the river country. I mentioned it to Ruth. "Yes, that's the only way they can preserve it, dry." I picked a cupful and dried them in a large paper grocery bag hung in the airy shade of the *vatto*, or ramada. The juicy berries dried slowly to a raisinlike consistency, tasting even better when dried. Sealed from insects, they probably would last indefinitely in a storage olla.

One fine spring morning in 1991, I stopped to visit Francis Vavages. We sat outside, talking for half an hour or so; then he looked out across the unpaved street to several wolfberry bushes growing in a vacant lot. The fruits were so dense the plants looked mostly orange (and this after a series of drought years). "See that *kuávul* over there?" he asked, pointing. "Now they're good. Put about that much water over it," he indicated about two inches, "boil it, then it gets mushy. Don't take long. Then put a spoon[ful] of thickener in it."

"What kind of thickener?"

"Flour. A kitchen spoon. Now is the time when they're best. Later on they get worms in there. My grandmother used to get it. Got to pray first—take one, make the sign of the cross with it, *then* she takes it—she prays to the plant. They pray to the food first, those old people. Same way with the *iivakĭ*. That gray one, *onk iivakĭ*

[Wright's Saltbush, *Atriplex wrightii*] grows on the canal banks. They get the leaves, the tender parts on top, make the sign of the cross. Then they can pick it."

"What do they call that when it's cooked that way?"

"*Kuávul hidoḍ*. Can drink it. It's not as thin as water. You can put it out in the sun to dry up. I saw my grandmother. Get a whole lot. Spread out. When you put it back in water, it'll turn like fresh ones. Same way with the *onk iivakĭ* they dry. Well, now that you know how to fix it, do you want to go pick some to take?"

"Sure," I said. Francis went into the kitchen and returned with two small containers. In less than 10 minutes we had picked nearly a quart of the plump juicy berries from the densely covered branches.

"Do you know the difference between *kuávul* and *toota kuávul?*" I asked.

"Yeah. *Toota kuávul* comes later, way later. But these you have to pick now."

"I heard some cook it with whole wheat."

"Yeah, some do that way, too."

A granddaughter about six years old, freshly scrubbed and with her pigtails just braided, came out to help us pick, more curious than helpful.

"Do you like those?" I asked her. She shook her head shyly, with a slight grimace.

"My kids don't even know how to fix these things," Francis said sadly. "But sometimes one of the ladies will bring me some."

Apparently, eating *kuávul* has the effect of temporarily blackening the teeth; this was the consensus in Russell's time (1908:160). I asked several Pimas about this. George Kyyitan said, "They say when you eat it raw, it'll do that. But if you cook it, it won't do that." Ruth Giff responded similarly: "When we eat them raw, without cooking them, it's supposed to do that to our teeth. I don't know. But we were told [that] when we were kids. I think we ran round eating them. I don't know if it really does."

Various birds are fond of the berries, especially Common Mockingbirds (*shuug*) and Phainopeplas (*kuigam*). In the spring, mockingbirds become territorial, singing day and night. Interspersed throughout their repertoire of mimicked songs is a sharp double-noted repetitive phrase, as if linking the other songs. To most, it is an

innocent enough phrase. But to the Pima women in the morning, out gathering baskets or cans of the orange *kuávul*, the bird seems to be admonishing them: *Keliv! Keliv! Keliv! Keliv!* (Pick! Pick! Pick! Pick!). But one time there was a lady whose husband was much older than she was. She went to pick *kuávul*, and she thought the *shuug* was making fun of her, saying, "*Keli! Keli! Keli! Keli!*" (Old man! Old man! Old man! Old man! Your husband is an old man!).

Kuávul occurs in the Pima Creation Story. Early in the narrative, I'itoi (Se'ehe, or Elder Brother) made Handsome Young Man, who ultimately caused the Flood. "And to this man whom he had made, [I'itoi or Elder Brother] gave a bow and arrows, and guarded his arm against the bow string by a piece of wild-cat skin, and pierced his ears and made ear-rings for him, like turquoises to look at, from the leaves of the weed called *quah-wool*" (Lloyd 1911:36). The leaves of many *kuávul* bushes are succulent and flattened, resembling the pendant turquoise beads the Pima found in the Hohokam ruins along the rivers. These were popular for earrings, especially among the menfolk. The word *kuávul* appears to be derived from an archaic verb, *kuavigid*, which means 'to string hanging objects along one string'. One might use this word to indicate attaching dangling feathers to the cotton string in making prayer sticks. We lack this concept in English. One look at a heavy-laden branch of *kuávul* in spring will show at once the appropriateness of this name.

Technical Notes: *L. torreyi* A. Gray, reportedly the species with the most palatable fruit (Benson and Darrow 1981:198), has not been collected on the reservation since 1930 (ARIZ). This is essentially a riparian species that has disappeared with the loss of the riverine habitat. One Sacaton specimen is annotated "uncultivated along fence." I suspect this species may prove to be the folk species *toota kuávul*. *L. parishii* A. Gray appears to be a plant of somewhat higher elevations. It has been collected between Casa Grande and Gila Bend, and I have found it near Vekol Wash and Interstate 8, but never on the reservation. Daniel and Butterwick (1992) reported both *L. andersonii* A. Gray var. *deserticola* (C. Hitchc. ex Munz) Jeps. and var. *wrightii* A. Gray from the South Mountains. This is a narrow-leaved species. *L. berlandieri* Donal, another narrow-leaved species, is dark stemmed, often spiny, with a short calix, a flared small corolla, and spheroidal red fruit. *L. exsertum* A. Gray is best distinguished from *L. fremontii* A. Gray by its densely pubescent lower half of the filaments (usually galbrous in *L. fremontii*). Flowers are essential to identification, which is complicated by staminate- and pistillate-form bushes in several species from Pima country. *L. fremontii* is the most widespread species here. Several old, ashy gray stemmed specimens (AMR 19, 21, 22) Sylvester called *chuchk kuávul* (I think this was an error). Reservation specimens have been determined by Richard S. Felger and Fernando Chiang.

For Tohono O'odham, Gary Nabhan (in Felger and associates 1992) reported *kuávulĭ* (unmarked) as *L. fremontii*, *s-chuk kuávulĭ* as *L. macrodon* A. Gray, and *s-toa kuávulĭ* as *L. andersonii*; *L. californicum* Nutt. did not occur in the area being reported on.

Vouchers: *19, 21, 22* (SM, *L. fremontii, chuchk kuávul [sic]*); *20* (SM, *L. exsertum, chuchk kuávul*); *23* (SM, DB, *L. californicum, sisipuḍ kuávul*); *258* (SM, *L. macrodon, chuchk kuávul*); *313, 314* (SP, *L. fremontii, regular kuávul*); *GPN 417* (RB, *L. fremontii, regular kuávul*); *LSMC 49* (*L. fremontii*).

Nicotiana glauca Graham
Tree Tobacco
Solanaceae
sanwán

Tree Tobacco is a strong, pioneering, nonnative, large shrub that may become quite tall and woody with age. Originally from xeric portions of southern South America, it has spread in arid and semi-arid areas of North America. It produces prodigious quantities of seeds, which ensure its establishment on the denuded soils of roadsides, ditch banks, and other places that few plants other than Russian-thistle (*Salsola australis*) might consider tolerable.

Tree Tobacco is unmistakable, with its large, glaucous green leaves about a finger long (or several times that in new growth). The long, yellow, tubular flowers emerge from slender grass-green cups (the calyces) and terminate in five short outward-opening lobes. These bunches of yellow tubes on the tips of branches produce abundant nectar that keeps the hummingbirds swarming, particularly during migrations.

At Komatke, Ruth Giff told me this plant was called *sanwán*. She supposed this was because they had always used these flowers to decorate the altars for San Juan's day, June 24 (St. John the Baptist being the patron of the village church). I was content with this derivation.

Then one day I was out with Pedro Estrella, a Pima Bajo from the lower Rio Yaqui in Sonora, 300 miles south of the Gila. A new road into Onavas had been punched out on higher ground. The bare banks had grown up quickly with Tree Tobaccos taller than my pickup. I stopped and asked Pedro about these. "*Shan shuan*," he pronounced. I checked what Pennington (1980) had recorded here a bit earlier: *sa'usuwam* or '*palo San Juan*'.

Some years later, when Ruth and several other Gila Pimas were visiting Onavas with me, I picked some Tree Tobacco in the yard. Ruth asked Pedro in Pima what he called it and was surprised at his response: "I thought that was just our name at St. John's!"

When we were discussing the Pima name *hiósig vaḍagaj* 'flower nectar', Sylvester used Tree Tobacco as an example. "The *vipismal* sure go to that wild tobacco flower, suck in there. I don't know why they call it *sanwán*. Mostly hummingbirds come to that flower."

"Is it used for anything?"

"Just for decoration. It really take[s] quite a while before it get dry. Put [it] in a vase, [with] a lot of water. Other flowers last a day or two and it gets withered."

Why did these two different speech communities, separated by hundreds of miles, have a common name for this non-native plant? Could it have arrived early, along with other Hispanic introductions? Was the name passed from one Piman community to the next in colonial times? As with so many other naturalized species, there are no good answers. Tree Tobacco was established in California by 1848 (Frenkel 1970). Janice Bowers found herbarium specimens (ARIZ) indicating that it was cultivated in Tucson by 1891 and escaped from cultivation at least by 1904. J. J. Thornber collected it at Sacaton in 1916. Possibly both Piman groups derived their name through the localized Spanish name Palo San Juan. But Maximino Martínez (1979) lends no support here, and Howard S. Gentry (1942) recorded only the name Cornetón in his Rio Mayo studies.

I have no information that this plant was ever used for smoking among Gileños, as it was evidently at Onavas (Pennington 1980:174).

Vouchers: *1655* (SM).

Pluchea sericea (Nutt.) Shinners
   [= *Tessaria sericea* (Nutt.) Coville]
   [= *P. borealis* A. Gray]
Arrow-weed
Asteraceae (Compositae)
u'us kokomagĭ

From a distance, a dense patch of Arrow-weed might appear to be a cloud of smoke or a patch of fog. The stems and leaves of this shrub are so covered with microscopic hairs that the entire plant appears grayish except the tender new growth. Arrow-weed colonies occur frequently throughout Pima land, always in wet and saline places such as lower river terraces, ditch banks, field edges, and pump sumps. You might find Screwbean (*Prosopis pubescens*), Seepweed (*Suaeda moquinii*), or Quail Brush (*Atriplex lentiformis*) growing nearby, but mostly the Arrow-weeds are in thickets with no other plant.

Arrow-weed stems are straight and tall—growing to head-height or more, branching mostly near their tops. The lance-shaped leaves, a finger joint or two long, clothe the slender branches, pointing upward. During a long flowering season, starting in midwinter, the flowers, pinkish to rose-purple, cluster in heads at the tips of the slender wands. These tell you this plant belongs to the aster or composite family, as does Seep-willow (*Baccharis salicifolia*), a somewhat similar plant found in the drainages. When the mature flower heads disintegrate, the seeds are blown away on short white fluff.

Drying Arrow-weeds have a distinctive herbal odor, somewhat like that of douglas-fir but sweeter and milder. From birth to death, the riverine Pima were surrounded daily by this plant, its gentle aroma permeating the home. To them it is known as *u'us kokomagĭ*, meaning simply 'sticks gray'. By a curious twist, the Tohono O'odham call the same plant *kokomagĭ u'us*.

George Kyyitan, Sylvester Matthias, and I stood in a traditional Pima *olas kii,* or round house, that had recently been built at the Desert Botanical Garden in Phoenix. The men were explaining to me names for the various posts that supported the structures—central mesquite posts, horizontal mesquite or cottonwood logs, a loose wicker of slender willows running vertically and horizontally like longitude and latitude lines on a globe of the northern hemisphere.

George reached up to tug at the barbed wire lashing the crossbars. "In the olden days these should be tied with mesquite or willow 'peelings' [bark]."

"How about *kuujul* [Screwbean]?" I asked.

"Yeah, that's okay. Screwbean or *che'ul* [willow]."

Each part had its name, the white-haired men agreed. Over the branched structure was piled or tied the *u'us kokomagĭ* from ground level to the zenith, with no smoke hole. Then almost everything was covered with fine alluvial silt. It had rained heavily the previous day and night, yet only a bit of floor was wet at the west side, along the more vertical part of the wall devoid of mud. We breathed in heavily, the sweet smell of mesquite timbers and Arrow-weed filling our nostrils.

"I grew [up] in one of these kind of house. I do [did]," commented George. "But not him," indicating Sylvester, "he's too young." I was surprised. Square adobe houses had begun to appear along the Gila in 1880 and in a few decades supplanted the round house. At the turn of the century, Russell wrote, "As an inducement toward progress, the Indian Department or its authorized agent has stipulated that a man must cut off his long hair and build an adobe house before he may receive a wagon from the Government" (1908:153). But George's village was a good distance from the agency, at least by horse-and-wagon standards, and harbored a pocket of conservatism. There were just three such round houses left in George's youth. Curtin located only a single *olas kii* in the 1940s. By the time of my arrival on the reservation in 1963 there were none.

"And I slept on a *main,*" George continued, pointing to a woven cane mat on the dirt floor to one side.

"What was it made from?" I asked.

"Ah, what do they call it?" George tried to remember.

"It's similar to a palm. They call it *umug,*" recalled Sylvester. "That's what they made the *main* out of." (Later they pointed out an *umug,* Sotol *[Dasylirion wheeleri],* along the trail.) Sylvester added that he had slept on a tarp on the ground, one of the inroads of modernism.

As we stepped out of the round house, the *vatto,* or open ramada, was facing us. Its roof was a thick layer of Arrow-weeds. Nearly every Pima house has at least one of these somewhere in the yard; it is the focus of many of the daily activities, especially during the long summer when the modern houses become unbearably hot. Many Pima families then cook, eat, and sleep out under their *vatto.*

Next to the *olas kii* was another structure that we entered: the roofless outdoor kitchen, made entirely of upright Arrow-weeds thrust into the ground in a large circle, held in place by a few horizontal stays. The walls were about shoulder-high, the space enclosed large enough to seat a dozen or more people easily.

"What do you call this?"

"*Uksha,*" said George. "I think almost everybody had it." The plural is *u'uksha.* They too had entirely disappeared by the time I arrived in the early 1960s.

"Eat and cook in there," added Sylvester. "Just sit on ground—no tables, chairs." Early in the century, Lloyd (1911:143) called this the *onum,* saying, "One or more may be found near almost any Pima hut." Both names are remembered.

Day and night these riverine people were literally surrounded by their *u'us kokomagĭ.* When I first went to the reservation, most houses (the so-called sandwich houses) still had roofs thatched with a thick layer of Arrow-weed covered with a still thicker layer of adobe mud. In summer these little square houses were far cooler than the modern frame houses, which turn into virtual ovens.

The next day I visited Ruth Giff at Komatke, mentioning our travels through the botanical garden. She said, "You know, on that *vatto,* I'll bet a lot of people never even notice the *homda.*"

"We did. Did you make it?"

"No, but I just sold one, finally, to the Arts and Crafts Center. They didn't want it. They want pretty things for tourists to see," she said, her frustration evident. "But I finally won, and now we're going to set it up the right way, over some logs."

The *homda* (pl. *hohomda*) is an Arrow-weed storage bin or basket that was once found atop the *vaptto* (ramadas), the *koksin* (storage houses), and standing on high spots of ground in the yard on a platform of logs. A family might have a dozen of these bins for the storage of mesquite pods, screwbeans, wheat, and other crops.

Like the *uksha,* the once ubiquitous *homda* had disappeared years before I went to Pima country. Joe and Ruth had told me about these and the *vashom* (pl. *vapshom*), the straw storage basket for grains. The weaving technique for the *homda* differed from that of all other baskets. I asked Ruth if she still remembered how to make one. She did, and she agreed to let me film the process.

We went to cut *u'us kokomagĭ* when the right time came, early in the summer— then the newest growth is at its most flexible for weaving. She selected the leafiest and thinnest shoots for harvest. We got a good load in my pickup from Co-op Colony. We also went down to the Salt River for some young willow branches, another item she needed.

The next morning Ruth began by weaving a circular base of willow branches about five or six inches tall; it looked like a large leafy Christmas wreath. Then she continued the spiraling walls upward by inserting withes of Arrow-weed branches, leaves and all, weaving them quickly in a braidlike fashion. Ruth explained that the usual Pima coil basket is woven from left to right (counterclockwise), but the *homda* goes in the opposite (clockwise) direction (see weaver in Russell 1908:134). In a matter of a few hours, the storage basket was finished, unlike the tray baskets, which may take months or even years to complete. When finished, her basket looked like a great leafy bird's nest, lacking both bottom and top. "They make *big* ones," she said, indicating a diameter of a yard or more. When used, the *homda* would be placed on a base of four small logs lying down and overlaid with sticks and finally a layer of Arrow-weeds. The top was covered with a roof of Arrow-weed and mud. The leafy lid rested directly on the contents. Then the bin looked more like a fat toadstool. "Never hold a *homda* from the rim," Ruth cautioned. "Hold it from the bottom." (Robert Brunig later bought this *homda;* it can now be seen in the ethnoecology exhibit at the Heard Museum in Phoenix.)

I had always assumed that this plant was selected for the weaving of storage bins merely because of its physical properties (long, slender wands) and the convenience of its growth along the rivers. Then one day when Francis Vavages was talking about wheat harvest and storage, he added, "It's usually stored in the *homda.* Reason they used that Arrow-weeds is because it has that smell—for bugs not to get to it. Things that you leave for a long time, [they] don't touch [it]. Mice won't get to it either. But the straw one, *vashom,* they will."

According to the Creation Story, when the Pima arrived in the area they now occupy, they were not irrigationists. Storyteller Juan Smith noted, "That farm [*oidag* 'fields'] that the Pimas made at that time didn't have any ditches in it; they were raising crops from the rains" (Hayden 1935:10). But they saw all about them a great canal system left by the Vipishaḍ, a Puebloan people. Each version of the myth tells how the people decided to try their skill at digging ditches to water their fields and how they had to resort to four *mamakai* 'shamans' to get the water to flow. Facing remnants of canals that showed great engineering skills, the Vupshkam, the conquerors or Emergenti (who are the riverine Pima), had to learn the tricks anew.

And they did, using brush weirs across the river. I asked Francis Vavages how the system worked.

"Getting ready for planting, they have a meeting. They always have somebody to sit and be the head of the people, appointed to get them all together, the *shuudagĭ ñuukuddam* ['water caretaker']. And they talk it over: 'Now it's planting time. Let us build that dam, right across there, and get some of that water.' After they all settle down, they give each one [a chore]. See, in those days they have wagons and horses. They say, 'You bring some Arrow-weeds. Get a wagonload of Arrow-weeds. And you, too. And you, too, over there, bring some wheat straw, *pilkañ moog.*' And they appoint so many.

"'You people cut some mesquite—long enough that we can put them down.'" Here he made pounding motions with his hands showing how the poles were pounded from the tops into the riverbed. "They put it in as far as they could. About three feet apart. They put a board over the top of that stick, and there's so many axes, hitting on there,

on the board. It'll go down. They go in, say, five feet. It's mostly sandy. They look for where it's kind of narrow. Sometimes somebody would drive their wagon and stand on there. And they wiggle it—it'll go down. They sharpen the bottom. It'll go down. It's loose down there.

"'And you bring some branches, just limber enough that you can kind of weave them between those sticks.' When they get through with that, they say, 'Alright, it's time to bring the *pilkañ moog.*' So that's what they do. They put the Arrow-weeds first, you know, like that. They put them right together, so it won't wash away. Then they put the straw in front [upstream side], and that's how they back it up, you know.

"Then pretty soon they'll say, 'Well, let's see, we'll get some of that dirt over there, and mix it in there, put it in there, to close some of those holes.' That's how they do. Of course, they let a little go by, from way under there. [Then] here comes the water [in the ditches]."

"What time of the year did they build that dam?"

"Right after the season, like in wintertime. Sometimes they say it's cold, [a] lot of [the] time, but they still go in the water. To get ready for [irrigating] the wheat [in March]. They appoint ditch riders to oversee. Sometimes the ditch needs cleaning. Some brush growing. I was just a young man then, but I know how they do [did it]. My father used to go to the meetings."

"How long would the dams last?"

"They'd last all season. They kind of let them go over the winter, because it's cold. Like I said, at the beginning of spring, that's the time they start."

"How about the people who lived in villages farther downriver? How did they get water?" I asked Francis.

"Oh, they have water too. It used to come out of the ground. The water level was so high around this place here. You can go down four feet and strike water. A spring—*vo'oshañ,* they call it, where there is springs coming out. They just have a big lake. And they open that thing to water there." Francis went on to describe five places he knew of where brush weirs were constructed for the Pima fields, with others for the Maricopa. (For another version of how weirs were built, see *Atriplex polycarpa* account.)

Pima war and hunting arrows were made from Arrow-weed stems. Sylvester

Matthias made hunting arrows, fletching them with the wing- or tail-feathers of *ñui* (Turkey Vulture) or *haupal* (Red-tailed Hawk) bound to the shaft with sinews. The finished arrow is called a *hapod.* But, as Sylvester explained, there were other Pima terms to indicate different stages of arrow making. When the *u'us kokomagĭ* is just cut and dried for arrow blanks, these are called *sheesha.* When they are peeled and smoothed, they are called *dadagiuma.* Finally, when they are notched and pointed but not yet fletched, the blanks are termed *hihivshana.* Arrow-weed grows so straight that the Pima had no need for arrow straighteners; they could just select the straightest stems and smooth these a bit with a coarse stone such as lava.

Joe told me that Pima boys formerly had to practice their skills with archery and shield to defend themselves during

u'us kokomagĭ   *Pluchea sericea*

Apache raids. They practiced shooting at each other and dodging arrows. They also played a game where they folded Arrow-weeds into a bundle and bound it with bark to make a *wulivga* or target. (*Wuliv* means to bind something tightly, he explained.) Curtin (1949:131) noted the game also, saying that the target might be made from Sweet Clover (*s-puluvam, Melilotus indica*) or long grass (*vashai*) as well. Ruth said a *wulivga* was also made for use as a torch to light one's way at night.

Sylvester also spoke of two kinds of *wulivga*. "A game for the boys—target practice. Somebody be dragging [it], somebody try to shoot it. Made out of bark, willow bark or Screwbean bark."

"What was inside?"

"Anything fine. But for a torch they bundle up dry Arrow-weeds. They have to use green bark or fine limbs, tie it up so they won't burn—pull the torch together."

"How long do they burn?"

"About 20 minutes. By then, they be [have] another torch ready."

Francis Vavages also explained the dual role of the *wulivga*. "Kids make it out of willow, sort of a target and bind it with something. Put it way down there and try to hit it with arrow. Sort of practice, too, when they hunt with their bows and arrows. They usually do that torch with Arrow-weeds. They burn for a long time. They go out at night for quail in the wintertime. They switch them back and forth. They use a mesquite branch with a lot of branches to hit them. It's called *wepegi* [nocturnal torch hunting], to get their food. It's sort of mean to the poor birds, but they have to get their food. [Use] dry Arrow-weed—bundle, tie up—[then] waving [them] back and forth—*ha vepegi; to vepegi kakaichu*." (For Joe Giff's description of this mode of Pima hunting see Rea 1979.)

Sylvester knew other traditional uses for Arrow-weeds. One of these was in the construction of quail and dove traps, called *eḑpa*. I asked him to make one. He had rods of *u'us kokomagĭ* but needed the binding material. So one day we went to the Salt-Gila confluence at 115th Avenue, and he trudged out across the sand with a small ax and knife until he came to a grove of good-sized willow saplings. He chopped away a strip of bark an inch or two wide, then peeled out strips of the fine inner bark about a yard long—one

strip from a tree. When I visited him a few mornings later, he was building the trap, laying the rods up log-cabin style, parallel sets crossing parallel sets to make a square. At his feet was a bucket of water in which willow bark was soaking. This remarkably tough material bound the four corners. The roof was finished off with a layer of parallel rods, bound on top with a carrying handle. Even the trigger was two notched pieces of Arrow-weed. Provided with a floor, the same structure became a cage for doves or other birds (see Russell 1908:102). The Pima name for the cage is also *eḑpa*.

Formerly an all-purpose screen was made of parallel rods of Arrow-weeds bound at each end and usually the middle with bark or rawhide (see Kissell 1916: 140–147). These might serve as the door to the olas kii. Such doors were called *kiijeg kuukpaḑag* 'house-opening closer'. Smaller ones suspended from a beam of the *vatto, kosin* (storage house), or *kii* served as general-purpose shelves once called *kuukta* (in Spanish, *tepestes*). While I have seen these *tepestes* commonly in Pima Bajo country, they have nearly disappeared from the Gila villages; their place is now taken by various plastic or metal baskets. When made rigid with two or three crossbars, the shelf doubled as a bird trap. Russell (1908:101) reported, "When used as traps they are tilted at an angle of 20° or 30° from the ground and supported in that position by a short stick to which a long cord is attached. Wheat strewn under the trap lures small birds, which are caught when a jerk on the line removes the supporting stick. One of these traps was seen at nearly every house during the winter of 1901–02." Sylvester said this type of trap was called *eḑpa*, just the same as the cage-shaped one, and George Kyyitan concurred.

U'us kokomagĭ served as the spindle in the two-piece fire drill, called the *iivdakuḑ*. (For an illustration of the drill, see Russell 1908:103.) Lloyd (1911:105) left the most detailed account of how this was used:

There were two parts to the apparatus. Gee-uh-toe-dah [*gi'atod?*], the socket stick [hearth or fireboard] was of a soft dry piece of giant cactus rib, and a notch was whittled in one side of this with a small socket at the apex, that is on the upper side.

This was placed flat on the ground, with a bit of corn husk under the notch, and held firmly in position by the bare feet. The twirling stick, eev-a-dah-kote [*iivdakuḑ*], was a hard arrow weed, very dry and scraped smooth. The end of this was engaged in the little socket, at the top of the cactus rib, and then, held perpendicularly, was twirled between the two hands till the friction rubbed off a powder which crowded out of the socket, and fell down the notch at its side to the corn-husk. This little increasing pile of powder was the tinder, and, as the twirling continued, grew black, smelled like burned wood, smoked and finally glowed like punk. It was now picked up on the corn husk and placed in dry horse dung, a bunch of dry grass, or some such inflammable material, and blown into flame.

It looked very simple, and took little time, but I never could do it.

The Arrow-weed rod served some other purposes as well. With a Saguaro crossbar it functioned as a simple spindle for spinning cotton thread. On the Pima loom, the lease rod, heald rod, heddle, shuttle, and part of the temple were of *u'us kokomagĭ* (Russell 1908:148–152). A long, pointed Arrow-weed rod once served, Sylvester told me, to turn over or remove fry bread from boiling grease. It was called an *uskonakuḑ*. This is apparently the same as Russell's (1908:101) "doughnut" fork. The *vaa'o*, or stirring stick, was a utensil used when cooking things in an olla. Sylvester made me a *vaa'o*. It consists of two smoothed rods about 16 inches long, tied closely together through grooves 3½ inches from their upper ends. When used, they are held somewhat like long cooking chopsticks. George Kyyitan reminded me of another use for the *u'us*—the *iágta* or decorated sticks held in each hand by dancers in many of the Pima ceremonies.

Still another ceremony, the lustration or 16-day purification of an enemy killer, utilized *u'us kokomagĭ*. In one of our last conversations, Joseph Giff told me about how it was used: "When a person is going on the warpath and kills an enemy [enemy killer, *oob muakam*], he'll have to purify himself—purification ceremony. My father said that they burn green Arrow-weeds—and they have to stand in the smoke—and stand in that smoke. He says it's unbearable, but he says you have to do it. Something like that—you have to be purified to run, too."

"And they didn't eat salt for so many days," added Ruth, his wife.

"There's something about salt," continued Joe. "I don't remember just exactly what that was. Just stand in the smoke till fire burns out. Then they have to go and jump in icy water, and then get out and stand before the fire. It used to get cold in those days. They said sometimes there would be snow [ice] on top of the water, but still they'd have to jump in there."

There were also joyous occasions where this common plant played an essential role. Russell (1908:160–161) told how both men and women painted their bodies: "Usually the face alone was painted, but during festivals and on other special occasions the entire body was painted. On dress occasions the lines on the face were made much narrower, and instead of being applied with the hands the color was laid on with a splinter or twig of arrow-wood 2 mm. wide by 80 mm. long." He also recorded that the Koahadk "were accustomed to gather the tufted ends of the arrow-bush branches and carry them southward into Papagueria [Tohono O'odham country] to be used as paint brushes" (Russell 1980:104).

There was even a medicinal use for *u'us kokomagĭ*. Russell (1908:79) recorded, "The bark of the arrow-bush root is separated by pounding between stones and then placed in water for a few hours to extract a liquid for washing the face and for sore eyes." Nearly 90 years later, I asked Sylvester if he had ever heard of any medicinal use for *u'us kokomagĭ*. "Yeah. The *root* . . for sore eyes. Also as deodorant [deodorizer]—in house. Just burn it. In the coach's house [at St. John's School] a skunk got in and they tried to get it out. It sprayed. Burn *u'us kokomagĭ*—it'll burn, even if it's green."

Another use for Arrow-weed was in burials. Traditionally, a Pima was buried in a side cavity at the bottom of a circular excavation. When the grave was filled, it was covered with a thick layer of Arrow-weeds, like a thatching, held down with heavy logs of cottonwood or mesquite removed from the dead person's house or shed. Clusters of these pit burials can be found at older village sites across the reservation, but

their Arrow-weed coverings have since decayed and only occasional traces of the beams survive.

Still another use for Arrow-weed—a sacred use—seems to have slipped entirely from riverine Piman memory: making the *omina*. Votive offerings—from candles to prayer flags to *pahos*—are found throughout the world. Throughout the Greater Southwest, prayer sticks or arrows were once prominent, as they still are among the Hopi and Zuni and certain other pueblos, where "cutting prayer sticks" is still an important activity of men in the kivas. These prayer sticks usually consist of a short stick, pointed at one end and bound with homespun white cotton string or yarn, often with feathers attached. Beyond this basic theme, the details varied with the different groups from the Cora and Huichol and Southern Tepehuan northward. Culver Cassa notes that this Piman name for prayer sticks, *omina,* is derived from the word *mulin* 'to break one thing' or 'to break off something'. The verb in the plural is *omin.*

Russell must have been the last to find the *omina* among the Gileños, and of the six sacred bundles obtained, he considered some to be very old. Most of these were in sets of four, some painted, others plain. He wrote (1908:106), "They are usually of arrow-wood; always bound together with cotton twine of native spinning, either with or without feathers attached to each separate stick." He (1908:265–266) noted that the prayer sticks were used by the *maakai*, or shaman, in curing certain diseases. In a rain song he recorded (Russell 1908: 331–333), the singers are in the fields of corn and squash throughout the night, singing before their prayer stick bundles:

The darkness of evening falls

As we sing before the sacred *omina.*

About us on all sides corn tassels are waving.

The white light of day dawn yet finds us singing,

While the squash leaves are swaying. (Russell 1908:332)

As with many sacred things, the origin of the *omina* is encoded in the Pima

Creation Story. The most detailed version (Lloyd 1911:133–142) ascribes the first manufacture of prayer sticks to I'itoi immediately after his resurrection. This is the prose account: "And he began then to cut up all kinds of sticks, four at a time, and to lay them down and look at them, but he liked none of them. Then he cut arrow weeds, four of them, and he liked their look. And he lit his pipe and blew the smoke over them, and spread his hand above them, and he liked the light of them which came thru his fingers. And he put those sticks away in his pouch" (Lloyd 1911:134).

Then follows, in mirror image, a formal oration:

He examined the sticks, but none suited him;

He eyed along the river, that green snake, which he had made, and found the sticks that pleased him.

And he cut those arrow-weeds, he found there, into four pieces, and blew smoke over them.

And out of them came sparks of light, that almost reached the Opposite World, the World of the Enemy, where things are different.

And when he saw the light from the sticks he smiled within himself;

He was so pleased he had found the sticks that suited him. (Lloyd 1911:136)

Then follow eight stanzas explaining how I'itoi endowed these prototypical prayer sticks with power.

For the Pima (at least formerly), the mundane, the commonplace, and the sacred were mixed. Every moment of the day or night, the Pima were surrounded with things made from *u'us kokomagĭ*. And one of their most sacred of votive offerings must be made from this bush. Other tribes might use homespun cotton string and feathers from various birds also, but the Arrow-weed was uniquely Piman. It comes only from along the desert river, "that green snake, which he had made." The river was a sacred thing because I'itoi had made it, and so were the Arrow-weed prayer sticks "that pleased him."

Today there is no green snake in Pima country. No one even recalls the *omina.*

Vouchers: *124* (?SM); *1748* (GK); LSMC *55*.

Salazaria mexicana Torr.
Bladder Sage, Paper-bag Bush
Lamiaceae (Labiatae)
siv u'us

Nearly all the older west-end Gila Pima knew a plant that was used in their youth for medicine, a plant called *siv u'us* 'bitter sticks [or] wood'. All remembered what it tasted like. But none know what it looked like. Older people than they had collected the roots, or peddlers had supplied them. None of the native consultants with whom I worked knew the plant itself.

Ruth Giff told me, "*Siv u'us* is a medicine stick; means 'bitter sticks'. Chew on it to get cured of whatever. Have to go up in the mountains to get it." Nearly two years later she told me, "*Siv u'us*—they suck on it and it's *siv* [bitter]! It grows up on the mountains."

Down at Santa Cruz village, Carmelita Raphael said she did not know if it was up in the Sierra Estrella or not. "Our parents would boil it when we got sick—it's sure bitter."

Sylvester Matthias echoed these words, adding that *siv u'us* can be found growing in the Sierra Estrella to one side of Tamtol Do'ag opposite Komatke. (The old village of Komatke used to be there until the flood of 1906 washed out the fields.) "It's good for headaches, laxative, or fever."

Some years later Sylvester told me, "There's a man that knows where to get *siv u'us*, Peter Fall. They call [him] Gakoḏk Piivlo, Crooked Peter. He move[d] up to Casa Blanca, Sweetwater. He's kind of a hunchback." I knew the name. The man had been dead for decades.

"He knew where to get *siv u'us?*" I asked.

"Right there—at Tamtol Do'ag on flats. [It's] something like a Mormon Tea [*Ephedra*] bush, only it's gray."

Sometimes it helps to check on a Pima folk taxon with a Tohono O'odham, because many names are shared. The next time I stopped to visit Tohono O'odham Frank Jim in Roll, western Arizona, I asked him about the plant. He had some dried root in the kitchen window above the sink. "*Siv u'us* is used for colds. Has purple flowers. Hard to find otherwise, when not in bloom. Dig out roots, scrape them. They're more or less white when fresh.

Boil them to use for medicine." The dried root he showed me was yellowish.

Over the next ten years he gave me more bits and pieces of information. "*Siv u'us* grows in the mountains, the Ajos, but not in the *akĭ,* arroyo. Can be maybe six feet tall. Purple [not bluish] flowers," he insisted, "with small leaves. Blooms in March, maybe found in April." And later he said, "Roots pretty bitter: gray-looking stuff. Found this side [west of] Gunsight. [Also] Ge Wo'o [Gu Vo or Kerwo on Anglo maps]. Flower purple and leaves small. If it's not in bloom, forget about it. You'll never find it." He was not sure about leaf arrangement but did not think the plant had thorns. At least with this I had a few concrete leads to pursue.

Thomas Van Devender has made a career studying the climates of the Southwestern deserts throughout the last 30,000 years by analyzing the types of plants that are incorporated at different places and times into packrat middens. A cemented midden accumulates materials over perhaps thousands of years. A researcher can radiocarbon-date different layers of the midden and identify the plants that were growing in the immediate vicinity of a nest at a given time. So unlike most botanists, who rely on whole specimens (usually with fruits or flowers) to identify a species, Tom has learned to identify whatever parts and pieces the packrats take home. From these identifications and his knowledge of the requirements and tolerances of different plant species—such as wet, hot summers and cool, dry winters, for instance— he builds pictures of past environmental conditions. When he was in San Diego to give an ethnobiology seminar, I asked him about *siv u'us,* giving him the verbal parts and pieces I had gleaned so far.

He thought for a while, and then said, "One time Paul Martin and I were searching the walls of Grand Canyon, and Paul threw me down a branch of something and said, 'Taste this.' It was terribly bitter."

"What was it?" I asked.

"*Salazaria mexicana,*" Tom said. The name meant nothing to me. "Well, it's a sage. And it has purple flowers, and it's pretty hard to find when it's not in bloom, just like that Papago told you. I've collected it as close to Pima country as the White Tanks." This range of mountains is just 18

siv u'us *Salazaria mexicana*

miles west of the present reservation boundary, within easy hunting distance for Gileños.

Andrew C. Sanders, herbarium curator at the University of California, Riverside, sent me an herbarium specimen in the fruiting stage with the diagnostic papery pods (really the expanded calyx) that give the plant its English names. Also, he dug out a number of roots and lower stems for me to use as specimens to check with the Pimas.

I gave some to different Gila Pimas; they thought this was the medicinal plant they had known. Meanwhile, Frank Jim and his wife Lela had moved to Sacaton, where I was writing. He stopped by for a visit at the beginning of April 1987. I gave him some of the root and stems.

He chewed into the stems and said, "It sure is bitter. Isn't this *siv u'us?* I think the plant is bigger than this. About as thick as your thumb." Then I gave him a root that size, and I guess he was satisfied. "You boil the root [and apparently stems] after scraping them clean. Can boil it over and over for medicine. Just take it out of the pot and dry it. Use it again next time until all the bitterness is gone. Drink it hot. Sweat it out of you. Now is the time to find it. In a few more weeks, will never find it when the flowers are gone." I asked Frank about the papery bladder pods. "We don't know about that, because after we dig it, we don't stay around to see what happens after the flowers are off. I couldn't find it the rest of the year. Just looks like all the other bushes then."

Salazaria mexicana is an intricately branched, dense shrub about waist-high. The branching is always opposite—one on each side of a main stem—and so is the arrangement of the small, simple leaves. The flowers are mostly purplish, but the tube is white and the lower lip is deep purple. The papery "pods" or inflated calyces are tan, about the size of a thumbnail. The plant occurs mostly to the north and northwest of Gila Pima country, but there are various disjunct populations. Tom Van Devender has traced the plant to the nearest range north of the Salt-Gila confluence. Tom Daniel and Mary Butterwick (1992) found it in a single canyon bottom on the north side of the South Mountains, a range once known thoroughly to the Gileños. Botanists have not yet discovered *S. mexicana* in the Sierra Estrella.

Vouchers: *A. C. Sanders 6517* (Frank Jim, Tohono O'odham, RG, SM).

Salsola australis R. Brown
   [= *S. iberica* Sennen & Pau]
   [= *S. kali* L. var. *tenuifolia* (Tausch.) Aellen]
   [= *S. pestifer* A. Nels.]
Russian-thistle, tumbleweed
Chenopodiaceae
vopodam sha'i

Kittie Parker (1972) justifiably calls the Russian-thistle "one of the most prolific and obnoxious weeds throughout the state

vopodam sha'i   *Salsola australis*

[of Arizona]." Perhaps more people know this plant by its vernacular name tumbleweed, a term to which biologists object because it is generic for any number of round, bushy weeds that break loose from the soil at the end of the growing season and roll along in the wind. Russian-thistles are not thistles but members of the chenopod family. They are early colonizers; fallow fields may be completely covered by seedlings of this Eurasian plant. They begin innocently enough as small, somewhat succulent plants, bluish green except for a few pinkish ridges along the stems. The leaves are linear, hardly distinguishable from the stems, and tipped with a small spine. The plants continue to grow until they form globular bushes about waist- or even shoulder-high. As with many other chenopods, the flowers are not much to write home about, but they make up for this inconspicuousness by prolific seed production. As the bushes dry the following winter, they turn pale straw-tan, uproot, and go rolling along in the wind, spreading seeds, until they are caught in barbed wire fences or mesquite trees, or pile up against people's houses.

The Gila Pima must contend with this plant, particularly when they leave a field fallow for a year or so. They named it *vopodam sha'i* 'rolling bush'.

Sylvester Matthias told me one day, "It's not a native. Like the Salt-cedar. They said

vopodam sha'i comes from Texas. Wasn't around when I was a child. Not good for anything. Stock don't eat it."

"Do you remember when it came?"

"I think it came recently, in the forties. No good for anything!"

In her fields one August morning, Ruth Giff mentioned tumbleweed as one of the weeds she remembers coming in, along with Salt-cedar, which I was helping her pull out.

Francis Vavages also mentioned *vopodam sha'i* while discussing Salt-cedar. "Came in about the same time, during the floods, from up that way, anyway," vaguely waving toward the east. "Some say they come from Texas."

The association of events and species is not a casual one. After Anglos came to dominate the state, irreparable damage was done to the Gila watersheds, and the delicate soils all along its course were disturbed by overgrazing the desert and semidesert vegetation. One immediate effect was devastating floods that ripped out riparian vegetation, turning the once channeled river in the Pima homeland into a broad sandy waste. Two invading plants that proliferated in these unstable and disturbed environments were Russian-thistle and Salt-cedar.

On an outing one morning to Black-water, Francis Vavages pointed out some Russian-thistle growing densely along the road edge. He called the plant *memḍedam sha'i*, meaning 'running [or] flowing bush'.

"Is that the same as *vopodam sha'i*?" I asked.

"Yeah. That's the plural."

I thought he was joking. But Mathiot's (n.d.:59, 66) dictionary gives *meḍdam*, *memeḍadam* as 'running', 'runner', 'one who runs repeatedly', the singular of *vopodam*. I heard no one else use the plant's name in the singular form.

In July 1991 I took Sylvester Matthias to Bapchule to visit with George Kyyitan: between them, they had 159 years of oral history. I had been asking about grasslands when their conversation drifted to intro-duced plants: Johnson Grass *(Sorghum halepense)*, Salt-cedars, and tumbleweeds, now-abundant plants that each of them remembered seeing invade their lands.

Then George turned to me and said in English, "I heard one guy, one old man said, it's self-thrashing seeds, *hejel kaihundam*—they roll out and spread out—the *kai*, seed, *kaipig*. Spreads out. But I never heard anyone else [use that name]. He's the only one." This is similar to one of its Tohono O'odham names, *hejel e'eshadam* 'something that sows itself' (Saxton, Saxton, and Enos 1983:20; Felger and associates 1992:25).

Vouchers: *425* (?sp); *906* (?rg); *1728* (fv).

Sarcobatus vermiculatus (Hook.) Torr.
Chenopodiaceae
Greasewood
kavk u'us, kauk u'us

In the saline flats and swales where Pickle-weed *(Allenrolfea occidentalis)* might be found is an even more common, usually shorter shrub that might pass as wolfberry *(Lycium* spp.) that has recently leafed out; all three of these bushes may be found growing rather close together. Greasewood has slender succulent leaves about ¾ inch long that give the plant an almost feathery appearance, out of keeping with its harsh habitat. In late spring a few coral-pink male catkins appear, harmonizing with the fragile blue-green leaves. Concealed stiff, spiky branches deter familiarity.

Kavk u'us is the name commonly used in the eastern part of the reservation; *kauk u'us* in the west.

Joseph Giff told me, "*Kauk u'us* is used for making corn huskers; plant has little needles like; and is used for making *ovij* ['awl'], too. They are in the alkali by Felix Enos' place." Ruth Giff added that *s-kavk* means 'hard'. Sylvester Matthias elaborated various additional uses for this special dense wood: "*Kavk u'us* means 'hard wood' or 'hard sticks'. Grows in saline areas with *iitañ* [Pickleweed]. Wood is used for making corn huskers; pegs to put into ground to stretch sinew for bow strings; pegs to put into ground for shaping a bow, if double-curved; also for wedges. Not eaten." The corn huskers he made for me, as well as the handles for some small awls for basket weavers, were fashioned from *kauk u'us*.

After discussing the *iitañ* growing in the *ongam*, a salt flat, near Felix Enos's place (at the edge of the Komatke feast grounds), Ruth Giff added, "*Kavk u'us* grows there, too. They make *ovij*, to make the *big* basket, the straw basket. I got some *[kavk u'us]* to use for mine, and I just ended up using an ice pick [instead]. And they use it, too, when they gather corn, ripe corn (and I use a knife!), because in those days they didn't have knives. And they'll get polished, *shining*, just like hard wood. I guess [that's] why they call it *kavk u'us*."

After killing an enemy, a Pima warrior had to be purified for 16 days. During this time he could not touch his head, particu-larly his hair. He kept a scratching stick, which might be made from Greasewood, for that purpose (Russell 1908:204–205; see also Group E, *Acacia greggii* account).

Hrdlička (1906:41) recorded a special use for *s-kauk u'us*. "A few of the old men about Casa Blanca at least occasionally assume the role of shamans. During the long dry spell which ended in the spring of 1905, they resorted to the ceremony of 'rain-calling,' a part of the preparation for which consisted in obtaining some sticks of *kauf-ku-us*, hard-wood (Sarcobatus vermiculatus), which were cleaned and notched on one side; they were then rubbed along the notched portion with a smaller plain stick during incantations to the rain deities."

Russell (1908:167) also remarked on these scraping sticks: "So important are these instruments in Pima rain ceremonies that they are usually spoken of as 'rain sticks.'" He collected four sets of scraping sticks. Though the materials are not iden-tified, two of those of local origin appear to be made from *Sarcobatus* wood.

The Pima certainly seem to have associ-ated this rigid bush (with its delicate leaves) with rain, as further indicated by another religious item Russell (1908:108) obtained: "There are two wands or cere-monial sticks in the collection. The longer (760 mm) is of grease-wood, Sarcobatus vermicularis *[sic]*, the material prescribed for iákita, or ceremonial paraphernalia of this class. It is spotted with black and red paint. The shorter wand (280 mm) is of willow, spotted with red. Both were made to be held in the hand during ceremonies intended to bring rain, to cure disease, and for kindred purposes."

George Kyyitan, still living in the vicinity Hrdlička wrote about 80 years earlier, told me about the *jujkida*, or 'rain-making', as he called it. "I've seen it once when I was a young child. It was on St. Peter's Day [29 June]. I didn't see what they were equipped with. They used to have it down there—in Anagam [on the Tohono O'odham Reservation]—every few years."

I asked George Kyyitan if he knew *kauk u'us*. "I know they use it for that singing—[to make] *hivkuḍ* [scraping stick or rasper]. I was looking for one of those *kauk u'us*; there's not much any more. Used to be lots over here." He went into a back room and brought out a scraping stick he had made from Salt-cedar. "It's good because it's got no 'brain' in there: *oág*." (In Pima, 'brain' and 'marrow' are the same word as the pith of a plant). I agreed to take George out the next morn-ing to look for *kauk u'us* to make *hivkuḍ* if it did not rain. But it poured. Some months later we scoured the east end of the reservation for suitable thick straight branches of Greasewood. We found a few.

Francis Vavages remarked, "*Kavk u'us*—it's hard. They use that to beat their grain, wheat [for thrashing]. Make healing sticks too, and touch the person with that stick, too."

"What do they call it?"

"*Hivkuḍ*. Little notches on there; makes noise. Sometimes they put it over a basket [turned] upside down, and they use that [scraping stick]. Singing. Dancing." What Francis was describing is a healing ceremony where the Greasewood scraping stick is not only a musical instrument but a blessing stick on affected parts after sacred songs have been sung. This is a widespread gesture among indigenous peoples of the Greater Southwest.

George said a digging stick made from *kauk u'us* was used to loosen the ground when making canals. Men and women worked together, the men digging the soil and the women carrying away the earth in baskets on their heads.

In 1901 the botanist J. J. Thornber of Tucson helped Frank Russell with Pima plant identifications. Russell (1908:84) quotes him as saying, "The native saltbushes . . . are eaten by range stock. . . . The true greasewood, Sarcobatus vermicularis *[sic]*, a species closely allied to the saltbushes, is also browsed to a considerable extent."

Some misnomers die a hard death. Greasewood in the Southwest is *Sarcobatus vermiculatus* and has nothing at all to do with Creosote Bush *(Larrea divaricata)*, to which plant the name is sometimes misapplied.

Vouchers: *161, 164, 1503* (SM); *1364, 1970* (GK).

Solanum elaeagnifolium Cav.
White Horse-nettle
Solanaceae
vakoa hai, vakoa hahaisig, vakoa hahaiñig

White Horse-nettle relishes the soft alluvial soils of Pima yards, fields, and ditch banks, particularly when they are slightly moist and saline. The erect semiwoody (suffrutescent) stems may reach almost to your knees. You can mistake this for no other plant in Pima country. The stems and even the leaf midribs are armed with slender spines, giving rise to the "nettle" part of the English name, a misnomer. The finger-length leaves have undulating margins. The entire plant is pale whitish green and somewhat scruffy, hence the "white" part of the English name. In the spring and early summer, the plant bears

attractive purple flowers that resemble violet starfish, the kind with webs between their five arms. Each flower has a central cluster of contrasting yellow stamens. Later in the season, fingernail-sized fruits develop, looking quite like green-striped whitish tomatoes; they turn yellow in the fall. These are quite tough, persisting like dangling Christmas tree ornaments even after the plants dry or freeze back in winter.

The Pima name literally means 'broken gourd' (*vakoa*, bottle gourd, *Lagenaria siceraria*), but the reasons for this seem to be lost in the mists of time.

Everyone who farms in Pima country is familiar with this plant. For instance, Sylvester Matthias said, "It spreads in fields. Not used for anything *[sic]*; awful pest; hard to dig out of ditch banks."

Ruth Giff had it in her fields, too, but took a more benign attitude: "I use that to

---

vakoa hai, vakoa hahaisig, vakoa hahaiñig

Solanum elaeagnifolium

make cottage cheese. Just crush some of those berries in there. You know, some people use something out of the cow's stomach. Save it and put a little in there. But we just use the *vakoa hai*."

"How many of those berries do you use?"

"Oh, a lot. I have to break them. About a dozen, I'd say, for a gallon of milk."

"Do you leave them in there?"

"No, I get them out . . . put them in something—a cloth."

"Do you keep them?"

"No. I just use them once and throw them out."

"But what if you want to make cheese at a time when those little yellow balls aren't on the *vakoa hai*? Do you store them?"

"Oh, yeah! I get a lot of it. And I use the dry milk."

"Do they have any flavor or anything?"
"No."

Use of this plant to curdle milk during cheese making was well known among older Pimas. Curtin (1949:88) 40 years earlier recorded a similar usage, as did Kearney and Peebles (1960:758), who

added, "A protein-digesting enzyme resembling papain exists in this plant." Cheese making is, of course, a post-Hispanic activity. The seedpods also contain a toxic alkaloid called solanine (Parker 1972:262).

The only name for the White Horsenettle I had ever heard in the western villages was *vakoa hai* 'broken gourd'. While writing at Sacaton, I was surprised to hear a variant, which at first I took as idiosyncratic. But with more questioning around the eastern villages, I found that *vakoa hahaisig* 'gourd broken into pieces' and even *vakoa hahaiñig* 'cracked gourd' were the usual forms of the plant's name among Ta'i O'odham. Variable names are most unusual for a native plant, but in this case, each name is a variant of the same basic meaning—a bottle gourd is somehow broken. They are better considered geographical variations of a single name rather than three different names for the same plant.

Vouchers: *28* (?JG or RG); *158* (SM); *LSMC 32*.

Sphaeralcea spp.
globemallow, Desert Mallow
Malvaceae
haḍam tatkam, haḍam tatk

Globemallow grows throughout the floodplains of Pima country, but especially in people's yards, at the edges of fields, along roadsides, and on ditch banks. The plants are mostly root perennials reaching from about knee- to hip-high, each plant with many erect stalks. These resemble a small version of the common garden hollyhock, a related mallow. The native desert mallows have a greener leaf that is more triangular than round, showing much variation from species to species and even from plant to plant. The wild plant has five small apricot-orange petals, each indented, and a central stamen cluster that is yellow or lavender or brown, according to species. An annual desert mallow that is probably included in this folk taxon grows out in the washes in early spring and has rather pale orange flowers; it is only a hand or two tall. Regardless of biological species, the herbage of these plants is covered with fine hairs, and therein lies a tale.

The Upper Piman name (both Akimel and Tohono O'odham) is *haḍam tatk* 'sticky roots' or occasionally *haḍam tatkam* 'sticky roots it has'. People know to avoid this common plant and teach their children to do the same. Ruth Giff explained, "*Haḍam tatkam* apparently is the real name. We knew we were going to get sore eyes. *Pi ñendag* means that it's going to give you sore eyes. I don't think that's the name, but every time they'd warn us not to pick it, because it's going to give us sore eyes and the new generation just *named* it 'sore eyes'. But [before] they didn't."

I had heard somewhere that the plant was also used as a cure for sore eyes, and I asked Sylvester Matthias about this. "Don't rub [it] in your eyes; not for any kind of medicine—just a weed. Its name, *haḍam tatkam*, means 'sticky roots it has'. Would be *sha'i* family because they grow big. This *gives* you the sore eyes," he continued. "It's not good. It will give you the sore eyes."

"What can you use for a remedy?"

"You can use *shegoi* [Creosote Bush, *Larrea divaricata*]—wash it off [out]. Or cottonwood leaves—make a tea of them."

But others knew of medicinal use. For instance, George Kyyitan pointed out some of the plants and said, "*Haḍam tatk*—you know—this one—you get the roots there, and pound it and boil it, and get the juice, and put it in the eyes—it looks pretty good [it makes you see well]."

Early in the 20th century, Russell (1908:79) had included the plant among the Pima medicines, calling it simply *hatam:* "The leaves are boiled and used as a remedy for diarrhea. Another informant states that the root is boiled, and the liquid extracted is used as a remedy for biliousness."

Early one April, Irene Hendricks and I were talking in the yard at Sacaton. At our feet were some globemallows. "Why don't you use some of the roots of that *haḍam tatkam*. Just dig them up. Wash the dirt off, they're dirty down there, and boil it. It's sure good for diarrhea. It's *s-haḍam*, real sticky!"

"How much?"

"Oh, any amount. A good little handful. That's all it takes and it gets real *s-haḍam*. Then it's ready. It sure tastes good. It's not like those other medicines. They're bitter. Oh! You'll like it."

"How long do you cook it, Irene?"

"Oh, not very long—about an hour and a half—and it'll be real *s-haḍam;* just

throw away those sticks, and drink the juice."

Sylvester told me of a more duplicitous use to which the plants were once put by a kind of *maakai* 'medicine man' who cured by sucking on the patient. "The medicine man used the root of the *haḍam tatkam* to chew on it, then he would go and suck the patient's disease out—and they used that—the roots."

Technical Notes: Species of *Sphaeralcea* taken in Pima country include *S. ambigua* A. Gray ssp. *ambigua*, *S. ambigua* A. Gray ssp. *rosacea* (Munz & I. M. Johnst.) Kearn., *S. angustifolia* (Cav.) G. Don, *S. coulteri* (S. Watson) A. Gray, *S. emoryi* Torr. var. *variabilis* (Ckll.) Kearn., and *S. laxa* Wooton & Standl.

Vouchers: *254* (SM, RG, *S. coulteri*); *524, 569* (SM, *S. emoryi*); *977* (GK, *S. emoryi*); *1836* (*S. ambigua* var. *ambigua*, SM "in the family of *haḍam tatk*"); *1837* (*S. ambigua* var. *rosacea*, SM "maybe also in the family" but IH and GK were doubtful); *LSMC 20* (*S. emoryi*).

Suaeda moquinii (Torr.) E. Greene
[? *S. arborescens*]
[≠ *Dondia suffrutescens*
 (S. Watson) Heller]
[≠ *S. suffrutescens* S. Watson]
[= *S. torreyana* S. Watson]
Seepweed, Quelite-salado
Chenopodiaceae
chuchk onk

In disturbed saline places such as the edges of fields, roads, and ditch banks is a common semiwoody (suffrutescent) shrub that might easily be overlooked. On first glance, it is not an attractive shrub: scruffy and unkempt, usually only knee- to hip-high. The minute flowers of this prosaic plant, like those of most other members of the chenopod family, are not likely to be noticed by anyone. But a closer look at the young growth shows a pleasing color combination: the stems are pinkish maroon and the slender, almost cylindrical succulent leaves are a dull aquamarine.

The bush is called *chuchk onk* '(pl.) black salty'. In the spring, when cholla cactus buds have been picked on the bajadas, they are usually brought down to the flatlands where there are roasting pits around peoples' houses. These pits are lined with stones, and a mesquite wood fire is built in them. After the fire has burned down and some of the coals raked aside, the very hot rocks are covered with a layer of *chuchk onk* branches, then tubs of *hannam* (cholla buds) are dumped in. Over these goes another layer of *chuchk onk,* then an old cowhide, canvas, or even a sheet of corrugated metal; finally everything is covered with a layer of dirt and left for a full day to roast. (This is the way I have always seen it done, but Russell [1908:71] gave a variation; see Group H, *Opuntia acanthocarpa* account.)

Joseph Giff told me that fresh *onk iivakĭ* (Wright's Saltbush, *Atriplex wrightii),* if available, can be used for this purpose as well. It is another chenopod with lots of moisture in the leaves. When the buds are roasted up in the bajadas where the *hannam* grows, Carmelita Raphael said there is no *chuchk onk,* so *shegoi* (Creosote Bush, *Larrea divaricata)* branches are used instead. She thought this would give the *hannam* a bad flavor, but she tried some that had been prepared that way and they tasted okay.

This use of Seepweed is undoubtedly an old one. From his work in 1901–1902, Russell recorded that *chuchk onk* was "added to greens or cactus fruit to give flavor."

Occasionally one hears of Pima children experimenting with food sources. This seems to have been the situation in a story Sylvester related from his wife when she was a girl at boarding school. "Irene said that they use it for greens, just plain. And they get an old lard can from the kitchen and there's some lard left yet in there, and they fried that *chuchk onk*— fresh [tender] ones."

"Fry it in lard and eat it as *iivakĭ?*"

"Yeah. Sacaton girls used to do that. They go around Keli Akimel [the main stream of the Gila] and swipe bread from the dining room [to eat with it]. *Chuchk onk* isn't used for anything, except for that *hannam chuámai.*"

"How about when you *chuáma* meat, beef?"

"No, no, no. Just the rocks and maybe a tin on the bottom. Then you put the meat on top."

Hrdlička (1908:183) gave some additional information. "The Pima say that if a stalk of the bush *chul-ick-un-ek* (Dondia suffrutescens) wounds a man and is not promptly removed, it is liable to give rise to blood poisoning and may have fatal results."

People I asked in the western villages seemed unfamiliar with this bit of lore. But George Kyyitan from the middle of the reservation told this story: "One of my uncles died of *chuchk onk.* He got it in his foot. He always goes without shoes. He didn't use even those rawhides *[shuushk].* He never got well. He died."

"How long did it take?" I asked.

"A long time. I don't know how long. But he died."

Hrdlička (1908:264) summarized: "*Chu-hch-kun-ek* ('black salty': Dondia suffrutescens), a small bush growing along the Gila, is considered poisonous. Nevertheless, as mentioned before, the Pima use the leaves and stalks to line the holes in which they roast the fruit of the *hanami* cactus. The purpose of lining the holes with this plant is to give the cactus fruit a salty taste and also to keep up a moist heat."

I had assumed that the Piman name *chuchk onk* 'black salty' was derived either from where the plant grows (in black alkali) or from its common use in pit roasting. However, Mathiot (n.d.:465) offered a different possible etymology. For Totoguañ Tohono O'odham (Santa Rosa district), she noted that *onk* is a saltbush, apparently in the generic sense. Most saltbushes (*Atriplex* spp.) are whitish. *Suaeda,* a similar but dark bush, is distinguished as 'black saltbush'. For riverine Pima, only one saltbush (*Atriplex elegans)* carries the base name *onk,* and in growth form it is quite unlike Seepweed.

Vouchers: *1* (?); *1459* (IH); *1502* (SM); *1572* (SM et al.); LSMC *62*.

Ziziphus obtusifolia (Hook. ex. Torr. & A. Gray) A. Gray var. *canescens* (A. Gray) M. C. Johnston
    [= *Condalia lycioides* (A. Gray) Weberb. var. *canescens* (A. Gray) Trel.]
    [≠ *Condalia obovata* Hook.]

Graythorn
Rhamnaceae
u'us chevaḏbaḏ

The only shrub on the middle Gila floodplains armed more viciously than Graythorn is the greener Crucifixion-

thorn *(Castela emoryi).* But maybe the two cannot be ranked, because Graythorn— being more slender branched and slender spined—is even more impenetrable. Graythorn, as both its English and varietal names imply, seldom shows any green. It is usually taller than a person and variously shaped, but it is almost always twice as tall as broad. It thrives in fence rows, along with wolfberry, mesquite, and Catclaw Acacia *(Acacia greggii).* Graythorn bushes reinforce a fence, making it much better than the original barbed wire by itself, while providing excellent habitat for wildlife. Most Graythorns you will find decorated with the bulky nests of the Verdin, or Cactus Wren. Various birds unwittingly help build these living fences when they perch on barbed wire fences after eating *Ziziphus* or *Lycium* berries, voiding the seeds below. At one time these living fences could be seen throughout the floodplains from one end of the reservation to the other.

The Gila Pima name for Graythorn is *u'us chevaḏbaḏ.* Mathiot (n.d.:489) gives *uusbad* as the name of an unidentified bush with berries, the name meaning literally 'deceased bush'; this probably is the Tohono O'odham name for one (or both) of the related species of *Condalia* found in Tohono O'odham country. I have always heard the suffix of the word as *-paḏ* as Russell, Hrdlička, Curtin (as corrected in the manuscript by Fr. Antonine), and Castetter and Underhill (1935) have written it. However, Culver Cassa notes that this suffix is really *-baḏ,* meaning 'the late' or 'deceased'. That leaves the middle lexeme, *chevaḏ,* which probably means 'tall' (from *cheveḏ* 'to become tall' [said of a plant]). If this etymology is correct, the implication is that *Ziziphus* is a tall, dead-looking bush.

The Piman name is not restricted to Pimería Alta. As I was driving out of Onavas pueblo in Sonora, I spotted a *Ziziphus* growing along the barbed wire fence together with the much more common wolfberries *(Lycium* spp.). I jumped out of the pickup, struggled to hack off a piece without getting too badly impaled in the process, and took it to Pedro Estrella, who was riding with me that day. He fondled the branch briefly, and his eyes lit up as he smiled in recognition. In Pima Bajo it was called *duwastbaḏ uus.* He said, "*Se comen las frutas de éstos*" ["They eat the fruit of these"].

The globular purple-black berries are scarcely as large as your smallest fingernail. They ripen in summer. "We eat *u'us chevaḍbaḍ* raw—don't cook it." said Ruth Giff. Sylvester Matthias told me, "Chew on the berries. Seeds inside are very tough; just spit them out."

U'us chevaḍbaḍ is not the kind of plant you want to tangle with if you can help it; it is hard to imagine anyone picking the fruit by hand. I asked George Kyyitan about harvesting them.

"We used to go across the river with my mother when I was young. There's lots of them by the highway, there. They go over there. Have to carry the canvas. She carries the things to put them in there and a box and a water [container]. So we have to go over there early in the morning, put the canvas under there and beat this one, this *u'us chevaḍbaḍ*. Just beat them up with a stick. Then shake them up. They fall on that one [canvas]. Then we gather up [the berries] and put [them] in there [a container]. It's pretty hard to get because they're small; take time. Then do some more. And when we get enough, go back."

"Did you eat them raw or cooked?"

"Just raw; just the way it is. But I don't know how they do [it]; they make a *sitol* out of it. Have to cook it. Same way with the watermelons. They make a syrup out of it. Those that are really ripe—ripe too much. Just put 'em in, cook 'em. It'll be just like that *haashañ sitol* [Saguaro syrup]. They use like that."

"Will it keep like *haashañ sitol?*"

"Yeah. Sure."

Russell (1908:76) wrote, "The black berry of this thorny bush is gathered in the basket bowls after it has been beaten down with sticks. It is eaten raw and the seeds are thrown away."

Hrdlička (1908:262) gave a more detailed account of its preparation as a food: "Of berries the Pima relish those of the *u-us dji-wuht-paht* (Condalia obovata), a bush growing along the lower Gila. These black berries are eaten raw by the Pima, and also by the Maricopa, roasted, or sirup is made from them. When eaten raw the solid parts are thrown away. The roasting is done in a frying pan and the berries are then eaten without additions. To make sirup the women cook and strain the berries, boiling the juice to the desired consistency. The sirup is used

on bread or otherwise, as in the case of saguaro sirup or honey."

Curtin (1949:50) recorded an additional use: "A syrup was made by boiling the berries in water; they were then squeezed out by hand and the seeds thrown to the chickens. The sweet juice was allowed to cool and thicken." And she was told at S-totoñig that "the Pima made suds of the roots for shampoo."

"Did you ever have a syrup of *u'us chevaḍbaḍ?*" I asked Sylvester.

"No. I just *heard* that. They didn't do that in my generation. They must have done it long ago. They make syrup now, all kinds of syrup—they just forgot about it." Then he added, "But they make a cure for dandruff—from *u'us chevaḍbaḍ* roots. They used it recently—in the forties were still using it. Now I don't think anybody uses it now, this generation."

Castetter and Underhill (1935:26) said that the Tohono O'odham fermented Graythorn fruits for a beverage, but not extensively. There was no ceremonial significance to this drink as there was with the cactus wines. No one with whom I worked among the Gila Pima recalled this use.

Sylvester noted that the wood was prized because it is very tough. "*U'us chevaḍbaḍ* [has] a very hard wood, used to make notched musical stick, played over basket, called the *hivkuḍ.*" The main branches are sometimes straight and quite long as a result of the initial growing stage. Although digging sticks had fallen into disuse by Russell's day, they were previously made from Desert Ironwood (*Olneya tesota*) or Graythorn. Russell (1908:97) noted, "It was used in planting maize and other crops, as a lever to pry out bushes when clearing the ground, as a pick when digging irrigating ditches, and in case of surprise it made an effective weapon of defense." One made for Russell (1908:97) reached from the ground to the top of the hip: "The short handle necessitated a crouching or sitting position by the operator." According to Sylvester, this implement was called an *eskuḍ.*

U'us chevaḍbaḍ has several medicinal uses. Russell (1908:79) recorded one: "The root of this bush is pounded up in mortars and boiled, the liquid extracted being used as a remedy for sore eyes."

Sylvester told about a Piman medical practice called *goi.* Whips of any very thorny plant such as the slender branches of a young mesquite or the Graythorn

were used. "Just beat it [lower legs] and let the blood drain out till you think when it's clear and then apply that solution there, from [boiled] inner mesquite bark. They call that [procedure] *goi.* They use glass now. I guess [it's for something] in the blood—arthritis, trouble with the legs, inflammation. Or used by runners before the race." Apparently this is the treatment that a Salt River woman told Curtin (1949:50) about in connection with *u'us chevaḍbaḍ:* "To prick the skin over rheumatic pains, after which treatment the blood is washed off with cold water."

I was at the church feast house in Blackwater, the far eastern Pima village. Mary Narcia Juan started talking about plants over her dish of *toota bavĭ* 'white teparies' and fried chicken. Mary was originally from Komatke but moved east when she married in the 1940s. She had just cooked the last of her dried *hannam* (*Opuntia acanthocarpa*) and *onk iivakĭ* (*Atriplex wrightii*) to take to the meal after services, and she complained that she would be making no more *hannam* because she did not trust her legs out picking chollas. I wondered at this; nothing seemed the matter. After she had her leg amputated, she explained, she was cautious about picking the cactus on the rocky slopes with her artificial limb. The subject turned to Graythorn. "When we were little, our mother would spread out a canvas under that *u'us chevaḍbaḍ* and knock those black berries down on the canvas. Then put them in a basket and put them on the table for us to eat."

"Raw?" I asked.

"Just like that. Chew it and spit out the seeds. But we were told never to eat [swallow] the seeds or we would grow up to be bald—like those seeds." This was the same as the better-known taboo about killing a tarantula.

"When would that *u'us chevaḍbaḍ* be ready to pick?"

"Should be right now [mid May], but I don't see any around anymore."

With a bit of effort, Mary got to her feet and left, seemingly none the worse for wear. One of her grandsons had just won an essay contest—the prize being a trip to Disney World, Florida. So she decided to fly there with all her grandsons she was rearing, and she was organizing food sales to help pay expenses. These old ones are hard to stop.

Vouchers: *LSMC 38.*

Group E, U'us Chuuchim
trees

Alone, *uus* (pl. *u'us*) may mean wood, stick, branch, post, or pole. And in some archaic narratives and songs, it means tree. But to refer to a tree in modern Pima, *uus* is used in conjunction with *keekam.* (The suffix *-kam* is an attributive.) *Keek* 'standing' appears in many modern northern Piman place names, such as Haashañ Keek ('Saguaro standing', the Pima name for Sacaton Flats); indeed, the Pima name for Blackwater village, Uus Keek, could mean 'tree standing' or 'pole standing'. But *uus keekam* as a life form means wood whose property it is to be standing upright, or simply tree. The plural of *uus keekam* is normally *u'us chuuchim* (although Ruth Giff used *u'us kekekam* for the plural). The Pima word for firewood, *ku'agĭ,* is not cognate.

Many other groups elevate the name of the most salient or conspicuous local tree to label polysemously the life-form category "tree" (Berlin 1972). For instance, among some Great Basin desert groups,

the word for cottonwood has been extended to name 'tree' as well (Trager 1939). Indeed, in Tohono O'odham, *kui* labels both mesquite specifically and tree in general: thus, the Tohono O'odham call the tamarisk *onk kui,* literally 'salt tree'. All the Pimas with whom I worked denied this being the case in Akimel O'odham. For them, *kui* means mesquite, nothing more: *onk kui* is a meaningless combination.

The referential expansion and marking of the 'wood/stick' term as seen in the language of the riverine Pima is another pattern of encoding the "tree" life form (Witkowski, Brown, and Chase 1981). Words that mean 'tree' do not reconstruct in proto-Uto-Aztecan or even proto-Tepiman, suggesting that they are relatively recent. At the opposite end of the Tepiman language continuum, Mason (1917) found *uush* meaning either 'stick' or 'tree' for the Tepecano, apparently without a marker.

Is the overt marker *keekam* obligatory with modern speakers? In a war speech, Russell (1908:367, 371, 373) gave both the plural *u'us chuuchim* and the terms *uuskaj* and *uus keek* in singular contexts. In some of the orations *uus* appears alone, suggesting that marking is not obligatory. War speeches, a highly formalized genre, are generally recognized as archaic. So the Gileño life form "tree" has a historical time depth of at least several centuries.

In most cases, there should be little debate over the species included in this Pima life form: ash, walnut, pine, cottonwood, mulberry, and oak are clearly trees to speakers of many different languages. But there are borderline cases where my own categorization might differ. Catclaw Acacia *(Acacia greggii),* for instance, is often shrubby. But where left undisturbed in fields and fence rows, it develops into a tree equal in form to Velvet Mesquite (*Prosopis velutina*), which also may be rather shrubby outside its prime flood-plain habitat. Desert-lavender (*Hyptis emoryi*) I thought of as a bush growing in the arroyos where the bajadas and mountains conjoin; Sylvester classified it as a tree.

Greater height and larger size distinguish Pima trees from shrubs. Drawing the line between Graythorn (*Ziziphus obtusifolia*) as a *sha'i* and the spiny Desert Hackberry (*Celtis pallida*) as an *uus keekam* may seem like a fine distinction in most of Pima country. But at higher elevations, the hackberry grows taller, and southward in Tohono O'odham country, another, more treelike species (*C. reticulata*) occurs. Apparently the Pima name *koom* applies to both species.

Folk classifications are not always unambiguous, particularly at higher levels. Ruth Giff, the first person to call to my attention the *uus keekam* category, debated the status of Soaptree Yucca (*Yucca elata*). "I guess *takui* will be included in *u'us kekekam* because they are so tall, like the Joshua Tree [*Yucca brevifolia* Engelm.] in California. You call that a tree."

The classification of the mistletoes, *Phoradendron* spp., is interesting. These are parasitic on various desert trees. Sylvester had no problem with them. Quite simply, they grew on trees and only on trees, therefore they were classified with the trees.

Only one species of palm occurs natively in Arizona: Northern Fan Palm (*Washingtonia filifera*). It is tall and trunked and therefore is classified as an *uus keekam*. The date palm (*Phoenix dactylifera*)—which occurs commonly now on the reservation—is classified as *e'es* 'planted things', in the category *ha'ichu iibdag* 'planted fruit trees'. The *e'es* category supersedes or is superimposed on the strictly morphological classification of plants. And even though the fan palms may be people's plants, they are known to the Pima to be naturally occurring as well, and they are probably most often dispersed by birds, so are more likely *hejel vuushñim*, something that grows voluntarily, rather than *e'es*, something deliberately planted.

Acacia greggii A. Gray
Catclaw Acacia
Fabaceae (Leguminosae)
uupaḍ

Quite similar to a mesquite in growth form, good-sized Catclaw trees are occasionally found growing in fence rows in the flatlands. More frequently they are small trees or shrubs of the arroyos cutting through the bajadas of the desert mountains. Their leaves are more finely divided than the mesquite's, and their similar flowers are in cylindrical, fluffy, light yellow clusters. Anyone attempting to hike through the arroyos will be forced to recognize the Catclaw; its slender branches are armed with numerous short, stout, clawlike spines or prickles that catch clothing and tear through skin and flesh.

One of the most rigorous ceremonies among the Gila Pima formerly was the lustration or 16-day purification required of a warrior following the killing of an enemy, usually an Apache. "During the period of sixteen days he was not allowed to touch his head with his fingers or his hair would turn white. If he touched his face it would become wrinkled. He kept a stick to scratch his head with, and at the end of every four days this stick was buried at the root and on the west side of a Catclaw tree and a new stick was made of greasewood, arrowbush, or any other convenient shrub. Then he bathed in the river, no matter how cold the temperature" (Russell 1908:204–205; see also Group D, *Sarcobatus vermiculatus* account).

The *keskuḍ* or scratching stick has its origin and explanation in the Creation Story. Specifically, Catclaw must be used. Its symbolism is given in Thin Leather's version recorded by Lloyd (1911:178; see also p. 94) where a *maakai* (shaman) is instructing a widow's young son: "He went and got a little piece of the oapot [*sic*], and tied a strip of cloth around the boy's head, and stuck the little piece of wood in it, and . . . this stick which the doctor had put into the boy's hair represented the *kuess-kote* or scratching stick which the Pimas and Papagoes [*sic*] used after killing Apaches, during the purifica-

tion time; and the doctor had made it from cat-claw wood because the cat-claw catches everybody that comes near, and he wanted the boy to have great power to capture his enemies."

Earlier in this same account, the young son lacks a bow and arrows to provide game. He is sent to his uncle, who says, "'That is an easy thing to do. Let us go out and get one.' And they went out and found an *o-a-pot*, or cat-claw tree, and cut a piece of its wood to make a bow, and they made a fire and roasted the stick over this, turning it, and they made a string from its bark to try it with; and then they found arrow-weeds and made arrows, four of them, roasting these too, and strengthening them; and then they went home and made a good string for the bow from sinew" (Lloyd 1911:173–174).

In the last episode of Thin Leather's version of the Creation Story, two *sisvañ* (Vipishaḍ chieftains) get together to gamble their wives: "Tcheunassat Seeven [Che'enashat Siivañ], after his bath [in the Gila near Blackwater], cut a piece of oapot [*Acacia greggii*] wood and sharpened it, and split the other end into four pieces, and bent them over and tied the ends of crow's feathers to them and stuck it in his hair, and dipped his finger in white paint and made one little spot over each eye, which was all the paint he used, and then he went and watched his wives dancing" (Lloyd 1911:236). Russell (1908:163) was told of similar hair ornaments, though their use had been abandoned by the turn of the century.

The broad papery pods of the Catclaw have no flesh. The flat seeds are rich brown and elliptical, about as broad as your little fingernail. None of the Pimas I asked knew anything about their use as food. This must have dropped out long ago, because even in 1901 Russell (1908:76) recorded under *o-opat*, "The beans of the cat's-claw were eaten in primitive times, but no one of the present generation knows how they were prepared."

According to Kissell (1916:147, 184), Catclaw was one of the materials used for making the circular rim of the burden basket and the U-shaped foundation for cradle boards. (It is unclear from her account whether the root or the branches were used for this purpose.)

Considerable information about the use of this tree survives. Joseph Giff told me, "The roots of *uupaḍ,* Catclaw, make the best bow, they say." George Kyyitan said that the limbs of *uupaḍ* are used to make bows, but he had not heard of this use for the roots. (He considered *gōhi* [Desert Mulberry, *Morus microphylla]* the best wood for bows.)

Sylvester Matthias said, "Yaquis made a bow of *uupaḍ,* but they have to boil it first." He gave this additional information: "It's medicine—use the root—for ulcers of the stomach—boil the root and it comes red like a strawberry pop—flat, no taste; but recently, I heard, those who have kidney troubles from drinking, it's good for the kidneys, and they pass [it] through, pissing. And then the tree itself is used for fuel. They say it will burn, no matter how green it is, it will burn like dry wood. Not eaten, seeds. Yaquis make bows out of Catclaw, but the Pimas don't know about it *[sic].* There's a lot of them by the old pumphouse [on the South Mountains bajada]."

Awls for weaving granaries or storage bins were whittled from either Catclaw or Greasewood (Kissell 1916:183). These were considerably larger awls than those used in conventional coil basketry. I asked Sylvester Matthias if he had ever heard of this. "I've *seen* it. My grandmother used one when she made the *homda* [granary; large storage bin]." I happened to have several Catclaw branches handy. He selected one of these and carved a fine nine-inch-long awl from the beautifully striated rose and blond wood. He called this a *vashomtakuḍ,* the instrument for making the *vashom,* or straw and bark storage basket. Kissell (1916:188) shows one in use.

A second species, the White-thorn (*Acacia constricta* Benth.), with quite different flowers and straight spines, grows to the south of Pima country. It is called *giidag* in Tohono O'odham. I mentioned this word once to Sylvester. "*Giidag?* I've heard the name, but I don't know what it looks like. Does it grow in the mountains? In Papago country? I've heard of it."

Vouchers: *455* (SM); *LSMC 58* (not extant).

Bursera microphylla A. Gray
Elephant-tree, Copal
Burseraceae
ko'otpidam

While hiking in the Sierra Estrella or in the South Mountains, you may be surprised to come across a stately bonsai-like tree that seems strangely misplaced from some foreign desert. You know that you are intruding on something centuries old. This is the Elephant-tree at the northeastern limit of its range.

There is something compelling about these anchorites, these small trees posted here and there up on the hot, rocky slopes of the mountains, never indulging in the kinder conditions of the bajadas or even the luxury of the lowlands. In their voluntary solitude they grow out of or even upon the great boulders; often they are taller than a person. The fat, elephantine branches spread tortuously from the base, forming a plant several times as broad as tall, each one different, shaped with such individual character to match its own rocky perch that it would bring approval from a Japanese gardener. On a mature tree, the main trunks (there are usually several) are whitish with shedding papery bark. At each turn there are fatty folds. These trunks are as thick as your thigh, if not thicker. If you run your fingers and eyes up along their softness, you will notice that suddenly they branch into smaller limbs that are lean and shiny maroon or wine-brown. These dark branches are covered with small compound leaves that resemble those of a miniature mesquite. The two rows of little leaflets down the flattened midrib are bright green. These leaves, when brushed against or crushed, have a potent, almost overpowering aroma—if not unpleasant, at least distinctive.

In winter the fruits are first green, then turn to a plum color. And indeed they look like a miniature plum borne on a short arched stem, usually two or three at a node. Later in the spring, some of the fruits pop off their frosted purple coats (capsules) in three segments, transforming the whole fruit into a bright orange-red, strongly three-angled fruit that continues to hang on the petiole. What disperser are they trying to attract? Some bird, no

doubt, that can forage about the pungent leaves and equally pungent immature purple fruits. But these now-red fruits have none of the repelling aromatic oils of the rest of the plant! This is invitation: solicitation, not repulsion.

The little fruits have one more startling metamorphosis. Scrape the bright fruits, and the waxy red coat (the aril) slips off, revealing a lightly mottled gray seed, flattened on two sides and broadly rounded on the third. These seeds look like a small gravel particle, the color of the rocks among which the trees grow.

Sylvester Matthias knew this tree, which he called *ko'otpidam.* I took him some branches from the Estrellas. "It grows in these mountains and the south side of South Mountain, where the pass is [Telegraph Pass]. No use for it. Wild animals don't bother it. Deer, they won't eat it. I guess it's just decoration for the mountains. It smells, too."

But once it did have a medical use among these people. Even though he apparently had nothing to identify botanically, Russell (1908:79) fortunately recorded the native name, *kâkpitam,* noting, "The leaves of this bush are boiled and the extract used as an emetic."

In folk taxonomic studies, when one gets down to just one or two people who remember the name for a plant, there is always the problem of recognizing collapsing or telescoping taxonomies or even outright misidentifications. So it is always a help to get some outside verification: an earlier document, or the folk taxon from a related speech community. And so it happened with the various thick-stemmed trees and bushes of the southern deserts variously called torotes among the Spanish-speaking peoples of Mexico.

There seemed to be an endless assortment of torotes on the lower Rio Yaqui where the lowland Pima Bajo live deep in Sonora. Over the years I collected various names for these *Bursera* and *Jatropha* species (and even one treelike *Fouquieria;* see species account in Group H), sometimes contradictorily named by don Pedro Estrella. Finally, I got a local cowboy to show me all the different torotes, which he found in the space of a single arroyo and canyon. Somehow all these different thick-trunked trees had merged in my consciousness to just a few species. There were four different species of

ko'otpidam　*Bursera microphylla*

species of *Bursera* was verified outside the Gileño language.

> Technical Notes: John Bates (1992) has found that wintering Gray Vireos are important dispersers of the seeds of this species of *Bursera* farther south, even holding winter territories among the stands. The two Pima populations and six other sites are north of this bird's wintering range.

> Vouchers: 572 (sm).

Castela emoryi (A. Gray) Moran & Felger
[= *Holacantha emoryi* A. Gray]
Crucifixion-thorn, Corona-de-Cristo
Simaroubaceae
a'agovĭ

Crucifixion-thorn completely measures up to expectation. These erect or spreading shrubs of head-height to about twice that are composed of stout branches that angle sharply from main stems and end in very sharp points. The entire plant is formidable. From a distance, what may appear as small clusters of mistletoe are on closer inspection seen to be persistent clusters of old fruiting heads. Though tightly packed, the individual carpels in fives, sixes, or sevens form a wheel pattern. Fresh clusters are greenish ochre, but older ones, from previous years, are dark brown to black, depending on age. Crucifixion-thorn is partial to the upper terraces of the alluvial flats, and with extensive developments for agriculture, this curious tree has almost disappeared from the reservation.

To the Gila Pima this shrub or tree is *a'agovĭ*. Sylvester Matthias considered it a tree because it has a main trunk and is taller than a person. He said that *a'agovĭ* is a plural form.

Russell gave no ethnographic information for Crucifixion-thorn (not even its Pima name) but commented that in his time (1901–1902), "It grows abundantly upon the mesas between the Gila villages and the Salt River Pima settlement, 30 miles northward." (One would be fortunate to find several plants along that same route today!) Hrdlička and Curtin do not mention it either. Fr. Antonine (1935) gives it in his list of trees, spelling it *a'agof*. (He used a double vowel to indicate a glottal stop.)

Bursera. Pedro called one of these *s-toa koopitkam,* another *s-veg koopitkam,* a third *s-tuk koopitkam,* and a fourth *kopaar,* a loan word from its Spanish name, copal. One more, *vii kootpitkam,* I still have to confirm. Two other torotes proved to be *Jatropha* species (see Group D, *Jatropha cardiophylla* account). Pedro called these *vaas* and *s-tuk vaas.* This nonanalyzable monolexeme turns up again as *waas* and *komagĭ waas,* two

Jatropha species, in southern Tohono O'odham country. The large Torote Verde (*Fouquieria macdougalii* Nash) he called *neliog,* clearly cognate with *melhog,* the Upper Piman name for Ocotillo, *F. splendens.*

So there was the confirmation. Between three different Piman speech communities, the 3 biological generics had sorted out as 3 folk generics, and Sylvester Matthias's name for the single northern

a'agovĭ   *Castela emoryi*

Celtis pallida Torr.
Desert Hackberry
Ulmaceae
koom

On mesic slopes above about 2,000 feet in the ranges surrounding the Gila Pima villages is a plant that superficially looks like the common Graythorn of the flood-plains but is much brighter green and retains its leaves year-round. This is Desert Hackberry, an erect but tortuously spiny large shrub or small tree usually well over head-height. The fruits, when ripe, are a pale orange and not as large as your smallest fingernail. The slightly sweetish flesh encloses a stone.

The Gila Pima (and Tohono O'odham) name for Desert Hackberry is *koom.* But there must be a change in referents in Tepiman languages from north to south. At the southern Tohono O'odham oasis village of Quitovac, Luciano Noriega gave us *kuwavul* for specimens of both *Celtis pallida* and wolfberry (*Lycium* spp.; see Group D account). At the desert Pima Bajo village of Onavas, Pedro Estrella consistently identified Garumbullo, or Desert Hackberry, as *kuwavul,* but *Lycium* he called *at wusha'i.* He did not recognize the word *koom,* even when Sylvester Matthias questioned him about it in Pima. The lexeme occurs in three Tohono O'odham dictionaries (Mathiot 1973; Saxton and Saxton 1969; Saxton, Saxton, and Enos 1983).

In Gileño country, hackberry is generally restricted to the several mountain ranges, which probably explains why most of the people with whom I worked were unfamiliar with the plant and only a few recognized the name as a tree. (Hunters would probably have been familiar with it.) Much of Tohono O'odham country is higher, so hackberry bushes are more frequent on the flats, particularly along washes. Castetter and Underhill (1935) noted that a closely related desert species is likewise called *koom,* and they described the collection of hackberry fruit: "If hackberries (*Celtis reticulata* Torrey) or mulberries (*Morus microphylla*) are found near the village, carrying baskets are filled with whole branches of them by the women." They also noted, "In the

When David Brown of Santa Cruz said the word, it sounded like *aago,* and indeed he noted how similar the plant's name was to *aaḍo,* which he gave as 'peacock' although it originally meant 'macaw'. Mathiot (n.d.:370) wrote it *aago,* without a glottal stop.

"*Aago* is the Crucifixion-thorn," David said. "You boil for tea; for diabetes, use dry seeds; it's worse than *shegoi* [Creosote Bush, *Larrea divaricata*]."

"What do you mean, 'worse'?" I asked, remembering that David often nursed a tin can of Creosote Bush brew at the back of his wood stove.

"It's more bitter!"

When Sylvester Matthias says the word, it is *aagovĭ*—with the entire last syllable, -*vĭ,* nonvoiced. "Its blossoms used as medicine, to cure headache or pain; make tea." He said he did not know exactly how and had heard of no other uses.

Joe Giff said, "They have these things that grow, they call it *a'agovĭ.* They call

them Crucifixion—stickers about four inches long. Those blossoms or the seed or whatever it is, they hang in big clumps and in that—(I had one but I guess something happened, or maybe I still have it over there)—but it's a star-shaped thing. Perfect! I think it has 6 points—5 or 6 points—like a star, little bitty things. . . . Now [they] dip that in white clay or red clay or something and you stamp your face [to make star designs]. It's just perfect."

Years later Ruth Giff explained this a bit further: "When you go dancing, I guess . . . [they use that for] their makeup. Very few of that *a'agovĭ* grows here now." She never heard of medicinal uses for it.

George Kyyitan commented on this tree: "*A'agovĭ* wood is very hot . . . the fire . . . and burns very bright. [But] it splits when you chop [it]. Few left."

Vouchers: *561* (SM).

Baboquivari valley people used for [sandals] the bark of the hackberry tree *(Celtis reticulata)* which comes off in smooth slabs."

Sylvester Matthias, apparently one of the last Gileños to remember this plant, said he was unaware of any uses for it. "There's a place in the Papago country called Koom Vo'o, [and] Koom Keek," he said. The names mean 'Hackberry Pond' and 'Hackberry Standing'.

Technical Notes: I have found Desert Hackberry growing somewhat commonly on shady north-facing slopes in the Sierra Estrella, particularly around cliffs and outcrops *(AMR 615)*. There are PHK specimens from the Santan Mountains and others from the White Tanks. In 1917 J. J. Thornber (ARIZ) collected a specimen marked simply "Maricopa, Ariz.," but I know of no others from the lowlands on or adjacent to the reservation.

Vouchers: *1068* (SM, *koom*; Pedro Estrella, Pima Bajo, *kuwavul*).

Cercidium floridum Benth.
[= *C. torreyanum* (S. Watson) Sarg.]
Blue Paloverde, Blue-green Paloverde
Fabaceae (Leguminosae)
ko'okmaḍkĭ

Two green-barked trees that in English are called paloverdes occur commonly throughout Upper Piman country. Both Gila Pima as well as Tohono O'odham use the same folk taxonomy to distinguish them. When the paloverdes bloom yellow and, a bit later, the ironwoods bloom pink, their masses of blossoms color whole sections of the desert, suggesting the seasonal phenomenon that is prominent in the thorn forests farther south in Sonora.

Though perhaps easily confused in English (indeed, both species are the state tree of Arizona), they are not linguistically associated in Pima. Blue Paloverde is the denser and bluer species found in lower washes or along riverbeds where it taps the more abundant underground moisture. It grows commonly to two or three times the height of a person. Its leaves are in Y's, like the two mesquite species, but are much smaller, with only three or four pairs of rounded leaflets on each arm of the Y. The flowering branches, especially,

may have short spines along the main stems. Blue Paloverde is the first of the two species to bloom, starting about late March. The entire tree may then appear to be bright yellow. The anthers (the pollen-producing parts of the stamens) are brown. Once the seedpods form, there is no reason to confuse the two trees. These are broad and flat, usually with only one or several seeds with only slight constriction between them. The flat seeds are brown and rather large, the size of your smallest fingernail.

The northern Piman name for this paloverde is *ko'okmaḍkĭ*. The word means something flat *(komaḍk)*, in the plural. The derivation is the same as for the village Kómaḍk, bastardized in English to Komátke. Apparently the tree has no use, not even for fuel.

Sylvester Matthias pointed this tree out to me, saying, "The *ko'okmaḍkĭ* is the knotty one, especially under bark. Has spines; it's found in arroyos." Later he said *ko'okmaḍkĭ* has "knotty wood; bark is rough—limbs not straight."

Because the Tohono O'odham make the same distinction between the two species of *Cercidium* as do the Akimel O'odham, I thought it would be good to verify the folk taxa with Tohono O'odham friends. Frank Jim, from the western part of the Tohono O'odham Reservation, confirmed the classification given by the Gila Pima: "There are two kinds of paloverdes— one in arroyo with spines is *ko'okmaḍkĭ* [*C. floridum*]. Some guys will eat it but we [from his village, Charco 27] don't because we don't like the taste. And seeds are flat and bigger; not very long pod. In *kuk chehedagĭ* [*C. microphyllum*], long pods; seeds when young good to eat (boiled like garden peas) but when dry, too damned hard—but old people [that is, people a long time ago] pounded them in rock mortars with long pestles. Mix the flour in water and drink like *haakĭ chu'i* [wheat pinole]." Another time when I asked Frank about the two he told me that *ko'okmaḍkĭ* seeds can be eaten only when dry and ground in a stone *chepa* [mortar], while *kuk chehedagĭ* can be eaten just green, uncooked. Juanita Norris, a Tohono O'odham from Ak Chiñ, confirmed Frank's information. "*Kuk chehedagĭ* they eat raw. But *ko'okmaḍkĭ*, they cook it when seeds come out, dry. Roast it like nuts. It's

ko'okmaḍkĭ *Cercidium floridum*

sure hard. It smells. I hate that smell."

No Gileño I talked to had ever heard of eating the dry seeds of Blue Paloverde. If they ever did, it is now lost from the oral traditions.

Another opportunity came for verification of the two paloverde names outside the Gileño community when Gary Paul Nabhan and I were afield in northern Sonora with Delores Lewis, a Tohono O'odham from Ge Oidag (Gu Oidak on Anglo maps). We stopped between Altar and Santa Ana, at a wash crossing the open desert where both *Cercidium* species were growing. Here Delores straightened us out: *kuk chehedagĭ* is Foothill (or Little-leaved) Paloverde (see *Cercidium microphyllum* account). He told us it is distinguished by shiny end branches, small leaves, light green (actually yellow-green) branches, small branches ending in points, and a lack of spines on the branches. The seeds, but not the whole pods, of these were once eaten when green and tender, with a bit of salt. The word *kuk* refers to the blotchy trunk. In contrast, *ko'okmaḍkĭ* is Blue Paloverde (*Cercidium floridum*). It is distinguished by dull bark on the end branches, larger leaves, darker-colored (blue-green) bark, branches that do not end in points, and spines present at the base of small branches.

I wondered why the green seeds of *kuk chehedagĭ* but not *ko'okmaḍkĭ* were considered edible. When I was out walking one afternoon in Sacaton where *ko'okmaḍkĭ* grows rather thickly, I picked a green pod and bit into it. Immediately I salivated copiously, spitting to try to rid my mouth of the lingering bitterness, and no longer wondered why.

Although common, this tree seems to have been little used. George Kyyitan said, "Can use it for firewood, but no coals in it. All right in summertime, just to cook with, but no coals." Russell (1908:101) noted that the wooden ladles "are commonly made of mesquite, though the Papagos make them of paloverde wood." Francis Vavages was probably one of the last spoon carvers on the Gila. His spoons were usually made from mesquite, although occasionally he used mulberry *(Morus alba)*. I asked whether he ever used paloverde. "No. They'll break—and they smell, too."

Not only Pimans have noticed these qualities. Botanists Kearney and Peebles (1960:407) remarked on the genus *Cercidium*, "The wood is soft and brittle and burns very quickly, giving off an unpleasant odor and leaving few coals." They added, "The Indians ate the seeds [of both paloverde species] in the form of meal." We are not told which Indians did this.

Another species called a paloverde in English is Bagote or Mexican Paloverde *(Parkinsonia aculeata* L.), sometimes known as Jerusalem-thorn. It grows naturally in some southern places in Arizona and is widely cultivated in towns—even in Pima villages. This species is fiercely spined on the main stems at each leaf node, and the minute leaflets grow from a flattened rachis (midrib) that may be as long as your arm. The flowers are bright yellow with some burnt-orange spots. The Tohono O'odham call this *oobgam* (see Saxton and Saxton 1969:170; Saxton, Saxton and Enos 1983:95; Mathiot n.d.:466). The name signifies something to do with the Apache. Even as far south as the Pima Bajo pueblo of Onavas, the plant is known as *ooba'igam* and in Spanish as Guacaporo. But I never have found a Gileño who knows this name or any other name for *Parkinsonia*. When pressed, they usually say it is just some kind of *ko'okmaḍkĭ;* they know it is different. Apparently the tree has reached this

far north relatively recently, without its southern Piman name.

Vouchers: *115* (Delores Lewis, Tohono O'odham); *994* (SM).

Cercidium microphyllum (Torr.) Rose & I. M. Johnston
[= *Parkinsonia microphylla* Torr.]
Fabaceae (Leguminosae)
Foothill Paloverde, Little-leaved Paloverde
kuk chehedagĭ

A greenish yellow tree, usually leafless, grows from the lowest bajadas right up to the crests of most of the mountains in Gila Pima country. (Some parts of the Sierra Estrella reach above the altitudinal limits of this tree.) These trees are commonly twice as tall as a person and are abundant, but their pale color makes them unobtrusive. After winter rains, fine green leaves appear on the usually bare twigs. Instead of the leaf being branched in a Y, it is branched in a V, and the leaflets are small, in four to seven well-spaced pairs. They do not last long, the burden of food production or photosynthesis being taken over by the greenish twigs and branches. The spines on Foothill Paloverde are pointed tips of branches rather than short points on the main branches themselves. Foothill Paloverde blooms a bit later in the spring than Blue Paloverde (*C. floridum;* see species account), though they overlap somewhat. Its flowers are a pale, slightly greenish yellow that complements the yellow-green bark. One petal is creamy white. The anthers are dull orange. When the seedpods form, this tree looks quite different from its cousin. The pods are rounded and strongly constricted between each seed (there may be up to four seeds), and the pod terminates in a long tail or beak. The two-toned brown seeds are only slightly flattened.

The northern Piman name for this tree is *kuk chehedagĭ.* The second word means 'green' or 'blue' (these colors are not distinguished; see chapter 8). Sylvester Matthias was puzzled by the word *kuk.* Ten years later I asked him again about it. "I don't know what that *kuk* means— maybe it's the name of some kind of tree,

which is green." I found the word in no dictionaries, but Tohono O'odham Delores Lewis said *kuk* was a reference to a blotch on the bark.

One May when the paloverdes were blooming on the hills all around us, Francis Vavages told me this story: "There were two boys who were always bad, who always fought with each other. And their mother got after them and spanked them, but it never did any good, and eventually their mother died and their grandmother had to take care of them. And still they fought with each other.

"One day the grandmother made a pot of food, an olla sitting on the fire, and warned the boys not to go near it, to be careful; it was all they had to eat. But they fought and busted the clay pot. And she got after them and spanked them, so they went off into the woods. And she followed them. And she could hear them talking there in the *s-uusig* [thicket], but didn't see them. So she followed their tracks. And the boys went up into the hills there.

"And the older boy said, 'I'm going to turn into a *haashañ,* a Saguaro. And I'm going to live up here. And every year I'm going to make fruit so the People will have fruit. They will have something to eat.'

"So the younger boy said, 'Well, up here there is nothing but rocks, so I'm going to turn into a *kuk chehedagĭ,* a paloverde. And I'm going to live up here in the foothills and make them beautiful. There is nothing but rocks, so I will make it beautiful when the People look up here.' So he turned into a paloverde.

"When she got there, the grandmother saw nothing in the hills but Saguaros and paloverdes, so she turned back."

Sylvester and I were out in the South Mountains where both species were growing close together on the short bajada above Broad Acres. "There are two kinds of paloverdes; branches are different; one is knotty, other straight," he said. "*Kuk chehedagĭ* when dry, they say you can roast them straight like *kui* [mesquite branches]. Used for beans, like green peas to eat, when they are tender. [But] *ko'okmaḍkĭ* is the knotty one, especially under the bark. Can't chop it. Has spines; found in arroyos."

Joseph Giff told me, "We use *kuk chehedagĭ* seeds for rattles; in the *ko'okmaḍkĭ,* the *kai* [seeds] are flat; don't use them in [gourd] rattles. We also eat *kuk chehedagĭ* when green and tender; eat

kuk chehedagĭ   *Cercidium microphyllum*

them raw. And we make spoons out of the wood, called *uus kusal* 'carved wooden spoon'. I think that word *kusal* comes from Spanish, from *cuchara*. They can be made from mesquite or paloverde, from *kuk chehedagĭ*, not the other paloverde." Then, in Joe's typical fashion, he mentioned related words. "The *ha'u*, that means to dip water to drink, as with the *uus kusal*. And *biikuḏ*, dip something to eat; [means] 'scoop'."

Sylvester mentioned these wooden spoons also: "The *uus kusal* or *uus biikuḏ ha'u* [is] made from *kui* or *kuk chehedagĭ*. A good-sized trunk of *kuk chehedagĭ* [is] used for *chepa* [wooden mortar], carved, set into the ground." He spoke further of this paloverde. "It's the one they ate

the seeds of; smooth branches; regular [= common] paloverde with straight wood. Doesn't make any coals, only ashes; fire smells bad." Russell (1908:172–173) noted that kickballs were carved either from paloverde (presumably this species) or from mesquite. They were covered with black gum, or lac, from Creosote Bush (*Larrea divaricata*).

Ruth Giff remembered the use of this paloverde for food. "We ate the pods of *kuk chehedagĭ* when tender—like eating [sweet] peas. They're small. When picking wheat in the field, someone would go out on horseback and gather a bunch. Can eat them raw, but they must be very young— or maybe eat whole pod, if they're really young." Three years later, she told me again of this. "*Kuk chehedagĭ*—comes yellow early in May. I remember when we were kids, years ago, they'd be cutting wheat, [and] someone would get on horseback and they'd bring so much in a sack. We'd be chewing on them. No one eats this anymore. These young ones don't even know it!"

Although Curtin misidentified *kuk chehedagĭ* as *C. floridum,* the ethnographic information George Webb supplied her does belong with this folk genus: "The green pods are gathered in summer and eaten raw, and the trunk and larger branches are made into ladles" (Curtin 1949:91).

At the turn of the century, Russell (1908:75) recorded the following about the use of this species as a food: "The paloverde bean was formerly eaten either as gathered or after being pounded in the mortar. It was not eaten as pinole, but was sometimes mixed with mesquite meal."

One day in late spring I was out on the bajada and noticed the *kuk chehedagĭ* were covered with still-tender young green pods. The soft seeds inside were light green and about the size of young garden peas. They were delicious—like eating fresh peas. And they were equally good cooked as a fresh vegetable. I was astonished that such an excellent and readily available food that requires nothing more than picking and shelling should be ignored by the thousands of Pimas who live on the reservation! They would make a delightful reprieve from the modern dietary trinity of sugar, flour, and lard.

Vouchers: *114* (Delores Lewis, Tohono O'odham); *1189* (RG or SM?).

Chilopsis linearis (Cav.) Sweet
Desert-willow
Bignoniaceae
aan

Desert-willow is a will-o'-the-wisp tree, a graceful production that may be from head-height to several times that. Its branches are slender, devoid of spines or thorns. The leaves are mere arced ribbons dangling in the hot breezes, their margins smooth, toothless. But it is the flowers—big tubular, irregular pink flowers crowding the tips of the branches— that somehow seem not to belong in the desert, where most living things seem formidable, armed to the teeth to protect precious water reserves.

The Piman name for Desert-willow is *aan. Aan* captures the hot breezes that flutter the long leaves on an afternoon, that waft a soft fragrance out to pollinating bees; the word *aan* summarizes the essence of the plant.

Aan almost misses the Pima reservation entirely. But old men always mentioned to me places where it would be found: the Superstition Mountains, Queen Creek, Higley, Kyrene—places just north of the reservation. In these localities, botanists had documented Desert-willow with specimens. It also skipped to the other edge of the reservation. There was once a thicket on the lower Salt River. In the 1970s on the south bank, one grew near 91st Avenue and another downstream in the V of land where the Salt and Gila Rivers join. They are probably both gone now, as part of the "flood control" program that the tribal council sanctioned.

Although skirting the reservation proper, *aan* is known to many elderly Gila Pima, though a few confuse it with *s-i'ovĭ che'ul,* Coyote Willow (*Salix exigua*). *Aan* comes up several times in the Creation Story. For instance, early in mythic time, after Toobĭ 'Cottontail', the first victim of Rattlesnake, died, the council decided to cremate his body. Coyote stole Cottontail's heart from the pyre and fled south of the Sierra Estrella. He paused, putting the heart on an *aan* bush. The halting place was called Aangam Chekwuañk, Place of the Uprooted Aan Bush (Russell 1908:216–217). From there he ran on to Giho Do'ag.

Gila Pimas are quite familiar with the Tohono O'odham village of Aangam. Joseph Giff gave me the following account of these *aan* people: "Aangam, they used to talk very slow, but not now. *Aan* is a plant, a weed. They say there is plenty from where they migrated from—not where they are [now]. They are not Papago. I don't know what they are. Somewhere back on Queen Creek some-where—that's where *aan* was—that's why they call them Aangam T-am O'odham. You see, they're different people than the Tohono O'odham. They say they were back in there somewhere along Queen Creek east of the Superstition Mountains. [There were] twin ponds—*gook vapchĭ*—water always there; a *vachkĭ* is a pond.

"And that's where the man turned to Eagle, and flew up on top of Weaver's Needle, and the rock standing up there—that's where they killed him, see. When he was up there, he swoops down, killing the people, and that's why the Aangam moved, they moved away from there. He was one of them. They say you can still see those ponds—those depressions—water standing all the time, year round. Weaver's Needle is close to it. They went over to Vav Giwulk [the Baboquivari Mountains]." George Kyyitan knew the story of the Desert-willow People, too, and was familiar with the place of their origin. He helped me locate Gook Vapchĭ east of Higley at the end of Williams Field Road. The area is now a golf course.

Apparently the Gila Pima made little economic use of the Desert-willow, which is more frequently encountered to the south in Tohono O'odham country. Frank Jim, a Tohono O'odham, related the fol-lowing: "You can smell it if you're any-where near the tree. Flowers are pretty; grows around water holes and charcos. It's used for covering the *vatto* [ramada]. When it rains and the water drips from the roof [with *aan* thatching] into the water olla, it gives the water a distinctive taste."

Sylvester told something similar. "We used to live in Queen Creek and we go chop those *aan*, Desert-willow, to make our *vatto* in camp. It grows there in Queen Creek."

Some years later I asked Sylvester if he had ever heard of Gook Vapchĭ where the Aangam people were supposed to have originated from. He had not, but added, "By Chandler and Queen Creek was *aan*.

aan   *Chilopsis linearis*

We used to cut it for *vaptto*. It's as good as Arrow-weeds *[Pluchea sericea]* or maybe better because it's leafy. Arrow-weed kind of shrink . . . sun can shine right through."

Mathiot (n.d.:370) misidentified *aan* as 'desert broom, a plant (Baccharis sarathroides *[sic]*)'. Could any two bushes be more different?

There are several fine mature Desert-willow trees growing in a yard in Sacaton and some young ones planted on the free-way interchange on the south side of the reservation. This species deserves a more widespread planting.

Eucalyptus spp.
eucalyptus, gum
Myrtaceae
chev uus, (pl.) che'echev u'us

I had worked for years, asking about every plant I thought the Pima might possibly know, might possibly have named, check-ing and rechecking everything countless times. And then another folk taxon appeared, right out of the blue—com-pletely unsought.

Ruth Giff came through the house, talking, taking a brief pause from work at her home, which was not far away. I think she was speaking mostly Pima, and not even to me, but my ears pricked up midsentence when I heard her say *chev uus* 'tall tree' in passing. Could this be still another name for the Lombardy Poplar *(Populus nigra)*, called *kapichk auppa* 'slender cottonwood'? I inquired if this *chev uus* she mentioned might be the poplar.

"No, 'tall trees'; it's those eucalyptus."

The next day I went to her house to find out more.

"*Che'echev u'us* used to be on 19th Avenue [in Phoenix]—whole row of them. And between Baseline and Dobbyns Road. We would see them."

"And the Pima have a name for them?"

"That's how we always call them. I guess that's their name."

"Are they used for anything?"

"I know somebody used them for medicine—boil the leaves. But I don't remember what it was for."

"When did they come in?"

"The first one was in 1938 at Gila Crossing Day School. Now it's a big tree."

The next step is to check a newly discovered name with others. I saw Sylvester Matthias a while later and asked if he had heard of *che'echev u'us*.

"Yes, that's eucalyptus. There used to be something like a nursery. I forgot where about—was it on 27th Avenue between Baseline and Southern . . . ? I don't remember because it's a long time [ago]. And they call it *che'echev u'us chuuchk*, means something like 'a grove there.' *Chuuchk* means 'standing' or 'there,' like a little forest there."

"Have you ever heard of the Pima using them for medicine or anything?"

"No, but I *did* hear that somebody used them for medicine. Maybe the Mexicans."

Eucalyptus was limited to plantings around public buildings on the reservation, but it is becoming more common as a shade tree around the newer private homes.

Fraxinus velutina Torr.
  [= *F. pennsylvanica* Marshall ssp. *velutina* (Torr.) G. N. Miller]
Velvet Ash, Arizona Ash, Fresno
Oleaceae
bitoi

Fresno, or Arizona Ash, is one of three broad-leaved trees of the permanent desert riparian communities, an honor it shares with the cottonwoods and willows. Unlike these, the ash is partial to higher elevations, preferring those Upper Sonoran/Lower Sonoran desert mergings where Saguaros often keep company with junipers on the adjacent hillsides. Arizona Ash usually has an elegant aloofness about it. Its erect is dark gray, often slender. The leaves are composed of two, three, or four pairs of elliptical, pointed leaflets, with an unmatched terminal one. These are never too many, being artfully arranged. They turn yellow in autumn and drop. The flowers are small, and the bunches of hanging winged fruits, each about as long as a thumb joint, are more likely to attract attention.

The middle Gila was one of the few places where Arizona Ash

descended from its higher-elevation strongholds along small streams into the lower hot desert. It is difficult for one to imagine what the now-bare banks of the Little Gila near Blackwater looked like just 80 years earlier when the botanist-ornithologist-schoolteacher M. French Gilman (1909) wrote, "Between the two streams, on the 'Island,' as it is called, are groves of cottonwoods, and a few Arizona ash trees *(Fraxinus velutina)*." The soil there is rich, and there are some plowed fields, a few mesquites, but not a trace of the riparian vegetation Gilman spoke of. He also mentioned (1915) finding migrant Pacific coast Red-naped Sapsuckers "in cottonwoods, willows, and occasionally Arizona ash."

It is quite possible that the ash did not make it all the way down the Gila to the lower western villages. But the word did. I showed Sylvester Matthias some specimens, asking if he knew anything about them. He did. "Ash, it's *bitoi*— I saw some when my father took me in to Phoenix, at Grant Street between First and Second Street, alongside a school. Maybe it's still there. Called *bitoi ch-eḍ mashchamakuḍ* 'it's surrounded [the school] by ash trees'. And there's a Papago village where they used to grow down there in Fresnal Canyon called Bitoikam 'It Has Ash Trees'." (The Spanish word *fresnal* has the same meaning.) He saw the tree along canal banks in Phoenix but never heard of any growing along the Gila.

But Russell (1908:22) noted an indication of local occurrence. Among the villages known to the oldest Pimas, Russell gave one called Petâi Kek, "Where the Petai (ash tree?) Stands." Though hesitant about this identification, he was right. (Russell's *â* = *aw* is an allophone of *o;* and the *b/p* substitution is not uncommon throughout his text.)

Culver Cassa puzzled over another place name, Pitaitutgam, on the upper (southern) San Pedro River, that Fr. Kino had recorded on his 1695–1696 map (Bolton 1936). Culver interpreted this as *bitoi chuuchkam* '[the place of] standing ash trees', or ash tree grove. In colonial Pima the *t* had not yet changed to *ch*, so that modern *chuuchk* was still *tutk*, as it still is in modern Pima Bajo.

George Kyyitan recalled that during his childhood the tree had grown along the Gila, but as far as he knew, it was not

bitoi   *Fraxinus velutina*

sought out for fenceposts or tool handles. Unlike that of other ashes, the wood of the Velvet Ash is reportedly not especially hard (Lanner 1984).

The word appears to be pan-Piman. The Tohono O'odham use it (though oddly Saxton and Saxton [1969:170] and Saxton, Saxton, and Enos [1983:7] gloss *bitoi* as oak, *Quercus!*). The 17th-century Névome vocabulary gives both *pitai* and *sopqui [sopki]* for Aliso (sycamore), the Spanish Jesuit compiler evidently lumping two native folk taxa; there is no separate entry for Fresno (Pennington 1979:7). In his work on the Northern Tepehuan, Pennington (1969:61) gives *petai* for Fresno (*Fraxinus velutina*).

There are Western Soapberry (*Sapindus saponaria* L.) trees growing as ornamentals in Sacaton and Bapchule. This is a tree native to the streams of the higher desert country. Francis Vavages said these are known as *bitoi*, too, the same as the ash, adding that they grow readily from seed. This is an odd extension of the name, because the leaves of the soapberry look more like a walnut or hazelnut than an ash.

Vouchers: *1550* (FV; SM "similar to").

Hyptis emoryi Torr.
Desert-lavender
Lamiaceae (Labiatae)
viopal

True sages (*Salvia* spp.) are members of the mint family. There are several species of *Salvia* in the Sierra Estrella, but one could wander around there for years without discovering them. However, there is one member of the family that no one is likely to miss in the sandy arroyo bottoms or up among the rocky slopes of this and other local desert ranges: a large bush or brushy tree, usually about head-height or more, appropriately called Desert-lavender. Its branches are gray, with a weathered look. If you have any doubt about its identification, just crush the small, broad, woolly leaves and smell the pungent lavender aroma, or look for the small lavender-shaped and lavender-colored flowers borne in profusion during the winter and spring months. These are a

haven for bees. Desert-lavender and its frequent associate in the Estrellas, Beloperone or Chuparosa, make it possible for the Costa's Hummingbird to breed here practically in the dead of winter.

Sylvester Matthias seemed to be the only one who knew the name for this fine gray bush, and he consistently gave me the same identification time after time when we were out skirting the Estrellas. He said there was no use for it, not even in cooking.

The word *viopal* is not the figment of one man's imagination, even though it does not occur in any of the dictionaries. In a footnote under tobacco, Russell (1908:119) mentions the various native tobaccos. One, *Nicotiana trigonophylla*, he said was known as *biopal viov(u)*, which he glossed 'like tobacco'. Russell added that this was one "gathered near Baboquivari by the Papagos and brought to the Pimas." No one seems to understand this translation, and I suspect that he got it wrong. My theory is that this tobacco, which grows well in the arroyos, was named for the Desert-lavender found in the same places; hence it should be *viopal viv*, the tobacco that grows in arroyos under the Desert-lavender. Perhaps with time and the loss of the more mesophilic tobacco, *N. attenuata*, the folk taxonomy collapsed, leaving in the Pima lexicon just one wild species, *N. trigonophylla*, now called *ban vivga* (see account, Group G). (The loss of this folk taxon is not quite complete. One day, after years of going over plants with me, Sylvester said, "There's some [tobacco] in a wash behind Santa Cruz Church—that's *viopal viv*." He had only heard of it, so he was not sure what the plant looked like.)

Desert-lavender always seemed like an overgrown bush to me. But when Sylvester arranged cards with different plant names into groups, he always put *viopal* with the trees. Perhaps, I thought, he had forgotten how small the plant really is; it is not something he would see very often. Late in the study, I drove him down from Santa Cruz village on the sandy road that cuts across the bajada. Near the south end of the Sierra Estrella, the road approaches the mountain. I turned off on a side road that took us right to the great masses of granitic boulders that form this mountain. Sylvester got out of the pickup and hobbled around on his crutches. We wandered over to several Desert-lavender bushes

viopal   *Hyptis emoryi*

growing between boulders twice the size of my pickup. The bushes were not much more than 6–7 feet tall. He crushed some leaves and smelled them, confirming that this was *viopal*.

"How would you classify this?"

"*Uus keekam*, because it grows straight—canes," he said, pointing with a crutch to the various rather straight branches springing from the base of the plants. "See how they grow up—with trunks?" Indeed, the larger ones did have several trunks, though not well-defined ones as on the paloverde species.

"But why not *sha'i*?"

"It's too tall for *sha'i*."

Carl Lumholtz (1912:53), traveling among the Tohono O'odham in 1909–1910, wrote, "The leaves of *viopóli*, a bush in the foot-hills, are also sometimes smoked." It would seem that he meant the Desert-lavender itself rather than *viopal viv*. I asked Sylvester if *viopal* had any use.

"Yeah—they use it as tobacco. They have to dry it first."

"Would they smoke it straight, by itself, or would they mix it with something?"

"Straight."

"Would anybody smoke it, or just the *maakai* [shaman]?"

"Just anybody who wants to."

Although Desert-lavender grows in the region of the lowland Pima Bajo, Pedro Estrella, the last speaker of that language, could give me only the name *shavir,* a loan word from its local Spanish name, Salvia. He did not recognize the word *viopal.*

Vouchers: 15 (SM).

Juglans major (Torr.) Heller
   [= *J. microptera* Berlandier var.
    major (Torr.) L. Benson]
Arizona Black Walnut
Juglandaceae
uupio

Arizona Black Walnut is a fine hardwood tree found in the cooler canyons and streamsides above the deserts but below the pine forests. The spreading trees rise on single black trunks. The leaves are compound with five or more pairs of long leaflets and a terminal one. Their odor is strong and distinctive. The hard black nuts, the *nogales* of Spanish, are each encased at first in a yellow-green hull, which later turns black and separates from the grooved nut. The shells are quite thick, making these walnuts difficult to crack, but the small amount of meat that can be picked out has an excellent flavor. With the fall change of weather, the leaves turn yellow. Sometimes there will be a whole grove. At least at this season the trees stand out.

So far as is known to botanists, Arizona Black Walnut descends no lower on the Gila drainage than near the mouth of the San Pedro, in what was formerly Sobaipuri country. It reaches Tohono O'odham country in the Baboquivaris. Both the Gila Pima and the Tohono O'odham call the tree as well as its nuts *uupio.* This is a curious polysemy. "The 'skunk' and 'walnut' have exactly the same name," said Ruth Giff. You must tell the words apart by context. In either case the name is derived through *uv-* 'smell'.

But the nuts were not sought as food. Ruth Giff explained, "This *uupio* [means] the black walnut itself, because they don't know the tree at all, only the nuts [that] washed down in the [annual] floods. Used for *kams*—only by women; the men

didn't." *Kams* can be used to mean any round object that is kept in one's mouth, such as an apricot pit. The noun *kams* or *kamsh* comes from the verb *kamshch* 'to have or to put an object into one's mouth' and is related to the word *kaam* 'cheek'.

Sylvester Matthias said, "*Kams* [is] something [used] in the mouth; Black Walnut used. They used to find a lot of them when the river was flooded. We used to gather up a lot of them; some of the ladies would come and grab them. Also used a peach seed [pit]. Could make a noise with walnut. [Trees] grow up towards Globe; they didn't grow here. Haven't any flood, they'd go without them."

Sally Pablo, much younger, confirmed this: "They make a noise—clicking the rough edges against their teeth and the women didn't stop—while breathing in as well as out when talking."

Ruth Giff said that she used to use the *kams.*

"Only women did that?" I asked. "Why?"

"They say you didn't get thirsty using that. And they'd make *noise,* rattling in their mouth. Oh, everyone would have it. I guess they'd put them away and pick it up again later. And we used to get them in the ditch . . . when the flood [came through]. One time I was out there waiting for it. And there's a little bridge, and I'd sit there. They were still green. Some were already black when they get them. Some were like that when they get it in the summer. And they'd fall off the tree and wash down. And we'd just keep a lot of them."

"How many could you find that way?"

"Oh, a *lot.*"

"You never cracked them open to eat?"

"No, huh-uh. We didn't even know. All we wanted was to make *kams* out of it." She laughed at this in retrospect. "I like that ice cream—you know—black walnut. But when we came to school, the Apaches, they used to bring them [to school] and eat the nut."

In April 1988, I took a break from writing and went up to the spillway on the Gila between Olberg and Sacaton

uupio *Juglans major*

Flats. On the sand I chanced to see a black walnut, washed down in the irrigation water where it was turned back into the riverbed. So I can vouch that the river still brings an occasional nut downstream.

There is always the temptation to think of the Pima in terms of a community of people restricted to the middle Gila in southern Arizona rather than as just one end of a long ribbon of people dispersed more than 900 miles across northwestern Mexico, barely reaching the United States. But folk taxonomic studies of the Pimans often jolt us out of this provincialism. Near the south end of the Piman continuum in Durango, Pennington (1969:7) observed, "Padre Rinaldini noted in the eighteenth century that *upivagaquer* meant *lugar de nogales;* the term is clearly related to the current place name *ïpïvïkïrï*, which is derived from *ïpïvï* or *ïpiokai* (nogal) and *-kïrï* . . . 'place of'." Assuming that the *ï* and northern *u* reconstruct in proto-Piman, we thus have the same word for walnut. (*Kai* is 'seed'.) Pennington (1969:125) also notes that for the Northern Tepehuan "*uúpai* is the term used for any of the striped, hognose, and spotted skunks."

While working in a shady canyon in Sonora with don Pedro Estrella, desert Pima Bajo, I came across a small tree with large leaflets arranged along a seven-inch rachis.

"*Uupio,*" he said. "Same word as for skunk."

This specimen was later identified as Western Soapberry (*Sapindus saponaria*). Although the walnut and soapberry are not botanically related, there is a similarity in leaf structure. Pedro evidently confused the two, because he commented that my specimen had nuts that were eaten.

At any rate, the Gila Pima undeniably had a concept and a word for the plant *uupio,* even though these riverine people are outside even the historical ranges of *Juglans,* as far as we know. The now-extinct Sobaipuri would have known walnuts from their local flora. Pollen found at archaeological sites near the Salt-Gila confluence probably originated from the spring floods, but I wonder whether the walnut might occasionally have established down to 1000 feet in elevation when there was relatively cool gallery forest along the rivers. One day George Kyyitan, discussing *kams,* said that the seeds would sprout but would die before they got very big. Sylvester, a few days

later, told me, "My late wife Irene tells me *uupio* used to grow along the irrigation ditches at Bapchule."

Vouchers: *1906* (SM, RG, GK, Dorothy Kyyitan).

Maclura pomifera (Raf.) C. K. Schneid.
Osage-orange
Moraceae
s-ho'idkam koli

Driving out of east Sacramento toward the southeast (on what turned out to be the "wrong" road), I soon found myself out of the urban sprawl, in the flat, dry prairie country John Muir once wrote about, with nothing but barbed wire and an occasional old stone wall. Half-absentmindedly I passed a tall hedgerow separating blacktop from grassland. Something clicked in my mind; perhaps I was not on the wrong road after all. I turned the truck around and drove back to the trees—tall tough saplings with slender branches studded here and there with stout spines. Among the dark green leaves I found what I was looking for: small grapefruit-sized hard fruits with knobby surfaces: Osage-oranges, as I had suspected. This was a plant I had failed to find in the Phoenix region. I braved the thorns and tried tearing off some fruits to take back to Takashi to illustrate, but to no avail. I went back to the pickup and got a knife and then managed to hack off a few fruiting branches.

Old Pimas for years had told me about this plant, which they called *s-ho'idkam koli* or *kolhi* (*s-* is an intensifier; *ho'i* means 'thorn', 'spine', or 'sticker'; *-kam* is an attributive; and *koli* or *kolhi* means 'corral' or 'fence'. It is an interesting name in that part of it, *ho'idkam,* occurs in the Piman name for Desert Ironwood (*Olneya tesota*) and another part, apparently a loan word from Spanish, is one of two different Gileño terms for fence. When the Pima started building fences to protect their crops from introduced Eurasian livestock, they used what was readily at hand: thorny branches. And when Anglos began settling the Salt River Valley, they brought with them from the southern Great Plains this fine plant for hedgerows. So Curtin's gloss (1949:83) of Osage-orange as 'thorny

fence' is both literally and factually correct.

Both Joseph Giff and Sylvester Matthias told me about *s-ho'idkam koli.* The Pima, traveling into Phoenix markets in the early days, quickly learned of the superior qualities of this wood for bow making. The prized *gōhi* or wild mulberry wood for making bows came only from lands held by Apaches, while their new plant could be collected in only a day's wagon trip into the new Anglo settlement. Both Russell (1908:95) and Lloyd (1911:190) record the prominence of Osage-orange by the turn of the century. Four decades later Curtin (1949:114) wrote, "Osage orange wood is preferred [for bows], but they are also made of willow (*chi ul*), and of catclaw (*oopat*) from the mountains. The trunk of the young osage orange tree about twelve inches in circumference is cut down and while green is split into halves and then shaped to the desired size with a butcher knife. The bow is bent to the proper arc and fastened with wire until thoroughly dry. Another method was followed by Lewis Manuel, who also used osage orange wood, but while the bow is green it is bent between the forked branches of a tree and allowed to dry in that position."

Curtin (1949:83) also noted that "the inner wood and the large roots are bright-orange in color, and were formerly used in dyeing." The statement is ambiguous. There is nothing in Curtin's field notes about such a use among the Pima. I asked George Kyyitan if he knew *s-ho'idkam koli* and if it had been used in dyeing. "Yeah, they make bows out of it. I saw some way down there—Pecos and Cooper [Road, in Chandler]. Think they're still there. Yeah, it is [bright orange] but never heard of dye use. Wood white with yellow middle. *Koli* means 'post' or 'barbed wire [fence]'—or 'wooden corral', like we saw over there [at the Desert Botanical Garden]."

The Pima name for bow, *uus gaat,* is often rendered 'wooden gun', the assumption incorrectly being that *gaat* is a loan word from the English *gun.* But as H. L. Mencken said, for every phenomenon there is an explanation that is simple, plausible, and wrong! This is a classic example of marking reversal gone to completion (see Witkowski and Brown 1983). The original Pima name for 'bow' (in Spanish, *arco*) was *gaat,* an unanalyzable monolexeme, as one would expect with ancient and culturally important items. A 17th-century Jesuit recorded, at Onavas in southern Sonora, *gato* and *gata* for *arco para flechas* (Pennington 1979:10). Centuries later, I was told by Pedro Estrella, the last speaker of the Pima Bajo language at Onavas, that the bow was *gaat* (see also Pennington 1980:208). While in the Mountain Pima country of Chihuahua in 1984, I was given *ga'at* at Nabogame for *arco.* Somewhat lower in the sierras at Maicoba, Pennington (unpublished notes) recorded *gat* and *ga'at* for *arco.* With the arrival of Europeans, firearms—which serve the same function as the aboriginal bow—became *vainam gaat* 'metal [or] iron bow'. This combination was in use at least until 1903 when Thin Leather told the Creation Story to Lloyd (1911:129). In time, as the Pima acquired guns and both gun and bow were in common usage, the bow became secondarily marked *uus gaat* 'wooden bow'. In more recent times when guns became almost the exclusive method of hunting, they became simply *gaat,* and the remembered but no longer used bow was left with the marker.

Vouchers: *1551, 1872* (SM); LSMC *56* (not extant).

s-ee'ekagkam, ee'ekagkamchu   *Melia azedarach*

Melia azedarach L.
Chinaberry, Chinese Umbrella Tree
Meliaceae
s-ee'ekagkam, ee'ekagkamchu

Though Chinaberry is an Old World species, there are good things to be said about its introduction. It has been planted commonly as a shade tree in yards across the reservation. In the spring the shiny, dark green leaves appear on the umbrella-like skeletons of this tree. The graceful leaves are double compound, the entire leaf being as long as your forearm and palm. Then for about two weeks the trees are covered with great clusters of oddly structured flowers—the petals being very pale pinkish, the central reproductive structures purple. These load the warm spring evenings with a gently pleasing fragrance. Next come the clusters of round green fruits, like marbles, which mature into the tan "berries" that persist through the following spring. These fetid-smelling fruits are toxic. The tree is native to Asia and seems to thrive here in the heat with only a minimum of watering. Most important, it does not spread into natural areas.

The Gila Pima name for Chinaberry is *s-ee'ekagkamchu,* meaning 'shady thing', a word related to *eekakuḍ* 'umbrella'. There is perhaps a double meaning here in that the tree is planted to provide yard shade and the tree is umbrellalike in shape. Sylvester Matthias said of this, "'Shady'— *s-ee'ekag,* 'shade'—[we used to] take them in class and hit the girls on the head. That's all they're good for [the berries]. Around since the late 1800s. They had some across the river [at the old Komatke village washed out in the 1906 flood]. They last[ed] a long time, even when the river went dry. Still there in 1929. Cottonwoods still in very good shape then."

I asked Ruth Giff about Chinaberry. "Yeah, *s-ee'ekag,* means 'shady', I guess."

"Have you heard of them being poisonous?"

She had not. "Not very many people had them."

"When did they come in—how early?"

"Let's see. . . . When did I see them? One man had one there in Santa Cruz. People didn't buy trees to plant. I guess what they plant is *siipuḍ s-wulvañ.*" This is the domestic mulberry tree *(Morus alba),* at least among speakers from Santa Cruz, Gila Crossing, and Komatke villages.

Technical Notes: The fruit of the Chinese Umbrella Tree, especially when ripe, can be fatal to humans as well as animals and is not to be taken lightly. According to Perkins and Payne (1978), "Symptoms differ, depending on the plant part ingested. They may occur in less than an hour, or be delayed for several hours. In man, overdose of a leaf decoction has caused burning in the mouth, scant urine, bloody vomiting, lethargy and occasionally a death. The symptoms of fruit poisoning in man are nausea, abdominal pains, vomiting, (bloody) diarrhea, thirst, cold sweat, incoordination, weak pulse, labored and irregular respiration, paralysis, (convulsions) and death due to respiratory failure in 12–15 hours. Gastric lavage or emesis should be performed as quickly as possible."

Vouchers: *721* (GK and FV?).

Morus microphylla Buckl.
Desert Mulberry
Moraceae
gōhi

gōhi   *Morus microphylla*
The painting was made from specimens
(AMR 1608) *collected in the Gila riparian*
woods near the mouth of the San Pedro
River. This was formerly the northern
extent of the Sobaipuri Pima.

A small tree grows at the bases of the higher mountain ranges forming a huge arc surrounding Gila Pima country—from Prescott through Superior, the Graham Mountains, Oracle, the Catalinas, to the Baboquivaris. Just a touch of its rough, dark green leaves should instantly suggest that it is a *Morus,* as indeed it is—a native wild species, Desert Mulberry. I have found it only as scattered trees and bushes, usually solitary.

Once while discussing bow making, Sylvester Matthias told me about a tree he thought was "in the family of ocotillo" called *gokoi.* (He later changed this to *gōhi,* the usual form.) This made the finest bows, but the Pima had to make excursions deep into Apache country, such as in the Chiricahua Mountains, to get it. Of course, the Sobaipuri on the Santa Cruz and San Pedro Rivers once put the Gila Pima into closer contact with the places where *gōhi* grows. Still, it is surprising that a plant growing so far away has remained in the oral tradition.

There are historical references to the Gileño use of *gōhi* in bow making. Lloyd (1911:190) wrote from his 1903 fieldwork, "Bows were made from Osage orange, cat-claw, or *o-a-pot;* or, better still, from a tree called *gaw-hee* [Desert Mulberry]. Arrows from arrow-weeds. The Apache

arrows were made of cane. The Pimas were formerly famous for archery, and the shooting of bird on the wing and of jackrabbits at full run while the archer was pursuing on horseback, were favorite feats." Russell (1908:95) also discussed the Pima bow: "Warriors made their bows of mulberry wood obtained in the Superstition and Pinal mountains. . . . Hunting bows are frequently made of osage orange wood, a material that is now obtainable from the whites along the Salt River. When mulberry wood was not available, willow was used, and most of the hunting bows which men as well as boys continue to make for hunting hares and similar small game are of that wood."

Later Russell (1908:202) described the Gileño retaliatory force venturing into Apache country with little more than some pinole, an eating container, and some tobacco: "A portion of each band was armed with bows and arrows; the former of the elastic mulberry wood from the same mountains in which the enemy found refuge, the latter of the straight-stemmed arrow bush, whose tufted tips waved in billowy masses on the Pimerían lowlands."

But not all knowledge of this higher-elevation tree is shrouded in the mists of history. I asked George Kyyitan about *gōhi.*

"I know it—[used] for bows. I had one. I don't know what they did with it. I like that one. It's sure good. It belonged to my late brother. A big one—about five feet. I used that to kill jackrabbits. It's sure good. I killed one, a buck jackrabbit, running. I shot him right here [through the chest]."

A year later, George was telling me about going out and hunting doves with bows and arrows in the cottonwood groves when he was young. He noted that bows could be made out of willow or *s-ho'idkam koli,* Osage-orange *(Maclura pomifera).* "There's another kind, too, that, I think, they grew up here in the mountains, deserts. It's that *gōhi.* It's good. I used to have one. My brother had one and he gave it to me. I used to have that one and sure like it. And one time they have a parade, over at the Melchuda, and they asked me, they going to use it. And I don't see it no more. No more."

The word is used among the Tohono O'odham as well. Mathiot (1973:298) gave *gohi* as "mulberry bush formerly used in making bows; strawberry bush; strawberries"; *s-gohig* was to be full of *gohi* in a single location. The Saxton dictionaries (see Saxton, Saxton, and Enos 1983:17) give *goohi* and *gohui* as variants for the Desert Mulberry. Pennington (1969:121) reported that the tree was important in bow making among the Northern Tepehuan, who call the tree *kóji* (=*kohi?*). Pennington (n.d.) recorded *kohi* for the Mountain Pima of the Yepachi area.

Mathiot (1973:298) also listed *gohi-meli*, which she described as the "name of a rain dance (from *gohi-med:* to go and [to] gather mulberry bush for bows?)." I was puzzled by this and asked George Kyyitan.

"Yeah, I heard of *gohi-meli*. Last year [1987] they had one over here at Jack-rabbit—that side of Chuichu. But they just used it as a social dance. I think it is connected with the medicine man when they want to have rain."

"Are there special songs?"

"Yeah. They call that 'cripple-dance' because they step one foot over the other."

Sylvester said it was a social dance, not a rain dance, and that it was difficult. To learn it, the dancers have to start when young.

Olneya tesota A. Gray
Desert Ironwood, Palo Fierro, Tésota
Fabaceae (Leguminosae)
ho'idkam

After the bright yellow wave of blooming Blue Paloverdes and the even more exten-sive but subtler yellow wave of Foothill Paloverdes across the reservation uplands, there comes a third wave of color, this one an almost ethereal pink, from the Desert Ironwoods on the bajadas.

"Ironwoods" grow in tropical and sub-tropical regions almost worldwide. They come from various plant families and share only the characteristic of having extremely dense wood, too heavy to float.

Desert Ironwood grows practically throughout the Sonoran Desert. It is a good-sized tree, generally taller than the paloverdes often found with it. On the better-drained soils of bajadas, and espe-cially along the arroyos, it forms a xeric "riparian" canopy. It is topped only by the Saguaros. The branches of this thin-barked tree are gray and generously sup-plied at leaf nodes with pairs of short, slender, and very effective spines. The broadly rounded (spatulate) leaflets are pale bluish green; these stay on through-out the year. A legume tree that has leaves in midwinter in Pima country is Desert

Ironwood—not mesquite, Catclaw, or either of the paloverdes. After summer rains, Desert Ironwoods along the washes may become quite verdant, lending the impression of the thornscrub forests farther south in Sonora.

In spring, certain trees become almost bald where masses of flower buds are developing. About mid May in Pima country, these trees transform into smoky pink clouds of flowers shimmering in the intense heat that marks the beginning of summer. The cloud is composed of pink and lavender wisteria-like clusters of flowers. These last through about the first week of June, when they begin to turn into brownish green masses of developing pods. When mature, in about seven weeks, the tough woody pods are brown, each enclosing usually two or three rufous brown seeds about the size of a Spanish peanut, though oddly ridged and com-pressed. One tree can produce prodigious quantities of seeds, which fall later in summer.

The Gila River Pima, a rancheria people like their Piman relatives for nearly a thousand miles southward, had a simple material culture quite distinct from that of the Puebloan peoples north of them. Their stone tool kit included the metate and mano, the pestle, and the ax (Russell 1908:95). But ultimately it was wooden items that were constantly in the hands of every working person. Pimería Alta was supplied with five hard, workable woods: the bushes Graythorn (*Ziziphus obtusifolia*) and Greasewood (*Sarcobatus vermiculatus*), and the trees Velvet Mesquite (*Prosopis velutina*), Screwbean (*Prosopis pubescens*), and Desert Iron-wood. In Desert Ironwood, the people had almost a substitute for both metal and stone. All one needed was time and patience to turn the rich dark wood into an enduring implement.

Throughout the Pima country of Sonora and Arizona, the name of the tree is the same, *ho'idkam*. "'It has the stickers," Sylvester Matthias said, glossing the word: *ho'i* is a thorn or spine, and -*kam*/-*gam* is attributive. "It makes good firewood but you better watch out! It'll burn out the inside of your stove." Its Cahitan name, *teso* (hence the Spanish name Tésota), is not cognate with Piman (Sobarzo 1966).

The earliest agricultural implement for the Pima was the straight digging stick,

about a meter long, made of either the heartwood of ironwood or a piece of Graythorn. "That's the *eskuḍ*," Sylvester reminded me (see Group D, *Ziziphus obtusifolia*). This indispensable tool was used to dig ditches, turn soil, make borders, and plant seeds (illustrated in Castetter and Bell 1942:135; Russell 1908:97).

One other item supplied the aborig-inal Piman field tool kit: a weeding hoe called the *giikĭ* (see chapter 4). This was like the blade of a machete, about three inches wide and about as long as a man's forearm. The farmer kneeled or sat on the ground to use his tool (Underhill 1938b:45), singing the plants to produc-tivity while loosening the soil or cutting away the roots of an undesired plant with his burnished brown ironwood hoe.

In the days when corn and teparies and beans and squash were the mainstay of Pima life, these two simple tools were apparently all that were needed, along with hard labor, to dig the canals and open the ditches and turn the fine alluvial soils along the Gila into flourishing fields. Then came wheat, revolutionizing this centuries-old pattern. Wheat, a winter crop, added a third planting to the Gileño pattern cycle. Exactly when plows were first used is not known, but these were simple wooden affairs called *giikĭ* also. The share and handle were fashioned from a branched piece of ironwood or mesquite. The tongue might be of some lighter wood, usually cottonwood. (In 1984 I found similar wooden plows, called by the same name, still in use by Mountain Pima in Chihuahua.)

While out hunting, gathering, or work-ing their fields, the Pima had to keep an eye out for raiding Apaches from the north or east. The calendar sticks record many episodes such as this one in 1843–1844: "A woman went with her daughter to gather cactus fruit for drying. She was accompanied by her husband, who went as a guard. While she was busy gathering the spring cactus, she heard a step and, turning, saw an Apache. She screamed for help and told her daughter to run to the village and give the alarm. The husband was hunting near at hand, but was too far away to rescue his wife" (Russell 1908:42). Formal retaliatory raids usually followed, if an immediate search of the vicinity did not reveal the enemy. In addition to bow, arrows, and shield, the heavy "potato masher"–style club was thrust into every warrior's belt. These were made from dense mesquite root or the still heavier ironwood and weighed about two pounds (Russell 1908).

Castetter and Underhill (1935:68) recorded the use of Desert Ironwood among Tohono O'odham to make the *hivkuḍ* or rasping stick to accompany songs intended to avert evil or to bring rain. The *hivkuḍ* fell into disuse among the Gileños so long ago that this cannot now be checked. At least for rain-making ceremonies, the Gileño rasper had to be made of Greasewood (see Group D, *Sarcobatus vermiculatus*). Russell collected one scraping stick that appeared to be of ironwood but noted that the workmanship was so unlike that of the others that he suspected it might have originated from the Yaquis.

The Creation Epic tells still another use for the *ho'idkam* that has been expurgated from English versions. In the flood episode, Handsome Man goes about ravaging all the young women of marriageable age. South Doctor, a powerful *maakai,* or shaman, instructs his grieving daughter to prepare for Handsome Man's visit by getting some tips of Jumping Cholla and the fine dark heartwood of *ho'idkam,* from which he fashions a *viha,* or artificial phallus. In the course of the night the tables are reversed, and Handsome Man himself is impregnated, bearing a child in the morning (see Group H, *Opuntia bigelovii*).

After the luxuriant display of pink and lavender blossoms, which sometimes almost cover a tree in May, the pods develop during June and are dry for harvesting in late July or August. "In July they're still hanging on the tree," Ruth explained, "but most would be on the ground in August. I always get them from the ground." Another time Ruth told me, "The *ho'idkam* comes in July—the *kai* [seeds]—almost the end of July. This past year [1987] there was *lots* of it! So much, we just picked a little bit of what was there."

I noticed that Ruth had a quart jar of the seeds next to her stove. Ironwoods do not grow on the floodplains with mesquites, but rather along the *a'akĭ,* the washes dissecting the bajadas. However, she did not have far to go to the South Mountain bajada to gather seeds. "We used to soak the seeds in water, but they aren't bitter. Soak, dry, *then* roast; then just eat them like nuts." She explained that they are supposed to be parched in an olla, but she just roasted these in a clay pan. Ruth gave me a handful of the rich dark brown seeds. They tasted rather like dry roasted peanuts.

"Why would you need to soak them?"

"Because they say they were bitter if you didn't soak it. (And I heard they [also] used to treat the screwbean, *kuujul.* And yet they're just *sweet,* so I don't know why they do that.) I heard some white people [archaeologists] when they dig the *vapaki* [ruins] and they dug some *[ho'idkam]* out, and they thought it was coffee, and yet it was the ironwood bean—I guess they put it away for the dead to take— they used it in those days, and I guess it stayed [in the *vapaki*] because it was buried. I'm sure I've eaten them—long time ago—roasted them."

Ruth had prepared *ho'idkam* for the past several seasons, gathering the fallen pods, *koshaj,* from the ground late in the summer. I had never been along when she harvested. "Do the seeds stay in the pod," I asked her, "or do they fall out?"

"They stay in the pod, just like the *vihog* [mesquite pods]. They're hard to crack them open—have to beat them with a stick."

"And you just roast them in the frying pan?"

"No. One time I did that, but I just burned some. This time I used my *muta* [clay parching pan]. So they came out better, because the frying pan is too hot."

"So you use your *muta* and live coals?" I asked, remembering how Rosita Brown used to parch wheat.

"No, . . . on the stove. I tried that way, too, and I burned it. So I use the stove. There were just a few of them that were kind of burned and I took them out."

"How long does it take to roast them that way in your *muta?*"

"Umm . . . about fifteen minutes."

Russell (1908:70–71) found people at the turn of the century doing much as Ruth was doing today: "The nuts of the ironwood tree are parched in an olla or, what is more usual, the broken half of one, and eaten without further preparation." George Webb, formerly Ruth's neighbor from Santa Cruz village, told Curtin (1949:93), "The Pima used to gather the fallen beans and roast them in a pit or in earthen bowls over a fire. The beans were parched and eaten whole, or ground and mixed with water to make pinole, and the flavor resembled that of acorns or peanuts."

But Sylvester Matthias recalled a more complicated leaching: "Lucy [Jackson] had to dig a hole, dampen the sand—has to be fine sand—put in the ironwood seeds, covered with sand, [and leave them] overnight, then dug them the next day. When dry—I don't know how long— then roast them. Bernardine's [Bernardine Vance's] grandmother just put it in a gunnysack and put it in the Santa Cruz River overnight, then dry it up. When it's dry, roast it. [Her name was] Josepha Tashquinth." He called the pods *ho'idkam vihogdag.* (The same term appears in Mathiot 1973:381.)

Sylvester's description is verified by Hrdlička's (1908:263) account written 80 years earlier: "The Pima used to eat also the seeds of the ironwood (Olneya tesota). As these seeds are bitter, it was the custom to put them into deep baskets which were hung overnight, each from four poles, in a swift current in the river. In the morning the seeds were dried and then preserved for future use. Before being eaten the seeds were roasted, and ground coarse. Another way of preparing these seeds was to grind them coarse, putting the meal into a clean hole in the sand near the river; here water was poured over the meal for a long while until all the bitterness was washed away.

ho'idkam *Olneya tesota*

The final preparation and mode of eating the seeds were the same in both cases."

Elisabeth Hart worked as a home extension agent among the Pima during the 1930s. This is her description of the preparation of ironwood meal: "Roast nuts whole. Break on a metate. Soak in water. Change to fresh water every day (every other day) for a week. Dry and use as nuts or grind and use as a meal. The meal is usually packed into an earthen pit and kept full of water to leach out the bitterness, and is then spread and dried for use" (Hart 1934:10).

Although all tribes living within the range of Desert Ironwood appeared to have eaten the seeds, it seems that preparation varied, even among the Gila Pima. Perhaps in earlier times, when eaten in greater quantities (not just as a snack food), antinutritional or toxic substances had to be removed by processing. This was the only food the Seri cooked in a second water (Felger and Moser 1985:331).

In wandering about the Pima desert for the past three decades, I have noticed that many ironwoods have been cut at their bases and have regenerated. The cuts are usually smooth saw cuts, quite old, rather than ax cuts, as one might expect had they been cut for firewood. Curtin (1949:93) in the 1940s remarked, "Palo-de-hierro has been known to reach thirty-five feet in height, but this is unusual, as the Indians prize its wood highly for firewood, tool handles, etc., and they do not spare the tree long enough for it to reach maturity." These beautiful trees have mostly recovered from this earlier overexploitation. At least today they are seldom bothered in Pima country.

Vouchers: *1190* (RG or SM); LSMC *73*.

Phoradendron californicum Nutt.
Desert Mistletoe
Phoradendron tomentosum (DC.) A. Gray
Yellow Mistletoe
Visaceae
haakvoḍ

Two species of mistletoe grow in Pima country. Both are parasites, deriving their basic nutrients from their hosts. Desert

Mistletoe is usually found in great masses in older Velvet Mesquite trees. The leaves of this species are reduced to mere scales, the plant being masses of dangling flexible branches, sometimes covered with bunches of small berries, some whitish, but most pinkish to coral.

Various birds, particularly the slender glossy black or gray Phainopepla, seek shelter in these mistletoe clumps. These infested mesquites serve as the Phaino-peplas' primary home during the winter and early spring months; in addition, their nests are placed there and the pink berries are their staple food. Joe Giff told me, "That bird's called the *kuigam,* the owner of the mesquite." Individual Phainopeplas hold territories, defending their Desert Mistletoe stores against others of their own kind. Phainopeplas (as well as Gila Woodpeckers and Common Flickers) also help spread the mistletoe to new trees. The berries inside are strongly mucilaginous, and sometimes they stick to the bill of a bird eating them. When the bird wipes the hitchhiker off, the enclosed sticky green embryo may become established in a new location.

Another, quite different, species found in Pima country is Yellow Mistletoe. This one, which is a parasite on cottonwoods and occasionally willows, has great ovate leathery leaves the size of a soup spoon. Its olive-green masses conceal clusters of watery white berries.

Cottonwoods were once common, even in the early 1960s, along canal banks on the reservation. When I realized that I had never yet collected a specimen of Yellow Mistletoe, Sylvester Matthias and I one day in 1986 made a special trip in search of it. Starting at Komatke, we drove along the canal banks, crisscrossing the river, scrutinizing every cottonwood until we reached Blackwater. We traveled more than 65 miles and were amazed at how few cottonwoods were left: we never found a clump of mistletoe! Nor could I find any among the newly established cottonwood groves along the lower Salt River bordering the reservation. Finally the next spring I discovered some among a few ancient cottonwoods along the main canal that runs through Sacaton.

Both kinds of mistletoe, though very different in appearance, have the same name in Pima: *haakvoḍ.* Everyone gave me the same answer. Joe Giff thought about the trees on which it grew: cottonwood,

willow, mesquite, ironwood, Catclaw, and, he thought, occasionally paloverde. But the mistletoe was always called *haakvoḍ.* Sylvester Matthias reiterated this: "You could say *kui haakvoḍ* [mesquite mistletoe] or *auppa haakvoḍ* [cottonwood mistletoe] or *uupaḍ haakvoḍ* [Catclaw mistletoe]. That's just what they're growing on. That's not different names." The lack of a formal folk taxonomic distinction is a surprise in view of the different cultural value as well as different appearance of the two.

Where the *haakvoḍ* joins the *kui* branch, Sylvester told me, a round ball or joint often forms. This is one of the materials that can be shaped and polished to make the *shoñigiwul,* or kickball.

I have often eaten raw the pinkish berries of Desert Mistletoe. Depending on the season, they are mildly sweet. According to Castetter and Underhill (1935), the Tohono O'odham ate them: "Clusters of mistletoe *(Phoradendron californicum),* which often grows on the mesquite bush *(Prosopis velutina),* are gathered, their branches laid on mats, dried and the berries rolled free. These berries form a sticky mass which is further sun-dried and stored." At the turn of the century, Russell (1908:71) recorded the Pima use of *P. californicum:* "The berries of the mistletoe that grows on the mesquite are gathered and boiled without stripping from the stem. They are taken in the fingers, and the berries stripped off into the mouth as eaten. Various species of mistletoe are very abundant on the trees along the Gila, but this one only is eaten."

The contrasted species here is *P. tomentosum,* Yellow Mistletoe. Why were they not eaten, too? I tried some of the milky-white berries from cottonwoods near Sacaton. They tasted awful!

Customs must have varied, however. No contemporary Pima with whom I worked recalled eating *haakvoḍ.* At the west end, Sylvester said, "We were told not to eat the berries of the one on the mesquite, even though they taste good. So everybody was afraid to eat." But 40 years earlier, Curtin had been told by Annie Thomas of nearby Co-op Colony, "Mistletoe berries were mashed or boiled and, when eaten, 'tasted like pudding'" (Curtin 1949:83).

Haakvoḍ apparently came into play both by the shamans and by the native herbal healers, *ha-i'ichudam.* Russell

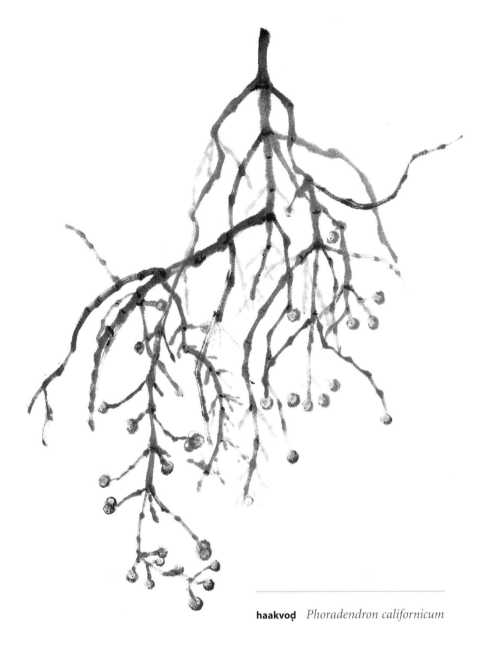

haakvoḍ *Phoradendron californicum*

u'us chuuchim "because it belongs to trees." Where white culture tends to view things as dissected parts, the Indian is more likely to see something as a relational whole. In this Pima hierarchy, mistletoe has no existence apart from "treeness."

> Vouchers: *1307* (?); *1841* (SM); LSMC *35* (*P. californicum,* not extant).

Pinus spp.
pine
Pinaceae
huk

You will not find pine trees in the desert ranges that immediately surround the Pima villages; but at higher elevations in the major mountains that are more distant, such as the Santa Catalinas and Santa Ritas near Tucson, several species of pine occur in extensive forests. The Pima were great travelers, and various series of ancient songs mention the names of these mountains.

Some years ago I had heard from someone, probably Joe Giff, a story about turkeys who were cutting down a pine tree (both turkeys and pines were outside the altitudinal range of Gila Pimas, at least in recent historic times). I do not think he ever told me the whole story, only the gist of it. One day when Sylvester Matthias was visiting me in San Diego, completely on an outside chance, I asked him if he had ever heard the story. With his characteristic instant recall, he began this story without further deliberation:

> The story is a Pima myth. *Totova ch huk,* turkeys and the pine tree. *Sh hek sh gaḍ hú.* A bunch of turkeys was going to chop the pine tree down and they gathered around and started singing:

> he-uk vii
> he-uk vii
> tova
> he'uk shónchagá
> he'uk shónchagá
> he'uk vii
> he'uk vii
> he'uk vii
> totovaga kwi'uk shónchagá
> he'uk shónchagá

(1908:257) was told that a teaching *maakai,* or shaman, "coughed up" white balls the size of mistletoe berries, called *chechaka,* and rubbed them into the breast of the novice in his process of "getting power" (*va'igida*). Curtin's information was more specific: "In olden times the medicine-man placed the berries in his mouth, leaned over the patient, and sucked at the afflicted places, then spat out the berries and claimed that he had removed them from the body—an old trick among many Indians. 'As long as you believe in anything, no matter what religion you have, you can be cured,' said George Webb" (Curtin 1949:83).

The berries were used medicinally by herbal healers as well. Curtin (1949:83)

was told, "The berries were boiled in water until reduced to a thick mush, which was cooled and a cupful given for stomachache; and it also acted as a purge, on the authority of Dean McArthur. Nowadays, at Stotonic, the plant is soaked in warm water and sores are washed with the infusion."

The etymology of *haakvoḍ* is no longer remembered, but I suspect that it may be derived from the kinship term *hakit* (pl. *haakit*) 'father's younger brother' or 'younger uncle', as an expression of the parasite-host relationship.

Perhaps the most interesting thing about mistletoe is its classification in Pima. Even though it is a parasite, Sylvester consistently put it in the category

Just then Coyote was passing by and heard them singing. And he came and said, "Oh, Cousins, what were you saying [singing]?"

And the turkeys answered, "Oh, we're not saying anything. We're just here and we're doing this, cutting down the tree."

"Oh, I heard, '*he'uk vii / he'uk vii*,' and I don't know how you end up."

Then the turkeys begin their song again and Coyote joined in. It was just about ready for the pine tree would fall down. So everyone get out of the way and poor Coyote was in the way and the tree fell and fell on his tail. And then he asked for help and nobody came back to help him, so he pulled and pulled and finally just skinned off his tail and ran after them, but they're too far ahead of him. But the one on the tail end he caught.

And then he said, "All right, I'll strip you out," and he stripped [the turkey], feathers and all, and then he said, "Now, go tell my wife to cook you so I'll have turkey for supper when I get home—I'm going this way—through the mountains over there, and make a circle over there, and I'll be home by sundown."

So the turkey obeyed his orders and went to Coyote's wife and told his wife, [but he] just peeped in the house because he had no feathers on, and asked: "Oh, Coyote said that you could cook his sinews—(used for arrows and bowstrings). Cook that and others that he's got around, and he'll be here around sundown."

And the turkey went [left] and there's another lake with a tall snag, there's a dead tree there, I think a pine, and so he climbed up there, because he can't fly up there without any feathers, and he was way up there on the top of the tree because it was getting too close [to the time] when the Coyote gets home.

Then Coyote came back saying, "Hey, Old Lady, Old Lady, have you saved me a turkey neck?" And he gets to the house there and he found nothing but his sinews cooking!

[Coyote's wife said], "Yeah, Turkey came and said for me to cook your sinews."

"Aaugh!"

Then Coyote tracked the turkey down to the lake and he saw (it was his reflection) that Turkey was way down in the bottom of the lake. And Coyote went back to his wife. "Come here. Let's go get the turkey and we'll have turkey [for supper]." So they both went and saw the turkey, because it was a clear lake, so they could see the turkey very clearly. [They could see the turkey's reflection.] He said, "Now, you dive first and get it." So his wife dived in and couldn't reach the bottom. So she came up. It's her husband's turn, so he dived in. No success. Then he came up and said, "Oh, let's get something that has a weight, and we'll both hang on and dive right in." So they finally got a piece of good-sized rock, a *machchud* [grindstone], and hang on and dived in. They both drowned.

At hoadag! And that's the end of the story.

In this story, Sylvester explained, the *he-uk* is song language for *huk* 'pine'.

Another legend speaks of the forest. Russell's (1908:224) version of the tobacco origin story says that the people "followed the tracks of the boys, who had gone toward the east with their parrots. The pursuers raised a cloud of dust as they went along, which betrayed their presence on the trail to the boy, who exclaimed, 'What shall we do!' At length they set free the parrots, which flew up into the mountains, where they concealed themselves in the forest. Following their example, the boys hastened to the same place, where they successfully eluded their pursuers." Presumably this refers to the pine forests atop such desert ranges as the Santa Catalinas, but the English does not specify. Thick-billed Parrots were once found in the pine country of southeastern Arizona.

In spite of their homeland being in the low desert, the Pima of the past century had contact with pines. One day Sylvester explained, "They would use *huk* when they go out and search for Apaches; they use that for firewood."

"What about timbers coming in floods?" I asked.

"Yes, they can use them."

Joseph Giff had told me likewise that in earlier days, before the dams, the Pima salvaged timbers from the floods once they

started building square adobe houses instead of round houses.

Huk is a pan-Piman word, with the same meaning among the desert Pima Bajo, Tohono O'odham, and Mountain Pima; the cognate *ukui* among the Northern Tepehuan designates the most common species of pine in their area (Pennington 1969:135, 177). Sylvester did not think there was a name in Pima for piñon, though he knew what they are. He would not include it or junipers in the category *huk* 'pine'. Ruth Giff, however, thought that pines, piñons, and junipers would all be included in the term *huk*. George Kyyitan also thought so. If this is the case, the label *huk* serves at two taxonomic levels: for the folk genus "pine" and for a more inclusive category of evergreens. (But this may well be a collapsed taxonomy, with the most salient genus, pine, surviving.) Where *uus* has come to mean 'wood' in Gila Pima, *huk* has come to have the extended meaning of 'boards' or 'lumber' among the Tohono O'odham.

Sylvester explained this use of the word: "*Huk* can mean 'lumber' or 'pine.' That's short. Really [should] say *huk komalkdam* 'flattened pine'. That's east end. That's better. At west end we say *uus komalk* 'flat stick/wood', for lumber."

The Pima name for the town of Prescott in northern Arizona (elevation 5300 feet) is S-hukak, indicating a place with numerous pine trees.

Populus fremontii S. Watson ssp.
fremontii
[≠ *P. deltoides* Marsh]
Frémont's Cottonwood
Salicaceae
auppa

Frémont's Cottonwood was once the ubiquitous streamside tree marking the course of the Gila, the "green snake" of the Creation Story stretching across the desert from east to west. In time, as the watersheds were damaged and streamflow fell and the defined channel of the Gila turned into a broad expanse of shifting sands, the great groves and stands of cottonwoods disappeared from the Gila proper. But cottonwoods were not gone. They survived along the canals and major ditches that crisscrossed the upper terraces of the floodplains, bringing water to the agricultural communities.

When I arrived in the early 1960s, rows of massive cottonwoods still shaded field workers around Laveen and in the eastern reservation villages of Upper and Lower Santan, Sacaton Flats, and Blackwater. These cottonwoods are nearly all gone now. The fields bordering the reservation south of Chandler once had ditches lined with miles of old cottonwoods. Today there is not a one; they have given way to slender palms, oleanders, and eucalyptus.

Fremont's Cottonwood is named after Captain John Charles Frémont, who once traveled through the Pima villages, practically the only "civilized" outpost in those days. No doubt he rested along the river in the shade of these great trees.

Old cottonwoods have light gray, furrowed bark. The massive trunks may be 3 or more feet across. Heights of 75 to 100 feet are not unusual. The crown is spreading. The waxy green leaves hang on slender petioles so that they rustle in the slightest movement of the hot dry air. Lucky is the Pima who still has one of these giants for a shade tree.

Midwinter frosts eventually turn the triangular leaves yellow. It might be after the New Year when the leaves are finally shed, leaving white skeletons. Scarcely six weeks go by before the brown resinous buds pop out green again. It was during this bare-branched season that Sylvester Matthias and I set out east across the reservation to search for Yellow Mistletoe (*Phoradendron tomentosum*) specimens. We drove more than 65 miles and found none among the occasional cottonwood colonies growing on canal banks. We did find a dead giant east of Bapchule, and Sylvester stripped off great shreds of its brown inner bark. He uses some of these to make the side braids on gourd masks.

On our return, we detoured from Gila Crossing down to Santa Cruz village to visit Carmelita Raphael. Cottonwoods are gone from there, too.

"We have been looking all over the reservation for that kind of *haakvoḍ* [mistletoe] that grows in the *auppa,* but we can hardly find any *auppa* anymore."

"They pulled them out when they put the ditches through here," Carmelita told us.

"Who?"

"The BIA. Long time [ago] when the [Pima] Indians do their own [ditches], they just let them grow." The BIA is the Bureau of Indian Affairs, a federal agency.

"The Pima let them grow along the ditches?"

"Yes, they just let them grow."

Ruth Giff grew up in this village and remembered the trees: "Cottonwoods grew all along the ditches there—seems like they just grew all over. Santa Cruz [village] used to have *big a'uppa.* Mostly cottonwood trees, along the ditches, . . . maybe a few willows. It's funny, on the [lower] Santa Cruz [River], they had no cottonwood trees, just willows."

Did the Pima just wait for nature to provide? At Santa Cruz village one day I proposed this to David Brown and Sylvester Matthias. No, they agreed, the Pima planted cottonwoods as well as willows along the ditches for shade, both for the People and for their horses.

Once when we were discussing hawks, Sylvester observed: "There used to be *lot* of *haupal* [Red-tailed Hawks] when the cottonwoods grew along the river. All the way down to the Salt. But the [white] farmers didn't like them [*auppa*] because they said they take too much water. But the Pima farmers like them because they enjoy the shade. Where there were willows, we got the *kumul* [edible mushrooms]. Then the river went dry and the cottonwoods died."

After the elegant native cottonwoods went, Athel Tamarisks were planted around nearly every Pima mud house on the reservation. An Asian species, they were easily propagated from cuttings and stayed put, not becoming naturalized like their cousin the Salt-cedar (see *Tamarix ramosissima* account). But the Pima were less pleased with the new tree. Sylvester explained, "You just sit in the shade of those, and you aren't satisfied. You just feel some way." The Pima have not deigned to give this introduced tree a name, even though it is everywhere on their land.

A few Pimas have tried to make a reversion; you will see an occasional young cottonwood planted in a yard. But usually these do not do well. When the water table was only a dozen feet down, cottonwoods thrived. But with the table now fallen to a hundred or two hundred feet below the dry alluvial soils, cottonwoods stay pretty scrawny.

Riverine Pimans everywhere in the desert were familiar with their beloved cottonwoods. The Tohono O'odham, like the Gileños, call the tree *auppa* (pl. *a'uppa*). At the Pima Bajo village of Onavas in southern Sonora, Pedro Estrella could remember no other name for the tree except *álamo,* which is Spanish, but María Córdoba called it *oopa.* The mid 17th century vocabulary composed in her village gives *álamo* as *hopo, opo* (Pennington 1979:7).

The wood of the *auppa* is white, light, and even-grained; the Gileños use it for specific purposes because it is strong. At Santa Cruz I watched Rosita Brown pounding dried mesquite pods in an upright wooden mortar, a *chepa.* Hers was made from a cottonwood log standing about 16 inches high. She explained, "My *chepa* is made from *auppa,* but it could be *kui* 'mesquite'. They're made like this, standing, or either on the side." Her *cheepidakuḍ,* or pestle, was an oblong stone.

The making of pinole is a difficult skill that older Pimas could appreciate. A quarter-century after Rosita's demonstration, Francis Vavages was discussing traditional foods at Sacaton. "*Auppa* [wood] is best for coals for making *haak chu'i* [wheat pinole] unless you're real good—then use *kui.* But it's real hot. Have to be very careful!" Cottonwood, being less dense than mesquite, produced coals that the less skilled parcher found easier to work with.

There were other uses for *auppa* wood as well. Joe Giff said, "The four outer and the four inner posts [of a round house, *kii*] are always *kui;* anything—cottonwood, willow—can be used for the others." Once cottonwood timbers played a more important role in Pima house construction. At the turn of the century, Russell (1908:154) wrote, "The central supporting framework is usually entirely of cottonwood, though other timber is sometimes used. When finished the *kii* is very strong and capable of withstanding heavy gales or supporting the weight of the people who may gather on the roof during festivals." Even the *vatto* 'arbor' uprights Russell (1908:155) said were made of crotched cottonwood posts (mesquite is now the material of choice for the four or six uprights).

The entrance to a storage shed, the *kosin,* was once closed simply by piling a stack of cottonwood logs, restrained by two uprights, before the door. The relatively light but bulky logs were unstacked to gain entrance. Upon the death of the owner, these logs were piled over his or her grave. When I first went to Piman country, traces of a few of these old-style burials still existed. They were for people not buried in holy ground, I was told.

The Pima wooden plow was called *uus giikĭ*. The share and handle were fashioned from a single-branched piece of some hardwood, usually mesquite or ironwood, but the tongue was cottonwood (Russell 1908:98; Castetter and Bell 1942:138). Joseph Giff explained, "The *yeevo* is the yoke [for plowing]. It's made of cottonwood because it's light. Rawhide is tied from it to horns of oxen; [there are] two places to fit on necks." Russell (1908:98) illustrated such a yoke.

Joe said the *shoñigiwul*, or kickball, was made of cottonwood or paloverde, never of mesquite or ironwood. It was cured with *kui ushabĭ*, mesquite gum, and polished down to the correct weight with a stone. The weight had to be just right for the individual user.

Sometimes cottonwood would be used for firewood, though it is not a long-lasting or hot fuel. Once in a while, the posts for a barbed wire fence may be made of *auppa*, but Velvet Mesquite (*Prosopis velutina*), Screwbean (*Prosopis pubescens*), and Catclaw (*Acacia greggii*) are the preferred and longer-lasting woods for posts. Sometimes Athel Tamarisk (*Tamarix aphylla*) is used, too. In spite of preference, the Pima are practical and often use whatever is most readily on hand.

Until 1987, masses of cottonwood logs and stumps were on the lower terraces of the Gila immediately east of Maricopa Road and Pima Butte. Sally Pablo said that until about the 1950s, there was a great grove of cottonwoods there, and on Sundays wagonloads of people from Komatke and Gila Crossing and Santa Cruz village went upriver to this spot on the river to swim and fish.

Ruth Giff told me more of the secrets of the *auppa*. "You know, when we were kids, we'd go looking in the riverbed for those roots—those *real* light ones, and we'd be smoking those. Suppose they are *auppa*. They're not *kui* because that is *hot*, will burn your tongue. Every kid did that in those days—look for that *vakolu*—real light driftwood—must have been cot-

tonwood—for smoking." She paused, then added, "But that's what the Hopi use to carve their kachinas.

"And you know," Ruth continued, "we call Florence [a town just upriver from the reservation] S-a'uppak 'many cottonwoods' or 'cottonwood grove.' There must have been a lot of *a'uppa* there."

As with the willow, the male and female flowers of cottonwoods are borne on separate trees. The male, or staminate, catkins are elongated. These are the pollen producers. According to Joe Giff, they are called *hiotamdam*, from the verb *hiot*, meaning 'to pop open', such as a flower bud. The female, or pistillate, flowers are better known. These are the *haupuldag*. Ruth Giff explained, "The 'berries' are called *auppa haupuldag*, but the fluff that blows out later from this or *che'ul* [willow], or from any other tree, is *tataindag*."

The Pima had good reason to know the pistillate flowers, as Ruth explained: "They say the right time to get *a'uppa* is in the spring. That was food for hungry people then. They say the worst time of the year was in the spring when their stored stuff was gone. Then they'd start eating their green wheat when it's real small." Joe told how they were used: "You get the strings of those berries, that *auppa haupuldag*, crack [them] with the teeth, and take the cotton out and chew it like gum, though it's not gummy; just chew it anyway. Best when they're young. Have a pleasant taste, near sweet." To be chewable, they must be young, about the size of your thumbnail. Ruth said a little lard helps hold everything together. The practice was widespread, and everyone among the Pima except the youngest generations seems to remember chewing *auppa haupuldag* when the trees were still at hand.

Russell (1908:69) stated that the male flowers were also once eaten. Under *aopa hiôsik* 'cottonwood flowers', he recorded, "In February and March the women send some of the barefooted boys into the tree tops to throw down the catkins, which are then gathered into baskets and carried home to be eaten raw by stripping them off the stem between the teeth." Only the staminate flowers could be called catkins, because the pistillate flowers are like round green balls, called berries by the Pima today. Russell made no mention of these being used as

chewing gum, the only thing Curtin (1949:109) noted 40 years later. Still, Russell's observations were in keeping with several other types of flowers eaten that way, such as Velvet Mesquite.

For instance, Francis Vavages told me, "Those blossoms—of the *kui* [Velvet Mesquite]—pull them through your teeth. They're sweet, the flowers. And the *auppa*, before they form berries, when young. You can eat that *auppa hiósig*, the fuzzy-looking ones, like a tail [of an animal]. *Auppa*, when they have berries, the *haupuldag*—the hard green berries—they use for chewing gum." The staminate (male) catkins appear a bit ahead of the pistillate (female) flowers and are on separate trees.

I asked George Kyyitan about this. He also called the female flowers *auppa haupuldag* and the male flowers *auppa hiósig*. "*Auppa haupuldag* are for chewing gum. They are the first ones that got the green on them. You'll see next month [February]. . . . In a few weeks, they'll be ready to chew. Also *che'ul* [willow] *haupuldag* can be chewed. And *auppa hiósig* [are] good to eat, too. Put salt on it . . . hummm," he said approvingly. "Also *kui hiósig* [mesquite flowers] . . . some are sweet. But not *che'ul hiósig* [willow flowers]."

"What about *tataindag*?" I asked him.

"That's nothing. It just blows off, when they [female flowers] get old, hard," said George.

About April this downy *tataindag* goes floating out from the mature *auppa haupuldag*. (The name for these green, cotton-packed fruits is the plural of *havuldag* and appears to be derived from *hauk/haupk* 'to be lightweight'.) To germinate, all they need is to fall upon the right surface. Once there was the right surface, as Sylvester remembered: "There is, especially after the flood—there would be little cottonwood trees, becoming . . . little trees so high. They called it *a'al a'uppa* ['little cottonwoods']."

After the floods on the Lower Salt, I would see dense patches of cottonwood seedlings coming up on the scoured riverbed. The Pima had a use for even these. Ruth knew how to use them in weaving baskets. "Those young shoots, seedlings of *auppa*, are stripped like willow shoots and used in weaving [baskets]. Smooth, without the 'knotholes' of willow. But they say that *auppa* turns

yellow sooner than the *che'ul* [willow],
but they used it." Later she told me,
"Sometimes they just come up—young
ones. And they are *smooth*. They don't
have those . . . where the leaves come out.
But they are hard to find. The only trouble
is they turn yellow. I did get some, but I
don't think I ever used it."

Nearly 90 years earlier, Russell (1908:
134) was told, "Wood from the slender
branches of cottonwood is sometimes
used to take the place of willow, but it is
less durable and soon becomes yellow. It
is prepared in the same manner and kept
in the same sort of coils as the willow."

Cottonwoods as well as willows root
readily from cuttings: young branches set
in damp soil in winter or spring. In places
in Sonora, careful manipulation of narrow
floodplain soils was achieved by the judi-
cious planting of cottonwood living fences
(Nabhan and Sheridan 1977). I have been
unable to determine for certain whether
this was a deliberate practice with Gileños,
or just the unintentional result of using
the right kind of posts at the right time of
the year for them to sprout.

I asked Sylvester about the deliberate
planting of willows or cottonwoods.

"I've *seen* it. My father and my half-
brother used to go cut cottonwood for
fence posts. Then they'd take root. They'd
have leaves on it. When they stop irrigating, they'd die. Some have good roots,
then they'd grow. Then what happened,
too, in my aunt's field, they had a cut—
they build a fence right here [with] those
cottonwood posts, and two of them
growed up, and they were still there for
a number of years. They were still there
a few years ago. Then, I guess the water
level went down, then they went dry."

This was almost certainly intentional
erosion control. I asked him if they
planted cottonwoods to hold the bank of
the river where it was cutting away.

"No. But Salt-cedars did hold. It
changes after the flood—but that
Tamarisk—it holds. In 1906, *both* the
Santa Cruz and the Gila flooded. [Dried]
cottonwood fences didn't last very long.
If it rooted—well, it's going to be there."

"If it didn't root, how long would it last?"

"Maybe two years. But these [Athel]
Tamarisks—they won't last. Maybe a year,
it'll be gone. It makes nice posts—but
don't last. *Auppa* make[s] good beams for
a *vatto*—because they're straight and not
much weight on it."

auppa *Populus fremontii*

"How do you get *auppa* started?"

"Just get them—by the roots, about January—when the sap is down. And you set them right into the ground, water it, and by the time the sun comes, then it starts leafing. Just like I said, they used to grow a lot of *auppa* on that little island next to the creek, and the people who cut for fenceposts, they start growing, because it's irrigated."

"A piece of the branch will grow?"

"Yeah, it'll grow. These *auppa* can start pretty good. I guess willow can, too."

The great cottonwoods served even medicinal purposes for the Gila Pima. One day Sylvester was explaining how the "sucking" medicine man once used to pretend to suck the root of *haḍam tatkam* (globemallow, *Sphaeralcea* spp.) out of his patient, saying this mallow was the cause of the illness. Then he added, "It's not good, *haḍam tatkam*. It will give you the sore eyes."

"What can you use for a remedy?" I inquired.

"You could use *shegoi* [Creosote Bush, *Larrea divaricata*]—wash it off. Or cottonwood leaves—make a tea of them."

"How much?"

"A little [few] leaves. Some pound it a while."

This echoed what Curtin had recorded 40 years earlier: "A handful of *auppa haahag* (cottonwood leaves) is boiled in a pint of water and sores are washed with the decoction. Like greasewood [*sic*, Creosote Bush], this is very healing." She recorded one additional use: "For hair-dye a brew was made from cottonwood leaves, strained and mixed with tea from mesquite bark."

Sylvester told of another use for cottonwood leaf tea: "Also good for broken bones—if you wash it. That helps, too. As strong as you can make it and as hot as you can stand it. It's just the same as when you burn it—*goi*—it kills the pain."

Along the lower Salt River, the northern boundary of the reservation, new growths of cottonwood-willow stands have been allowed to develop in the last 20 years or so. These are watered by permanent effluent discharged into the riverbed from Phoenix sewage treatment plants. There, Green Herons and Yellow-billed Cuckoos breed in numbers, as they once did along the middle Gila. And Song Sparrows, Yellowthroats, and Bell's Vireos advertise territories throughout the summer. But

only on the north side of the river. The tribal council was prevailed upon by whites a few years ago to let nearly all their woodland vegetation be removed as a flood control measure (*Phoenix Gazette*, 7 April 1980). It has never returned.

Vouchers: *579* (RG, SP); LSMC *41*.

Populus nigra L.
Lombardy Poplar
Salicaceae
kapichk auppa, (pl.) ka'akpichk a'uppa

Settlers throughout the West often planted rows of the tall, slender Lombardy Poplar around their houses and farms to serve as windbreaks, particularly where the water table was high or where the trees could keep their roots wet in a creek or ditch. The trees, all males, were easily started from cuttings or runners and grew rapidly. Their leaves, which look like those of the congeneric Frémont's Cottonwood (*Populus fremontii*), turn deep yellow in late fall. Unlike the broadly branching cottonwoods, the poplars are single-trunked spires.

When the Mormons settled in the Mesa area east of Phoenix, they planted the poplars about their farms. The Pima associated the poplar with this group and called the tree *moomli auppa* (pl. *momoli a'uppa*). This is the form Fr. Antonine recorded in his 1935 manuscript. Ruth Giff did not recognize the name but said it must be from Momoli (literally, 'Mormon, place of these'), the Pima name for Mesa.

But this name was not used uniformly among all Pima speakers. Sylvester Matthias knew the poplar as *kapichk auppa* (pl. *ka'akpichk a'uppa*), which he glossed 'slender cottonwood tree'. "There used to be a lot of them around Glendale, right in that area, not anywhere in Phoenix. Guess it's all just town around there now. They used to grow [there], just straight up! Unlike cottonwoods, that branch out."

I asked George Kyyitan of Bapchule about them. "*Ka'akpichk a'uppa*," he called them, "more than one. They grow straight up. Used to be some down here— Hamilton Corner. Straight up, not spread

out like a tree does, like limbs going out, spread out. *Ka'akpichk* goes straight up— just narrow."

When Phoenix was still in its infancy, John J. Gosper (1881:923), acting governor of Arizona Territory, penned this account of the new Anglo settlement wedged between Pima lands: "The first settlement was made in this valley a little over ten years ago. It was then a barren desert, covered with coarse grass, sage, and cactus; to-day it is one of the loveliest spots on the Pacific coast. Fields of golden grain and blossoming alfalfa; extensive vineyards and orchards; beautiful gardens, brilliant with their floral adornments nearly every month in the year; groves of cottonwoods and lines of the graceful Lombardy poplar diversify the landscape in every direction."

Prosopis glandulosa Torr. var. *torreyana* (L. Benson) M. C. Johnston [?]
[= *P. juliflora* (Swartz) DC. var. *torreyana* L. Benson]
[≠ *P. juliflora* (Swartz) DC. var. *glandulosa* (Torr.) Ckll.]
Honey Mesquite
Fabaceae (Leguminosae)
kui chepelk, (pl.) kukui che'echpelk

Through quirks of geological and evolutionary history, the Pima of the middle Gila are sandwiched between country occupied by Honey Mesquite to the east and to the west. All of riverine Pima country is populated by a distinctive species, Velvet Mesquite, that grows from the middle Gila southward in a broad tongue at least to Pima Bajo country (see *Prosopis velutina* account).

The leaflets of Honey Mesquite are more elongated and more widely spaced on the rachilla or stem (there is more space than leaf) than are those of Velvet Mesquite. Honey mesquite also has a more grassy green color and a much more prominent "beak" at the end of its pod, and it tends toward a sprawling brushy growth form, most noticeable when growing in dunes.

Older Pimas were astute observers of their surroundings. During a travel narrative, George Kyyitan made these comments: "The time we went to Mescalero [New Mexico]. On this side, too, the mesquite trees, they have large leaves on there. It's different from here. This one is smaller, leaves. I notice it. I kind of watch it when we were on the way there. Then when we got to Deming [New Mexico], we sleep there. And next morning, I saw there was a few of them standing out there. And I saw it was very different. All the way."

Is this what the Gila River Pima call *kui chepelk* 'dwarf mesquite'? There are various suggestions that it is, and some that it is not.

In 1979 Joseph Giff was telling me about various forms of mesquite. There was the "loco" mesquite and the "red-podded" mesquite and the "green-podded" mesquite. And then there was the "dwarf" mesquite. "*Kui chepelk* is short, wide, and low," Joe said. "We usually say *kukui che'echpelk*. They have hard wood, hard to chop; don't have big limbs on them; if one, there's more. That's why they use the plural. [Grows] like Bermuda Grass; the roots spread out and spread around. Roots are big, like they are all on the same root; maybe *kukui che'echpelk* don't have *vihog* [pods]."

They used to grow locally in the sand dunes, but he did not know if there were any left. He had a daughter-in-law, Cindy, whose father was from Pàlm Springs. "I think the *kukui che'echpelk* is the mesquite that grows in the Coachella Valley of California," Joe said.

The southern California and lower Gila mesquite is *Prosopis glandulosa*. Could it have once reached Komatke? The growth characteristics he described strongly suggested it. But Joe was then blind and could not take me out to look for them.

I asked Sylvester Matthias if he had heard of them. He had: they spread at the base and grew to no more than six or seven feet high, in sand dunes.

Later in the spring, when the leaves were on and the mesquites flowering, Sylvester took me out north of St. John's Cemetery and west of Sally Pablo's house, where he remembered the *kukui che'echpelk* grew. Along a little wash he pointed out three, and I collected specimens. They were nothing more than scrubby specimens of Velvet Mesquite, at best a growth form, though I would not have picked these individuals out as being much different from their neighbors. Perhaps the real

kukui che'echpelk was gone and these were the nearest approximation that Sylvester could come up with. Perhaps he had forgotten what they really looked like. Perhaps. But I have never found any mesquite other than Velvet Mesquite anywhere in Gila Pima country.

Years later I asked Francis Vavages of Sacaton about this form. He called it *che'echpelk kukui* and said it was the same as *kui* except smaller.

I then drove from Francis's house 35 miles across saltbush flats studded with wolfberry (*Lycium* spp.) bushes and mesquites to Ruth Giff's house at the west end. I asked her about *kui chepelk*. She brightened as if hearing the name for the first time in decades.

"Hey! Where is it? Somewhere in California? They have little tiny. . . . "

"Or Yuma?" I asked.

"Yeah. In that sand. And I guess we have some around here. They're not good for firewood. They're not big enough to cut down for firewood. They're midgets. I've seen them. Maybe they grow in the mountains, too."

"Do they look any different?"

"No. The *vihog* I think is the same."

But the mesquite origin myth, where Coyote steals the pods from the beautiful tree that grew from Bat's semen, clearly indicates that the Pima were aware of at least a qualitative difference in the mesquite growing downriver from them. When they discovered Coyote had stolen all the *vihog* from the first mesquite tree, they chased after him, beating him with their war clubs. Coyote sprinted toward South Mountain, defecating as he went. They chased him downriver, still beating him. And the seeds became fewer, poorer in quality.

And that is why the good mesquite grows here in Pima country and poorer mesquite grows down in Yuma country (see *Prosopis velutina* account for the more detailed story). And to botanists, who speak in taxonomic terms rather than archetypical mythical terms, the Pima have Velvet Mesquite (*Prosopis velutina*) and the Yuma, or Quechan, have Honey

kui chepelk, (pl) kukui che'echpelk   *Prosopis glandulosa*

Mesquite (*Prosopis glandulosa*). Still I wonder, could Coyote have shat a few of the bad seeds out up here on the sand dunes, just to confuse ethnobiologists many generations later?

Technical Notes: AMR 125, 126, 127 (SM)—supposedly representatives of *kukui che'echpelk* or dwarf mesquite—are actually *P. velutina*.

Prosopis pubescens Benth.
 [= *P. odorata* Torr.]
 [= *Strombocarpa pubescens* A. Gray]
Fabaceae (Leguminosae)
Screwbean, Screwbean Mesquite, Tornillo
kuujul, kuujuls

At a glance, one might not realize that Screwbean Mesquite is related to Velvet Mesquite, because their growth forms are different. Screwbean grows quite upright, aspiring early to freedom. The bark on smooth younger branches is gray, becoming blackish and fissured on old growth. The leaflets are fewer and more compact but, like the more common species of mesquite, are arranged on the two branched prongs of a Y-shaped leaf stem and are deciduous. The tiny flowers, appearing in late spring, are arranged in the same elongated clusters but are deeper yellow than Velvet Mesquite flowers. After pollination, all resemblance ends. Instead of a slender pendulant pod, the fruit begins to twist until there is a spring-tight ochre-colored coiled pod. These often cluster at right angles to the original flower stem. Screwbean pods are persistent. They become richer brown or even blackish with age. Frequently the crops from several successive seasons can be told from the deepening colors. Tiny bruchid beetles, departing their nurseries inside the very small seeds within, leave burrow holes on the coils.

The Gileño name for Screwbean is *kuujul*, exactly the same as Mathiot (n.d.:13) transcribed it for the Tohono O'odham. Fr. Antonine (1935) wrote *kujil*, Hrdlička (1908) *ku-u-dje*, and Curtin (1949) *koo-ejil*, all basically orthographic variations. The way Frank Russell (1908: 79) wrote it, *koitcilt*, suggests that the word might be etymologically related to *kui* (Velvet Mesquite), which he wrote just above as *koi*. But earlier (1908:75) he rendered it *koûtcilt*, which corresponds to the present form, given that Russell's *o* usually indicates *u*. Still, there might be an ancient derivation. In Pimería the tree occurs only in Akimel and Tohono O'odham country, so there are no clues from other Piman-speaking groups surviving to the south. A variant occasionally heard in the middle district of the reservation is *kuujuls*. Screwbean pods are called *kuujul vihogdag*.

This legume was important to all native peoples living within its range. Spier (1933:50–52) considered the two kinds of mesquite even more important a staple among the Pimas' neighbors, the Maricopa, than was corn. At one time, certain nomadic Piman bands, the so-called Sand Papago, depended heavily on Screwbean as well as the other mesquite (see Días 1774 in Bolton 1930, 2:260–263). But most Tohono O'odham, living away from permanent waterways, probably were unfamiliar with this tree.

Although she never fixed any for me because they were no longer available, Ruth Giff remembered the earlier processing of Screwbean pods for eating. They were soaked, dried, pounded to make flour, then sifted. Her recollection was part of a tradition stretching back centuries through documentary history. (Spanish chroniclers distinguished between *péchita* [Velvet Mesquite] and *tornillo*, so we know which species of mesquite was meant.) For instance, in 1774 Fr. Pedro Font noted the Gileños eating both (Bolton 1930, 4:44).

Both Hrdlička (1908:261) and Curtin (1949:69) tell of a "sweet and nourishing beverage" made from the pounded pods. Only Russell (1908:75) detailed its preparation method: "[Screwbeans] are cooked in pits which are lined with arrow[-weed] bushes *[Pluchea sericea]* set on end. The beans are placed in layers alternating with cocklebur *[Xanthium* sp.] leaves, the whole covered with earth and left to stand three or four days, after which they are taken out and spread to dry. They are then ready to use or store away in the arrow-brush basket bins on the housetops. They are further prepared for food by pounding up in a mortar, the fine flour then being ready to be eaten as pinole. The coarser portion is taken up in the hands with water, the juice sucked through the fingers, and the remainder rejected." This latter is called *kuujul vau* or *vaō*.

After this elaborate preparation of the pods fell into demise, young Pima still found other nourishment from the Screwbean. "*Kuujul ushabĭ* [sap, pitch] is not as sweet as the *kui ushabĭ*," recalled Ruth Giff, "but we ate it. Not sticky. Ah, we used to go out just to pick it. We'd have a little stick and we'd put it on the stick until we had big balls of *ushabĭ*. Then we'd go home. We'd go in between the dams [brush weirs] along the Gila River during the early summer. There was a lot of *kuujul* in there."

In 1979 I was talking to Joe and Ruth Giff about the kinds of foods made from *kui*, Velvet Mesquite. I mentioned the *kuujul* processing in pits. They had never heard of this. Ruth said she would ask Adolph Juan, who was then in his 90s. He had not heard of it either. All three had known of the pods being pounded directly as picked, just like the *vihog* from *kui*. At any rate, *kui* was much more important to the Pima of the Santa Cruz,

Gila Crossing, Komatke area, and they thought that maybe *kuujul* was more significant only when there was a poor *kui* crop.

However, Castetter and Bell (1942:57, 64) considered Screwbean so important in the Pima economy that they ranked it with cholla buds, only Velvet Mesquite and Saguaro being more important among wild gathered foods.

One time, Ruth Giff was asked to make a *homda*, or rooftop storage bin, for an exhibit in the lobby of a new hospital at Sacaton. I agreed to take her out to gather fresh Arrow-weeds to make it. A week later, when I phoned, Ruth had decided to make a *vashom*, or storage basket, as well, but she needed more *kuujul* bark for the binding material around the coiled straw base. "I have used *kui* before, though it's not the right color." I told her that I had just seen some *kuujul* trees in the El Centro area and would stop by in a few days on my way back to San Diego to get some for her.

I found a nice straight limb about two meters long and removed a great strip of the silvery gray bark. Double-pronged thorns defended the bark of even the stout limbs. The wood beneath was moist, slippery, and rich creamy yellow. Before long, I had enough ropelike strips. The Pima had certainly selected an excellent binding material for making their huge granary baskets.

But the *kuujul* was more than just food, drink, and fiber for the Gileños. Joe Giff, who had built many a fence in his day, said, "Screwbean seems not as hard as mesquite. But [it] has a core that stays. Dig hole and set post in place and put special dry soil around it; don't want it wet." Francis Vavages said *kuujul* posts were used to build the *kosin*, or storage shed, which was once found in every Pima yard. And, of course, *kuujul* made good firewood. Sylvester Matthias told me, "The frame for the cradle board is made of young screwbean." Also the *eskuḏ* or digging stick and the weeding hoe might be made of *kuujul*, though Desert Ironwood *(Olneya tesota)* was preferred. George Kyyitan said Screwbean was one material used to make a bow. Francis Vavages agreed. "Use *kuujul* for bow, *uus gaat*, because it's got a lot of spring to it. Sometimes they make it out of mesquite, too, when it's young. But *kuujul* is springy; they're hard; can't break it."

This once abundant tree formerly served as medicine for the Pima. Russell (1908:79) noted, "The bark of the root of the screw bean is pounded up in mortars, dried, and again ground into a fine powder in the metate; or it may be boiled without pounding or grinding and the liquid used as a dressing for wounds. After a few days, as the wound heals, the dry powder is substituted." Four decades later, Curtin was told the same. In addition, she noted, "To cure a woman who was having trouble with her menses, Lucy Howard gave tea made of the roots of this plant" (Curtin 1949:96).

Kuujul also finds its way into the version of the Creation Story that Thin Leather told to Russell (1908:231). Coyote, one of the four original beings, made two more creatures like himself: Sandy Coyote, the younger brother, and Yellow Coyote, the older. They went to live near the mouth of the Grand Canyon.

When they went to gather the screw bean the elder brother took the beans on the south side of the trees and the younger brother took those on the north side. One day the elder said to the younger, "How do the beans taste on that side of the tree."

"They are very good," replied the younger, but when they returned home in the evening he was taken sick.

"It is caused by the beans you ate," said Sandy Coyote. "The beans on the north side are not ripened by the sun as are those on the south side. To-morrow you shall see the difference." And so the next day they went again and found the screw beans sweeter on the south side of the trees.

Every evening they sat and split sticks with which to build bins, log cabin fashion, for the screw beans that they gathered. One day the elder brother said, "Let us play some kind of a game and bet our screw beans, and then we will not sleep too soon." (Russell 1908:231)

The younger brother lost his screwbeans, but the elder brother decided the next day that they would not get more screwbeans. The younger brother went hungry for many days, for the gambling continued until the beans became unfit to eat.

Screwbean is more restricted along the desert riverways than the almost ubiquitous Velvet Mesquite, although it was noted throughout the early part of the 20th century for its abundance along the Gila at the upper villages (Castetter and Bell 1942:64; Hrdlička 1908:261; Russell 1908:75). Curtin (1949:96) characterized its requirements well: "It loves moist, heavy saline soil." The eastern limit of the natural range of Screwbean was the Coolidge-Florence area on the Gila (Hastings, Turner, and Warren 1972:175). Today, this has changed. During the years I lived at Komatke, I found the tree occasionally on terraces of the Gila between that village and Co-op Colony, but I have never encountered it on the middle or eastern reservation portions of the Gila, along the lower Santa Cruz, or anywhere on the lower Salt River, although older Pima told me it was once common throughout all these areas.

George Kyyitan and I were discussing fenceposts and the various kinds of fences that the Pima used to make. He said, "I used to go over there and chop [*kuujul*] for posts. But nowadays you can't find nothing. Lot of them on that [west] side of Maricopa Road. Used to be lot of them there. But I don't know what became of them—just all got dry there. *Nothing* is there no more. Some, way right by the bridge, there. Used to be lot of it there, too. Lots of *kuujul* there. But there's nothing now." Apparently the Screwbean is more sensitive to modern patterns of land use and to lowered water tables than the Velvet and Honey Mesquite. It is now rarely encountered in Pima country.

Technical Notes: Although *P. pubescens* once occurred along the Gila to Blackwater and beyond, it is now apparently restricted to the far western part of the reservation (Santa Cruz village and downstream).

Vouchers: *38* (RG); *1553* (CR); LSMC *69*.

Prosopis velutina Wooton
   [= *P. juliflora* (Swartz) DC. var.
     velutina (Wooton) Sarg.]
   [≠ *P. chilensis* Stuntz.]
Velvet Mesquite
Fabaceae (Leguminosae)
kui

Trees seem few in the Sonoran Desert. Velvet Mesquite is an exception: it is common on the floodplains of southern Arizona and Sonora. Other mesquites live in similar arid areas from southern California across to Texas and south to Patagonia, in southern South America. They are members of the pea family, Fabaceae (Leguminosae), the third largest plant family in the world. Legumes are the major woody plants in the New World deserts except the Great Basin. Throughout the lowlands occupied by Tepiman groups, these trees dominate.

The mesquite tree reaches heights of 30 to 50 feet, though the only large ones found now on the reservation are dead skeletons of former giants in New York Thicket. Over most of Gileño country, the trees are much younger and only half this tall. Below the mouth of the San Pedro River, where the Gila still runs, I have found mesquite trees with trunks several feet in diameter; these were so tall that I could not reach their lowest branches.

Velvet Mesquite has coarse black bark on the upright trunks and spreading main branches. Younger branches, armed with long stout spines at the nodes, have dull greenish bark. Each leaf stalk (petiole) divides into a Y. (Sometimes there are four arms on a leaf instead of two.) A dozen or more pairs of leaflets crowd each arm of this Y. When mature, they are velvety, hence the species name, and slightly bluish green. (Leaflets of the Honey Mesquite are longer, greener, and more widely spaced on the leaf rib.) Mesquite flowers are minute, borne in slender catkinlike clusters.

Not that the mesquite was the only tree in the desert. The creeks and rivers were once lined with willows and great cottonwoods—the major components of the former desert riparian communities. Away from the floodplains, the bajadas (slopes) are studded with "forests" of

Saguaros *(Carnegiea gigantea)*, the tall columnar cactus that was much more important in the economy of the Tohono O'odham than that of the Gileños. The arroyos or dry washes that dissect the bajadas are lined with paloverdes *(Cercidium microphyllum* and *C. floridum)* and gnarled Desert Ironwood *(Olneya tesota)*. But probably all these trees combined did not equal the importance of mesquite in Pima culture.

The Pima name for the mesquite is *kui,* a primary unanalyzable lexeme. (The Pima Bajo in Sonora and the Tohono O'odham call the tree *kúi,* with two distinct but joined vowels; the Gileños say it much more rapidly, as if with just a single vowel, *kwi.)* No other plant affected the daily life of the O'odham to such an extent. It is so important, in fact, that two months in the Pima calendar refer to the life cycle of this species: *kui i'ivagĭdag mashad* 'mesquite leafing-out moon' (around April) and *kui hiósig mashad* 'mesquite flowers moon' (around May). For the Pima, Velvet Mesquite was the tree of life. They recognize at least five stages in the annual cycle of this tree: *kui i'ivgig,* when it leafs out; *kui hulkadaj,* when its flower buds are still green; *kui hiósig,* when it blooms; *kui vihog,* when it fruits; and *vihog baakam,* when it is ready to harvest.

Mesquite bosques or continuous-canopy woodlands, in Pima called *s-kuik,* in some places extend for miles along the river bottoms of Pimería Alta. One such bosque above the confluence of the Gila and Santa Cruz Rivers was about six miles long. A mesquite bosque has its own characteristic smell, a rich and distinctive odor that one can never forget or mistake. To reach underground water, the mesquite sends down a tap root that is 30 to 90 or more feet long. Lateral roots tap water nearer the surface.

The annual cycle of the mesquite and Pima cultural ecology meshed well. In temperate regions, most cultures appear to have undergone an annual period of stress when available resources were at their lowest—usually in late winter and early spring. Agricultural Piman tribes were no exception (see chapter 7). This was the time when stored resources were nearly depleted and game was scarce. Joseph Giff explained, "Now the People brought out their last jerked meat—dried lungs—to cook with late winter greens. This is the time when the Roadrunner rattles its bill

in courtship." In April, when the mesquite begins to leaf out, it is time for the seeds of beans, squash, and cotton to be planted in newly worked fields. Because resources are still scarce, the Tohono O'odham call May *ko'ok mashad* 'painful month'. Everyone in the desert, man and beast alike, awaits the mesquite crop.

Some Pima did not wait to start eating. Francis Vavages of Sacaton said that people would pull the catkins *(kui hiósig)* through their teeth and eat the flowers, which were sweet.

Sylvester Matthias told me of a food prepared from *kui hiósig* and mud. "There is a certain kind of clay. The one when the river is high. Not just earth. But it has to be *s-haḍam bid* [sticky mud], they call it. The floodwater overflow, or a puddle there. They always leave some kind of [mud that] becomes like [adobe] bricks. Some ladies chew on that—that *s-haḍam bid.* That's the one that you make [mix with] that *kui hiósig.* I don't know how much. I never seen it, but my grandmother told me that she used to eat it that way. It's sort of sweet, too, at that time. Now it's too dry. But at that time, there must be a lot of sap [nectar] in it. They have to wet it, mix it up, make something like a soup."

"Who ate this?" I knew of geophagy only among Pima girls and women.

"Anybody who wants."

"Did men do that? Or women and children?"

"Everybody."

This preparation apparently had no special name. Sylvester said it was called simply *kui hiósig.* He thought it may have been a starvation food.

As the desert heats to a pitch before the summer rainstorms arrive, the light-yellow mesquite blossoms fade and fall. If bees have done their job of pollination, light green pods resembling long, thin string beans soon cover the trees. Even these *vihog moikdag* 'mesquite pods something tender' are roasted on coals when other food is lacking. In June the pods thicken and turn pale yellowish white. Through photosynthesis, sunlight energy is stored in the pods as sweet carbohydrates (there is 25–30% sugar in the pulp). The fleshy pods have only 5–9% protein, the tough, hard seeds 32–40% protein. In most years, the trees are festooned with masses of these 6-to-8-inch pods—an abundant food source in a supposedly barren desert.

By midsummer the task of harvesting a year's supply of mesquite began in earnest. The pods are ready when they fall to the ground. There were so many that the Pima selectively harvested from those trees that produced high-quality fruits: the best trees yielded thick, sweet pods. Plenty remained in the bosques as food for animals.

The mesquite crop was one of the most dependable in the desert (Felger 1977). Because the roots could reach the water table even at considerable depths, the tree was not affected by the surface rains of a particular season. Russell (1908:66) was told, "About every fifth year in primitive times the Gila failed in midwinter." Then most wild resources were unavailable and farming impossible. In those years, "the fruit of the saguaro and the seed or bean of the mesquite were the most abundant and accessible resources. When even these failed the Pimas were driven to make long journeys into the Apache country—and whenever they got a mile from their own villages they were in the land of the Apache—in search of animal food, roots, berries, and especially the edible agave." (See also Russell 1908:74.)

I questioned Joe Giff about the failure of the mesquite crop. He took exception to this: "The drought never hurt the mesquite; there was a well 10 or 15 feet deep in New York Thicket. When it went dry, the *kui* stayed green until they couldn't reach the water table any longer. Now this Sally Hiinoi [the name is a species of fish] or Sally Fish [one of Frank Russell's "informants"] is Koahadk. She doesn't know what's happening here. May happen around Blackwater, where water started drying up. But look at Komaḍk [Komatke]—that's where there will be water for a long time. Because the South Mountains and Komaḍk [Sierra Estrella] joined [underneath], rock bottom, like a lake, so there's water to use for a long time. When they were trying to get the people to move down here [in 1873] to found this village [Komaḍk Wecho], one man kept trying to get his wife to [agree to] move, but she wasn't interested in moving from up there. Finally he told her, 'Look, the *vihog* on the mesquite are this thick up there' [indicating the size of his thumb or big finger], and then she decided to move up here, because the *vihog* was so good on the mesquite."

Joe and Ruth Giff said that bad wind interferes with the pollinization of pods, when young pods are setting at the end of flowering; it was not lack of water that made a mesquite crop fail.

The Gila Pima distinguished several different types of *kui*. Two of these variants are based on qualities of the pods, two on the growth form of the trees. Joseph Giff described one of these he called *cheedagĭ vihog:* "Even though they ripen, they still stay green-colored, and they're very sweet, very sweet; there's one down across the river at Santa Cruz [village]. Maybe one other. [We] stop when rounding up cattle and get some, but [otherwise] looks like any other *kui* when no *vihog* [on it]. You can't tell the difference then." I have never encountered one of these.

Another kind is the *vepegĭ vihog*. In this case, the normally whitish pods, when dry, are tinged or striped with pink or purple. This is a rather common form, but it is still considered merely a variant of "regular" *kui*. Sylvester Matthias said this about it: "*Vepegĭ vihog,* it's kind of pinkish, with stripes. *Vihog* is usually light yellow—sometimes almost white."

"Does it taste any different?" I asked.

"It's a little bit of sourness in it—a *little*. They like it. Sometimes they're really after [it] when it's green—fresh. They're really after the *vepegĭ vihog*." I questioned Margie Sabori and Ruth Giff about this one day. These red ones may be sweet, but they agreed that you really must taste the *vihog* from any tree first to find out its quality.

Everyone seemed to know the *loogo kui* (from Spanish *loco* 'crazy'). David Brown characterized this as "a mesquite that just goes up some way without growing up; branches go everywhere." For 25 years I have seen one of these individuals near a canal bank between Gila Crossing and Komatke. During all this time, the tortured growth seems never to have achieved any greater stature. David also mentioned a *leḍosh kui,* but I was not sure if this is another name for *loogo kui* or something he contrasted with it. I asked Sylvester Matthias, who knew *loogo kui,* if he knew *leḍosh kui*. "Yeah. I don't know what it looks like. Means 'silly' or 'crazy' or 'nasty'."

A fourth form, *kui chepelk,* or more frequently in the plural form *kukui che'echpelk,* appears to be a different folk generic (or at least folk species), as well as a different biological species (see *Prosopis glandulosa* account).

Although the Gila Pima were agricultural people, wild mesquite played a significant dietary role among them. Indeed, as one chronicler noted, "millions of pounds are gathered annually by the women of the tribe" (Grossmann 1873: 419). Another wrote, "The *principal* article of food was the bean [*sic*] of the mesquite, which still grows abundantly all over the desert" (Whittemore 1893:54). At the turn of the century, Russell (1908:74) noted that mesquite "formed nearly if not quite the most important article of diet of the Pimas in primitive times." A few years later, Hrdlička (1908:261) wrote, "The mesquite beans [*sic*] are still one of the most favored of the Pima native foods. They are dried in quantities and preserved in the storehouses." Later scholars such as Castetter and Bell (1942:57) likewise considered mesquite to be the principal food on the Gila. Ezell (1961:131), too, considered mesquite the most important of wild crops. A Gila Crossing Pima woman, born about 1869, told Curtin (1949:95) that she considered mesquite the most important of all wild foods.

Enormous quantities of mesquite pods were gathered every summer for storage in granaries. Lloyd (1911:123) noted, "The gathering of them was a tribal event, large parties going out," implying that both men and women harvested. As we were driving north one day into Gila Crossing, I asked Sylvester Matthias, "Did everybody pick the *kui vihog?*"

"Yeah—everybody picks, too, like *hannam* [cholla buds]. Sometimes they'd haul two wagons—over here under Lone Butte, where there used to be a *lot* of mesquite. They'd *all* go and gather the *vihog*. Everybody picks. They come back as soon as they get their wagons loaded."

"How long would that take?"

"Just one day. Load up, get it on a sheet, and just dump it on the wagon."

"Did your family have a particular place to go?"

"They just go in bunch, [different] families. They set a day when they go after *vihog*—they do that. I guess they help each other to fill their wagons up. Of course, they have to have a place to sit."

"How far away from Komatke would they go to pick *vihog?*"

"As far as Lone Butte, on this side. Because it's only mesquite forest up to that butte from here—maybe two miles from

the Presbyterian church. There's a wagon trail running all along the hill there."

"Did they ever get it off the tree or just off the ground?"

"Off the ground. They do that when it's fresh. They have to use the *ku'ipaḍ*—to get everything, when it's *fresh*." The *ku'ipaḍ* is the Saguaro-rib stick used to knock the fruit off tall cactus. Ruth Giff and Francis Vavages both said that the pods were picked from the ground, but some had to be knocked down first.

When George Kyyitan was a child, his family went on picking excursions. "We used to go out—camp for a week picking that *vihog,* filling those gunnysacks. *Lots* of them! Bring them back. Store them in the big *homda* [rooftop storage bin]. Then go out another week, pick some more. Come back on Friday, with my mother."

"How many *hohomda* [storage bins] would you fill?"

"Just one, one great big one. Lift the top to take them out."

Mesquite is less common in Tohono O'odham country than along the desert floodplains occupied by riverine Pimans. Sylvester explained, "Papago know where to go. They go to certain area[s] and there's good *vihog*."

"How about the Pima?"

"Naw. They just go where they can find [it]. They know where the good *vihog* they can get." This nonterritorial pattern among the Gileños held true for other gathered plant resources as well, such as wolfberry, cholla, and Saguaro.

There are two distinct mesquite crops each season, the second linguistically marked. Joseph Giff was the first to tell me about this: "After the first [summer] harvest of mesquite, there's a second harvest, ready in October, called *toomdaj*. Very sweet! You also say *toomdag* for second crop of figs, and you can get a second *toomdag* of watermelons—cut off vines— and new ones come, called *iipoñigaj* 'sprouts'. You can get a second harvest of tomatoes in fall, but it doesn't have a name." Then he added, "The second crop of mesquite is 'not very good' [productive] but very sweet, but smaller pods and smaller bunches [of pods]. But these are sweeter."

Ruth Giff looked at Vorsila Bohrer's Pima ecosystem chart (1970:424) and said it was not exactly correct. "The first *vihog* is ripe by *dia de San Juan* [24 June] and for about a month following. (*Haashañ*

[Saguaro] comes same time, but it doesn't last so long.) But the second mesquite harvest is too late [on the chart]— by about a month. Anyway, by that time everyone has all the *vihog* they need, unless something went wrong with the early fruiting." She added, "The *toomdaj* is the 'second crop'. *Now,* that seems like the only thing we can pick. There's no name for the first crop."

Others knew the word as well. Fr. Antonine (1935:11) listed *tawmdach* as "the second crop of mesquite beans." Mathiot's dictionary (n.d.:199) lists the verb *toom* as "to bear fruit out of the regular season, to give a second crop," and the noun *toomdaj* as "its second crop."

I asked Sylvester Matthias when *kui* was ready the first time.

"About June 20—anytime in the 20s. That's the first crop. *Vihog baakam* 'in season' [harvest time]. The second crop in the fall—somewhere in mid September, the *toomdaj.*"

"Which was the main one?"

"Both!"

Grossmann (1873) said of the mesquite, "These beans are gathered when nearly ripe, then dried hard." The reason for this is not clear, unless it was to avoid the bruchid beetle larvae, *vihog maamaḍ* 'mesquite pod babies'. Everyone with whom I worked told me that *vihog* was gathered after it had dried and fallen to the ground. This was common practice in the Southwest, according to Felger (1977); then the pods apparently have the highest carbohydrate content.

Great quantities were stored in granaries. There were two kinds of storage bins. One type, called the *homda,* was made from the long branches of Arrow-weed, leaves and all, roughly woven in thick coils spiraling upward until the desired height was achieved (see Group D, *Pluchea sericea*). The bottom layer was of fresh willow branches. These bins might be a yard across and about half as high. Most were placed on rooftops, but some stood free on the ground on a base of more Arrow-weed, protected from livestock by little fences. They were quickly constructed (I watched Ruth Giff make one in a morning) and served their purpose for several years before the elements destroyed them. The end product, looking like a huge bird's nest, was topped with a horizontal layer of Arrow-weed and mud. They were

kui   *Prosopis velutina*

entered by lifting the roof. Mesquite pods were most often stored in these bins.

The other storage basket, called the *vashom,* was a large affair woven of coils of wheat straw bound with strips of bark peeled in spring from Screwbean, mesquite, or willow trees. These jug-shaped baskets might be six feet tall. The weaver might stand within her basket as she completed it. These baskets stood under the *vatto* (ramada) or in a storage shed, but not in the open, exposed to the weather. They held corn, wheat, beans, or already prepared mesquite flour (*vihog chu'i*).

The imprecision of English has led to confusion as to what part of the mesquite *vihog,* or pod, was actually used (Felger 1977). Most accounts say the "beans" were used. But does this mean the floury mesocarp (pod) or the actual endocarp (the leathery coat and seed of the mesquite)? The pulp and the seed differ widely in nutritional content, as noted earlier. Although the dried pulp is easily converted into flour, the stone-hard seeds had to be crushed in a special apparatus, the gyratory crusher (Felger and Moser 1985:340;

Hayden 1969). These crushers are unknown from Pima country.

The confusion with regard to Gileño practice stems from an ambiguously worded statement by Russell (1908:75): "The beans are prepared for use by being pounded up in a mortar with a stone pestle, or, if a large quantity is required, with a large wooden one. The pods may be ground with the beans. Another method of preparation is to separate the beans from the pods, parch them by tossing them up in a pan of live coals, and reduce them to meal by grinding, whereupon they may be eaten as pinole. This has a sweetish taste and is reputed to be very nourishing." The parching and grinding in the second method implies the use of seeds, while the first singles out the "beans" (seeds) in contrast to the pods for grinding. Another statement by Russell (1908:66) appears to reinforce this interpretation: "The fruit of the saguaro and the seed or bean of the mesquite were the most abundant and accessible resources."

Pfefferkorn (1949:72) observed two ways colonial Pimans dealt with *péchita* pods. "They roast the pods, grind them between two stones, and mix this powder with water and drink it. Or they pulverize the unroasted pods in a wooden mortar, add water to it, and cook a very sweet porridge, which they deem unsurpassed." The second practice has continued to the present. But the first, now unknown, would have utilized both the mesocarp and the hard seeds.

During the 1970s I checked this point with various people in the western villages. David Brown said, "[They] throw away seed that is in the *vihog* pod and just keep the powder." Sylvester Matthias said just the pod flour was used to make *vihog chu'i.* Joseph Giff and Ruth Giff were strongly of the same opinion. "Let's ask Adolph Juan," suggested Ruth. "He's in his nineties." Indeed, Adolph was born before the time of Frank Russell's fieldwork! The next time I saw Joe and Ruth, they had their report: mesquite seeds, even at the turn of the century, were definitely discarded, not used. This practice is in conformity with all other outside accounts and with Pima oral tradition. Use of parched mesquite seeds must have been abandoned long ago; not one of the consultants with whom I worked knew about (or remembered?) a mesquite-seed pinole.

Julian D. Hayden, who worked among the Gileños in the 1930s, told me that the Pima specifically avoided cracking mesquite seeds by always pounding the pods in a wooden mortar with a stone pestle or in a stone mortar with a wooden pestle; stone on stone would have produced sharp-edged seed fragments. However, the seed meal would have yielded a higher protein content than the mesocarp pulp that the Pima used.

Infestation apparently was one of the problems with stored mesquite pods, but there may have been some recourse against infestation. Pfefferkorn (1949:72) said, "The Indians also gather the *[péchita]* pods, dry them in the sun, and keep a store of them for their housekeeping." Ruth Giff also noted that freshly gathered *vihog* were spread on the hot ground to cure for a few days before being stored in the granaries. The heat was believed to kill insects and their eggs already in the pods.

One day I said to Sylvester, "After that *vihog* stays in that storage basket for a while, there's those little worms that get into that."

"No, they won't get no worms," he corrected, "but it'll get something like bugs; and they flew out at dusk, just about sunset. So the *ñeñepoḍ* [nighthawks]

would be around eating the bugs that came out from the *vihog.*"

"What do you call those bugs?"

"*Vihog maḍ,* [pl.] *vihog maamaḍ* 'mesquite pod babies.'"

"Did they damage the *vihog,* or did anyone worry about that?"

"No, they didn't worry. They don't mind it. Whenever they wanted it, they just go and pound it. If they want to, they take it out and pound it, sift it, and make that cake. And they mix it up with *haakĭ chu'i* [roasted wheat flour]. Ah! I like that."

Ruth Giff showed me how the dry pods were converted into *vihog chu'i.* She used a mesquite log *chepa* (mortar) and an oblong stone *cheepidakuḍ* (pestle) to pound the mesocarp to a pulp. The seeds remained intact in their tough endocarp envelope. The resulting flour or meal was then sifted in a special basket *(giigdakuḍ)* to remove the fibrous portion and the seeds.

Mesquite flour was a versatile staple. To make a loaf or cake, a cloth was placed in a basket; then a layer of dry meal was added and sprinkled with water until moist. Then another layer was added and moistened. This was continued until a loaf six inches or more thick resulted. When dry, these unbaked cakes, called *cheeg,* would keep indefinitely. In this form they were resistant even to the ever-present bruchid beetles, *vihog maamaḍ.* The *cheeg* was incredibly hard, and one ate it by breaking off pieces with a stone. Some call these loaves *komkĭcheḍ* 'turtle' or 'tortoise', in reference to their shape.

The mesquite flour could be cooked directly with whole wheat flour to make a pudding *(vihog hidoḍ)* that looks like butterscotch but is richer, tasting more like carob. Another mesquite dish is called *hoohi-wichda* ('Mourning Doves done in imitation'). Dried mesquite pods were boiled until very soft, then cooled. The pulp was then mashed manually and strained. The juice was saved, the fibers and seeds discarded. Finally, whole wheat dumplings (the "doves") were dropped into the *vihog* juice and cooked until the dish thickened somewhat.

A similar dish called *ka'akvulk* or *ka'akvulkada* ('hills made') was prepared with thick wheat tortillas *(chechemit)* the size of small pancakes alternately layered with dried mesquite pods. The layers were covered with water and slowly cooked for several hours, until almost dry. The pods were then discarded

and the mushy *chechemit* (the 'hills') eaten.

Apparently these could be dried for use in travel. Sylvester Matthias said, "When they went hunting or on the warpath for Apaches, they take a little bag of deer hide and tie it on their belt. Take *ka'akwulkada,* like little flat tortillas-look-like, but thick, about an inch or more, and about this big [five or six inches] across. Made from wheat flour. I don't know how they fix it. They have to take that because it keeps."

Another food called *vihog weenagim chu'i* consisted of wheat pinole sweetened with mesquite flour. The *haakĭ chu'i* (pinole) is wheat or other cereal that has been toasted and ground into flour. The descriptions of all these mesquite dishes come from Ruth Giff. In many cases she prepared them for me. Historical descriptions of these dishes can be found in Curtis 1908 (p. 5), Grossmann 1873 (p. 419), Hrdlička 1908 (p. 261), and Lloyd 1911 (p. 123).

A sweet drink called *vau* (occasionally pronounced *vaō*) was prepared simply by soaking crushed mesquite pods, seeds and all, in cold water. The mixture was taken in the hand and the juice sucked out. Although it is no longer made, all the people I worked with said they esteemed *vau.* One told me that it was as ubiquitous at Pima meals as coffee is today.

Formerly, this mash was fermented by women to make a sort of beer called *o'oki navait* (women's wine), Sylvester Matthias once said, but no one remembers the process today (see Harvard 1896:37; Kearney and Peebles 1960:402; Palmer 1871:410). Felger (1977:162) notes that this drink, made by various Southwestern groups, "was probably very low in alcoholic content" and doubtfully intoxicating.

Why has the abundant and nutritious *péchita,* or mesquite crop, fallen almost entirely from the Pima diet in the last half-century? Several reasons are suspected. Wage earning and ready access to store-bought Anglo foods is certainly a factor in the decline in all native foods among Pimans. William Doelle thought that the timing of the winter wheat harvest largely eclipsed the mesquite harvest (see Felger 1977:165); but strong reliance on mesquite lasted two centuries after the Gileños began heavy wheat cropping, and the second crop of mesquite seems to have come at an opportune time in the annual cycle for gathering.

Another reason may be cultural, as Rosita Brown told me when we discussed mesquite dishes: "You better sleep outside," she warned me. "*Vihog* makes plenty of *uivĭ* [flatulence]." Others said the same thing. I asked Sylvester about this. "Anything [in the] line of the mesquite, the *vihog*. That would make you gas up. And, *oh!* the smell." He broke out in laughter. Apparently the rich mesquite produces even more flatulence than do beans.

Perhaps the strongest factor leading to the demise of mesquite as food stems from a characteristic that Bell and Castetter (1937:22–23) portrayed: "When stored in the form of whole or dry pods, partially pulverized, they soon became a living mass, since an insect, a species of *Bruchus,* was present in almost every seed. To the Pima or any other tribe of Indians, this made little difference. The insects were not removed but accepted as an agreeable ingredient of the flour, subsequently made from the beans. If reduced to a fine flour soon after gathering, the larvae still remained within the beans and became a part of the meal, forming a homogeneous mass of animal and vegetable matter." But as Felger (1977:163) points out, "More recently, because of changing prejudices, bruchid-containing food has become unacceptable."

Pima use of mesquite was not limited to food. The pounded roots of mesquite were boiled with a bit of salt to make a laxative. For a purge or for dysentery, the white inner bark of the tree was boiled. Joseph Giff once told me, "A medicine was used by Don Thomas, made from the thin skin [found] under the *kui* bark; this is peeled, boiled—makes one quart (usually for youngsters)—for weakness and dysentery." Ruth Giff later confirmed that Don cured children's diarrhea with the white inner bark of mesquite. Sylvester Matthias said, "[You] pound roots of *kui* [and] boil for laxative. My father added a little salt." In the late 1800s, Rev. C. H. Cook's wife was advised to give their weak newborn water to drink in which mesquite flour had been boiled (Cook 1976:113).

Another medicinal use was the treatment of intestinal worms. Parasitic worms in the intestines are called *vaptopad,* Sylvester said.

"What can you do about those?" I asked him.

"I guess they have to drink that mesquite bark. My father knew how to mix that. It's bitter . . . the inner bark."

According to Hrdlička (1908:244), there was one taboo associated with mesquite: "The Pima believe that anyone eating beans from a mesquite tree struck by lightning would have sores on the skin (herpes zoster?) beginning in the region of the liver." I asked Sylvester Matthias if there was anything special about a mesquite that had been struck by lightning.

"Avoid that tree. And avoid the wood from it. Avoid anything if it's struck by lightning."

"Just mesquite?"

"Any kind of a tree. Any—willows, cottonwood trees."

"Have you heard anything about getting sores on your skin?"

"Yeah. Yeah, especially right across here," he motioned across the shoulder and down to the waist. "They call that *honovad.*" He gestured to show how it went over the left shoulder and down across the belly. "Yeah, all around," he continued, indicating an area that ran like a girdle across the belly and around the back. "It's itchy, and something like a burning stuff. Mostly itchy, they say." Although Sylvester himself had never had *honovad,* his description of the disease fits "shingles" excellently, the herpes disease Dr. Hrdlička suspected. The name *honovad* comes from 'scarf', an allusion to how it encircles the body.

The mesquite exudes several types of sap. During the summer, injuries to the tree cause accumulations of clear or amber-colored resin called *kui ushabĭ* 'mesquite sap'. The globules are easily collected when they harden. They have a delicious, mildly sweet taste. This was candy or "gum" for the Pima children. It dissolves slowly in the mouth. Russell (1908:74) reported that the white mesquite gum was "used in making candy"—apparently by roasting, in analogy with other tribes. He also wrote, "The inner bark is employed as a substitute for rennet"; but the only substitute for rennet I have found contemporary Pimas using is the berry from the White Horse-nettle (see Group D, *Solanum elaeagnifolium*).

According to community healer George Kyyitan, *kui ushabĭ* is also used in the treatment of fallen palate or fontanelle, *koá i geesig,* a common affliction of young children but frequent in Pima adults as

well. "They kept it there. Get it up and kept it there. Some of them, but I never used it. Some of them, they use sugar too."

"How would they use the *ushabĭ?* Mix it with water?"

"I don't know. I don't know how they use it because I never used it."

"I've heard of them using sugar, salt, or flour."

"I don't know about salt. I never use anything like that. And I know some of them use eggs too."

"The yolk or the whites?"

"*All* of it. The whole thing. No, I never use anything. But it comes [out] alright." The purpose of these various substances, apparently, is to dry on the hair and scalp, making a sort of cast to hold the critical spot in place so it does not fall again.

Another kind of mesquite exudate *(kui chuuvadag)* is thin, black, and bitter. It has several uses. It was harvested by chipping off pieces of bark where the ooze had soaked. The chips were then boiled in water. The boiled liquid was a medicine used on open sores or cuts. It was also painted on pottery before the final firing to produce the black design, and is still used for that purpose by the Maricopa and Tohono O'odham.

Francis Vavages said, "The *kui eldag,* the black part, gather that and boil it; keep the liquid; strain it. Put a drop in the eye. Supposed to remove your cataracts. I guess it eats [them] up, or something like that. It's just like acid, I guess. That's why they have to dilute it."

George Kyyitan told me his personal experience with this remedy: "Mesquite juice—you know—the black one. It's pretty good, too, when you have sore eyes." He had just been talking about *haḍam tatk* (globemallow, *Sphaeralcea* spp.), a plant that causes eye irritation. "I used [it] once. It hurts, but your tears will be coming out; lay back, until tears wash out. Use a piece of it [the dried black sap]." I winced at George's story, knowing how potent, how bitter the black sap is.

Hrdlička (1908:245) had written of this use 80 years earlier: "*Kwei-chou-wa-te* (mesquite sap) is also used for sore eyes. The patient places a small piece of it in the eye and keeps it there as long as he can bear it. The tears dissolve part, coloring the eye brown. Occasionally this remedy is of assistance."

Sylvester Matthias added that *kui chuuvadag* was once used to cure venereal disease. It was applied externally as a liquid "when it's real sore."

The desert sun bleaches the black hair of the Pima to a dark reddish color. And age turns raven strands white. To remedy this, the Pima (especially the women) made a plaster of this black mesquite sap mixed with a black mud *(s-chuk bid).*

Sylvester told me about *s-chuk bid.* "Both sexes used the black dye, but mostly the females. The black mud came from the irrigation ditch or the river, but not just anywhere—but [from] little springs where the river runs slowly so as not to wash it [the mud] away. A soft mud from where the water is hardly moving, or standing at the edges. The mesquite gum adds something."

Not only the women had a streak of vanity. George Kyyitan, perhaps noticing my own graying hair, said, "Also, old time, if you have gray hair, use that *kui chuuvadag.* It will make you a young guy. Smells good. You get that black mud, you know, and put it in there, mix it, to wash your hair. Put it overnight."

Joking was the interminable pastime of the older Pimas. Sylvester told this story of Joseph Giff's father not allowing the vanity of an older coworker to go unnoticed: "When you mix up mud—black mud—and *kui chuuvadag,* they call that *veenda,* the *process.* It smells, of course. Like I said, we were working on a dam. That old man Giff said, ʻ*Si s-veenda uuv*' (I smell that *veenda.*) He knew who it was; that old man Thomas—José Thomas— had long hair, braids. He's the only one who uses it, in modern [times], in thirties." Women were allowed their cosmetic indulgence, but a man who not only kept his hair long decades after the government tried to stop the practice but also kept it black and shiny could expect some ribbing.

I asked Sylvester if the treatment was only for old people with gray hair. "No. Young people [too]. It's for the lice *[aa'ach].* Off goes the lice, nits, everything. The modern, well, I guess they got no lice, because they're well taken care of." (See also Russell 1908:159.)

There was yet a third kind of *ushabĭ* found on mesquite trees during the summer. Joe Giff told me the name and characteristics of it: "Also there is *si'al hikbin* ʻmorning cutting'. This one [sap] is white and sweeter than the regular *ushabĭ.*

It comes on the end of branches about as thick as your finger. Made by some animal; I don't know what animal makes them. They shine in the morning or in the evening as the light shines through the droplets on the small branches. The plant dies past [distal to] where these wounds are."

Eight years later, Sally and Ruth Giff met Sylvester and me at Tumacacori, between Tucson and Nogales. We were on our way to Pima Bajo country in Sonora. The women were a bit late; they had stopped to cut some mesquite branches with *si'al hikbin.* There were small circles of bark gnawed around the thin young branches. Ruth said this is very hard to get because it is usually on the ends of thin mesquite branches. "In the evening, you can just see it shining."

Sylvester knew it also. "I don't know what . . . who makes that ring around the branches of a *kui.* That's where they got *si'al hikbin.* It's real sweet, too."

The missing link with this *ushabĭ* is the culprit that does the chewing. David Faulkner, entomologist at the San Diego Natural History Museum, told me that these wounds are made by a cerambycid or long-horned beetle called the Mesquite Girdler *(Oncideres rodosticta).*

Mesquite wood, as well as the pods and the sap, played a major role in Pima culture. A Pima rancheria or settlement consisted of groups of houses loosely clustered, the families of the group related on the male side. Each family had a house *(kii),* an arbor or roofed working area *(vatto),* an outside cooking enclosure *(uksha),* and a storage shed *(kosin).* Mesquite trees in the yard provided additional shade and places to hang things, as well as places for children to play.

The Pima house was not a very elaborate affair. Originally, it was round and low, but it could support a dozen people sitting on the roof. The center was supported by four stout, forked mesquite posts *(chettonḍag),* which held horizontal vigas *(vavanaḍag)* of willow or cottonwood. Thin willow poles made a framework for the walls and roof, which was thatched with Arrow-weed, willow, cattails, or even cornstalks, covered with a thick layer of adobe mud. This *olas kii,* as the traditional round house was called, was used just for sleeping during wet or cold weather. The rest of the time, activities were carried out under the *vatto.* This

is a wall-less structure made of four or more upright mesquite poles, forked at the top, supporting a flat roof of willow or cottonwood poles, or perhaps the ribs of dead Saguaros, and a thatch of Arrow-weed. The thatch might be covered with adobe. *Hohomda,* ʻbins' for the storage of mesquite and other crops, were placed on top of the *vatto* as well as on the housetops. Under the *vatto* was always a three-branched mesquite stump, the *ha'a daikuḍ,* for supporting the large pottery water olla, or *ha'a.* (This might also be called the *va'igĭ daikuḍ* ʻdrinking water sitting place'.) Even on the hottest summer day, evaporation from this shaded water jar kept the contents very cool. Most Pima homes today have at least one *vatto,* though the storage bins are gone and the olla is usually replaced with metal or plastic water containers. In warm weather, a table and beds are under the *vatto.*

The Gila was the agricultural focal point for these riparian people, who learned to farm by irrigation. Some distance above each settlement, sometimes several miles, a brush weir or *kuupa* was built across the river so the rising waters would enter the headgates to be conducted into miles of ditches feeding into the rancheria fields. Constructing the *kuupa* was a highly organized community activity. Mesquite was the main material for building weirs. In early spring, long mesquite poles were driven into the bed of the river and thin mesquite branches were woven between these. Against this structure, on the upstream side, wagonloads of smaller plants and soil were put in place. (See Group D, *Atriplex polycarpa* and *Pluchea sericea,* and Group I, *Triticum aestivum.*)

After the Pima took up livestock raising as a serious pursuit, it became necessary to protect fields from the animals. Mesquite was the principal component of fences. There were several different types of fencing. The old-time fence, *sha'i koli (koli* is also pronounced *kolhi),* was made by standing poles upright in a double row about a meter or two apart and filling the intervening space with spiny branches, particularly from young mesquite growth. As the branches settled, more were piled on top. Because fallow fields quickly reverted to second-growth mesquite thickets from seed dispersed in horse manure, the cleaning of fields always produced plenty of fencing material.

George Kyyitan recalled these fences: "*Sha'i koli* [were made of] thick mesquite branches. That's what they used to have around here in the days when they don't have no [barbed] wires. I've seen them when I was a small kid. I remember. The field were just small pieces of land that they owned. There's *lots* then that they used at that time. In those days more mesquites had longer limbs."

These fences were substantial affairs, fortifying the field against human as well as animal intruders. An episode from the calendar stick recorded by Sharlot Hall (1907) gives us a picture of what these were like in 1840: "While a Pima man named Mewh was in his field, Apaches came. The fence was very high and the man could not see how many, so he ran home. Two other men were working in the field, and they ran home and told the rest of the Pimas. The Apaches could not get in the field."

An account by Joe Giff indicates how strong these fences were:

One time a man was missing melons in his field and he did not know who was stealing them. Searching, he found a tunnel someone had made through the bottom of the fence, concealing the opening on the inside and outside surfaces with brush plugs. So the man decided to spend the night in his field to find out who would come. Late that night, when everything was quiet, he heard some noise outside his fence as the melon robber removed the outer plug of brush and started crawling through the fence. When the thief pushed the remaining brush from the inside edge of the fence, the owner began throwing dry dirt and dust into his eyes. The interloper, unable to turn around, had to slowly crawl backwards through the tunnel, all the while getting dirt in his face. The next day the owner of the field looked around the village for someone with red eyes. (Rea 1979)

A simpler but nonetheless effective type of fence was still in use occasionally when I lived at Komatke. It consisted of stoutly spined young mesquite branches stacked in a row about chest-high, sometimes along an existing fence that was falling into disrepair. George Kyyitan called these *wiis koli*, "something we will drop there,

loosely piled." Another time he said, "*Wiis koli* is rough, mesquite branches just piled." Apparently the name is from *vuis* 'thrown, piled'. Usually Sylvester Matthias at the west end considered *wiis koli* and *sha'i koli* to be the same thing. Joseph Giff also was not sure of the distinction. "Same one. Some call it one way or other," George said.

Sylvester described one of the mesquite fences: "To make the *wiis kolhi,* they put a post, a post, a post—every [he indicated about two feet apart], just keep on going till you want to get an opening. Then you put a forked one that would leave an opening, or either have a place that you can crawl under—*wiis kolhi*—that way no animal can get into a field with a fence like that."

"How tall were they?"

"Oh, they got to be pretty high. Some cows can jump. So it's enough the cow won't jump over."

"Is there a difference between *wiis kolhi* and *sha'i kolhi?*"

"I think that *sha'i kolhi* is the modern fence—barbed wire and brush—just piled up. *Wiis kolhi* has posts, stick in here [and there], [a wall] built up with heavy mesquite branches—just pile it up."

Another kind of fence called the *uus koli* was made of upright mesquite posts with walls of horizontal mesquite limbs, often ones of good size. These were more for enclosing livestock. These rustic fences give a characteristic flavor to the Southwest. Sylvester described this fence: "*Uus kolhi* is just about the same, only it's for the cattle *[haivañ].* They have to put [mesquite branches] all around. Call them *haivañ kolhi,* so the cows won't get out once they're penned in there. It's all died out here. Nobody [has them]. Used to, years ago."

Barbed wire was introduced between 1892 and 1900. This new fence was called *vainamĭ koli* 'metal fence'. Mesquite was the preferred wood for fenceposts. Mesquite posts would not sprout as would those from willow or cottonwood. But in time, a wire fence would turn into a living fence row, with thorny shrubs such as Graythorn, wolfberry, and mesquite reinforcing the original fence. In the mid 1960s, these living fences were quite common across the reservation; only a few traditional farmers mostly in the western villages still maintain fence rows of living mesquite and other thorny shrubs.

Cutting mesquite posts was a lore in itself. When not cut at the proper season or if not properly treated, the wood turned into a powdery colony of buprestid beetles, greatly shortening a fence's life span.

Joe said, "If posts cut in spring, they turn powdery, with worms under the bark. To avoid that, I cut the bark off with an ax and strip [it] off. If sap up then they don't get wormy or powdery." Sylvester said that the big mesquite limbs for the *koli* were cut when the sap is down (in the winter), whereas the *chettoņḍag,* or upright forked house posts, were cut in summertime. But posts from *kuujul* (Screwbean) could be cut at any time of year.

Ruth told me, "When they used to chop posts, they'd soak them in water; they say that cures them. They peel it [bark] off—so it won't turn to powder."

At Sacaton I asked Francis Vavages, "What are the best trees for making fence posts?"

"Mesquite. They're hard. Hard for worms to dig in. Tamarisk will rot."

"When do you cut them?"

"Wintertime. When it's solid. 'Cause after it warms up, they start to get green and get looser and the worms start to get in. So it's better to cut it in winter for fencing. Yeah. Because the worms wouldn't bother, you know."

On my next visit to George Kyyitan, I asked him what he thought was the best wood for posts.

"Mesquite, because it's hard."

"When is the best time to cut?"

"I think before April—in April their peelings [bark] will be thicker and get loose, and the staples will come right out, won't do any good. But some of them, they burn them [to cure]."

"Have you heard of soaking mesquite posts?"

"Yeah, they do that too."

"Pima?" I asked, just to make sure.

"Yeah, some. Before they put them in. But most heat them."

This processing kills woodboring beetles in their various life stages.

Often old photographs of Pima fields show a mesquite tree growing in the field itself, not just in the fence row (for example, Hoover 1929:42; Russell 1908:Pl.XI). This was by intent, not accident, as Joe Giff explained to me (Rea 1978). A few large, productive mesquite trees might be allowed to stand in the fields—the Pima farmers plowed around them. The pods were a rich source of food for both the Pima and their animals. Workers hung their water containers under them. Especially during the hot months of harvest, men and women cooked their midday meals in the shade of these large mesquites, saving time walking back to their houses. After harvest, when the livestock were turned in to pasture, these trees provided them with shade. George Kyyitan said that the tree that would be left in the field was one that produced a superior *vihog.*

But were there other, perhaps even more important reasons? When Wendell Berry (1981:61), himself a farmer, visited the fields of one of my former Tohono O'odham students, he "noticed that the wheat was greener and of somewhat better quality" where a mesquite tree grew. Gary Nabhan (1985:72), who was showing Berry through Tohono O'odham country, has summarized recent research on this phenomenon: "Mesquite trees have long been known to be nitrogen pumpers. . . . Recently, however, scientists finally confirmed that symbiotic bacteria also associate themselves with mesquite rootlets as they do with other legumes, forming nodules which fix nitrogen as an additional source for the trees. In an otherwise nitrogen-poor desert, mesquite and its entourage of herbs sit in a pocket of riches." So these desert trees of life brought vigor to other plants as well as to the People.

But did the Pima realize this? In 1989 I brought the subject up cautiously to Sylvester. "Why did they leave a mesquite in those fields some time?"

"Yeah. They do that just a few years back, when I was a little boy. They used to have it lined, the fence, with the mesquite tree. And it serves, too, sometimes, to camp under it. You could see *all* over the field, they had mesquite trees lined along the fence."

"Did that help the field in any way?" I asked.

"The fertilizer!" Sylvester replied. "The leaves, when it's fall. When it's irrigated, you watch—there goes your fertilizer from the mesquite leaves."

"And the People knew that?"

"Yeah."

"How about in the middle of a field? Sometimes you see one tree growing in the middle of a field."

"Yeah, that's also used as a fertilizer. When it sheds [its leaves]. When it's

irrigated, well, there goes the fertilizer— being fertilized when it's irrigated."

"Is there any other way of fertilizing a field?"

"Oh, from the floodwater. When it's about clear. Not when it's real muddy. When it's, oh, about clear, ready to clear, then that be good fertilizer."

The preferred firewood, *ku'agĭ,* is mesquite. Paloverde is too soft and ironwood generally too hard; willow and cottonwood no longer grow along the river. But the hard, fine wood of mesquite burns with an even, intense heat. In winter, Pimas have a delightful aroma from their mesquite-wood fires. For nearly a century, cutting mesquite wood for sale has been an important livelihood for Pima men (see chapter 5). Pima George Webb (1959:122) noted, "The price for wood was small, but wood saved them from starving."

In the mid 1960s, I often made a weekly drive from Komatke up the Gila to Bapchule to spend an evening visiting with the Kyyitan family. At that time they lived in a house of mud and Saguaro ribs. It was an old house, with rooms added as the family grew. There were plenty of holes in the walls. On winter evenings, mesquite wood was burned outside in an old washtub until nothing was left but a bed of glowing orange coals. Then the tub was dragged inside to heat the house for the rest of the night. This was a common method of heating homes in earlier days because it reduced the amount of smoke indoors.

Many Pimas still heat their homes and cook with mesquite wood. In the 1980s, it was not an uncommon sight to find an old woman in the yard chopping her firewood. But the wagons and teams of horses that I once used to see in regular use by men cutting wood have largely been replaced by old pickup trucks.

One must not underestimate the importance of firewood in this desert country. The winter nights are often below freezing, and the dry air makes one feel even colder. Indeed, I know of one case where a tribal judge presiding over a divorce case sentenced the husband to supplying a monthly load of mesquite wood to his children and former wife.

Joe Giff told of additional uses for mesquite. "*Kui chuudag* is 'mesquite charcoal'. The men tattooed a zigzag line or lines on their forehead; the women,

three lines on chin, from lip edges down, and one in middle." Russell (1908:161) said that the tattooing material was the charcoal from either willow or mesquite wood. Joe added, "For a quirt, take top branch of mesquite very thin and long—so long— to make horse go."

Mesquite articles were numerous around the rancheria. The dibble or digging stick, *eskuḍ,* as well as a short weeding stick, *giikĭ,* were usually made of mesquite but sometimes of Desert Ironwood (see *Olneya tesota* account) or Graythorn (see Group D, *Ziziphus obtusifolia* account). Aboriginally, these were the sole instruments not only for preparing the soil for planting but also for digging miles of irrigation ditches. After wheat and draft animals were introduced, the Pima fashioned a plow, *uus giikĭ,* of a crooked mesquite branch share attached to a tongue of cottonwood.

The mortar or *chepa* for pounding wheat or mesquite pinole was made from a section of mesquite or cottonwood log about two feet long. Occasionally, Foothill Paloverde was used. The concavity, made either on a side of the log in a horizontal position or in the upturned end, was initially burned out with coals. Both types of *chepa* are still in use. Usually a stone pestle (*cheepidakuḍ*) was used. The mesquite-wood pestle is called the *chepa maḍ* 'mortar's baby'. Large serving spoons or ladles (*uus kusal* or *biikuḍ ha'u*) were carved from mesquite or from Foothill Paloverde (*Cercidium microphyllum*); Blue Paloverde (*C. floridum*) was never used for carving, the Pima say, because it is knotty and too brittle. During most of the last century, the Pima waged bitter defensive war against both the Apache and the Colorado River Yumans. In battle, the Pima carried, in addition to bow and arrows, a short but very heavy war club, the *shonchkĭ,* carved from mesquite root or ironwood.

Although this list is not complete, it shows that mesquite implements were in daily use by both men and women. Whether working in the fields or at home, whether cooking or eating, whether sitting in the shade or warming around the coals in the cold desert nights, the Pima were surrounded by the mesquite and its products. It was their tree of life.

Given the importance of *kui* in Gileño life, it seems surprising that this tree does not have a more significant role in the Piman Creation Story, in which it is scarcely mentioned. But during our sessions on the origins of certain plants in Pima mythology, Sylvester told me a story of how *kui* came to be.

Some knowledge of the girls' puberty dance is necessary for understanding the mesquite origin story (see Russell 1908:182–183; Underhill 1946:253–260). The ceremony lasted anywhere from four nights to a month. Parents of substance were expected to feast the dancers each midnight. Men and women dancers stood together in two tight lines facing each other. The lines of embracing dancers were covered with blankets. The emerging girl and the lead singer stood at the end of one line. "Anyone, but particularly the song leader, had a perfect right to sexual liberties," explained Ruth Underhill (1946:254). "Men [singers] who heard of the performance often came from great distance" (Underhill 1946:256). The only requirement for offering their services was that they know a cycle of songs appropriate to the ceremony and sufficient to last at least one night.

Sylvester's tale about this *wuágida* ceremony and the origins of *kui* began:

There were two beautiful daughters of a medicine man. Somehow Nanakmel (Bat) and Ban (Coyote) found out it's that time. From there on, they carried out [their plan]. Nanakmel was after the girls. Bat appeared as a Handsome Man. Ban sings beautiful songs [is the lead singer for the ceremony]. Once they get started, that would be excitement for the young people, both boys and girls, because all took part in the dances. Everybody wants to enjoy.

So Coyote said, "*Vatto keekiwua. Vatto keekiwua.*" (Let's get started.) That's the ceremony where they stand in a certain formation; *keekiwua*, that means 'to stand', but it's the beginning [of the ceremony]. That means they have to stand in formation to start the *wuágida*.

They came, every evening, just about dark. While Ban carries on with the *wuágida*, Nanakmel would take one of the [two] girls into his cave and have intercourse. And he get through about dawn. Each night he would take one.

Bat insulted the girls in public, saying, "A lot of hair! A lot of hair! [*muuspō*]." The people want to see what the person

[Bat] look like, because nobody sees the face. And they used those torches made out of Arrow-weeds, *wulivga*. And they say, "Let's see how he looks like." But [when they approached him] Nanakmel said, "Oh! My *dakōsha* [nose ring] dropped," then he dived down to hide his face [so nobody could see]. He keeps his head down [as if looking for his nose ornament].

But they were carrying [on] too often. That's when the girls' father, the *maakai*, told the girls not to go [out]. But the girls disobeyed their father. And every time Ban said, "*Vatto keekiwua*," to start the dance, they would run out. They would ignore their father's orders. That's why he got mad.

Finally the father got tired [of this and said], "I'm going to show whoever is running this deal." And he asked the daughter, "At *dawn*, you go look for a *kotḍopĭ hiósig* [*Datura* flower]—I'm going to fix that." So they put it in her cunt. And he said, "Let him have it. And then you squish it around [close the flower off] and bring it to me." And she did. He's intercoursing with this flower! And his semen got in there, so she took as his order, and twist [it] around [closed off the *Datura* flower]. Took it to her father. [He took it to] the community house, called *jeeñ kii*. Men of the village come to hear the news [at night], sit in circle. So he planted it right in there, right by where he tells his story or has his news meeting. Everybody was anxious to see it, too—how it would turn out.

And the first day, the semen came up a *beautiful* tree.

The second day, it was covered with yellow blossoms.

The third day, it produced green *vihog*, mesquite pods.

And the fourth day, they matured, turned yellow, and dropped to the ground.

The people had never seen *vihog* before; there was never a mesquite tree at that time. They watched the tree. But nobody touches it; they avoid it. They are afraid of the *maakai* who planted it. When Coyote came in, the other men said, "Sit down. Sit down. Sit down over there. And keep away from here."

Coyote is cunning. Without anyone noticing, he got [close] to the tree by the time the *vihog* dropped. Ban started chewing them. Because there was no light and it was dark. And nobody's watching. Of course he knows he is doing something [wrong]. He ate the *vihog* and chewed and chewed. He was very careful but finally the last one he made that sound. [Here Sylvester makes a sound sucking air in at the sides of his molars, imitating Coyote gobbling down and trying to chew the *vihog*.] That's the time the people know [what Coyote was doing]. When the father, the keeper of the *jeeñ kii*, found out he's eating that *vihog*, he yelled, "Hey! I knew you would do something." And they [all] jump and grab their clubs. They carry their own clubs. And they tried to block the door. But Coyote is so fast that he got by. He ran. They know how fast a runner is Coyote. He tries to outrun them. The first runner gets to him and hits him on the back and he shits out. They all ran after him, beating him with their clubs. All the way through Muhaḍag [South Mountain] and everything. [And they chase him, beating him.] All the way, along there, there's mesquites all the way. Everybody hits him clear down [the Gila] to Yuma. When [Coyote] gets to Yuma, there's very poor mesquite. Shits the last one. It's not very good; but along here [Pima country] they have very good mesquite trees. But Yuma [it's] not very good, because the tail end, they beat all the shit out of him. [These last ones are called] *kukui che'echpelk* [dwarf mesquite]. They don't grow up and branch off.

My grandmother heard the story. An ugly person, Nanakmel, appears as a Handsome Man. [Later] they chased him, Bat, back to his cave, and beat him with a stick, and threw him around his cave. He won't dare come to the village [in the day] after that, because they find out how he looks.

Sylvester told me that the Pima once knew the mesquite origin story well, and he added a related piece of folklore: "Slang [name] for mesquite beans—*nanakmel kekelda* 'bat's semen'. . . . Everybody back in them days knew what it means."

Vouchers: *1760* (SM); *LSMC 30.*

Quercus oblongifolia Torr.
Mexican Blue Oak
Quercus turbinella A. Greene
Desert Scrub Oak
Quercus spp.
oaks
Fagaceae
viyóōdi, toa

In the steepest, highest portions of the Sierra Estrella, on its north-facing, most moist slopes, grow thickets of Desert Scrub Oak. These can be found, too, in the White Tank Mountains, a bit to the northwest of Pima country proper; as you travel to the north of Pima country, oaks are usually the first indications of Upper Sonoran vegetation. These oaks are rather scrubby, scarcely trees, with small blue-gray leaves.

Few Pimas ever reach the tops of the Estrellas to see these thickets of oaks, gooseberries, and other plants so rare in this desert. This was the domain of the solitary hunter. Pima contact with oaks in more recent historical times has been primarily through trade of acorns with the Tohono O'odham. Nearly all modern information came to me in that context.

Sylvester Matthias said: "Acorns. They know what it is: *viyóōdi*. They didn't use it—maybe some people used it, like we use peanuts now. Papagos use a lot; they go after it just like the Apaches do." Another time he told me, "The Pima must have traded [these] with the Papago because they [the Papago] could get it on the Baboquivari—Papago go after them [also] at Fort Huachuca. Shell it, grind it, and make it into a gravy." The Pima, he thinks, used them just as nuts.

Ruth Giff also told me, "We didn't eat *viyóōdi*; the Papago children brought them to school." In another discussion, she said, "The *first* time I think I ever had *viyóōdi*, I think the Papagos brought it. Then the Ó'ob [Apaches]. I didn't care for it . . . too bitter."

I asked Irene Hendricks if she knew *viyóōdi*. "There's no tree around here. It's way out. I got it in stores—a little bag full. Crack and get the shells out. Sometimes they got little worms." She did not recognize any of the alternative terms for oak or acorns.

David Brown and a Tohono O'odham friend of his who was visiting from Menager's Dam, Jim Acuña, agreed that the word *viyóōdi* applies both to the acorn and to the oak trees on which the acorns grow.

But this is one of those curious cases where a loan word has come to supplant the original Piman name, *kus:* the Tohono O'odham have taken over from Spanish the word *bellote* 'acorn'. Since acorns have been coming up from the Tohono O'odham, the Gila Pima abandoned their native term and now know only the loan. This is rare; as Joseph Giff pointed out to me in word after word, in cases where the Tohono O'odham use a loan word, the Gila Pima have usually kept the original term or invented one from pre-existing Pima words. Among the desert Pima Bajo and the Mountain Pima, *kus* is the generic base word that is modified to signify specifically the many different kinds of oak trees these people know from the Sierra Madre.

Yet a third term has been used to designate the oak. Under *toa (Q. oblongifolia),* Russell (1908:78) recorded, "The acorns of this oak are traded from the Papagos. After the hulls have been removed they are parched and ground into meal." This word has been recorded from the Tohono O'odham as well. Saxton and Saxton (1969:170) give *tooa* as an 'inedible acorn' and later (Saxton, Saxton, and Enos 1983:94) give *toa* as 'oak tree'. In their ethnobiology of the Tohono O'odham, Castetter and Underhill (1935:18) give *toa* as *Q. emoryi* and *Q. oblongifolia*. Possibly this word has arrived among the northern Pimans as an abbreviated form of one of the Lower Pimans' specific names for oak: *toa kus* 'white oak'. More likely it may be cognate with Mountain Pima *tua* 'live oak' and Northern Tepehuan *tuai* (Pennington 1969, n.d.). Fowler (1983) found *tua* to be a widespread southern Uto-Aztecan root word for oak. But no one now among the Gileños recognizes the lexeme Russell found at the turn of the century. (This may well have been *tua* instead of *toa*, given Russell's transcription of *u.*) Mathiot (1973:260) notes, under the lexeme *doa*, "acorn; oak (old word, not as common as *viyóōdi*)." Saxton, Saxton, and Enos also list another beautiful name for the oak tree: *wiyoodi je'ej* 'acorn's mother'.

Salix exigua Nutt.
Coyote Willow
Salicaceae
s-i'ovĭ che'ul

Under the heading Gooding Willow, Curtin (1949:108) had recorded somewhat cryptically: "The Pima distinguish between the full-grown tree and the young shrub-like shoots, insisting that they are different species." How these are linguistically distinguished was not stated, and her field notes provided no additional clues. This statement puzzled me, because I had never found a case where the Gila Pima were overclassifying a wild biological species, that is, classifying a single Linnaean plant species as several folk species.

Perhaps the Pima were distinguishing between the male and female willow trees, which look quite different in spring when they are blooming. I asked people if these had different names. They did not.

Another kind of willow came to mind—Coyote Willow—that grew along the desert streams and rivers. Could that have been what the Pima were trying to tell Curtin about? I had searched the reservation streams for years but found only the common Goodding's Willow. The historic herbarium materials at the University of Arizona helped; in June 1926 the botanists Peebles, Harrison, and Kearney had collected a specimen of Coyote Willow in the Gila River bottoms near the Sacaton Diversion Dam. There once were two species in Pima country.

Then I remembered another incident. I was teaching some years ago at Prescott College, the original one up in the mountainous pine country northwest of Phoenix. One of my students knew Ruth Giff; Ruth had been housekeeper at one time for the family in Phoenix and helped raise their several children. Ruth arranged a trip to Prescott to visit the now-grown girl; she even took the girl a little willow and devil's claw Pima basket she had woven. The girl took us out to her hideaway, an old miner's cabin. The Hassayampa River there is a mere stream heading south, thickly lined with a bushy willow just a bit over head-height, with exceedingly narrow leaves and rather conspicuously reddish twigs. Ruth, the gardener and basket maker, always watches for plants wherever she goes, but especially for weaving materials. She remarked on the scrubby willows, mentioning their Pima name and saying that they once grew all the way down into the desert.

A dozen years later I began pulling together materials for this ethnobotany. I read about the other willow that Curtin's "informants" were trying to tell her about. Ruth's casual remark on the Hassayampa popped into mind. I telephoned her, and after some pleasantries, I asked if she remembered the willow from that visit to Prescott.

"That other willow—with the narrow leaves? I remember that one, called *s-i'ovĭ che'ul* 'sweet willow'. It has red stems. It used to grow around here at one time. Maybe along the ditch banks. But you can't use it for baskets because they're like weeping willow, they break too easily, even though they're *real* long, maybe a yard long, but brittle."

Now the Coyote Willow is almost completely gone from the middle Gila River. I found one growing in an irrigation canal in Sacaton, but it has since succumbed to herbicides. Its English name has no connection with the Piman name, though it would appropriately fit the Coyote's plants paradigm because of its inferior qualities. Perhaps, too, because it has managed to elude ethnobiologists for nearly a century.

I went back over Curtin's field notes once again to see if I might be able to sort out what the Pima in early 1940 were trying to tell her. Under Specimen No. 57 "Willow (Tree)" she had typed several pages of notes, all of which seem to fit legitimately Goodding's Willow. Then under Specimen No. 64 "Willow (Shrub)," I discovered something very interesting among her three brief sentences. Lewis Manuel, a Blackwater man then living on the Salt River Reservation, told her, "This shrub is called 'sweet' and the catkins are eaten raw." Ah! *Che'ul* does not mean 'sweet', as Curtin (1949:108) wrote when combining her two sets of notes. But *s-i'ovĭ che'ul* means 'sweet willow', and this is what Lewis was trying to tell her. Perhaps the catkins of this species are sweeter than the other, which is why they were eaten. (Catkins of Honey Mesquite [*Prosopis glandulosa*] and Frémont's Cottonwood [*Populus fremontii*] were likewise eaten raw.) I read on: "In making

the outdoor storage baskets, *hawmda [homda]*, the work is commenced with wands from this willow." These were the great bottomless granaries of Arrow-weed [*Pluchea sericea*], whose walls were always started with a ring or two of willow. And finally, "Slips are planted to form fences." Whether these were for erosion control or just living fences is not stated, but it is important to know that the Gila Pima did deliberately plant willows. Perhaps this is also why Ruth had recalled seeing Coyote Willow growing along the ditch banks that paralleled the fields.

Apparently when Curtin collected her voucher specimen of "Willow (Shrub)" she got by mistake a young specimen of the larger and usually more abundant Goodding's Willow, for she penciled into her field notes: "Mr. Peebles [the botanist] states that this shrub and the tree are both the same willow, but the Pimas make a distinction."

There is a lesson here, and that is that we must listen carefully to what our native consultants are trying to tell us. In a time when ecosystems are being drastically altered, when elements from the native flora and fauna are being lost, the native person may know many things that scientists do not know and will never discover, particularly in the pitiful remnants of modern desert riparian ecosystems.

In January 1990 I was discussing crop harvest and storage with Francis Vavages at Sacaton when he offered some interesting information on the big jug-shaped granaries: "The *vapshom* were made out of straw and sweet-willow bark, *s-i'ovĭ che'ul*. They don't grow as big as a willow [*che'ul*]—that's what they make bows with. Or Screwbean [*Prosopis pubescens*], *kuujul*."

The next year Francis said, "I know *s-i'ovĭ che'ul*, only they don't grow as big as the other *che'ul*. The bark is kind of sweet, alright. That's why they call it *s-i'ovĭ*. Peel bark off. Use the bark for *vashom*, to make the straw basket. Make holes in [a piece of] tin and pull it through." This was to pare the bark strips to a uniform width. "Take *kavk u'us* [Greasewood, *Sarcobatus vermiculatus*], make like a needle [*ovij*, awl], to make that *vashom*. *S-i'ovĭ che'ul eldag* [bark], put it in water [when it dries out] and it gets soft again when they want to use it." He demonstrated with an imaginary bundle of straw how the bark was laced around through the holes punched with the

Greasewood awl.

That afternoon George Kyyitan told me, "*S-i'ovĭ che'ul* is different from that other willow. Willow spreads out, but this *s-i'ov* goes straight [up]. Used to be a lot of them at that old store, along the canal, by the John Deere Tractor [west of Sacaton]. Used on the *vatto* [ramada] once, because it's long. And on the *vashom*, with straw."

Salix gooddingii Ball
Goodding's Willow, Western Black Willow
Salicaceae
che'ul

These are water people, these Pima. Through the centuries, they have moved through the hot country from one river valley to another. And wherever they went, there was always the willow, the *che'ul*.

Earlier they would have said *te'ul*, because the change from *t* to *ch* is recent (after the coming of the Spaniards). On the lower Rio Yaqui the remnant Pima Bajo still call it *tewer* and *te'evur*. Higher in the Sierra Madre the Mountain Pima say *te'egyi*. Pennington (1969) recorded it as *tubuli* [= *tuvuli*?] for the Northern Tepehuan.

The importance of willow in the everyday life of the Gileño in 1900 can be gauged by the number of items Russell recorded as being made from this tree (Russell 1908). For instance, the wood was used in the construction of round houses, yokes, saddle stirrups, and rope twisters, while smaller willow boughs were made into cholla bud pickers, calendar sticks, ceremonial wands, bows, and support hoops of the *gióho* (carrying basket). The inner bark of willow is stringy and strong. It was used in lashing the willow stays to the horizontal poles of the round house; shredded, it padded the cradle board; and flat, dyed pieces were woven into the cradle shield. Head rings the women used in carrying burdens might be woven from willow bark. Even in tattooing their faces, the pulverized charcoal was from either willow or mesquite.

Perhaps the major use has been, and continues to be, in weaving baskets. Ruth

Giff was busy at the kitchen table working on a basket, inserting thin white splints of willow she had trimmed exactly to uniform width, coiling these over the cattail-stem base. She explained that she had to get this basket finished in time for the Pope's visit to Phoenix the following month; she had been asked to make a tray basket to be used in the Mass. I hope it ended up among the Vatican treasures, as a basket by Ruth Giff should be. While she worked, keeping her splints wet in a tin can, she talked of the willow harvest. "The second picking is ready in June. Until mid July. If it is standing in water, it's still good by September. If it's on dry land, wouldn't be any good—you can't peel it.

"Use the longer ones to make their *uso* [a special basket used to remove the spines from cholla buds].

"Watch the buds in the spring . . . if a cold winter, they'll be late. Long time ago, they used to be after March first, but now it comes on earlier. When they're *just* budding out. When the leaves come on, it's too late. Naomi White used to go get her willows up by Lake Mary [near Flagstaff]—they're late getting ready. One time we went to Wickenburg after ours were too late, but they didn't have any buds on yet there. *Just* that close, but it makes a difference.

"Other kind, *s-i'ovĭ che'ul* [Coyote Willow, *Salix exigua*] [is] no good. They break *very easily*. It used to grow around here. Too dry now."

Even though her cattail and willow splints seemed meticulously uniform, she still trimmed and smoothed each piece more, holding one end in her teeth, while she worked. She paused to flatten the coil with a hammer on her cement floor. "My mother didn't use a hammer [to flatten the coils]. She used two rocks," Ruth explained.

"Did they use the same ones every time?"

"Yes, smooth ones. They got them in the river."

She pulled a new splint through one of several holes in a tin can lid. "Some white man said he invented this! He makes baskets! But it was already in use when I was a little girl. You know, one time this Pima boy asked me to teach him to make baskets. But I wouldn't. Turns out a *uikwuaḍ* [transvestite] if a man touches a basket."

When the Salt-Gila confluence at 115th Avenue (southwest of Phoenix) still had its groves of cottonwoods and willows and reeds near the paved highway, it was easy for elderly Indians to get in to gather materials they might need. Basket weavers had ready access to young willow shoots and acres of cattails in water shallow enough for them to pick. In the early 1970s Sylvester Matthias needed willow bark, so I took him there; I always needed to check the birds as this lush riparian community matured. Sylvester brought along a little hatchet. Before floods scoured this whole area, the confluence was a maze of sand banks and channels, cottonwood-willow groves and Salt-cedar (*Tamarix ramosissima*) thickets. Sylvester hobbled through the sand—crutch in one hand, hatchet in the other—looking over the willows, evaluating them, until he found a grove of saplings that pleased him. With trunks only four or five inches in diameter, they had not yet developed the deeply furrowed gray bark of the more mature trees. It was the flexible inner bark he was after. He chopped out a piece a couple inches wide, stripped it down to the ground, then severed it. He ended up with a belt of fiber. He took a strip from about a dozen trees, then tied up his load with bark.

A few mornings later, I learned what he was using the strips for. When I arrived, Sylvester was out working at a table, building a bird trap by laying up Arrow-weed (*Pluchea sericea*), log-cabin style. At his feet was a bucket of water with the willow bark, which he had soaked overnight. As he laid each alternating rod, he secured it firmly with bark at two corners. I took a strand out of the water and was amazed at how tough it was—just like twine.

Willow bark was the original source of acetylsalicylic acid, better known as aspirin. Many people in the world were aware of its analgesic qualities. Contemporary Gila Pima have forgotten its medicinal use, but at one time, the beneficial effects were indeed appreciated by the Pima: Hrdlička (1908:245) wrote, "A decoction prepared from the leaves of the willow is given in fever."

Some Pima tried willow leaves as a coffee substitute, Sylvester related. "They say it don't taste good—it's kind of bitter and it's too thick [tasting]. It's not like Mormon Tea [*Ephedra* sp.], but anyway they drink it. Just use it for tea or coffee. My father don't like it—it's sort of bitter on the side. But that Mormon tea, it's *tea*." He never heard of using *che'ul* for medicine.

Sylvester mentioned another use. "The *che'ul haupuldag* [catkins] can be eaten when young. When someone [is] picking willows, get[s] hungry, they can eat that, because it's just at the right time—for food."

"Did they use it for chewing gum like the *auppa* [cottonwood]?"

"No, for food." Only the female flowers were used, he explained, not the subsequent seedpods.

Ruth Garcia introduced herself to me at George and Dorothy Kyyitan's golden wedding anniversary. Dorothy is her sister. Ruth is another of the few remaining basket makers.

"Where do you get your materials?" I asked. "Isn't it hard to find them anymore?"

"I buy the *ihug* [devil's claws, *Proboscidea parviflora*] from this man who raises them, but the other two are hard to get. Some man gets my *uḍvak* [cattail] at the tribal farms. The *che'ul* is all gone now. There are some new *che'ul* along this canal on the Akĭ Chiñ Reservation. I'm going to ask the tribal council for permission to pick them."

Akimel O'odham, River People. Now a people without a river. Once a Pima woman was never more than walking distance away from two of the plants she needed for her baskets: cattails and willows. The third, devil's claw, she had plenty of water to raise in her own garden. And now a 77-year-old woman worries about picking willows on lands that once belonged to all Pimans. And she pays $50 for a wheel of devil's claw someone else grows.

Vouchers: 5 (RG); LSMC 57 (not extant); LSMC 65 (includes information on *S. exigua* also).

Sambucus mexicana Presl.
Mexican Elder, Desert Elderberry, Saúco, Tapiro
Caprifoliaceae
dahapdam

Elderberry is now gone from the reservation, or nearly so; I used to find it occasionally in dense, overgrown brush fences a quarter-century ago, especially those

dahapdam   *Sambucus mexicana*

Some years later I quizzed Sylvester on the uses of elderberry. "I think for medicine—dry it up—for open sores, just like greasewood [*sic*, Creosote Bush, *Larrea divaricata*]."

Four and a half decades earlier, Curtin (1949:75) gathered considerably more information on *dahapdam*. A Salt River Pima told her that it was one of the old Pima foods, later used for jams and jellies. On its medicinal uses, she learned that "to reduce fever, one-half cup of dried flowers are steeped in hot water which is then drained off and the liquid drunk lukewarm. The flowers, either fresh or dry, are boiled in water and the decoction is taken, while hot, for stomachache, colds, and sore throat."

Russell (1908) did not mention *dahapdam* either as a food or as a medicinal plant. Perhaps it was already becoming rare in the Sacaton area where he studied. Watershed and riparian destruction affected this part of the reservation first, and the tree likes plenty of water.

In the 18th century, the German Jesuit Fr. Juan Nentvig, while stationed in Sonora (which then included southern Arizona), penned extensive notes on plants, including the elderberry: "The Pimas, particularly the upper ones, still indulge [in drunkenness]. Their intoxicating beverages made of corn, mezcal, wheat and Indian figs [a cultivated prickly-pear, *Opuntia ficus-indica*] are bad enough, but the elderberry wine is worse because of its lasting effects" (Nentvig 1980:60–61). In another place he added, "The Mexican elder is a fairly common tree in both Pimerías. From its berries the Altos [Upper Pimans] make a beverage so potent that those who drink it to excess get so drunk that it takes two or three days to sober up. Believing that such a vice is the cause of many evils, the missionaries have tried unsuccessfully to stop the Indians from making it" (Nentvig 1980:41).

The Pima settlement of Attí was located on the Rio Altar, about 50 miles south of the present international boundary. In the Attí vocabulary compiled sometime before 1774 is an entry on elderberry that translates to "The elderberry tree: big *tohapitam* (little *tuhipitan[m]*) is lower [Pima] language: *tohapit navat* is elderberry wine." Although I have no idea what the distinction between the large and small kind of elderberry might refer to at Attí, the reference to its wine is clear. It

bordering moist fields. But these fields are likewise almost all gone, the victims of modern agriculture. Great thickets of elderberry trees and bushes once lined the stream connecting parts of Barehand Lane marsh. Elderberry flowers smell like allspice. After the rich cream-colored umbrellas of petals fall, they are replaced with heavy masses of small berries, at first covered with a whitish bloom, but later in the summer almost black. There, escaping the heat on the torrid flats, great flocks of delicate Phainopeplas used to congregate, eating the purple berries (see Rea 1983:218–219).

I have been unable to learn much about this tree in Pima culture. It seems scarcely a memory among the old people, and few of these even remember its name. Ruth Giff told me, "Old man Tashquinth used dried blossoms, stored away— [to make] tea, for fever, or for colds, maybe. He knew how to use it."

Sylvester Matthias knew the name also and identified the trees in the field with me. He said, "There's a place across the [Salt] river they call *dahapdam chuuchk* 'elderberry there', 'grove'. [It's] not planted, but it's there, voluntarily. Around 91st Avenue, where Santa Maria is now."

"Is there any use?"

"Dahapdam used for fever. Berry tree. They ripen in May. Old Man Tashquinth, every spring, he used to gather the blossoms and dry them and put them away; and boiled them when he needed them. But they never know they can use the berries. They think it's poison. Never used to be a lot of it."

would be *dahapdam navait* in contemporary Upper Pima.

I asked Sylvester if he knew anything about *dahapdam navait*. "I didn't know they could make [it]. I guess the [Gila] Pimas don't know they could, otherwise they would be making it. Nobody ever talks about *dahapdam navait*. If they know it, they'll be having it." George Kyyitan likewise knew of various fruits used to make wine but had not heard of elderberry wine, even in English. Apparently elderberry wine was one of the early victims of cultural erosion.

Vouchers: *149* (SM); LSMC 42.

Tamarix ramosissima Ledeb.
  [= *T. chinensis* auctorum]
  [= *T. gallica* L.]
  [= *T. pentandra* Pallas]
Salt-cedar
Tamaricaceae
vepegĭ u'us, s-vegĭ uus

Salt-cedars are attractive bushes and small trees with delicate light bluish green branches covered with leaf scales, suggesting the name cedar. These are not conifers of any kind, though. Throughout the warm season these trees sport masses of fine whitish through bright pinkish flowers. Their abundant seeds are windborne. Salt-cedars often form dense thickets at the edges of standing water such as charcos, borrow pits, and sump ponds. Once, even, I found Salt-cedar saplings at the edge of a tinaja high up in the Sierra Estrella, an indication of the dispersal abilities of their airborne seeds. They become most dense along the greatly disturbed river channels, almost to the elimination of any other plant, taking advantage of conditions detrimental to the native broad-leaved trees such as willows and cottonwoods (see Rea 1983). The middle Gila River channel is now almost exclusively covered with Salt-cedars.

In the heat of summer, Salt-cedars exude from their drooping branches a salty-sweet fluid. Perhaps this is why these thickets are generally devoid of most summer birds. The excess salt load picked up by a small bird attempting to clean its sticky feathers may exceed its physiological tolerance.

When the cold weather sets in, about in November, the leafy branches begin turning orange, and in a few months they fall, leaving just the woody branches and stems with their reddish bark.

I asked Sylvester Matthias how long this Eurasian exotic had been around the Komatke area, where the Gila and Santa Cruz rivers had surface water longer than any other part of the reservation.

He pondered a while. "The first Salt-cedars were in 1927. My father first cut one for a fencepost—very hard. Called *s-vegĭ uus* 'red branch [or] twig.'" Ruth Giff called it *vepegĭ u'us,* using the plural form. I found no one on the reservation who even recognized the name *onk kui* 'salt tree', a name used in Tohono O'odham country to designate the Salt-cedar. But this is not surprising in view of a basic difference between these two groups in designating the concept of tree. The Tohono O'odham use *kui* to mean both 'mesquite' and 'tree' in general; the Gila people use *uus keekam* 'woody thing standing' to mean 'tree'.

In the eastern villages I have occasionally heard Salt-cedar referred to as *kauk uus*, which properly indicates Greasewood (*Sarcobatus vermiculatus*). But users of this name for the exotic bush seemed always to recognize *s-vegĭ uus* when I inquired.

The plant has little cultural significance. Sylvester noted, "The young [shoots] make good arrows." But arrow making has been a low priority for almost as long as the tree has been there, and all the arrows I have seen Sylvester and others make have been of *u'us kokomagĭ*, Arrowweed (*Pluchea sericea*).

The culturally significant related tree is Athel Tamarisk (*Tamarix aphylla* (K.) Karst.), which has no Gila Pima name. It is a large tree with drooping nondeciduous branches, grayish green scales, and small whitish or very pale pinkish flowers that are hanging. This tree rarely if ever self-propagates, yet it has been planted as a shade tree and windbreak around houses across the reservation, replacing broadleafed trees as the water table fell. At Komatke Sylvester said that the first Athel Tamarisks were planted at the agricultural extension agent's house. "After that the Pima started planting them. They were planted around St. John's School in 1929. But you're never cool sitting under them. When [Franciscan] Father Matthias first came to St. John's he said he didn't enjoy the shade of the tamarisk. No coolness, except the cottonwood and willow. You get under a cottonwood or willow— oh boy! You feel the coolness. Tamarisk is shady, alright, but no coolness."

At Bapchule George Kyyitan told me, "I don't think there's a name for this [Athel Tamarisk] because it just came, and not very long ago. And that's why everybody likes it, because they'll bud easily [sprout from cuttings]. You just cut a piece and put it in there. It'll come up. Everybody likes it at that time, but later on, nobody likes it because they make it salty around, and the roots go under and spoil things. I think that one [Athel] comes [in] later than the tamarisk [Salt-cedar] because it was [19]23 when that school [St. Peter's] was started. And we planted those trees right in front of the church. Now—fifty-some years [later]—big trees!"

According to Benson and Darrow (1981:97), "Most of the [Athel] plants in cultivation in North America are part of an enormous clone derived from half a dozen cuttings secured at the beginning of the twentieth century by J.J. Thornber, Professor of Botany at the University of Arizona, from a correspondent in Algeria." The Salt-cedar followed shortly as an exotic species (Harris 1966).

Vouchers: *204* (?).

Washingtonia filifera Wendl.
Northern Fan Palm
Arecaceae (Palmae)
hevhodakuḍ

Bright green Northern Fan Palms are planted here and there across the reservation but are nowhere common. They have a single stout trunk topped by huge fanlike leaves with armed petioles or stems. Though these palms occur naturally at some desert oases in the state, those on the reservation have been intentionally planted. These are the favorite nesting place of the few remaining pairs of Hooded Orioles on Pima land. These birds suspend their nests, woven from the white palm-leaf fibers, from the undersides of the great fronds.

The word *hevhodakuḍ* in Gila Pima can mean 'fan', or this tree, or just one of its fronds. It is related to the verb *hevhogid* 'to cool [something] off' and ultimately *hevel* 'the wind' or 'to blow'. Asked about natural stands, Sylvester Matthias related, "They grow wild at Salome and Quartzite, [Arizona,] according to the Mexicans."

Joseph Giff said that three things could be used to make the sound inside the gourd rattle: seeds from Northern Fan Palm, seeds of *kuk chehedagĭ* (Foothill Paloverde, *Cercidium microphyllum),* or the fine water-ground gravel that certain ants bring up to the earth surface. Later, shot from a shotgun shell became another option.

Sylvester grew a grove of *hevhodakuḍ* outside his mud house. Each spring he is occupied cutting and cleaning the great fronds for the Palm Sunday liturgy.

I asked George Kyyitan if he knew the word *hevhodakuḍ.*

"Fan," he responded.

"Have you heard of the palm tree being called *hevhodakuḍ?*"

"Uh-huh. Hevhodakuḍ Tash . . . that's Palm Sunday."

None of the Gila people recognized the word *maahagam* as a name for the fan palm, though it is listed as such by Mathiot (n.d.:32) for the Tohono O'odham. The word is an old one that can signify 'hand' or 'palm and spread fingers' as well as the tree (see Group G, *Lupinus sparsiflorus* account).

I took a large fruiting branch to George one day. It was covered with purple-black fruits, each one small with a large hard seed inside.

"We used to eat *hevhodakuḍ.* But not in the olden days; just now when we were in school. We pick them up ourselves. Not these people. Some kids over at Goodyear have been using it. We never use it until that time. Then after that, we use [it] and they tasted sweet. Like *u'us chevaḍbaḍ* [Graythorn, *Ziziphus obtusifolia*]."

Pima Bajo Pedro Estrella supplied Onavas with brooms he made from *maahagam* fibers. He also collected them for the few remaining basket makers. On one of our visits, he showed Sylvester all about his local industry. The late Carolina Humar wove me a palm sleeping mat or *main* from *Sabal uresana,* a local species that is used extensively. This was rolled up and stood in the corner of the bedroom Sylvester occupied during his visits to San Diego. One day he commented, "My grandmother said she can make *main* out of palm—*hevhodakuḍ*—because it's similar to that *umug* [Sotol, *Dasylirion wheeleri*]."

Technical Notes: Mexican Blue Palm, *Glaucothera armata* Cook (= *Erythea armata* (S. Watson) S. Watson), was being grown at the Sacaton Field Station in 1924 as an ornamental. I have found no trace of this beautiful fan palm surviving anywhere on the reservation. A Sonoran Desert species that does not naturally reach the United States, it deserves broader cultivation now that it is commercially available.

Vouchers: *1642* (GK).

Group F, Iivagĭ
eaten greens

Iivagĭ (var. iivag, iivakĭ) is a utilitarian category that is superimposed on any other classification based purely on the morphological characteristics of the plants. It subsumes species that are currently eaten as greens, either cooked or raw, as well as those that are known to have been formerly eaten by the Pima. For instance, wild-heliotropes (*Phacelia* spp.) were once eaten. Although the use of these and several other greens ceased early in the 20th century, they are still classified as *iivagĭ*. The category *iivagĭ* is sometimes glossed as 'Indian spinaches' even though this is something of a misnomer; the term literally means 'new green shoots/buds', from the verb *iivagi* 'to put forth new buds'. The phrase *iivagim s-cheedagĭ* can be used to specify 'grass green', the color of new shoots.

I wondered if cooking was an obligate part of this special-use category. I asked Ruth Giff, "Does *iivag* have to be something you *cook,* or can it be something you eat green?"

"Just [green], like the *huai hehevo* [Annual Sowthistle, *Sonchus oleraceus*] that we call 'wild celery'. We eat them raw. We call that *iivag.*"

"Now, *s-ho'idkam iivagĭ* [Spiny Sow-thistle, *Sonchus asper*]. You cook that or eat it raw?"

"No, I think it's eaten raw. Have to peel them."

"So lettuce in the grocery store, you just say *iivagĭ?*"

"Uh-huh. Yeah." She wondered, though, whether *kumul* 'mushrooms' might be classified here. Although they are wild plants that are eaten, they are not greens.

This category, common to both Akimel and Tohono O'odham, is an old one and is probably pan-Piman. For instance, the 17th-century Névome vocabulary gives *hierbas que comen en general* as *hibaqui,* which in its Spanish orthography comes out to *ivaki.* In contemporary lowland Pima Bajo, potherbs or quelites are *ivak* (Pennington 1980:227; Rea field notes). The Northern Tepehuan also call collected greens *ivagi* (Pennington 1969:138).

The word *iivagĭ* forms the obligate base lexeme in the names of several greens, such as *onk iivagĭ* (Wright's Saltbush, *Atriplex wrightii*) and *s-ho'idkam iivagĭ* (Spiny Sowthistle, *Sonchus asper*). And in one case, a newly introduced Old World crop, lettuce, is called polysemously just *iivagĭ* (see Group I, *Lactuca sativa*): lettuce is an anomaly in that it falls into two special-use categories, the greens and *e'es* 'planted things'. But in most cases, the word *iivagĭ* is not part of a plant's name. As Joseph Giff once explained, "*Chuuhuggia* [Palmer's Amaranth, *Amaranthus palmeri*] is an *iivagĭ,* but we don't say *chuuhuggia iivagĭ,* although it is. It's understood—that it's an *iivagĭ.* You don't have to say it."

Greens are still very much a part of the Gila Pima diet, at least among certain middle-aged or older Pimas from different villages. As explained in the folk generic accounts, collecting and preparing pot-herbs is women's work, and consequently their knowledge of these plants is better.

The annual cycle of *iivagĭ* covers most months of the year. If winter rains come, *opon* (Patota, *Monolepis nuttalliana*) may be ready for gathering sometime in December, January, or February. After that the plants begin to get too tough and go to seed. The early spring *iivagĭ*, coming in the starvation period, have fallen into disuse. These included *ñiádam* (Star Mallow, *Eremalche exilis*), *havañ taataḍ* (wild-heliotropes), *vakoandam* (dock and sorrel, *Rumex* spp.), and perhaps others that are now completely lost from oral history. Later come various warm-weather spring *iivagĭ* such as *chuális* (Narrow-leaved Goosefoot, *Chenopodium pratericola*), *huáhi* (Pit-seeded Goosefoot, *C. berlandieri*), and most important, *onk iivagĭ* (*Atriplex wrightii*). Throughout the hot summer, *ku'ukpalk* (Verdolaga, *Portulaca* spp.) is one of the principal greens

chuuhuggia *Amaranthus palmeri*

gathered from cultivated lands. Later in the summer, great stands of another important *iivagĭ*, *chuuhuggia* (*Amaranthus palmeri*), begin to appear, and some young plants or tender shoots can usually be found until early in the fall. I know of no greens that are gathered during the period when the fields cease to be watered and before winter rains appear once again (October and November, more or less), but this is the height of the second harvest of cultivated crops for these traditional agriculturalists.

Amaranthus palmeri S. Watson
Palmer's Carelessweed, Palmer's Amaranth
Amaranthaceae
chuuhuggia

One of the most abundant weeds of Pima fields is the amaranth or carelessweed. Field edges, particularly those getting the most water and nutrients at the lower ends, often become pure stands of this rhubarb-red-stemmed rank plant by late summer and fall. Later the plants produce prickly heads loaded with minute black seeds. These are relished by doves and sparrows in fall and winter. If irrigated enough, the plants can reach a person's height. The diamond-shaped leaves may bear a whitish green or even purplish V pattern.

To the Pima, this is the summer *iivagĭ* par excellence. (It is so for other tribes as well, and for Mexicans, who call the plant *bledo*.) They pick the tender leaves and whole tips of stems and boil them. Of the various greens available to the Pima, this one is the most widely used and I suppose will be the last one forgotten.

Curtin (1949:48) gave the derivation of the name *chuuhuggia* as 'night carrying'. The word *chuhugam* means 'night' or 'darkness'. The word is probably pan-Piman. *El bledo* appears as *tucugusa* in Névome and as *tungia* among modern speakers of desert Pima Bajo. The Mountain Pima of western Chihuahua use *tukgya* as the base word for any species of amaranth (Laferrière n.d.). (The change from *t* to *ch* among the northernmost Pimans has occurred within historic times.)

Hrdlička (1908:264) wrote, "*Chu-hu-ki-ia* is a small plant the leaves of which the Pima use for food in the fall *[sic]*. They usually eat them cooked, with the addition of salt, in the same way as spinach, but occasionally they chew the leaves raw." And they are still cooked this way (I have not heard of anyone eating the leaves raw).

Russell (1908:78) did not identify the plant, but recorded the name *tchohokia*: "The leaves are gathered in spring *[sic]* and sometimes baked in tortillas. In summer the seeds are gathered, ground on the metate, mixed with meal or squash, or they may be parched and ground to be eaten dry." Lena Meskeer of Gila Crossing, who was about 72 years old when Curtin interviewed her in 1941, gave information on the preparation of *chuuhuggia* seeds: "The seeds are dried and ground, and one or two handfuls of meal are thrown into a pint of boiling water, a small teaspoon of salt is added, and the whole cooked until done" (Curtin 1949:48). By my time, however, use of the seeds had entirely dropped from Pima culture. Castetter and Underhill (1935:24) considered the seeds an important food source for the Tohono O'odham. There is no longer any tradition among living Gila Pima that they once cultivated any species of amaranth for seed, but Gary Nabhan and I have found their relatives in Mexico still growing white-seeded domesticates (see Group I, *Amaranthus hybridus* account). These can be parched, the tiny seeds exploding like miniature kernels of popcorn.

Reverend Isaac Whittemore (1893) noted that the Pima were raising (perhaps aboriginally) "a small round seed which they ground and boiled as mush." This might have been the cultivated amaranth or possibly *Panicum sonorum*, a Southwestern cultivar (see Nabhan 1985).

Chuuhuggia is mentioned in Thin Leather's version of the Pima Creation Story as recorded by Lloyd (1911:115–117). A young lady conceived by sitting over a boy's kickball to hide it. The part-human offspring, a girl called Ho'ok, grew up to be a witch who developed a taste for eating Pima children. The culture hero I'itoi was finally implored to kill her, which he did after four attempts. Her soul escaped, after a while turning into a "green hawk" (probably the Peregrine Falcon) that killed but did not eat people.

A woman was firing pottery, leaving one tilted on her kiln. The hawk swooped down but missed her, "and went into the hot pot in the fire, and so was burned up and destroyed. And one day they boiled greens in that pot, the greens called *choo-hook-yuh*, and the greens boiled so hard they boiled over and splashed around and killed people." The malevolent pot was at length smashed, but an old man and his orphan grandson ate the *chuuhuggia* and became, respectively, a black bear and a brown bear, which were likewise people-killers. At last this fourth manifestation was overcome: "There is a kind of palm-tree, called *o-nook,* which has balls where the branches come out, and the people burned the trees to get these balls, and threw them at the bears. And the bears caught the balls, and fought and wrestled with them, and while their attention was taken by these balls the people shot arrows at them and killed them."

And so the People's problem was solved. But mine was not; I am still wondering what the *onook* tree is.

Vouchers: *382* (sm); *1468* (multiple); *1750* (gk); lsmc 19. Excluded specimens: *511* (sp, *A. fimbrianthus* (Torr.) Benth.), *1751* (*A. albus* L.).

Atriplex elegans (Moq.) D. Dietr.
  [≠ *A. nuttallii* S. Watson]
Wheelscale Saltbush
Chenopodiaceae
toota onk

In spring and summer a bushy and quite erect annual saltbush grows in the alluvial soils of ditch banks, field edges, and roadsides. This saltbush is quite leafy, the upper surfaces being greenish but the undersurfaces a scruffy silvery white. Sometimes several annual species of *Atriplex* will be found growing together. Although these saltbushes may be confusing, Wheelscale Saltbush can be distinguished readily when fruiting because two bracts deeply toothed all around the margins enclose the seed. (In the related Salton Saltbush, *A. fasciculata* S. Watson, these margins are shallowly and indistinctly toothed.)

The Pima name for this saltbush, *toota onk* 'white salt', refers to the plant's farinose or mealy appearance, particularly on the undersurface of the leaves, as well as to its salty taste. Relatively few people today seem to know this *iivagĭ,* even those who esteem the similar but leafy Wright's Saltbush *(Atriplex wrightii), onk iivagĭ.* (Possibly a few Pima women may gather both species without making any distinction.)

I asked Ruth Giff about *toota onk.* She was not very sure about it, but then she said, "Does it have white under the leaves, like *onk iivagĭ?*"

"Yes."

"Well, I remember it, but its leaves are different, sort of broad below and pointed at tip, not like *onk iivagĭ.* Once, when we couldn't find any *onk iivagĭ* growing in the fields, we picked that one and cooked it, and it tasted okay, like *onk iivagĭ.* But you know, there's a third one." She laughed. "You can't eat it. When I was first married, I didn't know it wasn't *onk iivagĭ,* and I went out and gathered that one and cooked it, but someone told me it was the wrong one, so we didn't try eating it. It's different. On the ditch banks. Has some red things on it, kind of like berries. Don't know any name for this one." This third saltbush species is Australian Saltbush, *A. semibaccata* R. Br., an introduced species not named in Pima. As a child, Sally Pablo recalled eating the red fruits of this saltbush.

One of the problems of studying any folk taxonomy is that the native consultant may be using criteria to distinguish a folk taxon that are quite different from those used by the Linnaean taxonomist. Such is the case with the various annual *Atriplex* species. The Gileños use the plants when they are immature and still tender, but I usually need the fruits of the mature plants to make my identifications. By this time the Pima have no interest in the plants. So throughout the years I collected the plants over and over again, often with the same person, working later in the season so that I could find the diagnostic fruits on some of the more matured individuals. Eventually the differences in growth form and leaf structure that the Pima were looking for became apparent to me as I associated these with the older plants that are useless in the eyes of the cook.

Why has Wheelscale Saltbush fallen into disuse, while Wright's Saltbush is still eagerly sought by all Gileños who still gather wild foods? *Toota onk* may never have been the saltbush of choice, compared to *onk iivagĭ,* Wright's Saltbush. Russell (1908:77) said of *toota onk:* "The stems of this saltbush are boiled with wheat. They are cut in short lengths and used sometimes as a stuffing for roast rabbit." It was described as being "boiled . . . with other food, sometimes with the flower buds of *Opuntia*" (Kearney and Peebles 1960:258). Once when Sylvester Matthias was showing me how to distinguish various annual and perennial saltbushes, I asked him, "Can you eat *toota onk?*"

"Yes," he responded, "when you can't find no [other] greens."

Salton Saltbush (*A. fasciculata*), although sometimes considered only a subspecies of *A. elegans,* is not considered *toota onk* by the Pima, nor is it ever gathered, I am told. This smaller relative has no Pima name, according to everyone I asked. Perhaps Pima cooks and folk taxonomists know something botanists have yet to discover.

Vouchers: *101, 376, 701* (sm); *776* (rg).

Atriplex wrightii S. Watson
  [≠ *A. bracteosa* S. Watson]
  [≠ *A. coronata* S. Watson]
  [≠ *A. elegans* D. Dietr.]
Wright's Saltbush
Chenopodiaceae
onk iivagĭ

The largest annual *Atriplex* species growing in spring throughout Pima country is Wright's Saltbush. Like Wheelscale Saltbush (*A. elegans*), this bushy plant enjoys the moisture of roadsides, field edges, and ditch banks. Both have leaves with silvery undersides, but Wright's Saltbush produces male flowers on tassels at the tops of the erect leafy plants, and the fruits resemble a pair of clasped hands rather than a pair of fringed pancakes. This is the slightly greener of the two and is definitely more "leafy" appearing.

But it is long before this flowering and fruiting stage that the Pima women become interested in their *onk iivagĭ* 'salt greens'. These greens are ready for gathering after the season for Patota *(Monolepis nuttalliana)* has passed. When the plants are only about 12 to 15 inches tall, in March or April, they are picked and boiled like spinach. They are ready to serve after draining, or they may be fried at this time. They are still prized in the 1980s, as they were in the 1940s (Curtin 1949:69).

Pima women may travel a considerable distance to pick Wright's Saltbush where they find it growing profusely. Ruth Giff said, "*Onk iivagĭ* goes into June, if lots of water, irrigation. When you pick, new leaves come on and you can pick again. Starts early, in spring." I have seen women picking it in quantities and drying the leaves in the shade. These are usually stored in glass one-gallon jars.

Joseph Giff said, "*Chuchk onk* [Seepweed, *Suaeda moquinii*] is used for pit roasting. You can also use, if available, *onk iivagĭ.*"

Sylvester Matthias told me as we stood on the roadside in Santan overlooking a lush growth of saltbushes in early April: "*Toota onk* [Wheelscale Saltbush, *Atriplex elegans*] is more gray [than *onk iivagĭ*] and fixed the same way. [*Onk iivagĭ* is] taller than *toota onk*—during winter; good feed for horses in winter."

Several years later, I asked Sylvester Matthias again how one distinguishes the two saltbushes. "*Toota onk* has narrow leaves, more gray than *onk iivagĭ*—bigger leaves, greener. Both have these tassels."

"Does *toota onk* have red in the stem, like this?"

"No. Gray."

The botanists Kearney and Peebles worked at Sacaton for several decades in the early half of the century. They remarked, "This is one of the species held by the Indians in particular esteem as a potherb" (Kearney and Peebles 1960:258).

Russell (1908:69) said of *onk iivagĭ,* "These saltbushes . . . are sometimes boiled with other food because of their salty flavor. They are cooked in pits with the fruit of the [cholla] cactus, *Opuntia arborescens [O. acanthocarpa].* The young shoots of some of them are crisp and tender." Russell's account might imply that this rather succulent saltbush was sometimes used in the roasting pits in place of Seepweed, as Joe explained.

Hrdlička (1908:263–264) added more details on the preparation of Wright's Saltbush: "Of greens, the Pima use the *onch-ki-ie-wak* ('salt green'), a plant growing in the spring along the Gila. The leaves are cooked without seasoning or other addition, and the water is pressed out. Meantime there have been roasted and ground together some small beans [teparies] and maize; these are mixed with the leaves and thus eaten. Sometimes the greens are eaten with pinole or with the cooked fruit of the *hánami* [hannam, cholla buds]." *Hannam* and *onk iivagĭ,* rehydrated and cooked together, are still served as a Pima dish.

In *Pima Cookery,* Elisabeth Hart (1949:7) noted a "salt grass" with pink stems, probably this species: "Both stems and leaves are cooked in two waters to remove salt." Although this preparation might be appropriate for either *toota onk* or *onk iivagĭ,* the stem coloration applies better to *onk iivagĭ.*

One May Sylvester Matthias and I collected some particularly robust saltbushes from the road edge in Santan. Sylvester called these *ge'egeḍchu huáhi* (large *huáhi*), an implicit comparison with *Chenopodium berlandieri.* Ruth Giff looked at these
specimens and said they were probably just *onk iivagĭ,* and indeed they proved to be Wright's Saltbush.

> Vouchers: *33* (RG); *200* (SP); *377* (RG; SM called these *ge'egeḍchu huáhi,* in error?); *725* (SM).

Chenopodium berlandieri Moq.
[≠ C. album L.]
Pit-seeded Goosefoot
Chenopodiaceae
huáhi

This chenopod is far leafier than Narrowleaved Goosefoot *(C. pratericola),* a native species the Gileños call *chuális.* Pit-seeded Goosefoot likes disturbed areas also, and the two may be found growing together, although *huáhi* is much more partial to wetter areas, such as irrigation ditches and field edges. Its stems are bright green with red tinges rather than dull grayish green with maroon tints. Its leaves are broadly shovel-shaped and the marginal indentations deeply wavy. The only other goosefoot on the reservation with which *huáhi* might easily be confused is the foul-smelling Nettleleaved Goosefoot (*C. murale* L.), a common field weed, which the Pima try to ignore by not even naming.

Of the various *iivagĭ* that the Gila Pima still harvest and enjoy, the two chenopods are less well known and used, perhaps because they grow toward the end of the *onk iivagĭ* (Wright's Saltbush, *Atriplex wrightii*) harvest and the beginning of the *chuuhuggia* (Palmer's Amaranth, *Amaranthus palmeri*) crop—these plants are leafier greens. Its intermediate seasonal position between these may once have promoted its use, but fewer people today know the various greens that still are readily available. The preparation of this *iivagĭ* is the same as for the others.

I was given a number of variations on this plant's name: *huáhi, huáho, huáhai, huáhei.* I heard *huáhi* most consistently. Fr. Antonine (1935) and Curtin (1949) recorded it *hwahai.*

People tended to contrast *huáhi* with *chuális.* Sylvester Matthias called Narrowleaved Goosefoot *huáhi* (or "just *huáhi*"), contrasting it with this much larger leaved species, which he called *ge'echu huáhi* 'big *huáhi*'.

Ruth Giff said, "*Huáhi* seems like it doesn't have the salt on it." She was contrasting it with the annual *Atriplex* species, which are indeed salty. Then she added, "*Huáhi, huáho*—same thing." She considered it an *iivagĭ* of spring and early summer, together with *A. wrightii* and *C. desiccatum.*

However, among the Mountain Pima at Yepachi, Pennington (n.d.) recorded *waha* for *chual,* although the biological species is not given. Laferrière was given *waxa* (= *huáha*) for *Chenopodium neomexicanum* Standley.

Technical Note: The correct identification of some chenopod species is especially difficult. Richard Felger advised collecting *Chenopodium* voucher specimens extensively, which I did over the years. Curtin (1949:70) had called *huáhi* Lambsquarters (presumably *C. album*), noting its European origin "now naturalized almost throughout North America." James Henrickson, who studied the material, wrote, "Your specimens bring up some questions about the taxonomy of southern Arizona material. All your *[huáhi]* sheets are *C. berlandieri* Moq., which is a very widespread and variable species. Overall your specimens appear to conform to two intergrading taxa that have been named both as varieties and subspecies within the taxon. The two taxa are *C. b.* var. *zschackei* (Murr) Murr (synonym ssp. *zschackei* [Murr] Zoebel) and *C. b.* var. *sinuatum* (Murr) Wahl (synonym ssp. *pseudopetiolare* Aellen). All the *berlandieri* can be distinguished as it has a coarse reticulate pattern on the fruit walls and a style base. In the Great Basin flora, they recognize no varieties under *berlandieri* and this may be the best thing to do as there is such a continually mixing mass of variation within the taxon that it is not possible to maintain these poorly defined specific taxa." Unlike Lambsquarters, these are native forms.

A colony of *C. berlandieri* with conspicuously heavy seed heads and stems branching from the base (*AMR 1540*) was excluded from *huáhi* by Irene Hendricks and Albina Antone of Sacaton. They knew no Pima name for it.

Vouchers: *673, 699* (SM); *974, 1734* (GK); *1455, 1548* (IH); *LSMC 52* (DES, sterile); *JEL 1048* (Mt. Pima, Yepachi).

Chenopodium pratericola Rydb.
[= *C. desiccatum* A. Nelson var. *leptophylloides* (Murr) Wahl]
Narrow-leaved Goosefoot
Chenopodiaceae
chuális

The chenopods (goosefoot) and annual *Atriplex* species (saltbushes) can be confusing, particularly in young growth. But Narrow-leaved Goosefoot is the easiest of this group to identify. It grows from late winter through spring. In dry areas, such as on the bajadas, the slim plants may reach only halfway to one's knees, if even that. But along irrigation ditches and field edges where moisture is plentiful, it will reach to waist-height or more, forming an open bushy plant, branching from the main stem. In either case, the plant has a grayish green appearance and the leaves are narrow, rounded at the tip,

slightly lobed on each side near the base. The flowers are in gray mealy tassels at the tips of the branches. As the plants mature, rub the tassels between your hands to find the shiny black seeds, shaped like the clay spindle whorls left behind in some local archaeological sites. Of the various plants the Pima classify as *iivagĭ*, this one is the least leafy of those that grow erect. (The ground-hugging *opon* [Patota, *Monolepis nuttalliana*] is a much smaller winter green; it too is small-leaved.) And this is the least well known of the *iivagĭ*. It is harvested in spring and early summer.

Most Gila Pimas know this plant as *chuális,* which they contrast with the larger-leaved chenopod, *huáhi* (Pit-seeded Goosefoot, *Chenopodium berlandieri*). Carmelita Raphael and Ruth Giff both associated the plant with the flats along the river. Ruth added, "Maybe that name is from a Mexican word." George Kyyitan of Bapchule also had reservations about the name. He identified the narrower-leaved chenopod along the ditch bank as *chuáli:* "I *think* some of them know it that way because I think it's a Papago word, or something like Yaqui or something; I don't know what that word is, that *chuáli,* but I've heard that from the late Isabel Kisto. I think she learned this word when she was teaching over at Akĭ Chiñ [Reservation]. I think that's where she learned this word."

Chuális is used by other Pimans to the south and is apparently derived from the Mexican-Spanish *chual,* but the word originated from the Nahuatl *tzohualli.* Because this is a species native to the Southwest, it is odd that the local name would be a loan word. In the 17th-century Névome dictionary, *chual* is given as *cobu [kobu, kovu],* and modern Pima Bajo from the same pueblo give *chual* as the Pima equivalent for Spanish *chuale,* applying this name to a very similar narrow-leaved chenopod (*AMR 812,* apparently *C. pratericola* Rydb.). The Ópata name is supposedly either *zoale* or *cogue (kogei).*

Sylvester Matthias's taxonomic distinction differed from that of most other Pimas: he called the Narrow-leaved Goosefoot *huáhi* and the larger-leaved Pit-seeded Goosefoot *ge'egeḍchu huáhi* 'large *huáhi'.*

Vouchers: *364, 428, 672* (RG); *365* (SM; *672* SM called *huáhi*); *787* (RG, Frank Jim, Tohono O'odham).

Eremalche exilis (A. Gray) E. Greene
[= *Malvastrum exile* A. Gray]
[= *Malva* sp.]
[≠ *Malva borealis* Wallman]
White Mallow, Star Mallow
Malvaceae
ñiádam, ñeadam

In the late winter or early spring, a delicate wild mallow grows in the desert bajadas and washes or sandy riverbeds. It resembles the widespread European Cheeseweed (*Malva parviflora*), but the introduced plant is much more coarse and weedy. The native mallow has more intricate and deeply incised leaves. The flowers, with five whitish or pink petals, are no bigger across than your smallest fingernail.

I have never found this native species on the reservation. Joseph Giff, in explaining to me the meaning of the Pima name for Cheeseweed, *ñiádam chu'igam* '*ñiádam* it looks like', said that *ñiádam* was a plant that grows in the mountains, but he did not know it. Sylvester Matthias knew this plant and helped me search for it. He classified it as an *iivagĭ.* "Regular *ñiádam* is a smaller one from under the mesquite trees. Unlike other greens, it's flat, no taste, [but it's] a form of *iivagĭ.*" Sylvester's evaluation is reflected in the statement by botanists Kearney and Peebles (1960:548) that it was used by the Pima "in times of scarcity." It is harvested in late winter and spring, a time of dietary stress for desert natives.

Fr. Antonine (1935:12) listed *nyiadam* as "a weed similar to four o'clocks."

Hrdlička (1908:264), identifying the plant in the then current nomenclature, said, "*Ñi-a-tam* (Malva borealis [*sic*]) is another plant growing in the Gila valley, the fresh leaves of which serve the Pima as food. The leaves are cooked, mixed with white flour, again cooked, and eaten without further preparation." Russell (1908:76) recorded *nyiâtam* as a *Malva* species, noting, "This plant is boiled and the liquid used in making pinole in times of famine." Curtin (1949:80) mistook Russell's statement, apparently because of the scientific name *Malva,* as a reference to the introduced Cheeseweed rather than to this native species.

"Do you know what that name [*ñiádam*] means?" Ruth Giff asked me.

"No."

She laughed. "When you're awake—*ñiá* 'awake'. '*Napto ñiá?*' somebody asks you, 'Are you awake?' That means it's always *ñiádam,* always awake."

Both historic and contemporary accounts suggest that this was a famine food plant, which would explain why it has fallen from the Pima dietary. But why has the plant itself nearly disappeared? Has it failed to compete with the rank Cheeseweed, which is now found on the floodplains wherever moisture is suitable? Has it been selectively grazed out by European herbivores? The only recent specimens are from Sierra Estrella Regional Park (Eric Sundell 1974), where the plant's rarity was noted. On our many jaunts across the reservation, Sylvester and I have never found the plant.

Monolepis nuttalliana (Schult.) E.
 Greene
 [= *M. chenopodioides* Moq.]
Patota
Chenopodiaceae
opon

It is difficult to imagine such an inconspicuous and insignificant weed as the opon having a lasting dietary importance. During winter rains the tiny plants emerge, hugging the disturbed soils of yards, field edges, ditch banks, and river terraces. If there is sufficient moisture, the spreading plants are more erect and may reach ankle-height. The succulent leaves are linear, usually only 2–3 mm wide, and the stems often tinged with red, as in other chenopods. As spring progresses, microscopic greenish flowers appear at the bases of the leaves, followed by multitudes of minute dark brown seeds in grey coats.

But it is before this flowering and fruiting stage that the Pima gather *opon,* their main winter greens. *Opon* must be picked while young and tender, usually in January and February, perhaps through mid March, depending on rainfall. In early April 1978, Ruth Giff noted, "These are late this year, then they went to seed rapidly." At the beginning of February in 1985, she told me, "Not

quite ready yet, but we're using it anyway." In late October 1985, Ruth showed me the little plants just starting to come up behind her old house in the fields because there had been fall rains. "Have you pressed one of those yet?" she asked me. The Pima women recognize this plant in its very immature stages.

The picking of *opon* is a lot of work because the plants are tiny, unlike all the other *iivagĭ* the Pima still eat. Yet nearly everyone says this is their favorite cooked greens. The plants must be washed thoroughly, because they tend to pick up lots of sand. They are then boiled briefly, drained, and served as is or salted and fried with a bit of oil or lard, the same method as described by Curtin (1949).

Probably most people in the world today still cook without the benefit of timers and other measuring devices. One day I asked Ruth, "With *opon* and those other greens, how do you know when they're cooked?"

"Not long. They're tender. I never timed it. You just cook it till it's done. Probably the same time as Milgáán greens [spinach]. I think it's the roots—takes a little longer."

Hrdlička (1908:264) recorded the preparation: "*Oh-pon* . . . is a low spreading plant which grows in abundance near the Gila all along the Pima reservation. The green tops are boiled by the Indians and when cooled are drained, mixed with lard and occasionally with salt, and eaten with tortillas; sometimes the green tops are chewed raw." Russell (1908:70) made similar observations: "The roots are washed, boiled in an olla, and cooled in a basket. The water is squeezed out, and they are again put into the olla with a little fat or lard and salt. After cooking for a few moments they are ready to serve with tortillas." This is similar to what Sylvester Matthias described: "*Opon* [is used] for *iivagĭ,* or make a cake out of it like a bread. Boil it, boil it. Strain the water out.

opon   *Monolepis nuttalliana*

Then cook on the stove like a tortilla, and you make bread out of it, called *iivagĭ chemait*."

But Russell recorded an additional use that is no longer practiced and seems to have been forgotten: "The seeds are boiled, partially dried, parched, ground on the metate, and eaten as pinole." These small plants are highly prolific seed producers, but the grinding must have been difficult, because each seed is just a millimeter across.

Vouchers: *14, 463* (RG, SP); *295, 952* (RG); *674* (SM); *975* (GK); LSMC 22.

Phacelia spp.
wild-heliotrope
Hydrophyllaceae
havañ taataḍ

Early in the spring, before the heat has dried out the wildflowers of the bajadas and arroyos and other sandy places, you will likely notice some bluish flowers arranged on scorpioid heads. The leaves and stems are often quite bristly with stiff hairs, and the plants grow just tall enough to irritate your lower legs and ankles. The flowers are often showy, ranging from pale blue to deep violet, depending on species.

The Creation Story (Lloyd 1911:62ff.) tells a story of an orphan boy being raised by his grandmother. The boy was mischievous, often taking things literally that were only figurative.

And one day his grandmother sent him to get some of the vegetable called "owl's-feathers," which the O'odham cook by making it into a sort of tortilla, baked on the hot ground where a fire has just been. And he went and found an owl and pulled its feathers out and brought them to the old woman, and she said: "This is not what I want! It is a vegetable that I mean!" And so he went off again and got the vegetable owl's-feathers for her.

After that she sent him for the vegetables named "crow's-feet" and "blackbird's-eyes," saying to him that they were very good cooked together. And the mischievous orphan went and got the feet of some real crows and the eyes

of real blackbirds and brought them to her. And she said: "This is not what I mean! I want the vegetables named after these things!" And the boy went and got what she wanted and she cooked them.

Two of these plants—'blackbird's-eyes' and 'owl's-feathers'—so far have never been identified in contemporary Gila Pima; presumably these names have fallen into disuse (see appendix B). But the third plant, *havañ taataḍ* 'crow's feet', is still rather well known.

Ruth Giff told me, "*Havañ taataḍ* may be picked when tender—boiled just like *iivagĭ*." Sylvester Matthias said, "I've never eaten it. I've heard they used to eat them. They're not very good." Another time Ruth said, "Phillistina [Thomas] said she didn't like *havañ taataḍ* because it's not smooth. When we're picking *hannam* [cholla buds, *Opuntia acanthocarpa*] there would be a lot of them."

"*Havañ taataḍ* has purple flowers," said Carmelita Raphael. "They ate it a long time ago—when they first came out, the leaves. They cook it just like any kind of *iivagĭ*." Ruth Giff considered it an *iivagĭ* of late winter and spring.

Whether *havañ* refers to the Common Crow or to the somewhat larger White-necked Raven is still unresolved. Both are flocking species. Either might have visited the Gileños when native agriculture was practiced and there was a riparian corridor from the east. The much larger and usually more solitary Common Raven has a different name, according to Sylvester (see Rea 1983).

I was not sure how *Phacelia* should be classified in the Pima scheme of things. Sylvester tended to put it in the covert category with the other wild annuals. Then one day I overheard Ruth telling her daughter, Sally Pablo, about *havañ taataḍ*.

"What's that?" Sally said.

"That *iivagĭ* they used to eat," Ruth replied. Undoubtedly it had been a few generations since the plant was regularly used, but still it retained its classificatory position among the edible greens.

One spring as we were cutting through the Sierra Estrella bajada south of Santa Cruz village, Sylvester asked, "Oh, did I ever show you the 'poison ivy' we have? If it gets into your ankles at this time of year—oh boy!—it sure itches." I was puzzled at this one, knowing that neither

Poison-oak nor Poison-ivy grows in the low deserts. He found the plant for me— a *Phacelia*.

Hrdlička (1908:183) made medical observations among the Pima and reported this also: "The *hā-van tā-tat* ('crows'-feet': Phacelia, probably infundibuliformis) is a plant growing on the flats along the Gila, contact with which is followed by inflammation of the skin. The Pima say that when it touches the naked legs or arms it produces sores which, though they do not extend beyond the parts that came in contact with the plant, will last from three weeks to a month before they heal."

I am not sure how many species of *Phacelia* are grouped in this folk taxon. Sometimes Sylvester included the sprawling pale blue *P. distans*, but at other times he excluded it. The deep purple *P. crenulata* is sometimes excluded; it has a very foul odor.

The mystery to me is how anyone could eat a plant that is both fuzzy and ill-smelling. Probably it was less obnoxious when picked quite early in the spring. (Some basal rosettes and young growth I prepared lost their odor during boiling but were rather tasteless.) Or perhaps it was food fit just for poor widows trying to raise their orphan grandsons.

Vouchers: *P. crenulata* Torr.: *11* (RG, SM); *252* (SM); *1908* (GK). *P. distans* Benth.: *24* (SM, RG); *274, 1835* (SM). Excluded specimens: *728* (SM, *P. distans*); *756* (CR, *P. distans*). The diminutive *P. rotundifolia* Torr. grows in the Sierra Estrella; it has never been checked ethnographically.

Portulaca oleracea L.
Portulaca retusa Engelm.
[≠ *Trianthema portulacastrum* L.]
Common Purslane, Verdolaga
Portulacaceae
ku'ukpalk, ku'ukpaḍ

Of the various *iivagĭ*, the most succulent one is the Common Purslane, which grows during the hottest months, a plant sometimes found in some Southwestern supermarkets when in season. This species superficially resembles the inedible Horse-purslane (*Trianthema portulacastrum*), but there are a number of significant differences. Both grow on fully exposed, disturbed soils, particularly in fields and gardens, and both have soft, watery stems that tend to be prostrate, radiating out from the central root; but the edible plant has small bright yellow (rather than pink) flowers appearing at the leaf bases. These flowers close rather early in the morning. The leaves are alternate (that is, only one at each node) rather than opposite (one across from the other at the joint). The thick leaves are almost stemless, definitely elongated (spatula shaped), and flattened. When the plants mature, little seedpods open by means of tiny pointed caplike lids that pop off, revealing numerous black seeds. But native gatherers try to harvest the plants before they reach this stage. (For my early misadventures in seeking this plant, see Horse-purslane account.)

The Gila Pima as well as Tohono O'odham call this plant *ku'ukpalk* or *ku'ukpaḍ* or occasionally even *ku'ukpal*. The name appears to be derived from *kupal* 'upside-down', in the plural form *ku'ukpal*, as someone explained to me, but the reason for calling it this seems lost in ethnolinguistic history. (Another meaning for the word, 'face down' or 'prone', describes this plant's growth form well.)

Use of the plant is certainly not lost; *ku'ukpalk* remains one of the better known *iivagĭ*. George Kyyitan stood on a canal bank in Bapchule with me one spring, sorting through the abundant greens. He mentioned one that was not up yet. "*Ku'ukpalk* [is] something like a weed. They got round leaves. I think there [is] a lot of that—what we don't use now, that we're supposed to. It's good when you mix [it] with cheese and onions and maybe chiles."

Ruth Giff gave me similar information: "*Ku'ukpalk* is the Verdolagas. When Margie [Sabori] fixes it, she puts tomatoes, cheese, and onions [in it]. Drain that sticky juice, then fry your onions, then throw your *iivagĭ* in it." The plant is somewhat mucilaginous.

At an earlier time, more of the plant was used. Russell (1908:75) caught just a glimpse of this. Under *ku'ukpaltk*, he noted, "According to tradition the seeds were eaten in primitive times, but no one now knows how they were prepared. The plant is now boiled with meat as greens." This fits a fairly well defined pattern that is evident with a number of other species. Apparently with the perfection of wheat agriculture (which was certainly well

developed by the mid 19th century), alternative sources of wild seed to supplement cultivated maize dropped from the Pima diet as nonessentials.

The name is probably pan-Piman. At the Sonoran oasis village of Quitovac, Luciano Noriega told Gary Nabhan and me that *ku'ukpalk* is eaten more frequently than *kashwañ*, Horse-purslane. It is cooked with chiles and lard. At Onavas in Pima Bajo country, don Pedro Estrella gave the name *ku'umpurim* to both *P. oleracea* and *P. umbraticola* H.B.K., without any distinguishing markers. The 17th century Névome vocabulary compiled at this pueblo gives *kukpuriga, kuppurikka,* and *kapurhiga* as three variations for Verdolaga, while Verdolagas Grandes is *gegerpurha* (Pennington 1979:119). An Upper Piman vocabulary compiled between 1767 and 1774 at Attí in northern Sonora gives *kukpurik* and *guouorch kukpurik*, a larger form. In either case, what the large purslane is biologically remains unknown. Gila Pima do not distinguish a second folk species in their genus *ku'ukpalk*.

Vouchers: *147* (SP, Helen Allison); *372, 1719* (SM); *1716* (GK).

Rumex crispus L.
Curly Dock, Yellow Dock
Rumex violascens Rech.f.
Mexican Dock
Polygonaceae

vakoandam

The very green and extravagantly leafy sorrels or dock grow at the edges of rivers, small ditches, and charcos, and in frequently flooded marshy areas such as those that form from irrigation tailwater. Most are root perennials, growing in spring and summer to between knee- and hip-height. Spiky clusters of winged greenish seeds top these plants like tassels. When the marsh dries out or when the plants die back in cold fall weather, they stand in great rusty brown masses, with their rough seeds brightening up the flat winter desertscapes.

The Pima know this plant as *vakoandam*. "It's not a regular *iivagĭ*, but some do eat it," said Ruth Giff. "It's not *siv*

['bitter'], more *s-he'ek* [sour]." *Vakoandam* or *vapkoandam* (the plural form) means 'washing'. The verb *vakoan* means 'to wash', the suffix *-dam* indicating an agent. The plant must once have been used for washing something, although no one today could verify this.

Although dock and Canaigre are put in different folk families, Pimas usually compared the two when discussing identification. Joe Giff told me, "The species with small root is *vakoandam*. Has roots like carrots and grows in the water, that's why that name. They're yellow. Doesn't grow so tall [as Canaigre, *Rumex hymenosepalus*]. The *shashañĭ* [Red-winged Blackbirds] make nests in these when they bloom. Have rust-colored flowers—sort of rough."

During my work on the reservation, an Old World species, *R. dentatus* L., has completely replaced the earlier naturalized *R. crispus* and the native *R. violascens*. One time I drove Sylvester Matthias down to Barehand Lane marsh, where the aggressive new species had taken over the wet lower meadows. I showed him the plant and fully expected him to tell me it was *vakoandam*. He studied the plant intently. Finally he said, "This is something that looks like *vakoandam*, but it's not . . . see . . . too much space in here [on the stems], not leafy enough. It's related [but] they have small leaves. *Vakoandam* have wide leaves. The stems and tassels are the same as *vakoandam*, [but] this one is solid stems. Usually they have hollow [stalks] like bamboo [cane]. So it's in the family of *vakoandam*." He was not fooled, but he never once mentioned the very differently shaped wings on the fruit, the one obvious character to the botanist!

Ruth Giff considered *vakoandam* a summer crop, along with Verdolaga and Palmer's Carelessweed *(Amaranthus palmeri)*. She said, "Some people *do* cook it, but I don't like it. The old-time People never used it—just later. It's not an old food."

I have never found *vakoandam* being used by anyone during the past quarter-century, but apparently it was a regular food in the early 1940s when Curtin (1949:51) visited these same settlements: "Dock leaves are used for greens throughout the [two Pima] reservations. Women at Salt River told me that 'the leaves already have vinegar in them and we don't need to add any.' At Co-op it is also mixed with other greens."

vakoandam *Rumex crispus*

Curtin recorded one additional use not mentioned in the 1980s: "On the Salt River Reservation, to obtain a yellow dye the roots are pounded and then boiled. Mason McAffee [a 90-year-old Pima at Co-op Colony] said the yellow edges of ancient Pima cotton blankets were colored by this method." Russell (1908:150) said these yellow edges were made from "a dark buff ocher bartered from the Papagos."

On a visit to my home in San Diego, Sylvester pointed out that *vakoandam* is smaller than my cultivated Garden Sorrel, *Rumex acetosa* L., with pinkish petioles. "My sister said she ate it, and I know the Yaquis eat it as spinach, cooked. Name refers to 'wash-er', but Pima don't use it for washing." I put some sorrel in the salad that night. "Now I can really say that *vakoandam* is an *iivagĭ*, because I've eaten it," he commented.

Technical Notes: The invasive Asian species *R. dentatus* was already abundant below the Salt-Gila confluence by the early 1970s (Sundell 1974). In 1981 I first found it growing with *R. crispus* and native *R. violascens* at Barehand Lane marsh. By 1984 it was widespread in other parts of the reservation as well. Its tolerance of alternating dry conditions has apparently facilitated its replacement of earlier species in the few stressed marsh habitats remaining in Pima country as well as throughout much of the Sonoran Desert. In recent years I have found no native *Rumex* surviving on the reservation, except *Rumex hymenosepalus*, and that is now rare and localized (see Group G account).

Vouchers: *150* (RG, SM); *1046* (RG); *1746* (GK); LSMC *23*.

Sonchus asper (L.) Hill
Spiny Sowthistle
Asteraceae (Compositae)
s-ho'idkam iivagĭ, si'imel iivagĭ

In the spring, one of the coarse weeds you are likely to encounter in the alluvial floodplains where Pimas farm is Spiny Sowthistle. It prefers open, sunny places such as ditch banks, field edges, and fence rows. The erect plants may be knee- to hip-high. Like Annual Sowthistle *(S. oleraceus)*, the thick, sharply angled stalks of Spiny Sowthistle are hollow, about an inch thick, and often marked green and purple. Both have cone-shaped flower heads while maturing. But this one is definitely a rougher plant. The spiny leaves have narrow pointed tips, many lobes, and bases that always circle the main stem.

Its Gila Pima name, *s-ho'idkam iivagĭ*, means 'many spines it has greens' or 'spiny greens'. George Kyyitan identified some we found along a canal bank, saying, "I eat it raw. Just peel the stem when it's young. If it can peel easily—just eat it. Don't soak it in water. Just eat it same way as with that *huai hehevo* [Annual Sowthistle]. Eat the stems, not the leaves on *s-ho'idkam iivagĭ*."

Ruth Giff, however, told me, "We peel it and have to wash it—I guess the milk, to get the milk out of it. Grows along the ditches, but I don't see it now. Eat it raw." She added, "This is what some white people call Indian celery."

Sylvester Matthias said it was "like celery, but not as tough." He glossed the name as 'rough edges greens'. He confirmed that both *Sonchus* species can be eaten raw, but only *huai hehevo* was cooked. Curtin (1949:106) was told that both species could be eaten either raw or cooked.

Russell (1908:77) did not identify the plant botanically but recorded under *s-ho'idkam iivagĭ*, "The leaf of this thorny plant is eaten raw or boiled." Later, he mentioned that this "thorny weed, *sâitûkam iavik*, [is] said to blacken the teeth" (Russell 1908:160).

Some people from the eastern part of the reservation called this plant *si'imel iivagĭ* 'milk-running [lactating] greens', in reference to the milky sap of the broken stems. Mathiot (n.d.:168) noted that the verb *si'i* means 'to suck the breast'; *mel* means 'to run'. This is definitely a synonym, as people who used this folk taxon did not know its alternate and more widespread name, *s-ho'idkam iivagĭ*. This is another case of an introduced plant receiving several local names (see appendix D).

The densely spined Milk-thistle *(Silybum marianum)*, with large purple flowers and mottled leaves, is found occasionally in Pima fields and on ditch banks. It is also nonnative; I have not found a Pima name for this plant, although I have inquired of many farmers. But it is not included in the folk taxon *s-ho'idkam iivagĭ*.

Vouchers: *744* (SM); *991* (GK); *1325* (?).

Sonchus oleraceus L.
Annual Sowthistle
Asteraceae (Compositae)
huai hehevo

A late-winter to spring weed of disturbed areas such as yards, gardens, field edges, and ditch banks, the delicate Annual Sowthistle is rather common around Pima dwellings. These erect plants grow to about knee-height or so. Its delicate leaves are deeply incised and only softly spiny on their margins. Once the lettuce-like yellow flower heads have closed, they form green cones while the seeds mature. Then the little white seed tufts blow away in the wind.

Annual Sowthistle and Spiny Sowthistle *(Sonchus asper)* may be found growing together and may reach similar robustness, though Spiny Sowthistle (as its name implies) is the coarser plant, definitely spiny rather than lettucelike to the touch. On Annual Sowthistle, each side of the leaf base clasping the upper stem projects downward to a point like a shirt collar, but on Spiny Sowthistle the base is earlike, recoiling completely around the base of the leaf.

The Gila Pima used both species and had no trouble distinguishing these two introduced Old World plants. Annual Sowthistle is *huai hehevo* 'Mule Deer's eyelashes'. One day I saw some big, tender plants in Ruth Giff's kitchen. She had picked them in her yard and was about to add them to a salad. "Just peel it

s-ho'idkam iivagĭ, si'imel iivagĭ   *Sonchus asper*

[stems] and eat it green. Or use as greens, like lettuce. If there's not enough *opon* [Patota, *Monolepis nuttalliana*], can add *huai hehevo* and cook it."

After discussing *s-ho'idkam iivagĭ* (Spiny Sowthistle), Sylvester Matthias added: "There is another kind, *huai hehevo*—can eat like lettuce. It's tender—can eat *any* part."

Occasionally a well-informed Pima makes an error in an identification, but oftentimes they are surprisingly astute, not lumping even subtly different biological species into a known folk taxon. This was well illustrated one late February day when Sylvester and I drove across the east bajada of the Sierra Estrella. I collected a specimen of *S. tenerrimus* L. *(AMR 272)* and handed it to him. He looked the plant over carefully, saying

that it had no Pima name, then added, "Maybe it's a *do'ag* [mountain] form of *huai hehevo*."

Vouchers: *296* (RG); LSMC *51*.

Group G, Covert Category
wild annuals

The plants put into this category are categorized primarily by negatives: they are nongrassy, nonwoody, nonaquatic wild plants that are not planted and are not eaten for greens. For the most part, these are annuals or root perennials. Although Desert Broom-rape *(Orobanche cooperi)* was eaten, it has no chlorophyll, and no one ever suggested that it might be an *iivagĭ* 'eaten greens'. The two bulbs from Pima country (Bluedicks *[Dichelostemma pulchellum]* and Desert Lily *[Hesperocallis undulata]*) are included here.

The native Arizona Grape *(Vitis arizonica)* seems anomalous in this group, but perhaps Sylvester did not know what to do with it. When I asked him again years later, he looked puzzled for a long time, then said, "It doesn't quite fit." There is no Pima category for vines. Yerba Mansa *(Anemopsis californica)* is the most aquatic species included here.

Abronia angustifolia E. Greene
Sand-verbena
Nyctaginaceae
Centaurium calycosum (Buckley)
    Fernald
Canchalagua, Centaury
Gentianaceae
huhulga, huhulga ee'eḍ

Early Hispanic chroniclers referred to a cienega or marshy areas in the vicinity of what became known in the Anglo period as Maricopa Wells. There is nothing left there today to suggest a cienega, only a few straggling bushes of Seepweed *(Suaeda moquinii)* and some dead mesquites. Antisell (1856:137) remarked that the wells rested on yellow clay, the water table there reaching within 2½ feet of the surface, providing a "marked difference of vegetation." In addition to grass, rushes, and mesquites, the geologist remarked on one other plant there: Canchalagua, a little pink gentian. *Centaurium calycosum* is a tiny, fragile annual plant, just ankle- to knee-high, with simple opposite leaves and bright pink flowers spreading trumpetlike from a slender tube in five lobes, like a perfect star, emerging from a slender green calyx.

Richard S. Felger, an authority on Seri ethnobotany, asked if I had any plants in the Pima ethnobotany described or named after female body parts. I had not noted any, but I also had not asked specifically. I posed the question to Sylvester Matthias, and he thought for a long time, then finally said that he knew of none. "But did we ever get *huhulga ee'eḍ?*"

"I never heard of it. What does that mean?"

"*Huhulga ee'eḍ* 'menstruating blood' is a vine; grows on the Salt River, like *kashviñ* [Horse-purslane, *Trianthema portulacastrum*]; not exactly a vine, like *kashviñ*—it spreads—red flowers—more like maroon—grows where rocky and sandy. No use—just an ornament." He said the medium-sized flowers were single.

It seemed that he might be describing Sea-purslane *(Sesuvium verrucosum),* a rather rare sprawling plant of saline soils that had been collected just once along the middle Gila (and that still grows at Picacho Reservoir). But when I took him some specimens, he said that was not it.

"It's more of a vine [but] larger leaves."

I went to Ruth Giff next. She knew the plant only as *huhulga,* and she said it had "bright red flowers, very small; about six to eight inches tall; we used to see it along the ditch. [The name means] 'menstruation'. When the river and ditches were running. Blossoms like little stars with real pointed petals." She described the flowers as having no tube in back; they were more or less bunched or placed with just one at the top. On our next visit a bit later she gave some additional information on *huhulga:* "In summer— we would see it when we went swimming. It's red—not purple."

It seems quite clear that her concept applies to Canchalagua. It too had been collected on the reservation earlier in the century. (There is even a Sacaton specimen from 1926 at the U.S. National Herbarium.) But, like Sea-purslane, it is no longer found on the reservation lowlands. As another possible species, I tried the little Sand-spurry (*Spergularia marina* (L.) Griseb.), but this was not *huhulga,* nor was it known to have a name.

In the summer of 1990, Sylvester added more clues for the identification of this plant. "*Huhulga ee'eḍ* [is] like a *kashviñ*— but larger. Little bunches [of flowers] on each stem. Green leaves. Has to be damp."

I began to suspect that Sylvester's plant must be Sand-verbena *(Abronia angustifolia).* I showed him a single plant on the lower Salt, but he said it was too tall. He conceded, "This is in the family of *huhulga ee'eḍ.*"

But there was still another possibility. Ruth's *huhulga* and Sylvester's *huhulga ee'eḍ* might be two different folk taxa with different botanical referents, the first being Canchalagua, the other Sand-verbena.

In 1991 I took George Kyyitan some Sand-verbenas, along with several other plants. He crushed and smelled several composites, then came to the masses of tiny pink flowers.

"Where did you get this one?" George asked.

"In the riverbed, growing with the *vippi si'idam* [evening-primroses, *Oenothera* spp.] on the sandy flat."

"I know it. It's going to sound bad when you hear it [but] I'm going to say it: *huhulga ee'eḍ.* Do you know what that means?"

"Yeah. Why do they call it that?"

"From this color, I guess." he said, fingering the bunch of flowers.

"Does it have any use? Medicine?"

"I don't think so. Just to decorate the land. Spreads out a lot. Used to be a lot around that riverbed, where that *chuuvĭ taḍpo* [Owl-clover, *Castilleja exserta*] grows. Goes up in the middle, then spread[s]."

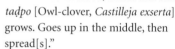
huhulga, huhulga ee'eḍ   *Abronia angustifolia*

huhulga, huhulga ee'eḑ *Centaurium calycosum*

The next day I took the plants to Francis Vavages in Sacaton. He looked briefly at them but did not identify any.

"Have you ever heard of a plant called *huhulga ee'eḑ?*" I asked.

"Oh, yeah. I think it's that one."

"Which one?"

"That one," he said, indicating the Sand-verbena.

I asked Sylvester if there was any use for *huhulga ee'eḑ.*

"They use it for sore[s]—make tea and wash it off. Boil whole thing."

"Where does it grow?"

"On the sandy place, where it's damp. Remember, we found it at the [Salt-Gila] junction."

There remained one more possibility to be checked. Sylvester, George, and Francis all mentioned the prostrate growth habit of *huhulga ee'eḑ.* A critical possibility in the contrast set was Trailing Four-o'clock (*Allionia incarnata* L.), a plant in the same family that grows in sandy places; it is even more supine and has flowers of the same color. I found some east of Gila

Bend and took them to Pima country, but everyone dismissed them outright, as they did a pink-flowered species of *Mirabilis.*

Some while later we took George to the Gila riverbed near Sacaton, where there was a profusion of blooming plants from a spring rain. Among the plants he found and identified there were Sand-verbenas, *huhulga ee'eḑ.*

I showed Ruth the same bag of Sand-verbena I got from the Sacaton riverbed that I showed to Sylvester, George, and Francis.

"No. *Huhulga* is not like this. Red, red—maybe fuschia. Just like little stars. On the ditch banks. Just little tiny plants. . . . This high [10–12 inches]. Usually a lot of blossoms toward the end of the plant, . . . just *red.* You could notice it from a distance."

In 1992 while working at tinajas in the southern Sierra Estrellas, I found Canchalagua growing at the edges of several of these rock tanks together with Yellow Monkeyflower *(Mimulus guttatus)* and Rabbitfoot Grass *(Polypogon monspeliensis).* I took some down to Ruth, and she confirmed the identification as *huhulga.*

Vouchers: *1656* (SM, "in the family of"); *1707* (GK, SM, FV); *1740* (GK); *1803* (RG, *C. calycosum*); *1863* (RG, *C. venustum* (A. Gray) Rob.).

Acourtia nana (A. Gray) Reveal & King

[= *Perezia nana* A. Gray]

Desert-holly

Asteraceae (Compositae)

ban auppaga, baaban auppaga

Few people ever encounter Desert-holly. (This plant is not to be confused with the little saltbush *Atriplex hymenelytra* (Torr.) S. Watson of lower elevations, nor the barberry [*Berberis* spp.] of higher elevations, all of which sometimes share the same English name.) The Desert-holly of Pima country grows in dense colonies under and near mesquite trees in soils that have not been disturbed in a long time. In the spring the plants emerge from a perennial rootstock. Some plants may be less than a finger tall while others are more than a

hand tall. The prickly leaves are broadly ovate, up to a finger-joint long, pale green, and prominently veined. Each plant may bear a slender, lavender, thistlelike flower head. If the papery leaves do not identify the plant, the powerful violet scent of the flowers surely will. By the time the floodplains warm in mid May, the open flowers turn into a powderpuff of winged seeds that eventually are carried away by the wind, dandelion-style. The skeletons of old leaves, clinging on dried stalks, may last through the next winter.

The only place I have discovered this plant in Pima country is between Holy Family Church and the cemetery at Blackwater. There a rather extensive colony grows under several mesquite trees. The area seems not to have been cultivated even at the time of Southworth's (1914) detailed mapping of fields. Perhaps not surprisingly, several other species also seem relictual there, including *Setaria leucopila* (Scrib. & Merr.) K.Schum. and a female specimen of Virgin's Bower (*Clematis drummondii* Torr. & A. Gray). M. French Gilman took a Desert-holly specimen *(MFG 123)* at Sacaton in 1908. Land use and loss of old mesquite groves seem to have made this a very rare plant on the reservation.

When I showed this plant to Irene Hendricks of Casa Blanca, she immediately called it *ban mauppa,* or so I thought; I dismissed this as an error, because *ban mauppa* is a common small Fishhook Cactus *(Mammillaria microcarpa)* of this region. Some years later I took some specimens of Desert-holly to George Kyyitan. He at once began to talk about *baaban auppaga* 'Coyotes' cottonwood'.

"How does it grow?" I asked.

"Just straight up. Little round leaves, smaller than an *auppa.* Light green leaves, about this tall [6–8 inches], under the *kui* [mesquite]."

"Does it grow just in the *s-kuik* ['mesquite bosque']?"

"Anywhere—under a mesquite—where there's lots of leaves and fertilizer—out in the desert." He had never seen the flowers or seeds on it. He emphasized again the pale green leaves but did not recall whether they had spiny margins.

The next morning I went to visit Francis Vavages, who was at his workbench burning the figure of a flute player on a flute he had just finished. I asked if

ban auppaga, baaban auppaga   *Acourtia nana*

he had ever heard of *baaban auppaga*. He looked puzzled for a while, then said, "*Ban auppaga,* yeah. 'Coyote's cottonwood'—only about this high," indicating about six inches. "All in bunch," he said, and he waved an arm wide to indicate how the plant spread out. "Under a tree, especially after a big heavy rain. The only place they'll grow is under a tree. It has leaves like a cottonwood."

"Have you seen any flowers or fruits on it?"

"I didn't see any flowers. Maybe they had some and I didn't notice."

"Is there any use for it?"

"I don't know. That's all I know—*ban auppaga.*"

Whenever possible, I like to have a folk species identified in the field, because looking at a plant out of context can be misleading. One morning I took Francis out to see the Desert-holly colony, without mentioning the plant's name. At first he called it *ban vivga* 'Coyote's tobacco', then settled on *ban auppaga* as the correct name. It had been a long time since he had seen it.

The next morning I took George to Blackwater, not telling him what plant we were looking for. We had scarcely reached the edge of the colony, where the outermost plants, growing in full sun, were only as tall as your little finger, when George exclaimed, "There it is!" He asked, "Where's Coyote? He's going to see us looking at his plants." He went in under the mesquites, where half a dozen Mourning Doves had their flimsy nests. "See, over there," George pointed to a spot under some dry branches where the plants were matted down. "There's where Coyote has been sleeping with someone." Then George pointed to a nearby *koson kii*, a packrat nest, where a couple of crumpled cans were stashed. "And see, there's where Coyote threw his beer cans."

Culver Cassa, who was with us, wandered over to the dance floor at the side of the church and brought back a cottonwood branch. I held up the two plants for comparison. "Ah," said George, pointing to the leaf shape, then the margins. "This is our *auppa* and this is Coyote's *auppa.*"

Some years later we were again in the field, and I stumbled into a second colony of these plants on a patch of land that had not been disturbed in a century or more, according to our maps. I called George over and pointed to them. "Ah! *Baaban a'uppaga!* Can't make no house with it, . . . no beams with it. It's good for nothing, just like the owner."

It is not difficult to see how this little plant fits the paradigm of Coyote's plants. In the spring, when the down drifts off the cottonwoods, whole "nurseries" of seedlings may emerge on the wet sandy flats along the river. The Pima call these *a'al a'uppa.* They grow rapidly; if not dried out or grazed off, a few will reach knee-height by late summer. Coyote's cottonwoods are also in dense colonies. But his plants are in the wrong place—out under a mesquite on the flats, away from the river. And they never get more than about a hand tall. The broad leaves have margins covered with spines, rather than just smooth notches. Late in the spring each plant bears a tuft of white pappus, like the cottonwood's downy seeds, that blows away in the wind. Unlike

the great cottonwood, Coyote's little *a'uppa* are deformed and never mature correctly.

Russell (1908:80) noted, "A composite Perezia [= *Acourtia*] wrightii is used as a styptic." This is a related species, with similar thin, papery leaves. But the leaves are longer and the flower heads numerous on these much taller (about knee- to hip-height), more bushy plants that still grow commonly in the desert mountains of Pima country. Russell gave no Pima name for this entry; perhaps the information was from Edward Palmer's manuscript, as is Russell's next entry. Irene thought the roots were used to wash baskets a very long time ago, perhaps in her grandmother's day. She knew of no medicinal use.

Ruth Giff and Sylvester Matthias at the west end had not heard of *ban auppaga* and did not recognize the plant. Nor have I ever found the plant there.

Vouchers: *1454* (IH); *1710, 1733* (GK); *1732* (FV).

Amoreuxia palmatifida DC.
  [= *A. schiedeana* Sprague]
Saya, Amoreuxia
Cochlosperaceae
shaaḍ

One of the most exotic-looking little plants growing in parts of the Sonoran Desert is Saya. The small leaves are palmate, but it is the flowers and seedpods that are unique among desert plants. The large flowers are somewhat poppylike. The five yellow-orange petals are lopsided, each with a purple basal spot. Even the stamens are asymmetrical. Later the kidney-shaped seeds are borne in a large pod, enclosed in "windows." The root is like a slender parsnip. New growth sprouts from these roots during the summer rains. Saya reaches its northern limits somewhere in Tohono O'odham country to the south of the Pima. But the Pima know it, too.

In the 1980s, Sylvester Matthias still recalled details of this plant as a trade item: "*Shaaḍ* doesn't grow here. Grows something like a sweet-potato. [Comes from] Chuichu, in Koahadk country. [The Koahadk are] a tribe [*sic*] of Papago—and [from] White Horse Pass. [Root] about two inches long and two or one-and-a-half wide; smells something like chewing gum—mint, but not as strong. Eat it raw; about size of *ihug* plant [devil's claw, *Proboscidea* spp.]." A few years later, Sylvester told me again of this plant: "Someone from Chuichu [a northern Tohono O'odham village] used to bring *shaaḍ* to my father—it seemed [tasted] rather flat to me, but others just ate it."

"What parts did they bring?"

"Just the root."

"Very much?"

"Oh, about half a burlap sack."

This was part of an old trade between these peoples. Among the items the Tohono O'odham took up to the Gila to barter at the time of the June wheat harvest, Russell (1908:93) mentioned "*rsat,* an unidentified plant that grows at Santa Rosa." (Elsewhere [1908:76] he confused *shaaḍ* with *haad;* see *Dichelostemma pulchellum* account.)

In the 19th century, Palmer recorded *shaaḍ* among Pimans: "*Amoreuxia schiedeana; Himajins* of the Papajos *[sic]; Saya* of the Pimos *[sic].* It furnishes to the

---

shaaḍ *Amoreuxia palmatifida*

Painted from plants grown from seed Wendy Hodgson collected in southern Arizona.

Indians of Arizona just named, an edible root. They eat it roasted to baked in hot ashes. It is quite palatable, with a slightly bitter tang" (Palmer 1878a:601). (I have no idea where Palmer got these names; both are doubtfully Piman.)

On the lower Rio Yaqui in Sonora, Sylvester and I went out looking for Saya with don Pedro Estrella, a Pima Bajo. There the plant is called *sha'adch* (Pennington 1980:232). With help from Noé Yuqui, we found a single plant (*A. gonzalezii* Sprague & Riley), dug the root, and collected the seeds for Wendy Hodgson of the Desert Botanical Garden in Phoenix. She is making a study of the two species. Noé apologized for his inability to find additional specimens, saying that the cattle had eaten all the rest.

Amsinckia intermedia Fisch. & C. A. Mey.
Amsinckia tessellata A. Gray
[≠ *A. spectabilis* Fisch. & C. A. Mey.]
[≠ *Lithospermum* sp.]
fiddleneck
Boraginaceae
cheḍkoadag

Fiddlenecks spring up on the floodplains, on the bajadas, and even in the mountains after the winter rains. If the moisture is light or the location dry, the plants may be minute, a single stalk just a finger or two tall, arising from a basal rosette of leaves. But where the moisture is more dependable, such as along the river, the plants may grow thickly and reach to knee-height or more. Every part of the plant is bristly—the leaves, stems, and flowering heads. These heads, rolled into characteristic scorpioid shapes, bear small tubular yellow-orange flowers. Two species grow on the reservation and in the low deserts in general: *Amsinckia tessellata* and *A. intermedia.* But it takes at least a hand lens and seeds to tell them apart.

Is it possible for a folk taxon to drop from the living lexicon of folk taxa, just to fall through the cracks of folk knowledge? This seems to be the case with fiddlenecks. Though these plants are very common and widespread spring annuals, try as I might,

I never found anyone who knew a Pima name for them or who even suggested that they did have a name that is now forgotten.

Russell (1908:77) recorded a plant called *Tci-itkwatak* (which we would now write *cheḍkoadag,* according to Culver Cassa) as a species of *Lithospermum.* This genus, however, occurs at considerably higher elevations. Surely the similar *Amsinckia,* also yellow flowered, was intended. And the plant was used: "The leaves are eaten without preparation."

Hrdlička (1908:264) similarly said, "*Djeh-t-ka-tak* (Amsinckia tessellata) is a plant which grows near the Gila. The young leaves are eaten raw without preparation." Essentially this is the same folk taxon Russell had recorded, allowing some variation in the transcription due to Hrdlička's Hungarian ear. Fr. Antonine (1935) did not know what to call the plant he listed as *chitkwadak,* but he described it as "when dry produces an itch." Nabhan (in Felger and associates 1992) recorded the southern Tohono O'odham name as *chetkom.*

If you are wondering why *Amsinckia* should fall here in the Pima higher classification of plants while the similar *Phacelia* is included in the category *iivagĭ,* the answer is simple. The eaten greens category is utilitarian: the People still retain a tradition that *havañ taataḍ* was once eaten, but no one remembers *cheḍkoadag* or its former use. When the plant is pointed out, they say it belongs in this category.

The mystery that remains, at least in my mind, is how people could eat the rather harsh leaves of *cheḍkoadag* and the similar *havañ taataḍ.* These two plants may have sprouted at a time when other greens were scarce and resources low. Or the very young leaves may not be as repulsively defended as they are later in the season.

Anemopsis californica (Nutt.) Hook. & Arn.
Yerba Mansa
Saururaceae
vavish

In cienegas and other swampy, usually saline places throughout the Southwest may be found a plant with broad leathery, almost fleshy leaves coming from a thick rootstalk mostly buried in the mud. Great coral-colored stolons may shoot out from these rosettes, seeking new places to establish their clones. What appears to be a flower is really an inflorescence of many tiny flowers bunched on a stalk, like those of *Plantago* species, with a ring of showy white to pinkish bracts appearing to be the petals. These are the famous Yerba Mansa, sought after by Indian and non-Indian alike.

Often Yerba Mansa grows densely packed with other saturated soil–loving plants such as certain grasses, rushes (*Juncus* spp.), bulrushes (*Scirpus* spp.), spike-rushes (*Eleocharis* spp.), and flat-sedges (*Cyperus* spp.). But with overgrazing and falling water tables, this valuable medicinal plant is often the victim. *Vavish* disappeared early on the Gila. Of all the botanical collectors on the reservation during the first half of the century, none found the plant except Curtin (1949:78), who noted, "The plant thrives in wet, alkaline soil, and my informants claim that it is rapidly disappearing from the reservations because the water-level has been lowered by pumping for irrigation by whites. On my long trips in every direction, the strictest vigil failed to reveal a single specimen. Only one locality where it still grows is known—St. John's Mission [Komatke] at the junction of the Gila and Santa Cruz rivers."

It was from this area, also, that Joe Giff described the dense growth. "Up in New York Thicket were many open meadows, covered with *mu'umk vashai* [Desert Salt Grass, *Distichlis spicata*] and *vakoandam* [dock, *Rumex* spp.] and so much *vavish* that the whole area there, whole patches were completely white with *vavish,* so you could smell them from a distance. You could hear the voice of the *o'okokoi* [White-winged Doves] early in the morning—all calling—like one voice. They nest in there, in the *kui* [mesquite] of New York Thicket, but go up in the *haashañ* [Saguaro] at the foot of the Estrellas to eat." Carmelita Raphael, who lives not far from there, echoed this: "There used to be a lot of *vavish* up at New York Thicket, but they're all dry now."

In 1979 at least one plant had been taken back to the reservation. Joe Giff told me, "*Vavish* has reddish, succulent runners, white flowers the size of violets. Ruth grows it. Margie Sabori brought new plants back from Bishop [California]."

The Gila Pima had many uses for this astringent plant. Sylvester Matthias told me, "*Vavish* grows in something like a swamp; like a little vine that grows where it's wet. In New York Thicket. Use for cold, sore throat; will take fever out just like aspirin. Use mostly root [the underground stem], fresh or boiled when dry. Put some into mouth, strong fumes [*sic,* flavor]!

"They can even bathe the baby in it, when it's born, if its mother is infected by syphilis—you can bathe the baby with the *vavish.* That's why one of my relatives' name [is] Vavish; that's what happened to him. He was bathed in *vavish*—*vavish* tea. That's how come they call him Vavish."

"His mother had . . . ?"

---

vavish   *Anemopsis californica*

"Yeah."

"So the Pima knew that it can go from mother to child?"

"Yeah. Yeah, I just think about it. I never thought that was used for syphilis. I know *kuupag* [Mormon Tea, *Ephedra* spp.] was, and also *shegoi* [Creosote Bush, *Larrea divaricata*], and [in this case] *vavish.*"

"How often did they have to use it, do you remember?"

"I think it's just occasionally until [the baby] gets to be maybe two months old. Then there's no danger of catching syphilis from the mother."

"Have you ever heard of *vavish* being used for menstrual pains, Sylvester?"

"Ah . . . yeah. . . . I guess they do. . . ."

Another time Sylvester related, "My father used to get *vavish,* sell it to those Mexicans. It sure goes fast."

"What part did he sell?"

"The root. I guess everybody use[d] it: the Papagos, and the Apaches. I used to take it to school—to day school—and those Apaches would ask for it, [would] be chewing on it."

"What for?"

"For medicine, colds. Things like that." Sylvester said that the dried *vavish* roots were one of the few medicines the Pima kept on hand in the home for year-round use.

In the 1940s the plant was one of the most important medicines used on both Pima reservations. Curtin records its use for stomachache, cough, colds, and to prevent a cough, and the infusion used both internally and externally for syphilis. Additionally, "George Webb asserted that wounds are treated by washing with a decoction of the plant, then a sprinkling of powdered roots, followed by the application of green leaves and a bandage" (Curtin 1949:79).

Vavish, which once must have grown in all the marshes and oxbows along the Gila, has long been an important plant for the Pima. Russell (1908:80) said, "The roots are crushed and boiled. The extract is used as a tea for consumptives, according to one informant, and as an emetic according to another." Hrdlička (1908:245) recorded, "The root known as *wávish* (yerba mansa) is reputed to be very effectual in syphilis. The Pima boil it and give the tea to the patient; they also powder the root and apply it externally to the sores. It is said that if a mother affected with syphilis while pregnant is given this tea the child will be free of the disease. Apparently the people are acquainted with the hereditary nature of the condition." For the Tohono O'odham, he (1908:242) recorded, "The root of the *wa-wish* is used in colds. They break it up, boil it, and drink some of the decoction hot. It makes them sweat. After drinking the tea they cover up and remain indoors. A strong decoction of the root is also used as a wash in case of scorpion bites [*sic,* stings]. Women drink some of the hot tea in painful menstruation." (This agrees with what a Pima woman told Curtin [1949:79], that the use of the hot tea to induce sweating was more frequent among the Tohono O'odham than the Pima.) Frank Jim, a Tohono O'odham, told me that "*wawish* is more a woman's medicine, for their problems; men don't know about it."

This folk taxon is probably pan-Piman. At the Pima Bajo village of Onavas, I found some people growing Yerba-del-manso in their yards for medicine, but no one knew a Pima name for it. The Névome vocabulary compiled at this village in the 1660s gives it as *vabis,* using a Spanish orthography with interchangeable *b* and *v.* Near Yepachi, Mountain Pima Cruz Castellano showed me the dried plants he keeps for medicine. One root he called *vavisa (vavaisa?)* has not yet been botanically identified, but its growth form, use, and even smell are similar to the lowland plant. Laferriére (personal communication) collected a *Cosmos pringlei* here that was called *bavish.* Even though *Anemopsis* does not reach these elevations, the name does.

Vouchers: *1187* (SM); *LSMC 7* (not extant).

Boerhavia erecta L.
Boerhavia intermedia M.E.Jones
spiderling
Nyctaginaceae
makkom jeej, makkum jeej

During the late summer rains and into the early fall, a leafy green plant grows abundantly in alluvial soils, especially at the edges of roads and fields. The lower part of the plant may be spreading, with entire (solid) deeply veined leaves, often with a purple tinge. The flowering stems are tall, branched, and very slender, giving the plant its English name, spiderling. These spidery flower stems often have several sticky bands along the stalks below the very small pink flowers. In Santa Cruz village I once found a Black Phoebe (a flycatcher) that had entangled itself in a spiderling and, being unable to extricate itself from the sticky stems, had died.

One of the foods that the Pimans relished was the *makkom* or *makkum,* a greenish yellow caterpillar with a horn at the tail, the larva of the White-lined Sphinx Moth, *Hyles [Celerio] lineata.* These appear during a brief period if there are good summer rains, feeding on either *Boerhavia erecta* or *B. intermedia,* plants the Pima call *makkom jeej* 'caterpillars' mothers'.

A number of *Boerhavia* species bloom on the reservation during spring, summer, and fall. I was long puzzled by why some were and others were not considered 'mother of the *makkom*'. For instance, once I collected a purple-flowered *Boerhavia coccinea* in June, and Sylvester Matthias said only that it was "related to *makkom je'ej.*" We found another in May, and he said, "It looks like *makkom je'ej,* but isn't this too early?" One time he told me, "*Makkom je'ej* 'mother of the *makkom*', the edible caterpillar; plant itself not eaten by people, but sooner or later the *makkom* [eat them all]. [It grows] close to foothills [bajadas] in washes, close to the mountains, more earth, not on sandy places [*sic*]."

Then one late August, as we were going down to Ruth Giff's fields, she asked me if I was still having trouble finding the one that was the right plant. I said I was. She took me to the lower edge of her field and showed me some spiderlings. "This is the one. It has sticky stems. And pink flowers." The others I had been finding had either deep reddish or purple flowers; these have no name in Gila Pima.

I wanted to verify the species of *makkom,* but somehow the caterpillars never appeared when I was there. Summer rains this far north are not dependable, and surely the caterpillars were not, either. One day Sally Pablo was going home from her office in Sacaton. The *makkom jeej* was growing lushly at the road edge, and she noticed *makkom* on the plants. She knew I needed specimens of the larvae for identification, so she stopped to pick some, along with some of the plant as vouchers, putting them into the only container she had available, her Styrofoam coffee cup. On she drove with the precious specimens. When she stopped at a gas station just off the reservation, she noticed that all the *makkom* had crawled out and were going different directions on the floorboard. Getting out, she got down on her knees and tried recovering the caterpillars among the foot pedals. The station attendant, sensing that the woman evidently was having some kind of mechanical problems, went over and asked, "Can I help you with something, Ma'am?"

Sally looked up, not knowing what she should say, finally decided on, "No, thanks.

I'm doing okay," and resumed her task of stuffing the fat caterpillars back into the cup. The man looked at the well-dressed Indian woman a few moments, then walked away puzzled.

Sally phoned me long distance to tell me that she had not only my specimen of *makkom jeej* but the *makkom* as well. What should she do with them?

"Better put them in the freezer, labeled, before they get away again!"

Technical Notes: The two botanical "species" lumped into the single folk taxon *makkom jeej* are sometimes considered by botanists two distinct species: *B. erecta* L. and *B. intermedia* Jones. Others consider the latter only a variety, *B. erecta* var. *intermedia* Kearney & Peebles. A larger and coarser species, *B. coccinea*, I have taken throughout the warm season. It is quite common. Occasionally *B. coulteri* (Hook.f.) S. Watson is found on the reservation. But neither is included in this folk taxon.

Vouchers: *361* (?); *859, 887* (RG); *1301* (SP); *1976* (GK).

the table, they will hold salt, pepper, Chiltepines, and sugar; if there are only two, they will most likely hold Chiltepines and salt. Apparently the word *ko'okol* is derived from the same root as the verb *s-ko'ok* 'to hurt, to give pain'. The Piman equivalent to Spanish *picante* (in English, usually called "hot" or "spicy") is *s-ko'ok,* which means something is hurtful or painful.

In Pima the plant is classified with the nonplanted, nonbrushy plants, apparently because it is suffrutescent, on that borderline between being herbaceous and brushy *(sha'i).* The source of the plant today is the store, but formerly, wild gathered *ko'okol* were one of the trade items that the Tohono O'odham took to the Gila Pima for exchange. (See Russell 1908:78, 93.)

Sylvester Matthias told me, "*O'olas* grows wild in Mexico. In Papago country, in the Baboquivaris, on the way to

[Magdalena], just before Sasabe, there's a mountain. That's the paradise [for it] there. They call it Ko'okolgam ['chiliness']; that's where the Papago go to get their chiles. . . . They pick them green, bundle them up, and dry [them]. I think *o'olas ko'okol* is the favorite of the Pima. That's my favorite."

For many years Ruth Giff had a *ko'okol* bush at her front door—until it froze during a severe winter. Such intentional cultivation is exceptional. Castetter and Bell (1942:121) stated, "It was never cultivated by either people." However, various other types of chiles were grown. These are in the Pima category *e'es* (see *Capsicum* spp., Group I).

Sometimes strong reliance on only one type of evidence, such as archaeological remains or notice in historical documents, leads to incorrect conclusions. For instance, Ezell (1961:34–35) wrote that the Chiltepín (*chiltepiquin*) was listed as one

olas kokol, (pl.) o'olas ko'okol   *Capsicum annuum*

Capsicum annuum L.
   [= *C. frutescens* L. var. *baccatum*
   (L.) Irish]
Chiltepín, (pl.) Chiltepines
Solanaceae
olas kokol, (pl.) o'olas ko'okol

On the lower Rio Yaqui, in the great bosques and fields along the river where the remaining Pima Bajo still tend fields, grow waist-high bushes with small bright green leaves and minute white stars of flowers with yellow centers. The plants are intricately branched, growing under mesquites, tepeguajes (*Lysiloma* sp.), and Desert Ironwoods *(Olneya tesota)*. In summer and fall, little round fruits appear on long slender stems. At first these are dark green, but they turn orange and red as they mature. These are the much sought after Chiltepines, which reach their northernmost natural distribution in Tohono O'odham country.

To the Gila Pima these are *o'olas ko'okol* 'spherical chiles'. If just the word *ko'okol* is used, this is the chile that is usually meant. In almost any Pima home you will find on the table a little jar of these potent Chiltepines: if there are four containers on of Kino's introductions into Pimería Alta and was considered to have been raised by the Gila Pima,

although questionably, because, as stated, no reference to it in connection with the Gila Pimas was found and because its Pima name, *tcil'tipin,* was recognizable as Spanish by Russell (1908:78). He noted that the Gila Pima obtained it from the Papago in trade, believing that they cultivated it—Castetter and Bell (1942:121) have since shown that it grows wild. Because its range lies well outside Gila Pima territory, and because if it had been used since aboriginal times, the Gila Pimas would be more likely to have their own name for it rather than an Hispanicized version of a Nahuatl name, it probably was introduced during Spanish times.

But why not ask the Pima? Even today, if a Pima youngster knows but a few dozen words in his or her native language, one of these is likely to be *o'olas ko'okol.* And the generic *ko'okolĭ* is known throughout Tepiman country, nearly a thousand miles from northern Jalisco to Phoenix.

Pimas of all ages love their chiles. (I was amazed at how young the taste is cultivated.) But they do have a proper respect for them. Often I have noticed someone at the table tearing off a little piece of paper, then crushing these little *ko'okol* between the protected thumb and forefinger over the dish. This is so one does not accidently rub one's eyes or some other sensitive part after handling these potent chiles.

And no account of Pimería would be complete without Pfefferkorn's (1949) first encounter with chiles:

No dish is more agreeable to an American, but to a foreigner it is intolerable, especially at first, because of the monotonous hotness of the peppers. The constant use of this hot sauce is at first an unbelievable hardship for the European. He must either be content with dry bread or burn his tongue and gums as I did when, after a difficult fifteen-hour journey, I tried for the first time to still my hunger with such a dish. After the first mouthful, the tears started to come. I could not say a word and believed I

had hell-fire in my mouth. However, one becomes accustomed to it after frequent bold victories, so that with time the dish becomes tolerable and finally very agreeable.

Vouchers: *1236* (SM).

Castilleja exserta (A.Hell.) Chuang & Heckard ssp. *exserta*
[= *Orthocarpus purpurascens* Benth.]
Owl-clover
Scrophulariaceae
chuuvĭ taḍ, chuuvĭ taḍpo

Many of the winter-spring annuals found in the Sonoran Desert have their centers of abundance on the Pacific Coast, their ancestral homes. These include Desert Lupines *(Lupinus sparsiflorus),* Mexican Gold Poppies *(Eschscholzia mexicana),* evening-primroses *(Oenothera* spp.), monkeyflowers *(Mimulus* spp.), and wild-heliotropes *(Phacelia* spp.). (These are balanced by annuals that appear here primarily during the late summer rains; these genera originated farther south.) Owl-clover is one of the spring ephemerals dependent on good winter rains. When these come at the right time and in sufficient amounts, whole patches of the uplands will be carpeted with these reddish purple flowers, often mixed with Mexican Gold Poppy. Then there are years (sometimes whole series of years) with none at all.

This purple annual could be confused with nothing else except Indian Paintbrushes *(C. integra* and *C. lanata),* which are bright scarlet, with less dense heads, and grow only higher in the desert mountains (see "Technical Notes" below). Owl-clover has one or several erect stems thickly clasped with finely divided leaves, the upper ones (the bracts) tinged with purple. The tubular flowers, densely crowded around the upper stem, have an irregular shape. Sometimes an inflated yellowish lower lip is visible. This interesting little plant is a hemiparasite, deriving part of its nourishment from the roots of surrounding plants.

To some the plant suggests a little whisk broom, hence its Spanish name, Escobita.

But to northern Pimans it is either *chuuvĭ taḍ* 'jackrabbit's foot' or *chuuvĭ taḍpo* 'jackrabbit's foot-hair'.

For about 20 years I had heard only the name *chuuvĭ taḍ* among the western Pima and only *chuuwĭ taḍpo* among the Tohono O'odham. But in talking with more people throughout the Pima villages, I have come to think that the two forms are used about equally by different Gileño speakers. For instance, Carmelita Raphael of Santa Cruz village used the *-po* form in talking to us, but Sylvester Matthias, who was listening, said he had never heard it that way. Ruth Giff said, "*Chuuvĭ taḍ*—lavender [colored]—[is] real pretty. On the hills when we go up to pick *hannam* [cholla buds; *Opuntia acanthocarpa*]." Irene Hendricks and George Kyyitan both called the plant *chuuvĭ taḍpo.*

Technical Notes: Two perennial species of *Castilleja* occur in the Sierra Estrella. E. Linwood Smith collected *C. integra* A. Gray in the higher elevations, and *C. lanata* A. Gray is widespread in the sierra: these are known as Indian Paintbrush because of their showy scarlet floral bracts. Although they are conspicuous in the bajada arroyos where they might be seen by Pimas harvesting cholla buds, Indian Paintbrushes are not included in the folk taxon *chuuvĭ taḍpo* and appear to be unnamed in Gila Pima.

Vouchers: *1892* (GK).

Convolvulus arvensis L.
Field Bindweed
Convolvulaceae
biibhiag

In the Pima cotton and alfalfa fields and along their borders may be found a sprawling weed that looks much like a morning-glory. Its tubular flowers expand to flat disks. Usually they are white, but some clones with smaller flowers are definitely pink. The flowers open in the morning and close in the heat of the day. The flowering mats may be quite conspicuous in places such as Upper and Lower Santan.

The arrival time of this European plant is not known for sure, but there are Arizona specimens going back to 1843 and reservation specimens at least as early as 1926. Its Pima name, *biibhiag,* comes from the repetitive verb *biibhiag* 'to coil or wrap around something'. Mathiot (1973:166) gives *biibhiag* as 'morning-glory, a creeper', and notes that it is a reduplicative form of *bihag* 'to coil, surround'. This word is used for naming various vinelike plants among other Piman groups as well. For instance, Batanene *(Mascagnia macroptera)*, a common vine growing about the village of Onavas in southern Sonora, is called in Pima Bajo *biibiogam*; this folk taxon is constructed a bit differently, in that *-gam* is an attributive, so the name might be rendered 'entwineness'. At the Mountain Pima village of Nabogame, perched in the Sierra Madre Occidental, I was given *bibiokam* as the name for the wild grape *(Vitis arizonica)*. In a country without many vines and without a life-form category for vines, this folk taxon is the closest various Pimans come to a word for 'vine'. Let me add in passing, though, that the most common and useful Gileño vine is *Sarcostemma cynanchoides*, Milkweed Vine. Its Pima name, *viibam*, comes from *viib* 'milk', as does *vii'ipkam*, the name for the various spurges *(Euphorbia* spp.). These have nothing to do with the superficially similar verb *biibhia* or *bihag*.

Before the arrival of the European species, the name *biibhiag* undoubtedly was applied to a native species such as Hoary Bindweed, *Convolvulus aquitans* Benth., which is now exceedingly rare in riverine Pima country.

Field Bindweed is considered an annoying weed by Pima farmers. One hot late spring day, Sylvester and I found the plant growing at the edges of fields in the Santan area. "*Biibhiag*," he commented, "that wraps around stuff. They can climb on a corn stalk. [Have to] get rid of it."

The true morning-glories, in the genus *Ipomoea*, are all higher-elevation species in Arizona except *Ipomoea purpurea* (L.) Roth, a locally naturalized tropical species that is well known in cultivation. Robert H. Peebles (*PHK 10255*, ARIZ) had collected it in the early 1930s around Sacaton, noting that it was a "troublesome weed in cotton fields." Ruth Giff said that the morning-glory grown in people's gardens would be called *biibhiag* just the same as the wild weed. I have heard the same usage among some Tohono O'odham. George Kyyitan used this name for a blue- or purple-flowered plant, excluding the white-flowered European plant, which he also noted was too prostrate.

Vouchers: *375, 381* (SM).

Cucumis melo L.
Feral Melon
Cucurbitaceae
baaban ha-miiliñga

At Sacaton, Francis Vavages told me about *baaban adavĭ* 'Coyotes' wild gourd', a plant I have never been able to identify, though I have asked nearly everyone about it. No one else seemed to know anything about it. I had one more person to ask, though, so I drove over to Bapchule one evening to check it with George Kyyitan. After some pleasantries, we started talking about cucurbits.

George said, "I've just heard [the name] *baaban aḍavĭ*. I don't know what it is. But I've noticed and I've seen this one that they call *baaban ha-miiliñga*."

"Where did you see that?"

"Queen Creek. Yeah."

"How big is it?"

"It's about this big," he said, indicating the size of an orange. "They were hanging on a tree there. About this big. They got no taste. It was like that—like them cantaloupes [*Cucumis melo*]. But they got no taste. Maybe it's frozen, that's why, why it doesn't. And about that devil's claw—you heard about that one, too?"

"The *ihug?*" I asked.

"Yeah, *ban ihugga* ['Coyote's devil's claw']—they're the little ones." (See *Proboscidea altheaefolia*.)

"Yeah, out in the wash."

"Hah! Baaban don't make no baskets!" laughed George.

"This *baaban ha-miiliñga*—is it soft?"

"Yeah, just same size as that *aḍavĭ*."

"And what color?"

"Yellow—when it's ripe. It's just the same as those cantaloupes, melons."

"And the inside, what color is it?"

"It's white."

"And the seeds?"

"The seeds is like the cantaloupe—but they're *small!*"

"In the middle or throughout?"

"In the middle. It was like a cantaloupe, but they had no taste. Nothing! They got no taste—just throw them away."

"Up by Queen Creek?"

"Up by Queen Creek, this side of Williams Field Road. You know where Boys' Ranch is? Just right across the railroad track there. There was a mesquite tree there—it was—they just climb up all over—just hangs on there."

"How did you know what that was, George? When you saw it, how did you know what to call it: *baaban ha-miiliñga?*"

"Because they call it that—all these people around here—they knew it. A long time [ago]—they knew it, they told me that—that the Baaban had a melon." He laughed. "He never plant[ed] that!"

"What about its leaves? How do they look?"

"I think it's got round leaves."

"Not like the *aḍavĭ?*"

"No!"

"But you never saw it down here in the villages?"

"No, no, I never seen that. We were working over there and I saw that one."

"The fruit would be like a tennis ball?"

"It's big—about this size—like a big orange. There was a cotton field on one side, the other side idle land with the mesquite tree. Now they're all fields there. Now there's nothing but cotton."

"So I can't go look for it?"

"No, you can't find it. Unless it might be by the railroad track, where some bushes are there."

"What time of the year did they have those fruit?"

"It was in wintertime—but I think they have same times as these melons—but all froze at that time. That's why it's got no taste. Maybe it's got a taste some way. But it's all froze at that time and it's got no taste."

Gary Nabhan had found this name applied to a cucurbit farther south in Tohono O'odham country. And much farther south, in Pima Bajo country, I had found a little melon (*AMR 1239, Cucumis melo* L. var. *dudaim* Naud.) growing out in the hills; don Pedro Estrella called it *ban me'eronig, melon de coyote* in Spanish. It was an attractive vine with broad fig-leaf-shaped leaves. The small flowers were deep yellow, the fruit about the size of a lime. But was this what George and his friends were calling *baaban ha-miiliñga?*

Gary Nabhan has made a special study both of cucurbits in the Southwest (especially the relationship of wild forms to domesticated ones) and of Coyote's plants in O'odham. I told him of my discovery.

He said, "There's some wild *Cucumis* species around this area, but it's not native and it's not been around too long. Find out how long ago he saw it."

The next time I saw George, I asked about this. He said his own encounter was "late—at least late [19]50s or probably [19]60s, but not before."

Then I stopped to see Francis Vavages one morning. I asked him again about *ban aḍavĭ*.

"It's just a little bit smaller, the gourds, than the regular *aḍav*. Hard. Leaf almost the same."

I asked if he had heard of *baaban ha-miilĭnga*. He paused, then started this story:

"Yeah. You know, Ban [Coyote] can steal. He gets into the field. He eats the *miiloñ* [muskmelon] in the field. Then he done his business out under a tree, a mesquite. Then that *baaban ha-miiloñga* comes up under a tree. A man was telling me one time he found them *way* out in the desert, growing up a tree; sure good. That's it. They're sweet. Just like *miiloñ*. You can eat them. We say *miiloñ*. Here. At the west end they say *miiliñ*."

George's account and Francis's are similar except that Francis said that the *baaban ha-miilĭnga* is sweet, the same size as the domesticate, and comes from domesticated seed, while George said it was wild, smaller, and tasteless. Perhaps only Coyote knows the truth about what this plant really is.

Cucurbita digitata A. Gray
 [≠ *C. foetidissima* H.B.K.]
 [≠ *C. palmata* S. Watson]
Finger-leaved Gourd, Coyote Gourd
Cucurbitaceae
aḍavĭ

As you wander up arroyos in Pima country, you may come across a rare vine that sprawls out across the sandy soils from a single central root. Sometimes it will climb a nearby tree or shrub. The stems and leaves feel very coarse, an almost certain identifying character. The tubular, yellow, finger-length flowers are unmistakably those of a cucurbit. As with its near relatives, the cultivated pumpkins and squashes, male and female flowers are separate (though on the same vine). The rough leaves have five lobes, each cut almost to the base or petiole, which gives this wild gourd one of its English names: Finger-leaved Gourd. Vines may flower

from June to October or somewhat later if cold weather is delayed.

What is more likely to catch the attention of a casual observer than either the mottled silvery green leaves or the flowers are the softball-sized fruits. At first these are boldly striped, end to end, light and dark green. As they season, the stripes change to pale yellow and deep golden. And when the gourds are completely dry in winter or spring, the pattern is lost on the uniform buff-yellow spheres. These remain attached to the tough vines radiating five or ten feet out from the tap root until a storm washes them away, some animal steps on them, or bacteria and fungi work through the shell.

If you wonder why animals do not eat these tasty-looking gourds, crack one open, put your finger on the white placenta supporting the seeds, then just touch your finger to the tip of your tongue. Carefully! That is why many Southwestern tribes refer to the various wild cucurbits as Coyote's melons or Coyote's gourds. Coyote plays tricks like that.

The name *aḍavĭ* is widespread among Piman speakers for various wild gourds. The Gileños say *aḍav* or *aḍavĭ*, the Tohono O'odham *aḍaw* or *aḍawi*. In Pima Bajo country, don Pedro Estrella showed us the abundant wild *chichi coyota* (*Cucurbita argyrosperma* var. *palmeri* (L. H. Bailey) Merrick & Bates), calling it *adaav* or *ada'av*. The Mountain Pima at Yepachi told us the name was *ára*, cognate with the lowland term.

Even though Finger-leaved Gourd has become quite scarce in Gila Pima country, everyone with whom I worked knew it well. Once Sylvester Matthias mentioned, "[They] use *aḍavĭ* for deodorant *[sic]*, to boil clothes, to keep lice [out]. It's not a soap, more like a detergent." One day we encountered the plant in the field and he elaborated on this: "*Aḍavĭ* has pointed leaves, like fingers; gourds just normal-sized. It's real bitter when they are green. The people who had *hi'opch*, body lice, would cut the *aḍavĭ* into boiling water along with their clothes, and prevent

aḍavĭ   *Cucurbita digitata*

the lice from coming back because it's real bitter. Use whole fruit. They [anybody] used to make rattles out of them, but they are too thin—they break. Modern, is used for Christmas ornaments, or a charm string [is made] out of them [to hang] next to door, both [ways] painted."

The Pimas with whom I worked denied any knowledge of the seeds being eaten, but Russell (1908:70) noted, "The seeds of this wild gourd are roasted and eaten." Selection for the oily seeds seems to have preceded selection for the sweet flesh we esteem today (Nabhan 1985:174).

Likewise, there seems to be no contemporary information on the plant's curative properties, though at the turn of the century it played an important role in Piman materia medica. The greatly thickened root of wild gourds is like an enormous daikon (giant Japanese radish). Hrdlička (1908:246) recorded this veterinary use: "*A-taf* (*Cucurbita palmata* [*sic; C. digitata*]) is a plant the root of which ground is used by the Pima as an application for all kinds of sores on horses." Sylvester Matthias, who told me of a similar use for Common Cocklebur (*Xanthium strumarium*), said he had never heard of this use of *aḏavĭ*. Russell (1908:79) recorded more detailed medicinal use of the root: "The root of the wild gourd is pounded up in mortars, boiled, and the extracted juice put into the ear to cure earache. It is poured into a hollow tooth to stop aching. 'It kills maggots in open sores.'" (Incidentally, the name Russell recorded for this gourd, *átaftak*, is really two words: *tatk* 'root', being the part of the plant that is used medicinally.)

The number of Gila Pima folk genera and the number of biological species formerly in their country remains uncertain. One day while discussing wild gourds, Francis Vavages of Sacaton asked me if I knew of *ban aḏav*: I did not. "Well, it's a little smaller than [regular] *aḏav*. About dollar-sized fruits, hard and striped. It spreads out—on the desert under a tree—in the flats. It's hard to find now." Fr. Antonine (1935) recorded, in addition to *adaf* or *a'adaf*, a second species of wild gourd: *sekufchu adaf* (see appendix B).

Hrdlička recorded *aḏaf* as *C. palmata* S. Watson, which is probably just a subspecies of *C. digitata;* in Arizona, *C. palmata* is found only along the Colorado River. Russell (1908:70) recorded the plant as *C. foetidissima*, but a footnote (1908: 79)—perhaps added after Russell's death—mentions *C. palmata* and *C. digitata*, evidently on the advice of Edward Palmer, and on page 91 is written, "There are three species of wild gourds that are quite common along the Gila, namely: Cucurbita foetidissima H.B.K., C. digitata Gray, and Apodanthera undulata Gray." If all three were really once found there, it seems likely that the word *aḏavĭ* has always applied to Finger-leaved Gourd, Fr. Antonine's name *sekufchu adaf* to *C. foetidissima*, and *baaban aḏavĭ* (which

Francis Vavages described) to the round-leaved *Apodanthera undulata.*

Laura Merrick (personal communication) notes that *Cucumis melo* var. *dudaim* is another possibility for Francis's *ban adavĭ*; it has a tough rind and is distinctly striped. Perhaps these were once here at the margins of their ranges. But there are no herbarium specimens (in ARIZ, US) from Pima country substantiating this possibility.

Coyote Gourd, which 25 years ago I encountered fairly frequently, I would now consider decidedly uncommon on the reservation, restricted to a few gently sloping bajadas in remote areas. Changes in land use and perhaps other factors have extirpated it from most of Pima country.

Vouchers: *514, 1508* (SM).

Datura discolor Bernh.
Desert Datura
Datura wrightii Regel.
   [= *D. meteloides* DC.]
Sacred Datura, Jimsonweed, Toloache
Solanaceae
kotḏopĭ, kotoḏopĭ

The daturas are likely to be noticed even by the casually observant desert visitor, even though the knee-high plants are not especially noteworthy. Depending on species, they are partial to the more sandy areas such as arroyo mouths, riverbeds, and roadsides. The stout branches and large grayish green leaves have an unpleasant odor. But the flower is a different matter: the huge white trumpet emerging from its elongate calyx will catch the attention even of those speeding by on paved highways. On a warm summer evening a bush may be covered with these delicate, sometimes lavender-washed flowers. By midmorning the following day, the flowers wither and the plant merges again unobtrusively with other desert shrubs and bushes. Weeks later, a prickly green fruit—the source of one of the plant's names, thorn-apple—is all that is left to mark the evening's satiny extravagance.

Throughout their ranges the various daturas once played significant roles in the religion of aboriginal peoples encountering them. Although the Pima name of

this plant was known to all the consultants with whom I worked, its use in consciousness alteration was either poorly known or denied. But everyone regarded *kotḏopĭ* with a certain respect, a rather vague awe—and with good reason. Daturas contain solanaceous drugs that block the parasympathetic nervous system; clinical effects include "rapid heart rate; dilated pupils; flushing, warmth, and dryness of the skin; dryness of the mouth; constipation; and difficulty in urinating and swallowing. In males, the tropanes also interfere with ejaculation. Toxic doses of these alkaloids produce fever, delirium, convulsions, and collapse. Death may occur in children, the elderly, the debilitated, and any persons unusually sensitive to the antiparasympathetic effects" (Weil 1980:168).

The plant's medicinal uses seem to have been forgotten, and much of what Curtin (1949:85–87) recorded came from her Maricopa consultants. However, Mrs. Pablo Chiago of Sacaton told Curtin, "One must cut the top of the bud and pour the liquid [presumably the nectar from the flower] into sore eyes." Another Pima, José Henry of Salt River, told her the same, adding, "The buds are gathered early in the morning and, to keep them fresh, they are wrapped in a wet cloth." This appears to be the only medicinal use among the Pima, and I am unable to find any vestiges of this treatment today. Edelmira Linares and Robert Bye tell me that even this nectar contains the dangerous alkaloids.

At first Sylvester Matthias was rather vague about datura. But on one occasion he said, "It's poison. It's a drug. If they chew on the root, it's just like marijuana—makes you see visions. When the effects wears out, their heads were about to crack! Oh, boy—makes them sick! The stock has more sense—they never bother it." Here Sylvester broke out in laughter over cattle having more savvy than some people.

A few years later, when we happened upon a blooming datura plant, Sylvester told the story again. "*Kotḏopĭ* [is] something like a lily [flower]; nobody used it; never heard of anybody being killed by it; maybe afraid of it. They *do* know it as a drug, though. Some fellow chewed on leaves [long time ago and] he goes crazy. He saw visions—lots of visions—they tried to catch him. When they get over it, oh! the headache. [It grows] along ditch banks. Even the cattle or the stock won't touch it."

Perhaps this is the same incident that Russell (1908:300) recorded: "A native thornapple, Datura meteloides D. C., is popularly believed [presumably by the Pima] that if one eats an undivided root it will render him temporarily insane, but if the root be divided or branching it is innocuous. There is a tradition that a man at Blackwater ate of the root and directed that he be locked in an empty house until the effects should wear off. He was locked in at noon and toward evening he was seen running through the thickets toward the river a couple of miles distant. He recovered his senses when in the middle of a thorny thicket of mesquites. His limbs were scratched and bruised, yet he had been unconscious of any injury until the moment of recovering his wits." Andrew Weil, a physician who has specialized in the cultural uses of drugs, noted that the Blackwater man was apparently showing a typical symptom of datura poisoning: "Many persons who eat solanaceous drugs develop burning thirst, fever, and the sensation of the skin being on fire. These symptoms often lead to visions of hellfire. In fact, a common cause of death under the influence of *Datura* is drowning—the result of stumbling into deep bodies of water while disoriented, in an effort to quench the thirst and fever" (Weil 1980:171–172).

Not all cases of datura intoxication are deliberate. Once, in the context of discussing how sacred things can cause one to sicken *(taach)* or have behavioral disorders *(iibad)*, I asked if *kotḍopĭ* might be among them. Sylvester thought not, strictly speaking, but then told this story: "It'll give you crazy actions just like *maliwáána* [marijuana] or any kind of a dope. Because there's one time that, used to be there is a little something, like a frost. And it is sweet. And it fell on the *kotḍopĭ* leaves, just the leaves. And he [someone] was licking—was licking the *kotḍopĭ* leaves. (And he know that it's a drug at the bottom, the root.) And he was licking, and finally there comes a vision. A *lot* of fishes crawling in the shallow water. And he goes and tries to grab them." I thought again of the connection Weil noticed with water during datura poisoning.

"What do you call that, on the leaves?"

"*Komvad*—like a little syruplike."

Museum entomologist David Faulkner tells me that this condition on desert plants is caused by scale insects, similar to those producing the honeydew mentioned in Exodus (see also Ebeling 1986: 116–118).

Both Joseph Giff and Sylvester Matthias told me the legend of a maiden who was instructed to use surreptitiously the *kotḍopĭ hiósig* 'datura flower', as a condom in an attempt to identify her unknown suitor (see Group E, *Prosopis velutina*).

Francis Vavages stopped his work on a new flute for a moment to discuss the big white-flowered weed I was showing him. "*Noḍagam*," he said. "If you take a little part of that, it'll affect your mind; that's 'crazy weed.'"

"Is that the same as *kotḍopĭ?*" I asked. His name was new to me.

"It's the same thing."

"What part do you use?"

"The root—a little piece of the root."

"Isn't that pretty dangerous?"

"Oh, yeah! Oh, yeah! It'll make you go crazy. Visions? Well, that's what they say—you see things different. Yeah. You see things different."

His name apparently is derived from a word meaning to make one either rabid or dizzy. In either case, the idea is that one becomes crazy.

Formerly, datura played a significant role in Pima culture. It was associated with both deer hunting and staying sickness, *kaachim mumkidag*. In both cases, its status parallels that of Peyote *(Lophophora williamsii)*, another psychoactive plant. Russell (1908:299–301) recorded two deer-hunting songs: one "Kotḍopĭ Ñe'i" (Datura Song), the other "Pihul Ñe'i" (Peyote Song). These songs are prayers to bring success in hunting. The following is Russell's free translation of the Datura Song:

At the time of the White Dawn;
  At the time of the White Dawn,
I arose and went away.
  At Blue Nightfall I went away.

I ate the thornapple leaves
  And the leaves made me dizzy.
I drank thornapple flowers
  And the drink made me stagger.

The hunter, Bow-remaining,
  He overtook and killed me,
Cut and threw my horns away.
  The hunter, Reed-remaining,
He overtook and killed me,
  Cut and threw my feet away.

Now the flies become crazy
  And they drop with flapping wings.
The drunken butterflies sit
  With opening and shutting wings.

Bahr, Giff, and Havier (1979) have analyzed this piece, treating it as a set of four songs. In the first three the hero, Mule Deer, tells in the first person what he did and what in consequence happened to him at the hands of two hunters. The last song shifts dramatically to the insects in a placid post-denouement.

Both datura and Peyote, as well as the Mule Deer, are dangerous objects, capable of bringing sickness to someone who abuses them (see Bahr and associates 1974). The characteristic symptoms of datura sickness are vomiting and dizziness; the primary cure is to sing the "Kotḍopĭ Ñe'i" (see Bahr and Haefer 1978; Russell 1908:299).

Undoubtedly at one time the plant was used more frequently by Pimans as a mind-altering agent, but unlike its twin, Peyote, datura is exceedingly difficult to control, with death of the user being a distinct possibility. Fr. Pfefferkorn, working with the Pima at Attí in northern Sonora between 1756 and 1763, reported on this use and its very real dangers:

Toloache leaves applied to a swelling will open it in a few hours. Many Indians prize this plant highly, not only because of its efficacy in healing, but for a disgraceful misuse which they make of it. The juice of this herb has the property of dulling the senses and robbing one of reason, as does intoxication. The Indians have the foolish notion that by means of this juice they are transported into an ecstasy, during which wonderful things and pleasant fancies occur to them. They also firmly believe that from it they receive the power to cure sicknesses and also the strength to defeat their enemies. Therefore, they drink this damnable juice with nonsensical eagerness and soon thereafter become so inebriated that they fall to earth like lifeless blocks and only recover their faculties again after some hours. It also happened more than once that they paid for this over-indulgence with the loss of health or with sudden death. Among the converted Sonorans, because of the earnest admonitions

kotḍopĭ, kotoḍopĭ   *Datura wrightii*

of the missionaries, this impious practice has been rooted out for the most part, though not yet entirely. I myself learned of two who drank themselves to death with toloache. As soon as these unfortunates had taken a good draft of it, they were seized by a general palsy; immediately afterward they pitched speechless to the ground and manifested no further signs of life. I was notified and hurried immediately to them, but on the way I received the report that both had died. (Pfefferkorn 1949:63–64)

But there are even more local and more recent examples of young Indians, perhaps lured by the fictive writings of Carlos Castañeda, who have been misled by the hopes of a cheap high. Weil gives a definitive thumbs-down on this plant: "*Datura* is not a nice drug. Although sometimes classified as a hallucinogen, it should not be confused with the psychedelics. It is much more toxic than the psychedelics and tends to produce delirium and disorientation" (Weil 1980:166).

The name for datura is nearly the same for the two surviving northern Piman groups. Curtin wrote it simply *kodop*, but Fr. Antonine gave *kododophi*. It is recorded as *kotobĭ* by Saxton and Saxton (1969:172); Mathiot (1973:497) gave *kotdobĭ*; and Bahr and Haefer (1978:95) wrote *kotdopĭ*. With the modern Pima Bajo at Onavas, don Pedro Estrella identified the plant as *hakatdam* or *hakandam*, saying that "it does something to your head." This name, not cognate with the Upper Piman word, is perhaps derived from an alternate name found in the 17th-century Névome dictionary (Pennington 1979:115), where *gugur ha'agama* 'big leaves' was used sometimes for *tokorhibi*. Its use as a poison (for murder or suicide) was known there (Pennington 1980:276). The Northern Tepehuan name for Toloache is *tokorakai* (Pennington 1969:185), cognate with the Névome name.

Two species of datura occur in Gila Pima country. They do not appear to be distinguished linguistically, though Curtin

was told that the smaller-flowered species, *D. discolor*, was the stronger and "when one inch of the root is chewed, 'ants look like horses and butterflies like airplanes'" (Curtin 1949:87).

Older Pimans have learned the dangers of datura and treat the plant with respect. Enjoy the blossoms as they burst forth in Georgia O'Keeffe splendor during the cool hours in the desert. But leave the potent alkaloids to serve the defensive purpose for which they evolved.

Technical Notes: *D. discolor* is an annual, attaining a height of 1–2 feet, bearing green leaves about 2–4 inches long. The flowers are trumpet shaped, with a purple band in the throat; the corolla may be 2–4 inches (rarely 6 inches) long; and the deeply angled, five-ribbed calyx is about 2¼ inches long. The fruit pods bear viscid, stout spines and contain seeds that when mature are black and pitted. *D. wrightii*, by contrast, is a perennial 2–3 feet tall, with leaves that are grayer green and longer (about 3–10 inches) than those of *D. discolor*. The flowers are funnel shaped and have no band in the throat; the corolla is 6–10 inches long; and the non-angled, sheathlike calyx is 3–5 inches long. The fruit pods have slender spines and contain light brown, smooth seeds when mature. A third species, *D. stramonium* L., a native of the tropics, has been naturalized in parts of Arizona and might be expected on the reservation.

Vouchers: LSMC 9.

Delphinium parishii A. Gray
[= *D. amabile* Tidestrom]
Desert Larkspur
Ranunculaceae
chiinō hiitpa

In spring out on the bajada flats and the banks of the arroyos cutting through them, and even in the lower parts of some of the ranges, you might find a larkspur of an incredibly delicate blue, seemingly out of place on these sunbaked flats. The single flowering stalks may reach to the knee. Root perennials, they grow in the protection of some Foothill Paloverde (*Cercidium microphyllum*) or other shrubby plant. Most of the divided grayish leaves are at the base. These are so reduced as an adaptation to their very dry microhabitat that the plant seems to be just ethereal flowers floating in the desert air. The bluish sepals have a greenish spot near their tips. The upper sepal is modified to produce the "spur" that gives the plant its name. Larkspurs contain alkaloids toxic to livestock.

Carmelita Raphael of Santa Cruz village goes out in spring to pick *hannam* (cholla buds; *Opuntia acanthocarpa*) on the bajadas. Here she would find the little blue flower she calls *chiinō hiitpa* 'Chinese queue'. She told me once that she thought it was all gone now, but when we went out later in April to pick *hannam,* I found some to show her.

Because Carmelita is part Tohono O'odham and when young used to spend some time down in that country, I wondered if she had perhaps picked up this name elsewhere. (She was not sure.) But the Tohono O'odham have their own, entirely different names for this larkspur. Saxton, Saxton, and Enos (1983:90) give *chuchul-i'ispul* 'chicken spur' and *kuksho-wuuplim,* larkspur; Mathiot (n.d.:451) gives *chuchul i'ispul* and *kuksho vuupulim* (*kuksho* 'nape of neck'; *vuupulim* 'tied together in a bundle'—hence, 'queue'). Neither source gives Carmelita's version, *chiinō hiitpa,* which must be local Pima after all. It is curious that two of the three Piman names for this native plant use a nonnative comparison as a marker.

I found no one else who knew a name for Desert Larkspur, but Sylvester Matthias was with us during the discussion at Santa Cruz and picked it up. During our later discussions of plant higher categories, he called the plant *chiinō hiitpaga,* putting it into the possessive form.

Vouchers: *757* (CR).

Descurainia pinnata (Walter) Britton [= *Sophia pinnata* (Walter) T. J. Howell]
Tansy-mustard
Sisymbrium irio L.
London Rocket
Brassicacae (Cruciferae)
shuu'uvaḍ

Sometimes folk taxonomies at a point of time capture transitions in cultural saliency and, consequently, naming. Such appears to be the case with two wild mustards, both locally abundant in late winter in Pima country.

Tansy-mustard is an altogether elegant plant, a native desert annual. If there are winter rains, the erect stems spring up in moist places on bajadas, along arroyos, or even on the floodplains, blooming there mostly from January through March. Its delicate, finely divided leaves are grayish. The flowering heads are typical of a mustard, with minute cream-colored or greenish yellow flowers at the tips, and tiers of relatively short but elongated seedpods below. Each pod is supported by a slender thread angling off the main stem. When dry, each side of the pod is packed with double rows of tiny rufous seeds. This native is partial to obscure places with sandy soils, but I once found it spread abundantly across the Gila River bed at Maricopa Road.

The introduced wild mustard, called London Rocket, grows in everyone's yards from winter through early spring, lasting longer where there is more shade and moisture. Both the stems and the leaves are bright grassy green, but the leaves are simply divided, with considerable leaf surface, particularly at their tips. The erect plants may grow to between knee- and hip-height and are crowned with tiny bright yellow flowers. As blooming progresses up the stems, slender and quite long pods mature below. These are packed with dozens of minute ochre yellow seeds in two back-to-back rows.

The two plants have a similar overall growth form, bearing spikes of small seeds in slender pods. And not surprisingly, some people call one or the other *shuu'uvaḍ.* (I have never yet found anyone among the three Piman groups who applies the native name to both the native and the introduced wild mustard.)

Sally Pablo and her mother, Ruth Giff, called the native Tansy-mustard *shuu'uvaḍ.* London Rocket was not "real" *shuu'uvaḍ* but something unlabeled. At Sacaton I pulled up some London Rocket in the yard to show to Irene Hendricks. She said, "That's not *shuu'uvaḍ.* But it looks something like that." I went through my plant press. When I came to a single specimen of Tansy-mustard we had found several days earlier in the Santans, she said, "There, let me see that one?" I handed it to her. She studied the plant carefully, then said, "*This* looks like *shuu'uvaḍ.*"

But then there is the other school of folk taxonomy, perhaps the more prevalent one. Sylvester Matthias, for instance, always identified London Rocket as *shuu'uvaḍ.* He told of its food use: "Tiny seeds—when they gather enough they grind it and make soup out of it—or boil it right and grind—just like *ga'ivsa* [parched corn]. When plants are dry—pull them [up] by roots and then pound them with stick and winnow in a basket and let the wind blow out—and just keep the seeds." Later, describing the prepared dish, he added, "Don't need to chew it—it just goes down."

At Sacaton, Francis Vavages also told of how pinole was produced from this European introduction. When he was young, his mother parched *shuu'uvaḍ* with coals, ground it on a stone, and cooked it as a mush.

George Kyyitan had participated in *shuu'uvaḍ* harvesting when he was young; for him the appropriate plant is Tansy-mustard. "We used to go up there with my mother and pick them, the beans *[sic]* on there, put them on a canvas, then . . ." He makes the motions of thrashing between his palms. "And they're *really* small seeds in there. Can't hardly make a cup[ful] on these. It takes too many. But it tastes good! I think I have drank from it two times, that's all, when I was small."

"Did you eat it or drink it?"

"Drink it. They got to roast the seeds. Then grind them. Then put in water like *chu'i* [flour]; they taste good. It's kind of greasy. It's just like that *haashañ kai* [a dish made with ground Saguaro seeds], but *haashañ kai* is a little thicker, because they use flour in it, sometimes."

Curtin (1949:84) told of both species being parched, ground, and mixed with water, calling *Sisymbrium shoo uvat* and *Descurainia such'iavik.* This latter may be a faulty transcription for the same word, as I could find no one who understood what it meant.

My theory is that *shuu'uvaḍ* originally meant the native *Descurainia* throughout Piman country and has been in the process of being transferred to the European introduction. Although I have had some Tohono O'odham identify the new plant as *shuu'uwaḍ,* Castetter and Underhill (1935:24–27) identified this folk taxon as *Sophia pinnata* (the earlier name for Tansy-mustard), as did Russell (1908:77) and Hrdlička (1908:263) for the

Gila Pima at the beginning of the century. In southern Sonora, Pima Bajo don Pedro Estrella identified roadside specimens of *Descurainia pinnata* (AMR 333) as *shu'awut.* Curtin's "informants," like my consultants, discussed either the native or the introduced wild mustard, but not both, indicating that the naming dichotomy was occurring before midcentury. *S. irio* arrived in the Phoenix area as early as 1909 and was collected at Sacaton by 1912. It soon became abundant in disturbed areas (ARIZ).

Once when we were traversing the Sierra Estrella bajada, I collected another mustard, *Thelypodium lasiophyllum* (H. & A.) E. Greene. I thought it was just a London Rocket with white flowers. But Sylvester was not deceived. "No Pima name [for this]; probably just call it *shuu'uvaḍ* anyway." Later, entomologist John Brown, who had studied certain mustards that were host plants to his butterfly larvae, pointed out the differences that Sylvester must have seen.

Early and contemporary records indicate that two different items were made from *shuu'uvaḍ* (apparently from *Descurainia* originally and *Sisymbrium* subsequently). One of these was a pinole from the roasted, ground seeds, the other a cold drink of the raw seeds. Hrdlička (1908:263) recorded the preparations: "Another seed used for food is that of a plant known as *řú-u-waht* (Sophia pinnata [= *D. pinnata*]). It is parched, ground, and eaten mixed with cold or hot water. Both the Pima and Papago use as food the seed of a grass [sic] known as *show-ou-wat.* The grass is gathered and rubbed on the concave part of a basket so that the seeds come out. These are thrown up and down, causing the bran to fly off into the air. The seeds are then ground and put into cold water and sugar is added; the liquid is used as a drink. The Papago use it much in summer, saying it cools them off." Both terms Hrdlička recorded are variants of *shuu'uvaḍ,* not two different folk taxa.

A number of other Old World mustards have appeared on the reservation this century as weeds, but none has entered the Gileño lexicon as a named plant.

Sylvester told me how this plant is used medicinally: "If something went into your eye and bothered you and you couldn't get it out, you use that *shuu'uvaḍ* and that catches the eye-butter and catches it and takes it right out. Yeah."

"At nighttime or any time?" I asked.
"Anytime. Both Pimas and Papagos do that."
"Does it hurt, sting?"
"No, it won't hurt."
"Do you use one or several little seeds?"
"*One* little seed. It goes in there and catches on the eye-butter—and here it comes out."
"Eye-butter?"
"That white stuff that comes out."
"How do you say that in Pima?"
"*Wuhi biit.*" We both laughed. The English phrase he was using was a euphemism for 'eye feces' in Pima!
"How long does it take to catch anything like that?"
"It won't take but a few minutes. That's why they always keep *shuu'uvaḍ*—just in case something happens—in a family—they always have it."
"Where do they store it?"
"Oh, just in some kind of a container. Keep it in the house."

Irene Hendricks noted this use also. She pointed to the pods, saying that when someone has something irritating the eye that will not come out, one of these seeds is placed under the eyelid when going to sleep, and the next morning the seed will be in the corner of the eye encased in mucus, along with whatever was causing the irritation.

Technical Notes: Several other invading Old World mustards in Pima country are unnamed. Wild Turnip, *Brassica tournefortii* Govan., is now common on roadsides, in fields, and in other disturbed places. Also recorded are Black Mustard, *B. nigra* (L) Koch; Field Mustard, *B. campestris* L.; and Indian Mustard, *B. juncea* (L.) Cosson.

Vouchers: *S. irio: 156, 260, 556* (SM); *1917* (GK, FJ); LSMC 4. *D. pinnata: 279, 281* (RG, SP); LSMC 60.

Dichelostemma pulchellum (Salisb.) Heller
[= *Brodiaea capitata* Benth. var. *pauciflora* Torr.]
Bluedicks, Covena
Liliaceae
haad

As you climb about the boulders of the Sierra Estrella in the spring, you are likely to notice growing among the mosses, liverworts, and spikemoss a slender, onionlike lavender-colored lily about 12 inches tall with a single sprawling leaf 16–18 inches long. If you dig the little bulbs out of the fine soil (as Javelinas do, and maybe Desert Bighorns as well), you will find them rather tasteless.

"This is not the wild onion," Ruth Giff told me, "but kids used to eat these (maybe everyone ate them), but no taste—just mushy." Carmelita Raphael told me, "They eat them—*haad;* don't cook them, just eat them." Sylvester Matthias said, "The *haad* is not a wild onion, but similar; only they have flat leaves—[grows] in the washes—and in the Estrellas; has bulb like [an] onion, but they don't eat them."

Later Sylvester told me of another plant he called *haad* also, that grows in the foothills of Table Top Mountain; this is a locality on the northern part of the Tohono O'odham Reservation, a place frequently mentioned in Gila Pima hunting and gathering activities of the past. This plant, he said, is stronger than the regular *sivol* (onion), so he must have been thinking not of *Dichelostemma* but rather some wild species of *Allium,* most likely *A. macropetalum* (see *Allium* accounts, Group I).

Among the Tohono O'odham, Castetter and Underhill (1935:17) recorded the name as *ha'at,* which they called Papago Blue Bells: "The bulbs . . . are eaten raw. They do not have a pleasing taste and are eaten largely because they appear in early spring, before other crops are ready."

Russell (1908:76) confused two quite different genera: *Dichelostemma* and *Amoreuxia.* Under the name of *Rsat* (*shaaḍ,* which is *A. palmatifida),* he said: "The bulb of the wild onion is eaten. It is common on the slopes at the foot of the Estrellas." This supposed wild onion is *haad, D. pulchellum.* The two plants are etymologically if not botanically related.

Vouchers: *10* (RG).

Erodium texanum A. Gray
Heronbill
Erodium cicutarium (L.) L'Her.
Filaree, Alfilaria
Geraniaceae
hōho'ipaḍ

In late winter and early spring out on
the bajadas, often growing on the desert
pavement with minute white daisies
(Desert Star, *Monoptilon bellioides*), you
may find a tiny annual with entire leaves
about the size of your fingernail and small
pink flowers that develop into long green
fruits, supposedly resembling a heron's
head and bill, larger than the plant that
supports them. This is the native *Erodium
texanum;* it eschews the soil about human
settlements. In the disturbed soils of the
floodplains as well as up on the bajadas
grows a similar plant, but this one has
finely divided leaves beginning with a
basal rosette; if there is moisture, it can
become a conspicuous weed. This is
Filaree *(Erodium cicutarium),* a European
plant that was probably introduced early
with Hispanic crop seeds or in the hair
and wool of livestock. In both species the
green fruits split when dry into five seeds
with the characteristic corkscrew tails that
ensure that the seed is transported by
some animal. And when rains come, these
tails unwind, helping to drive the little
spear-shaped seed head into the soil where
it may germinate.

To the Gila Pima, the green *Erodium*
fruits suggest not a heron's head but needles, and so the plant is called *hōho'ipaḍ*
(*ho'ipaḍ* 'needle'; from *ho'i* 'spine' or
'thorn'). Sylvester gave me this name for
the native species but was not sure if it
applied as well to the more common
introduced one. Irene Hendricks and
George Kyyitan said that both kinds
were *hōho'ipaḍ.* Ruth Giff's name for this
plant is slightly different: "We call them
hōho'ipaḍ. That's because there's a lot of
it. If it's one, it would be *ho'ipaḍ.* But,
you know, when you see them sticking all
over, they call it *hōho'ipaḍ.*"

There is no known cultural use for the
plant.

Vouchers: *E. texanum: 16* (SM, SP); *761* (RG,
CR); *1334* (IH). *E. cicutarium: 1743* (GK).

Eschscholzia mexicana E. Greene
**Mexican Gold Poppy, California
   Poppy**
Papaveraceae
hoohi e'es

When the winter rains have been good,
the bajadas sometimes burst out in blankets of bright orange poppies. In poorer
years the plants are smaller and fewer.
The leaves are rather succulent and quite
grayish green, finely divided. Even the
stems and buds, with their tight pointed
caps, are grayish. But when the cap slips
off, the four brilliant petals unfold. After
the petals and numerous stamens fall, an
elongate slender fruit with grooves
emerges from the center.

It is strange how a plant's name can
drop out of a speech community. No
one in Komatke knew the name for this
familiar and conspicuous spring flower.
But I knew the plant's name from the
Tohono O'odham. I mentioned to
Ruth Giff that Fr. Antonine's plant list
described *hoohi e'es* as a kind of poppy,
and she said, "I've never heard of it—
and poppies grow so much up there
where we pick *hannam* [cholla buds,
Opuntia acanthocarpa]." In the field one
day, I asked Sylvester Matthias about it.
He likewise knew no name for the poppy.
When I asked about *hoohi e'es,* he at first
said it was some kind of grayish plant:
"But it is mentioned in Thin Leather's
Creation Story." After five or six years of
my asking, he decided that the plant is
something entirely forgotten, or perhaps
just metaphorical.

Finally, Carmelita Raphael of Santa
Cruz village told me that *hoohi e'es*
was a poppy. But she did not remember
whether she learned this from the Pima
side of her family or from when she
would go stay with her Tohono O'odham
relatives to the south.

Later, at the east end of the reservation,
when I was out looking for plants with
George Kyyitan and others, I found that
many there still knew the Pima name and
recognized the plant. It is still a puzzle to
me how an entire village could lose a folk
taxon; yet the same thing has happened
with *tash maahag* (Desert Lupine, *Lupinus
sparsiflorus*).

The name *hoohi e'es* means 'Mourning
Dove's plants'. Although *e'es* is part of the

plant's name, the plant is classified in the
Pima system with the nonbrushy wild
annuals; it is only allegorically *e'es,* a crop.

I asked George if *hoohi e'es* had any use.

"Not for medicine, not for food. But
[they] used it for Easter, too, with *tash
maahag.* For the throwing, *ha'ichu hiósig
ñeenchuda;* just scattering different kinds
of flowers. Outside [the church]. They
used to. Then those educated men came
and it stopped." On Easter Sunday morning, petals and flowers were thrown
during the procession. Apparently this
was something the Gileños borrowed
from the Yaqui Easter ceremonies (see
Lupinus sparsiflorus account).

Neither Russell's (1908) nor Lloyd's
(1911) version of the Creation Story
transcribed from Thin Leather appears
to mention this plant. I asked Sylvester

hoohi e'es   *Eschscholzia mexicana*

if he had any more information. "It's a grandmother, the Hoohi; it's from the son's side, *Hoohi ka'akmaḍ;* and they were killed by the enemies—the grandsons. The messenger went and told Hoohi and she said, 'Wait till my *e'es* [crops] are ripe. Then I'll cry. That would sound better, when my plants are ripe and harvested. I'll be there working. That would sound way better than right now if I do the singing.' So, when the harvest is ready, then she start the song, and cry there, for her grandsons. I heard the song but I forgot [it]. It's been a *long* time. There's a song she sings with the crying, a mourning song. That's what the *hoohi* end[s] up there, with the harvesting of *hoohi e'es.* I don't know what *hoohi e'es* look like, but it's a grain [seeds], like *shuu'uvaḍ* [Tansy-mustard, *Descurainia pinnata*] here. That's that story."

The Piman name for the Mourning Dove, as with other doves, is onomatopoetic, the name mimicking the call of the bird. Interestingly, both the English name and Pima mythology link the bird with mourning. The Pima apparently associated the bird's call with their own keening, repeating the kinship term of the deceased.

Vouchers: *997* (GK); *1321* (IH); *1338* (IH, *E. minutiflora* included).

Euphorbia spp.
Spurge, Golondrina
Euphorbiaceae
vii'ipkam

Euphorbia species are usually found in hot dry disturbed areas, where they hug the ground with a spreading mat of minute stems and leaves. The whole plant may be as small as the palm of your hand or may spread over several feet. The flowers are so small that you will need a hand lens to appreciate them, but they are usually white with red or purple spots. If you are uncertain whether the plant is a *Euphorbia,* just pick some of the stems; they bleed a white sap. And the seedpod tumbles lopsidedly out of the flower like a three-lobed green pumpkin. There are many species, and they are difficult to distinguish, but almost any time from

spring through fall one species or another may be found blooming.

"It's called *vii'ipkam,*" Sylvester Matthias told me, "because it has milk [*viib* 'to have milk' + *-kam,* attributive], but it grows in the fields; I didn't know you could find it out here in the desert." (Then I told him that I had found the specimen I was showing him in Felix Enos's fields at Komatke.) "Use it for laxative—just enough or it makes you sick. Just chew it. It's different from *viibam* [Climbing Milkweed, *Sarcostemma cynanchoides*]—a vine that grows on trees and is used to make chewing gum." I pulled some and asked him how much should be taken. He folded a few stems in his small palm, saying, "As a laxative not very much." He laughed. "Otherwise it gets you down. Just a handful—small amount, raw. If you use more, it makes you sick the other way. [We were] warned not to chew on it."

Spurge's main claim to fame through history has been as a remedy for snakebite. The Pima George Webb described this treatment to Curtin (1949:99) in the early 1940s: "The wound is lanced immediately, the poison sucked out, and the juice of the spurge plant is squeezed into the cut. The green plant is chewed and the juice swallowed, causing vomiting, followed by sweating." (Botanist Robert Peebles told her that the plant was "highly poisonous and that he would rather suffer from rattlesnake bite than take the remedy" [Curtin 1949:100].)

Although Sylvester's father had a reputation as a curer of rattlesnake bites, he did not use *vii'ipkam:* "Father treated rattlesnake bites by sucking—*viiñunam*—sucking the whole body, to keep swelling down. A real [powerful] *maakai* ['shaman'] marked with a [burnt] match—and the swelling doesn't go any farther. They were very good at rattlesnake bites. Some died—if way in the area where they can get no attention. But with a man that can perform that snake healing, they get alright. Some get crippled for a while. Don't think they used any plant for helping snakebite. *Some* used mud from a gopher hole—on the bite itself. *Some,* but my father never used anything."

The account of snakebite treatment recorded by Russell (1908:264) largely parallels that provided by Sylvester 85 years later: "The bite of the rattlesnake is cured by sucking the wound every morning for

four days. Others suck it one or two days, and also ligature the limb with horsehair, or draw a circle around it with charcoal to define the limit of the swelling. The Papagos and Mexicans use the plant Euphorbia [albo]marginata to poultice snake bites, and it is possible that some Pimas use it also, though the writer was unable to find anyone who knew of its being so used."

Hrdlička (1908:246) reported, "In cases of rattlesnake bite the Pima suck the wounds; the latest remedy, however, is to kill the rattlesnake, tear it open, and apply to the wound a certain 'fat' which is found along the middle of the snake. This application is repeated and is said to be a certain cure. It is efficacious even when the limb has already begun to swell. Occasionally it is applied even without sucking the wound." (He made no mention here of Golondrina, though he did elsewhere in discussing Ópata treatment of snakebite.)

Grossmann (1873) left a lengthy account of the Gila Pima from a critical time period when little else was recorded. His account is filled with so many half-truths and hearsay mixed in with the factual material that it is difficult to know what in the report to trust. He wrote, "The Pimas know of many herbs which they use as food at times when wheat is scarce, but they have no knowledge of medicinal properties of herbs or minerals *[sic],* with the only exception of a small weed, called colondrina [Golondrina] by the Mexicans, which, applied as a poultice, is a certain remedy for the bite of a rattlesnake."

Given all this variable evidence, it would seem that the Gileño use of *Euphorbia* in snakebite was an occasional, secondary practice evidently originating from outside influences. However, all older people I worked with knew *vii'ipkam* on sight, distinguished it from *viibam* (Climbing Milkweed), and were familiar with its powerful emetic qualities as well as those of loosening "tight bowels."

I asked Sylvester Matthias about intestinal parasitic worms, called *vaptopad* (a word related to *vaptopĭ* 'worms' and 'fish').

"What can you do about those?"

"I guess they have to drink that mesquite bark. My father knew how to mix that—it's bitter. The inner bark. Some go to the medicine man, but won't do them any good. Has to be a first-aider that performs the remedies. They're not really

Poinsettia, *E. eriantha*) would be included in the folk taxon *vii'ipkam*.

Vouchers: *253, 256, 566* (SM, *E. micromera*); *282, 1511* (SM, *E. pediculifera*); *1497* (SM, *E. albomarginata*); *864, 893, 901* (RG, *E. albomarginata*); *378, 1512* (SM, *E. polycarpa*); LSMC *66 (E. polycarpa)*.

---

vii'ipkam   *Euphorbia albomarginata*

The uihimal, *or female Velvet-ant, figures importantly in Pima culture.*

maakai. [Rather] the ones that really work."

"Is there anything else they can use? Any *iivagĭ* [eaten greens]?"

"They can chew on that milkweed [sic], the *vii'ipkam* [*Euphorbia*], but not very much, because it's bitter. Not the *viibam* [*Sarcostemma*]. But just a little. Will wash them right off [out], but it's dangerous. They have to know how much—the patient uses—because if it's over, they'll just get sick."

"What will it do if you get too much?"

"Just keep on running, running! You'll be sick for quite a while."

"Do you know how much to use?"

"No, I don't."

The word will probably prove to be pan-Piman. At Onavas, Pima Bajo don Pedro Estrella was quite familiar with *s-viipgam*, an erect Golondrina (*Euphorbia hyssopifolia*) that grew abundantly as a weed in his milpas.

Technical Notes: Species of *Euphorbia* that have been collected in Pima country include *E. abramsiana* Wheeler, *E. albomarginata* Torr. & A. Gray, *E. arizonica* Engelm., *E. capitellata* Engelm., *E. eriantha* Benth., *E. melanadenia* Torr., *E. micromera* Boissier, *E. pediculifera* Engelm., *E. polycarpa* Benth., and *E. setiloba* Engelm. ex Torr. Most are now put into the genus *Chamaesyce*. Probably all these (except perhaps Desert

Helianthus annuus L. var. *lenticularis* (Douglas) Cockerell
Wild Sunflower, Annual Sunflower
Asteraceae (Compositae)
hivai, hiivai

John W. Audubon (1906:158), the younger of the famous naturalist's two sons, passed through the Pima villages in October 1849 on his way to the goldfields. Travel along the Gila floodplain was miserable, but there was some relief from the dust, in addition to occasional patches of grass in the bottomlands: "In places the sunflowers are marvelously luxuriant, and cover miles of the country, and are five to seven feet high, the road cut through them being the only gap in their almost solid ranks."

Even the most unobservant cannot help noticing the tall annual Wild Sunflowers that grow in profusion across the reservation, particularly along moist field edges, roadsides, and ditch banks. Although an occasional plant may be found in bloom at almost any time during the warm season, Wild Sunflowers are really summer flowering plants. These coarse plants with multiple heads may reach head-height or more. The dark dull green leaves are triangular, as large as your hand, and very rough. The flower heads (a collection of individual flowers), though small, are showy. The disk flowers are dark brown, the ray flowers (the "petals") deep yellow, almost orange, darkest toward the bases.

Parker (1972:298) wrote that these are native to the Great Plains, introduced to the Southwest. But Heiser (1976, 1985:96) considers the wild plant native to the Southwest, spreading prehistorically as weeds and domesticates to the central and eastern parts of the continent. Archaeobotany supports this. Karen Adams and Thomas Van Devender (personal communication) both note *Helianthus/Viguiera*-type seeds (achenes) in Southwestern sites and packrat middens at early dates, including many in the size range of Wild Sunflower. Van Devender suggests a dispersal northward of this frost-sensitive species into both the Southwest and the Great Plains during interglacial times.

Both the Akimel and Tohono O'odham call this plant *hivai* or *hiivai*. I suspect that the name may prove to be pan-Piman. Among the Mountain Pima of Chihuahua, I found this name being applied by different speakers to various tall weedy species of composites. Although Sylvester Matthias was not primarily concerned with etymologies, one day, quite unsolicited, he volunteered this information: "That's why they call it *hiivai*, because it's rough. You rub on it, it'll scratch you—*hiv* 'it's rough'. When you scratch your body where it itches, that's *hukshana*, you scratch. And the file and the rasp, *s-hivkchu*. They don't use that word no more. They call it *mu'ukadakuḍ* 'sharpener'. And *hivkuḍ*, [the musical] scraping stick. When you *hiv*, [you] rub on something that is rough."

One day when Francis Vavages and I were in the field, he heard me say the name of the Wild Sunflower. "That's west end—*hiivai*—they kind of drag it—*hiiv*— that's western people. Here we say *hivai*— very short." The way Francis pronounced it, it was almost *hihfai*.

At Onavas in Pima Bajo country, Sylvester pointed out the sunflowers in Pedro Estrella's fields. Pedro did not recognize Sylvester's name *hiivai*, calling the plants instead *tash nehidam* or *neadam*, in reference to looking at the sun, perhaps a back-translation of *mira sol*, its local Spanish name. But the colonial name there was *hiba* (= *hiva*) (Pennington 1979:80).

Today there appears to be no use for the plant among the Gila Pima, but that was not always the case. Curtin (1949:103) recorded both the petals and the stalk pith being used as chewing gum on the Salt River Reservation. Others told her of a "very bitter decoction of the leaves" that is "strained, cooled, and a tablespoon or more is given for high fever" (Curtin 1949:104). Another said this same could be used for screw-worms on horses (see also *Xanthium strumarium*).

Sylvester Matthias said, "*Hiivai* is Wild Sunflower. The pitch on that, when it's dry, you can use as gum or as putty."

"Putty for what?" I asked.

"For a broken olla. Just heat it up and seal it." Later he said, "That is from the shegoi [Creosote Bush, *Larrea divaricata*] that is used as putty, for the ollas, the cooking pot. But the [pitch of] *hiivai* can be used for the water jug." (See *Larrea divaricata*, Group D.)

Ruth Giff said that when she was a little girl she remembered using the pitch from *hiivai* for chewing gum, but not the stalk or the petals.

Although the Indians of the southeastern United States, when first discovered by Europeans, were growing sunflowers for their tubers, there is no definite evidence of domesticated sunflowers from southwestern archaeological sites. The Pima claim they never ate the seeds or the roots of the *hiivai* growing in their fields. I have found no tubers on the roots of this desert form, unlike those of the Native American cultivar (*H. tuberosus* L.), which is often called Jerusalem artichoke. The sunflowers on the Gila, though related to the species aboriginally domesticated in the East, appears to reproduce entirely by seed. Heiser (1976:80) considered this plant "unique among the early American cultivated food plants, for it is the only one to have been domesticated in what was to become the United States." This is an honor that we now know must be shared with devil's claw (see *Proboscidea parviflora*, Group I), although this peculiar plant is raised primarily for fiber and is only secondarily eaten.

Apparently nothing remains of the Sobaipuri Pima language except for a handful of place-names on colonial maps. Among the villages that Fr. Kino recorded on his map of 1695–1696 was one named Muihibai, probably *mu'i hivai* '[the place of] many sunflowers'. (A less likely possibility is that the Tyrolese Jesuit intended *mu'i i'ibhai* 'many prickly-pear'.)

One morning George Kyyitan and I stood looking at some tall Wild Sunflowers on the dried Gila River bed.

"Which way do these face, George?"

"Just any way. No particular way." That seemed to be the case with the plants before us.

"Are they used for anything?"

"Just food for the birds."

I checked the plants again late that afternoon. They still faced in various directions, disregarding a popular notion that they follow the sun.

Vouchers: *1059* (SM and Pedro Estrella, Pima Bajo); *LSMC 12.*

Heliotropium curassavicum L.
Quail-plant, Alkali Heliotrope
Boraginaceae
kakachu e'es, kaakaichu e'es, kaakaichu sisivoda [?]

Along Pima ditch banks, in fields, and in other areas with disturbed alluvial soils can be found commonly sprawling pale blue-green, rather succulent plants. They seem partial to soils with an alkaline buildup. No one can miss them, especially on exposed areas. These root perennials tend to spread out, reaching only halfway to the knee or so. They have neither spines nor protective hairs but are as naked as a cabbage. The flowering heads are usually in threes at the branch tips. These tips are tightly coiled, unrolling as the very small flowers open; these are white, with a yellow or purple throat spot.

Botanists have coined the word *scorpioid* to describe the coiled flowering stalks. But to the Gila Pima these stalks suggest the curved black topknot of feathers, the *siivoda*, that bends forward out of the crown of the Gambel's Quail. There are several variations of the plant's Pima name. In the west-end villages I was always told it was called *kakaichu e'es* 'the quail's plants'; the plural is *kaakaichu e'es*. Though its name includes *e'es*, signifying a plant belonging to someone (in this case an animal), it is not classified as *e'es* (Group I), because it is really a wild plant rather than something deliberately planted.

Irene Hendricks, Culver Cassa, and I were looking for plants on the flats northwest of Sacaton. We picked up some Quail-plant, which she called *kaakaichu sisivoda* 'quails' topknots'. On our way taking her home to Casa Blanca, we stopped to pick up a road-killed Gambel's Quail that Culver had spotted. When he handed it to her she said, "See—these things, the *siivoda*, right here on his head."

"Aren't you afraid of touching that?" I asked, knowing that there is a Piman taboo about touching the head of a quail.

"No, I don't believe in that," she said, deliberately touching the crest of curved feathers.

One day I asked Sylvester, "Was *kakaichu e'es* used for medicine?"

"Yeah. For sore eyes."

"What part?"

"The bottom—the roots, just the roots— boiled; make tea out of it. Could use *shegoi* [Creosote Bush, *Larrea divaricata*] also for sore eyes."

Curtin did not mention the plant, but at the turn of the century Russell (1908:79), recording the plant in the plural form, *kaakaichu e'es*, wrote, "The upper part of the light yellowish root is dried and ground in mortars, dried again, and ground very fine upon the metate, when it is ready to be applied to sores or wounds after they have been washed." Hrdlička for some reason missed this plant in his discussion of Pima medicines.

Although this is a field weed, apparently it is not considered a very bad one. When I asked Sylvester Matthias which were considered the most troublesome weeds in Pima fields, he did not include it among those he mentioned.

Vouchers: *36* (SM, JG); *1341* (IH).

Hesperocallis undulata A. Gray
Desert Lily
Liliaceae
naank je'ejgashakuḍ

Although dwelling in some of the harshest regions of North America, Desert Lily is as beautiful and delicate as any lily that you might find in a woodland meadow. Late in winter the little onionlike bulb deep in the moist sand sends up a rosette of dull green leaves with wavy margins. Then a single central stalk appears, followed by the fragrant cream-colored to greenish white flowers.

This attractive lily is confined to the most barren windblown dunes of southeastern California and western Arizona. It reaches its extreme northeastern range on the Gila River Reservation, on a few stationary dunes between the Gila River and the Lone Buttes community south of Gila Crossing and north of Maricopa Road. The remaining dune field is bisected by Riggs Road. These rare plants and their tiny bit of surviving habitat deserve protection by the tribe.

"The name *naank je'ejgashakuḍ* means 'ear opener'," Sylvester Matthias explained. "They use the stem to poke their ears, if they itch—either green or dry. If it itches, or something, you run that in, and clean it out. They keep it [the stem]. Used to grow a lot on Elliott Road and Carver Hill [south of Laveen along 51st Avenue on the way to Komatke, but] all gone now." Evidently all the dunes that were once in that area have fallen to field clearing and road construction.

Palmer (1878b:600) reported that "the bulbs of this beautiful plant are used by food by the Indians of Arizona." But I found no information on this among the Pima, and apparently only those from the west end of the reservation were familiar with the plant. The plant was never discovered elsewhere by M. French Gilman or any of the botanists working in the Sacaton region.

Frank Jim, a Tohono O'odham, called Desert Lily *ge'egeḍ haad* 'big *Dichelostemma*'. *Haad* alone is the name for the slender onionlike Bluedicks of the desert mountains. Both are lilies, though not very similar ones. Sylvester said he had never heard of this name for Desert Lily. Some years later when I took Frank a specimen again he called it *hiá siivol* 'sand dune onion'. "*Hiá* is just a pile of sand," he explained. "Some like it—the bulb—but I don't. It's kind of sticky. Eat it raw, like onions, the ones who live down there." He was referring to the so-called Sand Papago, Hia Ch-eḍ O'odham, living on the lower Gila.

Mathiot (1973:409) gives *naank jeejegashakuḍ* as "black grama, a weed *[sic]* formerly used to pierce holes in ears for earrings." This is a grass, *Bouteloua eriopoda* (Torr.) Torr.

Vouchers: *661, 695, 1896* (SM).

Hoffmanseggia glauca (Ortega) Eifort
[= *H. densiflora* Benth.]
[= *H. stricta* Benth.]
Hog-potato, Camote-de-raton
Fabaceae (Leguminosae)
iikovǐ

At first glance they may look like mesquite seedlings. The double compound leaves, somewhat bluish, give these little plants a definite legume look. There will be thickets of them in the deep alluvial soils of field edges, canal banks, and second-growth mesquite bosques. But when the erect little plants bloom (which is throughout the warm season), you will know that they are not young mesquites. The central stalks are covered with somewhat irregular flowers marked with yellow-orange and red. Whole colonies in bloom are bound to attract attention.

Iikovǐ is well known to the Pima, who once ate the tubers that grow deep in the soil. This is one of the first plant names I learned in Pima nearly 30 years ago. Even middle-aged people know about digging out the little "potatoes," so the practice must have persisted for quite some time. But the horizontal roots grow well over a foot down, so the digging is rather labor intensive.

Sylvester Matthias told me, "The 'wild sweet-potato,' *iikovǐ*, [grows in] flats, adobe, heavy soil below St. John's [Komatke]. Boil or roast [the edible enlarged roots]. [Tubers] go way down, at least six inches—dig it out with plow when someone is plowing. Sometimes gophers save them in bunch for winter, and once in a while we hit it when plowing a field—then rob [the cache] in spring."

Another time he told me, "When we're kids, used to follow the plow to get the *iikovǐ* bulbs *[sic]*. Sometimes the plow would go through the *jevhō to'a* [gopher's cache]; [they] store the bulbs, and kids get them."

I inquired about these caches.

"*Jevhō to'a, to'a* means 'gather and put'. A girl wouldn't eat *jevhō to'a*, [it's] prohibited, to eat the *jevhō to'a*. Older woman, whose period's finished, it's okay." Mathiot (n.d.:204) gives *toa[k]ch* as 'to have objects put, placed somewhere'.

I asked Ruth Giff about her experiences with the plant.

"*Iikovǐ?*—Oh, women used to go dig up people's fields. Always in January. They'd go—dig lots of holes, and have to fill them up. Yet it's got no *taste*. But they'd fill up on it—it's food."

A few years later when the subject turned again to *iikovǐ*, Ruth laughed and said, "Oh! Women used to go after those wild ones in January. I don't know why in January. They used to dig *big* holes in people's fields." She laughed some more, adding, "But I suppose they buried the holes. They go down *real* deep. I guess these things go way down. *Big* holes."

"How many feet down?"

"So high—about four feet. Some would be the old ones, some new. New *iikovĭ* [are] very light, and those others would be dark. *No* taste to them when we get home and boil them, eat them."

Sylvester told me, "The potatoes are yellowish when young, turning black when matured. Can eat them either way: roasted, boiled, or raw. I like them best roasted. Seems like more flavor than when they're boiled."

Although Russell does not discuss *iikovĭ*, except to mention that Thornber made reference to it, Hrdlička (1908:262–263) recorded its use among the Gileños: "Only one native bulb *[sic]* is used as food. It is the *eix-ko-we* ('underground-bulb': Hoffmansegia stricta *[sic]*), a small bulb, nearly black on the surface, which is dug out of the ground with considerable labor. It is boiled and eaten without additions. Occasionally the bulb is eaten raw, but consumed in that state in quantity it may give rise to 'sickness of the stomach'."

I asked Sylvester about this etymology, but he thought the word *iikovĭ* could not be translated. Nor had he heard of the raw plant giving anyone gastric problems.

Words can have extended meanings. Sylvester, Carmelita Raphael, and Ruth Giff all said that the sweet-potato or "yam" (*Ipomoea batatas)* would be called *iikovĭ*, just the same as *Hoffmanseggia*, without any special marker. Ruth said, "Papago say *kamóōdi* 'sweet-potato', but we call it *iikovĭ*, like the wild one; but I guess some say *kamóōdi* here, even the Pimas. *Iikovĭ* is better Pima, though." *Kamóōdi* is a loan word from Spanish *camote* (see *Ipomoea batatas*, Group I). At the Pima Bajo village of Onavas in southern Sonora, Pedro Estrella gave me only *kámoote* for sweet-potato, but the 17th-century Névome vocabulary written in this pueblo gives *icobi* for *Ipomoea batatas* (surely the same word, *iikovĭ*, in Spanish orthography). The word designating *Hoffmanseggia* is *iikowĭ* in contemporary Tohono O'odham. This word is probably pan-Piman, though to which plant it may be applied in different areas is still unknown. In its extended meaning of a crop, of course, the higher taxonomic position of *iikovĭ* changes from the covert category of wild things that are not brushy and not aquatic to the category of *e'es* 'planted things'.

Vouchers: *247* (sm); *900* (rg); lsmc *79*.

Lasianthaea podocephala (A. Gray) K. M. Becker
[= *Zexmenia podocephala* A. Gray]
Pioniya, Pionilla
Asteraceae (Compositae)
a'al ge'egeḍ

In spring 1987 George Kyyitan mentioned to me a medicinal plant, *a'al ge'egeḍ*, that he had taken as a youngster. He did not know much about what the plant looked like and doubted anyone was still alive who remembered it; but he thought it was a local plant. I asked Ruth Giff, and she thought she had heard the name but did not know what it was. Sylvester Matthias knew the name only as a kinship term. The name literally means 'children big', but in modern speech the Pima would use *ge'egeḍ a'al* to refer to older children in a family.

Hrdlička (1908:242) wrote that "*a-'a-li gu-gu-li* ('big children') is a plant the root of which is used by the Papago in fever. The root is broken up into little pieces, boiled, and the tea drunk a small quantity at a time. The root is also used in toothache. For this purpose it is ground up fine, mixed with some fat, and put into the cavity of the tooth. It is further used in neuralgic pains, when a mixture similar to that for toothache is applied externally to the painful part."

This left me with very little information with which to identify the plant, so the name sat for years in the category of unidentified folk taxa (appendix B). Even my Tohono O'odham consultant, Frank Jim, who knew this medicine, was unable to supply any details on what the plant looked like. He said the plant had tubers and probably grew more on the Mexican side of the border.

On the feast day of St. Francis of Assisi, 4 October 1992, I was at Magdalena, Sonora, for the great fall festival celebrating St. Francis Xavier. Indians come from all over Sonora and the border area of Arizona for the week-long celebration. I stopped in at Casa Natura, the medicinal herb shop there, and talked to the proprietress, Delfina Bravo, about what Tohono O'odham were buying from her, explaining my problem.

"Well, they don't say very much when they come in. They usually speak in their own language to each other, pick out a few things, and leave. Like the Yaquis, they buy a lot of this—*chuchubate* [Chuchupate, *Ligusticum porteri*]—and ointments for the skin."

I waited while customers came and went. Then an older man and woman, both well dressed in bright purples, came in and sat quietly in chairs at one end of the shop. I asked them in their language if they spoke O'odham.

"Sure," they both responded. Mary Miguel was from Al Chuk Shon, and Felix Antone from Poso Verde.

"Do you know a plant called *a'al ge'egeḍ?*"

They both nodded.

"You do?" I asked, hardly able to believe my ears.

"Sure."

"Do you know what it's called in *jujkam ha-ñi'okĭ* [Spanish]?"

"Ah!" said Mary, "it's something that begins with *pi-*, I just can't think of it, *pi- pi- pi-* . . . Oh, it won't come!" Delfina, finished with a string of customers, came over to join us.

"Where do you get it?" I asked Mary.

"Oh, I've bought it right here." She began searching through a rack of dried herbs in plastic sacks. In a few minutes she handed me one filled with brown tubers a finger joint or two in length, sharply tapered at each end. "Here it is—Pioniya—I just couldn't remember it."

A chill like electricity ran through my body. I was holding *a'al ge'egeḍ* after all these years. But what was it? What plant were these roots from? Delfina did not know, but she suggested I go ask the herb lady with a stall on the plaza, who gathers her own plants in the field and knows them thoroughly.

Out I went through the packed rows of stalls. I followed the crowd down the cobblestone street and pushed my way through the human mass to one of the tables set up at the edge of the plaza. Presiding behind a table packed with bags of roots, tubers, stems, and leaves sat a lanky lady with a gypsylike smile and long wavy hair, just beginning to gray. Olga Ruiz was a person I liked immediately. As soon as she finished selling Chuchupate to two ladies, I introduced myself and explained my problem. It was as if I had uncorked a bottle of champagne: the information came bubbling over. The flowers of Pioniya are yellow; they would be blooming about now, or have loose seed heads; and the leaves are like Yerba Buena, or mint *(Mentha* spp.). The plant is found on dry hillsides under oaks or mixed oaks and pines. It grows in the sierras near Nogales and Agua Prieta, Sonora; on the Arizona side I might try the Sierra Huachuca, above the fort.

Back in Tucson a few days later, I told Richard Felger what I had learned. He looked at the tubers, but nothing diagnostic jumped out at him. He referred me to Martínez 1979, in which there were columns for pages listing and describing plants called Pioniya or Pionilla in the Mexican vernacular. From the several dozen entries, only one was from the state of Sonora: *Zexmenia podocephala* A. Gray, a tuberous rooted composite. I walked over to the University of Arizona Herbarium, just a few blocks away, where I found pressed plants with tuberous roots just like those in my packet. And the plants had very coarse, large-veined leaves that did indeed resemble Yerba Buena, just as Olga had described.

Most of the Arizona specimens were taken from the southeastern border ranges, but two were from the Baboquivari Mountains, the range that forms the eastern boundary of the Tohono O'odham Reservation.

Even more interesting were some of the annotations on the specimens, reflecting uses described by the O'odham and two herbalists at Magdalena. There were three sheets collected by Howard Scott Gentry in his Rio Mayo studies in the 1930s. One said, "Roots decocted and drunk for indigestion and other stomach ailments"; another, "The tuberous roots highly regarded for the medicinal properties; decocted for stomach ailments." The prize was one Timothy Dunnigan took near Yecora, Sonora, while studying the Mountain Pima: "Root cooked with salt. Stomach remedy, and particularly constipation. Spanish name: *Pionia (Pioniya);* Indian [sic] name *Mármatham.*"

Back on the reservation a few days later I related my findings to Culver Cassa. "But the funny thing is, the Mt. Pima name isn't cognate with the Pima-Papago name."

"The words aren't, but I think the meaning is the same, anyway."

"How's that?" I asked, puzzled.

"The Mountain Pima name he wrote, *malmatham,* is probably [cognate with] our *maamaḍ* 'offspring', the young or

children of something, referring to those little tubers under there, just the same as our *a'al ge'egeḍ* does." Following Culver's lead, I checked Mathiot's (1973) dictionary and found the entry *maamaḍkam* 'to be the mother of several children'. Each Pioniya plant has from a half-dozen to several dozen of these fleshy tubers growing just beneath the ground—its children.

Vouchers: *1878* (Mary Manuel, Felix Antone, Tohono O'odham).

---

a'al ge'egeḍ   *Lasianthaea podocephala*

Lesquerella tenella A. Nelson
Lesquerella gordoni (A. Gray) S. Watson
Bladderpod
Brassicaceae (Cruciferae)
kopondakuḍ

In late winter and early spring a small mustard with bright yellow flowers may be found growing in the *a'akĭ* (arroyos) and other sandy places. Large plants may reach to your knee, but most are much smaller, depending on the amount of available moisture. As the petals drop, the little round pods that give this plant its English name enlarge. They are small but distinctively shaped, like two hollow hemispheres united, with an antenna poking out of the end opposite the stem.

One April, David Brown, Sylvester Matthias, and I went out on the bajada of the Sierra Estrella near Santa Cruz village to look for plants. They found some Bladderpods. David said they were *kopondam;* Sylvester called them *kopondakuḍ.* Each heard what the other was saying, and each stuck to his own pronunciation. But the two agreed, "The pods pop when you squeeze them."

David's name means "something full of air." We finally settled on *kopondakuḍ* as the better name for the plant: this means "something with which to make a popping sound," from *koponda* 'made popping' and *-kuḍ* 'agent' or 'instrument for'.

Once Sylvester told me, "*Kopondakuḍ* [makes] about the same noise as squeezing a *kekel viipiḍ* [Wright's Ground-cherry, *Physalis acutifolia;* see account]. Sounds

like a .22. [Can] play a joke with it, in a crowd—go smash it on the forehead, and—oh!—makes a noise!"

Likewise, Ruth Giff said, "*Kopondadag*—the pods, the berries—we snap them." This appears to be its sole function.

A white-flowered species, *Lesquerella purpurea,* grows on the higher reaches of the Sierra Estrella above 2500 feet. Presumably this also would have been called *kopondakuḍ* by the hunters who once stalked big game in this range.

For the Tohono O'odham, Mathiot (1973:496) gives *ge'egeḍ-kopon-dakuḍ* as "locoweed or milk-vetch, a plant with seed pods that are full of air." This needs to be verified with specimens. Her name means 'big *kopondakuḍ*', but she does not give *kopondakuḍ* by itself as a plant name.

Vouchers: *18* (SM, DB).

Ligusticum porteri Coult. & Rose
Chuchupate, Oshá
Apiaceae (Umbelliferae)
jujubáádi

For a long time *jujubáádi* was listed in my notes under the heading "unidentified plant, medicinal." Many people had told me about it. David Brown in Santa Cruz village opened a baking-powder can and showed me two sticks (maybe they were roots), appearing yellowish under the bark. "It's *jujubáádi*," he said, "for constipation; make a tea. Comes from Mexicans. Takes away fever, too." The placement of the accent suggested it was a loan word.

Other older people knew the medicine as well. George Kyyitan said, "*Jujubáádi*— I used that one, too. Bite off, like that *sivijul* [Canaigre, *Rumex hymenosepalus*], and chew in your mouth. It's almost the same as *vavish* [Yerba Mansa, *Anemopsis californica*]."

Sylvester Matthias told me, "*Jujubáádi* is a root from Mexico, peddled by the Yaquis of Guadalupe village to the Pimas. Called Chuchupate in Spanish. It was really popular with the Pima." Here was a lead to work on.

When I was attending an annual ethnobiology meeting in Albuquerque, Takashi Ijichi and I spent some time there with two dear friends: ethnobotanists Edelmira Linares and Robert Bye from Mexico City. Failing to find any place open in Old Town, we returned to the hotel lounge, where a band made any attempt at conversation an impossibility. Edelmira had been studying the markets in Mexico. When there was a break between songs, I told her of my problem identifying *jujubáádi*. Her eyes lighted up. She was about to tell me something when the band resumed full force; I could not hear what she was trying to say. She motioned for me to hand her my little pocket notebook and pen. She wrote, "Oshá, chuchupate = *Ligusticum porteri* Coult. & Rose."

A month later, Sally Pablo, Ruth Giff, Sylvester Matthias, and I caravaned down into Sonora to visit the Pima Bajo on the lower Rio Yaqui. At Magdalena we stopped in at Casa Natura, the medicinal herb shop. I explained in Spanish to the lady behind the counter that I was looking for Chuchupate. Did she know it? She did. But she was out of it and had been for

jujubáádi   *Ligusticum porteri*

The specimen illustrated (TRV s.n., ARIZ) *was collected by Tom Devender in the Chiricahua Mountains of southern Arizona. We do not know the source of the medicinal plants that were once bartered or peddled to the Gileños.*

some time. It comes from the mountains east of Mexico City; or it could come from the sierras in Chihuahua, if the people would go out and dig the roots.

Four years went by, and I despaired of ever finding this Pima medicine. In January 1990, after we finished a few hours of taping, Sylvester and I went to Phoenix for a Mexican meal. At Seventh Avenue and Baseline Road, I spotted a bright new shop with a big sign proclaiming "Yerberia Gonzales." Inside, a lady in a white druggist's smock was tending a constant stream of Hispanic customers. I thumbed through several books while waiting my

turn. Finally, I inquired about Chuchupate. After looking over two walls covered with plastic bags of dried leaves, seeds, twigs, and roots, she took down one labeled "chuchupaste."

I took the little bag of brown, misshapen roots out to Sylvester, who was waiting in the pickup. He looked them over carefully, then sniffed at the contents. "This is it—this is *jujubáádi*. This is it."

"How do you use it?"

"Just like tea, a drink, mainly for fever. *Hot!* Boil it. I guess it can be used in hot water, just the same as to boil it."

"How long?"

"Oh, maybe just ten minutes, about; just so it get[s] soaked up. Then, hot. My late wife [got it]—from some Yaqui elderly couple, [who] were selling *jujubáádi*, peddling on foot in those days. Of course, now it's just the same, they're scarce. Everybody were after *jujubáádi*—don't know where to get it. But those Yaquis took advantage of traveling on foot, to sell [it]."

"How long ago?"

"*Way* back. Must be in early—around the 1920s, somewhere in there. That's why she was a good-sized girl when her grand-aunt bought that *jujubáádi*. And she have to be careful with it, too, not to let it go to waste. Of course, they don't know where to get it after."

"Can you use that root more than one time, or just one time?"

"I guess it can be used more than once. Can dry it up and boil again. I guess it goes by the taste. When you won't taste no *jujubáádi*, well, that's no good."

The roots have a sweetly medicinal smell. I tasted one. The astringent flavor resembled that of wild carrot. I thought the tea would be sweet also, but it is bitter and unpleasant.

On subsequent visits during the October celebrations in Magdalena, Sonora, I would visit herbalist Olga Ruiz on the plaza. One of the main items Indians of various tribes would be buying from her was this plant.

Vouchers: *1877* (SM; Mary Manuel, Felix Antone, Tohono O'odham).

Lupinus sparsiflorus Benth.
[≠ *Malva parviflora* L.]
Desert Lupine
Fabaceae (Leguminosae)
tash maahag

When the winter rains have been sufficient, the roadsides, particularly through the sandy bajadas, may be densely lined with lupines. In March and early April the flowers may be so dense as to attract the attention of even the traveler speeding past on the adjacent blacktop. For those who stop for a closer look, the erect spikes are seen to be masses of small pealike flowers that are blue and white. The pubescent leaves are composed of five or more elongate leaflets radiating from a common center.

The Pima name for this plant almost slipped into oblivion. Ruth Giff, David Brown, and Sylvester Matthias each assured me every spring when I took them specimens that they knew no name for it, although they were familiar with the plant.

Finally Carmelita Raphael of Santa Cruz village, who picks cholla buds on the nearby bajada each spring, told me it is *tash maahag* or *haahag*—she could not remember which. "See these little hands— they just follow the sun." Later George Kyyitan of Bapchule told me the same: "*Tash* means 'sun', and *maahag* means 'the palm of the hand with the fingers', like that," he said, spreading his short fingers like the leaflets of a lupine. I checked the flora. Botanists call this arrangement "digitately compound." The word is also related to another Piman word, *maahagam* 'fan palm', still used by Pima Bajo and Tohono O'odham but curiously absent from Gila Pima (see Group E, *Washingtonia filifera*). In one of those odd parallel evolutions of language, in English *palm* means both a part of your hand and the tree. However, the English *palm* includes both the pennate (featherlike) palms and the fan palms, whereas the Piman *maahagam* is restricted to fan palms, the only kind occurring aboriginally.

Fr. Antonine (1935) listed a plant, identified simply as "similar to four o'clocks" as *tash haahag* 'sun leaves'. This must be a variant or misnomer for *tash maahag*, as with Carmelita's usage.

The lupine tracks the sun throughout the day, closing its leaflets to the night's dew. There is some vague association here with sun worship, though I have never been able to pinpoint anything explicit. Russell (1908:250) summarized Gileño religion as "a belief in the supernatural or magic power of animals, and especially in the omnipotence of the Sun." One does not get this impression from the various versions of the Creation Story, where the exploits of the four original beings or culture heroes—Jeweḍ Maakai (Earth Doctor/Magician), Se'ehe (Elder Brother), Ñui (Buzzard), and Ban (Coyote)—are recounted. But the Akimel and Tohono O'odham (who use the same name for lupine) were not ones to overlook this little plant that seemed to give reverence to the sun throughout the day by holding out its hands. And so they named it.

Another time, I asked George again about the lupine. We were going through an *Arizona Highways* issue with photographs showing thick growths of *chuuvĭ taḍpo* (Owl-clover, *Castilleja exserta*), *hoohi e'es* (Mexican Gold Poppy, *Eschscholzia mexicana*), and lupines. "*Tash maahag*," said George, pointing out the indigo flowers. "The leaves goes up to the sun. When it's down, then they go to sleep. Then in the morning, they be just the same thing. Every time they just follow the sun. This one they use for decoration in church. When Easter comes, they gather up this, you know, when it's time to throw them. They use that one. Because they come on those days . . . *tash maahag*." Then George qualified himself on this bit of folk Catholicism. "*Used* to, long time ago. But it's gone. They say [call] that time *ha'ichu hiósig ñeenchuda* 'throwing flowers'. Way back." Like many other elements borrowed from the Yaqui, this appears to have originated from their elaborate Easter ceremonies. At the Yaqui "Gloria" on Holy Saturday, flowers, confetti, and leaves are thrown at the *fariseos*, who are finally overcome with these *sewam* 'flowers', symbolic of divine grace.

I mentioned this to Sylvester, who is from a western village. "We do it on Corpus Christi [June], *ha'ichu hiósig ñeenchuda*. It's really a Papago custom. Yeah, we do that. Of course, then there was plenty of rain. They'd go right along the washes at that Muhaḍag Do'ag [South Mountain]. Just any kind of wildflowers. They'd throw them wherever the repository is set up. It was a long ways at that time." It is difficult to imagine today any wildflowers persisting this late in the season, but the native consultants in this study were growing up during a wet period. An extended drought ended in 1906, and the pendulum swung abruptly to the opposite extreme.

This plant has no other practical use among the People, as far as is known today. I wondered why the Pima did not harvest the seeds of this often prolific annual: in other parts of the world, other species of *Lupinus* are processed for food. Then I learned that the seeds of certain Arizona species contain toxic alkaloids, and the common reservation species,

L. sparsiflorus, is one of the most poisonous ones known (Kearney and Peebles 1960:414).

Vouchers: *759* (CR); *978, 1901* (GK); *1902* (GK, *L. concinnus* Agardh.).

Machaeranthera arida Turner & Horne
Desert Blue Aster
Machaeranthera tephrodes (A. Gray) E. Greene
Hoary Aster, Ash-colored Aster
Erigeron lobatus A. Nelson [?]
Fleabane, Wild Daisy
Asteraceae (Compositae)
chehedagĭ hiósigam

From late winter to early spring, the very small Desert Blue Aster springs up where fine silt has puddled winter rains. The plants, common on the saltbush flats, may be no taller than your thumb or several times that. The finely divided leaves have a foul smell. The flower head is made up of a yellow disc with blue to lavender rays. This little xerophyte is a species of *Machaeranthera.*

Occasionally, later in the spring or even in the autumn, another *Machaeranthera,* Hoary Aster, will add a splash of lavender to streamsides or roadsides through fields. This is a taller plant, about knee- to hip-height, with many asterlike flowering heads. Fleabane, or Wild Daisy, with narrower pale lavender ray flowers, may be found here as well, though it is more common in the desert sierras.

Of Desert Blue Aster, Sylvester Matthias told me, "*Chehedagĭ hiósgam* 'blue flowers it has'. I used to see them all over—and again this year—at one time right in the village there—just there—not way out in the hills, but just around the village. No use—just decoration—they [the Pima] liked them around there." We were on the bajada of the Sierra Estrella below Santa Cruz, and there had been a particularly dense growth of this little composite. Over the years we found these plants in the field many times, and he always called them by the same name. When we found Hoary Asters, Sylvester said that they too would be included in the Pima concept of *chehedagĭ hiósgam;* Ruth Giff concurred.

Fleabane may be, as well. Once Sylvester included the tiny Woolly Eriophyllum, *Eriophyllum lanosum* A. Gray, even though it has white ray flowers.

We left Hermosillo on one crisp October morning, heading for the Pima Bajo village of Onavas. I stopped at a roadside shrine to pick a few caltropes (*Kallstroemia* spp.) and other specimens to show to don Pedro Estrella.

Sylvester noticed bright blue wild morning-glories sprawling over the grasses and small shrubs at the road edge. "*Chehedagĭ hiósgam,*" he volunteered. "It used to grow in sandy places [on the Gila], but guess it's all gone now. Maybe seeds got lost in dry years. Didn't grow in fields, but on desert. As long as it's wet." A small-flowered pale magenta morning-glory I found at the Salt-Gila confluence (*Convolvulus equitans* Benth.) he called by this same name.

The Pimas who contributed to this ethnobotany were all quite rigid in their definitions of folk taxa, but none was narrower than Sylvester. A plant that did not appear to him quite right, or was out of place, or was blooming at the wrong season he would only admit to being "like" such-and-such. I was puzzled (I still am!) that he would call these vining plants by the same name he had always given me for the little blue asterlike plants that appear abundantly some years with the late winter rains.

When we got to Onavas, Pedro called all the different species of morning-glory we found in his *milpa* and along the river edges and arroyos simply *trompillo,* the Spanish name. He knew no Piman name for them.

Vouchers: *25, 305, 366* (SM, *M. arida*); *503* (RG, SM, *M. tephrodes*); *705* (SM, *Eriophyllum lanosum*); *1764* (SM, *Ipomoea triloba*); excluded *741,* a robust specimen of *M. tephrodes.*

Malva parviflora L.
Cheeseweed
Malvaceae
ñiádam chu'igam, ge'egeḍ haahagam, s-tadañ haahag

From midwinter until the drying weather of spring, one of the most abundant "weeds" of Gila Pima country is an Old World plant known as Cheeseweed. It grows in disturbed areas throughout the floodplains, being very common in yards and roadsides. This erect and sometimes bushy annual has broad, rounded dark green leaves, with the veins spread from the leaf stem like spokes of a wheel. The flowers are inconspicuous but look like minute pink hollyhocks (*Althaea rosea*), a related plant. The fruits are more noticeable. These look like a tightly coiled green caterpillar (the immature seeds) resting in a green cup (the calyx). As they dry, the individual seeds fall away like the triangular cuts from the old-fashioned cheese wheels, a resemblance that provides the English name, Cheeseweed.

The Pima have not been able to overlook this invader. At the west-end villages, the plant is well known as *ñiádam chu'igam* 'Eremalche it looks like' because of its similarity to *ñiádam* (Star Mallow, *Eremalche exilis*), a native plant formerly used as an *iivagĭ* 'eaten greens' (see Group F). The two grew about the same season. The name was also glossed for me as 'like *ñiádam*' or 'false *ñiádam*'.

This was the name we thought that everyone knew this plant by. Then one January I was to give a talk at Bapchule on "The Other Pima" from Sonora and Chihuahua. I picked up George Kyyitan and drove him to the community center, where I was speaking. He said he had just started working again and would be pretty busy through the spring.

"What are you doing?"

"We're cutting down these *ge'egeḍ haahagam.*"

"The what?"

"These weeds," he said, pointing to the *Malva* growing thickly along the roadside.

Later that day Sally Pablo, who had arranged my talk, told me, "I was talking with Dorothy Kyyitan at the meeting and she told me that George is working now cutting the *ñiádam chu'igam,* but she called it *ge'egeḍ haahagam,* not what we call it! That seems almost childish, it's just descriptive: 'it has big leaves'."

"I know. I got the same thing today from George," I told her.

"I always thought that the way we learned it was the correct way. Everyone here knows *ñiádam chu'igam.*"

As it turned out, many people from other parts of the reservation knew *ge'egeḍ haahagam* (*ge'egeḍ* 'large' [pl.]; *haahag* 'to

ñiádam chu'igam, ge'egeḍ haahagam, s-tadañ haahag   *Malva parviflora*

Matricaria matricarioides (Less.)
Porter
Pineapple-weed, Te-de-piña
Asteraceae (Compositae)
u'uv s-puluvĭ

If the rains come just right, puddling
up on exposed, packed soils, there may
follow a thick growth of this bright green
feathery-looking annual. The most inhos-
pitable soil seems to be to its liking. I used
to find it commonly for a few years along
bare edges of Riggs Road between I-17
and Maricopa Road; then it disappeared.
The plant eventually produces numerous
cone-shaped greenish yellow flower heads,
lacking "petals" or rays. The whole plant
has a pleasant sweet odor, suggesting the
smell of pineapple, hence both its Spanish
and its English name. Both groups have
learned to appreciate the mellow tea that
comes from the dried filamentous leaves
and flower heads.

Mayweed (*Anthemis cotula* L.), naturalized from Eurasia, has been found
occasionally on the reservation and is
apparently included in the same Pima
folk taxon. It has white ray flowers, giving
it the appearance of a daisy. This is the
common chamomile of English herbal
teas, called in Spanish Te-de-manzanilla.

One day I noticed some Pineapple-
weed along the roadway and stopped
to pick a plant for Sylvester Matthias,
who was riding with me. He crushed it,
smelling the fragrant foliage. "*U'uv s-
puluv,*" he pronounced. "It means smell—
'smelly clover'. Not used by Pima. Just
pull it and smell it."

"What about drinking it?"

"They experimented with it for tea,
but they didn't like it."

On his next visit to San Diego, I pulled
the jar of Te-de-piña from our overstocked
shelves of seeds, teas, and spices, and fixed
him a hot cup of the tea, enhanced with a
bit of honey. He liked the tea.

"All we need is a good rain," was his
comment. The little plants appear after a
good rain.

Another time I showed him the plant
and he identified it again as *u'uv s-puluv,*
saying, "It smells like apples."

The Pima name is an interesting
one: *u'uv* means 'smell', as Sylvester had
explained. And *s-puluvĭ* can mean 'Sweet-
clover' *(Melilotus indica)* or 'alfalfa'

have leaves' or 'to be leafy'; and *-gam,*
attributive).

But we had not reached the end yet.
I asked Irene Hendricks of Casa Blanca
about this weed, thinking she would give
me the same name others had supplied
from Bapchule and Sacaton.

"*S-taḍañ haahag,*" she responded, "'it
has wide leaves'" (*s-,* intensifier; *taḍañ*
'wide', 'broad'; and *haahag* 'to be leafy').

And so I learned a third Gila Pima
name for an abundant weed. And
although there was no agreement on
what to call this plant, there at least was

a consensus that the plant was not used
by the People.

And lest we go for a fourth name, let
me add here that Curtin (1949:79–80) or
one of her consultants must have confused
the plant with the lupine, as she gives that
plant's name, *tashmahak,* for Cheeseweed.
And the use she quotes from Russell really
is for the native *ñiádam,* which has nearly
(if not completely) disappeared from the
reservation (see *Eremalche exilis,* Group F).

Vouchers: *4* (JG, RG, DB, SM, SP); *LSMC 3,*
confused biological referent.

(*Medicago sativa*) depending on which part of the reservation the speaker is from. Culver Cassa showed me that it is really three different words: *s-*, the intensifier; *pul,* a plant perhaps forgotten; and *uuv* 'smell'.

Technical Notes: AMR 789, Maricopa Colony, *A. cotula.*

Vouchers: *1846* (SM).

Medicago polymorpha L.
    [= *M. hispida* Gaertn.]
Bur Clover
Fabaceae (Leguminosae)
s-ho'iḍkam s-puluvĭ

At the marvelous desert cienega just north of the former Pima settlement of Imuris, Sonora, Sylvester Matthias and I stopped one spring day on our way back from Pima Bajo country. I have always been interested in learning more about water plants that might once have grown on the Gila. We had stopped there earlier to look for plants (see *vak,* Group B: bulrushes, *Scirpus* spp.). Sylvester picked one of the roadside weeds.

"Have we ever gotten *s-ho'iḍkam s-puluvĭ?*" he asked.

"No, not yet."

He was carrying a piece of the prostrate Bur Clover, a troublesome weed from the Old World. It has small yellow flowers and spreads its seeds by means of its coiled fruit or bur. The plant may radiate a foot or two from its root. He said it was already a common plant when he was young.

I had once sent him a small pressed piece of Bur Clover, but he had not recognized it until he saw it growing in context. This should caution us about relying on elicitations made with herbarium specimens.

The Pima name means 'spiny' or 'thorny' clover (*s-*, intensifier; *ho'i,* 'thorn' or 'spine'; *-kam,* attributive marker); *s-puluvĭ* can be Sweet Clover (*Melilotus indica*) or alfalfa (*Medicago sativa*) (see Group C); its etymology is given above.

Ruth Giff also knew the plant, which she described as "a kind of *s-puluvĭ* that grows flat."

Vouchers: *1188, 1648* (SM).

Mentzelia spp.
Blazing Star
Loasaceae
iks s-hoohoidam

There are about five species of *Mentzelia* that have been found on the reservation. These range from the tiny-flowered, weak-stemmed *M. affinis,* found in the washes and usually supported by another small bush, to the large, showy *M. pumila,* an erect species found on mountainsides or other well-drained sandy places. Whether large or small, the flowers are golden-yellow stars with a mass of yellow stamens in the center. They are aptly named Blazing Stars. The stems seem to shed their bark in white scales. Even if you are unobservant of their strikingly attractive flowers, the leaves and stems will catch you should you brush against them; they are covered with minute hooked hairs.

I was never able to discover a name for this plant in the west-end villages. Then at Sacaton, Irene Hendricks told me about *iks s-hoohoidam* (*iks* 'fabric', 'textile', or woven material of any kind; *s-hoohoid* (also *s-hooho'id*) 'it likes'; *-dam,* attributive). "When you get it and stick it on your clothes; it just sticks because they like it." Others in the east end also knew the plant. When I went back to west-end people with the name, some told me they had heard of it but did not know what plant it referred to. For instance, Sylvester Matthias said, "I heard [it] from my grandmother. I don't know how it got that name. I just *heard* of it. I don't know what it looks like or where it grows. The flower likes that cloth or rag or whatever it is, in the definition of an *iks.*" Here was cultural erosion at work, even among the people who seemed to have preserved the most knowledge of things ethnobiological.

No one knew a use for this plant, even though elsewhere the seeds were an important food resource, high in oils (Ebeling 1986:127, 248). Perhaps it did not occur in sufficient numbers in Pima country to have been important in the diet. More likely it has not been used for so long that it has dropped from memory.

Technical Notes: Species of *Mentzelia* collected in Pima country include *M. affinis* E. Greene, *M. albicaulis* Douglas,

M. involucrata S. Watson, *M. multiflora* (Nutt.) A. Gray ssp. *longiloba* (Darlington) Felger, and *M. pumila* (Nutt.) Torr. and A. Gray var. *pumila.* All are included in the Pima folk taxon without distinction.

Vouchers: *1314* (IH); *1889* (GK).

Nicotiana trigonophylla Dunal
    [≠ *N. bigelovii* (Torr.) S. Watson]
Desert Tobacco
Solanaceae
ban vivga

Ban vivga 'Coyote's tobacco' grows wild throughout Gila Pima country and the Southwestern deserts in general. Although it might be found in just about any disturbed areas, the best place to look for it is in sandy arroyos. Some years, when the winter rains are just right, these erect green plants, about knee-high, will be common. The leaves, clammy and foul-smelling, may be six or more inches long and two and a half inches wide. Most of the leaf bases clasp the stem. The plants are crowned with greenish white or greenish yellow tubular flowers.

One hot day in late May, we bounced down the sandy road that skirted the Sierra Estrella bajada from Santa Cruz to Hidden Valley, plunging in and out of sandy arroyos. Not much was in bloom. I spotted a Desert Tobacco plant, stopped to pull it up, and took it back to Sylvester, who was sitting in the pickup. He did not recognize it, but I asked him to smell it. He crushed the viscid leaves. Then his face lit up.

He said, "It's *ban vivga* 'Coyote tobacco'; my father told me it grows along washes. When tobacco was scarce, they'll have to use the *ban vivga.* They say it's way stronger than *viv,* the tobacco they used to raise, and [it has] kind of a bitter taste. Not a drug, but ordinary smoking tobacco, but strong and bitter. Don't know what it looks like exactly, but [it] grows in washes. It's a native plant."

Almost everyone with whom I worked knew *ban vivga.* Castetter and Bell (1942: 108) considered this the first tobacco used by both Akimel and Tohono O'odham, adding, "It is our firm belief that neither Papago nor Pima cultivated any kind of tobacco aboriginally . . . but that they did smoke the wild *N. trigonophylla* and *N. attenuata*." (I do not share their opinion; see Group I, *Nicotiana rustica*.)

Gary Nabhan (1982:75–86) has explored the concept of Coyote's plants among Piman peoples. These are species genetically related to cultivated plants but that somehow have been "spoiled" by the machinations of the trickster. Usually the product is inferior, being either smaller or of poorer quality. Although in the cultural concept the wild species is considered to be a resulting degenerate form of the People's cultivar, in actuality the Coyote's form is the older one—sometimes the actual progenitor of the cultivar, as with teparies *(Phaseolus acutifolius)*.

Joseph Giff told Nabhan the story of the origin of the *aaḏo* or macaws, later called peacocks. It is also the story of a tobacco. An old lady instructed her two grandsons to take their birds far away and release them because the People wanted to steal them; when the boys returned, they would find that the People had killed her. She instructed her grandsons to bury her body in the soft sand of an arroyo and watch it for four days. Some kind of weed would come up.

"So they did. They took her out and buried her and they watched her grave and then weeds started coming up. And when it got to looking like she had told them what it would look like when it's ready, before they got to pick it, why Ban [Coyote] was running [by]; this mischievous creature was moving around and I guess he found it and he picked some. He knew what it was and he picked some. And this, I guess, was some kind of tobacco, because they were already smoking. But these old people, the wise men, they have a place they call *jeeñ kii,* you know, where they gather, a smoke house you might call it. And Coyote came over and he never had any *viv* to smoke, and he asked them if they didn't have any tobacco. No, nobody has tobacco. They just didn't want to give him any, for he never had anything to give them to smoke.

"Then old Coyote said, 'Well, you don't have it,' he says, 'I'm gonna smoke; I have tobacco of my own here.' I guess he was carrying that. They use the thin corn husk, you know, the [one] next to the kernels, real thin like paper, and they use that to roll tobacco in, which is good paper. So he got [it] out, but he had to wet it; you

know, if they don't wet it, it'll crack and break. So he wetted it and put the tobacco [in] and started smoking, and that's how they got that name, *ban vivga*. It wasn't, it doesn't belong to him. He stole it. But he got the credit for the name of that tobacco, *ban vivga*."

Although the cultivated species of tobacco and *ban vivga* are not the same species, the concept of inferiority applies. Forty years ago Castetter and Bell (1942: 108) recorded, "Both tribes regarded it as inferior in strength and quality to common tobacco *(N. tabacum)*." Still, the Gileños seem to have long maintained some ceremonial use for *ban vivga*. But I know of no modern use, and it is doubtful that anyone living now even knows how to prepare it.

Ban vivga is apparently a pan-Piman name for *N. trigonophylla*. This is the name I have always been given for it by Tohono O'odham. And with the Pima Bajo of southern Sonora, where the plant grows commonly about the villages, I was given the name *ban vivaag,* which they glossed 'tobaco cimarron' or 'tobaco del coyote'. Curiously, Russell (1908:119) gave *viopar viof,* glossed 'like tobacco', for *N. trigonophylla,* saying, "It is gathered near Baboquivari by the Papagos and brought to the Pimas." (For the identification of *viopal,* see Group E, *Hyptis emoryi*.) But this species is so abundant along the Gila that there would seem to have been no need for such importation. Russell assigned the name *pan viofu [ban viv]* to *N. bigelovii* (Torr.) S. Watson, a wild species from the Pacific coast. Surely this is *N. trigonophylla*.

This leaves a second wild species of tobacco, *N. attenuata,* and it is not known how this was treated in the Gila folk taxonomy. A robust specimen was collected in 1926 from the Gila River bottomlands, but this species is partial to riparian situations and long ago disappeared from the reservation. As a result I have been unable to learn whether any contemporary Pima know it, though I do not doubt it was once in their lexicon. In a footnote, Russell (1908:119) called this species *shegoi wecho viv,* but I suspect there is some error here. This name means 'under the Creosote Bush tobacco': this is the Tohono O'odham name for cultivated *N. tabacum*. The Pima concept from the late 1930s, as recorded by Castetter and Bell, is interesting: "The

Pima and Papago smoked to some extent, but never cultivated *N. attenuata,* which has a wide distribution in western North America. . . . More recently, the Pima use of it has been confined to the boys when they wanted to 'sneak' a smoke. A Pima informant referred to it as the 'uncle' of cultivated tobacco. Russell wrote of the Pima calling this species 'under-the-creosote-bush tobacco,' a statement which we have been unable to confirm; in fact, our Pima informants had never before heard this name." (Unfortunately, these authors studiously avoided giving any Pima names in their classic study, so

ban vivga   *Nicotiana trigonophylla*

we are left with no clue as to what the People of this time period actually called *N. attenuata.*)

I asked Sylvester Matthias how he would distinguish *viopal viv* (see Group E, *Hyptis emoryi* account) and *ban vivga.* He thought they were about the same.

Francis Vavages said, "They used *ban vivga,* too, for smoking. They came out under a mesquite tree. They just came out like that." This apparently refers to *N. trigonophylla.* The other wild species, *N. attenuata,* is more partial to the riparian community.

This usage agreed with what was known from the Tohono O'odham and Pima Bajo, as noted. But apparently this was a floating taxonomy, a situation where different people at various times applied the name to different species, as with devil's claw (see Nabhan and Rea 1987). If only wild and cultivated tobaccos were being contrasted, it was simply *ban vivga* versus *viv.* But there were once two wild native species along the Gila. And eventually there were two cultivated species as well, distinctive plants with different-colored flowers (see Group I, *Nicotiana rustica*).

With the acquisition of a second domesticated form on the Gila, some people, I believe, began using the name *ban vivga* for the lesser crop *(N. rustica).* Joe Giff told me once that *ban vivga* was cultivated, as did George Kyyitan. George Webb (1959:17) wrote, "The tobacco they grew was called *ban-vivega* (coyote tobacco). Only a few tobacco plants were grown by elderly Pimas as only the old people and the medicine-man smoked." Manuel Lowe at Sacaton gave Castetter and Bell (1942:111) seeds of domesticated *N. rustica,* calling them Coyote's tobacco. Unfortunately, in all three cases we do not know how (or even if) these men contrasted the wild species with *N. rustica.*

Technical Notes: Michael Nee (in Hickman 1993) relegates the name *N. trigonophylla* to the synonymy of *N. obtusifolia* Martens & Galeotti. *N. bigelovii* (Torr.) S. Watson is now known as *N. quadrivalis* Pursh.

Vouchers: *307* (sm?); *1365* (?); *1273* (sm; Pedro Estrella, Pima Bajo: *ban vivaag*).

Oenothera avita (W. Klein) W. Klein ssp. *arizonica* (Munz) W. Klein [= *O. deltoides* Torr. & Fré. var. *arizonica* Munz]

Arizona Evening-primrose

Oenothera deltoides Torr. & Fré. ssp. *deltoides*

Dune Evening-primrose

Onagraceae

vippi si'idam

In spring, about April and May, the sandy lower terraces immediately above the desert riverbeds are often covered with two low-growing plants, the shorter ones hugging the sand and bearing deep purple-pink flowers, the more erect ones gleaming white. The prostrate purple-pink ones are Sand-verbenas *(Abronia angustifolia).* The white ones are evening-primroses. Each flower has four petals, a large central stigma, and circling stamens that produce webby strings of copious pollen. When not blooming, the plants are easily overlooked. The dull grayish green leaves are narrow and deeply serrated. But by the time late spring is beginning to flex its muscles at midday, preparing for the real heat of summer, these plants are covered with large fragile flowers. Opening in the cool breezes of the evening, they last though the cooler morning hours, collapsing finally in the heat.

Ruth Giff first identified this plant for me. "It's *vippi si'idam.* I don't know why we call it that. It means 'sucking at the breast.'" *Viip* is 'milk' and *vippi* 'breast'; *sii* and *si'i* are 'to suck at the breast', so *si'idam* means one who nurses at the breast continuously. Another time Ruth said, "There were white blossoms, *vippi si'idam,* [along the] sandy road—[means] 'sucking on the nipple'—*big* blossoms, the size of *biibhiag,* you know, morning-glory [Field Bindweed, *Convolvulus arvensis*]. I wonder why they named it that!"

No one remembered why this beautiful flower had such a name. I thought it might have come out of some legend, now long forgotten. One time I was talking with Frank Jim, a Tohono O'odham originally from the western part of his reservation.

"Do you know a plant the Pima call *vippi si'idam?*"

"Yeah. There are two kinds over there— a white one and a yellow one [*Oenothera*

primaveris A. Gray]. They grow in the sand. But I only see the white one here [in Pima country]."

"Nobody remembers how it got that name," I commented.

"Well, did you ever put one of those flowers on your *vippi?* Then you would know."

"No. I don't have a *vippi.* I have a *baasho.*"

"Well, [it] don't make no difference," he said, rubbing his hand unconsciously over a pectoral as he spoke. "You just pick one of those flowers and put them on there. They just stick on." A picture flashed into my mind of brown-skinned little Piman girls of some innocent time, long past, playing in the river sand among the Sand-verbenas and evening-primroses, pulling off the big white flowers to attach to their nipples, shouting, "*Vippi si'idam! Vippi si'idam!*"

One April day I was out hiking in a wash and picked some tiny yellow evening-primroses (*Oenothera leptocarpa* E. Greene [= *Camissonia californica* (Nutt. ex Torr. & A. Gray) Raven]). The flowers soon closed. When I got back to the house, I put them in water on the table, waiting for the next set of flowers to open before pressing them. They opened about seven that evening. Ruth Giff happened in, looked at the slender erect stalks, and said these were *vippi si'idam* also. "But the other kind spreads out at the base, with larger white flowers; this one looks like a mustard." After Geoff Levin identified my little yellow plant, I looked up the species in Kearney and Peebles 1960 (p. 599), which noted, "The cruciferlike aspect of this plant is striking." Another source (Beauchamp 1986:196) called it False Mustard.

The spectacular white evening-primroses are almost gone now from the reservation. I have found them regularly only around the Maricopa colony on the lower Salt River, never along the Gila where Ruth remembered them and only once on the sandy riverbed at Sacaton after some flooding.

Technical Notes: There are two species with large white flowers in Pima country. Leaf characters that distinguish the plants elsewhere do not hold with the local subspecies, and even botanists have usually confused them. *O. avita* has distinctly spotted sepals that are conspicuously hirsute. These may be so woolly as to obscure the maroon spots.

The petals become distinctly pink before closing. Its seeds are dark and smooth. *O. deltoides* has glabrous sepals with few or usually no spots. Its petals may be slightly pinkish. Its seeds are pale and pitted. The two often grow together on the lower Salt River, but one opens its flowers in the afternoon, the other in the evening, closing later the following morning. Apparently all *Oenothera* and *Camissonia* species are included in the folk taxon, but the large white ones seem to be the focal species.

Vouchers: *754* (RG, *Camissonia californica*); *769* (RG, *O. avita*); *1037* (SM, *O. avita*); *1038* (SM, *O. deltoides*).

Orobanche cooperi (A. Gray) Heller
    [= *O. ludoviciana* Nutt. var. *cooperi*
      (A. Gray) G. Beck and var.
      latiloba Munz]
    [≠ *O. multiflora* Nutt.]
Desert Broom-rape
Orobanchaceae
mo'otaḍk

In late winter your hike along some bajada road or up through an arroyo might be interrupted by thick brownish purple fleshy heads pushing their way up through the sandy soils, asparaguslike. These heads, usually single, are leafless. As they mature their sides are decorated with lavender and yellow tubular flowers. There is not a hint of green. Most of the thick succulent stem is underground, pale and yellowish. It ends abruptly. Roots attach the purple and yellow stalk to its food source, the roots of some green bush nearby—often one of the bursages—a plant with the magic for making food. This obligate association was not lost on the early Pima, who named one of the bursages *mo'otaḍk jeej* 'broom-rape's mother' (see *Ambrosia confertiflora* account, Group D).

All the old people knew the name for this plant and knew that it had once served Pimas as a food. "*Mo'otaḍk* is *so* bitter, don't know how they could eat it," remarked Ruth Giff. When Sylvester Matthias identified specimens, he added, "They say it's bitter," but he had never eaten them himself. And so on among all the others I asked.

Curtin (1949:49) obtained somewhat more definite information: "When broom-rape first appears in the spring, the tender young sprouts resemble asparagus and, although bitter, are used in the same manner. They are cooked by covering with hot ashes and baking them in the fireplace; and this is still done by a very few of the old people. The lower part is eaten, but the upper end is discarded, declared George Webb."

At the turn of the century, Desert Broom-rape seemed to be more widely used. Russell (1908:75) did not identify the plant but recorded its native name, saying, "This is described as resembling asparagus. The stems may be eaten raw or boiled or roasted in the ashes." And Hrdlička (1908:264) noted, "In the case of *mo-o-tatřk* (Orobanche multiflora), the entire plant is used for food. It is somewhat bitter. The Pima cooked it without the addition of salt or sugar or other substance."

The plant must have fallen into disuse about this time, however. Sylvester commented, "The elderly in my generation [his youth] sure get that *mo'otaḍk*."

"Cooked or raw?"

"Raw. But they say it's bitter. But they eat it."

When I was working at Sacaton one spring, Frank Jim, my Tohono O'odham friend, was passing by and stopped to look at my morning's collection of specimens. Coming to the Desert Broom-rape, he said, "*Mo'otaḍk*—some of them are good to eat. Have to pick them early, when small; dig them out of the ground when they just [start to] come up. Can just eat them."

"You mean they're eaten raw?"

"Yeah, both [ways]. Just clean it and eat it. When finish, your lips are all black. We used to go around washes, where soft [sand is], and dig out the young ones. Sweet. But if [it's] too old! Take a bucket. Just eat up to where it gets hard, the soft part; [the] rest [is] too bitter."

He pointed to some of my specimens that were already beginning to discolor in the midday heat. "Those black ones they use for medicine. They're good for stomach problems. Just go get them and put them in a bag. Let them dry. Then pound them up and keep them till you need them. Make a tea. Pour boiling water over it."

"But they're so bitter when they get dark!"

"Well, some men can take more, some less. Each can adjust the amount. Put in the tea. Some boil it awhile. Then drink. Sweat it out of you."

None of the Pimas I talked to had ever heard of *mo'otaḍk* being used medicinally among them.

Was this just a starvation food, resorted to only in hard times? The Tohono O'odham must have always lived a fairly marginal existence, and the Akimel O'odham were barely eking out a living late in the 19th century as their riverine resources were drying up around them. The references to its use seem too common for that. Perhaps it was a stress-season food, coming up in the hard times of late winter and spring. Or perhaps this was an acquired taste that the northern Pimans once cultivated.

Vouchers: *271*, *706* (SM); *304* (?); *LSMC 65*.

Pectis papposa Harvey & A. Gray
Chinchweed
Asteraceae (Compositae)
**oidpa, kopondakuḍ [?], koán-
   damkuḍ [?]**

This tiny marigold, actually a Fetid Marigold, may appear in Pima country during almost any season of the year after rains have fallen, be they spring or summer. It is partial to the sandy bottoms of *a'akĭ*, or washes. If the rains are only light, the entire plant may be no taller than a joint of your finger, with but a single flower. But after good rains, especially the monsoons, the plant may spring up almost everywhere and may be a hand or more tall, covered with dozens of flowers. The blossom is a flat-topped, miniature, deep yellow marigold with linear leaves dotted with dark oil glands. The bruised plant has a powerful and distinctive odor—not as unpleasant as the garden marigolds but equally pungent.

The Piman name for this plant has been something of an enigma. George Webb told Curtin (1949:104) it was *oidpa*, a name that no one with whom I worked, either Akimel or Tohono O'odham, recognized. However, Gary Nabhan and Culver Cassa tell me *oidpa* is the name they received from a Hia Ch-eḍ O'odham (Sand Papago) (no voucher specimen). Sylvester Matthias once called the plant *kopondakuḍ*, saying it was used for colds. But this is the name of a mustard, Bladderpod (see *Lesquerella* account). This might have been a lapse on Sylvester's part except that Frank Jim, a Tohono O'odham then living in Yuma County, nearly 150 miles away, gave me the same name for a specimen of Chinchweed.

George Webb said *oidpa* was used as a laxative: "The fresh plant is boiled in water, allowed to steep, and strained off; or the dry plant is prepared in the same way and makes a stronger tea" (Curtin 1949:104).

I have always wondered if *s-oám hiósigam* (Russell 1908:80) might be the name for Chinchweed in the eastern part of the reservation.

Vouchers: *LSMC 68*.

Physalis acutifolia (Miers) Sandwith [= *P. wrightii* A. Gray]
Wright's Ground-cherry
Solanaceae
kekel viipiḍ

Shortly after I arrived in Pima land in 1963, I discovered the delightful little village of Santa Cruz tucked away on a narrow tongue of floodplain between the Sierra Estrella bajada and the lower reach of the river that gave the village its name. This was hardly a village, really: a collection of little ranchos dispersed among small fields planted in crops or overgrown with mesquite thickets of varying ages. The only road left the pavement at Komatke and Gila Crossing and snaked through axle-deep dust. This nearly impassible road preserved something intangible that the other two villages, connected by blacktop to Phoenix, had already lost.

I soon met Rosita and David Brown, very open people who invited me to their little rancho hidden amongst mesquite bosques and Salt-cedar thickets. They were two of my first teachers about things

Piman, particularly about plants and animals. One fall day, while exploring their garden, I came across a curious weed with creamy white disklike flowers, each with a darker yellow star marked in its center. The long, light green leaves had wavy margins. But what attracted my attention was the green pods, hanging like Chinese lanterns of veiny parchment.

I pulled up one of the plants and took it back to Rosita, who was sitting under her *vatto* (ramada) parching wheat.

She started laughing. "That's *kekel viipiḍ*," she said. "That means, you know . . . 'old men's testicles'. It's okay to say that, *kekel viipiḍ*, because that's just the name of the plant. But be careful. Don't ever say *keli viipiḍ*. That would just mean some old man's testicles. It would sound real bad."

"Is that what everybody calls this plant?" I asked Rosita as she continued to flip her tray of wheat, deftly bringing the live coals to the edge and brushing them into a clay container.

"Oh, well," she said mischievously, "why don't you take some of that back to Sally and Aggie at school and see what they say."

I took one of the curious but attractive plants back to Komatke, put it in a little vase, and left it where Sally Pablo and Agnes Kalka would see it when they came to work in the morning. I soon heard great rolls of laughter, the tears-to-your-eyes kind, from the two women. They knew the name for the plant, too.

Gary Nabhan suggested that I find out about the food use of ground-cherry among the Pima. The round fruits, enclosed in the husk, are an excellent source of vitamin C and are popular and even semicultivated among some Pueblo tribes to the north. In fact, related species of husk-tomatoes (*Physalis ixocarpa* Brot. ex Hornem. (= *P. philadelphica* Lam.)) are important vegetables in Mexico and now regularly find their way even into U.S. supermarkets along the border. But all the women I asked assured me that *kekel viipiḍ* was never eaten, just as Curtin (1949:88) had been told.

When I started writing this Pima ethnobotany, I realized that I had no voucher specimen of ground-cherry from the reservation. Because there are quite a few different *Physalis* species in Gila Pima country (and even greater diversity southward in Pima Bajo country), I began

searching again for the plant. But land use had changed; there were few little farms and ranchos left, just great mechanized agribusinesses. For years I unsuccessfully searched the edges of fields and ditches, trying to find even one plant.

One warm day in early November, Sylvester Matthias and I set out across the reservation, searching all the villages for this field weed. At midday, still without luck, we stopped at the Olberg trading post on the north side of the Gila. Coming through the door on his crutches, Sylvester met another old man. After exchanging pleasantries, Sylvester said, "We're trying to find *kekel viipiḍ*."

"Ah! You won't find it anymore. It's all gone. There aren't any old farms anymore."

This was discouraging. But later that afternoon in Upper Santan, we found a plant sprawling along a barbed-wire fence on a field bank. The lanterns, not much bigger than your thumbnail, hung in profusion on a plant a meter and a half across.

Sylvester verified that this was *kekel viipiḍ*, a plant I had not seen in 20 years. He said that kids played with the fruits, popping them on people as they did with Bladderpod (see *Lesquerella* account). The berries of *kekel viipiḍ* are filled with fine, firm, flat seeds in a glutinous pulp. They must have made a nice mess.

"Did you ever eat them?" I asked.

"Yes, when they are real black, real old; they say they're halfway sweet, halfway sour. I tasted [them], but I don't like the taste."

"Did you cook the *kekel viipiḍ*?"

"No, just eat it like grapes."

In San Diego I grew out the seeds from this plant, and when Sylvester came to visit, we tried them. He took the ones with the darker berry inside and liked them. Also, we cooked the larger Mexican ones, the store-bought tomatillos, which are much more tart.

I took some Wright's Ground-cherry to George Kyyitan one spring day. He started laughing. "Do you know it?" I asked.

"Yeah, I know it. *Kekel viipiḍ*."

"Why do they call it that?"

"I don't know. I think because [it's] kind of wrinkled; I think that's what they call it for."

"Did you ever eat these when you were a kid?"

"Uh-huh. They're sweet. Some of them are sweet. The ones, in summertime,

[when] it ripens, it's kind of yellowish. Sure good. It taste good."

"How did you eat them—raw?"

"Uh-huh."

"Did anybody ever cook these?"

"I don't know. I don't think so."

Many years later, I was asking Francis Vavages at Sacaton about *kopondakuḍ* (Bladderpod). He said, "You know that other one, *kekel ha-viipiḍ*? We used to pop them when green. When *very* ripe, they're good. Have little seeds in there, sort of like *suuna* [figs]. They grow in wet ground. Very few now." He added, "It doesn't mean anything bad, the way this generation might think."

So they were eaten! I had not even asked. But why was Curtin's account to the contrary? Was it because she asked only women? I went back to her original field notes. The 1940 information from S-totoñik was correctly published. But a year later (1941), Mrs. Anne Thomas, a Pima from Co-op Colony, told her, "Ripe ground-cherries used to be eaten." This was somehow omitted from Curtin's book. So a good vitamin C source was utilized, as with adjacent tribes, at least as a snack food. Perhaps in a more prudish era, the plant's name became an impediment to the plant's continued use as food. But only men told me they had eaten it.

I have found no Pimans to the south who use the name *kekel viipiḍ* for a species of *Physalis*. At Onavas, Sonora, the names I heard were *tukuḍ wuupuhi* 'Great Horned Owl's eyes' and *tómar,* a Spanish loan word from *tomatillo,* the plant's usual supermarket name here in the southwestern United States.

Nor have I found any Gileño who elaborated on why this plant is called *kekel viipiḍ.* However, the membranous outer covering with its prominent venation, loosely enclosing a firm, juicy tomatolike fruit with a strong musky odor, left little ambiguity in the poetic imagination of some early Pima who named it.

Technical Notes: Usually *P. acutifolia,* the annual weedy floodplains species, is the biological referent for this folk taxon. However, when I showed Sylvester Matthias the suffrutescent yellow-flowered upland *P. crassifolia* Benth. (*AMR 354*), he said he thought it would be called *kekel viipiḍ* also. The specimens had rather large (thumbnail-sized) calices. *P. hederaefolia* A. Gray and

P. lobata Torr. have been collected in the mountains of Pima country also but have never been checked ethnographically.

Vouchers: *516* (SM); *892* (RG); *1941* (GK); LSMC *34.*

Proboscidea altheaefolia (Benth.) Decaisne.

[≠ *Martynia arenaria* Engelm.]

Coyote's Devil's Claw

Proboscidea parviflora (Wooton) Wooton & Standley var. *parviflora*

[≠ *Martynia louisiana* Mill.]

Black-seeded Devil's Claw

Martyniaceae

ban ihugga

Some day during the hot and humid monsoonal season when you are hiking out across the bajadas or through some arroyos, a powerful but delicate perfume may capture your attention. The source will not likely be obvious. But follow it along the bed of the arroyo, perhaps to its mouth, and there, growing with Desert Ironwoods *(Olneya tesota)* and Burrobrush *(Hymenoclea salsola),* you may eventually find a rank-growing plant, superficially like Desert Datura *(Datura discolor),* which you are also likely to find nearby. Both are viscid plants, with broad leaves and coarse stems. But the fragrance is coming from a plant with glorious masses of golden-yellow flowers thrust above its leafy canopy. Except for some pink or lavender lines in the throat, these look something like the little Yellow Monkeyflower *(Mimulus guttatus)* found on sandy wet streamsides. The rounded-lobed leaves are shiny green, but often the whole sticky plant may be covered with sand or dried pieces of neighboring plants. The spreading little bushes, scarcely knee-high, are root perennials, emerging each season from a thick taproot in the sand. Its fruits are completely unlike the prickly "golf balls" on the datura. The devil's claw fruits give the plant its name. At first they look like an oversized green fava bean pod. But then the horns or claws at the tip begin to enlarge until they are several times as long as the seed-bearing capsule at the base of the fruit. Later in

the season, the green leathery husk or exocarp peels off, leaving the distinctive black woody fruit that gives the plant its various names: Unicorn-plant, Cuernito, and many others. If you spread the two horns of the dried fruit, you will find two crowded rows of rough black seeds along each side of the main body or capsule. Of course, this is exactly what happens when the pod becomes attached to some animal's ankle. (It may even be your own.) At each step a seed or two is spilled out onto the disturbed soil of the animal's track.

The Gila Pima name for this plant is *ban ihugga* 'Coyote's *ihug*' (*-ga* shows ownership). *Ihug* is the generic term for all the devil's claws, a word undoubtedly derived from *huch* 'nail', 'claw', or 'hoof'. As such, it is one of a whole suite of Coyote's plants, which Piman speakers consider to be inferior to the "regular" or linguistically unmarked counterpart (see Nabhan 1982:75–86).

Everyone with whom I worked knew the yellow-flowered devil's claw, which we may as well call Coyote's Devil's Claw, legitimizing its Piman name. Its claws are very short compared to those of the two pink-flowered forms *(P. parviflora).* Ruth Giff, a basket weaver who cultivates her own *ihug,* once told me, "You know, my mother told me they used to use the *ban ihug* in their baskets when they didn't have anything else to use. But it's *so* short. It will go around [the coil] about one time." Another time she said, "*Ban ihugga* wait till July or August [to come up]. The yellow flowers smell strong. Other *ihug* pretty, but no smell; lavender [flowers]."

Sylvester Matthias confirmed what Ruth said. "*Ihug*—that grows in the field—are long. *Baaban ihugga* are short. But they use them for baskets, too." He uses the name in the plural, meaning 'Coyotes' rather than 'Coyote's'.

George Kyyitan and I were discussing *ban mauppa* (Fishhook Cactus, *Mammillaria microcarpa*) and *baaban ha-miiliñga* (a wild-growing melon, *Cucumis melo*), when George said, "And about that devil's claw—you heard that one, too?"

"The *ihug*?" I asked.

"Yeah, *ban ihugga.* They're little ones."

"Yeah, out in the wash."

"Hah! Baaban don't make no baskets!" He found this most amusing—that Coyote had his own devil's claw.

But like regular Linnaean taxonomy, folk taxonomy is not a good study for those who have difficulty tolerating loose ends and inconsistencies. With both biological taxa and folk taxa, certain things occasionally do not fit conveniently into either/or categories. This is because of evolutionary changes, in one case with the genetics of populations, in the other with naming categories. There are two wild species of devil's claw: a yellow-flowered perennial, *P. altheaefolia,* and a pink- or lavender-flowered annual, *P. parviflora* var. *parviflora.* Both of these have black seeds. But the Pima have domesticated a third form, now formally named *P. parviflora* var. *hohokamiana.* It has probably been around for a century or more (see Group I). It is called simply *ihug* by everyone and is the preferred form for basket weaving because of the very long and supple strips derived from the claws. But how is the original wild black-seeded, pink-flowered variety named? Nabhan and I (1987) have toyed at length with this problem. Naming is a process, and the middle species is in transition. For instance, Ruth Giff said *ban ihug* is only the little yellow one not the black-seeded, pink-flowered wild one, which is "just" *ihug,* like the white-seeded one. We think the original contrast was between the two wild forms, yellow-flowered and lavender-flowered. But with the domestication and broad cultivation of the pink-flowered, white-seeded form, the linguistic situation was no longer tidy. Ruth's taxonomy probably reflects the second step in the lexemic evolution. The next step is including both wild forms in *ban ihugga,* in contrast to *ihug,* the cultivar. Some Pimas reportedly have adopted this taxonomy.

Both wild lavender-flowered devil's claws and the domesticated form are used in weaving, but the cultivar has a number of superior qualities. The selection of white-seeded *ihug* is still an ongoing process by the few Piman women who still both weave and farm.

Vouchers: *551, 1619* (SM, *P. altheaefolia*); *1747* (GK, *P. parviflora* [?]).

Rumex hymenosepalus Torr.
Canaigre, Wild Rhubarb
Polygonaceae
sivijuls

Canaigre grows out on sandy places along the rivers, or at least it used to. It is a robust plant with large green leaves and thick, succulent stems and stalks. The papery winged fruits are as big as your fingernail, larger than those of any other *Rumex* species here. The plant reaches from knee- to hip-height. If you dig into the sand or loose moist soil, you will find a number of firm tubers that resemble sweet-potatoes. They have a pungent, puckering taste. Once the plant grew all along the rivers here when they were flowing. It still grows upstream, above the Ashurst-Hayden Dam that diverts the flow of the Gila, but it is nearly extirpated on the reservation. I found one plant in the deep, fertile soil of the "island" north of Blackwater, one at the Olberg diversion dam, and a few elsewhere.

To the Gila Pima this important plant is *sivijuls.* There are at least two possible derivations of the word. Gary Nabhan suggested that the name should really be *s-hiwijuls;* Mathiot (1973:370) gives just *hivijul;* and Saxton and Saxton (1969:172) provide the variants *hiwidchuls* and *siwidchuls.* The root word would likely be *hiv* 'to rub against each other', hence *hivkuḍ* 'scraping stick' (the musical instrument), and *hivshanakuḍ* 'scrubber'. Joseph Giff suggested instead a derivation from *siv* 'bitter'. Certainly the bitterness is an outstanding characteristic of its tubers.

Although put into a different higher folk taxonomic category than *vakoandam,* the congeneric sorrels or docks (*R. crispus* and *R. violascens,* Group F), *sivijuls* is nearly always contrasted with those species. In his yard in Santa Cruz village, David Brown was growing some that he got from the Salt River years ago. He dug up one of the tubers for me. "Boil it or just put it in your mouth raw and chew on it. This is different from *vakoandam,* which has the root straight down." Joe Giff said, "*Sivijul,* the one with the tubers, has bigger leaf and [is] taller [than *vakoandam*]." Sylvester Matthias once described the tubers as yamlike.

Sivijuls was an important Piman medicinal plant. Sylvester told me, "When [you have] stomach ailments you chew on it—bad stomach. They used to grow here on the Gila, but the water level is too low now; everything's too dry. Nothing can grow now. You could find it in the sand—where there's lot of sand." Three years earlier he said, "Grew a lot in the sand where it was moist—used for medicine; boil and use bulbs [sic] on sores, or dried up and pounded, used on open sores, or sore throat or sore mouth or ulcers of the stomach. It's bitter. Also use as a dye for leather—made nice color—tan-brown, about color of saddle harnesses, or any kind of leather. Works fine on leather. We used to get them down on 67th Avenue on Salt River." And still earlier he said, "*Sivijuls* has yamlike root—used for dyeing leather *lately* [that is, leather dyeing was not an original use]. Put slice in mouth for sore throat; also use it for open sores. [I] think *vavish* [Yerba Mansa, *Anemopsis californica*] is stronger than *sivijuls.* Like aspirin—[it] takes fever out when you have [a] cold; make tea out of it and drink it. It's not a laxative."

The transplanting of locally available plants to Pima yards seems generally limited to a few species of cactus, particularly prickly-pears (*Opuntia* spp.), barrel cactus (*Ferocactus* spp.), and Reina-de-la-noche (*Peniocereus greggii*). But this medicinal plant definitely was tended. Sylvester said, "Santos had some [planted], and David Brown."

"Any others?" I asked.

"Just those two. Around my mother's home town, Piliñ Keek—almost to Snaketown—they planted it. A sandy place around the [village] well. It likes a sandy place. Nobody takes care of it, [but] they come up every year. And those bulbs multiply." I suspect that one of the reasons Canaigre was tended, whereas so many other plants were not, is that its tops die back and the underground tubers in their natural locations along the river might be quite difficult to find in the off season. But people would remember where to dig if the plants grew at dooryards or other nearby places.

Although the tuber can be dried, dried tubers are not nearly as astringent as fresh ones.

Ruby Allen, a Pima woman from near Vahki, told Curtin (1949:52–53), "To dye willow withes yellow for basket-making, they are left in the liquid [made from these roots soaked in water] for a short time; but if a brownish color is desired, they are soaked for a longer period."

Hrdlička (1908:242, 245) said that powdered "*se-wi-dje* is employed as a remedy for sore lips and sore throat." Ruth Giff and Sylvester both agreed that these were important medicinal uses in their day, too. Russell (1908:80) called the plant *sivichilt,* saying, "The root of the canaigre is dried, ground, and the powder applied to sores."

Hrdlička (1908:264) gave details of the preparation of this plant as a food at an earlier period. "Of *se-wi-je* (canaigre: Rumex hymenosepalus) a common plant in the Gila valley, the Pima used to eat the stalks. They cooked these in pots, or roasted them in the ashes; then, after peeling them, they ate the inside. The root is often chewed by the children, and is also used medicinally in the tribe." Both Ruth and Sylvester said they had never heard of this food use by their time, only the use of uncooked *vakoandam* as an *iivagĭ* (see Group F).

Curtin gave a recipe from Hart's "Pima Cookery" for making pies from this Wild Rhubarb; I strongly doubted that this really had anything at all to do with Pima cooking but rather was a suggestion of what might be done with a local plant. Well, so I thought. Then I was at a feast at Blackwater, the far eastern village of the reservation, and the conversation somehow turned to plants. Mary Narcia Juan told me, "You know those *sivijuls?* We used to make pies out of them. Like that rhubarb from the stores—you know, that plant with the pink stems. I never heard of that at the west end, but when I moved over here in the early forties with my husband's family, his sister would go out and gather a whole big armful of those *sivijuls* leaves. Cook that with sugar and a little bit of flour to make a pie. Like rhubarb. It tastes like lemon pie." Thus, although it was not indigenous Pima cuisine (pie making is pretty recent), it was locally adapted.

Vouchers: *13* (RG); LSMC *1* (not extant); DF *48* (DES).

Salvia columbariae Benth.
Chia
Lamiaceae (Labiatae)
daapk

Out on the bajadas, mostly in the washes, grows an annual sage, the famous Chia. The plants are small, the leaves finely dissected. The stems are erect, often single, with spaced clusters of green flowering cups circling the stem at intervals, leaving most of the stem bare. Protruding from these clusters, the calices, you may find some delicate blue sage flowers with exserted stamens. Chia's claim to fame is in its seeds, small vermiculated seeds that mimic grains of gravel. You can shake these out of the mature heads. But if you want them in quantity, the easiest way is to buy them in a health-food store, because these nutritious seeds are being grown commercially. They are rather bland, having none of the flavor you might expect of a sage; the leaves rather than the seeds have the aromatic oils.

It took some time to identify *daapk*. The plant is winter-rain sensitive, and in some years there are almost none. Once when Ruth Giff was discussing *shuu'uvaḍ* (Tansy-mustard, *Descurainia pinnata*), another small-seeded plant that was harvested, she told me, "*Daapk,* little plant with big seeds and big flowers [compared to *shuu'uvaḍ*]. *S-daapk* means 'you have no clothes on'. They used the seeds to make something [to drink]." That got me off to the wrong start. I checked Curtin but found nothing. Then I found this in Russell (1908:77): "*Tapk.* These seeds resemble those of flax in appearance. They are eaten either raw or boiled and are yet extensively used." Not much help there, although it might have helped if I were familiar with flax seeds.

I went over to Sylvester Matthias's place. He knew the plant. "*Daapk* 'slick' or 'smooth'; when cooked, grains [seeds] make like a gravy that looks nasty—like when you have a cold." We went out looking for it, but he did not find any. Later he told me, "Its name means 'slippery' or 'naked'."

Two months later, the plant name came up again with Ruth. "You know," she said, "the name *daapk*—it doesn't come from *s-daapk* 'naked' but from the texture in

your mouth—how it feels in the mouth: 'slippery'." Another time Sylvester told me, "It's slippery, like *shuu'uvaḍ.* You don't have to chew it [when prepared]; it goes right down."

I tried Hrdlička (1908:263): "*Tá-hapk* ('smooth') is a kind of grass [sic] which has small black oblong seeds used as food by both the Pima and the Papago. They are prepared like the seeds of the *show-ou-wat* [*Descurainia* or *Sisymbrium*]". More misleading information, but at least I learned that it was a folk taxon common to both northern Piman groups.

I could not find the word in Castetter and Underhill's (1935) ethnobotany. Saxton and Saxton (1969:175) identified *daapk* as 'Tansy-mustard', but I knew that was not right: that plant's Pima name is *shuu'uvaḍ (Descurainia pinnata)*. Mathiot (1973:239) gave 'Indian Wheat', and I knew that was not right, either: that is *muumsh (Plantago* spp.). But both of these had something in common with what the Pima were saying about *daapk:* the seeds became mucilaginous when mixed with water.

Then something clicked. Initially, Ruth had mentioned something about purple flowers and naked places along the stems. Something came back from my initial botanizing days in the fourth grade, when my teacher encouraged me to press and identify everything within walking distance, which covered quite a few miles of florally rich Sierra Nevada country. One of the plants I had identified using Parsons's guide (1907) was a little annual sage that the California Indians found an extremely high energy source: Chia.

The details started falling into place. Russell (1908:77) had never made the connection between his *tapk* and another plant he listed separately as "*Tchia, Salvia columbaria* Benth. The seeds when infused in water form a pleasant mucilaginous beverage, very popular with the Pimas." Actually, these are the same, the latter name not Piman but derived through Spanish from Nahuatl, *chia, chian* (derived from Mayan *chihaan* 'strong', 'strengthening'). Probably the person who gave him the name *daapk* had only the seeds, as happened in a number of cases. These were to have been grown out the following year for identification, but there was a drought in 1902 and everything was lost (Russell 1908:18). I checked the species

daapk   *Salvia columbariae*

in Kearney and Peebles 1960 (p. 741) and found further confirmation: "The seeds were utilized by the Indians to make pinole and also mucilaginous poultices. A mucilaginous beverage prepared from the seeds was popular with the Pima Indians." The seeds of certain desert plants, such as Chia and Tansy-mustard, are hygroscopic, absorbing great amounts of water on their seed coats, apparently to help them through the critical germination period.

I often wondered how the Gileños found enough *daapk* to make harvesting worthwhile. I now suspect that there may have been some management going on. One spring while picking cholla buds (*Opuntia acanthocarpa*) with Carmelita Raphael, we found a burned-over spot on the Sierra Estrella bajada above Santa Cruz village. The plants in the area burned a year or two earlier were quite different from those of adjacent areas. Two of the plants that grew more thickly in the burned area were Chia and Indian-wheat. It is now known that the Pima conducted fire drives called *kuunam* to hunt small

game (Rea 1979); perhaps there were other cultural advantages achieved through this hunting practice.

In a paper on Chia, Palmer (1881) left some valuable verbal snapshots. He recorded the Pima name as *dak,* a close enough confirmation of the name. His paper explains a discrepancy in the ethnohistoric notes I gathered a century later. He (1881:140)described the preparation of pinole: "In preparing Chia for use the seeds are roasted and ground, and the addition of water makes a mucilaginous mass several times the original bulk, sugar to the taste is added, and the result is the much prized semi-fluid 'pinole' of Indians and others." He (1881:141) continued: "Chia meal is often mixed with the meal of roasted corn, or other grains. If used without further cooking, it is called as above, 'pinole.' If cooked in water as gruel or porridge, it is 'atole.' It is a very agreeable food, particularly if sugar or flavoring is added. The Pima Indians are especially fond of chia 'atole,' and consume large quantities of it." Older Pimas also mentioned an additional, simpler preparation that Palmer (1881:141) also described: "One of the most refreshing drinks known is prepared by infusing the seed-like nutlets in water. The mucilaginous drink resulting retains the aromatic properties, which are lost in the roasting, which is the preliminary step in preparing it for food; and when sweetened and flavored with lemon juice, is especially grateful in the hot days of summer, even to the sick, as it is easily borne by the most delicate stomach, and at the same time affords considerable nutrition."

Pimas born just 30 years after Palmer's pioneering work scarcely knew the plant or how to prepare it.

Voucher: AMR 1992 (RG).

Sarcostemma cynanchoides Decaisne. ssp. *hartwegii* (Vail) R. Holm [= *Funastrum heterophyllum* Engelm.]
Climbing Milkweed, Milkweed Vine
Asclepiadaceae
viibam

Vines are relatively few in the Sonoran Desert. Of the several I can think of in Gila Pima country, Climbing Milkweed is

the most conspicuous one. Most often you will find it in an arroyo or at least nearby. But it enjoys fence rows as well, particularly where fields are near an *akĭ chiñ,* or mouth of an arroyo. The plants are very delicate for a milkweed, with fine stems and small, opposite, usually arrowhead-shaped leaves. They grow in great masses, often covering a mesquite or Saguaro that is the nurse plant. Perhaps it does not look milkweedlike to those familiar with the widespread genus, *Asclepias,* but the cluster of cream-colored or purplish, intricately constructed flowers are clearly those of a milkweed, as are the pale tan pods that open when dry, releasing silky "parachutes" that carry the seeds away in the air. Usually perennial, the vines may freeze back during severe winters (as in 1986–87).

Everyone I talked with from the older generation knew the preparation of chewing gum from *viibam* 'milk it has'. (The same preparation was described by Tohono O'odham from Sonora and Arizona and by Pedro Estrella, a Pima Bajo.) Children, in particular, would go down into the fields or out into the arroyos to gather the sap into the stem of a *haal* (squash) leaf. If there were green pods on the vines, these were the best, but if not, just the tender shoots were broken to get a drop of the white latex that immediately exudes from the injury. This took considerable industry, because only a tiny drop was produced from each wound. "We would just hold up that hollow squash stem to the sun to see how much it was filling up; it was sure slow," explained Francis Vavages.

When it was filled, the stem was put in the hot ashes to cook. "One time we had to go out," said Ruth Giff, "and we started a little fire out in the desert. Just a tiny one to make the chewing gum. Pretty soon it's ready, and we'd take that gum out of there. It would be hard. It was *good* gum. I don't know *how* they figured out there was gum in there."

Hrdlička (1908:265) recorded the same technique for gum preparation but added that there were two plants: the *viibam* that grew on fences in the fields, and another, *ban viibam* 'Coyote's *viibam*' that grew in the arroyos. Curtin (1949:82) mentioned only *ban viibam.* I have never found anyone on the Gila who makes this distinction or who was sure about Coyote's gum. Perhaps these do refer to different biological subspecies. *Sarcostemma cynanchoides* ssp. *cynanchoides* has been taken occasionally on the reservation, and its leaves are quite different from those of the more common *S. cynanchoides* ssp. *hartwegii*, although there is considerable variability in leaf shape within each subspecies.

One day I brought down from the southern end of the Sierra Estrella a specimen of the White-stemmed or Wax Milkweed, *Asclepias albicans* S. Watson *(AMR 179)*. This is a leafless bush that grows in a cluster of thin wands about as tall as a person. Sylvester Matthias was waiting below at the truck, and I showed him some branches with clusters of flowers. He had never seen it before but said, "[It's] same family as *viibam* we talked about this morning."

On the Salt River Reservation, Pima Diane Enos called the leafless Desert Milkweed (*Asclepias subulata* Decaisne.) *viibam* and Climbing Milkweed *ban viibam* (see Deanna Francis's notes on specimens *DF 80* and *DF 87*, ASU). This would fit the paradigm of Coyote's plants.

Both *Asclepias* species in Pima country are sturdy erect bushes, in contrast to the weak and sprawling Climbing Milkweed.

Vouchers: *113* (Delores Lewis, Tohono O'odham); *1510* (SM); *1731* (FV; GK, maybe; nominate ssp.); *1735, 1975* (GK); *LSMC 39*.

Solanum americanum Miller
[= *S. nodiflorum* Jacq.]
Marsh Nightshade
Solanaceae
mo'oial

There are few marshy places left on the reservation, except for several botanically sterile ponds where water may collect for a few years following good rains. However, one good marsh had water collecting year-round in an ancient Hohokam irrigation system; it was at Barehand Lane between Co-op Colony and the Maricopa village. The marsh hosted breeding Virginia Rails, Coots, Common Gallinules, Yellow-breasted Chats, and some other birds found almost nowhere else on Pima lands. Here, too, grew some rare plants such as Celery-leaf Buttercup (*Ranunculus sceleratus* L.) and Water Pennywort (*Hydrocotyle verticillata*). And this was the only place I found Marsh Nightshade, a tall water-loving annual, with its small white star flowers and purple-black tomatolike little fruits. Without these fruits you might have mistaken the plant for a Chiltepín (*Capsicum annuum*). But the little marsh suffered the fate of many of the aquatic microhabitats in the Sonoran Desert (see Rea 1983).

One day I took Sylvester Matthias out to Barehand Lane and showed him a nightshade, not really expecting him to recognize it. He looked the plant over carefully and without further hesitation pronounced: "*Mo'oial*, that's this plant in Pima. Also that's the

Pima name for Superior, Arizona. I don't know why. And the color purple is called *s-mo'oialmagĭ*." Ruth Giff pronounced the color *s-mo'oialk*. She said that was the name for the color of cloth used to cover statues in church during Lent.

Several others knew the plant. George Kyyitan said *mo'oial* grows in fields. Irene Hendricks said it sometimes grows on canal banks. Very few knew the Pima name for purple even though they were familiar with various other color names.

On a trip to the swampy Salt-Gila confluence in June 1990, I brought a big branch of a Marsh Nightshade back to Sylvester waiting in the truck.

"*Mo'oial*," he said. "This one is green. When it's mature, use it for purple dye. For cloth. Just make spots, purple."

"Can you wash it?"

"I don't know. I think it stays."

"Did they ever eat these berries?"

"No, I don't think so. I never heard."

"Have you ever heard of their being poison?"

"I don't think they know, because they never tried it."

In the seventh stanza of the Butterfly Bird Song, Russell (1908:297) mentioned Ma-ayal (Mo'oial) Mountain but did not translate the name or locate it. According to Sylvester, this mountain is somewhere very close to Superior. There used to be an Apache camp there.

Vouchers: *417* (RG, SM); *1480, 1654* (SM).

Trianthema portulacastrum L.
Horse-purslane
Aizoaceae
kashvañ, kashviñ

During the hot summer months several common weeds grow on the exposed disturbed alluvial soils of the desert. One of these is what the Gileños and other northern Pimans call *kashvañ*. It is a rather succulent plant, related to the ice-plants (*Mesembryanthemum* spp.) that flourish along the California coast. The plant tends to spread out from its central root, with the succulent, sometimes reddish stems lying along the bare soil. The leaves are flat, only slightly fleshy, and as round as pancakes or ping-pong

viibam   *Sarcostemma cynanchoides*

paddles. They may be as broad as a finger joint. The flowers are tiny and tucked away in the "armpits" where leaf joins stem. The minute pink petals close as the day begins to heat up, and there is nothing left of the flower except a plump green ovary.

I was going down to look for birds in Joe Giff's fields one morning, when Sally Pablo handed me a couple of paper grocery bags, saying, "The *ku'ukpalk* [Common Purslane, *Portulaca oleracea*] are really good in the garden now. If you bring back some, I'll fix them." After I finished my bird work, I took the bags and went looking for the greens. I ran into Tom Jose, an in-law who lived at Joe and Ruth's place. "I need to find some *ku'uk-palk* for Sally, but I'm not sure which greens it is," I said. He took me to the edges of the corn and bean rows and showed me the thick bunches of succulent plants. In no time at all, we had filled the grocery bags.

When I walked in with the two big bags, one in each arm, Sally was all smiles. "Oh, that's great." She dumped one bag out on the table to start cleaning them. Her expression changed quickly. "*This* is not *ku'ukpalk!*"

"It's not? But I asked Tom to help me find it. We picked these together."

She poured more out on the table, spreading them around. "None of this is *ku'ukpalk*—it's *kashvañ*—all of it." She put the plants back in the bag and dumped them in the garbage.

"But can't you use them anyway?"

"No! You can't eat *kashvañ*. It's not *i'ivagĭ*. We eat only *ku'ukpalk*."

I promised myself then never again to take a man's word about *i'ivagĭ*, or edible greens. That is women's territory. Curtin (1949:64) also had the two species mixed up; she identified *ko okpat [sic]* as *Trianthema portulacastrum*.

Later I was collecting voucher specimens. Helen Allison of Komatke showed me one: "This one is *kashvañ*. Big leaves. It has pink flowers. Don't eat this one."

Gary Nabhan and I have traced the name *kashwan* south in Tohono O'odham country as far as Quitovac in Sonora. Here Luciano Noriega told us the local people once ate *kashwan* when rains were abundant, but they do not eat it now. But can we trust information from a man, even though he correctly identified the plant?

In the Pima Bajo pueblo of Onavas I asked don Pedro Estrella the name for the

Trianthema portulacastrum specimens I showed him. He could not remember the Pima name but said they were Verdolaga. His daughter María just laughed, shaking her head. "*No pueden comer ésta. Es verdolaga-del-sapo*," she said, indicating her contempt for the useless plant that only resembled Verdolaga: this is the toad's Verdolaga.

Vouchers: *145, 146* (SP; Helen Allison); *477* (?SP).

Tribulus terrestris L.
Puncture-vine, Bullhead
Zygophyllaceae
tool a'ag, toolō a'ag

There are some organisms that it is hard to say anything good about. Most of these are naturalized Old World weeds, aggressive and coarse, that proliferate in their new environment at the expense of native forms. The examples, unfortunately, are legion: English (House) Sparrows, Common Starlings, domestic cockroaches, Salt-cedars, foxtails, to name a few. To this list we might as well add Puncture-vine, from southern Europe.

This plant looks rather innocuous, stems radiating out from the center, hugging the exposed dry soil of yards, fields, and roadsides. Each pinnate leaf has half a dozen or more pairs of leaflets, lying flat on the soil. (These suggest the related caltrope, *Kallstroemia* spp., found rarely in Pima country following exceptionally good summer rains.) The stems are bristly. The little yellow flowers near the tips of the vines are quite like those of Creosote Bush (*Larrea divaricata*), another relative. But each fertile flower turns into a wheel of five seeds (nutlets) each bearing two stout sharp spines. The plant continues to spread during the summer, producing more and more of these vicious burs, which attach themselves readily to tires and the bare feet of people and animals.

The Pima name, *tool a'ag* 'bull's horns', was quite well known to native speakers throughout the reservation. *Tool* is a loan word, from Spanish *toro* 'bull'. In similar fashion, Bullhead is one of the plant's English names.

I asked George Kyyitan how long the plant has been around.

"It's been here a long time."

"It was here when you were a kid?"

"Uh-huh. Really, it grows over here on the white [men's] farms. Then they start growing this way because the seeds of them. Now they are trying to grow here. We cut them right away."

Vouchers: *298* (SM); *904* (RG); *1463* (FV, IH).

Verbesina encelioides (Cav.) Benth. & Hook.
Golden Crownbeard
Geraea canescens Torr. & A. Gray [?]
Desert-sunflower
Asteraceae (Compositae)
a'al hiivai

Golden Crownbeard and Desert-sunflower are both desert composites with large showy flower heads, usually found growing in dry sandy areas. They both resemble miniature Wild Sunflowers (*Helianthus annuus*). But the leaves are considerably smaller and grayish green, not bright grass green as in the sunflower. Nor do the crushed leaves smell like those of the more common sunflower. These two less common species grow only to about hip-height; they have yellow ray flowers with tri-toothed tips and yellow rather than dark disk flowers.

The Pima name *a'al hiivai* 'small [or] baby sunflower' has been mentioned to me for years by Sylvester Matthias and over the years seems to have bounced around as the name for various erect annual yellow composites we found. But some he was not satisfied with, saying that they should be growing out in sandy places, not in the fields where *hiivai* grows. Finally one October we found Golden Crownbeard, and he was convinced that this one really was *a'al hiivai*. (While this is the focal species, I think that the somewhat smaller but otherwise similar Desert-sunflower might be included in this folk taxon.)

When we found some Golden Crownbeard growing in the Santan area, Sylvester remarked, "*A'al hiivai* is a desert plant [in contrast to *hiivai*, which grows in field edges and ditches]. Up—of course

Phoenix was a little town then—way up towards the Squaw Peak, they say it's a paradise for them—this here. Real desert sunflower, 'baby sunflower', close to the mountains."

Later I had confirmation of this species (*Verbesina enceliodes*) from several others. Once in this same area of Santan while I was gathering plants with Irene Hendricks, she warned me not to pick *a'al hiivai*: "It's just going to smell up the truck. It sure *stinks.*"

I was also given the name *s-oám hiósigam* 'yellow flowers it has', which we never seemed to be able to pin down. Finally Sylvester said he thought that *a'al hiivai* might be the west-end name for what the east-end people call *s-oám hiósigam*. Perhaps, but I have not yet succeeded in verifying this. Russell (1908:80) lists *soam hiâseikkam* without botanical identification, saying, "An infusion made from the flowers of this plant is used as a remedy for sore eyes."

Vouchers: *371, 501, 693, 907* (SM).

Vitis arizonica Engelm.
Arizona Grape, Canyon Grape
Vitaceae
mischiñ uuḏvis, mischiñ huuḏvis

Along the little streams and canyons in higher country, above the low hot deserts, there is a native wild grape, Canyon Grape, that climbs through and sprawls over the walnuts, sycamores, hackberries, and other trees and bushes found at these cooler elevations. You cannot confuse this native because its leaves are rounded, no larger than the palm of your hand, and they are not lobed or cleft like the domesticated ones. The purplish fruits hanging in loose bunches turn purple-black in the fall before the leaves turn bright yellow and drop.

One time Sylvester Matthias wrote me, "And about the wild grapes, the only place I saw wild grapes is on the White Mountain Reservation—it be hanging down the cliffs up on the upper country. Also wild strawberries. Pima name for wild grapes is *Mischin—Hood-vis.*" Later he added, "They're small but they're sour. Oh boy, but they're sour! You can eat them." Francis Vavages agreed.

This is a postcontact name. *Mischiñ* is something that is wild but should be tame, such as a wild horse. And *huuḏvis,* from Spanish *uvas,* is the domesticated grape, given by Saxton and Saxton (1969: 71) as *uḏhwis* and by Mathiot (n.d.:487) as *'uudvis.*

Many species of plants otherwise unknown in the Sonoran Desert found their way here to Pima country after the Gila finished its tortuous descent from New Mexico and spread out on the floodplain just west of the Buttes. Could Arizona Grape have been one of these? Benson and Darrow (1981) mapped the species as following the lower Salt River all the way to its confluence with the Gila at the northwest corner of the reservation. I have never found specimens to document this premise, and the riparian area today has been destroyed and contains no clues. I asked Ruth Giff. She agreed with Sylvester's name, pronouncing it *mischiñ uudvis.* "I don't remember any growing around here. Maybe before my time. Too hot and too dry. [It grows in] higher-up country." Others agreed.

Later both Ruth and Sylvester verified the identification of *mischiñ uuḏvis* when we stopped at the lush verdant cienegas north of Imuris, Sonora. We took a sample of it with us to show don Pedro Estrella when we reached the Pima Bajo village of Onavas. He called it just *uuwa,* much the same as the domesticated grapevines he has growing on the north side of his adobe house.

George Kyyitan was talking about acorns when he related this story: "We were on our way to Grand Canyon. Then we stopped over there in Prescott, on that side, where the road goes to Jerome. There's a tree there, acorn tree there. And wild grapes there. Boy! They looks pretty good. Wild grapes. *Mimschiñ uuḏvis.* Yeah. I know it. Yet, I know there was no good in there. I told one lady, 'There's *lots* of them on the ground there—acorn. And the grapes are just hanging on there. Boy, looks sweet!' And this old lady I was telling about started grabbing the acorns, got a bag about this much in there. I said, 'You girls didn't eat any grapes.' [One woman said], 'Yeah, I know; it's not sweet.' [A man said], 'I'll try it. I'll get some.' Then he got one and tried to taste it." George starts laughing at his pulling this fast one on his companion. "He opened his mouth . . . spit!" The wild grape,

especially when not fully ripe, can be highly acrid.

"Did that ever grow here, or only in the high country?" I asked the prankster.

"No, I don't see it here. They grow down there in Giho Do'ag [Quijotoa Mountain]. That's what I heard."

Vouchers: *1045* (SM, RG).

Xanthium strumarium L.
[= *X. saccharatum* Wallr.]
[= *X. canadense* Mill.]
Common Cocklebur
Asteraceae (Compositae)
vaiwa

One of the more troublesome weeds in Pima fields is Common Cocklebur. Where left for several years, this tall coarse annual will take over moist areas with rich soils. Carmelita Raphael of Santa Cruz village

---

vaiwa *Xanthium strumarium*

has a problem with it because she maintains permanent pastures for her livestock and the dull rusty-colored burs get in her horses' manes and tails. The plants are also prolific in her vegetable garden; indeed, one year she had to give up on the garden because the *vaiwa* took over and she could not get it cut. Carmelita knew of no good use for cocklebur. But Sylvester Matthias recounted its medicinal properties: "Both my father and my uncle—they were stockmen—cattle dealers. They take care, they do the butchering and all that kind [of thing]. And my father castrates the horses and bulls. And they have bull service and stud service. And what they use for sores on horses is *vaiwa,* that cocklebur leaves. If worms get in after you castrate, use *vaiwa.* And they pound it up, pound it up. No problem getting it because you can find it just about in every field. Pound and pound. Then make a good strong tea out of it and wash them out. Boil it. Bitter!

"The burs—you can burn it—it's greasy, just like Castor-bean. You burn [roast] it in the fire. They are just about the same. Use them on open sores [for] people. As ointment. Doesn't sting. The burs [are called] just *vaiwa,* the leaves and whole plant, just *vaiwa.* Used interchangeably—*vaiwa* and *maamsh* [Castorbean, *Ricinus communis*]."

Russell (1908:80) recorded an additional earlier use (though none of Sylvester's): "*Vaiewa.* Cocklebur pulp is combined with soot as a remedy for sore eyes." Forty years later Curtin (1949:97) was told that "fresh cocklebur leaves [are] mashed and placed on screw-worm sores in livestock. George Webb says the burs are boiled and a half cup of the strong tea is taken either for constipation or for diarrhea."

Vouchers: *508* (?); *LSMC 8.*

Group H, Covert Category
cactuslike plants

Given the abundance and diversity of cactus in Pima country, one might expect a formally named life-form category for this group. But this is not the case. The classification *ho'i* 'thorn, spine' that Mathiot (1962) and Pilcher (1967) give for the Tohono O'odham was denied as a category of plants by all the people with whom I worked: "Well, maybe it means that for them," a Pima told me, "but for us, it just means 'sticker.'"

Nevertheless, there was no hesitation about grouping plants into an unnamed or covert category that includes cactus and other plants such as agaves and yuccas. Another liliaceous species, Bear-grass *(Nolina microcarpa)*, was included here even though it is hardly spiny at all. Apparently, there is a conceived continuum from the yuccas and agaves through the somewhat less rigid leaved Sotol *(Dasylirion wheeleri)* to the almost grassy growth form of Bear-grass, at least in Sylvester Matthias's perception.

Even the almost spineless gray-stemmed Reina-de-la-noche *(Peniocereus greggii)* was included in this category. Like Bear-grass, it is only weakly armed, but its distinctive flowers—larger than those of any other cactus in Pima country— and its fruits are well known to the Pima. These parts are undoubtedly sufficient to associate the plant with other true cactus in the category of cactuslike plants. Spininess in itself is insufficient for inclusion, however; both Graythorn (*Ziziphus obtusifolia*, Group D) and Crucifixion-thorn (*Castela emoryi*, Group E) are classified elsewhere.

I am not sure how Peyote *(Lophophora williamsii)* would have been classified originally when only the dried plants were received as trade items. In that state they might have passed for a fungus. When shown the living plants still potted, Sylvester Matthias included them here, although they are completely spineless. Like those of Reina-de-la-noche, the flowers and fruits are cactuslike, though much less conspicuous. Perhaps someday we will find out more about this categorization from the Tepehuan (the southernmost Pimans), who still use the plant.

Agave deserti Engelm. ssp. *simplex*
    Gentry
    [≠ *A. americana* L.]
Desert Agave
Agavaceae
a'uḍ, a'oḍ

In the xeric regions, various liliaceous plants in the genera *Yucca, Dasylirion,* and *Agave* grow in the form of dense rosettes of leaves. The leaves of agaves have strong terminal spines and often stout clawlike spines along the leaf margins. The plants are sometimes known as century plants or (in both the Southwest and Mexico) as magueys or mescal. *Mescal* may refer to the sweet baked hearts or to a liquor distilled from them after fermentation, as well as to the plant itself.

Although a dozen species of agaves are known from Arizona, only one species, a relatively small one, was within easy reach of the Gila Pima. I have found it only in the Sierra Estrella, not in any of the other desert ranges of the reservation. Here at the generally higher reaches, well above the bajadas, the tightly compact rosettes grow to about knee-height. The thick bluish gray leaves are variable, some being so short and broad-based as to be almost triangular, others longer and relatively more slender: this is just local variation. About April fully mature plants start sending up a single blooming stalk, putting all their effort into a single reproductive effort, after which they die. The yellow flowers appear in early June. The inflorescence is quite tall (four to six meters, or two or more times a person's height). The plants in the Sierra Estrella reproduce by seeds, by pups (offsets), or occasionally by bulbils (small plants borne on the flowering stalks). I have found the plants only as scattered individuals or clusters of twos or threes, though E. Linwood Smith told me he found colonies near the highest peaks during his month-long survey of Desert Bighorn.

Agaves were once an important food resource to Native Americans throughout the Southwest. Mature plants that are about to bloom produce smaller, narrower leaves in the center of the rosette. The center tissue, the plant's meristem, is the sweetest. Among the 136 North American species, Gentry (1982:379) considered *Agave deserti,* the species growing in Pima

country, "among the more edible of the agaves."

When the *a'uḍ* is in just the right stage, with all the concentrated sugars about to send up the big blooming stalk, the entire plant is cut and the leaves are chopped off close to the base. The resulting "heart" looks like a white pineapple (*Ananus comosus;* see account in Group I). These hearts are put into pits to bake for 48 hours, most people told me.

During the long period of Apache raids on the lowland agriculturalists, Pimans traveling away from the protected villages to gather agave hearts, cholla buds (*Opuntia acanthocarpa*), or Saguaro (*Carnegiea gigantea*) fruit were vulnerable. The calendar-stick records list numerous episodes of attacks during these forays. For instance, in 1843–1844, "Apaches killed a party of Pimas who had gone to the mountains to gather mescal. The Pimas had planned to go to the Kwahadk [Koahadk] camp, but changed their minds and camped opposite them. The Apaches sent down scouts from the hills to see how many there were at the place where the smoke from the mescal pits were seen. It was a night attack and many Pimas never wakened to see another day; only one escaped to tell the Kwahadk's of the massacre" (Russell 1908:42). Two years later a similar attack occurred: "In the spring the Apaches waylaid a party of Pimas who were returning from a mescal-gathering expedition in the mountains. Nearly all the party were killed and two girls were made prisoners. The Apache were followed, most of them killed, and the girls rescued by a party from the villages." (Russell 1908:43).

According to Castetter and associates (1938), oral history recorded that the Gileños harvested agaves near Superior and in the Superstition Mountains, an area considered to be the domain of the Apache. Wendy Hodgson told me that there the target species was probably the Golden-flowered Agave, *A. chrysantha* Peebles.

Sylvester Matthias was familiar with the *a'uḍ* harvest in the Estrellas: "Pima, Papago, and Apache used it. Pound the leaves for fiber; all they use is the inner parts [hearts] for sweets. They have to go *way* up there in the mountain; it would be too much to bring down the leaves to make fiber out of them. All they bring down is the heart—maybe they can carry

three on their back, then they start down. They know about the fibers [in the leaves] but it's too much to carry."

Ruth Giff told me, "The time to go pick *a'oḍ* is when the *kui i'ivgi,* [that is,] when the *kui* [mesquites] are just starting to leaf out. This time [period] is like a month to them [the Pima], [the moon] when certain things are planted. The last time I tasted it [roasted agave]—maybe it was a white man—left some at Stobby's store [in Komatke]."

"Is there a special name for the prepared agave hearts?" I asked her.

"I guess *edaj* 'inside'."

Another time Ruth said, "*A'oḍ* [is ready] in April—that's the time to do it. There's a lot of it up there [in the Sierra Estrella]." The calendar-stick entry for 1877 says,

---

a'uḍ, a'oḍ   *Agave deserti*

"While a party was gathering mescal just before the wheat ripened a mare gave birth to twin colts" (Russell 1908:56). This would indicate that an agave harvest extended into May.

Although George Kyyitan lives in Bapchule village, far from any agaves, his years of archaeological surveying have put him in touch with almost every part of the reservation. "There's plenty [of] mescal—*a'oḍ*—up there in the Estrellas. . . . It's *sure* sweet. It's *too* sweet. And they cut—square. There are those leaves—and they call it the 'head' in there.

"The Papagos come in from there, because there's nothing up there, anything like wheat. They'd come over here, in the summertime, to cut the wheat for the People here, [with a] sickle. They'd bring that mescal, *a'oḍ*. Bring it. We use it. Cut it into little squares. When you use pinole, use that one. Oh, boy! It's good. And they'd bring the salt, too. They got somewhere, in that Rocky Point. They'd bring that one. We never starve. They [the Pima] get their salt from those people. And when the next year comes, they'll come again. They know where they've been cutting [wheat], who they've been cutting for."

George clarified for me several terms and usages of the century plant. "*Shaavaidaj* is that shoot out of *a'oḍ*, when it goes up, the middle part. That yucca does the same thing, I think. *Mo'oj* ['its head'] on that *a'oḍ* [is] the part they eat. And the leaves—to get the meat out—it's sweet, too. And they get the strings—I think they do those separately—*chuámai* [baked in a pit]"

"What do they call the strings?"

"*Shaavaidaj*, same as the shoot. For rope, *shaavai vijina*, made from *a'oḍ* strings."

What Ruth and Sylvester called simply *edaj* 'its inside', George called *mo'oj* 'its head'; in Mexico these are similarly known as *cabezas* 'heads' or *corazones* 'hearts'.

George was the only one to note that the leaves could be roasted, a practice found also among the Mountain Pima and the Cahuilla (Bean and Saubel 1972:32–33; Laferrière and associates 1991:101): "*A'oḍ* leaves, they eat that one, too, the leaves. They do the same thing."

"Does it have to be the leaves of one that's just going to put up that center thing, or any leaves?"

"Naw. They cut them, they cut the leaves off, and roast them right there."

"Can you go up there at any time of the year and take more leaves?"

"I don't know. I eat them when those Papagos came here, that time they were cutting wheat by hand in the harvest times. Yeah. They bring them, and boy, they're sure good, with pinole."

Ruth Giff, a skilled basket maker, spoke of the cordage: "*Shaavai*, that's made out of *a'oḍ* leaves. You know, what they make rope out of. . . . Guess they make string with it, too. And I think that's what they use when they make the *giho* ['burden basket']; can't remember what they call the thing they turn [to twist the rope]. It has a name." Joseph Giff said, "Dead leaves of *a'uḍ* picked, beaten, get fiber out to spin ropes."

One day Joe sat telling me about *hovij* (Banana Yucca, *Yucca baccata*). Then he added, "*A'uḍ* is different; when you pick it, it doesn't come back. Up there somewhere [in the Sierra Estrella] is a bunch of holes, for *chuáma*, where they used to roast *a'uḍ*." David Brown explained, "*Chuáma* [is] pit roasting for *hannam* (cholla buds), *a'uḍ* (agave), or even meat." George Kyyitan had had experience excavating these pits during his days of archaeological fieldwork: "I've seen those where they roast *a'oḍ*, up in the Estrella, *way* on top."

One day while looking for plants at the top of the bajada skirting the Sierra Estrella south of Santa Cruz village, Sylvester and I found an outline of a roasting pit. Pimas used two different kinds: one for agave hearts and another for *hannam*. Agave pits are larger. A *hannam* pit, Sylvester explained, is only about 2 to 2½ feet across, but deep. Both are rock lined.

But the whole pilgrimage was fraught with rigorous sanctions. "If a person has relations with [his] wife, the *a'uḍ* will come out of the pit just as raw as it went in," Sylvester said. This was a widespread taboo, even among peoples unrelated to the Pima.

Felix Enos lived on the edge of the Komatke village feast grounds. Though his wife had died a few years earlier, the house and yard were still impeccably neat. He said that he is half Papago. His mother came from Caborca, Sonora, and he still has relatives there. He is a quiet, bow-legged man, a pillar of the church. He told me this story: "One time some men went up to get *a'uḍ*. One of the men didn't stay in camp with the others, but came back here and slept with his wife, and in the

morning the men at camp were sure mad at him, because it was all no good!"

Sylvester, who lived in the same village as Felix, also knew the story. "I know who he is. He *told* me. Nobody's supposed to have intercourse till *after* they come back. But he went and had . . . , done it. Then he went [back] up [to the agave camp]."

"Does that apply to intercourse with anybody or just their wife?"

"With anybody."

Apparently even among the desert Pima Bajo in Sonora, where nearly all traces of the Piman belief systems have been lost through three centuries of close Hispanic contact, some vestiges of this taboo exist. Don Pedro Estrella told me that men should not touch their women at this time because of some danger, but he was vague on exactly what the danger was. His son-in-law, who ferments and distills agave hearts to make the powerful mescal, or *bacanora* (a local quality tequila), did not believe in this taboo of the ancients.

The strict sexual abstinence enjoined during the mescal roasting has led some to believe that women never participated in the gathering. But the calendar-stick record for 1876 notes, "For a short time the Pimas were free from Apache attacks, and they ventured into the mountains to gather mescal. While there, a race took place between a man and a woman, in which the woman won" (Russell 1908:55; see also p. 70).

When Russell did his fieldwork, in 1901–1902, maguey or agave fibers were still very much a part of Pima daily life, finding many uses. Maguey cords were used to fasten just about anything: parts of a saddle, the cross piece on the Saguaro fruit-picking stick *(ku'ipaḍ)*. He collected a saddlebag of open netting made of maguey cordage. Rope was made from the fibers, but Russell (1908:114) noted, "They are strong, but the harsh and coarse fiber renders them disagreeable to the touch of any but a hardened hand." It is not always certain whether these articles were made by the Gila Pima or by the Tohono O'odham, who took fibers to trade with the Pima during the wheat harvest (Russell 1908:93). The Tohono O'odham had access in their country to additional species of *Agave*, some with much longer fibers than those from *A. deserti*. When Sacaton Grass (*Sporobolus wrightii*) was no longer available at the cienega at Maricopa Wells and along the

Gila, maguey fibers were substituted in making hairbrushes (Russell 1908:116).

The disjunct and isolated distributions of this and several other agave species in the Southwest puzzle botanists. Some may be anthropogenic relicts from earlier cultures. However, the Pima Creation Story has a version of how the *a'uḍ* came to be (Lloyd 1903). A woman who lived in the mountains fell asleep while making cane mats; she was impregnated by a drop of rain and bore twin sons. When the twins were old enough they went to see their father, Cloud. Cloud asked them to prove that they were his children. Both flashed lightning, and one "thundered loud" besides. That night Cloud tested them again by leaving them outside while it rained and snowed; the next night he took them to an icy pond and left them in the water all night. When the boys passed one test after another, Cloud acknowledged them as his sons. They stayed with him for a while but then wanted to return to their mother. Cloud told them that they must not speak to anyone on the way home. He shaded them during their journey.

> And after a while they saw a man coming, and the younger boy said, "We must ask him how our mother is."
>
> But the older brother said, "Don't you remember that our father told us not to speak to anyone?"
>
> The younger said, "Yes, I remember, but it would not be right not to ask how our mother is."
>
> So when the man came the boy asked; "How is everybody at home, and how is the old woman, our mother?"
>
> And then the cloud above them lightened and thundered, and they were both turned into century plants.

The Pima story of the origin of *a'uḍ* is a widespread mythological theme of the conception of twin gods from the union of sky (clouds) and earth (the virgin). Russell (1908:239–240) gives a very similar version of this story narrated by a different storyteller, Inasa, that concludes: "This is the reason why mescal yet grows on the mountains and why the thunder and lightning go from place to place—because the children did. This is why it rains when we go to gather mescal."

The Tohono O'odham carried on a brisk trade in maguey products; there

a'uḍ, a'oḍ *Agave deserti*

were more *Agave* species growing in their country. Among the items they took to the Pima during the June wheat harvests were "agave sirup, maguey fibers for picket lines, *kiahas* [*giho*, burden basket] and fibers to make them" (Russell 1908:93). Russell also mentions "baskets of agave leaf," but it is unclear whether this means baskets filled with the leaves or baskets constructed from agave fibers. The "agave fruit in flat roasted cakes" must have been either the roasted *a'uḍ* hearts or the dried fruits of Banana Yucca. (The actual fruits of agaves are dry and woody.) Russell (1908:147) said that the Tohono O'odham bartered to the Pima sleeping mats made of agave leaves, but I am sure that these would have been made of either Beargrass *(Nolina microcarpa)* or Sotol *(Dasylirion wheeleri).*

Agave use was widespread throughout the Southwest. Sylvester noted that in his youth the Hualapai, called Het Kuá'adam 'Ochre Eaters' in Pima, brought roasted *a'oḍ* to Bapchule and Sacaton. "It's the heart, in there. They bake it. They chew on it. They say it's sweet. The Hualapai used to bring them. They use it too, like that. They just bake them. Then all you do is just cut a piece out, chew it just like a sorghum, and you throw whatever's left over from chewing."

Even the juice from baked mescal found a use. Hayden (1935:49) noted its

use as a vehicle or binder for red body paint, presumably red ochre. (Uncooked agave juice usually causes dermatitis.) It was also used to make *sitol:* "Sirup is extracted from the prepared mescal by boiling until the juice is removed, which is then thickened by prolonged boiling until it becomes a black sirup, somewhat similar to sorghum. It is inferior to saguaro sirup" (Russell 1908:70). Apparently this was diluted, like *haashañ sitol* 'saguaro syrup', to ferment for wine feasts (Russell 1908:37). But I found no confirmation of this preparation.

In addition there were indirect uses for agave among the Gileños. A species of carpenter bee (*Xylocopa* sp.) deposits its eggs in the century-plant stalks, providing one of two sources of wild honey for the Pima. Joseph Giff told me, "Bees, when stalk dries—they burrow in there and put honey in there. They [Pima] used that honey. Other honey [is found] in ground. Stick [a] stick in so [you] don't lose [the] hole. Inside [is] smooth like an olla. Dig all around and take it out. [Found in a] bald spot on the ground. They seal it, probably after they lay the egg in there. Sometimes one pot on top of another, but usually only one. If you get there a little late, bee [larva or pupa] will already be in its nest."

Agave flowers are often prodigious nectar producers. In a discussion about *hiósig vadagaj* 'flower nectar', Sylvester said, "Century plant—you suck that out. It's sweet! That's the only one *[sic]* you can suck juice out of the blossom, *a'oḍ*. And that stem—the bumble [carpenter] bees, *huhu'uḍagĭ*, [put] their honey [in there]. We get that, too."

"There's honey in there?"

"Yeah. Something like syrup, but it's a little harder than syrup."

"What time of the year?"

"In the early spring, when it starts blooming, when everything is in bloom."

"How would you get it out?"

"Just peel it open."

"Is there much in there?"

"Sometimes." He paused, thinking perhaps of his childhood quests as a youngster in the Estrellas after this elusive sweet. "Sometimes the bumble bees get busy—or they go somewhere [else] to build their nest."

"How does it taste?"

"Like the taste of a chewing gum."

"*S-i'ovĭ* ['sweet']?"

"Yeah."

A large cultivated century plant *(Agave americana)* is frequently planted as an ornamental around Pima homes across the reservation. There is disagreement as to whether this plant should be included in the Pima folk taxon *a'uḍ;* most excluded it.

Ruth and George both say *a'oḍ* rather than *a'uḍ*. Mathiot (n.d.:404) writes *a'uḍ*, as do the Saxtons (Saxton and Saxton 1969:3). Mathiot gives *a'oḍ* as a dialectical variant in Tohono O'odham. This usage does not seem to sort out geographically in Gila Pima country.

Vouchers: *599, 600* (?); LSMC *28* (not extant).

Carnegiea gigantea (Engelm.) Britton & Rose
[= *Cereus giganteus* Engelm.]
Saguaro
Cactaceae
haashañ

The great fluted columnar Saguaros are characteristic of the Arizona portion of the Sonoran Desert, and they are nearly confined to this area, with only a few spilling across the Colorado River in the vicinity of Blythe, California. In northern Sonora, Saguaros are still common, but farther south, in Seri country along the coast, they give way to the equally impressive Cardones, *Pachycereus pringlei* (S. Watson) Britton & Rose.

On the better-drained bajadas and well up the rocky slopes of the mountains of Pima country, you will find Saguaros growing two, four, or even six times a person's height; some reach 50 feet. In some places, especially on more southern exposures such as the Santan Mountains behind Olberg, there are great groves of this cactus. Younger individuals are solitary erect columns, but older ones begin adding side arms about a third of the way up the shaft. In mature plants, even the side branches may have smaller arms until a single Saguaro has a dozen or more branches, each with an active terminal growing center.

At these tips, green spineless flower buds begin to appear late in winter. When spring weather begins turning hot, the stiff, waxy blossoms open. When fully unfolded, they resemble a white tire filled with a mass of cream-colored stamens.

Blooming time and the arrival of White-winged Doves and Long-nosed Bats correspond. Each flower opens late in the evening, closing the next afternoon. Insects, bats, doves, woodpeckers, and even flycatchers are attracted to these flowers for one reason or another, incidentally pollinating them. Blossoming time is staggered, with dozens of flowers in various stages of development crowning the tip of each arm. The egg-shaped fruits mature during late June and early July. The Saguaro then puts on a second show, this time to attract seed disseminators: the greenish fruits split open,

haashañ   *Carnegiea gigantea*

revealing the dark red pulp with its mass of black seeds. Even the inside of the fruit skin (the ovary wall) is bright red, signaling seed dispersers. The splayed fruit rinds adhere to the Saguaro, and some people have mistaken these for brilliant vermilion flowers.

Among both Akimel and Tohono O'odham, Saguaro is called *haashañ*. This Piman lexeme was probably widespread, referring to any tall columnar cactus. Saguaros do not grow at Onavas on the lower Rio Yaqui, so I did not get a name for them from don Pedro Estrella. Farther south, a cognate term, *aasáñi*—labeling the giant columnar Echo, *Pachycereus*

pecten-aboriginum (Engelm.) Britton & Rose—appears among the Northern Tepehuan of Durango (Pennington 1969:136).

Surely the largest plant that dominates the scenery of Pimería Alta must have entered the cosmological thinking of the People. How did they account for these massive green columns all around them, no two quite alike?

There are many stories of Saguaro's origin among the Pima. In Juan Smith's version (Hayden 1935:22) of the Creation Story, *haashañ* was created directly (see also Font 1931:40). In a story Francis Vavages told, two misbehaving orphan boys went up into the mountains and turned themselves into a Saguaro and a Little-leaved Paloverde (see Group E, *Cercidium microphyllum*). Russell (1908:247) recorded the same story. In another story Russell (1908:245, 250) narrated, a boy set out to find his mother who had been captured by Apaches. On their return home, they turned into Saguaros.

There are many versions of the origin of *haashañ* and *navait* that agree in essence (see Densmore 1929:149, 161–162; Saxton and Saxton 1973:211–219; Wright 1929:109–121). In a story told by the Tohono O'odham, a young mother who was inordinately fond of *toka* (women's hockey) abandoned her son, leaving him with some gourds of milk. As the child grew, he sought his mother in distant villages. When he finally found her, he waited outside the village with some children playing in an arroyo. When she did not come at the fourth summons, he found a tarantula hole and began sinking into it while the children circled around, singing. In four days, the sunken child reappeared as something green, a little *haashañ*, the first anyone had ever seen. The children abused it, so it sank into the ground and came out far from anyone. The animals searched for *haashañ* until Raven found it.

When Raven found the *haashañ*, it was covered with bright red fruit. He gorged on this *bahidaj*, returned to the village, and vomited the fruit into ollas. The People removed the seeds, then fermented the remaining pulp for four days, producing the first *navait*, Saguaro 'wine'. But there was quarreling even at this first wine feast, and some birds, such as *hikvig* (Gila Woodpecker) and *chikukmal*

(Western Kingbird), have "bloodied" heads even today. So the community leader gathered up all the *kaij* 'Saguaro seeds' and sent someone to carry the seeds far away, usually toward the east. Some versions say the seed carrier was *chuuvǐ* 'jackrabbit', others *kaav* 'Badger', and others just the chief's messenger. The People (this included the animals at that time) sent Coyote to rescue the *kaij*. Coyote circled the messenger, approached him from the opposite direction, and inquired what he was carrying in his fist. When Coyote finally tricked the seed carrier into opening his hand, finger by finger, Coyote hit the hand from below, scattering the seed into the air. *Hevel* 'wind' was blowing from the north and scattered the *kaij* all over the southern faces of the mountains. And that is why they still grow on that exposure today. Two important phenomena are encoded in this myth: the first is the origin of the annual wine feast, the second an explanation of why Saguaros are now found distributed as they are.

Sylvester Matthias and I were once discussing plants that could be dangerous to people, a very small category that might cause *kaachim mamkidag* 'staying sickness'. I mentioned Saguaro. "No. They say *haashañ* is a good friend, harmless. Want to serve people, like *kui* [mesquite]."

One time I came upon an engraving (Bartlett 1854) showing some Indians shooting at a Saguaro, called *pitahaya* or *pitaya* in Spanish. Bartlett commented, "It is a favorite amusement with both men and boys to try their skill at hitting the petahaya, which presents a fine object on the plain. Numbers often collect for this purpose; and in crossing the great plateau, where these plants abound, it is common to see them pierced with arrows." It is unclear exactly who engaged in this practice because the previous several paragraphs speak of both the Pima and the Maricopa. Browne (1869:78) referred to the same practice when he was on the lower Gila near Yuma. Whitworth, a young Englishman traveling with the Mormon Battalion, encountered Saguaros between the San Pedro River and Tucson. He sketched one pierced at the top by an arrow (Gracy and Rugeley 1965).

I mentioned this to Joe Giff. He took immediate exception. "Those wouldn't be Pimas. It would be wrong for us to do something like that, to injure, you know,

to shoot at a Saguaro, because that would be like shooting at a *person*. See, in our way of thinking, the Saguaro *is* a person. That's where Saguaros come from; in the stories it was somebody who turned into a Saguaro, so that would be forbidden, to do that. Maybe that was some other kind of Indians in that picture you saw. Not Pimas."

Joe's teaching is also found in a Tohono O'odham account of the origin of Saguaro: "This giant cactus was a very strange thing. It was just a tall, thick, soft, green thing growing out of the ground. All the Indians and all the animals came to look at it. Ah-ah-lee [*a'ali*]—the children—played around it and stuck sticks into it. This hurt Hah-shahn [*haashañ*] and he put out long sharp needles for protection so the children could not touch him. Then ah-ah-lee took their bows and arrows and shot at Hah-shahn. This made Giant Cactus very angry. He sank into the ground and went away where no one could find him and he could live in peace" (Wright 1929:116).

Perhaps, then, it was Maricopas who were illustrated in Bartlett's work. Originating on the Colorado River where the Saguaro is less common, they may have had a different mythology of the cactus that did not portray the sacred character that Joe described for the Saguaro.

As cosmic sacredness was rooted out of Pima culture in the late 19th century, traditions became "superstitions." When Herbert Narcia moved to Blackwater in 1940, he enquired of an old man how the neighboring village, Haashañ Keek, had gotten its name. "There used to be a single Saguaro standing in the village," the old man said. "But the boys shot arrows at it and it dried up."

So important was the ripening of the Saguaro fruit among northern Pimans that their new year started with the harvest. The first month of the year was called *haashañ bahidag mashad* 'Saguaro harvest moon' (roughly June), followed by *jukiabig mashad* 'rainy moon' (roughly July), then *shopol esiabig* 'short planting' (roughly August). The months were probably shorter lunar months, beginning approximately at the summer solstice, 21 June. In Southwestern folklore the summer storms are said to start after the feast day of St. John the Baptist, 24 June, though about a month later may be a more realistic expectation. The "short planting" is still a critical point in the year for all Gileños who plant: seeds must be in by 15 August, the Feast of the Assumption, if they are to mature before the first winter frosts in November.

The calendar sticks record many incidents associated with the Saguaro harvest. For instance, for the year 1835 the oral historian narrated, "The next summer we went over across the Gila River in the Estrella Mountains and camped there to gather giant-cactus fruit to make syrup. The Apaches came and killed some Maricopas. We stayed two weeks and went home" (Hall 1907). In 1836, "early in the morning [during the wine feast] a woman started toward the hills to gather cactus fruit" when she was overtaken by an Apache (Russell 1908:39).

Russell (1908:66) learned that about every fifth year the Gila failed in mid-winter and the People were forced far afield in search of native plants; Saguaro fruit and mesquite pods were the most abundant and accessible resources in the dry season. But spring rains apparently were not good for Saguaro productivity. Sylvester Matthias related, "If it rains, it ruins those that are already opened. But if they aren't opened, won't do any harm. But the ones that are opened, they damage that. If it's a wet spring, they won't have much *bahidaj* [fruit]. In some years, they don't produce very much. Then in another year, it would be in abundance." These observations are widespread among northern Pimans (see also Underhill and associates 1979:21).

Sylvester was explaining that Tohono O'odham families had specific areas for gathering mesquite pods when he added, "Same way with the Saguaro. The Papago had a territory where they can find good Saguaro. When it first ripens during the season, they go and camp. When their Saguaro fruit dried up, they put it away for later use."

"But the Pima didn't have territories for the Saguaro?" I asked.

"Naw. Just go anywhere to pick." Although the Pima lacked family territories for these natural resources, individuals did have favorite places to gather cholla (*Opuntia* spp.) buds, mesquite pods, and Saguaro that were accessible and produced the best-quality fruit.

Saguaro fruits were knocked down by means of a *ku'ipaḍ* made of several Saguaro ribs tied end to end with a small crosspiece (the *matsig, machchuḍ*) near the upper end to dislodge the ripe fruit. Culver Cassa pointed out the specific terms used in picking Saguaro fruit. To pull the fruit down by hooking the *ku'ipaḍ* over it is *oḍ*, while to knock the fruit off by pushing up below it is *hemchkwua*.

Ruth Giff explained: "People cooked it or just eat it raw. The *juñ* is the dried fruit. *Bahidaj* when it's ripe; means the fruit by itself. I guess it's the Papago word. But a Pima, when they go after the fruit, always say, 'We'll go pick *haashañ*'—but the Papago go after *bahidaj*. Say *oḍop* 'to go pick *haashañ*' or *to oḍo* 'let's go pick *haashañ*'. *Juñ* is the fruit picked up from the ground after they fall. If they really dry up there [on the Saguaro], won't pick up the sand [when they fall on the ground]. But the Papago don't care if they get sand because they just want the juice, boil them and strain out the seeds. Nobody goes for it now [in Pima country]. But right here, west of the church [St. Catherine's in Santa Cruz], they are so thick; we went there when we had to walk, to make *sitol* ['syrup']." The word *bahidaj* is derived from *bahidag* 'fruit' or 'crop', which in turn comes from *bahid* 'to ripen'.

There are some differences in terminology between the Akimel and Tohono O'odham, as noted already. Sylvester Matthias explained: "Papagos call *haashañ bahidaj*. *Bahidaj* is not Pima. *Gakidaj* is the fruit *off* the Saguaro, open, dropped on the ground. *Bahidaj*, the fruit still *on* the Saguaro, fresh; but us Pima just call it *haashañ*. They go pick it *[gakidaj]* and make it into that ball called *juñ*." Mathiot (1973:271) gives *gakidaj* as "dry saguaro fruit formed into a cake (also called *juñ*); *gakidag*, 'dried fruit; dried out plant'."

Joe Giff said *iá* was the *haashañ* fruit picked and put in a basket and *juñ* was the dry fruit picked up from the ground.

Juñ is both sweet and rich, because it contains all the pulp from inside the fruit. Sylvester said, "If you want to, you can break it off and combine it with pinole [or] just add water and make a beverage out of it."

George Kyyitan of Bapchule provided the following classification: *haashañ iibdag*, the fruit, not ripe yet, still up there; *bahi, bahidaj*, the fruit, ripe, still up there; *gakidaj*, the fruits when they fall on the ground; *iá*, what they take out *(iáḍa)* from the *bahidaj*, when they still have the juice

in there. Then he explained, "When the fruit still have the juice, just open up and put it in the can, [still] juicy. But those others *[gakidaj]* are dry. Then they put in a chunk like that, [make] a ball of it pressed together—[called] *haashañ gakidag*."

"What about *juñ*?"

"I think [that's] when they made that chunks and put it away."

The Pima have a few other terms for the Saguaro in addition to those referring to the fruit. The side arms are called the *mamaḍag*. (Mathiot [n.d.:44] gives it as *mamaḍog*.) This word does not apply to the central or main trunk. Both Gila Woodpeckers and Gilded Flickers excavate nest cavities into the soft tissues. The cactus seals off the newly exposed tissues inside with a tough barklike scar. Almost every Saguaro will have one or more of these round openings called in Pima *haashañ shuud* 'Saguaro's vagina'.

Sylvester mentioned two growth-form variants. "There were dwarf Saguaro, *jeejim haashañ*, about 12 feet, more or less, but they bore fruit. Don't see them anymore. [Name means] '*haashañ* with a mother'. Also there was a 'woolly Saguaro' near Del Monte's store called *voshḍk haashañ*. They made [a] postcard of it."

Injured or senescent Saguaros may succumb to a disease caused by a bacterium, *Erwinia carnegieana*. By-products important to the Pima result from the rapid tissue decay, which takes less than three years (see Alcock 1987). The *haashañ shuud* comes free as a great tan-colored "boot," looking more like a petrified moccasin. These were formerly convenient containers—for tobacco and perhaps seeds. Also these were once used, according to Sylvester, to make the ceremonial mask for the *s-kukpadkam*, one of the dancers who accompanied the Navĭchu. (He and I attempted to work scar-tissue boots into masks but found them quite brittle, even when soaked for days. Sylvester's masks are all of bottle gourd, *Lagenaria siceraria*.) But it is the vertical circle of outer ribs that the Pima found most useful. These light but very tough ribs are called *vaapai*.

Beginning probably sometime early in the Hispanic period, Pimans gradually started the process of abandoning the *o'olas kiikĭ* 'round houses', substituting rectangular houses of adobe. Europeans, both Hispanic and Anglo, pushed the new form, even though the new style had a serious drawback, as Russell (1908:156) pointed out:

> From a hygienic point of view it is a great pity that the Pimas are learning to build adobes, for the tendency is for them to live indoors and to abandon the healthful arbors, every inch of whose floors is purified by a burning sun that throws its sterilizing rays well under the arbor during the morning and afternoon. Tuberculosis is present in nearly every family, and it is difficult, if not impossible, for the agency physician to induce those stricken with it to remain out of doors; they invariably confine themselves within the bacilli-laden dwellings. The arbor is kept well swept and clean, as is the entire yard about the house, so that a more healthful habitation could not be devised.

While the Tohono O'odham, perhaps because of their closer contacts with Hispanic settlements, adopted the making of adobe or sun-dried bricks to build their houses, the River People instead improvised a *jacal,* or wattle-and-daub house. In its simplest and most original form, this consists of vertical timbers to which are attached horizontal bars, inside and out, of Saguaro ribs. Adobe mud is then packed into the intervening space. Successive layers of mud are piled up until the height of the eaves is reached. Then crossbeams, a thick layer of Arrow-weeds *(Pluchea sericea),* and finally a thick layer of adobe completes the roof. When I arrived at Komatke in 1963, this was almost the universal house. Some were plastered, inside and out, then painted white, but many were just bare puddled adobe and Saguaro ribs. The next innovation was to substitute thin pine lumber for the crossribs. People who built houses told me Saguaro ribs were preferable because they were rough and held the mud in place better than did the lumber.

These adobe mud dwellings on the reservation are universally called sandwich houses, supposedly for a fanciful resemblance between the *jacal* and a sandwich. In Pima they are called *eḍa wuisim kii* '[something] laid down house' or '[something] put in house', from the way the mud is packed between the ribs. Others call this type of house *wua kiita;*

in contrast, a house made from adobe bricks is called *shaamt kii.*

There were other uses for these *vaapai,* or Saguaro ribs, as well. Sylvester said that his father was good at making *vaapaidag* 'splints' or 'casts' to set broken bones. Hrdlička (1908:247) wrote, "Splints and bandaging are also employed [for broken bones]. The splints are generally made from the flat, elastic ribs of the saguaro."

Fetishes used by the *mamakai* 'shamans' were carved from Saguaro rib wood. Russell (1908:107) as well as Bahr and associates (1974:229) illustrate one representing a horned lizard and another said to be a lizard. Russell remarked that "they are either carelessly or clumsily made." Some Pima had the last laugh on a naive gringo. The horned lizard is a good enough representation of that animal. The lizard supposedly represented by the other is the *jusukal,* Desert Spiny Lizard, a common inhabitant of mesquite woodpiles around Pima houses. Women going out to urinate behind woodpiles feared this lizard would enter their vagina. The carved "lizard" fetish is not at all clumsily made; it is an erect phallus, anatomically correct in every detail. Furthermore, the curing song for this species of lizard is associated with "loose women" or harlots (Russell 1908:308–309; see also Bahr and associates 1974:319). Sylvester once brought us a *chukuḍ* (Great Horned Owl) fetish carved from Saguaro wood. Although there was an owl disease and its curing song, no fetish was associated with it. Apparently his owl carvings were intended as purely decorative.

During the Navĭchu ceremony, bull roarers were swung by the *s-kukpadkam.* Both the handle and the rotating blade were made from Saguaro rib because it is light but strong. The preferred wood for the hearth of the fire drill, called *iivdakuḍ,* was a flat, dry piece of Saguaro rib. Likewise, four similar-sized pieces of *vaapai* were marked on one side and used as dice in the gambling game called *kins,* presumably from Spanish *quince* 'fifteen'. Although Curtin mentioned Arrow-weed and mesquite root as materials for making these game sticks or *kinskuḍ,* a set Sylvester made is from Saguaro, as were those collected by Russell. Birdcages and various types of latticework were sometimes made from *vaapai* instead of *u'us kokomagĭ,* Arrow-weed, as shown by Kissell (1916:141, 144).

Saguaro ribs are expanded near the base of the plant where they are joined, so that occasionally they will pry apart, with the lower foot or so of the rib being oar-shaped. These sections are used to rake cholla buds over a screen so that the spines are broken off, either before pit roasting or after drying. Sylvester also made wooden kickballs from these thickened bases. They are between 2 and 2½ inches in diameter.

The Saguaro harvest was a rich food resource for the riverine Pima, coming after the spring crops were planted and the winter wheat harvested. *Haashañ* competed only with the first mesquite crop. The fresh juicy fruit could be eaten as is or dried, seeds and all, for use anytime during the year. Fresh fruit could also be used to make a nonalcoholic beverage called *iá.* This sweet and flavorful blackish red semiliquid mass that comes from the fully ripe saguaro fruit is an evanescent treat to desert dwellers, both human and nonhuman. Sylvester described this, "See, some are really juicy. We call that *iá.* That *iá* don't keep long. It'll turn sour. Have to drink it right away," he explained. "It will turn into wine, come sour, if you don't drink it. But it would taste awful, because it's kind of stale, old. Same way with that wine, in that *navaitakuḍ* [wine-making olla]. You have to watch it till it's just ready, then you drink it. If you have it too long, it'll get stale. They call it *gahi dahiwush.* That means it just come weak. It has to be [drunk] when it's ready. That's the time you drink it up."

Even more important in the long term is *haashañ sitol,* a syrup made from the pulp of the ripe Saguaro fruit. Both *iá* (fresh fruit) and *juñ* (dried fruit) were broken up in water, the seeds strained out, and the liquid boiled down to the desired thickness. The dark reddish brown *sitol* was once a staple item among Akimel O'odham families and still is among many Tohono O'odham. It is eaten with *haakĭ chu'i* (pinole) or on *chemait* (tortilla).

The Gila Pima have almost entirely abandoned harvesting Saguaro fruit during the last half of the 20th century. But in early July 1973 I participated in a Saguaro harvest along the bajada on the southeast of the Sierra Estrella. Most of the fruit was boiled down to make *sitol.*

Ruth Giff said in 1988, "[Someone] told me that this Papago lady who used to be married here—she separated and went back to Chuichu—made some *haashañ sitol* to sell—$5 for a little jar, maybe a pint. So I told her I wanted some. You know, that's a *lot* of work. You have to cook it so *long!* Some sell it for $10 a pint. I asked her what she did with the seeds, and she said she just throw them out. The Papago don't care much about that. So I told her, next year, if I'm still around and you're still around, to save the *kai* for me. You know, they keep for a long time."

"I know, I've had some for eight or nine years and it isn't rancid yet," I told her.

"It's because they cook them. But they get those little pebbles in there. They don't take them out. But I just put them in water, and they just settle, and the *kai* float, and I skim them off with a sieve. We haven't gone after it for a long time. It's ready in June, late June, about the 24th of June." This is the time when daily temperatures soar to 110°F and more.

George Kyyitan was explaining the preparation of *sitol* out of Graythorn (*Ziziphus obtusifolia*), sweet sorghum, and watermelons that are very ripe. "Just the same as that *haashañ*. Make chunks out of it. Store it away. Keep it for the year to meet the year. Sometimes they have enough to make wine out of it, they can make it. I think this month [August]; they're supposed to have it this month. They have lots of ways to make [it] out of that *haashañ juñ*."

Sitol could be kept for long periods. Sylvester told about this technology: "The *haashañ sitol* they seal it into little container, olla."

"What did they use to seal it with?"

"The clay. It's potterylike. Maybe half-gallon, I don't know. They're not too big. Then they make that lid, then pour the syrup in it. Then seal it. Then add some more clay on top. There, it's sealed."

"Not baked or fired?"

"No, but it holds."

"How long would it last?"

"I imagine several months." Others said it kept for years. Whittemore (1893:55) observed that the *sitol* was "stored away in small earthen jars hermetically sealed, a foot or two underground." But George Kyyitan told me, "Those *haashañ sitol* they use in those little ollas—cover up, and put it over there, and let it stay there. When it's really too long, and heat it up in the summer time, it's kind of, tastes a little like chile when you eat it. You could taste it

that way. When it stays too long, it'll go that way."

The *kai* 'seeds', high in proteins and fats, were very important—even more important, Ruth Giff said, to the Pima than to the Papago. "*Haashañ kai* 'Saguaro seed' is ground on the *machchuḍ* [metate]; makes it greasy, and after a few times you have to chip the surface of the *machchuḍ*. You can hear somebody doing that: ping, ping, ping. Then the wheat ground after that gets a little sand in it from the chipping."

Ground *haashañ kai* mixed with ground whole wheat makes a nutritious breakfast dish called *ku'ul* or *haashañ kai hidoḍ*. Russell (1908:118) said, "The seeds were available at any time, as they were always kept in store as an article of food." Earlier he noted (1908:71), "The supply is a large one and only industry is required to make it available throughout the entire year, as both the seeds and the dried fruit may be preserved." When Tohono O'odham came to the Gila to help with the wheat harvest, Saguaro seeds, dried fruit, and syrup were important trade items brought with them (see Russell 1908:93). We should not underestimate the significance of *kai* to the Gileños. Russell (1908:7) also recorded, "Seeds that have passed through the body are sometimes gathered from the dried feces, washed, and treated as those obtained directly from the fruit, though there would seem to be some special value ascribed to them as in the case of the 'second harvest' of the Seri." Needless to say, no vestige or even knowledge of this particular custom seems to have persisted into the generation I worked with!

But I have been served *ku'ul* (ground wheat and *kai*) several times in Pima homes. Ruth Giff said, "I miss that. We had a lot when we were growing up."

One day George was discussing how his mother used to prepare seeds of *shuu'uvaḍ*, a wild mustard. "It's just like that *haashañ kai*, but *haashañ kai* is a little thicker, because they use flour in it sometimes. I sure like that *haashañ kai*. When this late Ira Hayes' mother was living, she usually get the seeds from somebody. And do that, make gravy. She have to give me some."

"What do you call that?"

"I think they call it different ways. We call it *ku'ul*."

One time I asked Ruth about initiating lactation when there is no flow. "If no

milk comes to a mother, make *ku'ul*. That's the same as *átol*. Made with *hannam* and wheat. Or probably sometimes *kaij* ['its seeds']."

"What kind of *kaij?*" I asked.

"*Haashañ*. Don't have to say. If you're O'odham, you'll know."

Leather tanning was of relatively little importance to the Gileños, because rawhide sufficed for sandals and shields and the Tohono O'odham probably brought up for trade most of the tanned hides. However, Russell (1908:118) described the local methods. "Two tanning media were used—brains and saguaro seeds. The former were kept dried into a cake with dry grass until they were needed, when they were softened in water. The seeds were available at any time, as they were always kept in store as an article of food." Joe Giff and Sylvester Matthias thought that the Pima did not do any tanning; even in 1901–1902 Russell found "that Gileño women did little more than enough skin dressing to keep the art alive among them. At present there are very few who know anything about it."

The annual Saguaro wine feast among the Gila Pima is something of an enigma. What was its nature? Was it an occasion for personal indulgence in a social context, or was it a community prayer to bring rain, the ultimate necessity for desert agriculturalists? Historic records kept by Hispanic and Anglo outsiders of these "*tizwin* drunks" do not even hint at an underlying religious and life-sustaining ceremony (see Cremony 1868; Curtis 1908; Garcés 1900; Grossmann 1873; Russell 1908:72; Whittemore 1893). For instance, Grossmann (1873:419) wrote, "All Pimas are inordinately fond of this beverage, and old and young partake of it until the whole nation are wildly dancing about in a drunken frenzy, until at last they drop to the ground overcome by the stupefying effect of the liquor." Twenty years later Whittemore (1893:55) wrote, "The feast is kept up until universal intoxication ensues; and one or more are usually killed."

The combined Gila Crossing, Black-water, and Salt River calendar-stick oral histories record 25 festivals where liquor was brewed, and 24 people killed during drinking bouts between 1833 and 1902 (Russell 1908:38). It is not known, of course, how many of these were associated with Saguaro wine feasts and how many were simply drinking bouts.

An active effort to prohibit wine feasts began in 1890 (Russell 1908:61); however, the western villages seem to have been the center of traditionalism, with some groups clinging both to native ceremonies and their centuries-old folk Catholicism on a reservation rapidly being converted to Presbyterianism. Sylvester Matthias summarized the situation: "After the flood of 1906, old Komatke was right there near Gila Crossing. Catholics [were] forced out, so [they] started chapel where St. John's is now. They were forced out because they kept the old ways, having that drinking, not for rain, I guess just for the heck of it. It says in history that it's the government that stopped it, but my father says it's the Presbyterians that stopped it. They were going to put my father in jail for running that feast. And then Father Bonaventure Oblasser repeated the same story: that the *mimsh* [Protestants] stopped the wine feast; and I believe it."

On another occasion Sylvester related, "The last wine feast was in the 1890s. According to history, the first Mass was said in 1897 at St. John's. So it must have been earlier than 1897, when the wine feast [was held] at old Komatke, west of the [Gila Crossing] Day School. Joe Giff's father, the old man Giff, rounded up the kids, and keep them away from the doings. Oh, they had *fun* by themselves." Apparently the community recognized it was a dangerous affair. A few years later, Russell (1908:171) observed that the wine feasts were considered "drunken orgies in which, since the introduction of knives and firearms, men were sometimes killed."

According to Sylvester, at that time there were two ways of drinking: "Have it and share it, *navait i'i [dag]*, just social; or 'sit and drink', *dahiwuak ii,* is the little ceremony, social way, too. That's about the only ceremony they had in that wine feast."

There was no ceremonial roundhouse where the wine was carefully fermented by specialists who watched over it and sang, he told me. Rather, wine making took place in individual houses. I quizzed Sylvester further. "In Pima way, when you go to a house with the wine ready, what do you do?"

"Just drink."

"No songs?"

"No songs. Except for the ones who want to; they sing. There is a song they call the 'Navko'i ha Ñe'i.' They, maybe just a few years ago, a few people sing the songs, the 'Navko'i ha Ñe'i.' A *navait* song. Or any other songs, like the Devil's Songs. Some use that for the 'Navko'i ha Ñe'i.'"

"Where did they get their *bahidaj* for a wine feast?'

"*Jeweḍ haashañ* ['land with Saguaros'], off reservation, where all [now is] cotton fields, west of 51st Avenue. Had plenty of *haashañ* [there]. Or on Muhaḍag [South Mountain] or on [Sierra] Estrella. And they have vineyards across the river. That's where they go to get grapes and make it into wine."

"The Pima knew how to make wine from *huudvis* [grapes]?"

"Uh-huh. That wasn't ceremonial. They called that *navait ii, navait ii'e.* Just a drinking party; there's no ceremony; just a social gathering. It's different from the Papago. Salt River People came to join in."

Certainly there was little in Sylvester's account to suggest a profoundly religious annual event occurring early in the 20th century to call down the rain. Also, Saguaro and the resulting cider- or beer-like drink was at least partially abandoned and replaced with the much stronger grape wine.

Francis Vavages was a few years older (born 1905) but was raised near the tribal agency, under the watchful eyes of white officials, Indian police, and missionaries. I asked him if he knew of the Saguaro wine feast.

"*Haashañ navait*—that was once-a-year treat that they get in those days. I heard they used to do . . . during the summertime . . . each one . . . to bring the rain. I know they put it in some places where it's warm. I just heard about it, but they don't have it at the time when I was small. But people would know where they—Papago—would have it, up there. Take off [to] Chuichu or Santa Rosa. They'd go for that drinking party up there. They'd make it in big ollas, too. And so many persons will have it in their homes. And when it's gone, they'd bring some more of that. [Lasts] two or three days, I guess."

The Tohono O'odham wine feast or *navait i'idag* 'wine drinks' has been carefully investigated by some of the most thorough analysts of desert ceremonials (Crosswhite 1980; Densmore 1929; Underhill 1938b, 1946; Underhill and associates 1979). A summary of Tohono O'odham wine feasts, which persist on their reservation, will help us determine if the River People too might once have had a ritual wine feast.

The harvest for Tohono O'odham was a family affair. The picking of Saguaro fruit might last for several weeks sometime in June and July. Some *sitol* (syrup) prepared by each family is taken to the *olas kii,* or communal roundhouse. An old man who knows the ritual speeches and the fermentation process mixes the *sitol* with equal parts of water, setting the liquid out in great clay ollas in the *olas kii.* (Densmore [1929:153] said that about half a pint of *sitol* was mixed with a gallon or so of water.) Outside, people dance while singing formalized songs, to assist the fermentation. In two days and nights, the *navait* should be ready. Several men then go to neighboring villages, delivering a formal oration of invitation. A man seated before the *olas kii* of the invited village accepts with a parallel oration. At midday everyone arrives at the host village for the *dahiwuak ii* 'sit and drink'. There is another formal speech assigning the visitors to be seated at the north, east, and south, with the hosts to the west. Serving wine and singing songs occur in a specified order: east, south, north, and west. All the *navait* is consumed at that time. The drinking is concluded with the formal memorized Mockingbird Speeches, 'Shuugaj Ñi'okĭ.' Then people return to their own homes to consume the *haashañ navait* they have been brewing there as well.

The result is intoxication and vomiting, for the *navait* is an emetic. But the ceremonial function is drawing the wind and clouds and encouraging these to fall as rain. The planted seeds and wild growing things can then come forth. And if rains have already started, as is usually the case, the ritual is intended to continue them. (Villages hold the feast sequentially.) The People sing to "pull down the clouds" (Underhill and associates 1979:33).

But did the Akimel O'odham have such a ritual, whose songs and speeches and symbols were permeated with clouds and rain? Was there once a profoundly religious ceremony directed to the four rain gods (symbolized by the four drinking groups seated about the communal *olas kii*)? Or did the Gileños, dependent on irrigation from the rise and fall of their river rather than on tenuous *de temporal* agriculture at the mouths of arroyos, observe merely a drunken annual social event, as oral histories suggest? Was it as Frank Crosswhite (1980:50) surmised: "Due to their use of the water resource provided by the river, the Pima were probably not as preoccupied with rain-making as were the Papago"? Crosswhite continues: "Although there are many historic records of Saguaro wine used by the River Pima, there seems to be no clear indication that wine was associated with rain-making."

Would the riverine Pimas even have had a motive for a critical rain-bringing and rain-sustaining ceremony? I think the answer is yes. Although *akĭ chiñ* 'arroyo mouth' agriculture, using run-off from rains, contributed only a very small portion of their economy, it was an option that had not yet been entirely abandoned earlier in the 1900s, according to Joe Giff and Sylvester Matthias. The Gileños were double croppers, with a spring planting followed by *shopol esha* 'short planting' that had to be in by middle August to beat the autumn frosts. But the short planting was also timed to the second rise in the river, caused by the summer storms over the Gila watershed. Surely the People made a connection between the great thunderheads that gathered in the east each afternoon and the subsequent rising of their river. And various wild grasses and greens that were harvested locally depended on summer rains, which are even less reliable in Gileño country than in the higher Tohono O'odham country to the south.

But motive alone is not sufficient proof that there was a rain-making ceremony. There is good evidence in Pima culture for the same symbolic associations that are found among the Tohono O'odham. In the Creation Story that Lloyd (1911: 218–229) recorded from Thin Leather, for example, there is a marvelous story of Tobacco Man and Corn Woman, two interdependent deities in the Gileño cyclic scheme (see Group I, *Zea mays*). Their offspring, a pumpkin, when broken, sinks into the ground, coming up somewhere else as a *haashañ*. Both tobacco, when smoked, and *haashañ*, when fermented and drunk, bring clouds and rain.

In her studies of Tohono religion, Underhill (1946:59) saw parallels with Pima oral literature. For instance, she equates Russell's *jujkida*, or rain-ceremony speech (Russell 1908:347-352), with a

Mockingbird Speech. Like the Tohono O'odham, the Akimel O'odham fermented the *navait* for two days, "and in the meantime the people gather and dance in the plaza nearest to the spot where the large ollas are simmering [*sic*]" (Russell 1908:170). She even points out four songs from the Russell anthology that have the structure of *navait* drinking songs.

But the clearest association of the ceremonial aspects is found in Juan Smith's version of the Creation Story (Hayden 1935:22): four *mamakai* (medicine men) were seated at the four cardinal points, as in the Tohono O'odham ceremony, reminiscent of the four rain gods. "So when the sun was just coming up, two men put some of this wine in some vessels and took them and gave it to the medicine man that was sitting at the east side." He sang this song:

> Some blooming, blooming wind
> I'm sending my wind
> I'm making the earth turn yellow.

One of the wine servers next went to the north; the other went to the south. After he had drunk, the north *maakai* sang:

> They have come to their blooming,
> Wine,
> I drank it and I am drunk
> And everywhere it's turning green.

Next the south *maakai* drank, then sang:

> They gave me that red water to drink,
> And when I drank it,
> I am drunk,
> And I am making some rainbows.

The servers then both went to the *maakai* sitting in the west and gave him *navait*. After drinking he sang:

> You have given me
> Some crazy water [*nodagiam shuudagĭ*]
> And you have made me drunk
> The ground is getting damp [raining].

This is an exact description of the liturgical actions described by Densmore (1929:152–157) and Underhill (1946:58) for the *dahiwuak ii*, except that it is described as occurring at dawn instead of midmorning or midday. Here we find once again the basis of the most important Pima ritual encoded in its

most fundamental outlines in the Creation Epic. The songs that the four symbolic gods sing are associated with rain. East Doctor turns the earth yellow, just as the desert turns yellow with poppies and sunflowers after rains; North Doctor turns everything green with rain; South Doctor makes rainbows; and West Doctor wets the ground with rain.

The oblique reference is a characteristic form of Piman respect, particularly in public speaking or singing. For instance, an orator announcing the object of a communal hunt might refer to the "long-eared ones"; it would be much too direct and offensive to say, "Tomorrow we will hunt jackrabbits." So too we usually find *haashañ navait* not mentioned directly in song. For instance, Densmore (1929:161) recorded a drinking song in which it was called simply "a red water." As noted, Underhill (1946:63) identified four Pima drinking songs from Russell's anthology (Russell 1908:311, 319, 325, 337). Two of these obliquely mention "reddish water"; only one refers directly to *navait*. (Russell listed three of these as curing songs but none as drinking songs.)

George Kyyitan perceived the Pima wine feast quite differently from Sylvester and Francis, as noted earlier.

"When the Akimel O'odham did the *dahiwuak ii*, was it supposed to be a prayer?" I asked him.

"It *is*. They do that. I kind of see that when I was small. In the days when they don't have any water. *Nothing*. But they use it for rain, to their crops. Some they gather up to pray. That's what they do. Pray. They used to do that. They really get the rain. It won't be like this. . . . See the clouds—you see the clouds over there, all around it, but we wouldn't have any rain here." George pointed east, drawing my attention to some peaks on the horizon crowned with clouds in an otherwise bright blue desert sky.

"In those days, all clear like this. And there be a *little* lightning, *way* up there somewhere. Just one place there. Late in the night, the rain will come. It comes *all* the way. See. And so I may believe that they get what they ask for praying like that. But I don't think some of this new generation—they don't believe these things like that. Some of it, they do believe, but they don't ever do that [make wine feast]. So that is [the] way when I was a kid."

I asked George if he ever heard of the "Shuugaj Ñi'okĭ," the Mockingbird Speech. "It's hard to understand. Pretty hard to understand. It's pretty hard to put it up [recite it]. I know one guy does. He's my first cousin. Living down here. And he passed away. And he did one, he talked. He mentioned some of the words. But I couldn't understand what he meant. It's really hard talk. Long time [ago]. Different way [of speaking].

"I think it's two ways of saying that. Coming in or the home speech. That's the way it is—two ways. That mockingbird speaker. Then after, when he gets through, the home man, the home speaker will return—will answer his speech. Yeah, it's like that. But anything like that—a singer or anything, like that drinking, or *jujkida* or anything like that [is a prayer]."

Clearly George was describing the invitation oration and the acceptance speech of the guest village. These were formalized in archaic language, difficult to deliver as well as to understand.

I asked him about the *jujkida*. "Is it part of the rain ceremony?"

"Uh-huh."

"Is it associated with the *haashañ navait?*"

"Oh, yeah! That's why they are getting their rain. It happened here once, just for entertainment [demonstration]. But I didn't see it all. Because I was just a kid at that time. I didn't see it very good. I'd sure like to see that one [in Tohono O'odham country], to see all the parts in there. I'd like to see it before I go away."

If the Gileño wine feast was originally associated with rain making as was that of their Piman cousins living away from the river, what happened to secularize it? After the turn of the century, Russell (1908:170) wrote, "Of course all festivals partook somewhat of the nature of sacred ceremonies, but when this element was at a minimum, as in the saguaro harvest festival, its description may properly appear here with the arts of pleasure."

I suspect that the loss of the sacred aspects of the wine feast is just one facet of the complex of general secularization of Pima society that began with European contact. Corn, the temperamental crop that had to be enticed into good production with all the prayers and ceremonies that the people could evoke, was supplanted by wheat. Whereas maize found

its way actually or symbolically into nearly every ceremony, wheat was purely profane. With the advent of Anglo-Americans on their way to California, wheat production temporarily made the Gileños successful entrepreneurs. It grew in winter, a cool season when rains were less critical. With cash and access to other crops that could be fermented, Saguaro, once at the very heart of the Piman liturgical calendar, lost its sacred character. Beer, wine, and whiskey, generally unavailable during Spanish and Mexican times, became more readily available with the settlement of Anglo communities near the reservation. (Indians were not permitted to buy liquor until after World War II, and possession of alcohol on this reservation was legalized only in the late 1960s.) Drinking was no longer a communal celebration indulged in for two or three days in summer. By the time most people I worked with were growing up in the early 20th century, the sacred rain-making aspects of *haashañ navait* and the *dahiwuak ii* had been largely forgotten.

According to an account by Fr. Garcés, the wine feast was perceived to have a therapeutic effect on the People as well. On his return from the Hopi in late August or early September 1776, the Pimas Gileños were astonished that he was still alive and said they wished to celebrate all together. Garcés, always one to see the positive side of the Indians and their country, consented, provided that the festivities were held away from where he was lodged. But he wrote, "The next day I complained of these excesses to the governor, who told me that it only happened a few times a year and in the season of saguaro, and added that it made his people vomit yellow and kept them in good health" (quoted in Russell 1908:72n; see also Garcés 1900, 2:439–440).

The actual fermentation of the *navait* between the two Piman groups seems to have been similar. Russell (1908:72) mentions "a process of boiling and fermentation." Sylvester provided more information on this wine. "One time somebody from Sells, they took some of that cactus wine and analyze it and they say it's weaker than ordinary beer. Then my father said at the time that [when] it's ready, it should be used. If not, they call it *gahi dahiwush* ['overdue' or 'over past']; then it won't be good. *Navait* [is] not like [grape] wine; when it's wine, it's always

wine." Sylvester considered the analysis an invalid assessment.

The large olla for fermenting the wine is called the *navaitakuḍ* 'wine something to make with'. Its use was abandoned, apparently, early in the century. I asked Sylvester if anyone still had them when he was a child. "No, nobody. But I do seen it years back; that potter at Topawa, Laura Kermen, she had one of those *navaitakuḍ*. Stands about that high [about a meter tall]. Must have hold about ten gallons. It's a big-sized one." Saguaro wine ollas average about half a meter tall and wide, holding about 10 gallons.

The Saguaro wine celebration among the Yuman-speaking Maricopa Indians who lived adjacent to the Gileños appears to have had an entirely different symbolic framework (Spier 1933:56–58, 162–163, 269). Although there was formal communal fermentation, invitations, distribution, and singing, there was no rain invocation. The "red water" seemed to symbolize blood, and thoughts turned not to mounting summer thunderheads but to the enemy and war. "The intoxication and the incitement of song commonly ended in a decision to go on a raid" (Spier 1933:58). Perhaps some of this diffused back into their Piman neighbors on the Gila, where later historic *tizwin* debaucheries often ended in bloodshed rather than violent summer rainstorms.

Vouchers: *LSMC 28* (not extant).

Dasylirion wheeleri S. Watson
Sotol, Desert Spoon
Agavaceae
umug

Skirting Gileño country at elevations of 3000 feet or more in a great semicircle from the northeast to the southeast is the somewhat yuccalike Sotol. The dense rosettes of long leaves indeed suggest either yucca or some narrow-leaved agave until you come closer and see the structure of the bluish gray leaves. Unlike an agave or maguey, these are very narrow, of uniform width nearly their entire length, and not fleshy or succulent to any degree. And they lack the agave's strong spine at the leaf tip. Any attempt to handle Sotol leaves will immediately convince you that they are not yuccas, because the parallel leaf margins are ferociously barbed with short slender teeth that catch you either coming or going. (There are even microscopic teeth along the edges to discourage small predators from injuring the plant.) The leaf surfaces are rough. Whereas you might come to terms with yuccas and even magueys, Sotol firmly stands its ground.

It is an attractive plant in its rocky slope habitat. At blooming time it sends up a densely flowered spike to head-height or even twice that, with each plant bearing small flowers of only one sex. The bud of the emerging stalk is high in sugars. Unlike agaves, the plant does not die after blooming, but the old stalks persist for a season, at least.

The name for Sotol may be pan-Piman. Among the Gila Pima and the Tohono O'odham it is *umug*. At the Pima Bajo pueblo of Onavas in southern Sonora, I was told *húumug*. The Mountain Pima say *umoga*. (The Tepehuan names Pennington [1969] recorded for this genus are not cognate.) Very few Gileños know the name *umug* today, and some who do fail to recognize the plant even though they have some idea of its appearance. In earlier days they must have encountered it when making forays into Apache country, and the plant grew throughout Sobaipuri country (above the San Pedro floodplain).

But the lasting contact with *umug* came by way of the Tohono O'odham, who regularly gathered the plant to make the *main* 'sleeping mat' as well as a number of other specially plaited items that have also fallen into disuse. Kissell (1916:150), who called *Dasylirion* "palmea," wrote, "The leaf of the palmea is the useful part of the plant for plaiting, and is in perfect condition to be gathered at any time of the year. Its harsh spiny edge makes it difficult to collect, necessitating the use of a stick for breaking off the leaf; so that for the gathering the women travel afoot armed with long sticks for severing the leaves. When a sufficient number has been secured, they are carried home in bundles on the head, or in the kiaha *[giho]* carrying frame on the back."

Castetter and Underhill (1935) give a concise summary of the preparation methods:

Sotol, too coarse and rough to be used in making clothing, was utilized to make large, tough sleeping mats which

could be rolled up and carried about; also in making cradle mats, back mats for the carrying frame, headbands, headrings, and two kinds of baskets.

Sotol leaves are prepared for use by first scraping off the spiny edge with a knife, then splitting the leaves lengthwise in two with the thumbnail (the thumbnail in olden days was kept long to be used as a knife). These withes are triangular in cross-section, and the cushion of pulp on the inner margin is removed, making a flat strip of uniform thickness about an inch wide, with both a rough and a smooth side. The withes are dried in the sun for a few days, then wrapped and stored. When needed they are buried the night before in moistened ground in order to make them flexible.

The Gileños once made their own *main* from the split stems of Carrizo *(Phragmites australis)*, but this reed was one of the victims when the river declined, and the Pima had to turn to the Tohono O'odham for sleeping mats made of *umug*. In Russell's day there must have been a barter in both the plant's leaves and the already prepared mats. The one he illustrates (1908: 146) is of *umug*. The one figured by Kissell (1916:152) is the original Gila Pima *main* made from split, flattened Carrizo canes.

Russell, by the way, confused this plant with agave. The trinket basket and the scalp or medicine basket shown on p. 146 of his book (and mentioned on p. 93) are woven of *umug*, not agave. But the method of preparing fibers from the succulent maguey leaves Russell (1908:142) ascribes to "Tasylirion *[sic]* wheeleri" really are for agave. Kissell (1916:229) gives nearly the same description of the agave fiber preparation.

Vouchers: *Hodgson 7368* (GK, SM).

Echinocereus engelmannii (Parry ex Engelm.) Rümpler
Engelmann's Hedgehog Cactus
Cactaceae
iisvikĭ

The flat gray terraces of the Gila, dominated by Desert Saltbush *(Atriplex polycarpa)* and wolfberry *(Lycium* spp.),

burst into purple patches of flower about midspring. The sources of the color are scattered colonies of an elongate, heavily spined cactus: Engelmann's Hedgehog. The large trumpet-shaped flowers are usually bright purple, with pale stamens and a green stigma, but occasional plants bear almost pink or rose flowers. The flowers disappear almost as suddenly as they appear, and as the next few weeks build up toward the summer heat, the fertilized fruits continue to swell until they are the size of a plum. At maturity the whitish spines begin to loosen from the fruits, a concession to whatever animal would come eat them, thereby disseminating the seeds.

And here the Pimas obliged. The cactus they called *iisvikĭ*. Ruth Giff said, "Have to be right there or the animals get them ahead of you. It's something like strawberries inside: some are pink and some are white. Full of seeds. They're not cooked, just eaten raw."

Sylvester Matthias recalled, "Just eat the *iisvikĭ*. Don't make jam or anything out of it. Just eat it. We used to go after it. A bunch of kids would go after it. We all picked—everybody—however many boys. Then we divided it equally. Just about early May—the wheat would be just about ripening up then. Mostly white inside, some are pink. Not really too sweet, but when they get real ripe, they are sweeter. Sometimes they would taste a little sour. Even when they're dried, you can eat them, rinds and all. Except the thorns!"

Little has changed through time. Curtin (1949) was told, "When ripe, the scarlet fruit, which has a network of white spines, is eaten, the spines being removed with a stick." At the turn of the century Russell said, "The thorns of this cactus are removed as soon as gathered, and it [the fruit] is eaten without further preparation." But the present generation of Pimas have not learned to enjoy the late spring abundance of this plant.

Among the dialects of Tohono O'odham, two names are in use for the hedgehog cactus. Frank Jim, originally from the western part of the Tohono O'odham Reservation, called the Engelmann's Hedgehogs on the flats in Sacaton *giishul* while his wife Lela, a southern Tohono O'odham, called them *iisvik*. She used the *v* sound, saying she may have learned this name from Pimas while attending school at Bapchule. But this name occurs in Tohono O'odham dictionaries as well. On several occasions Sylvester told me of a cactus called *giishul*. He said that it is larger than *iisvikĭ*, has larger flowers but of the same purple color, and grows in quite large colonies. The only place you could find it was at Giishulik, 'place of the *giishul*', across the Santa Cruz River from Komatke where the stagecoach crossing was. Perhaps *giishul* was some species or variety of hedgehog cactus that was washed down the Gila from higher elevations, as occasionally happened with other plants. But the terraces across the river from Komatke were washed away with floods, starting in 1906, so Giishulik was probably washed away as well. With the Tepecano, the southernmost Pimans of northern Jalisco, *gisur* appears as 'pit[a]haya', a general Spanish term designating various taller columnar cactus. This term is probably cognate, indicating some cereoid cactus. Mathiot (1973:294) and Saxton, Saxton, and Enos (1983:16) use *gisokĭ* for a purple-fruited prickly-pear (*Opuntia* sp.). We have verified this ourselves at Quitovac oasis (Nabhan and associates 1982; *Opuntia violacea*). But this appears to be an unrelated name.

Vouchers: *592* (CR); *731* (SM); LSMC *17* (not extant).

Ferocactus spp.
[= *Echinocactus wislizeni* Engelm.]
[= *E. lecontei* Engelm.]
barrel cactus, bisnaga
Cactaceae
chiávul

The several species of barrel cactus form a distinctive and readily recognizable portion of the Gila Reservation flora. One species grows commonly on the alluvial flats; the other is found only on steep rocky slopes. Both are robust plants about knee- to hip-high when of flowering age. They have 20 or more deep grooves with clusters of stout spines along the edge of each ridge. The spine arrangements differ with the species, but the main spines are roughly ringed and the central one often hooked. These are the most stoutly spined cactus in Pima country. The heavy spines may be grayish or reddish. In spring or summer the heavy plants are crowned with rather stiff funnel-shaped flowers that are either yellow, orange, or reddish. After fertilization, the flowers develop into large, light yellow fruits about thumb length, filled with black seeds. There may be a dozen or more fruits lasting well into the following season. They have no pulp, but the fruit walls have a somewhat lemonlike flavor that will quench a hiker's thirst.

In Gila Pima this plant is *chiávul*, not to be confused with *jiávul* 'devil', Joe Giff jokingly reminded me. I checked this with Frank Jim from the western part of the Tohono O'odham Reservation. He said, "*Jiawul*—'devil' and 'barrel cactus'; same name exactly." I tested the two words on Sylvester Matthias. Invariably he heard the differences between *chiávul*, the 'barrel cactus', and *jiávul*, the 'devil', a loan word from Spanish *diablo*.

The fruits are seldom if ever eaten by Pimas, but the plants frequently find a place in their yards as ornamentals, at least at older, more traditional homes. These are probably picked up and transplanted from land that is being cleared.

Sylvester told me that the Pima were well aware of the use of the *chiávul* as an emergency supply of water. Under *tciavolt*, Russell (1908:77) recorded the following uses: "The pulp of the visnaga [Bisnaga] is considered valuable in lieu of water to those suffering from thirst. It is also eaten after being cut in strips and boiled all day. It is sometimes boiled with mesquite beans, a layer of each in the cooking olla. It is occasionally boiled with sugar."

I read to Sylvester Hrdlička's (1908:262) account of preparation of barrel cactus for eating: "The *biznaga* cactus, known by the Pima as *tsa-ult*, also serves occasionally as food. The top is removed and the inside pulp is sliced and cooked, usually together with the pods of the mesquite beans; the combination is said to be very agreeable to the taste."

"Yes," he agreed, "this is how my aunt, my father's sister, fixed it." Sylvester described the preparation of this robust cactus: "Peel it off, get the inside, slice it, boil it. My aunt did, not my grandmother. Use mesquite beans [pods] to flavor it. I don't know how long it takes to cook." Another year he told me, "My aunt cooks it. She'll get a little barrel cactus, a young one. She'd clean it off and slice it. Yeah. But I never tasted it. Usually when she fixed it, she'd give us a bowl full. But she got stingy with it."

"Did she have to cook it very long?"

"Not very long, because they're tender."

There is a specific name in Pima for this dish, but no one I talked to could recall it. In the early 1940s two Pima women in their 80s described the same preparation to Curtin.

I asked Sylvester what it tasted like.

"It tastes something like a *keli baasho*, a casaba [melon], when it's green [unripe]."

Sylvester added, "And there's a great demand on the barrel cactus by the Mexicans—they make that candy. There's a Maricopa man that, oh! he'd go out and get a truckload. There's a great demand lately—about the [19]20s, middle [19]20s. [He is recalling this in the mid 1980s.] And they [the cactus candy] used to come out in the markets—in Mexican stores, or they sell it right at the house—or somebody peddle around wherever they were camping."

Curtin was told that the recurved central spine of the floodplain species (*F. wislizeni*) was once used as a fishhook. Sylvester said he had never heard of this use, nor of the Pimas using fishhooks of any kind. (At least one tribe in central Baja California was reportedly fishing with these central hooked spines [Wagner 1966:49–53]; Palmer [1878b] described how an old Mohave prepared them.)

I found a barrel cactus near Lone Butte that had been uprooted and the top sliced off and cored. I had wanted to try one but did not want to injure a cactus just to find out. The liquid from the pulp had an insipid taste. George Webb told Curtin (1949:56) that the greenish liquid obtained from the pounded pulp "is neither sweet, sour, nor bitter." Let's hope that people will take our word for it and not destroy these magnificent slow-growing plants to find out for themselves; continuous land-clearing is destroying the habitat fast enough!

Barrel cactus occurs in a beautiful nature poem (the Swallows' Song, "Gigidval Ñe'i") recorded by Russell (1908:294). The eighth stanza mentions *tchiavor hyawsiga* (*chiávul hiósgga* '*Ferocactus* its flowers'). The free verse reads, "Thence I run as the darkness gathers, / Wearing cactus flowers in my hair."

One day I stopped with Irene Hendricks in Blackwater, near Holy Family Church, to measure the very large pads on

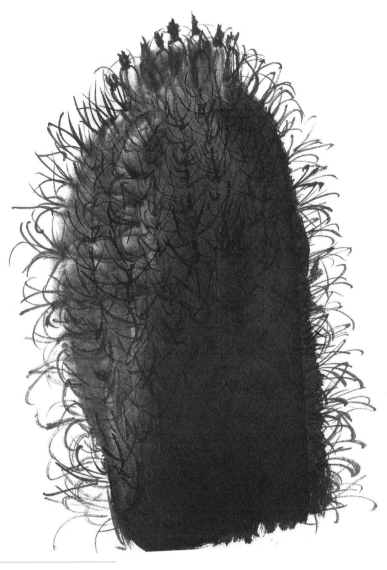

chiávul *Ferocactus cylindraceus*

Fouquieria splendens Engelm.
Ocotillo
Fouquieriaceae
melok, melhog

Ocotillos are one of the most distinctive members of true desert vegetation. Out on the well-drained bajadas or up on the rocky slopes of the mountains, Ocotillos wave their multiple, very slender arms in the wind. They occur on the floodplains only where people have transplanted them. Some may think of these as cactus, but they are not. The wands, a dozen or so growing from a single base, are completely woody, covered with a gray and greenish yellow mottled bark. The entire stem is spined, the sharp points left over from spring leaves during initial stem growth. The willowy wands may leaf out again at different times of the year when rains are sufficient. But these leaves have no spines and appear in the "armpits" at the base of old spines. Toward the end of spring and occasionally at other times, great masses of waxy red flowers cover the branch tips. These tubular tempters of hummingbirds exsert thick masses of stamens bearing pollen loads, dusting the heads of the active birds to ensure cross-pollination.

The northern Piman name for Ocotillo is *melok.* Both Akimel and Tohono O'odham use the upright poles, thrust into the ground and trimmed to an even height, for fences and as walls for the *vatto* 'arbor' and *kosin* 'storage house'. In Russell's day (1908:156), as early photographs attest, Ocotillo was more commonly used in construction: "Beside each dwelling will be found a rectangular storehouse *[kosin]* built with a framework of about the same shape and size as the arbor, but with walls of upright okatilla *[sic]* trunks or cactus ribs. The large bush, Baccharis glutenosa *[sic],* is often used for this purpose." The Gileños tend to use Arrow-weed *(Pluchea sericea)* for this purpose much more frequently, but just outside the door where I write at Sacaton, a wall of Ocotillo encloses parts of three sides of the *vatto.* In rainy seasons, many of the stakes leaf out, and some even bloom.

George Kyyitan said, "Sometimes Pimas make the fences of *melok,* like the Papago. It'll grow." He called such a fence *unma,* adding, "*Unamad* 'to do that', to put it

some prickly-pears (*Opuntia* spp.) I had seen there earlier. I suspected that people of an earlier day had been moving choice plants around, practicing some kind of semicultivation. Irene found a barrel cactus nearby. "Don't be afraid," she joked. "This is a *chiávul*, not a *jiávul*." I assured her that I knew the difference. She wanted to take the plant home, but I convinced her it weighed far more than two people could lift. And it looked quite contented where it was.

Technical Notes: Both *Ferocactus cylindraceus* (Engelm.) Orcutt (long known as *F. acanthodes* (Lamaire) Britton & Rose var. *lecontei* (Engelm.) Lindsay) and *F. wislizeni* (Engelm.) Britton & Rose are common on the reservation. *F. cylindraceus* (Mountain or Ridge Barrel Cactus) has a principal central spine 2.0–5.5 inches long that is red to pink with a yellow tip and is not hooked or recurved at the apex; this is a tall plant clinging to steep slopes or cliffs, flowering profusely in May or June following winter rains (sometimes also flowering a little again after summer rains). *F. wislizeni* (Saltbush Barrel Cactus) has a principal central spine 1.5–2.0 inches long that is red with a surface of ashy gray and is hooked or recurved at apex; it flowers in late summer and is barrel shaped to columnar. The seed coats of the two species are very different (see Benson 1969). *F. emoryi* (Engelm.) Orcutt (long known as *F. covillei* Britton & Rose), Emory's or Coville's Barrel Cactus, is a more southern species that must reach the very northern limits of its range on the reservation. It has major spines only, lacking the whisker-like ones; the pink or gray hooked or curved central spine is 3–4 inches long. It is red flowered, blooming in fall. Scattered individuals are found throughout the Sierra Estrella but are mostly uncommon (L. McGill with E. and M. Sundell *461,* ASU; *AMR 1887,* photos). This is the most bare looking of the three species.

Vouchers: *1481* (SM?); *1558* (IH); *LSMC 74* (not extant).

together, *all* the way around, like that *a'al hiha'iñ,* 'Children's Shrine' [at Santa Rosa in Tohono O'odham country]." Then George sang a song about this famous shrine, where two boys and two girls were once sacrificed to prevent the ocean from flooding out of a hole (see Underhill 1946:23; Underhill and associates 1979:140–146). In the Gileño Creation Story, an Ocotillo stockade (called a "cane cactus fence") was also said to have surrounded one Vipishaḍ pueblo, about four miles west of Santan, that the Emergenti conquered along with others in the Gila and Salt River valleys (Russell 1908:227; Lloyd 1911:156).

Out in the Sacaton Mountains one day, as we looked at the scarlet flower clusters, George told me, "Ocotillo blossoms, *melok hiósig,* [are used] for the basket dance; the song says they came with baskets of Ocotillo flowers."

I asked if he knew it. Then he sang this song:

siko li tonodho ma hiya vo ñeheda
melong ga hiosi ka mo to ka
siko la biñimim
ka i ya vo ñeheda
hayaha a'a hayaha a'a hayaha'a
ko ñeheda sialing
kio hoñegevam

"What do the words mean?" I asked George.

He thought awhile, then made this free translation:

There was a shining place there;
Here I sing; here I sing.
There's Ocotillo blossoms
That I hold up in a basket
And circle around.
Hai ya ha'a. Hai ya ha'a. Hai ya ha'a.

"Those last words," I asked George, "don't really have any meaning?"

"No. Just when things you really like, then you say, '*hai ya ha'a.*' Like a basket, when it's good, this is what it means. When basket made in springtime, when the young girls, they made baskets, to see who's going to be the best one to make [them]. See, that's where they have that basket dance. And that's why they mention all that about basket that they had the flowers in; good things they have in there. And circle around dancing. And that's why they have that kind of song—to see,

to see and to show the people—the men or whoever—to see who's got the best made baskets.

"Then they'll say, 'I'll get her' or 'I'll get her. She done the best!' That's the way it goes, when they [do the dance]. I might say this way again, that a young girl, a young man, if, when he is about matured to go, marry, like that, and look around to see who's be the best worker, a good worker, like that. That he's going to, that she's going to get. And that's the way it is. *Used* to be. Long time [ago]. But nowadays, it's really different. They just go out, meet somewhere, and join together.

"See, that's the way it goes, this song goes, why they mention baskets, like that."

"So it's almost more about the young ladies and their baskets than about the *melok hiósig.*"

"Uh-huh. Yes, that's the way it is."

Culver Cassa's literal translation of George's song is

Circle shining, here will sing
Ocotillo flowers carrying on head
    [in basket]
Around pass by
And here you will sing:
Haia! Haia!
And will sing morning
And will shift body

Another day, when the Ocotillos were blooming in the nearby Sacaton and Santan mountains, Herbert Narcia and I were discussing flower eating and the *vaḍag,* nectar, in flowers. I asked if he had ever heard of anyone sipping the juice from Ocotillo blossoms. He had not, but he said there was a song about the gray fly that comes like a hummingbird, drinks the juice, and gets drunk. Then he sang this song:

kokomagi mumuval
melonge hiosige tasho na
gamhiyo ke noḍa
himk huduñi wewecho
mo hime me e

I asked Herbert how he interpreted this song. "Well, it says that the gray fly went and drink the Ocotillo flower juice. And there's where all that juice is, in the flower. I guess it's done just like the *vipismal* [hummingbird]—you know the *vipismal?*—when he drinks the flower

juice. But this fly, he drinks the *melhog hiósig vaḍag noḍa.* He went crazy, I guess, and he went all over. It must have had some strength to it, to get him *noḍa* ['dizzy, rabid, intoxicated']."

The spoken Pima version, as reworked by Culver Cassa, is

koomagĭ muuval
melok hiósig todshag
am ii k noḍa
himk huḍuñ wecho
am o him

Gray fly
Ocotillo flower foam drinks
    and becomes dizzy
Under the evening he's going
Down there he goes and goes.

Later I played the tape of Herb's song for George, who listened attentively then said, "That's a 'Chechpa'avĭ Ñe'i' ['harlots' song']."

"How can you tell, George?"

"Because it says *ma . . .* or *me . . .* at the end [of each stanza]."

Modern Gila Pima seem to have forgotten the medicinal uses for this plant (see Curtin 1949:90). Earlier Russell (1908:265) had said, "The remolinos, or whirlwinds, that are so common in Pimería, cause pains in the legs, but not swellings. The remedy is to sing the wind song (p. 324) and to rub the limbs with the black gum of the okatilla *[sic],* Fouquiera *[sic]* splendens." Hrdlička (1908:245) recorded another medicinal use: "*Müh-lok* is a little bush *[sic],* the root of which, fresh or dried, preferably fresh, is boiled and the decoction given for coughs and in the beginning of consumption." Sylvester Matthias and others told me they had never heard of these uses.

The word for Ocotillo is widespread among those Pimans who live where there is some species of *Fouquieria.* The Tohono say *melhog.* The treelike *F. macdougalii* the Pima Bajo at Onavas call *nüliog,* while in classical Névome it was recorded as *murioga* (Pennington 1979:52).

Vouchers: *705* (SM).

Lophocereus schottii (Engelm.) Britton
   & Rose
Senita
Cactaceae
cheemĭ

At the very northern periphery of its
natural range, Senita just crosses the
international boundary into southern
Arizona in Organ-pipe Cactus National
Monument. There are fewer than 100
individuals here. But they become
abundant farther south, such as around
the Tohono O'odham oasis village of
Quitovac.

The plants grow in great clumps, of
dozens to several hundred arms, reaching
several times a person's height. They are
easily distinguished from the similarly
growing Organ-pipe Cactus *(Stenocereus
thurberi),* which has a dozen or more
shallow ribs: Senita's green stems are
deeply ribbed and in cross section resemble stars with five, six, or seven arms,
occasionally more. On most branches
the spines are sparse and very short. But
at the tops of older, taller branches the
spines become elongated, twisted, and
abundant, giving an unkempt appearance
to these otherwise neat plants. These
shaggy sections bear most of the small
greenish pink flowers that open at dusk.
The bright red fruits ripen in fall, reaching
the size of a thumb joint.

There is one Senita colony at Sacaton,
an ancient plant in Myrtel Harvier's
yard. This clone has many hundred arms,
growing amid the old-style fences *(uus
koli)* made of horizontally stacked
mesquite branches. Peebles first collected
from this historic plant in 1935, noting
that it was cultivated by a Pima. It was
then more than 12 feet high. Other
herbarium specimens and photos (ARIZ)
were taken in 1940, noting that the plant
was then old and had been brought from
Sonora about 20 years earlier. That made
the colony about 65 years old when I again
made specimens and photographed it.

Although it flowered abundantly, the
Sacaton plant produced no fruits. I asked
Myrtel about this. "I don't know," she
replied gently. "You know, I never studied
botany. We girls weren't expected to
study that sort of thing in those days. I
think maybe it has something to do with
the ants." Indeed, the ants were making

regular visits to the flowers. Perhaps the
clone needed cross-pollination.

Senita is known to the Tohono
O'odham as *cheemĭ* and to the Pima
Bajo as *teemĭ* or *teemis.* For the colonial
Eudeve, *témi* is given as *pitahaya chica*
(Lionnet 1986). Sylvester Matthias
consistently identified Senitas at the
Desert Botanical Garden in Phoenix
as *cheemĭ.*

But by one of those curious twists
in folk nomenclature, some Gila Pima
call the Sacaton plant *chuchuis* (see
Organ-pipe Cactus, *Stenocereus thurberi*).
There are other cases where the biological referent of a folk taxon becomes
transposed, including several cactus
examples among the Gileños. With only
one colonial columnar cactus reaching
Pima country as a tended plant, the
name *chuchuis* might have been inadver-
tently switched from Organ-pipe Cactus
to Senita.

Some Gileños still remember *sitol*
'syrup' and preserves of both *cheemĭ* and
chuchuis being bought north by Tohono
O'odham friends and relatives.

Vouchers: *1452* (Myrtel Harvier), *Hodgson
7369* (GK, SM).

Lophophora williamsii (Lemaire) J.
   Coult.
Peyote
Cactaceae
pihul, pihulĭ

Northern Piman scholars have long been
puzzled by the word *pihul* or *pihulĭ* (also
written *pihuri* or even *ikul*), the name for
a plant or perhaps a being. In the early
1930s, when preparing the ethnobotany of
the Tohono O'odham, Castetter and
Underhill (1935:27) wrote, "Peyote appears
not to have been known by the Papagos
unless it be one of the unidentified roots,
rumored vaguely as powerful magic, that
come from Mexico." But later, Underhill
(1946:295–296) observed, "Not only
animals caused sickness, but plants, winds
and undefined mythical beings. A very
common one was *pihuri,* spoken of as a
little old blind man with red-rimmed eyes.
He passed people on the road, and the
wind blowing from him to them caused

cheemĭ   *Lophocereus schottii*

trachoma and other eye troubles. A whole
series of songs describes *pihuri*'s personal
habits." She added, "Some informants
thought that *pihuri* was a plant which
took human form. It is supposed to be an
intoxicant, and there is some mysterious
connection between it and the deer. The
Pima are said to consider it a tree growing
in Mexico. These scraps of information
suggest a connection between *pihuri* and
the Huichol *hikuli,* peyote."

Belief in such plant-human metamorphosis is rare in the north. Underhill
(1946:297) noted, "Two other plants take
human form and cause hallucinations.
One is *Datura meteloides* (jimson weed).

pihul, pihulï   *Lophophora williamsii*

The other plant is the unidentified *tcïnacat*" (see Russell's [1908:229] *TchU-Unarsat,* or *cheenashaḍ* in modern orthography).

The linguistic evidence for a connection between northern Piman *pihul* and this cactus seems quite sound. The Huichol know this cactus as *hikuri.* The plant is *ikuri* among the Northern Tepehuan (Pennington 1969), *hikur* among the Southern Tepehuan (Mobley 1980), *hikuur* among the Tepecano (Mason 1917), *jikuri (hikuri)* among the Tarahumara (Bennett and Zingg 1935; Lumholtz 1898; Pennington 1963), *pejori (pehori)* among the colonial Ópata (Nentvig 1980), and *ikoli* among the colonial Névome (Pennington 1979). With the northern Pimans, several variants have been recorded, including *pihuri* for the Tohono O'odham. One of two hunting songs recorded by Russell (1908:299–301) is called "Pihul Ñe'i." (Although Russell spelled the word *pihol,* his *o* is *u* in current orthography.)

I cannot help but think some ancient pun may be preserved here. The northern and middle Piman word *keli (keri)* means 'old man'. The *hikuri/keri* pun (and symbolism) could be retained from a time before the sounds *u* and *e* diverged from proto-Tepiman. (And whether one writes *l* or *r* for the intermediate sound is purely a matter of orthographic convention.) This would be entirely in keeping with the fun-loving, joking, and punning Pimans, who did not separate humor from awe in their religion and ceremonies.

While the lexical connection throughout the whole of Piman country appears clear, is there an authentic conceptual and symbolic unity between the northern *pihuri* and the southern *hikuri?* That is, did the northern Pimans participate, perhaps in some diluted form, in the highly elaborate Peyote religion of the south? To answer this question, we need to look more closely at the Peyote religion among the Piman neighbors southward. Peyote symbolism is most strongly developed among the Huichol Indians of a mountainous region where the Mexican states of Jalisco, Durango, Zacatecas, and Nayarit interdigitate. Peyote is pivotal to understanding not merely the religion of these people but the very fabric of their culture. The Huichol make an arduous quest some 300 miles inland to the southern central plateau in San Luis Potosi in search of the plant. During the six-week trek on foot, all pilgrims are required to abstain from salt, sexual relations, bathing, full meals and drink, and full amounts of sleep.

The cactus is intimately associated with the deer, as Lumholtz (1898) long ago wrote: "The moving principle in the religion of the Huichols is the desire for producing rain; the deer and the *hikuli* being the chief factors in attaining this end. Once upon a time the all-important thing for the Huichols to eat—the deer—became god, and he is to them the symbol of life and of fertility. With his blood the grains of corn are sprinkled before being sown. The great god of *hikuli,* when he appeared the first time out in the country of the *hikuli,* showed himself as a deer, and every one of his tracks became a *hikuli* plant—the plant of life; and the life is that of the deer."

But this is only part of the symbolism, which has been made more explicit through research by Myerhoff (1970, 1974): a deer-maize (corn)-Peyote symbol complex is recreated and united during each pilgrimage. Peyote, in contrast to both deer and corn, is the symbol of essentially a private and beautiful nonutilitarian experience. In producing demonstrably safe but individually unstructured private experiences, it stands at the opposite end of the religious spectrum from things ritualized and priestly. Myerhoff (1970:72) concludes, "Peyote is neither mundane, like maize, nor exotic and exciting, like deer. It is that solitary, ahistorical, asocial, asexual, arational domain within which man [as well as woman] is not complete, without which life is a lesser affair." Taken together, these three symbols represent the Huichol unity. "The animal represents the past—hunting, masculinity, independence, adventure, and freedom; the [maize] plant is the labor of the present—food, regularity, domesticity, sharing between the sexes, routine, and persistent diligence; the cactus which grows far away is plant and animal at once—non-utilitarian, free of time, sex, and specifiable meaning, existing for its own unscrutinized, quiet gift of beauty and privacy. Truly such a vision is, as the Huichol often state it, 'beautiful because it is right'" (Myerhoff 1970:73).

The Huichol complex appears to be the prototype; elaboration to such a degree has not been found among surrounding native groups. Yet not only the word but also the use of Peyote seem well embedded in Tepiman culture, at least in the south where pockets of language and culture have survived. Various types of evidence suggest conceptual and symbolic affinities with what is known of Huichol Peyote religion, rather than an independent derivation.

Among the Akimel and Tohono O'odham there are six preserved oral texts (all of them songs) in which *pihuri* or *pihul (ikul)* occurs. The first is the hunting song recorded by Russell (1908:301), as mentioned earlier (this is the only one preserved in the original Pima); the other five texts were recorded by Underhill (1946), who published only her free translation in English.

The song recorded by Russell was provided by Vishag Voi'i ('Falcon Flying'), a Gila River Pima. In addition to the original form, which is in song language, the words are then given here as they would be in spoken Pima, and finally in the free translation from Russell.

Song language:

> Yali chuve maakai chukak et meḑedach
> yoai chukak yoapa, si'alim anta
> yoapa, si'alim anta yoapa-a.
> Yali kave maakai chukak et meḑedach
> yoai tatat yoapa, si'alim anta yoaka,
> si'alim anta yoapa-a.
> (repeat four times)

> Nañ pia hyewelik, nañ pia yewelik, nañ
> pia hayeweli-ika-a.
> Nañ pia hachevakik, nañ pia chevakik,
> nañ hacheva ki-ika-a.
> Kahova siyali wecha sahama Ikul
> kukach yamha ñe vaita, nañ pia yew
> lik, nañ pia hayeweli-ika-a.
> Kamhova hondoñ wecha sahama Ikul
> kekach yamha ñe vaita, nañ pia yew
> lik, nañ pia hayeweli-ika-a.
> Kamhova hondoñ wecha sahama Ikul
> kekach yamha ñe vaita, nañ pia
> chewekik, nañ pia hacheweki-ika-a.

Literal translation of spoken Pima:

> Younger Hare Magician blackness in
> running Black-tailed Deer meat
> bring, morning I will bring, morning
> I will bring.
> Younger Badger Magician blackness in
> running Black-tailed Deer feet bring,
> morning I will bring, morning I will
> bring.

> Had I no wind, had I no wind, had I no
> wind.
> Had I no clouds, had I no clouds, had I
> no clouds.
> Distant east under yellow (?) [Ikul]
> standing there me calling, had I no
> wind, had I no wind.
> There west under yellow (?) [Ikul]
> standing there me calling, had I no
> wind, had I no wind.
> There west under yellow (?) [Ikul]
> standing there me calling, had I no
> clouds, had I no clouds.

Russell's free translation:

> Young Hare Magician running
> Brings Black-tailed Deer venison.
> And young Badger Magician
> Brings the feet of Black-tailed Deer.

> Had I neither winds or clouds?
> In the east the Yellow Ikul,
> In the west the Yellow Ikul
> Called me. I had no winds or clouds.

In a footnote to the Datura Song, "Kotḑopĭ Ñe'i," Russell (1908:299–300) added, "This and the Pihul song are sung to bring success when setting out on a deer hunt." Both datura and Peyote are linked with Mule Deer. All three are associated with *kaachim mumkidag* 'staying sickness'. The first part of the Pihul Song relates the story of two young shamans, Jackrabbit Maakai and Badger Maakai, running in the darkness. In the morning they will bring back the venison and feet of the Mule Deer. The second part tells of Peyote (Yellow Ikul) standing at a distance in the east, then in the west, calling to the hunter (who is the singer).

This song is a departure from the usual northern Piman deer-hunting song in which the hunter and the deer alternately speak, the deer addressing the hunter. Here Peyote rather than the deer beckons the hunter. Clearly this song establishes a symbolic unity between Peyote and the deer.

The other five song texts were recorded by Underhill (1946:296–297). The first of these begins, "Am I not *píhuri, píhuri, píhuri!*" This song, from an unidentified source, is obviously of recent vintage; in the following line it describes painted playing cards. The remaining four songs were obtained from Juan Diego, a Tohono O'odham from Tecolote village.

In three of these songs, Pihuri is a little old man. The first one (like Russell's) describes Badger Shaman's deer hunt and then notes, "Lo, it is *píhuri*, the little old man! / The [deer] meat he is pounding." Underhill's accompanying notes say, "*Píhuri* is so old that he even pounds up fresh deer meat before cooking, so that he can chew it." Jerked venison was regularly pounded and pulverized before being cooked, but here Pihuri must tenderize fresh venison before he can eat it.

The second one is narrated in the first person: "Lo, I am *píhuri*'s little familiar." This is ambiguous because we do not know what or who Pihuri's "little familiar" is, except that he is the singer and a protégé of Pihuri, and the girls are afraid of him. The original Piman text would clarify the relationship between the two men.

In the third song, in typical Piman truncated style, Pihuri *is* the deer and the deer is drunk: "Around the great mountain he runs." Hunting songs often refer to the wounded deer as being "dizzy" (Bahr and associates 1979; Underhill 1946).

The fourth one is perhaps the most interesting in that it appears to describe an actual quest for the Peyote cactus and the behavior of the pilgrims in the field. The narrator runs toward the Peyote (on which a Sotol flower has fallen), which he says will make the pilgrims drunk or dizzy. Then he says, "In a circle we shall go singing," and in the next stanza, "In a row we shall go singing." Most likely, for these northern people, living far from the plant's natural range, the pilgrimage was re-enacted only in song.

There is some strong circumstantial evidence that northern Pimans once did have knowledge of Peyote hunts. Underhill was told, "Shamans owned love magic. A 'mushroom,' which corresponds to the Huichol description of peyote, was a strong love charm. A man stalked it like a deer, and shot it with an arrow before it had time to disappear into the ground. He must then give it to a shaman for magic treatment before he could carry it as a love charm" (Underhill 1946:264). This is such an apt description of how the first Peyote plant is stalked on the Huichol pilgrimage that it cannot be a coincidence. Unfortunately we do not know what Underhill's consultant meant by "mushroom." Was this an attempt to gloss Peyote in English without knowing what word to use for it? This seems most likely; the shape of the dried plant as well as the growth form of the scarcely green living plant is strongly suggestive of a puffball or unopened mushroom.

As noted earlier, the Tohono O'odham associated a little old man, a hallucinogenic plant, and the deer (Underhill 1946:296). There is no explicit association in northern Piman text or commentary between *pihul* and corn. But the deer-corn symbolic association is strong. Although deer hunting had less significance with the highly agricultural Gila Pima, busy tending fields both summer and winter, it was more important among the Desert People.

Ceremonial deer hunts are a part of the Tohono O'odham annual cycle, opening the winter season (Underhill 1946). A ritually killed deer (traditionally it was strangled) is cooked without salt, and corn is pit roasted. The purpose has nothing to do with a rite to increase game, as found in many other tribes. Rather, the focus is a cleansing ceremony. Venison, crop produce, and people are ritually purified with thorny branches of Ocotillo

(*Fouquieria splendens*) or cholla (*Opuntia* spp.). "During the dance in the night, women painted their bodies to represent corn and men carried hunting arrows. Again, the overall explanation is that the deer was killed to make the new farm crops safe nourishing" (Underhill and associates 1979:77–78, 81). At another village, eight dancers, two boys and two girls from each moiety, had their bodies painted with dots to represent corn kernels.

Like the Huichol narration where Maize is considered unpredictable and quick to take offense, northern Piman, too, preserves in the Creation Story an account where Corn leaves the People on several occasions when he is offended (see *Zea mays,* Group I; also Lloyd 1911:217–239).

Apparently only in the north is the association (or impersonation) of *pihuri* as a little old man found well developed; this person causes diseases that require the singing of the *pihuri* songs (Bahr and associates 1974; Underhill 1946). As noted earlier, Underhill was told he caused trachoma and other eye problems. All the songs she gave were used for curing sore eyes. Russell called him an evil spirit. Russell's song also was used for curing, though the disease is not specified. At least some *pihuri* songs sung for curing are also hunting songs.

In the single Gila Pima text, collected by Russell, Peyote is called S-oám 'yellow' Ikul (written as *sahame* in song language). Is this just a random color name, as in many other Piman songs? I think the symbolism is more specific than that. Whenever Peyote is used in Huichol ceremonies after the return home, the people paint designs on their faces to represent the plant. A specific root (*Mahonia trifoliata* Moric.), only recently identified, is used to make this paint (Bauml and associates 1990). The Peyote symbols are always yellow; these are small round flowers, said also to be the emblem of the deer (Myerhoff 1974:147, 173).

After Huichol Peyote pilgrims had carried out the ritually prescribed activities on the quest and had been purified on returning to their communities, the cactus was available for use by individuals who had not participated in the "hunt." Some Peyote might be sold to neighboring tribes that did not have a pilgrimage. Its subsequent use was primarily as a medicine; during a Tarahumara curing ceremony, all participants, including the

patient, ate dried ground Peyote mixed with water. The Peyote and the scraping stick or rasper both were believed to have curative properties. Bruises, snake bites, and rheumatism are some of the afflictions treated (Bennett and Zingg 1935). The Tepecanos, Huichol, Ópata, and Yaqui all are known to have used the plant medicinally.

Medicines traveled widely throughout the Greater Southwest. Northern Pimans had names for and knowledge of various plants that grew far from their indigenous territories. Some of these they knew only in their dried, prepared state. Did the northernmost Pimans share in the use of Peyote plants obtained from their neighbors to the south, or did they merely participate in a shared symbolism of maize, deer, and Peyote?

Hrdlička left three passages indicating that both Akimel and Tohono O'odham knew the actual Peyote plant. First, "*Peyote* is taken [consumed] among most of the Mexican tribes [he visited], above all by the Huichol, and also to a slight extent by the Papago and Pima" (Hrdlička 1908:173). Second, "The *peyote* is eaten in small quantities by the Papago, but its use is probably not entirely medical" (Hrdlička 1908:242). Third, "The Papago bring and sell to the Pima every year a little *peyote.* The Pima eat it in small quantities, but probably not for medical purposes only" (Hrdlička 1908:244). Hrdlička, who did his Piman fieldwork between 1898 and 1905, may have been catching the end of trade of Peyote to the Gila, though he emphasized it was still an annual affair. (He makes no mention of Peyote among other Southwestern tribes he studied north of the Pima.)

There remains the question of where the Tohono O'odham obtained Peyote. In trade, of course, but from which of the tribes engaging in a Peyote hunt? Of the three groups who ventured out into the dry interior deserts for Peyote, only the Tarahumara and the Huichol are known to engage in selling the plants. Earlier there may have been other sources. One possible source was the annual October trek to Magdalena, Sonora, where the northern Pimans encountered other tribes not only from Sonora and the northern Sierra Madre but also from far to the south. By the early 1930s when Underhill was studying the Tohono O'odham, even this source may have dried up; a

pan-Piman phenomenon disappeared as these people north of the border forgot they were part of a much greater Tepiman whole.

When Peyote returned to southern Arizona, it was not part of the traditional deer-maize-Peyote complex. It came as the Native American Church, a revivalistic movement whose origins are with the southern Great Plains tribes. Semi-Christian and to a certain extent nativistic, this Peyote cult is no more Piman than peace pipes and tipis and eagle warbonnets.

Mammillaria microcarpa Engelm.
Fishhook Cactus
Cactaceae
ban mauppa, ban maupai

In the well-drained soils of bajadas, hills, and sierras throughout Gila Pima country grows a small, erect species of *Mammillaria,* solitary or in small clusters. It is well shrouded in white radiating spines; the green flesh of the plant can seldom be seen. Each nipple is tipped with a dark hook that will ensnare the unwary small mammal (or human). In trying to get loose from one, the victim usually manages to get hooked on several more. This may be a one-trial learning situation that assures the little plant of safety most of the time from would-be predators.

At various seasons bright pink flowers encircle the upper part of the plant. These later develop into elongated red fruits. Inside, each has a bit of white pulp with minute black seeds. The fruits have a slightly acidic taste.

Fishhook Cactus is well known to the Gileños, who call it *ban mauppa* 'Coyote's paws'. *Mauppa* is the plural of *mavĭ,* a word seldom used today, signifying 'spread hand'. Ruth Giff called the little cactus *baaban mauppa* 'Coyotes' paws', using the plural form, but the singular, *ban,* is more frequently heard. Among some Tohono O'odham, Gary Nabhan recorded the name as *baaban ha-iiswikga* 'Coyote's Hedgehog Cactus' (Felger and associates 1992).

Although the hooked central spine might protect the plant from some predators, it was no defense against a thirsty Pima, according to George Kyyitan. "*Ban mauppa,* just one, stick straight [up]. We used to [have it] here, [at] that little butte there," he said, pointing to Gila Butte across the river from Bapchule. "And we'd find one, there—have to dig it out and put it on a rock there and get a sharp pointed rock—and pound it on there. And take out the center in there. Just open it. It's got a water in there. And chew it. It got lots of water in there."

"Did it taste good?" I asked.

"Yeah, it's alright. Not bitter—it's just like water—get the juice."

"The one up on the rocks?"

"Yeah."

"Did you people ever eat that little red berry on it?"

"No. Huh-uh. I never use it."

So far as I could learn, no one else ate the tiny but tasty fruits.

Curtin (1949:57) was told additional uses: "The small red fruit of *ban maupai* (meaning 'like coyote paws') is rubbed on arrowshafts to color them. For earache, the thorns are removed, the cactus is sliced, boiled, and placed warm in the ear. This is also a remedy for suppurating ears."

Curtin thought that the *Mammillaria* species growing in the heavy soils of the flatlands was the same as that in the uplands, but the two are quite different. A few Pimas thought that the Thornber's Fishhook Cactus (see *M. thornberi* account) of the floodplains would be called *ban mauppa* as well; others disagreed. I suspect this is a case of a collapsing folk taxonomy, in which several originally distinguished species are telescoped into a single more conspicuous, better known, or more useful folk taxon.

A third *Mammillaria* species is also found on the Pima reservation: Colorado River Fishhook Cactus, *M. tetrancistra* Engelm., a plump, somewhat less white pincushion cactus that seems to have grown outward rather than upward. I have found two individuals in the Sierra Estrella and one between sand dunes near Lone Butte. I took Sylvester Matthias out to the dune locality and asked him what he would call this plump pincushion cactus.

He looked at it for a while, then said slowly, "I don't know." It was not *ban mauppa.*

Technical Notes: Some botanists now include *M. microcarpa* in *M. grahamii* Engelm.

Vouchers: *1986* (GK); *LSMC 75* (not extant).

Mammillaria thornberi Orcutt
Thornber's Fishhook Cactus
Cactaceae
ban bisuldag, giishulĭ [?]

Whereas no one but the unobservant is likely to miss Fishhook Cactus (*M. microcarpa*) because it is out in the open, being white and conspicuous in elevated rocky places, few will ever find Thornber's Fishhook (also sometimes called Thornber's Pincushion) Cactus in its secluded spots down on the fine alluvial soils of the flatlands. I have discovered it in only a few places: once in a mesquite bosque, the others out in undisturbed Desert Saltbush-wolfberry (*Atriplex polycarpa–Lycium* spp.) flats, always concealed under protecting bushes.

If you do find it, you are not likely to confuse it with its much more common relative. Thornber's Fishhook Cactus forms colonies of long flaccid bodies, sprawling in a disorganized mass. The green of this plant is more conspicuous than the white radial spines. In fact, the plants look quite weak and naked, though each plump nipple is well armed with a central hooked spine and some radials. Like its cousin, it has showy pink flowers circling its crown at various seasons, but the reproductive parts are different. The stamens or male organs are all gathered to one side, and the stigma or pollen collector is bright purple rather than green. Both species have bright red elongated fruits.

I long wondered whether the Pima really distinguished this rarer *Mammillaria* species from the more common *ban mauppa.* Thornber's Fishhook Cactus should once have been the species more familiar to most Pimas because it grows down in the fine soils where the People farmed and had their villages. Before the wholesale leasing of lands and the inroads of agribusiness, this little cactus must have been common, too. Some Pima, as noted in the previous account, included both species in *ban mauppa.* Others did not.

When George Kyyitan finished telling me about *ban mauppa,* I asked him about the other cactus down in the flats, under the mesquite trees.

"I've seen one, too, that was like that, but it grows—[he spread his hands out, away from center, rising a bit at the edges,

ban bisuldag   *Mammillaria thornberi*
Painted from a plant near St. John's cemetery, Komatke, the northernmost colony discovered so far.

just as the sprawling *M. thornberi* grows]. Bunches."

"What do you call that one?"

"I don't know. I don't know that one. I thought this was the same as that, as the—that *ban mauppa*—but it does that . . . [he spreads his hands apart again, indicating how it grows], so it wouldn't be."

Fr. Antonine's list (1935) gives *ban bisuldag* as the "smallest variety [of cactus] (bears small red berries)." Because this is not *M. microcarpa,* well known as *ban mauppa,* it must be *M. thornberi.*

Bisuldag means, according to George Kyyitan, "something that will open—when it stands, then when it's dry—or will come up, like a peeling. I think that's when they say *bisuldaj* [the noun]."

At the west end, however, Sylvester Matthias told me about another plant that seems an exact description of Thornber's Fishhook Cactus: "*Giishulĭ* is about [the] height of *ban maupai*, in mesquite or open; small flowers. One patch toward the Sierra Estrella foothills, [at a place] called Giishulik, name of patch. Flowers purple and light yellow; have hooked spine just like *ban maupai*, and fruit similar—not good for eating—just hollow inside; not like *iisvikĭ* [Engelmann's Hedgehog Cactus, *Echinocereus engelmannii*] or *haashañ* [Saguaro, *Carnegiea gigantea*]." Ruth Giff and Sally Pablo did not know a name for this cactus, nor did they know the plant itself.

No one else could verify either *ban bisuldag* or *giishulĭ*. Among at least some dialects of Tohono O'odham, Saxton, Saxton, and Enos (1983:6) found *ban mawpai* used for Thornber's Fishhook Cactus, an interesting transposition if correct, and *ban chepla* for *M. microcarpa*. Gary Nabhan, among some group of Tohono O'odham, found *baaban ha-iiswikga* 'Coyotes' Hedgehog Cactus' for *M. microcarpa* and *ban ha-mauppa* for *M. thornberi* (Felger and associates 1992). My suspicion is that the Gila Pima name for Thornber's Fishhook was originally *ban giishul* 'Coyote's Hedgehog Cactus'. *Giishul* is an alternate name for *Echinocereus* among some northern Piman dialects. The transposition of names among Piman groups, particularly among cactus, appears to be a common phenomenon; there is a greater conservation of the word, the folk taxon, than of the word's biological referent. The naming of this contrast set of two *Mammillaria* species needs to be worked out among all groups of northern Pimans and documented with specimens.

This interesting cactus is found almost exclusively in Pima–Tohono O'odham country. The northernmost colony of Thornber's Fishhook Cactus yet found is between Ruth Giff's house and St. John's Catholic cemetery in Komatke. This species has the misfortune of growing in the *Atriplex-Lycium* plant community, a habitat shared with another rare cactus, Desert Night-blooming Cereus (*Peniocereus greggii*). These seemingly useless flats are prime targets for those who would turn the reservation into a monotonous sea of mechanized agricultural fields.

Vouchers: *1191* (?sm).

Nolina microcarpa S. Watson
Bear-grass, Sacahuista
Agavaceae
mōhō

Bear-grass, unlike all the other liliaceous plants in the desert, grows in loose colonies with such long, flexible leaves that the rosette configuration is not evident and the plant looks, instead, like some coarse grass. The leaves making up these grassy clumps are as long as your leg but scarcely a quarter-inch broad, grooved lengthwise, and barbed on their edges with almost invisible teeth that make them too rough to handle comfortably. In May or June the plant may send up a plumelike stalk, about head-height, bearing dense but minute cream-tan flowers. Bear-grass grows on rocky hillsides, not in Gileño country proper but in that great semicircle of higher country to the north, east, and south. Yet the Pima knew the plant, for if nothing else, they encountered it on frequent retaliatory expeditions when they drove Apache raiders back into rugged mountainous strongholds to the east.

The Gila Pima, and every other Piman-speaking group with whom I have worked, call Bear-grass *mōhō*. The word is undoubtedly derived from *mo'o* 'hair' or 'head'.

Once Sylvester Matthias and I drove up out of the hot desert to the cooler elevations northeast of Phoenix. Above Sunflower, we turned off on the Mount Ord road. On the hillsides were great disorderly masses of Bear-grass. Sylvester knew the plant. "The Papago women use *mōhō* the same way Pima use the *uḏvak* [Southern Cattail, *Typha domingensis*] when weaving baskets. They can't get any *uḏvak* in their country, so they have to use this."

Ruth Giff knew the plant, too. She had been out near Oracle, north of Tucson, to collect it. An innovative and resourceful person, when she was low on *uḏvak* stalk as the foundation for basket weaving, Ruth would substitute *mōhō*. But it had a disadvantage. "*Mōhō* tapers, because it's leaves. Have to keep adding more [to the coils]. But our *uḏvak*—it's even, all the way to the end. Don Bahr brought me *mōhō* from Santa Rosa [a Tohono O'odham village], and I used it all."

It is not known for certain whether the Tohono O'odham women once brought *mōhō* to trade with the Akimel O'odham, as they did *umug*, or Sotol (*Dasylirion wheeleri*). Presumably they did. Their gathering of this plant early in the century was described by Kissell (1916:198):

Summer is the season for harvesting beargrass, when the women generally go for it in groups, at the present time in wagons, but formerly on foot. Even when the trip is taken by wagon, an entire day is none too long for the journey, so when the time arrives for a particular group of friends to gather this basket material in the foothills, they must get an early morning start. Beargrass grows in great bunches from 30 cm. to 60 cm. in diameter, and from 60 cm. to 90 cm. in height. In the center of these clusters the grass stands erect, but around the edge it is dry and bends to the ground; so this outer portion is rejected by the gatherers and only the center cut away with axes and large butcher knives. Each woman collects for herself as much as she needs, some selecting with care the material in the best condition, others gathering more carelessly; when the beargrass is carried home it is laid on the ground to dry in the sun for four or five days, but it must be taken in during showers. When needed for basketry it is taken without moistening, and split by the teeth, fingers, finger-nails, or at times a knife, and worked into the basket, dry.

The gathering changed a bit with time. Castetter and Underhill (1935:57) observed: "Formerly the women severed the beargrass leaves one at a time and brought them home in the carrying net; the leaves were sun-dried and, for use, split into four, five, or six strands. At present the practice is for a man to go out and cut whole tufts of beargrass with an axe and haul home in a wagon a six-month supply."

Vouchers: *424* (sm).

Opuntia acanthocarpa Engelm. &
    Bigelow var. *major* (Engelm. &
    Bigelow) L. Benson
    [≠ *O. arborescens* Engelm.]
Buckhorn Cholla
Cactaceae
hannam

Up on the bajadas, among the Saguaros
(Carnegiea gigantea) and Little-leaved
Paloverdes *(Cercidium microphyllum),* you
will find one of the most important Pima
plant foods, a common shrubby cholla,
usually waist- to chest-high. It is about as
broad as tall, with rigid branches close to
the ground. The stems are covered with
slender raised tubercles as long as a finger
joint. A cluster of stiff spines grows out
of the upper end of each tubercle. If the
winter rains are good, two or three or
more buds begin emerging from the
branch tips as soon as the weather warms.
Later in the spring, Buckhorn Cholla may
be covered with showy burnt yellow to
burnt orange flowers (in western Pima
country; in the Sacaton Mountains and in
the mesa country east of Florence, some
are lemon yellow, others orange, and some
even deep purple-red—all in the same
area). The spiny fruits never get fleshy.
They seem scarcely differentiated from
the branch that bears them. The dry fruits
fall off later in the season.

A few simple field marks distinguish
Buckhorn Cholla from the other common
local species of cholla. Chain-fruit *(O.
fulgida)* and Teddy-bear *(O. bigelovii)*
Chollas have erect blackish trunks and
dense spines, giving the plants a whitish
or blond appearance from a distance.
Both shed branches readily (in contrast
to Buckhorn Cholla), earning them both
the name Jumping Cholla. The two lepto-
caulines, Christmas *(O. leptocaulis)* and
Pencil *(O. arbuscula)* Chollas, have sparse
long spines borne on slender, almost
smooth joints rather than on distinctly
raised tubercles. These two species have
a green, bare look about them. The fruits
of Buckhorn and Teddy-bear Chollas are
dry and fall off by late summer, while
the fleshy fruits of the other three are
likely to be found on the plants throughout the year. This leaves only the rare
Cane Cholla *(O. spinosior),* which has
bright red flowers in Pima country, and its
localized hybrid with Chain-fruit Cholla,

known as *O.* x *kelvinensis.* These have
greenish yellow, spineless fruit, and the
plants are more erect than spreading.

Except for the collecting of *iivagĭ,* eaten
greens, the only wild plant gathering that
still persists among the Gileños to any
extent is the spring harvest of *hannam,*
cholla buds. Perhaps as many as a dozen
women and their families (possibly even
twice that number) still go after *hannam.*
It is the one gathering in which I have
regularly participated. My trips have been
to the east slopes of the Sierra Estrella and
the Sacatons.

Gathering time, Pima women say, is
always around Holy Week. When Easter is
early, so are the buds; when late, the crop
is late as well. This means that there is
about a three-week period, sometime
during March or April, when the buds
are swollen but not yet opened. The crop
quantity varies considerably from year to
year, depending apparently on the winter
rains. Some years, the crop is hardly
worth picking because there are but a few
buds on each plant. In a good year, you
can
pick a gallon-bucketful just at one cholla.

Ruth Giff explained, "In March-April,
they pick *hannam;* in May, *vipnoi* [Pencil
Cholla] after *hannam* is gone. [They're]
called *iibdaj* 'fruit' even though they are
the flower. If there isn't enough *hiósig*
['flowers'] on *hannam,* then pick the
shoots to make enough to eat. They taste
the same." Actually, when the new shoots
start to appear on the tips of the branches
at the same time as the buds, it is easiest
to pick shoots, buds, and all.

In the third week of April 1987, Culver
Cassa and I drove to Santa Cruz village; I
had offered to take Carmelita Raphael to
the bajadas to pick *hannam.* She was ready
with buckets, tubs, and tongs of various
shapes and sizes. We went down to St.
Catherine's Church, drove up one of the
many roads that crisscross the bajada, and
found a spot that suited her. She took
from the truck a jar of water wrapped in a
white cloth, placing it in the shade of a
rear wheel.

Most of the plants were just starting to
bloom, though a few already had opened
flowers. Most are a clear yellow (and the
unopened buds greenish), but many are a
sort of burnt yellow (and their buds more
pinkish or reddish). An occasional plant
has flowers a pale salmon color. I asked
Carmelita about these differences. All

hannam   *Opuntia acanthocarpa*

are *hannam* and all are picked and used
indiscriminately, she said. But if the
flowers are opened or even beginning to
open that morning, they are not good
because they fall apart during the roasting;
the unopened buds (the perianth) are the
best part for eating.

The only other chollas growing on the
bajada were the diminutive Christmas
Cholla, with its oversized red fruit, and an
occasional arborescent Pencil Cholla, with
its buds just beginning to emerge from the
stems. It would be at least a month before
these would be ready for picking.

This was a good season. The tips of
individual branches were so loaded that
it was easy to get at least three buds with
each reach of the pickers. One grabs these
between the arms of the tongs, gives
them a lateral twist, then drops them into
the bucket. The mature buds break off
rather easily.

I timed my picking of a five-gallon plastic bucket; twice it took me 34 minutes to fill it. As we brought our buckets back to the truck, Carmelita would dump them into the tub and mash them down with a large stone. When compressed, they take up considerably less space. By midday we had filled the tub as well as three five-gallon containers, all packed down. And the bajada was beginning to get hot.

Back at Carmelita's cool rancho tucked away in the bosque, we unloaded the harvest under a large mesquite in the yard. Her *hannam chuámaikuḍ* 'roasting pit' was ready, but the fire had not been built. Her pit is cone-shaped, entirely lined with flat rocks all the way up to and including the edges. Carmelita assured us that all 25 to 30 gallons of packed buds we had picked that morning would fit, even with the layer of *chuchk onk* (Seepweed, *Suaeda moquinii*) that would be added. It was windy when we got back, and she said that was bad because when the mesquite wood was burned, the wind would blow away the heat too much, and the *hannam* that went in after would not get done all the way. So they would wait for a better day to fire the pit and roast the *hannam*.

When George Kyyitan and I finished a long interview under the shade of his *vatto* 'arbor', I said, "I want to have a look at your *hannam chuámaikuḍ*."

We walked around the back of the house. George said, "Well, let's see if *you* can find it." I looked around at the flat, almost colorless, tan adobe soil. There were a half-dozen pits and circular depressions, most evidently the remains of garbage pits or outhouses.

I pointed to a circle of green. "How about this with the *komal himdam vashai* [Bermuda Grass, *Cynodon dactylon*]?"

"Nope. That's where a *goks o'odham wuhiosha* [bush morning-glory, *Ipomoea carnea*] was growing."

I wandered over a larger area, still with no luck, then gave up.

"Here," said George. He pointed to a few rocks protruding haphazardly at his feet. To my eye there was nothing at all to indicate anything different on the pale soil surface, not even a depression. George started to etch a circle around the rocks with the point of a stick he was holding. As he scraped, an outline of pale orange soil began to appear—the burnt clay of the pit's edge. When he finished scraping, he stood over a circle two feet in diameter,

the same size as Carmelita's. I wondered how many archaeologists would ever recognize a *hannam chuámaikuḍ*.

Not all pits are this narrow. At the other end of the reservation, Ruth Giff and Sally Pablo roast their *hannam* in great tub-shaped pits more than three feet across. These can hold several washtubs of raw buds at one time.

Regardless of shape, the *hannam* roasting pit is lined with smooth river rocks. Over these is built a fire of mesquite wood. The wood is allowed to burn down completely. The rocks, too hot for direct contact with the delicate unopened buds, are covered with a thick blanket of *chuchk onk* before the packed buds are dumped in and covered with another layer of the succulent Seepweed branches. "We usually burn the wood for four hours, to get it real hot," explained Ruth. "Leave *all* the coals in there and cover with enough *chuchk onk* to keep the *hannam* from burning." While it is still daylight, the pit is closed. A piece of corrugated sheet-metal roofing covers the plant material (formerly, a hide or canvas was used), and dirt is shoveled over it all to hold in the heat overnight.

Sometime the next morning, about 12 hours later, the pit is reopened and the buds tested. "If they think it's not done," said Ruth, "they just leave them till evening and it's still hot." The buds should then be uniformly steamed and about the consistency of cooked asparagus but not so mushy as to lose their individual shapes. If the Seepweed layer is too thin, the buds nearest the edges may be burned or too mushy and thus will be discarded. And if the pit is not hot enough, the *hannam* will have to be cooked for several additional hours on the stove before it is used.

But if all has gone well, some of the steaming hot buds will be served fresh from the pit, while the rest will be spread on mats and grates to dry. Buds the size of a fat thumb are reduced when dry to a hard shriveled remnant—calyx, tepals, and all—scarcely the size of your little finger joint.

As far as I have been able to learn, sexual continence is not enjoined upon those practicing *hannam* pit roasting as it once was on those who roasted agave hearts. One day Sylvester was telling me of a man who left the agave camp, came back to the village, and violated the sexual taboo, ruining all the *a'uḍ* that was roasting in the pit (see *Agave deserti*).

"Will that affect the *hannam* or just the *a'uḍ*?" I asked.

"Just the *a'uḍ*. Oh, yeah, there *is* a saying, and it's still kept: Nobody's supposed to *fart* over, around [the pit], when the *hannam*'s being roasted."

"What will happen?"

"It wouldn't be done, either. This one woman did that. A heavy woman. She's the one that's doing it [roasting]. And stamping. Put some more *chuchk onk* on top, and then covered it. And she was stepping on it—stepping to pack the dirt down. And she fart[ed]. And it's not done [when they opened it]! These other ladies get after her. With all their work and it's only half done. Everybody has to keep away. Don't fart. Just keep away."

In the days before pickup trucks, some women chose to roast their *hannam* on the bajadas rather than carry their heavy, raw harvest all the way back to the village. Because *chuchk onk* does not grow up on the bajadas, *shegoi* (Creosote Bush, *Larrea divaricata*) was usually substituted in the roasting pits. Although this was common knowledge, apparently few modern pickers had ever tried it. Ruth said she thought *shegoi* would give the buds a strong and bitter flavor. "I'm going to try it next time!" she told me. In the 1988 season, Ruth tried some that had been prepared this way and told me, "You can't tell the difference, yet it's *very* strong, the *shegoi*." She was surprised that it did not impart a bad flavor.

I asked Sylvester Matthias whether both men and women picked *hannam* in the old days. "Yeah, they'd all go pick. In the washes [*sic*, bajadas]. My grandmother used to take out the stickers [with] something they call *uso* [a basket designed for despining cholla]. They get rocks [to support the *uso*] and pour *hannam* over [it] and get a piece of wood and stir them up, and the thorns would just [fall] to the ground. Then they'd have a tub full. They'd all pick *hannam*. My father used a pliers. Pick in a bucket, take to wagon, dump in tub."

However, in my experience, there does appear to be a division of labor in the *hannam* harvest. The woman is the principal organizer of the picking. The places on the bajada are mentioned as where this woman goes or that woman has picked, even if the actual picking is a family affair. This pattern is reflected in the Creation Epic: it was a woman whom Corn-man encountered near the Superstition Mountains roasting the *toota hannam* (Chain-fruit Cholla). The pattern is also reflected in the tribal oral history, such as in the calendar-stick records (see, for example, Russell 1908:42). Men seem to be primarily responsible for digging or re-excavating the pit, gathering the mesquite wood, and firing the pit. However, I have known women to help with all these activities as well.

Once the cholla buds have been pit roasted and dried, they can be stored indefinitely. However, they must be rehydrated before they can be eaten. Ruth Giff said, "Cook the *hannam* same as beans, about three hours." She mentioned some recipes: "I cook *hannam* with onions— [and] a little shortening. Sometimes if I have the *[onk] iivagĭ* [Wright's Saltbush, *Atriplex wrightii*] I cook it in there, but I never use *chuuhuggia* [Palmer's Carelessweed, *Amaranthus palmeri*]. Somebody told me she used canned spinach. Well—it's greens. You could put chile in it. Some do it just plain, but I mash mine, after [it's] cooked, before shortening. Then kind of fry it. There's a way where they grind it—dry—then mix it with whole-wheat flour. Guess [that's] what they call *atol [atole]*—same as *ku'ul* [porridge]. They do the *s-oám bavĭ* ['brown tepary'] like that, too."

Hannam can also be ground and cooked with ground *kaij* 'Saguaro seeds'. Ruth said that if a new mother is slow in getting her milk, the *ku'ul* of ground *hannam* and whole wheat helps stimulate her milk flow.

At Onavas in Pima Bajo country, *hannam* labels a large, thick-stemmed cholla, *Opuntia* aff. *fulgida*. Pedro Estrella explained that it was prepared in much the same manner as the northern Pimans prepared Buckhorn Cholla.

There are different schools of thought on preparing cholla buds. Most of the people with whom I picked removed the spines from the green fruits when the fruits were still uncooked. A shallow rectangular box is constructed with a bottom of hardware cloth (quarter- or half-inch soldered screen). This is propped up on four rocks, and a few quarts of buds are dumped in. A four-foot section of a Saguaro rib, one with a flat, flared base, makes an excellent utensil for raking the buds back and forth across the screen. The abrasion knocks off the spines, which

fall through to the ground, leaving the calyx and unopened flower intact.

Before screen was available, the Pima used a special despining basket called an *uso*. This was made from slender, peeled willow twigs, laid parallel and bound to an oval frame of willow. Ruth and Sylvester both said that the *uso* was similar to the sieve shown in Russell 1908 (p. 146) and Kissell 1916 (p. 143). Ruth used a different name once for this basket: "*Hannam ho'ipigakuḍ*—to take thorns off cholla. Hoop on edge, not round, shallow." She put her arms around in an oval in front of her, hands overlapping—"about this much. *Che'ul* [willow] stems all the same size, next to each other. Let the bark [stay] on." Both said the *uso* was the same as the *hannam ho'ipigakuḍ,* but these are different from the *giigdakuḍ* 'sifting basket' (see Mathiot 1973:382).

Other Pimas first roast and dry the buds, then remove the thorns in the screened box. I asked Ruth about the relative merits of the two methods. "I tried it [drying them first], but I don't like it. I wait *just* one day, then *ho'i*—clean—then the little ones come off without wrinkling. If longer, some thorns stay in there in the wrinkles, the little ones."

Hannam pickers, called *vaa'o* (the same as stirring sticks) were once made from a piece of willow branch split for most of its length. Today any metal tongs, such as those used in handling deep-fry foods, are used.

One day George Kyyitan was telling me of the almost endless ways the People had of preparing the *vihog* or mesquite pods, a staple food: "Same way with that *hannam.* Nowadays they're doing so many things out of it. They have to fry it. Yet, that wasn't that way [frying]—not *supposed* to be. But when this young generation, they start eating *lard* and things like that. They have tortilla—they have to mix lard with it. But it's good, too. First time, when I married over here, I like it myself. Just watch them [fix it]. But it taste good. When you fry it. And mash it up and fry it. And use tortilla and coffee. And I do that once in a while. When I have some. But now I never have any *hannam.* Trying to find where to get some. I was going over here to Santa Cruz. I know a lady over there that's always got some."

"Carmelita?" I asked.

"Yeah."

Hannam is one plant that is well entrenched in the oral literature of the

People, and well it should be, considering how it has withstood the many inroads of cultural erosion. One song sung by the late Blaine Pablo and transcribed by Sally Pablo, Don Bahr, and me tells of a Rock Wren, *vavas,* who makes wine from the *hannam* blossoms. It is sung by a Bullock's Oriole who gets drunk. Here is Blaine's version:

Koomagĭ Vavas Ñe'i

kokomagi vavachus
hannami yosike navaitoke
iyañe melivita

kuñama wemaji yo ke navamo
haiya vañe piñe mamache
gamu shakalano meneta

(in song language)

koomagi vavas
hannam hiósig navaitk
iiya at o ñ-meliew

kuñ am wemaj ii k ñ-navamch
vañ pi ñ-maach
shakal meḍ

(in spoken Pima)

Rock Wren Song

Gray wren
Cholla flower makes wine and
Here runs to me
[To invite me to drink].

And I with [him]
Drink and get drunk
I don't know,
Diagonally run [stagger].

Francis Vavages at Sacaton also sang the song. His version is slightly different:

hanama hiosigai'iyo ke'e navama
vañ pi ñmamache he
shakali memeda
kuñ ama wemaj haiyo ke'e navama

Here is how Francis explained his song:

[I] got drunk with cholla flowers
And I don't know what I [was] doing
Running side by side [or crooked]
With [them] there I get drunk

There is even a song Rosita Brown told me of a little squirrel, *cheerrkul* or *cheekol,* that comes to steal the pit-roasted buds the women have spread to dry.

Vouchers: *732, 734, 774* (SM); *765* (CR); *LSMC 47* (not extant).

Opuntia arbuscula Engelm.
[≠ *O. versicolor* Engelm.]
Pencil Cholla
Cactaceae
vipnoi, vipĭnoi

In Gila Pima country there are two slender-stemmed chollas that might be confused. Both have quite green, rather naked-looking joints with sparse spines up to two inches long. In both, these spines are ensheathed in straw-colored envelopes, as in other chollas. The slenderer of the two is the shrubby Christmas Cholla *(O. leptocaulis),* which usually has persistent red fruit.

But up on the bajadas, principally, grows the more treelike Pencil Cholla, whose joints are about finger-thick. The trunked plants may be freestanding or clambering almost vinelike in some tree. Pencil Cholla may reach shoulder- or head-height and has a more open crown than any other cholla in Gila Pima country. The cupped greenish yellow to deep yellow flowers, about an inch across, come on when the weather turns seriously hot in May. After closing they turn rust brown. Unlike those of Christmas Cholla, the plump and spineless fruit of Pencil Cholla remain green, with perhaps merely a wash of purple.

In Gila Pima this cholla is known as *vipĭnoi.* Joseph Giff contrasted it with *hannam,* Buckhorn Cholla *(O. acantho-carpa),* saying, "It's way bigger than *hannam,* tall, but has smaller buds; joints [are] smaller and break off more easily; but it's easier to get stickers off [the buds] than [off] *hannam.* The fruit are smaller—you have to reach *up* to get them—and the branches are smaller, with yellow flowers. Grows down on flats; also below the middle peak [of Sierra Estrella] way up close to the mountains [on the upper bajada]."

Ruth Giff, a veteran cholla bud picker, said, "They use the buds of *vipĭnoi* also. They come on later [in the season]."

I asked George Kyyitan, who is very fond of *hannam,* if he knew *vipĭnoi.* "I eat [it] once. It's kind of a little bit different from *hannam*—kind of hard. Or maybe it wasn't well cooked, that's why. It's kind of a little sour, or something like that."

One April when I was out on the bajada above Santa Cruz village picking *hannam*

with Carmelita Raphael, we discovered a Pencil Cholla with its old greenish yellow fruits still attached and the new buds just appearing. "The *vipĭnoi* can be used for roasting buds later in the spring," she explained. "They have a different taste. If you miss *hannam,* you still have another chance for *vipĭnoi.*" Sometimes the spring crop of Buckhorn Cholla was poor; then people would gather the early summer Pencil Cholla. Surely this was more work, because the plants are less numerous and not as easy to pick as the waist-high *hannam.* Use of *vipĭnoi* has waned.

Curtin (1949:59–60) included both *O. versicolor* and *O. arbuscula* under the name *vipe noi,* saying, "The fruit on both Deerhorn and Pencil cholla is always green, and in the spring young fruit ends are gathered, placed in a basket, and thorns removed with a stick. For storage, the fruit is dried, or when green it is boiled with *onk ivakhi* (salt-bush) and tastes more sour than *hannam* (cholla)."

As with several other cactus, the bio-logical referents of the Piman names change between various dialectical groups. In her dictionary Mathiot (1973:278) gave *che'echem vipnoi* 'small *vipnoi*' for Christmas Cholla and *ge'egeḍ vipnoi* 'large *vipnoi*' for Pencil Cholla, quite a different contrast set from Gileño but one that makes phylogenetic sense. At the Pima Bajo village of Onavas, Sonora, don Pedro Estrella on several occasions gave me the name *ti'órim* for Pencil Cholla, saying that both the flowers and the fruits were eaten. This is interesting because this word is cognate with Tohono O'odham *cheolim* (Saxton and Saxton 1969:175, 1983:8) or *chiolim* (Mathiot 1993:211), a name applied to one of the cane chollas in Tohono O'odham country. The word comes full circle, because the Gila Pima know *chiólim* (Russell [1908:77] wrote it *chíawldi*) as a trade product in earlier days from the Tohono O'odham who came up to work on the Pima wheat harvest.

Vouchers: *766* (CR); LSMC *16* (not extant).

Opuntia bigelovii Engelm.
Teddy-bear Cholla, Jumping Cholla
Cactaceae
haḍshaḍkam

Two reservation chollas share the common name Jumping Cholla and are superficially similar in that they both have abundant long whitish or straw-colored spines on short, easily detached joints. One of these, Chain-fruit Cholla *(Opuntia fulgida),* is a tall plant of the lower bajadas and flatlands. Its joints are conspicuously green, the spines on deep triangular tubercles well spaced on the stems. The other is Teddy-bear Cholla. It grows on rocky, usually south-facing slopes. It is an erect plant, with a single blackish and usually heavily spined trunk, growing to about hip- to head-height, with a thick crown of fairly short, upward-reaching, chubby blond branches. Nothing is pendulous about it. Even the flowers and fruits face upward. The flowers are small, with dull greenish yellow tepals, some-times with a hint of flesh color along the tepal midline. The fruits are infertile, producing no seeds.

Nevertheless, where you find one, you are likely to find an entire hillside choked with Teddy-bear Chollas. The secret to this success is in their short, readily detached, massively and intricately spined branches. The tubercles are short, packed on the stem at least twice as densely as in the local form of Chain-fruit Cholla. All this to transport themselves via some unwary animal to a new location, where they quickly root on contact with soil.

Smaller animals find refuge in these Teddy-bear Cholla forests. Packrats haul fallen joints to their nests, scattering them about. Some birds, most often the Cactus Wren, build their nests in the crowns of these chollas. Doves and thrashers may nest there as well. The adults seem to manage, but occasionally I have found a young bird that must not have learned the tricks and paid for its inexperience with its life.

To the northern Pimans this cholla is *haḍshaḍkam,* a name that is derived from the verbs *haḍshaḍ* 'to attack instant-ly' and *haḍshap* 'to stick to [something]'. So *haḍshaḍkam* is "it has something that sticks to something." The word is etymologically and conceptually related

haḍshaḍkam   *Opuntia bigelovii*

to *s-haḍam*, also meaning 'sticky', but in the sense of mucilaginous, as in glue (see *haḍam tatkam*, globemallow [*Sphaeralcea* spp.], Group D).

In discussing this cactus, Sylvester Matthias said, "The Papago put their dead in a cave. If no cave [is available], then [they] pile up rocks and put their dead body in there. Then mesquite limbs were cut and put it [them] on top. Then pile some more dirt on it, to make a mound like. Then they put that cholla cactus— *haḍshaḍkam*—all over to protect [it] from coyotes or anything. That's the way they protect their dead when there's no caves. Just make that pile of rocks, put their dead in there [and] as many *haḍshaḍkam* as it can hold. That's the way I heard, to protect them."

So far as is known today, *haḍshaḍkam* was never used in Piman culture as a food. But it played a very important symbolic role, which is critical to understanding certain legends. Russell (1908:92) aptly summarized this for the Gileño: "Night marauders [in the fields] were in olden times kept at a distance by the rings of the terrible cholla cactus, Opuntia bigelovii Engelm., that were laid up around the individual plants. It is recognized as the most effectually armed of the many cacti and is the symbol in Pima lore of impenetrability."

One of the most crucial episodes in the Creation Story is the account of the flood (Lloyd 1911:36–50). I'itoi (Se'ehe) decided to make a special person, "the most beautiful man yet made." The narrative goes on to describe how the "young man married a great many wives in rapid succession, abandoning the last one with each new one wedded, and had children with abnormal, even uncanny swiftness, for which the wives were blamed and for which suspicion they were thus heartlessly divorced. Because of this, Juhwerta Mahkai [Earth Doctor or Creator] and Ee-eetoy foresaw that nature would be convulsed and a great flood would cover the world." A powerful medicine man had a beautiful daughter, who realized that she would eventually be victimized. But her father had a plan, which included instructing the girl to gather some *haḍshaḍkam*. Lloyd adds in a footnote, "What the Pimas call the *haht-shan-kahm* is the wickedest cactus in Arizona. The tops of the branches fall off, and lie on the ground, and if stepped on the thorns will go thru ordinary shoe leather and seem to hold with the tenacity of fish-hooks, so that it is almost impossible to draw them out." But what exactly was this plan? Lloyd's version says only that the girl's father asked her to wear a piece of this cholla wrapped with her hair on one end, which would protect her. It appeared that the details had been edited out. Perhaps Sylvester Matthias would know the expunged parts. On a visit to Arizona I read him the Lloyd version and asked if he knew what really happened. He did.

"How does *haḍshaḍkam* come into the Creation Story? In this way: Handsome Man was coming around looking for young girls. The daughter of the *siivañ*, the medicine man, was crying. She heard the news that he's going to come [Handsome Man]. And he [her father] asked her, 'What's the matter? You've been crying?' And she said she's afraid of that man, because she'll have a baby overnight. The *siivañ* told her, 'When you go after [fire]wood'—she generally goes after wood [in the morning]—'you get a piece on the inside of the ironwood [tree; *Olneya tesota*] and you bring it to me. And then you bring me some *haḍshaḍkam*— not in a big bunch, but just enough.'

"So she did—and go after wood, and get that inner part. And they [he] carved it into an artificial cock [phallus]. She placed it in her cunt. And that *haḍshaḍkam* she placed as hair around it. And her father gave her instructions: 'If he tries to do anything to you—[he'll be stuck by the cholla, then]—the first thing you do is as soon as he turns around, you jump on his back and poke this thing [ironwood phallus] right in his *siipuḍ* [anus]. Have him do it, till you come.' So the next morning that man has this baby [whose crying caused the flood]."

Vouchers: *1494* (?).

Opuntia engelmannii Salm-Dyck
  [= *O. phaeacantha* var. *discata*
    (Griffiths) L. Benson &
    Walkington]
Engelmann's Prickly-pear
O. phaeacantha Engelm. var.
  major Engelm.
Desert Prickly-pear
Cactaceae
iibhai

Sally Pablo and I headed north of Phoenix on Black Canyon Freeway on our way to Prescott where I was then teaching. Suddenly the steep grade topped out and we leveled off onto a mesa top that was almost barren— almost, except for scattered mesquites and acres of prickly-pears.

"Look at all the fruit on the *iibhai*," Sally said, astonished at the density. "Let's stop and pick some." I pulled off at the Sunset Point rest stop and fished a paper bag out of the truck, while Sally found a trail out into the flats of cactus. In no time, she had broken off a handful of a fine dry bush, knocked down some deep-purple fruits, and was rolling them around on the sand, whisking off the tiny but insidious spines called *vii ho'i* 'glochids'. Soon we were munching on *iibhai*, the burgundy-colored juice dripping all over as we spat out stone-hard seeds.

Engelmann's Prickly-pear, the most common prickly-pear in Pimería Alta, is a sprawling plant reaching between knee- and hip-height but spreading out often several times that much. The pads, usually longer than broad, are bright green with few spines. The large and showy flowers are usually bright lemon yellow. While writing this description, I watched a bee climb past the green stigma and down the style; the mass of stamens closed slowly over the bee like a sea anemone engulfing its prey. The bee, unperturbed, emerged a few minutes later, thoroughly dusted with pollen from the arched stamens.

Although Engelmann's Prickly-pear is found primarily on the bajadas of the four mountain ranges of the Gila Reservation, clumps do occur here and there on the floodplains as well. These clones are so highly variable that I wonder if they may all have been planted there by earlier people. Agriculturalists (as both the Hohokam and the Pima were) are notorious for experimenting with growing plants of all kinds. A prickly-pear colony at Blackwater just west of Holy Family Church has exceptionally large pads (many are 9½ or 10 inches across and 12 to 13 inches long). Two colonies Sylvester Matthias took me to see northwest of the Komatke cemetery each have differently colored pads. I asked Sylvester about these colonies. A Tohono O'odham family by the name of Cabulla had settled there late in the 19th century, and they had planted the *iibhai* there. "They bring them down from South Mountains—up where the antennas are. They know the good ones." Sylvester mentioned a locality in Komatke called "place of many prickly-pears," S-navag.

In Gila Pima, by some curious twist of usage, the entire prickly-pear plant has come to be called *iibhai*. This is different among most if not all Piman dialects to the south. I checked with Frank Jim, a Tohono O'odham, to see how he used this term. "*Navĭ* is the prickly-pear plant itself, *ii'ibhai* the 'fruit' of *navĭ*."

I asked Ruth Giff. "This isn't the way in Pima; it's different up here," was her blunt reply. Others told me the same, and indeed this is what Curtin (1949:60) was told in the early 1940s and Fr. Antonine in the 1930s. But in the early 1900s Russell (1908:75) had recorded *O. engelmannii* simply as *naf,* saying, "The thorns are

brushed off the fruit of the prickly pear before it is gathered. It is then peeled and eaten, the seeds being thrown away."

I discovered some variation, however, among the Gila Pima. For instance, George Kyyitan reserved *navĭ* for the cultivated prickly-pear (Group I, *O. ficus-indica*), calling the wild ones all *iibhai.* Francis Vavages' usage seemed to correspond to that found to the south, with *navĭ* being all the prickly-pears and *iibhai* being their fruit.

Sylvester probably put his finger on the puzzling word change that we find today among most Gileño speakers. "There *is* a *navĭ* which I don't know. And that Christmas Cholla *[O. leptocaulis]* they call *a'aji navĭ,* because they are thin. I don't know which is [just] *navĭ.*"

"If there's an *a'aji navĭ,* then there has to be just a plain *navĭ,*" I reasoned.

"Uh-huh." He pondered for a while.

"Have you heard anybody call the *iibhai navĭ?*"

"Yeah, that's it. *Navĭ* and *iibhai* would be the same thing. That's what it is."

"What part do you eat, the *navĭ* or the *iibhai?*"

"The *iibhai.* Now, from my father, he called it *navĭ.* That *iibhai,* that's way back. But later on, I got that *a'aji navĭ* from my grandmother. She talks a lot about *a'aji navĭ.* And *iibhai* is *navĭ.* But later on, they just drop that *navĭ* and just call it *iibhai.* See, that's 1900, 1908." This would mean that the curious transposition of the fruit's name to the entire plant is a relatively recent phenomenon, and would explain why a few Pima speakers on the Gila still make a linguistic distinction between the fruit and the plant.

The various Piman names for *Opuntia* species appear to have gone through an interesting series of stages in their linguistic evolution. In proto-Piman, all prickly-pears were called *nav.* The southern Pimans have markers for the various species of prickly-pear: *vii nav, s-moik nav, kauk nav.* By the 19th century, the Upper Pima had come to refer to the plant of *O. engelmannii* as *nav* and called the fruit *iibhai; O. leptocaulis* was named *a'aji nav.* During the 20th century, the riverine Pima have begun to call both the plant and the fruit of *O. engelmannii iibhai; O. leptocaulis* is either *navĭ* or *a'aji navĭ.* The anomaly in this scheme is that the little *O. leptocaulis* is not a prickly-pear at all, but a cholla. (Folk taxonomic

evolution, like biological evolution, is not for those who like tidiness.)

I suspect that both the word *navait* 'wine' and *navich* 'friend' are derived from *nav.* (Wine was consumed only at the annual rain-making ceremonies, and in this context, the individuals exchanging the liquor addressed one another with some relationship term; where none existed, the word 'friend' was used.)

Russell (1908:75–76) noted, "The Papagos make a sirup from the fruit (which is said to cause fever in those not accustomed to its use) and dry the fruit as they do that of the saguaro, but the Pimas make no further use of it than to eat it raw." This, too, is what I was told more than 80 years later. I had even heard about the prickly-pear giving one the chills. Lloyd (1911:123) noted this also: "They warned me that a novice to eat freely of prickly pear produced a lame, sore feeling, as if one had taken cold or a fever. I noticed no symptoms however." I checked with contemporaries about the effects of eating prickly-pear fruit.

Sylvester Matthias differentiated between two growth forms of *iibhai:* "Ones so high—tall—make you sick, give you chills; also [have] maroon fruit. [But] the good-eating *iibhai*—just go down, spread—*s-keegchu iibhai*—ripe fruit appear very dark—dark maroon. Yellow or red flowers, like where the antennae are on South Mountains. Smaller, thinner fruits, I *think.*"

Five or six years later, I asked Sylvester again how the People avoided problems with *iibhai.* "They know which ones. They go up there—they know which ones are best. They got on top of Muhaḍag, South Mountains. They say if you eat too much of the local one around here, it will make you sick, give you the chills. When they're ready to eat, they're real dark red, the good one. But the local ones are just red."

"What time of year is that?"

"In the fall."

The word Sylvester used to denote the good one, *s-keegchu,* comes from *keeg* 'beautiful' or 'pretty' but also meaning 'good' or 'nice', the connotation here. Often when we were driving across the reservation he would point to some sprawling colony of prickly-pear, saying that that was the good kind to eat. Donald Pinkava (personal communication), who has studied the taxonomy of Southwestern *Opuntia* species, believes that Sylvester is distinguishing between the varieties *discata* (the better one) and *major* (which may make one ill). (Incidentally, the only truly trunked or arborescent prickly-pear native to the reservation is the tall blue-green *O. chlorotica*, or Pancake Pear, so called because of its very round pads. It grows only up in the mountains, above the bajada, and I have never been able to show it to a Pima.)

But not all agreed with Sylvester's distinction on morphological grounds. For instance, Ruth Giff said she had not heard of any difference between upright and spreading *iibhai*. "People just never know which kind makes them sick. It gives you the chills. I think you get a temperature. I don't know how many it takes." For her it was amount rather than growth form.

This seemed to be the same under-standing George Kyyitan had. He related this event: "I heard there's a man. The ladies had been out picking *iibhai*. This man came, and they told him to have some. He just started eating them. Never stopped. And later on, this lady said, 'Aren't you afraid you might get colds [chills]?' '*M añ ge cheḍhum.*' ('I got a blanket.') Finally, he did [get the chills]. Yeah. 'Where's your blanket?' the lady said."

"Was it the wrong kind, or did he eat too much?" I asked George.

"I think too much. They never eat too much. Just a few—three." George indicated that the cultivated species, *O. ficus-indica*, did not cause this problem.

At Sacaton, the next village upstream, Francis Vavages told of similar experiences. "Saguaro [*Carnegiea gigantea*], Giant Cactus, is good, the fruit. *Iibhai* is good, too, but don't eat too much. Or you're going to have the shivers. It comes right out of your body. You get so cold, you have to lay out there in the sun. Yeah. Start shaking. Yeah. It happened to me one time. I ate too much, you know. There used to be a trader up there at that four-miles post, they call it. Right down below were lots. Right around September. Oh, they were real *dark* purple! So we start to eat them. If you eat too much, why— they're sweet, too—if you eat too much, that's what will happen to you; you'll get the shivers."

"How many fruit can you eat safely?"

"Oh, you can eat about two or three of 'em. But if you eat too much, overeat, that's what will happen. Because it's sour and at the same time sweet, too. That's what makes that."

The discrepancy over causality seems not to be only recent, because in Curtin's ethnobotany (1949:61) we read of both explanations: "'If too much pricklypear fruit is eaten, it results in chills and the person shakes all over,' stated José Henry; and George Webb explained that there are several varieties of pricklypear— one has light-red fruit which is not poisonous, and another, of darkish-purple, which gives the 'shivers.'" (These colors are the opposite of what Sylvester said.) Nor was the idea of a more edible prickly-pear limited to the Pima: Castetter and Underhill (1935:22) recorded, "The Papago distinguish two kinds of edible prickly pears (*Opuntia* sp.), one of which produces chills and nausea in susceptible people."

Throughout the millennia, not only the prickly-pear fruit but also the cooked pads have served native people in the Southwest and Mexico as food. For instance, in the Mountain Pima country of western Chihuahua we were served quantities of cooked nopales.

In 1764, Nentvig (1980:37) referred to native use of the pads farther south, among the Ópatas and apparently the neighboring Pima of northern Sonora: "The tender sprouts, called nopalitos, are cut, boiled and dried. Because of their tartness, they are used as a relish for pinole." He then discussed the use of the fruit itself, "which when ripe is found in different sizes and colors. . . . The tuna, because it brings on fevers, etc., is not so healthful as the pitahaya." The translators and annotators of this edition, Pradeau and Rasmussen, added in a footnote, "The good father was misinformed. Both are healthful and nourishing." Well, Nentvig gets the last laugh on this one. It's the annotators who are misinformed, as they are on most of the modern plants' names. *Caveat lector!*

I asked Ruth Giff about eating pads. "I don't think the old Pimas ate the pads of the *iibhai*. They do now with the one they raise. The wild one is too small. We learned that from the Jujkam [Mexicans]." Four years later I questioned her about some young pads lying on the stove waiting to be cooked. She said, "They never did the leaves [*sic*] of the [native] *iibhai* in the old days. I never heard of pads being

called *navĭ*—just *iibhai haahag* 'prickly-pear leaves'. We started eating them when the *ge'egeḍ iibhai* [domestic prickly-pear, *O. ficus-indica*] came in."

One day on our way back from Phoenix, I stopped with Sylvester along the road where someone was growing *ge'egeḍ iibhai* as a fence row. He was telling me about how these were cooked like zucchini squash (see Group I). Then he added, "Regular wild *iibhai*—they can still use that as squash, too—Apaches, not Pimas. They [the Pima] know, but they never try it. Except cholla buds, *hannam* [*Opuntia acanthocarpa*]." In other words, the Gileños know that some neighboring tribes cooked prickly-pear pads, but they prepared only cholla buds until the arrival of the cultivated Mexican nopal. George Kyyitan likewise thought this was the case. He said, "I've seen them buy it . . . those tall ones. *Navĭ?* They're good to eat. Call all of them *iibhai* [the fruit] in O'odham."

I checked with Francis Vavages. "When you were young, did anyone eat the pads on the *navĭ?*"

"No. Never did. Only the ripe ones. Peel them off. Have to get a bunch of brush. Go like that [sweeping] and all the stickers will fall off."

Engelmann's Prickly-pear served the Gileños for other uses as well as a source of fruit. "*Iibhai* thorn [spine] is used for tattooing because they are strong and fine—only thing preferred," Sylvester said.

This anecdote may seem insignificant in a land where so many plants are armed with spines and thorns, but I remembered something similar from Russell's turn-of-the-century ethnography. Flipping to a section on tattooing (Russell 1908:161), I read, "A few lines were tattooed on the faces of both men and women. Thorns and charcoal were used in the operation. The thorns were from the outer borders of the prickly-pear cactus; from two to four were tied together with loosely twisted native cotton fiber to enlarge the lower portion to a convenient size for grasping, while the upper end was neatly bound with sinew. The charcoal, from either willow or mesquite wood, was pulverized and kept in balls 2 or 3 cm. in diameter." That was more than 80 years ago. The Pima have been very conservative and specific through time on how they use their plants, even in such a seemingly insignificant detail as this.

iibhai   *Opuntia engelmannii*

A Maricopa woman—who told Curtin (1949:35, 61) she was 98 when the ethnobotanist interviewed her—related, "To encourage the flow of mother's milk, heated pads of the pricklypear are placed on the breasts." No Pima I talked to knew of a similar practice among the Gileños, although Ruth Giff said that they might use a *ku'ul,* or porridge (*atole* in Spanish), made from cholla buds and wheat or the ground seeds of Saguaro to make a mother's milk begin.

Vouchers: *733* (SM); *1492* (IH); LSMC *16* (not extant).

Opuntia fulgida Engelm.
Chain-fruit Cholla, Jumping Cholla
Cactaceae
toota hannam

The trunked Chain-fruit Cholla is usually found in aggregates on the flats. For instance, a whole forest of them is tucked around the southwest end of the Sierra Estrella, and others are between the Sacaton Mountains and Blackwater at the east end of the reservation. These are probably clones of related individuals that became established in the dim past.

The chollas can be sorted out using a few field marks (see *O. acanthocarpa* and *O. spinosior* accounts). Chain-fruit Cholla has a single erect trunk that may be as

thick as one's thigh, either black or still bristling with old spines. It grows usually to well over head-height. At the top, younger branches spread and droop, loosely umbrellalike. The long spines are white or blond but well spaced; the dull green branches bear raised tubercles or nipples, each with an areole or spine cluster near its top. The flowers come late for a cholla, beginning in late May when most others have finished blooming. These are among the simplest cholla flowers: about half a dozen bright pink inner tepals and about an equal number of duller pink outer tepals, encircling a thick mass of bright pink stamens. (These tepals are arranged in sets of threes, each group of three going about its business of opening and closing separately, giving the flower an untidy appearance.) When fully open during the night the tepals curl back, leaving a flat radiation of stamens, like a pink and white sea anemone. As with the other *Opuntia* species, the stamens close back inward when disturbed by something such as a bee, insuring that the insect leaves with a maximum pollen load.

Instead of falling off, the plump green fruits, only about a finger-joint long, sprout additional green buds on their outer margins, until drooping chains of fruits accumulated from many years hang from the edges of these plants. If you slice a fruit open, you will find only several seeds. This cholla reproduces by means of fallen joints lying in the shade of any of these plants. These readily detached joints give the plant its other name (shared with *O. bigelovii*), Jumping Cholla.

Joseph Giff was one of the first to tell me about *toota hannam* 'white *hannam*' or 'white cholla'. "Has lots of stickers, white, very thick; breaks easily; the fruit hang down off of each other. It's very white."

Ruth Giff also told me of *toota hannam,* "the white one—I guess they call it that because there's more than one kind [of *hannam*]." She, too, mentioned how the fruits hang down in chains. Some years later she said, "It's not eaten—I wonder why, because it's a *hannam*. Maybe they're not the kind to be cooked." All the old people seemed to know *toota hannam.*

Sylvester Matthias said it grows on the flats near New York Thicket. We found it there on the lower bajada. We also went around the south end of the Sierra Estrella and found the plant growing there in a great thicket.

In the "Hevel Ñe'i" (Wind Song) recorded by Russell (1908:324–325), the sixth (final) stanza talks of this plant. The song, put into modern orthography and retranslated by Culver Cassa, is

toota yanam i'ivakime
toota yanam i'ivakime
kenda ñ melivekai
shoinge memelihi'i

Jumping Cholla is budding
Jumping Cholla is budding
Just then I came running
Miserable, [I] run and run.

Culver notes that in this song Wind is singing, probably about blowing through the cholla. This is one of the curing songs. "The remolinos, or whirlwinds, that are so common in Pimería, cause pains in the legs, but not swellings. The remedy is to sing the wind song and to rub the limbs with the black gum of the okatilla [Ocotillo, *Fouquieria splendens*]" (Russell 1908:265).

To the south, among other Piman dialects, this cholla is referred to simply as *hannam*. For instance, at Quitovac oasis village in Sonora, Luciano Noriega, a Tohono O'odham, called *O. fulgida* just *hannam*, saying it was used as a cooked vegetable. (Likewise Juanita Ahill of this tribe told Martha Burgess [personal communication] that *hannam* was Chain-fruit Cholla.) And among the Pima Bajo at Onavas pueblo far south in Sonora, Pedro Estrella called the local variety of *O. fulgida* (a robust-stemmed, very bare form) simply *hannam* also, saying it used to be cooked and eaten. No Gileño I worked with knew of this practice. On the coast of Sonora the Seri collected and ate the persistent fruit either fresh, cooked in water, or roasted on hot coals about half an hour; these fruits and the young stems of Teddy-bear Cholla (*O. bigelovii*), which were pit roasted, were the primary chollas eaten by this maritime tribe (Felger and Moser 1985:266–268).

Yet the Gila Pima also may have once eaten *toota hannam*, at least in starvation times. In Thin Leather's version of the Creation Story appears the episode of Corn and Tobacco. Tobacco was a woman, a vain and somewhat fickle one, quite aware of her critical role in bringing rain (see *Nicotiana* spp., Group I). She quarreled with Corn, ultimately causing his departure, along with Pumpkin. For years

the People were without these two most critical crops, and they were starving. One day a woman living near Kotkĭ Oidag (Gila Crossing) went to the Superstition Mountains to pick and roast "the white cactus." This must have been an unusual circumstance, and mostly likely the original Pima listeners knew full well how severe the situation had become. Although *hannam* (Buckhorn Cholla, *O. acantho-carpa*) was a dietary staple, it was available for harvest only during a few weeks in the spring. The persistent chaining fruits of *toota hannam*, by contrast, would have been available year-round as a food of last resort. I suspect its seemingly casual use here was a literary device indicating starvation. When Corn notices the woman and her brother and comes into camp, at length instructing the woman to open her roasting pit, the woman finds not *toota hannam* but corn and pumpkin together, "all nicely mixed and cooked."

Opuntia leptocaulis DC.
Desert Christmas Cactus, Christmas Cholla, Tesajo
Cactaceae
navĭ, a'ajĭ navĭ

Two related (leptocauline) chollas growing in Gileño country might be confused with each other, at least at first. Both Christmas Cholla and Pencil Cholla (*O. arbuscula*) have thin, very green, and almost smooth branches and sparse spines up to two inches long. In both species the spines are encased in loose straw-colored papery sheaths that slip off readily. Both have greenish yellow to yellowish flowers, followed by fruits that persist all year on the plant. But here the resemblance ends.

Christmas Cholla is unlike any other cholla in Pima country. It grows as a dense shrub, usually knee-high, on the flatlands and bajadas and frequently in the arroyos. The stems are so brittle that the plant often grows protected by some other woody shrub. It is in these concealed places that the unwary may first discover this cholla the painful way.

The stems of Christmas Cholla are scarcely as large as a pencil, and the few-petaled flowers a bit larger than a man's

thumbnail. In the evening the pale yellow flowers open wide. With their exposed brush of stamens, they look more like a Blazing Star (*Mentzelia* spp.) than a cholla. But the fruit will always identify this species. Though small, they are bright coral red and often abundant enough to be showy, hence the common names associating the plant with Christmas. (This desert plant is not to be confused with a popular houseplant, Christmas Cactus, *Schlumbergera [Zygocactus] truncata* (Haworth) Moran, a tropical epiphyte.)

One of the most interesting phenomena in Piman folk taxonomy is how Pimans (or proto-Pimans) dispersed, taking a word with them and applying it to the new biological referents they encountered. Presumably the Gileños, the northernmost Pimans, were the last to invade new desert country before the arrival of Spanish colonists in the late 17th century. In Pima Bajo and Tohono O'odham the monolexeme *nav* or *navĭ* refers generically to prickly-pears (see *O. engelmannii* account for more details on this linguistic evolution). This folk genus is variously modified to indicate different species such as *vii nav* and *s-moik nav* at Onavas (Desert Pima Bajo). But on the northern frontier of Pimería Alta, the lexeme as a terminal taxon with most speakers is preserved only as the name for this cactus, the tiniest of the chollas in their country.

But there is some variation in exactly how (and at what taxonomic level) the name is applied. Fr. Antonine (1935) and Curtin (1949:60) gave the name of Christmas Cholla simply as *nafyi*, or *navĭ*, as did most of the Pima with whom I worked. However, Sylvester Matthias called it *a'ajĭ navĭ*: "It's very slender with round red fruit. The name means 'skinny'."

"What about *navĭ* by itself?" I asked.

"Just prickly-pear," he said. "They look for it and plant it."

I asked Ruth Giff about this. "I don't know the *a'ajĭ navĭ*—just the *navĭ*. It seems like it has fruits all year round!" she added.

Earlier in the century when Hrdlička (1908:244, 261) was studying foods and medicines in the Sacaton region, he wrote of *O. leptocaulis* as *a-a-dji naf*, which he glossed both times as 'slender cactus'. So Sylvester's name appears to be an old one.

I never found anyone who ate the attractive but small red fruits, which are often covered with long glochids—those easily detachable bristlelike spines found on *Opuntia* species. But this was not always the case. Ruth Giff said, "We don't use the *navĭ* buds. We used to eat these red ones, the fruit, when we were kids. Got lots of stickers. You have to brush them off. But not [used] for *hannam*." What she meant is that the unopened flower buds were not pit roasted as were those of Buckhorn Cholla (*Opuntia acanthocarpa*) and occasionally Pencil Cholla (*O. arbuscula*).

Forty years earlier, Curtin (1949:60) had recorded a similar statement: "The coral-colored fruit is gathered in baskets and most of the fine thorns are removed with a brush-like branch. More thorns are rubbed off with a cloth. Then the fruit are eaten raw at any time of year."

Still earlier, Hrdlička (1908:261) had found the fruit being "used as food to a limited extent. . . . These small fruits are eaten raw, the seeds being thrown out." He also (1908:244) noted one medicinal use for the plant: "The root of the *a-a-dji naf* ('slender-cactus': Opuntia leptocaulis), ground up and boiled, is given as a tea in children's diarrhea when the excretions are whitish in color." This remedy must have fallen into disuse soon after, because Ruth Giff and others I asked had never heard of this treatment.

Vouchers: *767* (CR); *LSMC 77* (not extant).

Opuntia spinosior (Engelm.) Toumey
Cane Cholla
Opuntia spinosior x *O. fulgida* Engelm.
  [= *O.* x *kelvinensis* V. & K. Grant]
Gila Cholla
Cactaceae
kokavĭ hannam

In late April at the eastern end of the reservation an occasional and usually solitary cholla on the flatlands may startle you with its many deep rose to purple flowers open during the day. (Most other cholla flowers there are yellow or burnt orange.) This is the Cane Cholla and its hybrid with the Chain-fruit Cholla that has been formally named.

The variable plants bearing these garish colors are single stemmed, erect, and between shoulder- and well over head-

height. The branches are somewhat slender and tend to be drooping at their ends. The spines of the tubercles are relatively short and well spaced. The yellowish fruits are strongly tuberculate, that is, they have raised areas from which come a few spines; and they stay on the plant for well over a year. On some of these, the Gila Cholla, a hybrid, the fruits "chain" to a certain extent. There is only one other similar-growing cholla in this area, *toota hannam* (Chain-fruit Cholla, *O. fulgida*). Its flowers are pink, opening in the evening; its plump spineless green fruits form long chains; and true to its Pima name, its spines are conspicuously white, not gray and pinkish. Gila Cholla and Chain-fruit Cholla are easily distinguished where they are growing together, as along Blackwater School Road (see Technical Notes).

After Mass in 1987 at Bapchule village, I sat with four older men, and we began discussing my *hannam*-picking expedition to Santa Cruz village the previous week with Carmelita Raphael. "Were they good, or was it too late already?" one asked.

"Some were already blooming, but most were just right. There were so many buds together that it made picking easier—often three or four at a time. In a few hours we picked over 30 gallons."

I took the opportunity to ask the four about the deep-rose-colored cholla flowers I had been finding about Blackwater and a few other places on the surrounding floodplain. The discussion dropped entirely into Pima. But I could follow it, catching the words here and there for Pencil Cholla and Christmas Cholla and various other possibilities. I heard the word *chiólim* come up, but they dismissed it because it comes from down in Tohono O'odham country. Then I heard *toota hannam* discussed.

"How tall is it?" one of them asked, switching momentarily to English.

"About this high," I said, holding my hand stretched high over my head, "at least the one by Holy Family Cemetery near the Little Gila. But it's like the *toota hannam*, with a single trunk; grows up like that."

"And the fruit," interjected George, "are they big?"

"I think they are bigger than regular *hannam*. Looks like they would be good to pick. The old ones stay on and turn yellow."

They went back to discussing chollas and their properties in quiet Pima tones. Finally a decision was reached to which all four gave assent.

"It's called *kokavĭ hannam*," George announced in English.

"What's *kokavĭ* mean?" I had never heard that word before.

"That's what we don't know. We don't know what that word means," said George, summing up their discussion. "There's lots of regular *hannam* over there in the Santans, but this one, *kokavĭ hannam*, grows here in the riverbed. Has lots of meat on them [fleshy fruits]. They pick them. Taste like *hannam*, but a little bit different. They're good. They're not bitter and more meaty. Used to be here in the riverbed," he pointed over toward Snaketown, "but floods took them away, washed them out."

The four men assured me that *kokavĭ* was not derived from the word *kauk* 'hard' or 'tough'. Mathiot wrote, "*Kokav:* Cane Cholla cactus (a variety of 'chiolim')." In fact, she even had *s-kokavag* as "to be full of cane cholla cactus in one location." Culver Cassa believes that the word *kokav* means "hanging at angles," in reference to the pendent limbs.

Technical Notes: Two biological forms of cholla are included in the Gila Pima concept of *kokavĭ hannam:* a parental form, *O. spinosior,* and a hybrid, *O.* x *kelvinensis.* The other parental form, *O. fulgida* var. *fulgida,* is considered a distinct folk taxon by the Pima. The three forms may be distinguished by the following characters (see Grant and Grant 1971).

O. spinosior is moderately tall and shrubby and has slender joints less than an inch thick that are not readily detachable. Its spines are short and dark gray to pinkish. The flowers are red or purple, many tepaled, and diurnal. The fruits are nonchaining, yellow, and strongly tuberculate, with a deeply cupped apex.

The other parent, *O. fulgida,* in contrast, is tall and treelike in growth form, with thick (nearly two inches wide) branches that are easily detachable. The spines are long and white. Its flowers, deep pink, usually have only 5–8 tepals and are vespertine. The fruits are conspicuously chaining, green, and smooth, with only a slight or no apical cavity.

The hybrid *O.* x *kelvinensis* is trunked but intermediate in growth form, joint size, and ease of detachability. Spine length is intermediate, but the spines are gray to pink. Flowers, variably colored, are diurnal and many tepaled. The fruit is intermediate in color, tubercle size, and depth of apical cavity. To complicate matters, there are two clones in Pima country: one somewhat more like one parent, the second more like the other.

I have seen several individuals in Pima country that appear to be pure *O. spinosior.* Kearney and Peebles (1960:585) said the hybrids were found from Florence to Casa Blanca (or more exactly, Bapchule, where George Kyyitan noted them). By the 1980s they were more restricted due to flooding and agricultural clearance. Most *O.* x *kelvinensis* from the lower population are in unfarmed flats between Sacaton Flats and Blackwater, where *O. fulgida* may be found as well.

Vouchers: *1736, 1962* (GK).

Opuntia versicolor (Engelm.) [≠ *O. echinocarpa* Engelm. & Bigelow]
Staghorn Cholla
Cactaceae
chiólim

Chiólim does not grow in Gila Pima country but is one of the most abundant arborescent chollas found in the Tucson Valley (once Sobaipuri Pima territory) and in northern Sonora. Until recently it was quite well known to the Gileños as a trade item, one of the many products brought by the Tohono O'odham coming north to work in the Pima wheat harvest. Most Pimas did not know what the plant itself looked like but said it was different from *hannam (Opuntia acanthocarpa)*, which they themselves harvested, and that it had a distinctive flavor.

Martha Burgess, who has made a study of plant-gathering practices among the Tohono O'odham, has found that the name *chiólim* in Papaguería is most frequently applied to Staghorn Cholla, a taller, distinctly trunked cactus in the cane cholla group. Unlike Buckhorn Cholla *(O. acanthocarpa)*, Staghorn Cholla is treelike and openly branched. The tubercles on the stems are slender and prominent, the spine clusters well spaced. Flower color is variable, as its scientific name suggests. The fruits are usually spineless. In the vast expanse of Tohono O'odham country, it appears that different

speech communities apply the folk taxon *chiólim* to different biological species. The details of cholla taxonomy need to be worked out for each district or dialect.

Even though the wheat harvest exchange ceased early in the 20th century with mechanical harvesting and the decline of Pima farming because of water loss, *chiólim* occasionally still finds its way northward. As we were discussing the plant, George Kyyitan said, "Just a minute. I think I have some here." He took down from the top of the refrigerator a dusty jar full of the dried pit-roasted buds and spread some on the table. The shriveled hard dull olive-colored buds were scarcely the length of a small finger joint. From another dusty container, he poured some *hannam,* the steamed buds of Buckhorn Cholla. We compared the two, which were both now perfectly spineless. The *chiólim* were more slender, especially toward the base, with fewer tubercles. "Some Papago lady gave me these *chiólim* last year for my birthday. She teaches over here at the school. Coaches."

"Where did she get them?"

"I don't know. Somewhere around Vainam Keek. I'm not sure which side. Somewhere around there, anyway."

Sally Pablo had various kinds of roasted cholla buds: *vipnoi* (Pencil Cholla, *O. arbuscula*), the regular *hannam,* and the kind some Komatke ladies had recently begun collecting around Wickenburg.

"What about *chiólim?*" I asked.

"I've seen that one, finally, around Sells. It's like our *hannam* but the branches are longer and I think more slender."

I puzzled over why *chiólim* would have been an important trade item between Tohono and Akimel O'odham when the Pima already had a good supply of *hannam* growing abundantly on the bajadas throughout their own country. Could it have been that the nearly spineless-fruited Staghorn Cholla could be prepared more easily and in greater quantity than the spiny-fruited Buckhorn Cholla? Would the Tohono O'odham have had ample time in the two months preceding the Pima wheat harvest to devote to cholla picking, taking in a plentiful supply both for themselves and their northern Piman neighbors? Perhaps. But there may have been an additional reason as well. Perhaps the Gileños were such connoisseurs of pit-roasted cholla buds that they relished the subtle distinctions

between the different species and varieties. Much the same as a good grocery store will stock dozens of types of cheese or wine, the Pima may have added to their larder, stored in the Arrow-weed *(Pluchea sericea)* koksin 'storage sheds' various kinds of this staple food.

Vouchers: *1907* (GK, dried prepared buds from Tohono O'odham), *1988* (FJ).

Peniocereus greggii (Engelm.) Britton & Rose
Desert Night-blooming Cereus, Reina-de-la-noche
Cactaceae
Ho'ok vaa'o

Out in the flats dominated by Desert Saltbush *(Atriplex polycarpa),* Seepweed *(Suaeda moquinii),* and wolfberry *(Lycium* spp.), you may come across a pitiful lifeless-looking plant with a few slender gray stalks reaching haphazardly upward. You may have passed it on this trail a dozen times before without seeing it. Its ungainly branches may be supported by some shrubby mesquite or saltbush, or it may be standing alone, having outlived its nurse plant. If you get down and take a closer look you may notice as many dried skeletal arms as living branches, and even the living branches seem borderline. They are scarcely as thick as a finger, with four to six ribs edged with very short and innocuous spine clusters.

Then again, you may never find one. In that case wait for just the right summer night, one of those hot nights in late June or July when the rocks and floodplains and mesquite trees have sucked in so much daytime heat that no reprieve comes even with the setting sun. When the heat radiates outward like discordant music and you know no reprieve will come till two or three in the morning. When even the night, like some stern Zen master, will not permit you the escape of sleep. Then wander out among the saltbushes and Seepweeds and the now-leafless wolfberries. The still hot air may bring to your nostrils a heavy fragrance, a white fragrance, that will lead you to the plant when it bursts out with slender white trumpets of many delicate petals, trumpets as long as your hand out to the tip of the middle finger, sometimes longer. Within a few hours of sunrise the exuberant tubes hang limp and hairy on the bony gray plant, spent from an all-night quest for pollinators. At the very base of that dangling gray tube is an ovary that swells, slowly turning from green to bright red. In a few weeks it reaches the size of a small elongate pear and is filled with pink pulp and small rough black seeds. The fruits are inviting, sparsely spined, but with a peculiar, unsweet flavor. One wonders what animal out in these flats is responsible for dispersing these seeds. Perhaps Coyote himself.

How can such a pencil-thin, nearly dead-looking plant produce such an extravagant show? It has resources unseen to the casual observer. Down in the fine alluvial soil lies a thick jicamalike tuber, as big as your skull or larger. Near Sacaton the botanists Kearney and Peebles once photographed a plant with 24 flowers that opened in a single night. They dug out the tuber, which they found to weigh 87 pounds. But most are in the 5 to 15 pound range.

The Gila Pima, as well as the Tohono O'odham, call this straggling cactus *Ho'ok vaa'o* 'Ho'ok's tongs' or 'cholla pickers'. Ho'ok was a witch from the Creation Story who was conceived when a young man's wooden kickball rolled near a seated girl and she hid it under her apron to tease him. Although Ho'ok was half-human, she had long claws on her hands and feet, and teeth like a dog's. She developed a taste for babies and young children and eventually had to be destroyed (see Group F, *Amaranthus palmeri*). Pima tongs for picking cholla buds were originally made from a partially split willow branch; apparently Ho'ok's were instead made from this cactus.

The tubers of this sparse cactus were formerly used in the treatment of diabetes and were once in considerable demand by Mexican herbalists.

One afternoon I took George Kyyitan and Sylvester Matthias out to the Desert Botanical Garden in Phoenix. Gary Nabhan met us at the herbarium before we toured the new ethnobotany trail. Gary had long been intrigued by stories I had related from Sylvester about the use of *Ho'ok vaa'o* for controlling blood-sugar levels in diabetics. He produced a half-gallon potted *Peniocereus greggii* and put

it on the patio before the two elderly Pima men. They knew it, of course.

"Is this a medicine?" he prompted.

"The *bulb*," responded Sylvester without further coaxing, "for diabetes. Cut it up and dry it, and make a tea out of it and drink for diabetes. It's the *Mexicans*. My father used to get it for them."

"How often do you use it?" asked Gary.

"Not all the time," said Sylvester.

"Just once in a while," added George.

"Just for diabetes," said Sylvester. "I never heard of it used for anything else."

"For gonorrhea," George said.

Gary showed the men a potted specimen of *Peniocereus striatus*, a smaller, multiple-tubered species that just crosses the border into Tohono O'odham country. Both men were curious, but neither recognized it.

One day I asked Ruth Giff if she had heard of Mormon Tea *(Ephedra aspera)* being used for venereal disease. She had not, but she added, "You know, I heard they used the *Ho'ok vaa'o* for that, for the syphilis."

"The tuber?"

"Yeah."

"O'odham?" I quizzed.

"Yeah."

"How about for diabetes?"

"No. I never heard of that."

One spring morning in 1990, we got together to videotape some of Herbert Narcia's Pima songs, which he then translated and explained. The subject of Ho'ok came up. I told him I knew who she was.

"You know *Ho'ok vaa'o*, too?" he asked.

"Yeah. It has white flowers and grows out here on the flats," I assured him.

Herb continued, "*Ho'ok vaa'o* is for diabetes—if people stay away from sweets. Boil it. My grandfather at [the] west end used it. He was Papago, from Chuuvi-geeshk 'Jackrabbit Fell Down', Sonora." His grandfather was Keli Manol, a healer I had heard a great deal about.

I was curious whether his grandfather had learned this use from the Pima or from Mexicans.

Herb was not sure but added, "He hung around with the Yaquis at Guadalupe [village, south of Phoenix], so he might have picked it up from them."

Later in the afternoon, I stopped at Bapchule to talk to George. He mentioned *Ho'ok vaa'o* in the context of various plants used by diabetics but said, "I don't think there is anybody [among the Pima] who uses this one very much." Some cultural

traits are pervasively Piman, while others, such as the use of several herbal remedies, seem to have been acquired late from surrounding cultures; the use of Reina-de-la-noche in controlling diabetes seems to be one of the more recent acquisitions.

Occasionally *Ho'ok vaa'o* is grown in Pima yards because of its beautiful flowers. I went to visit Helen Allison in Komatke because I remembered that years earlier she had had one that her father had planted. It was still there.

No one recalls ever eating the fruits.

Formerly I found this unobtrusive gray plant among the sand dunes near Lone Butte, but in recent years, I have been unable to relocate any. Perhaps they were destroyed when the highway was put through here. I hope a few are still in hiding. In 1987 I came across a plant here and there in the drainage flats (called *vo'oshañ*) southwest of Blackwater as well as between Blackwater and the former Little Gila channel. The plant, overall, is exceedingly rare on the reservation, owing today to wholesale habitat destruction and perhaps formerly to commercial exploitation. Before more tribal land is cleared for industrialized agriculture (sadly, mostly for the benefit of non-Indians), the *Atriplex* flats should be searched for these marvelous creations and the plants moved to some other suitable habitat.

Vouchers: *1495* (?); *LSMC 78* (not extant).

Stenocereus thurberi (Engelm.) F.Baxb.
   [= *Lemaireocereus thurberi*
   (Engelm.) Britton & Rose]
Organ-pipe Cactus
Cactaceae
chuchuis

Organ-pipe Cactus, like Senita *(Lophocereus schottii),* barely makes it north far enough to qualify as part of Arizona's native flora. It is relatively common in western Pima County, especially in the national monument set aside for it. A few more hardy individuals wander a bit farther northward: at least one grows on the east slope of the Picacho Mountains. Although birds get implicated in some unlikely cases of seed transport, this cactus was probably a legitimate dispersal by a bird, possibly a White-winged Dove.

Organ-pipe Cactus may have several dozen branches of rather uniform girth. Each branch has many ribs, suggesting a slender-armed Saguaro *(Carnegiea gigantea).* Even the whitish nocturnal flowers, like those of Saguaro, have a stiff appearance, as if carved from wax. But Organ-pipe Cactus branches from the ground and each arm is only about as thick as your calf.

Flowering is staggered on an individual plant so that fruiting occurs over a long period from mid to late summer. (Saguaros are much more synchronous.) The plump bristly fruits are about as long as your index finger. When they mature, the spine clusters loosen. The pulp within is blood red and juicy.

In Gila Pima and Tohono O'odham this cactus is *chuchuis,* in Desert Pima Bajo *tutuis.* (In southern Sonora the plants become densely branched on short, pedestal-like trunks.)

Evidently syrup of both *chuchuis* and *cheemĭ* (Senita) was once traded regularly from Tohono O'odham to Akimel O'odham country. Sylvester Matthias pointed out the Organ-pipe Cactus at the Desert Botanical Garden and contrasted it with *cheemĭ.* "Eat both of them," he said.

George Kyyitan, who was with us, said that when he was going to school at Parker, Arizona, Tohono O'odham students would get the syrup of both *chuchuis* and *cheemĭ* from their parents and share it with the Pima students. Later he elaborated: "*Chuchuis,* over there, [is] something like a Saguaro, that little one, little, tall ones. They make syrup out of them. But it's *sure* good. Up in Papago country. That's why all them Papago people do that. We had it one time in Mohave. They sent it, they sent some of that *chuchuis sitol* ['syrup'] over there. And those *paan,* whole wheat buns. They sent it over there. Boy, we sure had a picnic, by the river there, river bank. And we had some of that *haakĭ chu'i* [wheat pinole] there. Mix it up in a gallon can, just pass it around, and around. I sure like that. Boy it's good! Mix it with that *chuchuis.*"

Other Pimas were familiar with the fresh fruits from their forays into Tohono O'odham country. The annual wagon pilgrimage to Magdalena, Sonora, in early October probably kept others familiar with Organ-pipe Cactus.

Vouchers: *Hodgson 7370* (sm, gk).

Yucca baccata Torr.
Banana Yucca, Datil
Agavaceae
hovij, hovich

Banana Yucca, or Datil, is a robust plant that forms a prominent part of the landscape in desert grasslands well above Gila Pima country. The rather thick and broad dark green leaves may be as long as your arm, with white fibers slivering along the margins and a stout spine at the apex of each leaf. The leaves grow from rosettes, themselves sometimes in loose colonies. About the beginning of summer, great masses of large creamy white waxy flowers appear on stalks above the plants, betraying their lily ancestry. The floral parts are in two sets of threes. If the yucca moths have carried out their coevolved task of pollinating the pleasant-smelling flowers, the sepals and petals fall and a

large bananalike green fruit as long as your hand develops. The fruits are thicker and shorter than the tropical bananas *(Musa* x *paradisiaca)* commonly found in grocery stores. But the comparison is apt; the yellowish tan flesh has the consistency of an overripe banana. Its flavor is reminiscent of very sweet applesauce or custard-apple, *Annona reticulata* L. Inside are stacked six rows of flat black seeds.

"*Hovij!*" Ruth Giff said. "Just remember it's almost like *ovij,* the 'awl.'" The *ovij* is the indispensable tool of basket makers such as Ruth, and I never forgot this association, though I still wonder which word was derived from which.

Joseph Giff told me, "*Hovij* [is] shaped like bananas, but not as long; sweet, they dry them. I don't know where [the fruit comes] from. *Hovij chu'igam* is [the] name for bananas: 'like *hovij.*'"

Sylvester Matthias agreed, saying, "Can call bananas just *hovij.*"

hovij, hovich   *Yucca baccata*

David Brown of Santa Cruz village told me, "*Hovij* [is] the one here in the Estrellas; the one they roast; like bananas. *Hovij* and *a'ud* [*Agave deserti*] are both plants; both should be up here." But on being asked further, he said he personally knew only *a'ud.*

Working among the Tohono O'odham, Mathiot (1973:378, 381) recorded *hoi* as the name of this yucca and *hovij* as its fruit; Saxton, Saxton, and Enos (1983) call the plant *howi je'e* 'Banana Yucca's mother'. They also use the phrase *utko jeej* for *Y. schottii.* Apparently the Gileños, familiar primarily with the fruit, use only the term *hovij* (or *hovich*) for both the plant and its fruit.

As far as I know, yuccas have never been collected in the Sierra Estrella, but they are there. One May, Gary Nabhan, Paul Johnson, and I climbed to Butterfly Peak near Montezuma Sleeping. There, above the forests of Crucifixion-thorn (*Canotia holacantha*), above the groves of Desert Scrub Oaks (*Quercus turbinella*) and Oak-belt Gooseberry (*Ribes querceto-rum*), perched on a rocky cliff on the ridge, grew some stout-leaved yuccas that appeared to be Banana Yuccas. But we could not reach the plants without ropes, and the plants remain uncollected and unidentified. During a Desert Bighorn survey, E. Linwood Smith (personal communication) also discovered yuccas, which he called Banana Yuccas, at the tops of these desert islands.

When the Tohono O'odham came north to the Gila in late spring to help the Akimel O'odham with the wheat harvest, one of the items they brought with them was dried Banana Yucca fruits and the cakes made from these. Castetter and Underhill (1935:23) described the production of the cakes:

The seeds are removed by hand and the pulp ground on the metate, some cornmeal being rubbed on the metate and muller to prevent sticking. The pulp is then cooked with cornmeal and the seeds are dried.

In preserving the fruit two methods are employed. One is to open the fruit in the field and remove the seeds and fiber, these being taken home in sacks to dry. In drying they are spread on racks of sahuaro ribs (*Carnegiea gigantea*) in the storehouse, and when perfectly dry are beaten on a mat to remove the fiber.

The released seeds form hard lumps which are stored in baskets and ground into meal when needed. The other method is to dry the pulp in the field. Women scrape it out with the fingers, pat it into cakes and dry these on a hot stone. This drying is not sufficient for storage so the cakes are taken home and spread on the roof, and when thoroughly dry are stored in a jar.

Probably these are what Russell (1908:93) referred to as "agave fruit in flat roasted cakes" that the Tohono O'odham brought up to trade.

Juanita Norris, a Tohono O'odham living on the Ak Chiñ Reservation, told me, "You can eat the fruit of the *hovij,* but they have to be prepared. Do it in the ground like the *hannam* [cholla buds]. Put it in the ground, cover it up. And in the morning, you take it out. It's real sweet." The Pima apparently were unfamiliar with this processing. They probably received most of their *hovij* already prepared.

At least until the turn of the century the Gileños were still quite familiar with *hovij.* Hrdlička (1908:262) wrote, "*Ho-wich* is the fruit of a yucca growing in the mountainous parts of the Papago country and used by both the Papago and the Pima as food. The fruit is brought by the Papago and sold to the Pima in a dried state. It looks somewhat like bananas halved and dried, and even in the raw state is sweet and agreeable. It is ordinarily eaten cooked, with the addition of white flour; but it is also eaten raw." Russell (1908:72) spelled the Pima name *havalt,* saying, "The fruit is boiled, dried, ground on the mealing-stone, and boiled with flour. It is also eaten raw as a cathartic. The stems are reduced to pulp and used as soap." When Big Saca-ton (*Sporobolus wrightii*) was no longer available along the river, Banana Yucca (along with maguey) was one of the fibers used as a replacement in making hair-brushes (Russell 1908:116). Various species of yuccas were used among different tribes as soap and especially as shampoo.

Technical Notes: *Y. arizonica* McKelvey and perhaps *Y. schottii* Engelm. might be included in the folk taxon *hovij.* This needs to be verified in the field with Tohono O'odham, the ones primarily responsible for collecting the plant and bringing it to the Gileños. The Pima would have encountered *Y. schottii* in their forays into the San Pedro Valley in search of Apaches more than a century ago.

Yucca elata Engelm.
Soaptree Yucca, Palmilla
Agavaceae
takui

This other species of yucca in the Gileño lexicon, Soaptree Yucca or Palmilla, is quite different from Datil or Banana Yucca (*Yucca baccata*). Soaptree Yucca is also a plant of the desert grasslands, often being the most conspicuous plant across the plains. Mature plants are trunked and have thin, narrow leaves with fine, stringy margins and a terminal spine. The waxy white flowers are smaller, borne in a cluster on top of a bare stalk. The fruits are dry capsules that persist on the stalks after the black seeds have dispersed in the strong desert winds. While the earthy Banana Yucca is a heavy plant rarely much more than head-height, Soaptree Yucca is an ethereal plant that often reaches two or three times that tall.

This yucca is called *takui.* Occasionally Ruth Giff uses *takui* for the binding element in making "Papago-style" baskets. On a trip to the Prescott region, she showed me how *takui* is gathered. During the growing season, the central cone of still tightly enfolded leaves is grasped firmly in the hand and is pushed back and forth strongly, until the whole growing center of young leaves breaks off. (If too many older leaves are clutched, the centers will not break free.) The young leaves are segregated into the pale green ones and the still yellowish innermost ones and are spread out to dry. The sun bleaches the central ones almost white, the usual ground color of the finished baskets. On Pima baskets the white is the split slender twigs of willow. "It's sure easier to use *takui* than *che'ul* [willow]," Ruth said, but she seldom strays from her tribe's traditional materials. After harvest the yucca plants look like shaven-headed recruits while the bald spots grow out, but otherwise the plants are unharmed.

"*Takui* can be picked from April to October," she said. "Somebody picked some in October, so you can still do that. I noticed there wasn't much this summer [1989] because of the dry weather."

A tall double-branched yucca was growing in Carmelita Raphael's ranch yard in Santa Cruz village in 1985. It was taller than the gable of her house. I photographed it, then drove my pickup under it so I could reach some leaves for voucher specimens.

"*Takui,*" Carmelita volunteered. "I used to gather the leaves each season, but I never learned to make baskets, never had time, so I quit saving the leaves."

"Where did it come from?"

"There was a flood one year, a flood in the Gila. Two of them came washing down in the flood. So I transplanted them here. This one is left. Sometimes it blooms, in May, but not every year."

"When was the flood?" This low-lying village between the Gila and Santa Cruz Rivers was always being flooded, it seemed.

"Ahh, let's see. About 30 years ago. Yes, 30, because that was the year my son was born when we planted that."

I asked her to save me some pressed flowers should it bloom again, but during the next few seasons it failed to bloom, and in 1987 I noticed that it was gone.

"Some horses knocked it over," she explained.

Sylvester Matthias was with me. He too had heard of *takui* coming in floods; for instance, his father-in-law had one that came in the Gila River. "Maybe from down by Tucson," he added. Other fine specimens grace older yards in nearby Gila Crossing and occasionally other villages.

Russell (1908:72, 142) recorded that the Gileños used *Yucca elata,* as well as *Y. baccata,* as soap and that Palmilla was one of the fiber plants for spinning the thread used in making the netting of the *giho* 'burden basket'. These, of course, were trade items brought by the Tohono O'odham. Apparently the nearly exclusive use of *takui* as the major coiling fiber in Tohono O'odham baskets is a relatively recent change stimulated by commercial sale of baskets. Kissell (1916:197) said that for domestic use the Tohono O'odham woman would use only those baskets made of "tree material," the more easily made yucca ones being almost entirely for sale.

Juanita Norris, a Tohono O'odham living at Ak Chiñ, told me that *takui hiósig* 'Soaptree Yucca flowers' were cooked and eaten. They taste like cabbage; some are bitter and others are not. Some have purple on them. Contemporary Gileños, at least, were unfamiliar with this dish. It was probably a delicate food that could not be preserved for trade, unlike the sweet fruits of Banana Yucca.

Vouchers: *771* (CR, SM).

takui   *Yucca elata*

Group I, E'es
crops, planted things

Like *iivagĭ* (eaten greens, Group F), *e'es* is a special-use or utilitarian category that is superimposed on any classification of plants based on their structure or growth form. This is a high-salience category, as one would expect from such a strongly agricultural people as the Pima. Sylvester Matthias called *e'es* "anything that's planted, that belongs [to someone]": both the planted or cultivated aspect and the ownership concept are included.

The category comprises primarily agricultural plants that are annuals but may include cultivated perennials as well. For instance, the cultivated prickly-pear *(Opuntia ficus-indica)* and bulbs such as the various onions *(Allium cepa)* are classified as *e'es.* Even one semiwoody bush, Castor-bean *(Ricinus communis),* is included; Joseph Giff used this species as an example of the normative status of the category *e'es* because in Pima country, it is always "someone's plant," but in southern California, he had seen it growing wild (naturalized). Another shrub, Yellow Bird-of-paradise *(Caesalpinia gilliesii),* is categorized as *e'es* because it occurs in Pima villages only where planted in people's yards. According to Sylvester Matthias, both *iivagĭ* when meaning 'lettuce' and *chichinō iivagga* 'cabbage' belong in the category *e'es*—not the category *iivagĭ*—because they are planted things. Others concurred. Closely related forms of plants may be segregated by the *e'es* category. For instance, the cultivated white-seeded devil's claw is *e'es,* but the wild black-seeded one is not.

The category name is derived from the verb *es* 'to plant seeds'. It would not surprise me if *es* and *uus* 'stick, woody thing, bush' share a common origin in proto-Tepiman.

When I asked doña María Córdoba of Onavas, Sonora, for a contrast with the life-form category *u'us* she replied, "*E'es, cosas que se siembran* ['things that are planted']." At the southern end of the Tepiman language continuum, Mason (1917) found the word *esh* (pl. *e'esh*) indicating 'milpa' or 'cornfield'. Whether it was used polysemously for 'planted thing' as well is not evident from the data. At the northern end of the continuum, among the Akimel and Tohono O'odham, the word *oidag* is used for 'field[s]' and is the base word in many colonial and modern place-names.

Allium cepa L. *aggregatum* group
"wild" onion
Liliaceae
mischiñ sivol

Occasionally found growing in Gila Pima and Tohono O'odham gardens is a clumping green onion, with the thin dried outer scales of the bulb brownish, the inner ones pink, and the still living ones white. The plants are summer deciduous: the leaves dry back with the coming of hot weather in May, to be regrown only with cool fall weather.

The plant is something of an enigma on several counts. Some I found growing in Sacaton were said to be of local origin, but they proved to be from Tohono O'odham country. Sylvester Matthias told me that wild onions "grew in the foothills of Table Top Mountain." This is south of Interstate 8 and east of Vikol (Vekol) Wash. "They said it was *very* strong; white flowers, about 10 inches, round leaves. Also grows in the Baboquivaris." The round leaves and the flavor clearly eliminate Bluedicks *(Dichelostemma pulchellum),* a wild lily sometimes confused with *Allium.*

Among the Tohono O'odham this supposedly wild onion is known as *I'itoi sivol* 'Elder Brother's onion'. Its origin is reputed to be the Baboquivaris. Yet it appears not to have been collected in the wild state there, and the cultivated ones rarely produced a flower, which is necessary for identification of the plant to species.

Another enigma is whether *mischiñ sivol* (pl. *mimschiñ sisivol*) should be classified as *e'es* 'planted things' or in the covert category of nonwoody, noncultivated wild plants (Group G).

The connotation of the Gila Pima name, *mischiñ,* is something that ought to be tame but has become wild or feral, such as a wild horse. But it could just label something that has tame relatives, the norm, with the less well known wild relative being secondarily marked (a marking reversal). Because the word *sivol* is a loan word from Spanish for the large cultivated *Allium cepa,* it would be appropriate to mark the local wild species with an adjective after its original name was forgotten. Calling it 'I'itoi's onion' likewise signifies that it is something wild, growing perhaps on I'itoi's mountain, the Baboquivaris.

I have long wondered if perhaps it is just some green onion such as those still readily available commercially or perhaps left from earlier Hispanic introductions of the Jesuit period. Or could it be some asexually reproducing hybrid? In 1988, plants in several locations bloomed. Sister Anne Fischer collected several flowers from a colony on the Ak Chiñ Reservation. That spring, Gilbert A. Voss of Quail Gardens saw those specimens as well as others at the Desert Botanical Garden in Phoenix. He considers them to be from a very old line of clumping onions of European origin. These well could be descendants of plants introduced by Fr. Kino or some other early Jesuit missionary three centuries ago.

Probably two different entities are included in this folk taxon: one of Arizona's native wild species that Pimans occasionally encountered in the hills, and their own shallotlike cultivar that appears to be "wild" when compared to the next species.

Technical Notes: The Ak Chiñ specimens are *AMR 1562,* collected 8 and 16 May 1988. The plants had passed through several hands but supposedly had been taken by the grower's brother in the mountains somewhere on the Tohono O'odham Reservation. None of the plants I have grown from any of the Piman reservations has ever bloomed.

We have found a wild onion, *Allium macropetalum* Rhydb., growing upstream on the Gila as close to the reservation as Cochran. The bulb scales of this species are distinctly different from the *mischiñ sivol* grown by the Akimel and Tohono O'odham but may account for some of the references to "wild onions" found naturally occurring.

Vouchers: *1562* (Juanita Norris, Tohono O'odham, Ak Chiñ Reservation); *AMR, s.n.* (Albina Antone).

mischiñ sivol   *Allium cepa, aggregatum* group

Allium cepa L. *cepa* group
onion
Liliaceae
sivol

It would be difficult to think of Pima cooking without *sivol,* but apparently this plant was a late addition to their cuisine. Even though Fr. Kino and subsequent Spanish missionaries introduced this European cultivated species to the more southern Pimans, and the current Akimel and Tohono O'odham name is derived from the Spanish word *cebolla,* the Gila Pima seem to have picked up this crop only within the last 100 years or so. Emory (1848) did not mention it among Pima crops (but surely he overlooked all but the most conspicuous ones); nor did Russell (1908) or Lloyd (1911:58). However, Grossmann (1873) included onions among the crops of lesser importance.

Sylvester Matthias put the introduction of *sivol* rather late: "When did they start using this? In the early times, when the Mexicans settled in the [Salt River] Valley; [use] in stew; or just fried like or with potatoes. Used for spice [seasoning] now. [But] when my father made tortilla soup, he always adds onions to it. Nowadays onions is used in *hannam* [cholla buds, *Opuntia acanthocarpa*] and in all those greens, *iivagĭ*. I think it is the main spice now. In my early days, I eat raw onions. I didn't like them cooked." Ruth Giff gave a similar account of how her family began using such seasonings (see garlic, *Allium sativum*).

The development of improved strains of yellow onions and White Bermuda onions was one of the more successful achievements of the Sacaton Field Station. Reporting on the 1922–1924 period, King and Leding (1926) said, "Much of the Indian land along the Gila River is well suited for onions, and should marketing conditions become more favorable for the Arizona crop, on account of the newly developed onion industry in the southern counties, the Indians would have several advantages in engaging in the commercial production of onions. Many of them already grow small patches of Bermuda onions for home use, and the station distributes large quantities of seed and seedling plants to them every year." Later, Castetter and Bell (1942) still found the Pima growing onions in small acreages, particularly for home consumption.

As far as I know, no wild species of *Allium* have ever been found in Pima country proper (see *Allium cepa* L. *aggregatum* group). There appears to be no aboriginal Gila Pima name for a wild onion. But the Gila Pima once hunted afar and were often aware of such plants, so they might have once used a native species. Sylvester Matthias related, "The *haad* [*Dichelostemma pulchellum*] is not a wild onion, but similar; only they have flat leaves—[grow] in the washes—and in the Estrellas; has bulb like onions, but they don't eat them [*sic*]." Sylvester is exactly correct in this distinction, but ethnographers and lexicographers have sometimes confused *sivol, haad,* and *shaaḍ* (Saya, *Amoreuxia palmatifida*), plants belonging to three different genera in two families. In each case, an underground part was eaten.

aahō *Allium sativum*

Allium sativum L.
garlic
Liliaceae
aahō

Although it may be presumed that the early Jesuits introduced garlic to Pimería Alta, there is no evidence that the Gila Pima picked up this crop until well after the Anglo period began. I have found no mention of the crop by Grossmann (1873), Russell (1908), Lloyd (1911), Castetter and Bell (1942), or any other writers from outside the culture. Minor crops, however, were usually overlooked by outsiders.

In January 1968 I was working at Komatke and got back to the house at midday. Nobody was home, but Ruth Giff from several doors away saw me come in and brought me over a steaming pot of delicious hot soup. I asked her about Pima seasonings, what was used, and when they learned about them.

She said, "Old Pimas never used oregano or anything. But my mother, I guess, when they used to park their wagons someplace in Phoenix, and I guess one cold morning, I guess this Mexican family asked her in to get out of the cold; you know, they'd bring their mesquite wood, and the men would go around town and sell it. So she was watching when they were cooking. She never had a chance to know about cooking; so she saw her [the Mexican lady] cook potatoes and put chorizo in there; so she learned that. And I don't know *where* she learned about garlic; she *liked* garlic, and she'd put it in the beans. And I guess from there on I liked garlic. Of course she uses onions. Like, a Pima stew would be potatoes and onions—that's all that goes in there. So there's no oregano."

Apparently no one knew to use Desert-oregano (*Aloysia wrightii*), which grows higher in the Sierra Estrella, or even Desert-lavender (*Hyptis emoryi*), which has a sagelike flavor.

The Pima and Tohono O'odham name *aahō* is derived from the Spanish *ajo* 'garlic'. The word *aahō* seems to be well known among Gileño speakers, but use of the plant does not seem widespread. Sylvester Matthias commented, "The Papagos like *aahō*. Pimas don't care so much about it." It seems not to be one of the plants raised in the small "kitchen" gardens of chiles (*Capsicum annuum*), onions (*Allium cepa*), and tomatoes (*Lycopersicon esculentum*).

Amaranthus hybridus L. var. *ery-throstachys* Moq.
    [= *A. hypochondriacus* L.]
grain amaranth
Amaranthaceae
ki'akĭ, giád

Grain amaranths, found throughout Mesoamerica, were once an important Native American staple. Street vendors still sell cookielike disks of the popped seeds in many parts of Mexico. They were once grown even in the northern fringes of Mesoamerica in Salado country, A.D. 1350–1400 (Sauer 1967) as well as in Hohokam country (Gasser and Kwiat-kowski 1991:441–442). And were it not for a single slim reference by Russell at the turn of the century, even the name for this crop, *ki'ak* or *ki'akĭ*, would have fallen from the Pima lexicon. No one I worked with had ever heard of a cultivated ama-ranth. Russell probably had no specimen but provided enough information for identification of *ki'ak*: "The heads of this annual are gathered and the seeds beaten out with the kiâhâ *[giho]* stick used as a flail. The seeds are moistened, parched, which makes it resemble popcorn, ground on the metate, and eaten by taking alter-nately pinches of meal and sips of water" (Russell 1908:74).

Whittemore (1893:52) made another possible reference to this crop: "A little over 100 years ago, nearly all the Pima Indians . . . resided . . . in seven villages . . .

west of the agency. Here they raised cotton, corn, melons, and pumpkins, and a small round seed which they ground and boiled as mush."

Gary Nabhan (personal communica-tion) has traced the name *giági* as well as the plant southward to the Mountain Pima at Maicoba and the Northern Tepehuan. Pennington (1969:54, 139) found the Northern Tepehuan still extensively growing *giági*, a domesticated amaranth.

Grain amaranths were once cultivated widely throughout the Southwest, judging from archaeological remains (Bohrer 1991; Sauer 1967). This cultivated form differs from its abundant wild relative *chuuhug-gia* (Palmer's Carelessweed, *Amaranthus palmeri*) in having larger seeds and a distinctive seed morphology caused by larger embryos (Miksicek 1983) and often pale rather than black seed coats. Sauer (1967:134) notes that Pennington's *giági* were pale seeded, as were specimens collected among the Guarijío. Recovered Hohokam seeds are apparently all black (Gasser and Kwiatkowski 1991:441; Miksicek 1983:697, 1987: 208). Seed color of the Gileño crops is unknown. Speci-mens should be watched for in historic and protohistoric Piman sites.

This is a hot-season crop. Perhaps it was grown in what Crosswhite (1981) termed second fields. The seeds are scarcely a millimeter across and are pro-duced in prodigious quantities, usually several tens of thousands per plant. These are similar in carbohydrate content to the true grains and even higher in pro-tein and fat content (Sauer 1967:104–105).

Like the domesticated chenopod *kovĭ* (*Chenopodium berlandieri*), grain ama-ranth appears to have fallen from the Gileño crop inventory during the stress years of the 1870s and 1880s when water-shed deterioration and drought reduced the flow of the Gila and Anglos and Mexicans upstream cut off what water remained. Those Pimas with whom Russell worked in 1901–1902 may have remembered *ki'ak* only as a plant from their past.

In June 1990, George Kyyitan was discussing one method of controlling weeds in fields by broadcasting *bavĭ* (teparies, *Phaseolus acutifolius*) very thickly. When I asked him what kind of weeds were a problem, he surprised me with a linguistic artifact.

"That—what you call it?—nightshade or whatever? That tall ones?"

"*Mo'oial?*" I asked, offering the name for Marsh Nightshade (*Solanum ameri-canum*).

"No, that *chuuhuggia*." The Pima word *s-chuhugam* means 'darkness' or 'night', so George associated *chuuhuggia* with the English name *nightshade*.

"I see it's already up, here."

"Those that they grow here, right along that ditch there [in Bapchule], the regular *chuuhuggia*, that one. You know it's two different kinds of *chuuhuggia*. These here are the real *chuuhuggia* that grows a long time [ago]. *Then*, later on, them others came up, those big stalk, tall ones. But this one here, their leaves is not so big and it's good to eat. But this one here, their leaves are thicker—thicker than these. And when it's kind of bigger, it doesn't any good to eat. Because it's too thick. But this one here, you can eat it. And this is the one coming up early in the summer. It's kind of got pink stalks on there."

"The early or the late?"

"The one now there." He was referring to Palmer's Carelessweed, an important greens for cooking.

"Do you call them the same thing in O'odham?"

"I think that they said that they call it *ki'ak* or *giád*—the big one. I know that one pretty well—*chuuhuggia*—because it's been here for a long time. And later, later, I found out that this one comes in. It grows over here in the Milgáán ['Anglo'] fields, then they come over here, start coming up. Same thing with that Johnson Grass *[Sorghum halepense]*. See, they grow way over here somewhere; then they come up in the water, then they come. Now it's getting *all* over."

"Can they use *giád* for anything?"

"Yeah, some of it for cooking, like *chuuhuggia*. When they were young— their leaves still young, soft. But little bit, when they grow up, their leaves will get thicker and heavier."

"Have you ever heard of anyone using the seeds of that?"

"Huh-uh, no. And that other one, the regular *chuuhuggia*, they used to grow *all* over in the fields there. Sometimes they, when it rains and the water will get in there, then they'll come [up]. Then they'll grow way up. We'll go out and cut 'em, cut 'em. Give it to the horses. Eat 'em up. Just like hay."

ki'akĭ, giád *Amaranthus hybridus*
*Painted from plants grown in San Diego from
seed supplied by Native Seed/SEARCH, Tucson.*

But apparently someone early
in the century still remembered
the robust cultivated species
(*A. hybridus)* and applied the
name *giád* either loosely or carelessly to the large wild plants
(*A. palmeri),* an example of
cultural erosion in folk taxonomy.
My surprise is that the name for this lost
cultivar has survived at all.

Vouchers: *1936* (GK, rank mature *A. palmeri).*

"Were the seeds black or white?"

"I didn't see the seeds, just the *muḍadaj*
['its tassel']."

"I heard a long time ago, last century,
they used to raise one, with white seeds."

"*Giád?*"

"Yeah."

"I think they do. Some things that we
don't know [now]. But later on we learn,
when we see it. It's the same way with
everything that was taught, back, how
things grow and how they do."

The second form of carelessweed that
George referred to here proved to be just
the taller, rangier plants of wild Palmer's
Amaranth growing later in the season.

Ananas comosus (L.) Merrill
pineapple
Bromeliaceae
a'uḍ, a'uḍ chu'igam

In summer, Sylvester Matthias would visit
me in San Diego. It was a respite from
the severe southern Arizona heat and an
opportunity for us to get stretches of
uninterrupted work done. During each
visit we would go to a grocery store with
a good produce section and study the
items. On one of our shopping excursions,
Sylvester paused to pick up a pineapple.
He said, "We call this *a'uḍ.* When we start
seeing them in Phoenix, we call them *a'uḍ.*"

"But that's a century plant! You didn't
call them *a'uḍ chu'igam* ['agave looks like']
or something like that?"

"Oh, that would be wonderful. But it's
just *a'uḍ.*"

As he held the plant, it became immediately obvious to me why the Gileños
should make such an association. This
new plant bore a nearly perfect resemblance to the "hearts" of the agave after
all the leaves had been lopped off. I had
seen agave hearts once piled under a great
mesquite tree near the Pima Bajo village
of Onavas, Sonora, waiting to go into
the roasting pit. They looked for all the
world like robust white pineapples, with
the tightly packed rhomboid leaf bases
appearing as the floral bracts of the
pineapple.

But the parallels did not stop there.
After the agave *corazones,* or hearts, have
been pit-roasted some 48 hours, they
emerge from the ground marvelously
sweet, with a flavor Pimas describe as
something between pineapple and sweet-
potato. Both agave and pineapple may
have a fibrous quid you must discard after
chewing. And were that not enough, the
new green plant crowning the pineapple
has tough, tooth-edged leaves that even
resemble some small agaves.

Despite the similarities, the pineapple,
a terrestrial bromeliad, is unrelated to
the liliaceous agave. And though we may
associate pineapples today with the
Hawaiian Islands, the crop originated
in tropical South America, perhaps first
cultivated by Tupí-Guaraní Indians of
northern Paraguay (Schultes 1984). In
pre-Columbian times, the crop spread
widely through the West Indies and
Central America as far north as Veracruz,
Mexico.

Apparently the Pima were unaware that
pineapples are commercially raised rather
than gathered wild. Sylvester at first said it
should be classified with the thorny or
cactuslike plants rather than with crops.
Later he changed it to the *e'es* category.

But one puzzle remained, a linguistic
one: why this plant was called simply *a'uḍ.*
When a new item is introduced into a
culture, it is often linguistically associated
with something preexisting and then
"marked" with a term to distinguish
it from the original, as in the case of
the pear being called *vakoa chu'igam*
'bottle-gourd it looks like'. With time,
this marking may be reversed as the new

item achieves greater cultural significance. This has happened, for instance, among some Tohono O'odham, who now call bananas *(Musa x paradisiaca)* just *hovij* (or *howij*) rather than *hovij chu'igam* 'Banana Yucca looks like'. The Tohono O'odham have completed the marking reversal and now refer to their native Banana Yucca *(Yucca baccata)* as *O'odham hovij* 'the People's banana'.

Then I read an interesting paper by Witkowski and Brown (1983) discussing the evolution of such linguistic marking reversals, and my puzzle was resolved. Presumably there is an intermediate stage in the process when both the new item and the old one are referred to by the same name or lexeme, a situation linguists call polysemy. As the cultural salience of one item rises, that item becomes the unmarked one. But with the pineapple and the agave, we have caught the rare intermediate polysemous stage. No Pima goes out into the Estrellas any longer to harvest the agave, and only the very oldest Pimas remember anything about it. Pima acquaintance with the tropical pineapple must have been relatively recent in the Anglo period. And pineapples are expensive produce items that rarely find their way onto a Pima table. So neither plant has reached the prominence necessary to push the linguistic process to completion: both remain unmarked.

A few years after our initial discussion, however, Sylvester consistently called the pineapple *a'uḍ chu'igam,* using the presumed earlier stage in linguistic marking.

Arachis hypogaea L.
peanut
Fabaceae (Leguminosae)
kakawáádi

Words may make long and circuitous journeys through time and space. Often there are permutations along the route so that meanings as well as forms may be changed. Sometimes a word may come to signify the opposite of its original meaning, like the word *manufacture,* whose Latin roots literally mean 'handmade' *(manuus + facere).* Other times a word may become a loan word with little change in form or meaning. The peanut, now worldwide, was confined to the New World in pre-Columbian times. Though found in South America and among the higher civilizations of Mesoamerica, it appears never to have made it into the American Southwest. At least it is unknown here archaeologically and was unrecorded at the time of Hispanic contact. But farther south the Nahuatl called the plant *tlalcaca huatl.* The Spaniards modified this to *cacahuate* and apparently brought the crop northward with them. The Pima and Tohono O'odham took the seeds and this Hispanicized version and further Pimanized it to *kakawáádi* and *kakawááda.* (Mathiot [1973:268] wrote it *gagavuadi.*)

Peanut growing appears never to have achieved much prominence among the Gileños. The crop is not even mentioned by Morrisey (1949) or Ezell (1961) (both summaries of documentary sources) or Castetter and Bell (1942). I found Ruth Giff growing *kakawáádi* in her field. But she and Sylvester Matthias assured me that peanuts were not grown in their youth. For home consumption, commercial outlets off-reservation have probably been the major source.

"The Mexicans raised them," Sylvester said, "between Central and Seventh Avenue. We bought them."

Everyone, even at the west end, seemed to know the Pima nickname of a Sacaton man with whom I worked, Kakawáádi Shuushkam 'peanut shoes he has'. Apparently in the early days when he worked on ditches, he wore oversized galoshes or rubber boots that looked to someone like giant peanuts on his feet.

Asparagus officinalis L.
asparagus
Liliaceae
mo'otaḍk chu'igam

Unlikely as it may seem, the Gila Pima formally named asparagus, a crop introduced early this century at the Sacaton

kakawáádi   *Arachis hypogaea*

Research Station. Some asparagus escaped, and the plant was collected in the Gila riverbed in 1926. But it seems not to have survived subsequent years of alternate flooding and desiccation.

Names—folk taxa—sometimes make curious circles. The Pima, seeing the more slender whitish purple and green shoots of asparagus erupting from the sand, were reminded of the similar-shaped but thicker yellowish and purple Desert Broom-rape *(Orobanche cooperi),* which they called *mo'otaḍk.* Using one of the conventions of the language, they named asparagus *mo'otaḍk chu'igam* 'Orobanche looks like'. But *Orobanche* is sometimes called wild asparagus by those writing on native foods, particularly by those not very careful to learn the real identity of the plant they are talking about. So by a circuitous twist, the Pima name for asparagus becomes 'wild asparagus it looks like'.

I have never found anyone growing asparagus today, and it is only an incidental part of their diet as a store-bought commodity. I asked Culver Cassa whether the Pima ate it. His blunt and prompt response was, "That's rich people's food."

Beta vulgaris L.
beet
Chenopodiaceae
Raphanus sativus L.
radish
Brassicaceae (Cruciferae)
s-vepegĭ daadaidkam

The Gila Pima name for beet is *s-vepegĭ daadaidkam* 'red roots they have'. They were being grown experimentally at the Sacaton Field Station in the 1930s as part of the development of commercial truck crops. I have never found beets among the vegetables some Pima grow for home consumption.

Sylvester Matthias told me, "*S-vepegĭ daadaidkam*—not a native here. Hardly anybody eats them. I learned to eat them at Phoenix Indian School."

Few Pimas recognized the name for this plant. Apparently the cultivated radish, *Raphanus sativus,* is sometimes included in this term as well, but Sylvester said that the name was first applied to beets, because the Pima knew them first.

The word *dahidag* means the thickened underground part of a plant, such as an onion or beet, growing straight down; it does not refer to other parts, such as sweet-potatoes, that grow laterally.

Vouchers: *1490* (SM).

Brassica oleracea L. var. *capitata* L.
cabbage
Brassicaceae (Cruciferae)
chichinō iivagga

I have sometimes wondered, if a loan word in a language were identical to its source, whether we would be able to recognize it as a loan rather than merely a substitution after the source has become the prevalent language. This is not merely a philosophical question. With Gila River Pima, words borrowed from Spanish are readily recognizable, but not necessarily those from English. Even native speakers quite often disavow a folk taxon whose source is readily perceptible to them: "We don't have a name for apple. We just call it *mansáán* or *aabals,*" overlooking that these names are now distinctively their own. More perceptive Pimas, though, will point out that the accents, the vowel lengths, the pluralization, and sometimes even the consonants have been modified to fit characteristically Piman patterns.

For years I asked about the Pima name for cabbage, an Old World cultivar that should have been introduced early in the Jesuit period but which perhaps had never caught on this far north. Or it may have been integrated early as a winter crop, then fell out of use. Had a loan word been acquired at this time, it should have sounded something like the Spanish *col* or *repollo* or perhaps *berza,* all cabbage varieties. Always the answer to my question was the same: "It's cabbage, just cabbage. We have no special word for it except cabbage." Was *kabij* actually the Pima name, or was there none, as I was being led to believe?

One day, when I thought this book was nearly finished, Ruth Giff and I were going over a list of products the Gileños would buy in Phoenix. She asked if I knew the

---

chichinō iivagga   *Brassica oleracea*

Pima name for cabbage, and of course my reply was negative. She said, "*Chichinō iivagga.* You know what that means?"

I thought for a moment. "Chinese lettuce or greens?" She nodded approval. *Chiino* was a loan word right out of Spanish, and it is pluralized in Pima with a reduplication of the initial syllable. *Iivagĭ* means 'eaten greens': generally wild greens, but it could mean 'lettuce' in a restricted and very recent meaning. And -*ga* is a possession marker.

She went on while these pieces flashed through my mind. "They must have eaten it at a Chinese restaurant," she said, bringing us back to our discussion of the Pimas going in to south Phoenix in their wagons to shop. "Tom Jose, Joe's brother-in-law, raised lettuce, cabbage, and carrots on 35th Avenue. His rows were so straight. It's the one they see in stores. Frank Matthias [Sylvester's uncle] used to raise it in a little garden at the school. All kinds of vegetables. Stew it. People didn't eat lettuce in those days. They had their own greens!"

Ruth told of an incident her husband recalled from his youth: "Joe, when he was a little boy, went out to the fields where

the men were plowing. He waited until they stopped, then asked, 'Do you like *chichinō iivagga?*' They said yes, so he ran home and his mother dumped the cabbages in the soup she was cooking for the men's lunch. Maybe some of the men didn't know what *chichinō iivagga* was—it was new here—but they said yes. Maybe it could mean other things, like brussels sprouts too, that they saw them using in Chinese restaurants. But they grew cabbage here [in Komatke]."

"How would you classify it, as *e'es?*"

"Yes, because it's a crop. They don't go out and gather it like the greens; they raised it. Did you ask Sylvester about it?"

"Yes, but he doesn't know it. He never heard of *chichinō iivagga*."

About a year later, when we were discussing local variations in Gila Pima names, Sylvester told me, "Those people from Santa Cruz [village] say *chichinō ha-e'es* ['Chinese crops'] or *chichinō iivagĭ*, but here [in Komatke] we just say *kabij*."

aadō o'ohoḍag, meek ha'ichu hiósig   *Caesalpinia gilliesii*

Caesalpinia gilliesii Wall
Yellow Bird-of-paradise
Fabaceae (Leguminosae)
aadō o'ohoḍag, meek ha'ichu hiósig

One of the small ornamental shrubs one encounters planted in more traditional Pima yards is this bird-of-paradise. Its wispy leaves are bipinnate or double compound, a characteristic of many desert legumes. The leaves have an unpleasant smell. The Pima cultivate it for its showy red and yellow flowers, which bloom from late April through summer. It is a native of southern South America.

Although most Pimas were familiar with the plant from the many years it has been grown on the reservation, no one was at first able to give me a name for it. But Sylvester Matthias told me this of its history: "My grandmother named it. It's not a native here. When the ladies went to Phoenix to peddle their baskets they saw it. Brought back its seeds to grow in Komatke. Called it something to do with 'peacock', *aadō*." But for many years he could not remember the plant's full name.

Finally, in 1989, when I stopped to collect a specimen from a colony of the shrubs west of Jacumba, California, he called it *aadō o'ohoḍag*, which he glossed 'peacock, the color of'. The dictionary definition of the second word is 'design' or 'sign'. However, *aadō* originally meant 'macaw' and has only recently been transferred to the bird brought in by Anglo farmers, after the macaw of tradition faded from Pima memory (Rea 1983:162–163). The original designation would be most appropriate for this plant, with its yellowish flower and enormous exserted red stamens, like the long tail feathers of the Scarlet Macaw.

Sylvester said this plant was also known as *meek ha'ichu hiósig*, a back-translation of its English name. "*Meek ha'ichu* means 'far away' because the Pima didn't know where paradise was. They left the bird part out of the name."

This is an atypical plant to be included in the *e'es* category because it is definitely a woody shrub rather than an annual, as are most other *e'es*. In Pima country I have never found it growing feral, as happens in some parts of the Southwest. In Pima perception it is always "somebody's plant."

The similar Mexican Bird-of-paradise, the *tavachín* (*Caesalpinia pulcherrima* (L.) Swartz), with more orange flowers, is replacing the earlier red-and-yellow-flowered species as the favored planting in some Pima yards, at least on the west end. It is unnamed in Pima.

Vouchers: *841, 1623* (SM).

Cannabis sativa L.
marijuana, hemp
Moraceae
maliwáána

An elderly man at Sacaton was telling me about the types of tobacco once raised there. He mentioned one kind the Pima were growing. Long ago someone from Mexico tried it and said it must be marijuana—in Pima, *maliwáána*—because it was so strong.

"What color flowers did it have?" I asked him.

maliwáána *Cannabis sativa*

I asked a number of elderly Pimas—people who remembered their first associations with the plant around the 1920s and 1930s by way of Mexicans in Phoenix.

One warm evening in late May I sat outside in the dark with George Kyyitan talking of *jeweḍ hiósig,* the mystical lichen. When we finished, I brought up the subject of marijuana.

"*Maliwáána* makes them crazy," George said.

"How would you compare it to *jeweḍ hiósig?*"

"Well, *jeweḍ hiósig* is not like that. It's a good smelling, good smoking—not go crazy." Nonetheless, that was a risk everyone took with this native lichen. But George meant something else. "A guy told me one time, 'Well, there's nothing to it when you smoke that *maliwáána*. It's just the same thing as you see things in your mind, like that. Like you just see funny things like that. That's the only thing that you can see that if you smoke that.' He said, 'Yeah, one time I *did* smoke. And I was sitting there. And finally I saw an ant coming—right there. A big ant.'"

"*Totoñ?*" I asked.

"Yeah. 'You could hear his stepping on there—his heels—you could hear.' 'How big is his feet, that you could hear his heels stepping?!' [It's] nothing like when you get drunk. In this you could just see something, funny like that, like you can see in the movies. And there's another guy—he's over at Ak Chiñ—he said, 'One time we were sitting over there and we had some of that *maliwáána,* and we were sitting there, and finally this man who was with me said we should go over to his place, so we both ride on one horse. I went over there. And I saw *lots* of ants about this big [he indicates about as tall as a horse], going through there, and you could see the horse tracks—and all those big ants there.' So you can just see things, awful things. So I don't know why they like to see that." I laughed at George's recounting someone else's experiences with smoking the plant, but George did not think it was a laughing matter. "Well, there's *nothing* to it, that you could really think it's funny, and something that is good to see—but that's just crazy things."

"Have you ever heard of Pima raising marijuana?"

"Yeah! I *think* he's a Pima—over here at S-totoñig. Ah—nobody knows it. And there was a house [abandoned] on this

"I don't know. I don't even know what that plant looks like."

I had to laugh inwardly, because not two yards from where he sat, one of his grandchildren was trying to raise a sickly-looking marijuana plant in a coffee can.

It was clear on listening carefully to him and to others that the Spanish name for marijuana had officially entered the Pima language as a loan word and was distinct from its Spanish and English equivalents. But when this happened was less clear. Older Pimas had considerably less familiarity with the plant than the current generation; what were their perceptions of marijuana and their attitudes toward it?

side. And he had it in that—the plants. And one time, there's a man that goes through there and went in that house. Then he find that one. He notice it, because he knows it. He threw them away! There's another one over here at Ak Chiñ that he raises that one. It's one of my—uncle, you might say. And he passed away. Somebody find that out. It was planted with something. I don't know what. Anyway, they find that out. I think they put him in."

"They what?" I asked.

"They put him in the can [jail]."

"In general, would you say not many Pima raise it?"

"Might, but I don't know how it smells. But some of them, they know it."

George told another story about his early contact with marijuana. "I know one guy, going around with me in Phoenix, back in 1928, I think. O'odham. And he got one roll there. Cigarette. Went over there on Third Avenue [Phoenix]. He got about two, three puff[s] out of there. Tried to give it to me. I said, 'No, I don't want it.' [Of] course, I don't know what it is."

Sylvester Matthias is always good at pinpointing arrival times of plants, so I asked him about his first recollections of marijuana.

"My late uncle, when he was a very young man—he's the one who mentioned about that. My father never mentioned it. It's mostly used by the early Mexicans. Nobody else uses it. Later on they found out it was a drug, then other people started using it."

"Does anybody grow it?"

"Yeah, they grow it, but the narcotic agents will surely spot it; no matter where, they'll spot it. Once on Buckeye Road (now it's all residential area, but there was farming area then, and out of town) somebody planted [it] with the cotton, in the cotton field. But *maliwáána* grows [taller], and they just grow over the cotton—and they found it!"

"What happened?"

"They have to go after whoever is growing it, or else pull them off."

"Do you know what it looks like?"

"I don't know what it looks like when they are fresh. They must be similar to cotton leaves. I saw it once, but it's already too dry, and crushed. It's that kind of green, like cotton leaves."

"Do you know anybody in your generation who would use marijuana?"

"No, not in my generation, but they *talk* about it, they talk about *maliwáána*. Seems like they don't know it. But I know it's really used by the Mexicans. It wasn't common. Now it's getting [so that] everybody's using [it]."

"Have you ever heard of anybody growing it on the reservation?"

"Yeah. But they don't know *who* they are—are they Indians or Mexicans? There's a patch over here, this end of Sun Lakes. Where the wastewater runs out, that's where they plant—in the seepage. But they found it!"

"It's a drug, huh? What is it supposed to do to you?" I asked.

"Oh, just make you . . . not drunk, but just kind of make you feel . . . happy."

I have heard stories from Pimas who have seen small planes landing on remote stretches of the reservation, dumping bales of something they suspected was marijuana. Older adults look on such activities as criminal and generally associate marijuana with a low class of people; but the younger and more acculturated Indians are less fastidious.

In all my wanderings around Gila Pima backcountry, I have never found a marijuana plant being grown. (My only specimen is from the lower Salt River, an area frequented by non-Indians.) The few individual potted plants I have seen have all been in Sacaton. I doubt that most of these survived long. They have to be watered!

Capsicum spp. L.
chile
Solanaceae
ko'okol

The Gila Pima love their *ko'okol*. It is hard to imagine a feast without great stews of beef cooked in a thick red broth of ground red chile, or a Pima table without a little jar of the tiny but potent round wild *o'olas ko'okol* (Chiltepín, *Capsicum annuum*, Group G). One fall I was going to visit Pima Bajo country in Sonora. "Be sure to stop and pick up a string of red chiles at Magdalena," Sally Pablo reminded me, "The one I bought a month ago is almost gone." She was in charge of cooking for funerals that year. That takes lots of chiles.

Although we might expect chiles to be broadly distributed in Hohokam and aboriginal Piman country, apparently no seeds have yet been recovered from archaeological sites in this region (Gasser and Kwiatkowski 1991; Gasser, personal communication). But then, information from protohistoric Piman sites is almost nonexistent. So we do not know when any of the cultivated varieties arrived in Gileño country. Gary Nabhan tells me he believes that all of these are cultivars of *Capsicum annuum* L., as is the wild Chiltepín from southern Arizona.

Only the potent little Chiltepines are known to have been used by the various northern Pimans and other prehistoric peoples of the Southwest. These wild plants grew in Tohono O'odham country and were traded to the Akimel O'odham (see Group G, *Capsicum annuum*). Undoubtedly at this stage in the folk taxonomy these were called just *ko'okol*. A 17th-century Névome dictionary (Pennington 1979) recorded *cocori* 'chile'. With Fr. Kino and the mission period, other chile varieties were introduced from southern tribes, and it became necessary to "mark" the original word. Now in Onavas, the pueblo where the Névome dictionary was written, the wild Chiltepín is *vii ko'okol* 'little chile'; with the Akimel and Tohono O'odham, it became *o'olas ko'okol* 'round chiles'. Unlike various other cases, the wild plant was not named *mischiñ* or *ban ko'okol*.

What kinds of chiles were brought northward during this period is unknown. The documents do not specify. Lloyd (1911:58) added just "peppers" to the list of crops the Pima were raising in 1903. But subsequently the Gila Pima have been eclectic in their acquisition of different varieties. Some they have raised themselves, and other kinds that they have named originate from off-reservation sources. Sylvester Matthias gave me seven folk specifics. One, *o'olas ko'okol*, he classified with the wild nonwoody plants (Group G). The remaining six are in the category *e'es* 'planted things'. These are

miitol vippi 'cat's nipples', an important cultivated type that was small, very potent, pointed, and stood up on the plant when ripe. This is also called *a'al ko'okol* 'little chile', although some Pimas use this as an alternate name for the Chiltepín *ge'egeḍ ko'okol* 'large chile', bell peppers

s-cheedagĭ ko'okol, 'green chile', the long Anaheim chile

s-vepegĭ ko'okol 'red chile', the one most frequently used. This is also called *s-ko'ok ko'okol* 'sting ["hot"] chile', *juukam ko'okol* 'Mexican chile', or *jujkam ha-ko'okolga* 'Mexicans their chile'

s-toota ko'okol 'white chile' (pale yellow, bought in grocery stores)

chabaniis ha-ko'okolga 'the Japanese their chile', Jalapeño

I sometimes wonder, particularly with cultivated varieties, whether one person's taxonomy is idiosyncratic or at least local. With current language loss, it is often difficult to verify this. Even people who still speak the language may rely largely on the folk taxonomy of the surrounding dominant culture rather than their own. One afternoon I ran into George Kyyitan in a Sacaton market. He was dropping some Anaheim chiles into his cart. I asked him what he would call them.

"Just *ko'okol,* I guess."

"What about the *cheedagĭ ko'okol?*"

"That's what I would call these," he said, indicating his Anaheims.

I asked about some whitish ones.

"*Oám* ['yellow'] *ko'okol.*"

"Which ones would be *toota ko'okol?*"

"The same thing. Either *oám ko'okol* or *toota ko'okol.*"

"Have you ever heard of something called *chabaniis ha-ko'okolga?*" I asked, thinking that I would stump him on this one. He went pushing through a loose section of assorted chiles in the display case. He finally came to two Jalapeños and said, "That's what these are."

"How about *miitol vippi?* Have you ever heard of that one?"

He pawed again through the mixed chiles but could not find one. There was a packet of narrow dark green ones with pointed tips. "It's sort of like that." I handed him a packet of dried red pointed ones. "Yes, something like that one."

"How about these?" I asked, showing him a little quarter-ounce packet of the precious dried Chiltepines.

"*A'al ko'okol.*"

"What about *o'olas ko'okol?*" I asked, using the name I always had heard.

"Either way, *a'al ko'okol* or *o'olas ko'okol*—it's the same." *Ali* means 'child',

ko'okol   *Capsicum* sp.

a'al 'children', but the word *a'al* is often used by the Gila Pima to indicate 'little', as in this case.

In a matter of several minutes I had independently verified nearly all of Sylvester's chile taxonomy. (The bell peppers were all bought out that day.)

One chile, though, has never entered the Pima folk taxonomy. I enjoy tossed green salads. A Pima lady once watched me put the little green pepperoncinis or "Italian Peppers" into a salad.

"What's the point of those?" she demanded.

"Well, I like the flavor," I replied somewhat defensively.

"But they're not even *hot!* They're . . . , they're like making love with your clothes on."

According to Sylvester, the only chile commonly grown in Pima gardens was *miitol vippi ko'okol,* more frequently called just *miitol vippi,* which he glossed 'kitty's titties'. "That's very common around here." He considered it always *the* domesticated chile. Ruth Giff knew the plant well and confirmed the characters that Sylvester had described: "*Miitol vippi—little* tiny ones—and when they grow they just stand up. Stick up. Very hot *[picante].* I guess that's what you get in Mexico. The Pimas used to grow them. I see them in stores—

the green ones, now." But she had no seeds of it, nor did anyone else she knew of. Apparently it has been lost.

Takashi cooks with a small Japanese chile, *to karashi*. It is small, slender, red, pointed, and quite potent. Sylvester thought it had exactly the right flavor and shape for the old Pima *miitol vippi*. I grew some seeds out the next spring, but when the little green pods came on in the summer, they hung downward in the foliage, to my disappointment. Toward fall, they started to darken a bit. And as they turned red, they turned upward, sticking out of the plant like so many erect little nipples! This was quite likely the same cultivar. Although native to the New World, chiles spread rapidly around the world after 1492.

Not many Gila Pimas raise any kind of *ko'okol* today. When grown from seed, chiles are planted during the *kui i'ivgidag mashad* 'mesquite leafing out moon', approximately March. Castetter and Bell (1942:208) found the Pima raising primarily the Anaheim chile. These were cultivated or hoed twice, first when about they were six inches high, then again at about a foot tall.

> It was sowed in small vegetable gardens and harvested in early November. The ripe pods were pulled off by hand and the green ones allowed to remain if weather conditions indicated they might still ripen. However, if frost were indicated, the plants were cut off at the base and piled in the field, and many of the green pods would ripen there. Green fruits were often roasted, then peeled and eaten. At times these dried roasted fruits were stored in sealed ollas and when utilized were steeped in hot water to soften. Those that had not dried readily were tied into strings and hung outdoors, then stored in ollas when dry. Another way of preparing the unripe fruits was to roast and peel them, then pound in a mortar and shape into a large cake, which was dried for future use. When needed, a piece of the mass was broken off and utilized as seasoning. (Castetter and Bell 1942:208)

At any feast in a Pima village, when you get in line to be served under the *vatto* (shade structure) you will see on the serving table at least one huge kettle holding perhaps 20 gallons of chile stew.

This is the central dish. It consists of hunks of beef that have been cooked with powdered red chiles until the stringy meat is falling apart and the chile has permeated every fiber of it. The Gileños may make trips into northern Sonora to buy it or may purchase it locally from Mexican peddlers.

Ruth explained to me how the red chile was prepared. "Boil it [the whole chiles]. When it's done, my mother would grind it on the *machchuḑ*, then gather it in a dish. Or you could do it in the colander and squeeze it out. Then the skins are in here and the meat *[sic]* gets out, like a *paste*. Or you can buy that in the store already done. Cook with beef with thickening—flour. Mexicans put in spice, garlic, but the Indians don't care for that. There's a different taste in the powder and in the pods. Pods are better."

The chile in Mexico has certain allegorical connotations, particularly among the womenfolk, and we find at least some echoes of that on the Gila. In discussing chile taxonomy, Sylvester happened to recall, "There is an old man—I think he's from Santa Cruz, I'm not sure—there is an old man named Ko'okol Vihakam" (literally, 'chile penis', with the attributive marker *-kam*). "And another man called S-chuk Viha 'black pecker' was also from Santa Cruz," Sylvester added. These are names women gave to men. Rather than being offended, men were quite proud of these special names, which were widely used.

Chenopodium berlandieri Moq. ssp. *nuttalliae* (Safford) Wilson & Heiser
[≠ *C. murale* L.]
huauzontle
Chenopodiaceae
kovĭ

When two cultures collide and one is technologically more advanced and therefore dominant, a number of disasters may befall the subordinate culture, ranging from language loss to the disease of alcoholism. One of these calamities is what cultural agronomists call genetic erosion. This may be a decrease in the

varieties of a particular crop, as has happened with Pima corn. Or it may be the loss of an entire crop. Spanish-introduced wheat, for instance, proved to be so suitable on the Gila and so economically lucrative in trade that it eclipsed all other grain crops.

Certain crops have now disappeared completely. One of these that long puzzled me was *kovĭ*. No one with whom I worked knew it, nor had Castetter and Bell or Curtin picked it up 40 years earlier. I was probably a century too late. In Juan Smith's version of the Pima Creation Story (Hayden 1935:10–12), Se'ehe first made cotton, followed by four food crops: corn, then squash or pumpkins, white tepary *(Phaseolus acutifolius)*, and speckled beans (probably *Phaseolus lunatus*). Songs are sung about these plants during the narration, so the sequence is critical. At this point the narrator paused to comment that four is the important number and that four food crops were made. The Pima were then given three additional crops: black-eyed peas, "some kind of a food that they call *koff*," and tobacco. Surely this *kovĭ* was a crop of some significance, but Juan said he never saw what it looked like. He knew only its name.

Two earlier accounts offered bits and pieces suggesting what this plant might be. One was written by Russell (1908:73), who said: "*kâf*, Chenopodium murale *[sic]*. The seed is gathered early in the summer and prepared by parching and grinding, after which it may be eaten as pinole or combined with other meal." Another account by Hrdlička (1908:263) is more detailed: "The people say that in former times they cultivated a certain plant for its seeds, which they used for food. The name of the plant was *khof* or *kopf* (as pronounced by different individuals). It had big pods [seed heads?], with many small seeds somewhat like those of the saguaro in color and size. This seed was roasted, ground, and eaten like mush."

Piecing these comments together, Gary Paul Nabhan has deduced that *kovĭ* is the domesticated chenopod huauzontle, cultivated in Mesoamerica as well as in the Mississippi watershed (see Wilson 1980, 1981; Wilson and Heiser 1979). He believes this crop fell out of the Piman crop inventory during the drought years of the 1870s that brought the Gila Pima agricultural boom to a halt.

kovĭ   *Chenopodium berlandieri*
Painted from plants grown in San Diego from seed supplied by Native Seed/SEARCH, Tucson.

A few years earlier, Russell may have collected at least some seed for his botanical consultant, Dr. Thornber; this is probably what led to the identification of *kovĭ* as *Chenopodium murale,* a foul-smelling introduced weed still found in Pima fields. (*C. murale* has no Piman name, as far as I have been able to discover.) Thornber had planned to grow out the still unidentified seeds in 1902. "Unfortunately, the season proved too dry for them to germinate" (Russell 1908:18). Probably Russell did not realize that he was dealing with the last remnants of an indigenous cultivated crop. Nonnative peoples entering the Southwest were often slow to realize that certain "weeds" growing in Indian fields were actually crops that had been selected and cultivated for centuries.

Some have been disturbed by Hrdlička's use of the word *pod* in reference to *kovĭ.* Most turn-of-the-century Pima still spoke no English. The Pima word that would have been used to describe the dense seed heads of this plant is *muḍadaj* 'its tassels'. This could easily have been interpreted as 'pod', and Hrdlička did not see the plant.

Charles Miksicek has paid close attention to the domesticated *Amaranthus* and *Chenopodium* seeds from southwestern archaeological sites. He finds them in Hohokam sites primarily from the mid-Sedentary (around A.D. 1000) through the Classic period (see summary in Gasser and Kwiatkowski 1991:441–442). These cultivars, of whatever geographic origin, have clearly been part of the Southwestern inventory of crops for a very long time.

Although several species of wild chenopods flourish naturally in Pima fields and along ditch banks and provide greens that are still eaten (see Group F, *Chenopodium berlandieri*), this domesticated form has a more compact fruiting head, with the seeds ripening simultaneously, and pale-colored fruits that must be stored and replanted each year. The seeds tend to stay on the plant rather than shattering and dispersing themselves as do their wild relatives. In selecting for the pale seed coat, a recessive trait, humans also eliminated the germination inhibitor that permits the seeds to lie dormant in the soil until the proper growing season. As a result, the domesticated form cannot reseed itself (Wilson 1981). But what the Pima were raising was apparently not a white-seeded crop: Russell thought *kovĭ* was the weedy Nettleleaved Goosefoot (*C. murale),* and Hrdlička compared it to Saguaro (*Carnegiea gigantea),* both of which have black seeds.

Cicer arietinum L.
garbanzo, chick-pea
Fabaceae (Leguminosae)
kalvash

One of the crops that Fr. Kino introduced into Pimería Alta was the garbanzo. It is not known for certain when this cultivar reached the northern frontier of Piman country, but I suspect that everyone experimented with the seeds in a very short time. Ezell (1961:35) noted that garbanzos were not recorded specifically as a Gila Pima crop until 1849, when several Anglo-Americans mentioned that the Pima were raising them.

They were still being grown at the turn of the century. For instance, Lloyd, recording crops he found in 1903, mentioned "a pea called *cah-lay-vahs.*" This must refer to the *kalvash,* the present Pima name, which is a loan word from Spanish *garbanzo.* Russell (1908:73) listed the garbanzo as *kaifsa,* saying, "The chick-pea is raised in small quantities and is also purchased from traders." *Ga'ivsa,* however, is the Pima name not for the garbanzo but for a dish made from large grains that have been parched; this may include garbanzos but most often means something made from roasted, crushed, and boiled maize. The lexeme *ga'i* means 'a roasted thing'; the root word is *ga'a* 'to roast' or 'to broil'.

Castetter and Bell (1942:196) gave this account of *kalvash* cultivation:

The chick pea is well adapted to arid and semi-arid regions and matures in about ninety days. Both Pima and Papago made a single planting, sowing

kalvash   *Cicer arietinum*

them as a winter crop from November first to February and harvesting in April or May, consequently they required little moisture. It has been reported that garbanzos will endure a temperature of 13°F without being injured. People began to harvest chick peas when mesquite beans were about ripe, and picked them progressively. The first pods to ripen were simply rubbed by hand. When the plants began to dry up, tops were cut off with a sickle or pulled up, piled and dried for a few days, then threshed by beating with a stick or trampling. They were then winnowed, dried further and stored in a sealed large gourd or olla. On good soil a yield of one thousand pounds of chick peas per acre was not unusual.

Chick-peas, they said, were planted three inches deep in rows of 1.5 to 2 feet apart.

Ruth Giff told me, "They used to grow *kalvash*—and nobody knows about it now. We soak it—for half a day. Rub with hands." She showed me how to prepare them by boiling, then pouring off the water with the loose skins, then continuing the boiling. "Cook *kalvash* with meat [beef] called *baashonaj*. It's that fat from the belly," she said.

Francis Vavages said, "*Kalvash* taste more like peas, only they got tails. *Kalvash* is a little harder to eat, a little tougher, than *u'us bavĭ* [cowpeas, *Vigna unguiculata*]."

Sylvester Matthias said, "*Kalvash* [is] planted in fall [or] about January. My father makes pretty good money on that. That time there's refugees here from Mexico. Oh! They go for that *kalvash*. Sell them by 50-pound flour sack." He added, "Pimas pound it, like split peas." Even though Sylvester uses the present tense, his father has been dead for decades; the growing of *kalvash* was during the Mexican revolutionary years from 1910 to 1917.

Except for a few rows in one Santan kitchen garden, I have never found any Pima growing the crop from the early 1960s onward.

Vouchers: 1538 (Albina Antone, Clement Antone).

Citrullus lanatus (Thunb.) Matsum. & Nakai
[= *C. vulgaris* Schrader]
watermelon
Cucurbitaceae
do'ig hugĭ, miiliñ, miiloñ

In 1698 Fr. Kino looked over the crops being raised at the O'odham village of Botum, probably near present-day Cocklebur on the Tohono O'odham Reservation. Among these he was surprised to find *sandías,* or watermelons, an Old World crop. That night he noted in his diary that these and several other introduced crops had arrived here ahead of him (in Bolton 1936). A year earlier, Manje had found watermelons already being grown at the Sobaipuri Pima village of San Agustín de Oiaur (now Tucson) on the Santa Cruz River. Two years later, Kino found the Quechan (Yuma) growing them near the mouth of the Gila. In parts of the Southwest, watermelons were 100 years ahead of the Jesuits. Watermelons proved such an immediate success among North American Indians—the seeds spreading so rapidly—that in several other places as well, the crop was being grown already by the time European explorers arrived (Blake 1981). Its presence misled some early ethnobotanists into thinking that the watermelon had originated here.

There is another perception, though: the native one. Agricultural specialists Castetter and Bell (1942:118) remarked, "The culture of watermelons long since has become so thoroughly established among the Pimans that all our informants regarded them as aboriginal." This is a curious statement, because all the people with whom I worked regarded this as a crop originating with the Spaniards at some point.

The one version of the Creation Story that is most explicit about the origin of crops, Juan Smith's (Hayden 1935), also exhibits the most obvious influence of the surrounding dominant cultures. This version acknowledges two beloved Hispanic introductions: "Se'ehe made the watermelons and the muskmelons (*Cucumis melo*)" (p. 16) after giving the Pima the four primary food crops (see *Chenopodium berlandieri*).

The generic name for watermelons is *do'ig hugĭ*. "That's watermelon in general," explained Sylvester Matthias. "It means 'raw-eat', something that was eaten raw." In contrast to all the other cucurbits the People were raising, this one and muskmelon could be eaten just as they came from the vine.

Several different varieties must have been among these early introductions. Shortly after I arrived in Pima country, I remember going down to Walter Rhodes's place in Co-op Colony. In the fall he had for sale piles of round melons with yellow flesh interspersed with a few large black seeds. This apparently was a form of *s-oám eḍakam* 'yellow inside it has'. Any watermelon with black seeds could be called *s-chuchk kaikam*. Sylvester Matthias said there was also a kind of watermelon with yellowish seeds. "People used to raise a lot." Sylvester told me of additional varieties the Pima cultivated, listed below with the names he assigned them.

miiliñ or *miiloñ:* similar to Klondike. This is the oldest variety among the Pima. The fruit varies from small to large and is oblong, with a green rind and red flesh. The seeds are large and black.

s-oám eḍakam: "Pine melon," according to Sylvester. This is another old variety. The fruit is medium or small and oblong, with a whitish or whitish green rind and yellow flesh. The seeds are large and pink.

o'olas miiliñ or *ko'okol miiloñ:* Chilean or "chillian." George Kyyitan called it *ko'okol miiloñ* (*ko'okol* is the Chiltepín, *Capsicum annuum*). This round, medium-sized watermelon has a greenish rind with stripes. The flesh of this watermelon is red; the seeds large and black. According to Sylvester, this variety "should come in the fall when it cools off. During the summer season here it burns." Its Pima name is a play on words.

vii kaikam: Klondike. This is the most recent of the varieties. The fruit varies from small to large and is oblong. The rind is green, and the flesh is red. The seeds are small and black.

Although watermelons apparently originated in Africa's arid regions, there may have been further selection for the extreme xeric conditions once the crop reached Pimería. Castetter and Bell

(1942:119) observed, "Today their cultivation is general throughout our area, and even the Papago are able to grow them under flood-water farming on the desert since they have varieties which are rather drought resistant."

King and associates (1938:55–56) at the Sacaton Experiment Station wrote, "The Pima Indians are heavy consumers of watermelons and are being encouraged . . . to plant at least an adequate acreage for their own use. Varieties that have produced well at the field station are Klondike, Striped Klondike, Angelino, Kleckley Sweet, Chilian, Irish Gray, Tom Watson, Stone Mountain, and Georgia Rattlesnake." Some of Sylvester's named varieties may have been obtained from this station.

"Watermelons quickly gained popularity here, taking second place, among the introduced novelties, only to the winter crops, wheat and barley," said Castetter and Bell (1942:75). "Watermelons are among the most important crops of the Pimas and are eaten for at least six months of the year," wrote Russell (1908:75) at the turn of the century. "When we planted them," said Ruth Giff, "we used to eat them from morning till night."

But why was the new crop so successful? Blake (1981) concludes, "Reasons for rapid acceptance of the watermelon by the Indians over a wide area, often with different cultures, appears to lie in the fact that methods of successful cultivation are similar to those of the native squashes (*Cucurbita* spp.) with which they were familiar. Also, the long, hot continental summers over much of the United States favored its growth."

The Mormon Battalion in 1846 passed through the Pima villages on Christmas Day. They were treated to watermelons the Pima had carefully stashed away. This was something to write home about, and several of them did (Bigler 1962; Standage in Hayden 1965:23). The secret to midwinter watermelons lay in rather sophisticated storage techniques that the Pima developed, probably an outgrowth of whole squash and pumpkin preservation. Castetter and Bell (1942:206–207) described the watermelon storage pit in detail: "Poles were placed on the bottom of the pit and covered with cornstalks, on which were placed alternate layers of wheat straw or bean vines, and watermelons. Sides of the pit were not lined. For a roof, poles were laid across the

opening and a layer of arrow-weed covered with dirt over all. Earth at one end of the pit was dug away to allow a vertical entrance about three feet high, which was covered with a framework of slender poles. Watermelons thus stored kept well for as much as three to four months." They added, "The Pimans stored a goodly supply of watermelons for winter use, but did not bury them in sand as did the lower Colorado River tribes, although informants had heard of the Yuma and Maricopa doing so." This information evidently came from the upper end of the reservation where these two ethnobiologists worked. At the west end, where the river ran a half-century longer, the Pima did indeed bury watermelons. But not in the sand. Sylvester recalled this storage practice

"Where did they bury them?"

"In the field, just where the vine is."

"How deep?"

"Just below the surface," he said, indicating about eight inches or so. They had to be deep enough that cattle walking over would not damage them or animals dig them out. "Sometimes they forget where they bury it. And they hit it when plowing. So they have watermelon at planting time. Of course, when they bury it, the vine is still there till the frost gets it. But still I couldn't understand: a melon is buried but it didn't rot." Watermelons were also kept in a haystack, covered with straw. He added, "Best way is in *kosin* ['storage shed'] with straw in pit, so they can get air." This is the method Castetter and Bell had described.

Even in the central part of the reservation, there is remembrance of watermelon-keeping in winter. At Bapchule I asked George Kyyitan, "How do you keep watermelons, George?"

"In straw, in [the] *kosin;* cover them up. Even if it's not ripe yet, they have to stay in there till ripe."

"How long would they last?"

"Year-round. Year-round. Meet the season again when they plant. I've seen that. They used to plant melons a lot, then when it froze, get frost on them, pick 'em up, all of it. Then before it's too frosty, they put 'em in there, put the straw in there and cover 'em in there. Whenever they wanted, they'd take one out, eat 'em. Then they do that."

"Did you people ever bury them in the sand, in the ground?"

"I did once, over here. Big one! Boy, it's sure sweet."

"How long did it last?"

"Not very long. I just keep it there for a while." Then George added, "Won't last because I like to eat it!"

Castetter and Bell (1942:206) also added this seemingly casual anecdote: "Pima informants said when the first planting of watermelons had matured the fruit was picked, then the vines irrigated and in about six weeks there was a second, although smaller, crop." Joseph Giff told me about second crops, which are formally named in Pima: "After the first [summer] harvest of [Velvet] Mesquite (*Prosopis velutina*) there's a second harvest, ready in October, called *toomdag.* Very sweet! You also say *toomdag* for second crop of figs— and you can get a second *toomdag* of watermelons—cut off vines and new ones come, called *i'ipoñigaj* 'sprouts'. You can get a second harvest of tomatoes in fall, but it doesn't have a name."

Ruth said, "We used to squeeze the juice of watermelon and boil it down and use the juice—*miiliñ sitol.* Doesn't last—about a week. Have to use it right away. *Haashañ sitol* ['Saguaro syrup'] keeps for years."

Castetter and Bell described the cultivation of this crop. Watermelons are planted when the mesquites leaf out (in late March and early April). These are ready for harvesting in late June. If there is a second planting, it is after the Saguaro harvest (in the latter part of July). These watermelons would be ready during October. Seeds are planted three inches deep and germinate in four days. Aphids are a pest, and gophers are said to favor the roots over those of any other plant.

The Gila Pima do not use the word *gepĭ,* the Tohono O'odham generic term for watermelon. Several times while we were talking about crop plants, Ruth Giff asked me if I knew what *gepĭ* meant. She had heard the word and wondered if it was some kind of pumpkin. George Kyyitan mentioned it one day when we were discussing crops: "There was a *gepĭ.* I don't know [it]. My father mentioned it. I think it might be a melon; *uuv haal,* they [the Tohono O'odham] would say: 'smelly squash'. They mean *keli baasho, s-heheviñ miiloñ, oámakam* [varieties of muskmelon]."

Vouchers: *862* (RG).

Cf. *Citrullus lanatus* (Thunb.)
   Matsum. & Nakai var. *citroides*
   (L. H. Bailey) Mansf.
Cucurbitaceae
citron, preserving melon
s-kauk miiloñ, s-kauk miiliñ

I had not asked Herbert Narcia of
Blackwater (originally from Komatke at
the west end) many things about plants.
While I was recording some songs from
him in 1992, I brought up the subject of
the Coyote's plants *baaban ha-miiliñga*
(feral *Cucumis melo*) and *ban aḏavĭ*, an
unidentified cucurbit. He thought some
while before answering.

   "Well, no. I don't think I ever heard of
either of those. All's I know is this *s-kauk
miiliñ*. Like white, striped melon, real
round. They're hard! Can get big, too."
He started to laugh. "Over here someone,
long time ago, had a *big* watermelon
patch. And he planted those *s-kauk miiliñ*
all along the outside of his field. So if
anyone goes in there and steals his watermelons, he's going to get that one. You
can't eat that one."

   "Was that O'odham planting them?" I
asked.

   "No! Miligáán ['Anglos']."

   A few days later, as the matachini
dancers processed out of church for the
feast of St. Francis of Assisi, I noticed
Frank Reed sitting in a wheel chair under
a tamarisk. He was an old acquaintance,
a Pima from Santan I had not seen in
many years. I was surprised to see him
here on the Ak Chiñ Reservation. I
stopped by and we began to talk. I asked
Frank about several of the Coyote's plants
I had recently discussed with Herbert, but
Frank did not know them either. "How
about *s-kauk miiloñ*?" I asked.

   "Yeah, I know that one. It was around
here. Last time I saw them was about 1920."

   "Can you eat them?"

   "No! You'll break your teeth."

   "Are they colored like a watermelon
inside?"

   "No, kind of pinkish, pale." Then he
spread his hands, indicating about the size
of a large cantaloupe. "I think they're
striped."

   "Did people plant them?"

   "They're just wild."

   About a month later when I saw
Sylvester Matthias again, I mentioned this

plant to him. He laughed. "*Kauk miiliñ* is
a foreigner. If you drop it, it'll bounce
around just like a basketball. Some use it
to make pie out of it. That's about all you
can do with it. You can't eat it—can't chew
it. But the hogs can eat it. They would
bring them out to St. John's for the hogs."

   "What color are they?"

   "Either like a regular watermelon or
like the Pine melon—yellow inside, but
the seeds is always black."

   "How about outside?"

   "Could be any color—like regular or
striped. It's hard! It looks like you can
chew on it, but you can't. I think it's native
to the Caribbeans."

   "How big does it get?"

   "The largest around basketball size.
They come in early August through
September."

   "Is it wild?"

   "No."

   "Is it an *e'es*?"

   "Yeah. It grows in the cotton fields
around Laveen. You could find it around
Cheetham's silo. It just comes up around
there, around Laveen."

   "Then maybe it's *hejel vuushñim* [a
volunteer]?"

   "I think so. It just grows wild. I don't
know who brought the seeds. Around
Laveen."

   "Were *kauk miiliñ* around when you
were a kid?"

   Sylvester knitted his brow and thought
for a long time before answering. "It was
after World War I when I saw them. In
the fields around Laveen. I don't know
how it got here."

   As nearly as I can tell from these
descriptions and from what I can remember myself of the insipid round watermelons I found decades ago in old fields and
waste places between mesquite thickets
around Komatke, *s-kauk miiliñ* appears
to be the citron, a variety of watermelon
used for pickling. This should not be
confused with the citrus fruit that is also
called a citron. Its classification status
remains anomalous. It is not grouped with
the covert category of wild annuals,
because it is clearly perceived as a plant
of human agency, normally found in
fields. However, in many cases mentioned,
it was not deliberately seeded. Rather, it
was a volunteer, something that came up
by itself, *hejel vuushñim*. Also, in more
than a dozen years of my fieldwork, *s-kauk
miiliñ* was never once mentioned as a *kind*

of watermelon, even though most elderly
people were familiar with it. Instead,
the name came up during the discussion
of Coyote's plants, a category of intrinsically aberrant organisms, encompassing
incongruent characteristics. But it is not
really one of them because its arrival time
and agency are too well known.

Cucumis melo L.
muskmelon
Cucurbitaceae
s-uuv miiliñ, miiloñ

There is no individual plant that the
Gileños call *s-uuv miiliñ* (or *miiloñ*). This
is a category of plants, the muskmelons,
each cultivated variety of which has its
own Piman name. There are at least six
such cultivars recognized linguistically.
All of these *s-uuv miiliñ* are distinguished
from another category of plants called
do'ig hugĭ 'eaten raw' (watermelons), each
variety of which is named as well. When
you go out to a Pima field, if you are lucky
enough to find one, you must ask for
specific kinds of muskmelons, such as
oámakam or *keli baasho*. I say "lucky"
because there are not many Pimas left
who grow traditional crops, and these
centuries-old, desert-selected races are
among the most sweet, aromatic, and
flavorful melons.

   The Pima people I worked with considered muskmelons Hispanic introductions,
as indeed they are. (The crop probably has
an African origin; see Whitaker and Davis
1962:4–5.) Among the six formally named
varieties, they recognize older and newer
varieties. The older cultivars may well date
from initial Spanish contact. Muskmelons
may have dispersed even more rapidly than
the Europeans themselves, as was the case
with watermelons (see *Citrullus lanatus*).
The only version of the Pima Creation
Story to mention these two nonnative
crops is a late one recorded by Julian
Hayden from Juan Smith in 1935. Here an
addendum ascribes the origin of both
species of melons to Se'ehe, who wished to
pacify the People. This late version of the
Origin Myth is notable for its heavy
Judeo-Christian interjections.

   According to Sylvester Matthias and
George Kyyitan, the generic term for all

the muskmelons among the Tohono O'odham is *uuv haal* 'smelly pumpkin'. It is an interesting comparison. The People already had the pumpkin and squash. The newly arrived muskmelons are in the same family but when ripe are sweet smelling. The Gila Pima generic name, *s-uuv miiliñ* 'smelly melon', also emphasizes this aromatic characteristic.

Six varieties of *s-uuv miiliñ* are known to the Gileños:

oámakam 'yellow' + attributive marker: crenshaw. This melon is oblong, with a smooth yellow rind and orange or white flesh. It is the oldest of the Pima varieties.

keli baasho 'old man's chest': Pima casaba. This onion-shaped melon has a green to greenish yellow wrinkled rind and greenish white flesh. It is another old Pima variety.

s-heheviñ miiliñ 'chapped melon' (from *heevĭ* 'chapped'): cantaloupe. This melon is round, with a rough tan rind and orange flesh. It is another old Pima variety.

kuushp hihikam 'sticking intestines' + attributive marker: unknown variety, but something like casaba. This melon is onion-shaped, with a greenish yellow, wrinkled rind and greenish white flesh. It is another old Pima variety.

chuchkuḍ ha-miiliñga 'ghosts' melon' or 'owls' melon': "banana melon." This spherical or oblong melon has a smooth rind speckled greenish and yellow; the flesh is whitish. It may be an old variety among the Pima, perhaps from the Tohono O'odham.

toota miiliñ 'white melon': honeydew. This round melon has a whitish smooth rind and whitish green flesh. It is the newest variety among the Pima.

S-heheviñ miiliñ is the same as the commercial cantaloupe. Indeed, seeds are often saved from store-bought cantaloupes. The same may be true of *toota miiliñ*, or honeydew. But others in the Pima repertoire are not likely to be found in stores: for instance, *chuchkuḍ ha-miiliñga*. Ruth Giff said it was a little melon, grapefruit sized: "I don't know if it [the name] means 'dead people' or the 'owl'." Sylvester Matthias called it banana melon in English. "It's muskmelon type—real

keli baasho *Cucumis melo*

tender. Can just peel it [the rind] right off. Ghosts' melon or owls' melon . . . I don't know which." *Chukud* means either 'ghost' or 'Great Horned Owl'. "A *chukuḍ miiliñ* is the size of an *aḍavĭ*, wild gourd [Finger-leaved Gourd, *Cucurbita digitata*]. They used to grow here. Papago raised it here. They turn yellow like *oámakam*; inside is white." The melon is gone now. Perhaps it is the unusual melon that Castetter and Bell (1942:120) once found among the Tohono O'odham.

And there was the *kuushp hihikam*. According to Sylvester, "*Kuushp hihikam* seeds stick to the inside. Hardly any seeds inside; [it's] mostly the part you eat. *Kuushp* means 'something kind of halfway burned'. The seed part we call *hihi* ['intestines']."

"What color were they?"

"Either plain white or kind of yellow."

"What about the skin?"

"Grayish yellow or grayish green—two colors. And they're real sweet. But it's the smell—if you smell, it's ripe."

"What about the size?"

"They are *big*. Larger than the average size of a *keli baasho*. You don't see it no more. Just like *chuchkuḍ ha-miiliñga*, it's out of the line now. I guess nobody has the seeds. I remember the last one. The old man Giff had a whole patch of that. People come and buy it. Somewhere along 1919 or [19]18. I was a little boy then."

Ruth said she heard of *kuushp hihikam* but did not remember what it looked like.

"What does the name mean?"

"It means that those seeds are not loose. They're sticking on—*hihi* means the insides."

A muskmelon I had long heard about was *keli baasho* 'old man's chest' (the name alludes to the deep wrinkles on the skin of this melon, similar to that of the casaba now found in grocery stores). Joe Giff used to raise it, but he was gone, and apparently so were the seeds. Ruth, his wife, had started gardening an acre or two but was unable to find seed. "*Keli baasho, keli bapsho* [pl.], is a dark green melon with very thick flesh, very sweet," she said. "It was lost about ten years ago. It had been lost before also, but the *kaij* [its seeds] were found when some old man died."

Gary Nabhan began searching among the Tohono O'odham for it. They call it *oks tooton* (pl. *o'okĭ tooton*) 'old woman's knees'. After five or six years, he said, "I think you're going to have to give up on *keli baasho*. I think it's all long gone. People know it, but haven't raised it in ages."

I asked Frank Jim from the western part of the Tohono O'odham Reservation if he had seen any. He looked serious for a few moments then answered, "Oh, you mean them melons they call *oks baasho?*" he said, pulling my leg. We both laughed. "No, I haven't seen any in a long time."

One day at Santa Cruz village, I chanced to ask Carmelita Raphael about it. "Just a minute. Maybe I have some," she said. She went inside and in a few minutes came out with a metal toolbox filled with little bottles of watermelon and muskmelon seeds, with cryptic pencil notes marked on bits of paper. "Here, I think this one or this one may be it." Carmelita told me the *keli baasho* should stay green on the outside when ripe; casabas turn yellow.

I distributed seeds for various people to grow, but none produced quite the right fruit. The next season, Ruth or Carmelita found more seed from an elderly woman in Santa Cruz and grew it out. This time Ruth was satisfied that these were the real *keli baasho*.

Oámakam (considered the oldest variety known among the Pima), unlike *keli baasho*, indeed did seem to be long lost. Sylvester knew this melon well. He said, "It has smooth skin, yellow skin. Green when not ripe. It smells [good]. White [flesh]. Can eat it even without teeth! Sweet, but bitter when [still] green. Call it *sivchu* when green. Then they call it *oámakam* when ripe." In a grocery store, he compared it to a crenshaw melon.

Ruth Giff said that *oámakam* was the only one they dried. "They turn yellow when they're ripe. And that's what they dry. When it's real ripe—peel it and dry it. Then I guess they just eat it like that. It's still soft, I guess."

Sylvester warned of a condition that can result from muskmelon consumption. "If eating *oámakam* in summer during [the] heat of day, [you] must use salt on it or it will cause this illness—called *hi'ivshkushdag*—the bladder gets sore. Remedy for it is to drink ice water or drink something cold or dive into cold water. When [it's] cold (night or early in the morning), won't do it; but during the hot day. *Keli baasho* without salt [is] no problem—never does it. I don't think cantaloupe would bother either."

Muskmelons made it quite early into Pimería Alta, perhaps as early as the *sandías,* or watermelons, that Kino and Manje noticed had preceded them into the fertile river bottoms of southern Arizona. For instance, Manje wrote in 1697 that the Sobaipuri Pima at San Agustín de Oiaur (now Tucson) were harvesting both *melones y sandías* (Mange 1926:256).

Centuries after the initial introductions, new varieties were introduced at the Sacaton Field Station. King, Beckett, and Parker (1938:55) said, "Muskmelon varieties that have proved well adapted for home consumption are Hales Best, Perfected Perfecto, Rocky Ford, Eden Gem, Tip Top, and Honey Ball. The Honey Dew, Persian, and Golden Beauty Casaba melons are good producers especially on fairly fertile soils."

The importance of the muskmelon through time is difficult to evaluate. At the turn of the century, Russell (1908:77) called it *siechu* [*s-i'ovĭchu* 'sweet'?], saying, "The muskmelon is extensively raised by the Pimas." Forty years later, Castetter and Bell (1942:119) wrote, "The Pimans have never been fond of muskmelons, and this, coupled with unfavorable keeping qualities, is responsible for their lack of popularity. They are little grown by either the Pima or Papago today." And here we are, another forty years later. I would say that the muskmelons are as well liked as the watermelon and that their local races, at least those that survive, are esteemed.

And why not? In about 15 minutes today, one can rush over smooth roads from any Pima village to buy a cool watermelon inexpensively. But tasty *oámakam* or *keli baasho* are not available anywhere except from a few garden plots of traditional Gileños.

Technical Notes: Cucurbit specialist Laura Merrick suggests the following identifications of Pima cultivars, based on the classification of Whitaker and Davis (1962:42): *C. melo* var. *inodorus* Naud. = *oámakam, keli baasho, kuushp hihikam,* and *toota miiliñ; C. melo* var. *reticulatus* Naud. = *s-heheviñ miiliñ; C. melo* var. *chito* Naud. or possibly var. *cantaloupensis* Naud. = *chuchkuḍ ha-miiliñga.*

Vouchers: *862* (RG).

Cucurbita spp.
squash, pumpkin
Cucurbitaceae
haal

The four major crops of the Pima have traditionally been given as corn, cotton, beans, and squash or pumpkins. Some Europeans may have overlooked the distinctions among the three species of beans—*bavĭ, muuñ,* and *havul*—but they seldom missed the cucurbits. In fact, many of the Anglo-Americans on their way to the California goldfields commented specifically in their diaries and letters on the hoards of pumpkins that these desert riverine farmers were able to break out from their storehouses in midwinter.

One of those so impressed was Henry Standage. Traveling down the "Ela" with the Mormon Battalion in 1846, he entered in his journal for 23 December: "Saw a loom and some Indians weaving blankets of cotton; see some spinning very curiously. Bought a [mesquite?] cake for a button. Called at an Indian hut, where they gave us some stewed pumpkin."

By 1871 Indian Agent F. E. Grossmann (1873:419) said of the Gila Pima, "Wheat, corn, beans, and above all, pumpkins and mesquite-beans are their principal food."

In the Creation Story (see Lloyd 1911:217–230), the story of Corn and Tobacco, Corn Man and Pumpkin Man are always associated. Corn is the protagonist, and Pumpkin, his silent partner. In the conflict with Tobacco Woman, both men retired to the east, leaving the Gileños for many years without these important crops. When at last Corn Man returned to Tobacco Woman and they spent the night in a special house covered with mats instead of bushes, the skies rained down both corn and pumpkins throughout the night.

So Corn lived there with his wife, and after a while Tobacco had a baby, and it was a little crooked-necked pumpkin, such as the Pimas call a dog-pumpkin.

And when the child had grown a little, one day its father and mother went out to work in the garden, and they put the little pumpkin baby behind a mat leaning against the wall. And some children, coming in, found it there, and began to play with it for a doll, carrying it on their backs as they do their dolls. And finally they dropped it and broke its neck.

And when Corn came back and found his baby was broken he was angry, and left his wife, and went east again, and staid there awhile, and then bethought him of his pets, the blackbirds, which he had left behind, and came back to his wife again.

But after awhile he again went east, taking his pets with him, scattering grains of corn so that the blackbirds would follow him.

And the Dog-Pumpkin Baby lay there broken, after corn went away, but after awhile sank down and went to Gahkotekih, and grew up there, and became the Harsan [haashañ] or Giant Cactus [Saguaro, *Carnegiea gigantea*].

I have found no one today who recognizes the English gloss 'dog-pumpkin' given by Lloyd (1911:225).

Although no trace of field ceremonies survives today among the Gileños, even as late as the late 1930s Castetter and Bell (1942) were reminded that "ritual was just as important as actual cultivation in producing a crop, and no crop could be expected without it." Song must accompany the growing of the aboriginally all-important crop, maize, but these invocations must be sung for other crops as well, especially teparies, pumpkins, and tobacco. Two such songs, Rain Songs, were recorded by Russell (1908:331–334). They are long and beautiful poems. In one of these songs, "Huuñĭ Ñe'i," Corn Man looks down from his home atop Taatkam Mountain south of Picacho and sees not only the corn standing but also squash.

The other song is to be sung in the fields during the night. The singers sing before the sacred bundles or *omina,* the prayer sticks they have planted in the fields as votive offerings. "And the white light of dawn yet finds us singing, while the squash leaves are waving." Each leaf of the Pima *haal* bears on it the white marks of clouds. At first just the humming of the *hivkuḍ,* the Greasewood (*Sarcobatus vermiculatus)* rasping stick, is heard reverberating on the overturned basket in the middle of the circle of petitioning singers. But the vibrations of the basket drums become the rumbling of the heavens, with a flash of lightning seen along the edges of the world where sky touches earth. Dancing stops. Only the deep voices of the men continue to the end of the song: "Earth is rumbling / Everywhere raining."

Pumpkins and squashes, particularly the winter squashes, can come in a bewildering variety of sizes and colors and shapes. Ruth Giff, who has raised squash and pumpkins throughout her life, supplied this classification of the Pima species:

haal ("just" *haal*): has spongy seeds like *daapk haal* but does not last as long.

daapk haal (sometimes called *O'odham haal*): has spongy seeds without dark margins.

sha'ashkaḍk haal: has seeds different from those of *daapk haal*—not spongy and with wavy dark margins. The skin of *sha'ashkaḍk haal* is light orange and has grooves; this variety will outlast the others.

Although his notes are difficult to unravel, the three folk forms recorded by Russell (1908:71, 77, 91)—*halt, tapkalt,* and *rsas'katuk*—are the same as those given by Ruth and others.

George Kyyitan and Sylvester Matthias called these three together *kiikam haal* (*kii* 'house' + *-kam,* attributive), which implies that these are the People's varieties, the indigenous forms. These are contrasted to another set considered nonnative or foreign (no Pima word marks this set). Two named forms are included in this unnamed set: *s-oám haal* 'yellow squash', known as "goose-neck" or crook-neck squash; and *s-hikijuls haal* 'scalloped squash', scallop squash. Both of these are forms of *C. pepo,* and both are summer squashes grown by the Pima and cooked while still immature and tender. A third variety of *C. pepo* sometimes grown is the Zucchini, which has no Pima name. These three have a compact, erect or bushy growth, while the three *kiikam haal* are trailing-vine types that require a great deal of space. Carmelita Raphael grew pepos in a small fenced garden near her house, along with onions, chiles, tomatoes, melons, and watermelons. But the *kiikam haal* were all interplanted in an acre rectangle with corn.

Russell (1908:91) noted, "According to tradition the first pumpkins, called *rsas'katuk [sha'ashkaḍk haal],* were obtained from the Yumas and Maricopas." One day in discussing the classification of *haal,* Sylvester said, "*Sha'ashkaḍk haal* is mentioned in the legends, so [it] must be the earliest. They're natives. Corn Man's baby was a *sha'ashkaḍk*." He explained the meanings of the names: "*Sha'ashkaḍk haal* means it's not in good form, just always kind of rough shape, shaped that way, not round. . . . *Daapk* means just smooth, instead of those grooves."

The English use of common names of cultivated cucurbits is almost hopelessly muddled. One man's winter squash is another man's pumpkin—completely interchangeable (Whitaker and Davis 1962:50). Ezell's attempt to make sense of this from documentary sources is futile (1961:34). So too with the names *calabash* and *gourd.* (We use here *bottle gourd* for *Lagenaria* to avoid confusion.) Summer squash in U.S. markets are *C. pepo,* although not all *C. pepo* are summer squash. In northwestern Mexico and the southwestern United States, *C. argyrosperma* is also used as a summer squash. A few of the English names are more specific, though: most cushaws are *C. argyrosperma,* the "regular" *haal* of Pima, and cheese pumpkins are *C. moschata.* We live in an imperfect world.

The Pima world was imperfect as well, but in a different way. One August morning, Ruth and I visited the acre or so field she was farming down at her old house. We sat under a big mesquite growing along the fence line, while she shelled young *O'odham havul* (Pima limas). I was going over the specimens I had collected in her field, mostly cultivars. When I got to the squash flowers, she said, "Every time we cook a *haal* we have to taste it. Some are very bitter. Doesn't look any different from the good *haal*. [Have to] taste each squash when you cut it up. Just like that *adavĭ* [Finger-leaved Gourd, *Cucurbita digitata*]—very *sivk* [bitter]."

"Why?"

"We have no answer for that. One whole plant would be like that. We find out, we pull it out. Some man didn't taste it first and ruined everything in the pot. We haven't had it in a long time now."

"Does the *adavĭ* grow near the fields?"

"No, it doesn't. Long way away from where we farm."

A few months later, the subject came up with Sylvester. Here is how he explained it: "Bitterness goes *into* the *haal*. Used to. Useless! Plants come from the same seed. *Siv haal*—you have to be careful when you pick them up for cooking. You have to taste . . . well. It's few in a patch that would turn bitter. I don't know why. As soon as they spot the *siv haal*, they can whack it off with the hoe. No use. Taste it right there in the field."

"That doesn't happen any more?"

"That seldom happens [now]."

"Would *sha'ashkaḍk haal* get bitter, too?"

"No, never, just the *haal*."

I checked this again with Ruth, and she said the same: "Only the *haal* gets bitter."

Later in the day I went down to Santa Cruz village and asked Carmelita Raphael if she had heard of this, but she said she never did! This was surprising because she raises a considerable amount of squash.

But *siv haal* was encoded even into the Creation Story. In the version Julian Hayden (1935:16) recorded from Juan Smith, when the pumpkin baby born of Corn Man was dropped and broken, Se'ehe became upset: "Se'ehe at that time made this bitter pumpkin, and he told them that that would be a sign that at that time that thing happened, that this corn's baby died."

In the Onavas region of southern Sonora, Gary Nabhan (1985) and associates have found the local wild gourd

(*C. argyrosperma* ssp. *argyrosperma* var. *palmeri*) back-crossing with the domesticated *haal*. Thickened peduncles, larger size, and a tendency toward pear shape are indicators of gene flow from the fields of *C. argyrosperma* into the surrounding wild population. The most immediate indication of gene flow in the opposite direction is a squash with a terribly bitter taste. There, Pima farmers must take precautions to prevent this from happening. But on the Gila, Gary tells me that the local wild Finger-leaved Gourd is not closely related to the cultivated *haal*, so back-crossing should not be the explanation for the occasional *siv haal* in Gileño fields. Perhaps, then, it is only a coincidence that the *adavĭ* or bitter wild gourd has become uncommon and the *siv haal* is now seldom encountered in fields.

The various species of pumpkins have had a long pre-Columbian history in the Southwest. Gary Nabhan has looked at archaeological squash remains from southern Arizona, mostly from Hohokam sites. He tells me that *Cucurbita pepo* appears earliest in the record, followed by *C. moschata* and *C. argyrosperma* together, though it is not always possible to distinguish their seeds due to loss of margins. There is no *C. maxima* (Hubbard or turban squash) until historic times. *C. argyrosperma* has greater drought resistance and appears to have been in greater production. The historic Piman *C. pepo*, similar to an acorn squash, was already disappearing by the time of Castetter and Bell's study in the 1930s. Gary feels that the decline in pepo cultivation was related to the introduction of watermelon.

Most of the time on the reservation we have eaten *haal* as a summer squash. It is a versatile vegetable. The young ones, boiled in water alone or with something else, are called *haal maamaḍ* 'squash babies' (*maḍ* 'offspring', 'child'). The immature fruits may be steam-fried with onions, tomatoes, and meat or cheese. But thoroughly ripened cushaws will keep in winter, even though they are not as durable as *daapk haal*. *Sha'ashkaḍk haal* stores the longest.

Late one January, when I stopped to visit Ruth, she was working on *haal* in her crowded little kitchen. She said she prefers the *haal maamaḍ* to the mature *haal*. Therefore she likes the light-colored ones, rather than the dark, more or less solid green ones, because they are easier to find among the leaves when young. She was

washing the *haal* seeds from two dark green ones, then drying them on her mesquite-flour-sifting basket. Ruth explained that usually she does not wash *haal* seeds; she washes seeds to save if they are messy (for instance, melon seeds, but not watermelon).

Castetter and Bell (1942:171) were given the following information on cucurbit cultivation, which is the same as Pima farmers in the mid 1980s told me: "Pumpkins were usually irrigated twice before setting fruit, after which plants were never watered heavily, as certain fungus would attack them, causing stem rot and fruit decay. . . . Care also had to be taken with young pumpkins. When the fruit was about the size of a baseball each one was raised off the ground and supported by a piece of brush, stones, etc., before irrigating, to prevent rotting from too much moisture."

The average Pima family had a crop of 60 to 100 pumpkins each year (Castetter and Bell 1942:188–189). These were piled up outdoors and covered with cornstalks. Or they were piled in a specially built storehouse of Arrow-weed (*Pluchea sericea*), willow, and cornstalks, where they kept through late winter or spring. Only the regular *haal* was sown in the second planting after the July Saguaro harvest. However Sylvester thinks *any* type could be planted at this time; it is always a gamble with killing frosts.

There were other ways of keeping pumpkins in addition to putting them whole in the outdoor storehouses. Pumpkins or winter squash were routinely dried. There were two methods, Ruth told me. *Simna*, produced by the first method, is chipped with a knife and dried. She thought *sha'ashkaḍk haal* was the best kind for *simun*. The other method, called *hikchulda*, is to cut them into strips or strings and to hang them up to dry on a pole or wire in the sun. Dried pumpkin is then stored in an olla or sealed glass jar. Ruth once told me, "*Simna* keeps forever. Once I kept some for five years—it was still good." She fixed some, and I found it incredibly sweet, more like a pudding than a vegetable dish.

Sylvester said *siimna* is "similar to potato chips; just dried in the sun and stored." George Kyyitan said he would use *daapk haal* for *simna* and *sha'ashkaḍk* for *hikchuvulidă*, which was "like jerky [cut] in strings."

I asked Sylvester how the dried squash was prepared for eating. "First you boil it, then drain the water out; then grease it up and fry it with chiles. But if you want to save it for dessert, put some aside to drain; put your sugar in it, according to your taste, and put a little canned milk. And then you have your dessert or cereal or whatever."

"How about *daapk haal?*"

"They bake it—sugared or salted—cut into quarters or eighths and baked in oven."

Those studying New World crop evolution suspect that the *Cucurbita* were domesticated first for their oily seeds and only secondarily for their flesh as nonbitter forms (mutants) happened occasionally and were selected for. The use of pumpkin seeds continued among Pimans. For instance, one day George remarked, "My mother used to put out roasted *haal* seeds, *haal kaij.* I guess most of them done that. But now nobody knows how." Joseph Giff related, "My mother told me that they used the squash seed inside. You know, it's oily. They used that for cream—face cream. They grind it, then they put it on their face." And Frank Jim, a Tohono O'odham, told me "The *haal kai* ['squash seed' is] roasted in [the] hull—very greasy; eat them seed by seed, like sunflower seeds. Don't mill and thresh them. Girls would sit around and chew on them at night, then when they went to bed, [they] had a mouthful of oil to smear on them[selves]. Can't roast watermelon seeds or anything else this way for food, just squash."

Sylvester had also heard of this use. I asked, "Where did they smear the *haal kai* oil?"

"On their skin . . . or hair; anywhere."

The peduncle is called *vaksig.* Sylvester added, "*Sha'ashkaḍk vaksig* is a girls' game. They burn it and tossed it around like a torch, because it's slow burning. Only the *sha'ashkaḍk* has it."

Technical Notes: Merrick (1991 and personal communication) believes that both *haal* and *daapk haal* are *Cucurbita argyrosperma* (long known as *C. mixta*) and that *sha'ashkaḍk haal* is *C. moschata.* She notes that the name *dapcahari* for a *calabaza de cuello* ('necked squash') appeared in the 17th-century Névome vocabulary (Pennington 1979:17). The colonial terms *aari* and *sarcarhcaari* may correspond to the remaining two modern Piman cultivars.

Merrick (1990, 1991) proposes *C. argyrosperma* subsp. *sororia* from southern Mexico and Central America as the wild ancestor of *haal.* In contrast, var. *palmeri* of the lowlands and foothills or northwestern Mexico south to southern Nayarit is most likely a weedy form of *haal.* She notes, based on information from don Pedro Estrella, that var. *palmeri* is called *siv haal* as well as *adav* in Onavas.

Vouchers: *866* (RG, *haal, C. argyrosperma*); *903* (RG, *sha'ashkaḍk haal, C. moschata*); LSMC *40* (not extant).

sha'ashkaḍk haal   *Cucurbita moschata*

Daucus carota L.
carrot
Apiaceae (Umbelliferae)
s-oám daadaidkam

Although the "humble umbels" as a family supply European cultures with such fine seasonings as caraway *(Carum carvi),* anise *(Pimpinella anisum),* dill *(Anethum graveolens),* and coriander *(Coriandrum sativum),* apparently none of these have found their way into either the Gila Pima's garden or lexicon. This is somewhat surprising in view of the great popularity of cilantro *(coriander leaves)* in Mexican cuisine throughout the southwestern United States as well as farther south. And the plant is readily and quickly cultivated. But all those excellent Old World members of the Lamiaceae (Labiatae) and Apiaceae (Umbelliferae) that Fr. Kino is known to have introduced to the more southern Pimans late in the 1600s seem never to have caught on at the northern frontier. (I have found mint *[Mentha spicata]* growing in Pima yards in several villages, but it has no Pima name, and I'm assured it is rather recent.)

There is one exception, of course, and that is the carrot, called by the Gileños *s-oám daadaidkam* 'yellow/orange root it has'. Apparently few, if any, Pimas raised carrots. I asked Sylvester Matthias about their role. "Yeah—modern, they buy. Some eat it fresh from the ground. That's modern. And they use it in stew," was his summary.

A native carrot (*D. pusillus* Michx.) grows wild in the desert mountains and upper arroyos of Pima country, but I have found no one who recognizes this little plant. It is a small spring annual that has no root enlargement.

Fragaria sp.
strawberry
Rosaceae
siipuḍ s-wulvañ, s-vepegĭ iibadkam

With strawberries being raised in the new town of Phoenix, the Pima developed an acquaintance with this crop more than a century ago (Gosper 1881). Sylvester Matthias said that the Pimas "didn't raise strawberries, but went to pick them. They [the Anglos of the western Phoenix area] had a patch over here, and some people from Santa Cruz [village] and a lot of other people, they went and picked strawberries. And by the time they got to the other end of the patch—and they looked back, and there's already ripe strawberries behind them." He laughed. "I thought we picked everything. I was working for an old lady here, and she had in her back yard, in the alley side, a few. And she asked me the first thing in the morning to go pick strawberries. So I went and pick them. And then again about ten o'clock she told me to go pick strawberries. Yeah, and [I] went over and I picked just as much again. Yeah, they ripen pretty quick."

Whenever we discussed strawberries, Sylvester would mention their origin. "I wonder where the strawberry comes from—whether it's native here or comes from some other country. I've never heard. But I see them in Apache country, up in White Mountains, growing wild. But I never heard if the crop is native here."

This crop has had a complicated history (Wilhelm 1974). The wild plant had been cultivated in Europe since Roman times.

siipuḍ s-wulvañ, s-vepegĭ iibadkam   *Fragaria* sp.

Another species, ranging along the Pacific coast of North and South America, was extensively cultivated and marketed by Chilean Indians. But our present large-fruited, long-producing crop is of mixed parentage resulting from hybridizing the European species with other forms from Canada, Virginia, Chile, and California. As a result of this combination and selection, Wilhelm says, "Strawberries are the first fruit to ripen in spring and the quickest to bear after planting. No other fruit yields more per unit of cultivated land in so short a time." The plant thrives from north temperate to desert climates and may yield up to 40 tons an acre each year.

The strawberry and the domesticated mulberry tree *(Morus alba)* provide an example of different applications of a folk name in different parts of the reservation. At the west end, mulberry is called *siipuḍ s-wulvañ* 'wrinkled anus' by everyone with whom I talked. I thought that was all there was to it. But George Kyyitan told me that *siipuḍ s-wulvañ* is "what we call strawberry."

For five or six years Sylvester could not remember the name for strawberry. Shortly after his 80th birthday (and near-fatal bout with pneumonia) we began talking about some of the loose ends still remaining on this book. He mentioned *s-vepegĭ iibadkam* (*s-*, intensifier marker + 'red' + 'fruit' + -*kam*, attributive marker). I was puzzled, not knowing the reference. Sylvester rambled on. "It's no name, but we just call them *s-vepegĭ iibadkam* because they're always red. I've seen them wild in White River, but they're small. But they look like strawberries and they smell like strawberries and they taste like them."

I went to see Francis Vavages at Sacaton. It was a hot mid May morning. He was sitting outside, putting the finishing touches on a flute. I sat on a stump, and we talked. A pile of stout mulberry limbs lay at our feet, material for carving spoons. He called mulberry *gōhi* and, like George, said that *siipuḍ s-wulvañ* was strawberry.

Gossypium hirsutum L.
   [= *G. hopi* Lewton]
G. barbadense L.
cotton
Malvaceae
toki

If the agricultural Pima Indians have been known for a single plant, it is surely cotton, a crop mentioned frequently by both Spanish and Anglo explorers. It is ironic that the cotton that has come to bear their name ("Pima Cotton") has nothing to do with the variety of cotton the Pima had been raising "from time immemorial."

At the end of the 17th century, Fr. Kino his companions were amazed at the productive fields as well as the resplendent cloth and blankets woven by both the Gileños and the Sobaipuri of the Santa Cruz and San Pedro Rivers. The Jesuit and Franciscan missionaries who followed were no less complimentary (see chapters 1 and 3). Anglo writers beginning in the mid 19th century commented not only on the quantity of the cotton the Pimas raised, but also on the excellent quality of the local landrace. Listen to their testimony:

> Upon the Gila River grows cotton of the most superior kind. Its nature is not unlike that of the celebrated Sea Island cotton, possessing an equally fine texture, and, if anything, more of a silky fibre. The samples I procured at the Indian village, from the rudely cultivated fields of the Pimas and Maricopas, have been spoken of as an extraordinary quality. (A. B. Gray 1855:33)

> Pimo Indians from time immemorial—certainly since they were first visited by Coronado, in 1542—have cultivated cotton of excellent quality. Specimens, which I showed to gentlemen in Texas, were pronounced nearly equal to the best Sea Island cotton of South Carolina. (Whipple 1856:18)

> They live in villages, raise luxuriant crops of corn, wheat, millet, melons and pumpkins, and also cotton of excellent quality resembling the sea island. It is from the black seed. They also spin and weave their cotton, by hand, into blankets of a beautiful texture, an art not acquired from the Spaniards. (Mowry 1858b:587)

Richard C. McCormick (then secretary of Arizona Territory), writing in the *New York Tribune,* 26 June 1865, noted, "They also raise a superior quality of cotton, from which they spin and weave their own garments." But for some reason, the same year Colonel Charles D. Poston (1865:1320), first delegate to Congress of Arizona Territory, reported, "Last year I had the pleasure of contributing something to those Indians by giving them [500 pounds of] cottonseed, hoes, spades, shovels, etc." Was he carrying coals to Newcastle? Commissioner William Dole (1865:164) commented that "[the Pima] were urged to renew their attention to this staple, in which they had formerly been successfully engaged." Was wheat production usurping cotton? One researcher contends that Poston's 1864 introduction began the genetic dilution of the native Pima strain (Shapiro 1989). However, others say the two are different species, Pima-Hopi cotton being *G. hirsutum* and Anglo cotton *G. barbadense* and that chance crossing is unlikely.

In any case, cotton cultivation had seriously declined by 1873, when Grossmann (1873:419) found the Pima growing only small amounts. Two factors led to the decline of cotton cultivation: manta, a type of cotton cloth of Anglo manufacture became the currency of exchange for Pima wheat and other supplies during the Civil War, and the volume of water in the Gila started to decrease drastically due to various factors beyond the Pimas' control (see chapter 5). By 1902 in the Sacaton area, Russell (1908:148) found that spinning and weaving as well as cotton cultivation were disappearing: "The Pimas no longer spin and weave, the art is dying with the passing of the older generation. It was with difficulty that enough raw cotton of Pima raising was secured to make the beginning of a piece of cloth on the small model loom." Lloyd (1911:58), carrying out his fieldwork a year later, confirmed that cotton had indeed dropped out on the middle and east-end villages as a Pima crop, apparently a result of loss of irrigation water. A few decades later, Castetter and Bell (1942:105) wrote, "We have found no trace of Pima and Papago cultivation of aboriginal cotton at the present day; in fact, it was impossible to find a single specimen of the ancient product among these Indians, and it is very doubtful whether any aboriginal cotton has been

grown by the Papago-Pima for many years." (Apparently things were different at the west end of the reservation, where the river ran the longest; during the mid 1980s, Ruth Giff and Sylvester Matthias recalled that the native cotton had been raised there in their youth.)

The U.S. Field Station at Sacaton, Arizona, was established in 1907 through a cooperative arrangement between what is now the Bureau of Indian Affairs and the U.S. Department of Agriculture. When its seed farm was established ten years later, its purpose was to provide the reservation Indians seeds of the "new and improved strains" of crops. Cotton breeding was one of the major areas of experimentation (King and Leding 1926). From the beginning of the cotton industry among the Anglos of the Salt River Valley in 1912, the Yuma variety of Egyptian cotton was the sole commercially grown crop. What became known as the Pima variety in 1910 had its origin as the selection of a single plant out of a field of Yuma cotton being grown at the station. As far as is known, Pima Cotton—the variety that was released to the public for commercial growing starting in 1916, completely replacing the Yuma variety— was unrelated to the cotton race the Pimas had been raising in their homeland for at least several centuries, the variety that came to be formally named Sacaton Aboriginal. In fact, the two forms are considered to be strains of two distinct species: the "Yuma/Pima" varieties of Egyptian cotton being *G. barbadense,* the Sacaton Aboriginal then being called *G. hopi.* The strain formally designated "Pima Cotton" was only bred and selected on the reservation, from parentage originating in a different land.

But then Sacaton Aboriginal dropped entirely out of sight. The research station had closed in 1958, and not even an herbarium sheet of native cotton was preserved among the vast collection acquired by the University of Arizona. I had hoped to find the *real* Pima crop to grow for Takashi to illustrate. My Pima friends remembered the plant but had not seen it in half a century. Then in 1984, my good friend Gary Paul Nabhan (who has an excellent track record for recovering obscure races of native Southwestern crops), found six seeds marked "Sacaton Aboriginal" with a code number from the University of Arizona cotton breeders (via Dr. John Endrizzi) that had last been grown out in 1964. He kept two, mailed me two, and sent two as a backup to another interested researcher.

The next spring I took my two 21-year-old seeds up to the roof of the Natural History Museum and planted them in a warm place. Nine days later the strangely misshapen cotyledons were springing vigorously from their little pots. Transplanted into large containers, they grew, bloomed, and by September began producing abundant bolls filled with the famous fibers—and plenty of seeds.

The word *toki* means cotton in all the Tepiman languages, attesting to an ancient association of the plant and this group. The word can signify the plant, the cotton still in the bolls, or the product after it has been picked. In Gila Pima there are two kinds of cotton. Ruth Giff recalls the plantings of her youth. "Just *toki* is the original Pima cotton. It was planted about the first of April and grew *tall*, about 7 feet. The pods were small, but the fibers real thick. It took lots of work to pick and fill the bags because they were so small. Then came the modern cotton, the short one, called *maskal toki* 'silky cotton'. They [non-Indians] must make the modern plant short so they can pick it with machines."

During World War I a few west-end Pima farmers were still raising cotton commercially, Sylvester Matthias recalled, "but all those people who raised cotton passed away. Enough material coming in [from Anglos]—gingham principally used. They buy their goods [off-reservation]." He thought that by 1930 cotton again became a significant crop for the Pima. There were only two kinds, he said: "The original Pima cotton was native here— small bolls, tall, called *che'echevĭ toki* ['tall cotton'] or just *toki*, tall, long staple. It's been around all the time. And the short staple, *maskal toki*."

"When did the Pima start raising that?"

"I think it came in around 1926."

George Kyyitan of Bapchule used the same taxonomy: "*Che'echevĭ toki*, that's long staple [Sacaton Aboriginal]. That's what *they* [the Pima] use; they *used* to: weaving, things like that. Nowadays, no more. They're lazy." I asked what he called the cotton in the field we were standing in. "*Maskal toki* 'silk[y] cotton'."

As so often is the case, the crop had several more purposes with the native

people than with the Anglos. "We used cottonseed as food—just roast them and eat them like peanuts; just pick 'em up and throw 'em in the fire and it's ready to be chewed on," Sylvester recalled. "Someone found out you could make cottonseed meal—grind [it]. They also make dye out of the tall [uppermost] Pima cotton leaves when they are *real* tender. Makes green dye. Boil it. [Used] when men wanted to color their G-strings."

History confirms Sylvester's comments on the seeds. At the turn of the century, Russell (1908:77) recorded "*Tâki*. The cotton plant is no longer raised but from pre-Spanish days down to the last quarter of a century it was cultivated both for the fiber and the seeds. The latter were pounded up with mesquite beans [*sic*, pods] in the mortar or they were sometimes parched and eaten without grinding." Among the various cottonseed dishes that Castetter and Bell (1942:198) were told about was "a sort of cake-like tortilla baked in ashes" made from fine-ground parched cottonseed and either mesquite-pod meal or cornmeal.

Another time Sylvester told me, "I just come to think: there *is* a cotton grasshopper. It's green. Only you can find it on cotton plants. You can always find a cotton grasshopper. They call it *tōki t-ab shōō'ō*. It didn't bother nothing. Just in the cotton patch."

The Pima were once famous for their fine woven blankets. As with the Hopi, weaving was generally the work of older men (Bartlett 1854, 2:229), but spinning seems to have been women's work (Emory 1848; Russell 1908:148). Palmer collected a loom for the National Museum in 1885, and his field notes give names for its various parts. The blankets, for wearing or for bedcovers, were woven on horizontal looms. Cremony (1868:106) wrote, "They manufacture a very superior quality of cotton blanket, which will turn rain, and is warm, comfortable and lasting." Although by the 20th century this art was nearly unknown, Russell (1908:150) had Thin Leather make a small model showing how the loom was set up. Only a few Pima blankets (*iks*) were saved in museums; these were characteristically white with an ochre edge. Among the items enumerated that the Tohono O'odham brought up to trade with the Gileños at the time of the June wheat harvest were "red and yellow ochres for face and body paint, and the

buff beloved by Pima weavers" (Russell 1908:93). In exchange the Pimas gave cotton blankets and cotton fiber, with the seed. The dry-farming Tohono O'odham were unable to raise enough of this water-loving subtropical plant to supply their own needs.

A second woven product was found throughout Tepiman country from the south end of the Sierra Madre to the Gila: colorful sashes and bands of tightly wound cotton yarns. Four of these have been preserved in the National Museum. Dr. Barney Burns, who has had extensive experience collecting textiles in northwest Mexico, commented to me how nearly identical the pattern and texture of these Pima items are to those made by the Tarahumara and other Pimans to the south. But the Gileño specimens are of local manufacture, the work of Pima women (Bartlett 1854, 2:224; Michler 1857). Also, Sylvester Matthias remembers that his grandmother was one of the last who knew the art of weaving bands. Presumably sashes were woven on a back-strap loom.

Other cotton items have been lost as well. Nothing is known today among the Pima of the *omina*, the medicine man's bundle of prayer sticks, but Russell (1908:106–107) at the turn of the century collected several sets from the last *mamakai*, or shamans. These are of Arrow-weed (*Pluchea sericea*), wrapped with homespun cotton, with or without attached feathers. These votive offerings resemble the prayer sticks of the Hopi and Zuni to the north and, less closely, those of the Huichol and others to the south.

With the pumping of groundwater and other sources of irrigation water, cotton cultivation once more came into vogue. By 1924 cotton was the second most important crop after wheat for the Gila Pima, and it was their leading money crop (King and Leding 1926). Today it probably ranks first in the areal extent of both fields being leased to whites and those farmed by the Pima themselves.

Technical Notes: Hopi Cotton was described as a new species by Lewton (1912), who included the aboriginal Pima strain in his new species on the basis of a single specimen E. W. Hudson obtained from Pimas at Sacaton. Thomas Kearney of the Sacaton Field Station wrote ethnobotanist Volney Jones (1936), confirming that Lewton's Pima specimen "was secured from the Pima and the strain has been kept from becoming extinct by making annual plantings at the Field Station." Castetter and Bell (1942:105) concluded, "It is evidently seed of this specimen [Hudson's] from which annual plantings have been made at the Sacaton Station under the name of 'Sacaton Aboriginal' and which was doubtless the kind of cotton cultivated aboriginally by the Pimas. It is a strain of *G. hopi* Lewton, a species found to be very early and genetically quite heterozygous." Both "species" are now considered to be in the broader species *G. hirsutum*.

Paul A. Fryxell wrote me the following letter (21 April 1992) clarifying the taxonomic status of the Pima cultivar and the confusing name in current use:

It is perhaps unfortunate that the phrase "Pima cotton" has come to be applied to three different entities, because this inevitably gives rise to some confusion. Firstly, it refers to the pre-Columbian cottons grown by the Pimas and Papagos *(Gossypium hirsutum)*, of which Sacaton Aboriginal is the surviving remnant. Secondly, it refers to the cottons introduced to Arizona about 1912 *(G. barbadense)*, originally native to South America but introduced from Egypt (hence called American-Egyptian cottons). Thirdly, the name refers to modern (i.e. post World War II) cultivars of *G. barbadense*, generally referred to as "Pima S-1" or S-2, up to whatever the current varieties are, which are highly modified versions of the earlier "Pima" cottons. About 10% of the current commercial crop of Arizona is *G. barbadense* (Pima S-?) and 90% improved varieties of *G. hirsutum*. Botanically they are very different, and commercially they are separate industries.

The Sacaton Aboriginal and the Hopi cottons are simply variants of the same thing, the Hopi having somewhat smaller stature and greater earliness, as is logical knowing that it comes from the northern limit of cotton agriculture. Together, they differ from other representatives of *G. hirsutum* in being precocious and prolific, with relatively small bolls.

We grew out Sacaton Aboriginal in San Diego and Tucson. The sepals are very long (45–50 mm) and the petals a pale greenish yellow with vinaceous veins. Characteristically the lateral branches from the tall main stem are double. The mature bolls are only 23–24 (rarely 25) mm deep, about one-third the size of the modern commercial varieties.

Vouchers: *870* (SM).

toki *Gossypium hirsutum*

Sacaton Aboriginal, painted from a plant grown out in San Diego from seed (HA 76 G8 A-5 S) obtained from the University of Arizona.

Helianthus annuus L. var. *macrocarpus*
D.C. Cockerell
Russian sunflower
Asteraceae (Compositae)
ge'egeḍ hiivai

The so-called Russian sunflower, a cultivated variety of the native Wild Sunflower (*H. annuus* L., Group G) that abounds in Pima country, is seen today infrequently on the reservation. Although Russian sunflower resembles Wild Sunflower, the cultivar has but a single flower head on top of a heavy stalk and the disk is many times broader than in the wild form, whose disk (the seed-bearing part) is only a finger joint across. The Gila Pima name *ge'egeḍ hiivai* means 'large Wild Sunflower', one of the ways the Pima had of naming newly acquired cultivars that were more robust than their wild progenitors or other associated familiar species.

Sunflowers were not raised aboriginally or even historically in Pima country (Castetter and Bell 1942:63, 74), and indeed, the large cultivar we know today did not return to the United States until the 1880s after selective breeding in Russia (Heiser 1976). Earlier in the 20th century, Russian sunflowers were raised occasionally in Pima country for chicken feed. They still maintain some popularity as a snack food. Sylvester Matthias commented, "My great-grandchildren—oh boy!—they're crazy about them." They never achieved the ceremonial and esthetic significance found among the Hopi (Whiting 1939:96–97; Nabhan 1989:129–132).

There is a tradition among various Euro-American cultures that the flower heads follow the sun during the day. I asked Sylvester if he had heard of this.

"We don't have that in Pima, but Pedro [Estrella, a Pima Bajo] does. Pedro called it *tash neidam* 'sun looking at.'" For the most part the ones on the reservation face east. But then, everything there faces east; that's Pima tradition!

Hordeum vulgare L.
barley
Poaceae (Gramineae)
sivááyu

Barley and wheat were the two major winter crops that Fr. Kino introduced to Pimería Alta in the late 1600s. The Pima name for the plant, *sivááyu* (or occasionally *siváá'iu?*) is a loan word from the Spanish *cebada*. Barley proved even better suited to Gileño country than wheat. It tolerates the heavy alluvial soils found in the lower terraces along the river, has fairly good drought resistance, and is the most salt-tolerant of the cereal crops (Chapman and Carter 1976).

Castetter and Bell (1942:75) considered barley second in significance only to wheat in the early post-contact Piman economy. In 1870 the U.S. government issued 10,000 pounds of barley seed to the Pima and Maricopa (Grossmann 1873). A few years earlier, these Indians had sold 2 million pounds of wheat and corn, along with "a large surplus of barley, beans, etc." (Rusling 1874). In the early part of the 20th century, barley ranked fourth in the number of acres planted, following wheat, cotton, and alfalfa, but by 1940 it had dropped to fifth place, with only 700

acres planted (Castetter and Bell 1942:76, 78). Some 700 varieties of barley were grown at the Sacaton Experimental Station, many from North Africa, but the Pima usually planted only two, Common and Vaughn, in the 1940s. "Used to when they have water. Not today," commented Sylvester Matthias. Castetter and Bell (1942:118) wrote: "Barley is usually planted from November to January, although sometimes as late as February, and is grown almost entirely for pasturing cattle; this begins about six weeks after the crop is planted." They added (1942:203): "Barley can be pastured three or four times and still give a crop, wheat only once, so it is considered superior to wheat for this purpose. Both are first pastured when about a foot high. Pima like to plant their barley in late October and harvest in late May, as there is little frost damage to this crop."

I asked Sylvester Matthias about *sivááyu*.

"Some of them [some crops] originated from the Spanish. It's feed for the stock. Wheat had a higher price than the *sivááyu*. It's not [used] for [human] food." Rolled barley, used for livestock feed, is called *sivááyu shoñvi* or *sivááyu ha-shoñvi* 'pounded barley'.

A clump of some grain I did not recognize was growing along the roadside near the canal below the Sacaton Mountains. I asked Irene Hendricks what it was.

"*Sivááyu.*"

"But how can you tell *sivááyu* from *pilkañ* [wheat]?"

"*Eeeeh!* See—" she pointed to the awns, or beards, "because this one has a long *chiñvo* like you!"

Vouchers: *724* (SM); *1340* (IH).

Ipomoea batatas (L.) Lam.
sweet-potato, "yam"
Convolvulaceae
kamóódi, iikovĭ

Sweet-potato is a sprawling morning-glory that seldom blooms but produces enlargements of the roots (unlike the tubers of the white potato *(Solanum tuberosum)*, which are stem parts with buds). Both sweet-potatoes and potatoes

were taken back to Europe shortly after the discovery of the New World, and the confusion between these two "root" crops began. Some forms of sweet-potato were known as *batata* in the West Indies, apparently a Carib name (Standley and Williams 1970; Yen 1971). This name somehow became associated with the crop from the Andes, *Solanum tuberosum,* and was transposed into English as the word *potato*. One more plant is involved in this ring of misnomers: if you go to a U.S. market, you are likely to find moist, sweet varieties of sweet-potatoes labeled as *yams*. But true yams are tropical plants of the genus *Dioscorea* (family Dioscoreaceae) that seldom find their way into temperate commerce.

Nor does the confusion end with names. Plant geographers have been intrigued by the distribution of sweet-potatoes, a crop that was spread at an early time across tropical Asia, the Pacific Islands, and tropical America. What was its origin? And how was it dispersed? Was there contact throughout this area before Europeans "discovered" the New World? The evidence is not all in yet (Brand 1971; Yen 1971). But the crop was absent from the northern frontiers of Mesoamerica, including Tepiman country.

The Pima may call the sweet-potato simply *iikovĭ*, the same as their own wild Hog-potato, or Camote-de-raton *(Hoffmanseggia glauca),* which they once dug for its tubers (see Group G). Because neither the wild plant nor the new cultivar was of high cultural significance, neither Pima name was secondarily marked to indicate which was which. Other Pimas (see Willenbrink 1935) and the Tohono O'odham (Mathiot 1973:455) prefer to use *kamóódi,* a loan word from Spanish *camote* (itself a loan word from the Nahuatl *camohtli*). Although vowel length and a consonant among both Indian groups have been modified to fit the Piman canons, the accent remains atypical, as in its Hispanic source.

When did the sweet-potato reach these northern Pimans? Documentary sources are unclear, but it would appear that the introduction occurred sometime during the Hispanic period (Ezell 1961:35); at least one Anglo on his way to the California goldfields noted potatoes and "yams" among the Gileños (Pancoast 1930:244). It is unfortunate that this account of his sojourn at the Pima villages seems to have been written from memory as it mixes cultural elements that cannot possibly pertain to the Gila Pima. Ezell (1961:35) interpreted Pancoast's reference as an Hispanic introduction. At this early date it is doubtful that Anglo-American crops had much influence on these desert people. (Apparently the introduction was largely unsuccessful for some reason; Castetter and Bell [1942] did not even mention the crop in their study of Pima and Tohono O'odham agriculture.)

In the early 1920s, King and Leding (1926) reported, "Special effort is being made at the [U.S.] field station [at Sacaton] to increase the interest of the Indians in the production of sweet-potatoes, especially in home consumption. Experiments have demonstrated conclusively that the soil along the Gila River is capable of producing large yields of sweet-potatoes, and there is apparently no reason why this crop should not occupy an important place in the agriculture of the Indians." Georgia "Yam" and Porto Rico were the varieties experimented with. Later at the station, King and associates (1938:56) noted that Porto Rico was the most popular variety among local commercial producers. "The larger plantings are made on light well-drained soils, and it has been found in recent years that the tubers keep well in such soils even after frost. As a rule they are allowed to remain in the soil and are dug from time to time in quantities to meet market demands."

Ruth Giff told me, "Somebody used to plant iikovĭ—the kamóódi—when I was a child. Has to be sandy soil. But not very many [people did]."

Sylvester Matthias said that only a few families used to raise sweet-potatoes in the western villages. These were planted in the house gardens rather than out in the regular fields. "Some use it as potatoes—fry it. Others bake or boil [it]; my late sister used to make turnovers out of them. Just [call it] s-i'ovĭ chemait; chechemait is plural. That's anything in the line of pastries in general." The making of these turnovers, or empanadas, was something learned evidently from the Mexicans.

"Why didn't people raise that more often?"

"Maybe they didn't know how. Another thing about iikovĭ—when these local farmers [both white and Pima] in our area there—they just let it [stay] there [in the ground] until the frost got them. Then

that time they harvest. Then it's real sweet. But now they get [them] too soon, and they don't have that flavor that I tasted when I was a kid."

Sylvester noted that the name iikovĭ could be used to designate either the wild plant or the new crop.

I asked him, "If I just said iikovĭ to you, how would you know which one I meant?"

"It's the same thing," he said, "but now they just say kamóódi; that's a Spanish word."

I have never found anyone growing sweet-potatoes on the reservation, but canned "yams" are served at feasts rather commonly.

Ipomoea carnea Jacq. ssp. fistulosa (Mart.) D. Austin
bush morning-glory
Convolvulaceae
goks o'odham wuhiōsha

On a late May afternoon I went to the Bapchule community building looking for George Kyyitan. He was sitting with a group of elderly men after their lunch but came over to talk when he saw me. We discussed plants for some while. Then my eye caught a series of bushes about knee-high planted in front of the building. The bright pink trumpet-shaped flowers from a distance looked like some datura, and the rather large triangular leaves suggested that as well, but on closer look they seemed more like giant tubular morning-glories.

George immediately followed my eye to the pink-flowered bushes and said, "Goks o'odham wuhiōsha. That's what they call those. You know what that means?"

"Yes, 'monkey's mask'."

"'Monkey's face'. It's the seeds. When the seeds down there develop, they look like a monkey's face." (Where the Tohono O'odham use the Spanish loan word chaango for 'monkey', the Gila Pima say goks o'odham 'dog person'; wuhiōsha can mean either 'face' or 'mask'.)

I cut one of the seedpods open, but the seeds were still white and fuzzy. "Have they been around for long?"

"Not very long. They were around, then I didn't see them for a while, then these

were planted here. They freeze back but come up again, about in February. See here where the old stalks were. This is the second year here."

"I thought they were kotḍopĭ [datura]."

"No. Kotḍopĭ is white. These just look good, nothing to use. Just for decoration for the yard." The plant, it seems, is a relatively new one. As far as George could remember, it was not cultivated in his childhood.

I asked Irene Hendricks of nearby Casa Blanca if she had ever heard of such a plant. "Um-hum. It makes a great big bush with purple flowers."

"Why do they call it that?"

"Because the seeds are little moomĭ ['heads']."

"Was it around when you were a little girl?"

"No."

"How about when you were a teenager?"

"Yes. About the [19]30s."

Later I saw some mature seeds. Their rounded outer surface is covered with quite long and silky dull brown hair, but the inner, slightly angled surface, dark brown, is nearly bare, with a little "mouth" depression on the midline at one end. The Pima name of 'monkey face' is most appropriate.

Vouchers: 1466 (GK); 1493 (GK, IH).

Lactuca sativa L.
lettuce
Asteraceae (Compositae)
iivagĭ

Although missionaries of Pimería Alta introduced lettuce to the missions, it is doubtful that any of the indigenes on the "rim of Christendom" had much interest in such a crop. Kino (1919) did list lechugas along with other garden vegetables. There is a passing reference by Nentvig (1980:23) to this and several other potential crop species from the Jesuit period: "Sweet potatoes, carrots, turnips, radishes, lettuce, and whatever else grows in truck gardens might be sown and abundant yields obtained were it not for the scarcity of workers for neither Indians nor whites are willing to apply themselves to such a task." (This section is missing from editions such as Viveros' [Nentvig 1971], made from the Mexico manuscript.)

Lettuce grows primarily as a cool-season crop in the desert, and Gileños already had several greens to choose from that did not have to be cultivated. Additionally, two introduced Old World weeds—the sowthistles *Sonchus oleraceus* and *S. asper*—came along and were incorporated into the dietary. Why cultivate a plant when there were already sufficient equivalents available just for the gathering?

Lettuce probably entered the Pima diet well into the Anglo period. By the late 1920s and early 1930s, lettuce was being grown at the Sacaton Field Station and had become a major crop in the state (King and Loomis 1932; King and associates 1938). After off-reservation supermarkets became the major source of food, head lettuce began to appear on Pima tables, but my general impression from the last quarter-century is that it has never been very popular. Still, lettuce has replaced the various wild *iivagĭ* (see Group F) as the greens of choice in all but the most traditional families.

This minor commodity is more interesting from the linguistic point of view. Though called polysemously *iivagĭ* 'eaten greens', should it be classified as a wild gathered greens in the category *iivagĭ* or as a crop, *e'es?* Both are special-use rather than morphological categories such as grasses *(vashai),* bushes *(sha'i),* or trees *(u'us chuuchim).* The consensus is that it is an *e'es* because it is planted. The literal meaning of the word does not affect its conceptual category. By default, everything in the special-purpose class *iivagĭ* is a wild plant.

Lagenaria siceraria (Mol.) Standl.
   [= *L. vulgaris* Ser.]
bottle gourd
Cucurbitaceae
vakoa

Bottle gourd vines have broadly rounded, rather foul-smelling leaves. They typically climb over some support such as a tree, holding on by tendrils. In summer, usually late, they bear white flowers, the united petals like crumpled silk. The following day these turn dirty white and close. As with other cucurbits, the female flowers are separate, bearing a green fuzzy miniature gourd at the base of the wrinkled petals. If this is fertilized by insects and fruit set occurs, the gourd begins to swell, developing slowly over the next few months. Frost limits the varieties of bottle gourd that can be raised on the Gila. Slow-developing strains from too far south in Mexico set fruits so late that the frosts ruin them before they mature.

Gourds were found aboriginally on both sides of the Atlantic, in Africa and the Americas. Their ability to maintain viable seeds during long periods in ocean water no doubt contributed to their transcontinental distribution, and which continent was their original home is still debated (Heiser 1979; Whitaker and Davis 1962). Various sizes and shapes of these hard-rined fruits have been preserved in archaeological sites in the southwestern United States and northwestern Mexico. The numerous cultivated strains of bottle gourd differ in seed shape and color as well as in overall size and shape of the fruit. These landraces were also adapted to local conditions in various other ways, such as fruiting time.

All the Gila Pimas with whom I worked knew and used the bottle gourd, which they call *vakoa* (pl. *vapkoa*). The name applies both to the plant and to the fruits it produces. This is a pan-Piman word. Among the Tohono O'odham and the Pima Bajo at Onavas (and the 16th-century Névome) it is called *vako* (pl. *vapko*). Mountain Pima at Maicoba say *vak* as well as *havo.* The Northern Tepehuan call it *vakoi;* they maintain four named varieties (Pennington 1969). All these forms of the gourd word contain the water prefix, *va-* (see discussion of *vak,* Group B), linking the plant with its early aboriginal use as a canteen. Use of gourds as water containers is so important that a canteen, even one not made from the gourd, is called *vakoa.*

The Gileño varietal taxonomy today includes only two folk taxa: the *vakoa* (a moderately large rounded form with an elongated "handle") and the double bottle gourd, called *giwuḍ vakoa* 'belted gourd'. The Pima are not alone in distinguishing these. "In a pattern exhibited across a number of unrelated language groups in Mexico," writes Merrick (1991:217), "two major classes of *L. siceraria* are distinguished, with one term designating fruits with a globose shape and the other designating fruits with a constricted neck." Perhaps there were once other named forms when more varieties were maintained by the People for various uses. Even today there is considerable experimenting among the Pima with different genetic strains, and I continue to discover new shapes and sizes being grown that I never saw 20 years ago. These are probably traded among different tribes. But these novelties are not named.

Sylvester Matthias told me, "*Giwuḍ vakoa* [is] something shaped that way, 'belted gourd'; gourds' main purpose is for canteens; also big gourds, several feet tall, used in the storage house. I don't know what they used it for."

"Did they raise lots of gourds?" I asked him.

"When the river was running, and irrigation ditch[es], they could just put, plant it along a ditch bank by a mesquite tree, and, oh, they would just go up; oh, they'd be covered with gourds."

This observation agrees with a more detailed account left by Castetter and Bell (1942:154):

> Gourds were planted differently from other crops. Among the Pima, about four hills, each three feet in diameter, were located and prepared near an irrigation ditch and close to a tree or bush if possible so the plants could climb; otherwise a forked pole was placed in the hill for support. Four seeds were placed in each hill at a depth of two to three inches, and the hills walled to retain water. When water in the ditch was high, it was conducted to the hill and walled in; when low, the channel connecting the ditch and hill was left open, allowing water to move to the hill at will. It was woman's responsibility to cultivate and care for gourds, as they were rather looked upon as her pet crop and easy to grow.

In emergencies, it was the gourd plants that received special treatment. "Neither Pima nor Papago carried water to plants (other than gourds, which were sometimes watered by hand) to tide over periods of drought" (Castetter and Bell 1942:172).

According to tradition, the gourd originated with Navĭchu, a deity who appears in the Creation Story after the original four culture heroes (Earth Maker, Buzzard, Coyote, and Elder Brother). The masked cult associated with Navĭchu has a number of unmistakable Puebloan characteristics and indeed appears to be borrowed from the north, most likely from the Hopi. Two types of the masked dancers who accompany Navĭchu wear distinctively kachinalike half-gourd masks that look like some found among the Hopi today, as does the Navĭchu's case mask, a canvas bag fitting over the entire head. The cult of Navĭchu was primarily a healing ceremony for swelling of the knees, but there is also a fertility prayer recited by Navĭchu (Lloyd 1911:215). (The comparable ceremony among the Tohono O'odham is called the Viigida.)

vakoa   *Lagenaria siceraria*

This painting shows the dipper gourd, the form most frequently grown by the Gila Pima.

Russell (1908:326–329) recorded two songs from the curing. He seems not to have had an entirely clear understanding of the cult (which he confused with the Pascola ceremony, pp. 108 and 186), but there are stray references to it scattered here and there throughout his ethnography. For instance, he (1908:174–175) said, "A woman might gamble away the family sleeping mat, the metate, in fact any household property, although she hesitated to wager the drinking gourd, probably owing to the fear of provoking

Navitco [Navĭchu], the deity who gave the gourd to man." And Russell (1908:91) or perhaps his interpreter, Jose Louis Brennan, said, "Cultivated gourds have been known to the Pimas for a long period—how long, it is impossible to say. The Papagos have a tradition that this plant was introduced by Navitco, a deity who is honored by ceremonies at intervals of eight years—or, if crops are bountiful, at the end of every four years—at Santa Rosa."

Sylvester Matthias, from his father's side, is in the Navĭchu line, but because he has no sons in his family, the line will end with him. He made both *kokshpakam* and *vipiñim* masks out of half-gourds. The mask is called *wuhiosha*. The *vipiñim wuhiosha* he often fitted with side bundles made from the fine inner bark of the cottonwood, like a wig. The Navĭchu ceremony was last performed at Komatke in 1923. Sylvester made a painting of a Navĭchu dancer wearing a simple case mask and a light-colored shift reaching to mid calf, fastened with a belt. "And he wears wrist bands, bands about three inches wide. He sticks the seeds under there to give to the People," Sylvester explained, showing me on his drawing.

"What kind of seeds?"

"Gourd seeds."

"Any other kind of seeds he distributes?"

"No, just gourd seeds."

A rounded gourd with a long neck was grown for making the *shavĭkuḍ* 'gourd rattle' (*shav* 'to shake something'; *-kuḍ*, agent). One might think that a rattle was a readily made object, but proper preparation of this important item, still used to accompany nearly all native songs, is an involved process if the *shavĭkuḍ* is to have superior sound qualities. Joseph Giff was a maker of excellent rattles for singing and dancing. Four things he said can be used to go into the gourd as resonators: paloverde seeds (from *kuk chehedagĭ* [*Cercidium microphyllum*] only); lead shot (#9 shot is best); the small round seeds from the fan palm *Washingtonia filifera;* and small, water-polished rocks that a species of ant brings up to the surface of the earth. Joe learned the technique of making rattles from the Cocopa. You cut off the tip near the peduncle, then bury the gourd in the ground and fill it with water. Soak it for seven to ten days. Then take rocks the size of the original gourd seeds, place them inside, and shake the rattle by the neck "handle" to work away the soft insides of the gourd. (Someone once used broken glass inside to clean it out instead of soaking and drying.) Rinse everything out. Finally, when all the soft material is out, make the holes to let the sound out and make the right tone. (I think Joe used an ice pick to make these holes.) Dry it for several days. Then it is ready for use. The handle that grows with the gourd, the neck, is discarded, and the gourd seeds are never used to make the sound. Finally, Joe smoothed the edges with a pocketknife and fitted the gourd with a Saguaro rib handle. (See also Curtin [1949:123] and Castetter and Bell [1942:201] for brief accounts of rattle making.) The wild gourd *aḍavĭ (Cucurbita digitata)* is considered too thin to make the *shavĭkuḍ,* except for play.

Still another variety of gourd (or sometimes just a small rattle gourd) was grown for the *ha'u* 'dipper'. The *vakoa ha'u* was made by longitudinally cutting a gourd having a small bowl and large handle. Early in this century Russell (1908:91–92) said, "Dippers and canteens are occasionally made of gourds, but the chief use of gourds seems to be in the form of rattles which contain a little gravel and are mounted on a handle. Gourds are never used as forms over which to mold pottery."

On a long trek across the desert, where water sources are few and far between, the light and durable bottle gourd had advantages over ceramic vessels that might break, leaving the traveler stranded. Gourds were precious. Russell said, "The gourd is used as a canteen, and if it becomes cracked a rabbit skin is stretched over it which shrinks in drying and renders the vessel water-tight again." Various California-bound itinerants in the mid 19th century were supplied with bottle gourds of water for the jornada between the Pima villages and Gila Bend. For instance, in November 1846, Emory (1848:87) wrote, "We were notified that a long journey was to be made without water, (to cut off an elbow in the river) and the demand for gourds [from the Indians] was much greater than the supply. One large gourd cost me four strings of glass beads, which was thought a high price."

Late in October Ruth Giff and I were walking through her fields. The satiny flowers of her two types of *vakoa* were still white in the early morning sun. "Some

Pima lady told me if you wipe off the fuzz [of the developing fruit] when they get the size you want, they won't grow any larger." Then she took a piece of paper nearby and wiped off a very small one. "I'm going to see if it works." But I think the *tasi'ikol* 'Javelinas' ruined her experiment by trampling most of the vines shortly afterward.

Vouchers: *902* (RG); LSMC *61* (not extant).

Lens culinaris Medic.
   [= *L. esculenta* Moench]
lentil
Fabaceae (Leguminosae)
laanji, laanjikĭ

Among the many new crops introduced to the Upper Pimans at the end of the 17th century were four legumes that seem to have been accepted: the garbanzo, black-eyed pea, garden pea, and lentil. The Pima invented their own names for the black-eyed pea (*u'us bavĭ* 'stick tepary') and the pea (*vihol,* akin to *vihog* 'mesquite pod'). But the Pima baptized the other two with names derived from Spanish: *kalvash* for the garbanzo, and *laanji* or *laanjikĭ* for the lentil, called *lanteja* in Spanish.

The pea and the lentil could be grown in the cool season (with wheat and barley), when Pimas aboriginally were apparently planting little or nothing. Kino's (1919:457, 1989:361) *Favores celestiales,* summarizing work in Pimeria Alta, lists the lentil along with beans (*frijol),* garbanzos, favas, and *alberjon* as productive legumes. In the 1764 *Descripcion . . . de Sonora,* better known to English readers as the *Rudo Ensayo,* Juan Nentvig wrote, "Similarly, in Ópata and upper Pima lands beans, favas, lentils, and other legumes yield abundantly. Except in a few places, garbanzos do not yield always to the expectation of the worker, nor do the *alverjon, alverjas* or *chicharos* and others" (Viveros 1971:73 [my translation]). Lentils were already being grown at San Xavier by the Sobaipuri between 1687 and 1716 (Velarde 1931; see González 1977:52) and did not have far to go down the Santa Cruz to reach Gileño country.

According to Castetter and Bell (1942: 197), "Pima and Papago planted lentils as early as February, and they were ready to

harvest in April or early May. Pima frequently made two plantings of lentils, the Papago but one; this second sowing was after the saguaro harvest. Like chick peas, they were fairly drought-resistant. Lentils were harvested and stored in the same manner as teparies, and were reported as returning a yield of tenfold."

No doubt there were more similarities between these two legumes than just the harvest techniques. The lentil looks like a small flat tepary and has a somewhat similar texture, though it cooks in only a fraction of the time that the little native bean does. But most important, it ripened at a critical time of the year—during the spring stress period.

The lentil seems to have fallen out of favor among the Pima sometime in the last 40 years, however. I have never seen it served on the reservation and know of no one raising it. All locally adapted races apparently have long since been lost. Castetter and Bell mentioned that a number of cultivars survived at least among the Tohono O'odham in 1940, although they made only two seed collections. Formerly lentils were planted about 1–1½ inches deep in hills and rows about a foot apart (Castetter and Bell 1942:153). None of the people with whom I worked knew much about its cultivation.

Lycopersicon esculentum Mill.
tomato
Solanaceae
tomáádi

Although a native of the New World, the tomato, a domesticated weed, was unknown aboriginally as far north as Pimería Alta. Its initial introduction was probably during the Hispanic period, and its Piman name is derived from Spanish *tomate,* which was in turn derived from Nahuatl *tomatl,* the green husk-tomato (*Physalis* sp.; see Group G). Nentvig (1971) mentioned both *tomate* and *jitomate* among medicinal plants in northern Sonora in 1764, so one may presume that it reached at least this far north in Pima country.

Very few of the Pimas grow tomatoes, kitchen gardens being rare. Those used now are purchased in supermarkets located at convenient distances from the reservation. Occasions calling for communal feasting or potlucks will usually have one or more green salads with tomatoes, but these never go as quickly as the heavy potato salads, which are usually made in considerably greater quantity. For 1931–1935, King and associates (1938:9) noted that "truck crops, consisting of beans, corn, tomatoes, onions, and melons, are grown by the Indians each year for home use and, to a small extent, for commercial purposes."

Sylvester Matthias once told me, "Make *tomáádi sitaḍoi* [stewed tomatoes]; can [also] make corn or chile *sitaḍoi.*"

In certain Pima families, a most excellent salsa is made from finely chopped tomatoes, green chiles (Anaheim type), and onions, served with *chuukug ga'i* 'steak roasted over coals' and fresh *chechemait* (wheat flour tortillas).

Nicotiana rustica L.
Nicotiana tabacum L.
tobacco (cultivated)
Solanaceae
viv

For the Pima, everything about tobacco is sacred: its origin, its cultivation, its use. Only with the coming of Europeans has tobacco, like Saguaro wine, been taken out of its ceremonial context and desacralized. As Underhill (1938b:22) wrote, "It must be confessed that the Whites, who have taught the Indian that smoking is not an act of worship but a daily pleasure, have also encouraged him to drink whenever he has money instead of only when he wants to 'pull down the clouds'" in the Saguaro wine rain-making ceremony. Russell (1908:118) explained, "[Tobacco] was to [the Pima] a plant of divine origin that in its death (burning) released a spirit (odor and smoke) that was wafted by the breeze to the home of the magic beings that shape men's destiny."

The various versions of the Piman Creation Epic record several different origins for this sacred plant. The first story recounts simply that after Se'ehe gave the People their four primary food crops (see *Chenopodium berlandieri*), then came three additional crops, one of which was tobacco. But the other and more widespread account is more complicated. Joe Giff told me the story, which is also recorded in various other versions (see Densmore 1929:29–30; Russell 1908:224, 248; Underhill 1946:83–84).

After the cannibal witch, Ho'ok, was killed, everyone was invited to a feast except for an old woman and her two grandsons. She instructed the young men to go look for remnants of Ho'ok, and they brought back all that they could find: only a few drops of her blood. She told the boys to check the blood in four days, and they found two bird's eggs; in another four days, two macaws had hatched. She told them that the people would be jealous of these beautiful birds and try to steal them, but that they were to flee eastward to the mountains with their birds. When the boys returned, she said they would find that the People had killed her, but that they should take her body and bury it in the sand of a wash, then watch the grave. These things happened as she predicted. In four days they saw something green emerging from her grave, and in four more days, they had a tobacco plant with large leaves they could pick. Se'ehe came to look after the boys and instructed them in how to roll these in corn husks and smoke them.

According to Joe, the grandmother explained the special nature of this plant to the boys by telling them, "You take that and the People, the medicine men, will use that. And they can see far ahead. They can see things and they can have the power to cure and the power to prophesy and all the things of the medicine men only. So they did."

Even the growing of tobacco was to remain a sacred activity. Sylvester Matthias told me that tobacco "had to be away from the People, off from the village; they say tobacco is *shy!*" The restrictions were stringent. Joe explained some of them to me. Normally tobacco was planted by old men where it could not be seen by a casual person, particularly by women. If a menstruating woman were to look on the plot, the whole crop would be ruined. Another time Sylvester said, "[At] Pima Butte there is a place they planted—*viv esikuḍ*—it's a place where they grow their tobacco." He continued, "They didn't want to plant it around women. Women didn't

smoke in those days. The first one I saw was a Mexican old woman that smokes. Then a Papago woman from across the border. They do now. They're not chain smokers."

The old men knew the songs that must be sung to the seeds, as well as to the germinating and growing plants. No one now remembers these, but Underhill (1938b:47, 1946:84) recorded the one song she found among the Tohono O'odham:

Am I not the tobacco shaman?
Here I come forth and grow tall.
Am I not the tobacco shaman?
The blue hummingbird finds my
    flowers.
Above them softly he is humming.
Hiro a'aro.

This is one of many tobacco-growing songs that the old woman in the tobacco origin story taught her grandsons. The last line (rendered *Hi'ilo'o ya'a'a* in Gila Pima) is also an important part of corn-growing songs among both Akimel and Tohono O'odham. It is untranslatable.

There is still a third origin tradition for tobacco from the Creation Story (Lloyd 1911:217–229). An unmarried woman, daughter of a *maakai* 'shaman', "grew tired of her single life and asked her father to bury her, saying, 'We will see then if the men will care for me.' And from her grave grew the plant tobacco, and her father took it and smoked it, and when the people who were gathered together smelled it, they wondered what it was, and sent Toehahvs [Coyote] to find out. But, although the tobacco still grew, the woman came to life again and came out of her grave back to her home."

Living there was Corn Man, who beat Tobacco Woman in gambling. She refused to pay up, and they quarreled. "She told Corn to go away, saying, 'Nobody cares for you, now, but they care a great deal for me, and when the doctors [*mamakai*] use me to make rain, and when they have moistened the ground is the only time you can come out.'" Thinking everyone was gossiping about her, she left for the west. Drought followed, and there were no crops. The People reproved Corn Man. "And Corn left, going toward the east, singing all the way, taking Pumpkin with him, who was singing too, saying they were going where there was plenty of moisture." In the second year of drought,

Giisobǐ (the Verdin), another powerful *maakai*, went searching for Tobacco Woman, taking her four votive offerings: a crystal, a Hohokam slate or palette, soft down feathers, and eagle tail feathers. But she was still in a huff and would not return. Instead, she gave him four balls of tobacco seed, with the instructions, "Take the dirt [feces] of the tobacco-worm, and roll it up, and put it in a cane-tube and smoke it all around, and you will have rain, and then plant the seeds, and in four days it will come up, and when you get the leaves, smoke them, and call on the winds, and you will have clouds and plenty of rain." And thus the essential ritual relationship of tobacco–smoke–winds–clouds–rain was established.

There were two different ways of smoking *viv*. It could be stuffed into tubes cut from *vaapk* (cane or Carrizo [*Phragmites australis*]) to make a pipe or cigarette called *oachkǐ*. These are six to eight inches long and were once important votive offerings at the *vapaki* 'ruins' or at other shrines in Pima country. The Carrizo tube was once the only acceptable ceremonial method of smoking. The other technique was to roll the tobacco in the fine white inner husks or shucks of the corn. These cigarettes were also called *oachkǐ*. "Wet [the] *huuñ eldag*. They have to roll it up and *hold* it and smoke it," Sylvester explained.

I asked him, "Have you heard of Pima using the *vaapk* for the *oachkǐ*?"

"Yeah. Way back—maybe a century ago—the bamboos [cane tube] were used for a cigarette or pipe or whatever."

"Did that burn too, or just the tobacco?"

"Just the tobacco."

Although Pimans smoked several different plants, tobacco was reportedly never smoked with anything else except occasionally the powerful lichen, *jeweḍ hiósig* (see Group L).

Joe told me, "In every village there was a round house, called the *jeeñ kii* 'smoke house' where the men came together every night to smoke and discuss important matters. You know, first they smoke, then they'd talk and talk until everyone reached a decision. They all had to agree on whatever it was, where they were going hunting or whether to go on the warpath against the Apaches or whatever. They'd all have to agree. They would pass that cigarette around, in the *vaapk* [cane tube], and take a few puffs, and then hand it to the next

one, addressing that one in his proper relationship, his kinship term, whatever he was to him. They'd pass it that way, using that greeting. It was something sacred. They didn't just smoke the way we do now. See, it was a kind of a ceremony, something sacred—before they would discuss their business for the evening." The person who took care of this communal house was given the title *jeeñ chekchim*, keeper of the smoke house.

The sacredness of the communal smoking Joe described also had its origin in the Creation Story. Verdin Shaman, Giisobǐ, who obtained seeds from Tobacco Woman, took the dried leaves one night to the *jeeñ kii*, rolled a coal toward himself and lit them in a cane tube pipe, and began to smoke. He then passed the pipe around the smoke house to the others. "When it came back to him he stuck it in the ground." The next night,

> the father of Tobacco said, "Last night I had a smoke, but I did not feel good after it."
>
> And all the others said, "Why we smoked and enjoyed it."
>
> But the man who had eaten the greens *kah-tee-kum [katikum]* the day before, said, "He does not mean that he did not enjoy the smoke, but something else troubled him after it, and I think it was that when we passed the pipe around we did not say 'My relatives,' 'brother,' or 'cousin,' or whatever it was, but passed it quietly without using any names."
>
> And Tobacco's father said, "Yes, that is what I mean."
>
> (And from that time on all the Pimas smoked that way when they came together, using a cane-tube pipe, or making a long cigarette of cornhusk and tobacco, and passing it around among relatives.)

There are various other accounts of ceremonial social smoking throughout Piman legends, but they always involve the same sequence: the men enter the communal round house and sit around the fire in silence. They may talk for a while. The person who brings the tobacco stuffs the tube, rolls a coal from the central fire toward himself, lights his *oachkǐ*, takes four puffs, then passes it to his right. This person acknowledges the giver with the

proper kinship term, takes four puffs, and continues the tube counterclockwise until it again reaches the first smoker, who then sticks the butt upright in the dirt next to him. "In warrior circles, it was moved along near the ground, with the words, 'It is traveling low—sneaking like a warrior'" (Castetter and Bell 1942:219). "The men (among both tribes) then discussed the topic of the evening for a time, and when all had agreed on the issue in question, the chief [*sic*] relighted the cigarette and passed it around the circle just as before; or, if the problem were a knotty one requiring lengthy discussion, the cigarette might be relighted and again go the rounds. The main reason for all smoking the same cigarette at the meeting was, in the view of these Indians, to develop a unity of thought, spirit and purpose" (Castetter and Bell 1942:219). (Those familiar with the Japanese tea ceremony will appreciate the parallels with Piman ritual smoking.)

Tobacco smoke was intended to enhance vision or deepen perception. In a formal speech, considered one of the oldest genres of Piman oral literature, Navǐchu, the masked Puebloan being from the north, says:

> It was after the creation of the earth,
> and there was a mud vahahkkee
>   [ruins],
> and inside of it lay a piece of wood
>   burning at one end,
> and by it stood a cane-tube pipe,
>   smoking,
> and we inhaled the smoke,
> and then we saw things clearer and
>   talked about them. (Lloyd 1911:214)

The smoke was a votive offering or prayer, much as a lighted candle is used today: "In ceremonies or meetings, each man blew the smoke upward to a Spirit in the heavens, upon whom he was calling for help or guidance in making decisions. Some informants, particularly among the Pima, asserted that the great power called upon was the sun. Smoking, both group and individual, was regarded as a sort of prayer—a medium of communication between the smoker and the Great Spirit [*sic*], while the smoke itself was a means of gaining an audience, as well as the bearer of messages" (Castetter and Bell 1942:219). The smoke also might help an individual to "see." For instance, in going to war the *chukuḍ namkam* 'Great Horned Owl meeter' might blow smoke toward the enemy to see where they were camped. The diagnosing medicine man *(saíjukam)* would blow tobacco smoke over the patient to help him "see" what was causing the illness. Then another healer is called in, one who knows the proper song to cure that particular disease. This curing *maakai* not only sings but may blow the curative tobacco smoke over the patient and apply certain fetishes.

In addition to this general use of tobacco smoke in curing, there was one specific curative use: "The eagle is also blamed for the lice that find refuge in the hair of the Pimas. The remedy is to blow cigarette smoke over the head" (Russell 1908:263). This would appear to be an empirically derived treatment. Nicotine in tobacco leaves is toxic to insects and is still used in some garden sprays. Sylvester Matthias told of another medicinal use: "Tobacco was used for toothaches. [Use] plug chewing tobacco."

To protect his unborn child, a man had to observe many taboos while his wife was pregnant. Most dealt with hunting and the handling of game, but one had to do with tobacco. In the Pima Creation Story, where men are being instructed on ceremonial smoking, the *maakai* says, "Of course you are all aware that if any man among you has a wife expecting to have a baby soon, he should not smoke it, but pass it on without smoking to his neighbor, for if you smoke in such case the child will not be likely to live very long" (Lloyd 1911:179).

Traditionally, smoking was considered proper only for hardened or mature men. Castetter and Bell (1942:218) were told, "It was considered injurious to young men, weakening them, causing a cough, making them lazy and fat, or unable to stand cold and preventing them from being alert. They did not believe it injured old men in any way." The negative attitude about youthful smoking is expressed early in the Creation Epic. Jeweḍ Maakai, Earth Shaman, made a series of people but ultimately had to destroy each world by pulling down the sky on them and starting over. "But these new people made a vice of smoking. Before, human beings had never smoked till they were old, but now they smoked younger, and each generation still younger, till the infants wanted to smoke in their cradles. And Juhwertamahkai

[Jeweḍ Maakai] did not like this, and let the sky fall again, and created everything new again in the same way, and this time he created the earth as it is now" (Lloyd 1911:30).

At the turn of the century Russell (1908: 119) recorded, "Boys learn to smoke at an early age, though the use of tobacco is not encouraged. The father's favorite saying in reply to a request for tobacco is, 'I will give you some when you kill a coyote.'" Of course, this would be said only by a member of the *ñui* (Buzzard) Moiety.

Francis Vavages said that when he was a youngster in the early part of the century, he and other boys would accompany their parents into a certain trading post where papers and tobacco were provided for the adult patrons. While their parents were busy shopping for flour, lard, and sugar, the boys would cut off a piece of tobacco and sneak around back of the post to smoke. "We would sure think we were big!"

At least early this century, it seems that young people began to experiment with smoking, often with various substitutes, some comic. Sylvester Matthias described one such early event, in which a Pima went away to school and learned about chewing tobacco. In the *jeeñ kii,* "He offered it, but nobody wants to chew! Alright. He picked up a broken pottery [sherd], that broken olla, and heat it up. And he took out his knife and sliced it off, slice it off, cut that chewing tobacco in little pieces, and dried it over the fire— heat the broken pot—and then slice it— cut up, just in small pieces—then he rolled—he rolled. And he smoked. Then he said, 'Alright, go ahead. You go ahead and smoke.' Then everybody grabbed it! *Eeeeh!* They were sure smoking with that chewing tobacco!"

Ruth Giff said, "When we were kids we used to smoke some soft root that we found at the river."

Sylvester told of this also: "[We also smoked] *vakolu* 'driftwood', smoked the root—found in the riverbed. Certain kinds of tree—maybe [from] aspen *[Populus tremuloides],* because it's real light. Not a native [local] tree." Others thought this was the root of *auppa,* cottonwood *(Populus fremontii).* He also told of youngsters trying the dried leaves of willow and Arrow-weed *(Pluchea sericea)* pulled from arbor roofs, and even dried horse manure.

There were two local wild species of tobacco, one *(N. trigonophylla)* usually found growing on the desert washes, the other *(N. attenuata)* mostly along the river (see Group G). Castetter and Bell (1942:111,113,212) were strongly convinced that there was no aboriginal cultivation of tobacco among the Pima. They noted that neither Emory in 1846 nor Bartlett in 1852 noticed the plant among the Pima crops and that the neighboring Maricopa Indians were not growing it themselves (Spier 1933:333). But I find this difficult to believe. With the exception of corn, no other plant has such a prominent role in the Creation Stories, always as a cultivated crop. The only nonaboriginal crops to be mentioned in the Creation Story are black-eyed peas, watermelons, and muskmelons—and these but in passing in a late version (Hayden 1935). Even Kino's wheat, which became the backbone of the Gileño economy, is absent from the myths. The fact that there was an injunction against planting tobacco publicly would easily explain why most earlier explorers failed to notice the crop. Surveyor A. B. Gray (1855:33) said that tobacco was being raised together with cotton, wheat, beans, and melons, but he did not specify which were Pima and which were Maricopa crops. Lieutenant A. M. Whipple (1856:28) listed tobacco as a major crop "cultivated by Pimas and Maricopas." However, Spier (1933:333) stated that, unlike the Pima, the Maricopa never raised tobacco. In 1869 Browne (1974:109) noted that the Gila Pima planted tobacco when the mesquites began to leaf out (late March to early April), and Lloyd (1911:58) recorded tobacco cultivation in 1903.

Both *N. rustica,* generally an eastern North American species, and *N. tabacum* from the south were being raised by the Gila Pima earlier this century. Castetter and Bell collected seeds of both species during their fieldwork (1938–1940), but I have been unable to relocate their samples (UNM). They note that Manuel Lowe near Sacaton called *N. rustica* (the smaller species) Coyote's Tobacco. Joe Giff and George Kyyitan likewise called *ban vivga* a cultivated species, though others restrict this name to one or the other of the wild tobaccos (see *N. trigonophylla,* Group G). It seems likely that this was a floating taxonomy, with one or more of the inferior species falling under the folk taxon 'Coyote's Tobacco'. *N. rustica* is the older species in the Southwest, and there is some evidence that it was part of the Hohokam crop inventory at a rather early date (Bohrer 1991; Setchell 1921).

One day when we were discussing smoking among the Pima, Francis Vavages mentioned the name of a tobacco I had never heard of before: *cheedagĭ viv* 'green tobacco', which people smoked while his *viikol* (great-grandfather) told stories. A year or two later Francis said, "This old man [called Vipismal 'Hummingbird'] plants all that. He saves the seeds. He planted every season and then he saves all the Bull Durham bags. Then he'll crush it up, then fill those bags, then go around and give it to *all* the old people. I have a great-grandfather [who] was so old that people come over there and ask him questions. Lot of things, that happened from way back. They sat around in a circle and burnt a log. And so, in those days, they don't have any rolling paper. They have corn husk, you know, they save that. And what they did is use their finger nails. They just cut it with their finger nails. *Huuñ eldagĭ.* That [is] the only thing they can do is hold it together to smoke. They light it right on that log, that burning log. They sit around and talk about this and that."

Francis described *cheedagĭ viv* as being about chest-high: "It grows about that high, I think. I've seen it, but I don't really remember. But it must be this tall." Francis again indicated to his chest. "They got purple flowers. Kind of light, light purple. This man [a Mexican] came and said, 'What do you do with that?' 'We smoke it.' 'Do you know what it is? That's marijuana [sic].' They didn't know. They just like the taste."

"Do you remember the shape of the leaves on the *cheedagĭ viv?*"

"Well, it's not too big. It's about that big, say two and a half inches long and about one and a half inches wide. They get them when they're kind of young, you know, and still green. And whatever is left forms the flower and the seeds. They thrash that and replant it again. It is like a rose flower, only it's pinkish looking—pink-looking flower. They're not too big, but I could see it when they were blooming."

Another time I asked, "What kind of flowers did it have?"

"Pink flowers. I've seen it, the blossoms. They're kind of pink. They [the plants] grow about that high [chest-high]. This

old man went to pick them, too, then spread them out, dry them out, crush them, then put them in a little tobacco bag—a Bull Durham bag—and gave it to each of the old people."

The scene Francis was describing took place about 1910. The old people could well have been reminiscing about the 49ers headed for California and the gold-

fields. The plant they were smoking was *N. tabacum,* the species of modern commerce (not marijuana, *Cannabis sativa*). It appears to have originated in South America or the West Indies, then spread through various tribes during historic times. Only *N. tabacum* has pink flowers. All other species are yellow or whitish flowered. Some tobacco cultivars from

viv *Nicotiana rustica*

Painted from a plant grown from Santo Domingo Pueblo seed supplied by Native Seed/SEARCH, Tucson.

the northwestern Mexican rancheria area are known to be very potent.

Francis told a bit more of the lore of tobacco. "Bull Durham, [in] bags, came when the white man start to come in. There's another tobacco they call *tosiv viv*. That's 'meadowlark [tobacco]'. That's kind of a brown package, wrap around. *Tosiv viv* they call it. [From] the cover. A little bigger than the Bull Durham." The Pima called this meadowlark tobacco because that bird was pictured on the wrappers.

When I asked George Kyyitan about *ban vivga,* he told me this story: "There's a man, right here. His name was Antone. His son was Francisco. His grandson, Henry Antone. And he used to plant that *ban vivga*—right there—by that canal that used to run through there, and there's a little garden on that side. He used to plant those *ban vivga* there. Then after that, he moved over on this side of the school. He plants this tobacco."

"Did they have yellow flowers?"

"I don't know what kind of flowers, but the leaves [were] still green yet. He always stays in that garden all the time—weeding it, weeding it, keeping it clean all the time.

"I heard that *ban vivga* sure is strong. You can't inhale it. Oh, boy! That's why most of them, when they buy it from him, they have to mix it with Bull Durham. That's not strong, not so strong. Just going to mix it. If you smoke that by itself—boy, it's just going to choke you up, make you cough."

"How tall were they, George?"

"About this tall," he said, indicating hip-height. "Almost like a *hivai* [Wild Sunflower, *Helianthus annuus*], *hivai* leaves, the size of your hand."

"How about *cheedagĭ viv?*"

"Maybe same. No . . . I've never seen it." George continued, speaking of old Antone: "He said he had some *[ban vivga]* seeds there, not much, just a little bit. And when he died, I don't know what he done with it. No more tobacco here."

"Have you seen the little wild tobacco out in the *akĭ* ['arroyo']?"

"I don't know. I don't know that one. Maybe I've seen it but didn't know it."

Already early this century, seed was being lost at some villages. Sylvester said that it was already gone when he was young. He never saw anyone raising it. They had to go to the store for it.

Joseph Giff told me that when his village realized that no one had any seed left because they had buried the last of it in an old man's grave, they dug it up, retrieving the seed, and had tobacco again for a while. Then they lost it for good.

Eventually the many traditional practices of Pima tobacco growing and use fell away, as they did with most other crops. It was left to the more conservative Tohono O'odham to preserve a sense of the sacred. Francis Vavages of Sacaton thought that tobacco could be sown just anywhere in fields, although those in the western villages recalled that it had to be in a secret place. In the early 1940s, people from the Sacaton area gave Castetter and Bell (1942: 212) this description of its cultivation:

Among the Pima, tobacco was always grown on irrigated land just like any other crop, preferably in sandy loam, never in saline soil as it would not do well. The ground was first loosened with the digging stick, then leveled. Tobacco seed was planted by an old Pima man holding it in the palm of his hand, placing the tongue in contact with a portion and more or less spitting or blowing it over the soil. Another man followed and covered the seed by brushing the soil with a branch of saltbush *(Atriplex)*. Holes at each corner of the field, and a brush barrier at the point where the ditch entered the field, trapped any debris brought in by the water.

The Pima practiced deliberate selection by preserving seed from the largest plants with the biggest seeds.

Even though the ritual context has been lost, smoking is still light among the relatively few Gila Pima men who do smoke. At the high school where I used to teach, there was a "smoking wall" where the boys could indulge. One or two cigarettes a day was their norm. The majority did not participate. Although I was told that there was no taboo against Pima women smoking, I must confess that in all my years on the reservation, I cannot recall ever seeing a pure-blooded Pima woman smoke.

Opuntia ficus-indica (L.) Mill.
domesticated prickly-pear, nopal, tuna
Cactaceae
ge'egeḍ iibhai, navĭ

Various species of prickly-pears, both North and South American in origin, have found their way into Pima gardens. One of these, the nopal, or domesticated prickly-pear, has large flat, nearly spineless jointed stems that form hedges taller than a person in many parts of Mexico and the Southwest. The tender pads in early summer are peeled, diced, boiled, drained, and fried to make the nopales of Mexican as well as Pima Bajo and Mountain Pima cooking. The large fruits, variably yellow through reddish, have a somewhat melon-like flavor.

Sylvester Matthias called these *ge'egeḍ iibhai* 'large prickly-pear'. They are found planted rather extensively in yards across the reservation. It is curious that these northernmost Pimans use the word *iibhai* in the generic sense, whereas the Pima Bajo, Mountain Pima, and Tohono O'odham use *nav* or *navĭ* as the generic, reserving the word *iibhai* just for its fruit (see, for instance, Mathiot n.d.:105, 436). But the Gileños have not lost the generic word entirely; it persists as the name for the tiny Christmas Cholla *(O. leptocaulis),* called *navĭ* or sometimes *a'aji navĭ*. Russell (1908:75) gives *naf* for the usual wild prickly-pear found on the reservation, so the change is recent. At the village of Bapchule, George Kyyitan made quite a different distinction, calling the large cultivated one *navĭ* and the wild ones all *iibhai*.

Ruth Giff grew a tall row of nopales in the yard, as do many other Pima. One day when I visited her, she had several pads lying on her kitchen stove. She explained that she first burned them over the open gas flame to remove any "stickers," then peeled and boiled them. She had heard that some pick them younger to avoid the need for peeling; as the pads mature, the epidermis thickens.

George Kyyitan was aware of the problems of eating too many *iibhai* fruit (see *O. engelmannii,* Group H). I asked if this was a danger with cultivated *navĭ* as well. "I don't know. Just tried it one time—[tastes] like apples."

I stopped the pickup at a tall row of domesticated prickly-pears growing along a fence and asked Sylvester what they were.

"*Ge'eged̦ iibhai.* Some call it *nav.*"

"Is it good for anything?"

"You can eat the fruit, just like *iibhai.* The Mexicans and the Apaches go for the young shoots. They make it just like the Mexicans—just like [summer] squash."

"Did Akimel O'odham do that?"

"No. Mexicans do use that and add chile, just like Dr. [Allan] Phillips' wife Juana. Add chile, onion, and cheese. They make a meal out of those—shoots of *ge'eged̦ iibhai.* And Apaches use it as squash."

"What about the regular wild one, *iibhai?*"

"Yeah. They can still use that, too, for squash."

"Did the Pima ever cook that, use that?" I asked.

"No, but they know. But they never try it, except cholla buds, *hannam [O. acanthocarpa].*"

Others I asked were of the same opinion as Sylvester. The pads of neither the wild local prickly-pears nor the cultivated species were cooked as food until relatively recently. Castetter and Underhill (1935:14) mention such a use among the Tohono O'odham, however.

Another curious twist is that the Spanish name for this plant or its fruit, tuna, appears to be the source of the pan-Piman word *suuna* for the domesticated fig, suggesting an early association between the two early in the colonial mission period. (Saxton, Saxton, and Enos [1983:55] derive O'odham *suuna* from Yaqui *chuuna* 'fig'.) The cultivated plant

is sometimes called *higo de Indies* or in English, *Indian-fig.* On the Gila it may be rather recent, perhaps dating from the settlement of the Phoenix region. At least in the late 1930s, Curtin (1949:61) noted, "Mr. Peebles told me that on the Salt River Reservation, Indians are growing an imported Mexican pricklypear for its abundant fruit, and the Mexicans in that locality are so fond of this species that they procure it from the Indians."

ge'eged̦ iibhai, navĭ *Opuntia ficus-indica*

Phaseolus acutifolius A. Gray var.
   acutifolius
   [≠ *P. vulgaris* L.]
tepary
Fabaceae (Leguminosae)
bavĭ

Two of the first new foods I encountered when I took up residence in a Pima village on the Gila were *hannam,* or cholla buds, and the little bean called the tepary. Both the white and brown varieties of tepary resemble familiar common beans, only smaller. But I thought just Pimans knew and cultivated the tepary; no one off-reservation seemed to have heard it. I was shocked out of my provincialism three or four summers later while touring the United States Department of Agriculture seed bank in Fort Collins, Colorado. The guide used as one of his examples of the bank's function the preservation of all the known cultivars of teparies, a desert-adapted crop that was slipping into oblivion. He was the first white man I had ever heard say the word *tepary.*

Since that time, the tepary has experienced a well-deserved renaissance. A considerable corpus of information now in print extols the merits and limitations of this amazing crop from the arid regions of North America (see especially *Desert Plants* vol. 5, no. 1 [1983]; Nabhan 1985; Nabhan and Felger 1978; Pratt and Nabhan 1988).

The tepary is special partly because even as a crop, it retains the strategy of a desert ephemeral—get up fast, reproduce quickly, produce seeds before drought has a chance to hit, then die. To accomplish this, the tepary germinates rapidly, sending down a deep root. Its leaves track the sun during the day, use plenty of moisture, and photosynthesize to the maximum. "Under irrigation, teparies can be done producing a crop in two months' time, whereas other cultivated plants may need as many as two more irrigations over an additional five to seven weeks to complete their seed yield" (Nabhan 1985:115).

These characteristics made teparies especially suitable for the nonriverine Tohono O'odham, dependent solely on rain-fed irrigation. When summer rains soaked the arroyo fields deeply, their *bawi* could be planted and in three out of four years would bring off a good crop before the soil moisture was depleted. The early Jesuits correctly noted that the *papavi*, as they called it, was the staple that made possible these marginal agricultural communities (see, for instance, Velarde 1931:128). But not only the desert Pimans exploited the tepary's unique qualities; the riverine Pimans did not abandon *akĭ chiñ*, or summer rain agriculture, until early in the 20th century. And if the river was too low for irrigation, Gileño teparies were likely to produce when the slower-growing varieties of *muuñ*, common or kidney bean *(Phaseolus vulgaris)*, were still just in the vegetative stages.

Other qualities in addition to the desert ephemeral strategy suited this bean to desert agriculture. Under drought conditions, as many Anglo-American agronomists have observed for 80 years, the tepary can outproduce the common bean fourfold. The tepary is more tolerant of salinity, an important factor in alkaline desert soils. Even in the searing summer heats of southern Arizona and northern Sonora, teparies can still set pods. Their roots have been shown to be resistant to several blight and rust diseases disastrous to other beans. And Pimans have noted that their tepary plants are less likely to be eaten by insects than are other kinds of beans grown together in the same field.

Moreover, the tepary fares well in nutritional comparisons with the now ubiquitous varieties of common bean. For example, teparies "have higher iron, calcium, and protein content and fewer antinutritional factors than your 'average' grain legume" (Nabhan 1985:120). As a result of acculturation and adoption of the nutritionally deficient modern diet, the Akimel and Tohono O'odham have become increasingly susceptible to nutrition-related diseases (see chapter 8).

Although there is a long archaeological record of these little beans in the Southwest, there is no evidence they were domesticated in this region; no semidomesticated forms have been recovered. Pratt and Nabhan (1988:419) suggest that the origin may have been the interface of arid North America and Mesoamerica in the central west-coast states of Mexico. (This area is the historic southern limit of the Piman or Tepiman corridor.) They continue: "We feel that there exists very strong evidence for the Jaliscan plateau as a region where beans and maize were domesticated, based on multiple criteria. There, in the west-central states of Mexico, early tepary domesticates may have been literally intertwined with early maize and common bean domesticates."

The Spanish word *tépari* appears to derive from the Ópata and adjoined Eudeve from northwestern Sonora. An Eudeve vocabulary of the 17th century (Lionnet 1986:117) gives *tépar* as "*frijolito silvestre comestible*" in distinction to *mun* and *muni*, which are given simply as "*frijol*" (see also Pennington 1981:124). Sobarzo (1966:319) gives the derivation through Ópata *tepa* ('frijol' in the genitive case) plus the suffix *-ri*. (Incidentally, that author's repetition of the idea that the *tepari* arises by degeneration from other bean species is erroneous in every respect. This idea originates from Juan Nentvig's comments in the colonial *Rudo Ensayo;* see Nentvig 1971:73, 1980:23.)

As with most unanalyzable monolexemic words in Piman languages, *bav* or *bavĭ* or *bawĭ* has an interesting distribution. In contrast to the upper and lower Piman dialects, both Northern and Southern Tepehuan as well as Tarahumara use the word in referring to small *P. vulgaris*. In colonial Névome, *babi (bavi)* is the base term for four kinds of frijol (Spanish for 'bean'): the word *mun* does not appear. Except for Tarahumara, the word *bav* does not occur outside Tepiman languages (Fowler 1994). This suggests that the desert-adapted tepary has long been the bean throughout all Piman lands, with the more mesophytic common or kidney bean that came more recently, together with its Uto-Aztecan generic name, *mun/muñi*. (At least among the Pima, Tohono O'odham, and Pima

Bajo, a cognate term, *bawui,* labels Coralbean *[Erythrina flabelliformis].)*

The Gileños inhabit lands outside the ranges of the two wild varieties of teparies (see "Technical Notes"). But the related Sobaipuri (now extinct) on the San Pedro and upper Santa Cruz Rivers and the Tohono O'odham to the west lived within range of wild teparies, the same species as those cultivated in their fields (see maps in Buhrow 1983). Our Topawa friend and renowned potter Laura Kermen told Gary of harvesting the wild tepary she called *chepulina bawĭ.* The modifying word at first appears odd, but in colonial times it would have been pronounced *tepurina,* the word that yielded *tépari* in Sonoran Spanish and *tepary* in English. Culver Cassa suggests an alternative derivation through *chuchpulim* 'being square', in reference to the angular shape of the wild tepary seeds. Others called these *ban bawĭ* 'Coyote's tepary'. They are difficult to collect because, once dry, the pods shatter or dehisce—that is, the pods explode into two halves, flinging the seeds some distance. The seeds, furthermore, are small and look like gray-speckled bits of gravel, matching the arroyo bed. These characteristics suit the paradigm of "Coyote's plants" (see Group G, *Nicotiana trigonophylla;* Nabhan 1982). From our evolutionary point of view, these are the very traits that early domesticators would have selected out of their wild plants. By the end of the 1940s, the Tohono O'odham gave up on the wild harvest (Nabhan 1982, 1985).

If the Akimel O'odham ever knew about wild teparies or if the Tohono O'odham ever brought up *ban bawĭ* to trade during the wheat harvest, that fact has been lost in the mists of history. Today the Gila Pima know only the cultivated varieties, which they produced in abundance until their rivers went dry.

Among the cultivated legumes, Russell (1908:92) gave *toota bavĭ* as the most ancient among the Gileños: "At least five varieties of beans are now cultivated. The first known, the *tatcoa pavfi,* 'white bean,' is said to have been brought in some forgotten time from the valley of the great 'Red River,' the Colorado. Considerable quantities are raised and the thrashing is done by horses driven in a circle on the same hard floor that is prepared for the wheat thrashing." The remaining four varieties are not enumerated.

bavĭ *Phaseolus acutifolius*

It is difficult to access the role of teparies historically because most Anglo-Americans writing on the Pima and their generous supplies mentioned only "beans," without regard to type. Browne (1869:110–111), however, noted that the Gila Pima produced "a large quantity of beans called *taperis* [sic], and a vast quantity of pumpkins, squashes, and melons." Three years later, the Pima sold 20,000 pounds of beans, presumably also teparies, to the California Column.

In Juan Smith's version of the Pima Creation Story recorded by Julian Hayden in 1935 (see *Chenopodium berlandieri*), the first two crops made by Se'ehe were corn and pumpkin. "The next thing that Se'ehe made was the little Indian white bean [*toota bavĭ*]. So the next thing that he made was another kind of bean, that they used to called speckled beans. You don't see them anymore now" (Hayden 1935). Whether this latter refers to a speckled tepary now lost or to the mottled Pima lima bean is unknown. Then there is the song of the four crops:

Hi'i yana héo(lt) vava heo(lt)
[They are] singing
In the fields are singing,
These little white beans and the corn
    are singing.

Then the speckled bean and the
    pumpkin are singing together.
Vava heo(lt) they are singing.
Haicheya. Ya'eena.

(This song [reworked from Hayden 1935] is quite similar to a growing song or rain song Russell [1908:331–332] recorded; see *Zea mays.*)

Teparies were once a major crop and, like wheat, were usually sown broadcast. Ruth Giff said, "They broadcast teparies, then plow—watered first. Some do it dry, then put water on it [plant then irrigate]."

Interplanting was a widespread if not universal Indian practice. When George Kyyitan was telling me about how he planted corn, I asked whether he planted common beans or teparies with the corn.

"Yeah, mixed sometimes. We do that sometimes. We just broadcast them like wheat does. On the corn and on the *bavĭ*, we do that too, that way." But there was an additional motive, which became evident as George continued. "If we wanted to try to control the weeds, in there, we'd broadcast the *bavĭ* thicker, a little thicker, so the weeds wouldn't come up on top of them. I know that there are some that don't have no *bavĭ* on it because they are too thick. But that does that, just control the weeds."

A few days later at Sacaton, I learned more details on tepary cultivation from Francis Vavages.

"Just sow it, not much at a time: *gantaḍ* 'broadcast'. Have to get all the weeds out [first]; water the ground. Any weeds come up. Then I plow." He makes signs of spreading the seeds across the prepared soil. "Then rake or drag a plank," he points to a railroad tie lying near our feet, "kind of covers it up. I used to plant [a] lot of brown tepary and white. Use tractor to soften the ground—then drag, cover the seeds, about so much" (he indicated about two inches deep).

"And on the harvest?" I asked.

"Just *vopon* 'pull them up', put them in piles. Until really gets dry. Have to turn them over. When it gets too dry, [the plants] just crumble up. What I used to do, I used to pile them on canvas, then drive over with a horse. All the leaves come off.

"If a big field, lots of people come, but they feed them. Pile them up [the *bavĭ*] on a hard place. The ground has to be taken care of, too. Put water on there, make it hard, then you put your cloth, canvas over it. Maybe run a horse, too. One horse. Then whatever's left, you just beat it with a stick. Old ladies like that—beat them. Then they feed them [the workers] again, before they go home. But they like to eat meat—stew meat and later they remember, the next time, when [someone else's] field ready.

"And they get every bit of it, too. They [the teparies] don't stay on the ground, they get every bit. Then they put it out in the sun or they [the teparies] rot—they have to get *dry*—until it really cracks. Few days. Bring it in every evening because of the moisture, till they know it's *really* dry. They really get it dry. *Baví.* Others they can just pick bunches, *u'us baví* [black-eyed pea], *muuñ* [common or kidney bean]. But with the *baví* they pick the whole bush."

There were two ways of thrashing teparies, explained Ruth. "We beat them with a stick unless [there's] a lot, then we use the horses like with the wheat." She then explained these terms: *vinwa,* to winnow from a basket, spilling them onto canvas; *geegev,* to beat with a stick (either beans or corn); *kehimun (kehivia),* threshing with a horse; *da'ichuda,* to throw wheat or whatever up in the air to winnow; *hiik,* to cut (you *hiik* the *pilkañ* [wheat]); *pokash esh,* to punch a hole in ground and plant with *eskuḍ* 'digging stick'; *gantaḍ,* to spread or broadcast seeds.

When George Kyyitan was telling about stationary thrashing of wheat with horses, I asked if they did the same with the *baví.* "Uh-huh. We did it in no time because it's too dry and it pops up [shatters] right away. It's easy that way when it's really too dry. Get a basket and let the wind blow— left the big ones in there. Then we can get a big pan and get the dirt out or whatever it is in there and clean it up."

During the past half-century, only two kinds of teparies have been regularly found among the Gila River Pima: the white and the brown. The brown one comes from the pod sort of yellowish, but then oxidizes slowly to a warm red ochre-brown. This is a rather flat tepary with sharp edges, while the white one is plump, without abrupt edges where the two cotyledons meet under the seed coat. Gary Nabhan maintains that the *s-oám baví* currently available was selected by Anglos and that the original northern Piman form was a plump red-brown bean, now lost to Akimel and Tohono O'odham.

These two are the only ones I have ever seen cooked during my years of eating at Pima events. I asked Ruth Giff how many kinds she knew. "Just *toota baví* and *s-oám baví;* once in a while we get a black one in the white. We had a black *u'us baví* [black-eyed pea, *Vigna* sp.; not a kind of tepary]; we had it, but then we lost it; don't know where we got them; all black; grew true." Occasionally a black individual appears among the *toota baví.* "Can come in the brown ones, too," Francis Vavages noted.

I asked Ruth, "Do black teparies come on the whole plant or just a few pods?"

"I don't know because there are *very* few. I leave them in. But we plant just the white ones." She did not think they grew true.

S-oám baví 'brown tepary' could be called *s-vegĭ baví* 'red tepary', Ruth said, but the former term is preferable. George Kyyitan said the same. I showed him and others a plump gray-speckled tan tepary that I had selected from white sports and that had grown true for five years. No one recognized it. Sylvester added a third kind to the tepary pantheon: "There are some blue ones, *chehedagĭ baví*—that I know, that I've seen, but it's rare. I don't know who's got it, maybe the Hopi. Just plain blue, just the same size as *toota baví, s-oám baví.* They just died out. There's a [Pima] man who grew *chehedagĭ baví* [who was] originally from Santa Cruz. He lived right over here [in Komatke] and he had a little patch of land and he raised them, but he wouldn't sell them."

But Ruth Giff said, "I've never heard or seen a *chehedagĭ baví*—blue, nor a yellow one."

Even in the early 1940s, Castetter and Bell (1942:97) found only the current white and brown:

Both Papago and Pima had only two staple varieties of tepary—the white (really greenish white) and the brown. Black and variously speckled teparies also occurred but were rarely selected or grown as distinct varieties, and many Papagos and Pimas would pull up and discard such plants when found among crops of the white and brown ones. Informants of both tribes, with a single exception, regarded the brown as the oldest of the teparies. A present-day factor which limits the growing of teparies is the commercial demand for

only the white and brown varieties. The one most commonly found on the commercial market is the white, and this is often brought to trading posts for barter. Both tribes regard the white one as superior in taste and general quality to all the others.

But this was not always the case. At an earlier time, more varieties (reportedly at least 46 in southern Arizona; see Nabhan 1985:119) were maintained among these desert agriculturalists, each with its particular characteristics of microenvironmental tolerances as well as differences in seed coat color, shape, and flavor. Early in the 20th century, Freeman (1912, 1918) found 15 varieties of teparies being cultivated by the Tohono O'odham but only 3 still in existence on the Gila: in addition to the white and brown, he reported a *s-bidmagĭ baví.* This word indicates a 'very dark muddy color' (see O'neale and Dolores 1943:388, 393). This variety, "Brown Speckled *spate mook pave [sic],* 'muddy tepary,' is somewhat rare, having been found only twice among the Pima and not at all among the Papago." Russell's description (1908:76, 92) of Pima beans is inadequate, not distinguishing the folk generics.

When I arrived in Pima country in 1963, only the white tepary could be found in reservation stores, and then it was in low supply and expensive: approximately two ounces of teparies cost more than a dollar. Commercially the brown tepary was essentially unavailable, although a few Pimas still raised some for their own use. During the 1980s, *s-oám baví* once again became available on the reservation, supplied by several growers—including Sacaton's tribal-subsidized new crop farm.

Although I prefer the brown tepary because of its richer, more nutty flavor, most Pimas prefer the white one. Ruth Giff explained at least one factor that has led to its popularity: white teparies cook more quickly. Brown and speckled tan teparies I usually soak 8 to 12 hours. During this time they triple in volume. But Ruth said, "*Baví* are not all the time soaked. Can just start boiling them. Soak maybe [for an] hour or two hours. Sometimes soak pintos for just a short time. The pink beans are harder than the pintos."

Francis Vavages said, "In cooking *baví,* you can put in beef cracklings or *kooji* ['pork'] cracklings—or green *ko'okol* ['chile'].'' The best tasting *bavĭ*, according to Pimans, are those cooked slowly in a round open-top olla. In more traditional homes, one may still see these red clay bean pots with their smoke-blackened bottoms, usually stored inverted. Today these are made only by a few remaining Tohono O'odham potters and cost forty or more dollars for even a modestly sized one.

At Holy Family Church in Blackwater one Sunday, I had some *poshol* made from whole wheat, similar to the wheat berries of health-food stores, and brown teparies. The lady who made it said she first pounded the wheat, then cooked the two together. I asked George Kyyitan if he knew about *poshol*. "That's a lot of work. You have to pound that *pilkañ*—soak it and crack it—before you cook it with *oám bavĭ*."

Then I visited Ruth Giff at the west end of the reservation. "*Poshol* is *pilkañ*," she said, "whole or pounded, cooked with *oám bavĭ*. It's called *ga'ivsa* when mixed with corn instead of wheat."

There were other ways of preparing teparies, such as parching and grinding to speed cooking. In speaking of grinding the dry *hannam*, Ruth Giff added, "They do the *s-oám bavĭ* that way too: crush it, not grind [it], just big pieces, to make it cook faster, I guess, but they don't use *toota bavĭ* [ground]. Brown takes longer to cook. I don't know why." She could not remember what the cracked tepary is called. I asked George Kyyitan. He called them *bavĭ chu'ikana*. When I returned to Ruth with the word, she said, "Oh, yeah! But it doesn't taste very good, though, because they want to do it fast." Then Ruth explained some culinary distinctions in Pima: "*Chu'ikad* means [to] crack the brown *bavĭ* on the *machchuḍ* ['grindstone'] to make them cook faster. Maybe the white ones, too, if they're in a hurry. When you do it fine, it's *chu'i*. For *atol* [atole], just *haak* ['parch'] in the coals, then grind it, the *bavĭ*. But I've never done it. Just eaten it."

Francis Vavages likened the storage and preparation of tepary to that of corn: "Store the *bavĭ* in the *homda* ['Arrowweed storage bin'] too, like corn, too, when they're dry. Roast them and make *ga'ivsa* out of it."

Another day George Kyyitan was explaining how a gravy could be made from ground Saguaro seeds or the parched seeds of a wild mustard, *shuu'uvaḍ* (*Sisymbrium irio*).

"What do you call that?" I asked him.

"I think most of them call it different ways. We call it *ku'ul*. Same way with the tepary beans, they do that, too. They have to roast, heat, then grind, then make a *ku'ul*."

"[With] *toota* or *oám bavĭ?*"

"*Oám bavĭ*. Roast it first, then grind it with [the] *machchuḍ*."

"Roast in the *muta* ['clay parching pan']?"

"Yeah."

"How fine do they get that *bavĭ?*"

"Not too fine. Not too thin."

"Like *huuñ* ['corn'] or *pilkañ* ['wheat']?"

"That's too thin, too fine. They do that, too, with the *hannam*."

As with all other crops, teparies are not the miracle crop to supply all the protein needs of vegetarians, even though sometimes touted as such. And, as Nabhan (1983a) points out, the crop has its disadvantages as well: it is sensitive to overwatering, it has the same deficiencies in key nutrients as other grain legumes, and some may object to the flatulence it might cause. (I soak mine well and cook them thoroughly and suspect this may be an important factor in the reduction or elimination of flatulence.) And teparies are several times the price of the more readily available kinds of beans. In addition, many of the more interesting cultivars simply are not commercially available. Through the efforts of seed collectors, once again seed of more than a dozen varieties can now be bought from Native Seeds/SEARCH, Tucson, by those who wish to raise their own. For me, these ancient beans are an acquired taste, one of those facets of variety that makes life that much more worthwhile.

The tepary's claim to fame is not that it differs significantly from other beans in nutrients, minerals, and fiber, but that it is so excellently adapted to the agricultural conditions of the Sonoran Desert. With megalopolises such as Los Angeles and Phoenix gobbling up ever more and more surface water, and the energy costs rising for pumping groundwater from everlowering aquifers, the little *bavĭ* is one of the secret alternatives to abandoning agriculture in these regions altogether.

Technical Notes: The taxonomic status of the three named varieties of *P. acutifolius* has long been disputed and treated differently by various authors (see Pratt and Nabhan 1988: table 3, p. 414). Gray's description of the species dates from 1849. The domesticated forms have often gone under the name *P. acutifolius* var. *latifolius* (Freeman in 1912). Because of various discrepancies and irregularities Pratt and Nabhan concluded that "var. *latifolius* is a *nomen confusum* and obsolete [later or junior?] synonym for *P. acutifolius* var. *acutifolius*." This latter name subsumes both the domesticated forms and the broader-leafleted wild individuals growing at lower elevations along streams, canyons, and arroyos. The other form, with narrow leaflets, is *P. acutifolius* var. *tenuifolius* (A. Gray in 1853). This form occupies more xeric sloping habitats, usually above 200 meters. It does not enter the equation of domestication. "Already adapted to natural disturbances which scarify their seeds and open the vegetation enough to reduce competition for light, wild var. *acutifolius* was pre-adapted to colonizing floodplain fields which prehistoric farmers opened up for planting. These trends provide a possible answer to the question, 'Why was one [botanical] variety domesticated and not the other?'" (Pratt and Nabhan 1988:423).

Phaseolus lunatus L.
lima bean
Fabaceae (Leguminosae)
havul

The Pima lima bean is a most attractive bean. The seeds have a tannish pink coat modulated irregularly with purple scrawls or splotches. No two are alike. A few are immaculate pink or nearly so, while others are marked almost solidly purple. In time, a year or so, the coats oxidize to a burnished brown blotched with black. Generally, two or three of these handsome seeds are borne in the broad pods. The bushes are freestanding, with some tendencies toward vining.

Something of a controversy has surrounded this bean for decades: Is it aboriginal? It has been found in Hohokam and Salado archaeological sites. But did the Pima have it before the 20th century? The early writings of Kino, Manje, and

O'odham havul   *Phaseolus lunatus*

Painted from a plant maintained by Ruth Giff of Komatke.

Velarde help little with beans other than *bavĭ,* tepary *(P. acutifolius).* Russell (1908:92) noted that five varieties of beans were cultivated, and he discussed the white tepary, but he did not even list the other four. Inexplicably, he listed the word *havul* in the food supply as Banana Yucca *(Yucca baccata).* Freeman (1912, 1918) made extensive collections of beans among the Akimel and Tohono O'odham but did not report finding any limas.

Yet all the people with whom I worked knew this bean in their youth (early in the 20th century). It is called *havul,* and it is not considered a subset of either *muuñ* (common or kidney bean) or *bavĭ,* even though all three are considered to be beans, in the English language. (*Havul* also means 'knot' or 'to tie a knot'

or 'to have a knot tied in something' [Mathiot 1973:314].)

But the folk generic *havul* pops in and out of Gileño history. It would seem to be absent from the recorded versions of the Pima Creation Story. Juan Smith's version (Hayden 1935:11) tells of the four original food crops created by Se'ehe for the People: corn, pumpkin, tepary, and "another kind of bean that they used to call speckled beans, you don't see them anymore now." Is this the pinto, a variety of the common bean that arrived late among the Pima, called *s-bibiolmagĭ muuñ?* Or is this speckled bean the lima? Given the time of the telling (1935), my own opinion leans toward the lima.

George Kyyitan knew both the narrator and the translator. One day when we were

discussing beans, I asked George whether he thought the "speckled bean" in the story was a pinto or a *havul.* He said, "I think that's that *havul,* there. *Havul* is like that."

"So that's considered an ancient crop?"

"Yeah."

Castetter and Bell (1942:99–100) opposed the concept that the Gila Pima have had *havul* all along:

A lima bean of the Hopi type was cultivated by the Pima for many years, although evidently not aboriginally *[?].* Between 1915 and 1930 it almost disappeared from the Gila River Indian Reservation and seldom was seen at the trading posts. About 1930 the Cooperative Field Station at Sacaton began to select Indian lima types for adaptation to the warmer areas of the Southwest. Beginning in 1931, several color types were secured from the Hopi reservation, and the selected strains of white Hopi limas were obtained from the Agronomy Department of the University of California. In 1932 some samples of the Pima strain were secured which have been designated Pima-Hopi lima. Varying somewhat in color, this strain is typically light tan with black longitudinal markings. Selections were made from the strains and some improvement obtained in their ability to retain flowers and young pods during hot weather and to resist spotting of the young beans.

This paraphrases what King and Beckett (1938:58) had reported just a few years earlier. However, where I inserted the bracketed question mark in Castetter and Bell's account, King and Beckett had said, "Among the older Indians and Indian traders there is some disagreement as to its origin and its importance as a crop." According to Whiting (1939:81n.12), "the commercial 'Hopi Lima' does not represent any aboriginal variety," being selected from a mixed collection from Second Mesa.

At any rate, the pure Pima strain was preserved at the University of California, Davis, and reobtained in 1932 for breeding purposes at the U.S. Field Station. This *havul* was a locally adapted landrace with superior qualities for hot country: "In

comparative plantings of unselected seed
the Pima strain ordinarily is more productive than the native Hopi strain" (King and
Beckett 1938:58). (For some reason, this
original strain was designated Hopi-Pima
lima, rather than just Pima lima; the logic
of this escapes me.)

The exotic white strains, according
to the field station experts, were more
susceptible to spotting and shriveling of
young beans. "Nearly all of the Hopi and
Hopi Pima lima (Indian lima) strains and
selections," they wrote, "show a greater
resistance to root knot nematodes than
do the various kinds of common beans
and teparies, but some strains of lima
show greater resistance than others and
selections have been made in the effort
to obtain a completely resistant strain"
(King and Beckett 1938:58). The selected
strains were distributed to the Pima,
"who are showing much interest in
making greater use of this type of bean
as a food crop."

George was explaining that pink beans,
the variety the Gileños had in his youth,
were difficult to raise but that the pinto
that came in later was easier.

"What about *havul?*" I asked.

"Which *havul?*"

"Which one did you use[d] to raise?"

"That *O'odham havul.* We didn't raise
those white ones. They didn't agree here.
I had [a] first cousin [who was] living
over here. And he tried it and plant[ed]
some of those white ones. Yes, it grows,
but nothing on it. Just get bigger and all
those stalks are *big.* Looks pretty good
though. But there's *nothing* on it. Nothing!
Then he tried again. And it's the same
thing, again. Then he didn't raise [it],
no more."

I asked Sylvester Matthias about the
kinds of limas he knew. There were three:
O'odham havul, the one grown by the
Pima; *toota havul* 'white lima', the one
bought in stores; and *al ha'as havul* 'baby
lima', a strain once grown by Gileños,
apparently long lost (*al* means 'small
[or] baby', *ha'as* is a qualifier: Culver
Cassa renders this 'smallish lima').

In July 1984, Ruth Giff told me, "All
my corn and squash haven't produced.
The rabbits are eating a lot. But the corn
doesn't have much kernels on the cobs.
Maybe I didn't plant it at the right time.
But the *O'odham havul* are really doing
good. We like them green. I've been
picking mine already and cooking them."

Some years later when we were discussing
tepary harvesting, Ruth said, "*Havul* is a
lot of work, because you don't pull up the
plants [like *bavĭ*], but just pick them, or
the seeds might fall out. It doesn't ripen
all [at] the same time."

Her limas are purple marked, at least
when fresh, suggesting that these have
descended from the field station stock.
One of the interesting things, though,
is these are invariably referred to as
O'odham havul, the People's (or, more
specifically, the Piman) lima bean. And
that is the only one raised. It is their
lima. It has dipped in and out of history,
but the People's lima is still being preserved by a few old people who will leave
their beds at four in the morning, before
the onset of the soaring temperatures, to
tend to a few ancient crops.

Vouchers: *894* (RG).

Phaseolus vulgaris L.
common bean
Fabaceae (Leguminosae)
muuñ, muuñĭ

The common bean is the bean species best
known to most English speakers. In our
markets it goes by a number of different
names, such as kidney, pinto, pink, black,
navy, string, and wax bean. These are all
varieties or cultivars of a single species, in
contrast to limas *(P. lunatus)* and teparies
(P. acutifolius). (The Pima folk generics
also segregate these out exactly along the
three biological species lines: *muuñ* or
muuñĭ in contrast to *havul* and *bavĭ.*)
Although each variety of common bean
has its own flavor, these are all subtle distinctions compared to the very different
tastes and textures of limas and teparies.

The common bean is a big-leaved plant
without all the fine-tuned desert-adapted
physiological characters that Gary Nabhan
and others have discovered in the tepary
(Pratt and Nabhan 1988). It was not a
good crop for *temporal* dry farmers but
was satisfactory for irrigation agriculturalists on the river.

It is difficult to assess the role of bean
species among the Gileños in historic
times. In Hispanic documents, the distinction was seldom made between *frijol*

'bean' and *frijol tepari* (or just *tepari*).
And Anglo-Americans usually recorded
everything simply as "beans."

But Velarde (in González 1977:50–52)
shed some light on the arrival time of
the common bean among the northern
Pimans. In one passage he established
that the bean raised by the Tohono
O'odham was *bavĭ;* in another passage
he stated that the indigenous bean of
Pimería Alta was the small tepary, with
various other kinds of beans (likely
P. vulgaris) coming in with Spanish
contact, presumably along with the name
muuñ, widely used by many rancheria
tribes in northwestern Mexico. (Velarde
also mentioned *haba,* or fava bean, but it
seems never to have caught on as far north
as the Gila; no one with whom I worked
recognized it.)

In discussing beans among Gileños at
the turn of the century, Russell (1908:76)
considered at least one variety of the
common kidney bean to have been grown
by the Gileños before Spanish contact.
However, he did not realize he was dealing
with different biological species as well as
folk generics, listing all under "*Pavĭ,
Phaseolus vulgaris.*"

Lloyd (1911:60) was more explicit:
"Perhaps the staple food of the Pima
even more than corn *(hohn)* or wheat
(payl-koon [pilkañ]) is frijole beans—
these of two kinds, the white *(bah-fih)*
[and] the brown *(mohn).*" Apparently the
important beans at this time were a white
tepary and a pink (or brown) common
bean—exactly what the ethnohistorical
accounts suggest.

By the 1940s, Castetter and Bell
(1942:78, 98, 100) found only the common
pink or Mexican pink bean being grown
commonly on the Gila. Despite Russell's
contention, they doubted the pre-Hispanic
status of the kidney or common bean
entirely, partly because Underhill found
no growing songs for *muuñ* among the
Tohono O'odham: "It was unthinkable!"
(Castetter and Bell 1942:98). While corn
and squash or pumpkins share a prominent place in the Creation Story (see
Chenopodium berlandieri) as well as in
growing and rain songs, the position
of beans is less secure. In Juan Smith's
version of the Creation Story (Hayden
1935:11), the first four crops were "corn,
pumpkins, little white Indian beans, and
speckled beans." The little white bean, of
course, is *bavĭ,* the tepary. By implication

the speckled bean is larger, but is it *muuñ* or *havul,* the lima, which is also mottled?

Archaeological evidence shows that early inhabitants of the Southwest had had *P. vulgaris* for millennia. And linguistic evidence shows that the Gileño *muuñ,* a primary unanalyzable lexeme, is a widespread Uto-Aztecan name, appearing in identical form with the Tohono O'odham. The word *muni* occurs in Tarahumara, Yaqui, Mayo, and Guarijío, apparently always for the common bean. Cognates occur in Cora and Huichol and to the north in Southern Paiute and Hopi (Fowler 1994). In Eudeve, the *frijol—mun* or *muni*—was contrasted with *frijol pequeño,* or *tepar* (Lionnet 1986; Pennington 1981). Mountain Pima use *bavi* as the term for their many varieties of common bean. According to Joseph Laferrière (personal communication), they do not raise the desert tepary, however.

Most Pima distinguish only two varieties of *muuñ:* the pink and the pinto. Sylvester Matthias provided the following amplified taxonomy:

> *vepegĭ muuñ* 'pink bean'
>
> *s-bibiolmagĭ muuñ* 'muddy bean', pinto bean
>
> *toota muuñ* 'white bean', navy bean
>
> *chehedagĭ muuñ* 'blue bean', a bluish-greenish Hopi bean, a little smaller than the pink bean.

Another time he pointed out kidney bean, calling it *ee'eḍ muuñ* 'blood bean' or *s-vegĭ muuñ* 'red bean'. Sylvester used both names as well, saying that the Pima did not raise them but bought them from the Chinese in Phoenix. He added, "There used to be a man called Ee'e Muuñĭgakam ['he has kidney beans']. They called him Muuñgaj for short." According to Sylvester, "Recently, maybe 30 years ago, everybody used *vepegĭ muuñ.* But then they all switched to *s-bibiolmagĭ muuñ.* I don't know why."

Ruth Giff said, "All there was was *vepegĭ muuñ.* Pinto came out maybe in the [19]40s, I'm not sure. It's easier to cook than *s-vepegĭ muuñ.* I don't think anyone cooks it anymore." Later she explained, "They [*s-vepegĭ muuñ*] take longer to cook than the pinto beans. The skins are tough."

Similarly, George Kyyitan said that when he was younger, the variety they raised was *s-vepegĭ muuñ.* "They didn't have no pinto beans in those days. I don't know where they were coming from."

"How do you say 'pinto beans' in O'odham?"

"*Pipinto muuñ.*"

"How about *s-bibiolmagĭ muuñ?*"

"I think the same thing." When talk turned to other beans, George said, "I only know those two, *s-toota bavĭ* ['white tepary'] and *s-oám bavĭ* ['brown tepary']. That was the main crop. We were also raising those *kalvash* [garbanzo], *u'us bavĭ* [black-eyed pea], peas, *muuñ;* now we raise some of those pintos. Those are good. They can grow anywhere. Anytime. But pink beans are kind of scared to grow. It'll come up. But sometimes it's not too sure to come up. But pintos are all right."

"What about cooking? Is pinto easier?"

"It is. Easier to cook, and faster."

The practice of soaking beans before cooking seems to vary among families. I asked George if he soaked teparies.

"I do myself. Some don't."

"How long?"

"Just any time . . . not make any difference how long. I think it's better to soak it a long time so it'll get big and soak up and cook good. I used to do that with that *bavĭ* and the white one [lima]. Besides, the water [here] is kind of salt[y], and you've got to have some of that [water] from in town [Phoenix]."

"How about *muuñ?*"

"No. If you wanted to, you could do it."

"*Havul?*"

"Same thing. Some of them doing that. Some don't."

Ruth's practices were similar. "I don't soak *muuñ.* Some do . . . overnight. *S-oám bavĭ* are very hard, harder than the white; *those* I soak. Don't soak *toota bavĭ.* Those are easy [to cook], and the *havul* you don't have to soak." Apparently at an earlier time, beans—at least teparies—were parched and crushed before cooking (see *P. acutifolius*).

Common beans are seldom if ever raised by contemporary Gila Pimas, but 50 years ago Castetter and Bell (1942:78) recorded some 450 acres of Pima country devoted to the three species combined. Pintos now are so readily available as surplus commodities and in area markets that it is hardly worthwhile to raise them.

Many older O'odham, both Akimel and Tohono, noted their preference for beans cooked in the traditional bean pot, or

ha'a. Sylvester Matthias said, "You know, cooking in that earthen pot, beans got more flavor than using aluminum or enamel." Yet only a few families still have their clay *haha'a,* which are now obtained from the several Tohono O'odham potters who practice this trade.

Certainly to this day, common beans remain a most important article of diet with the Pima. A funeral or feast-day potluck would hardly be complete without one or more pots of pintos, even if white and brown teparies are lacking. Concessions at fairs and bazaars will always have fry-bread and burros of thick mashed beans. This wonderful high-protein high-fiber food was and is important in the Pima diet.

Pisum sativum L.
garden pea
Lathyrus odoratus L.
sweet-pea
Fabaceae (Leguminosae)
vihol

Peas were one of the earliest Hispanic introduced crops. Though they never achieved the importance among northern Pimas that wheat did, they were another winter crop (apparently no other crops were grown at that season aboriginally). So the garden pea joined the lentil and garbanzo as legumes that could supplement the diet by growing in the cool season.

The northern Piman name for garden peas is *vihol,* which appears to be related to *vihog* 'mesquite pods', which hang abundantly at certain seasons, and more distantly to *viha,* the '*miembro viril*' of similar disposition.

Ruth Giff told me, "We had yellow peas, planted in October or November, ready by April. They used to grow free like a little bush, not as vines." Another time she said, "The Papago plant yellow peas and sell them. Pima have yellow, too. Guess that's the only one. We didn't see any green ones."

In the spring Ruth had great masses of various colored ornamental sweet-peas growing in her yard. She said these are also called just *vihol.* (Mathiot [n.d.:270]

vihol *Pisum sativum*

A Tohono O'odham strain obtained from Native Seed/SEARCH, Tucson.

gives *s-iovĭ vihol* for sweet-peas among the Tohono O'odham.)

In 1856, Emory (1857:117) considered peas to be among the seven major crops being raised on the Gila. By 1862, in the heyday of their agricultural economy, the Gila Pima sold to the U.S. War Department more than one million pounds of wheat, as well as other crops, including green peas, for the California Column (Browne 1869:111). In the early 1940s, Castetter and Bell (1942) found the Pima still growing garden peas on a small scale. A single planting was made in January; it needed good moisture to produce. But 40 years later the crop has nearly disappeared. I know of no one who maintains seed from early stock, although Gary Nabhan and his associates at Native Seed/SEARCH have found some among the Tohono O'odham, presumably selected over time for the particular growing conditions of the desert.

Vouchers: *770* (RG, *L. odoratus*).

Proboscidea parviflora (Wooton) Wooton & Standley var. *hohokamiana* Bretting
[≠ *Martynia louisiana* Mill.]
[≠ *M. fragrans* Lindl.]
White-seeded Devil's Claw
Proboscidea parviflora (Weston) Weston & Standley var. *parvifora*
Black-seeded Devil's Claw
Martyniaceae

ihug

A plant domesticated by Southwestern Indians, possibly right here along the Gila River, seems to have escaped the notice of almost all the Hispanic and Anglo chroniclers who commented on the excellent agricultural skills of the Gileños. Alamán (1825) referred to what must be this plant, but his sentence is worded so ambiguously that it is impossible to tell whether Alamán is implying that the

"cat's claw" was one of the domesticates (along with wheat, corn, beans, and cotton), or whether the baskets made with it, as well as buckskins and cotton blankets, were economic products that they traded. This plant has gone by a number of different names, including martynia (from its former generic name) and catclaw.

Ihug, as the Upper Pimans call this plant, came to the attention of the early ethnographers because of its essential role of producing the black fiber for the designs in coiled basket weaving (see Nabhan and associates 1981). For instance, Russell (1908:133) wrote, "The supply of wild plants is not large enough, and a few martynia seeds are planted each year by the basket makers. These are gathered in the autumn at any time after the plant has dried." Kissell (1916:202) observed that the cultivated plant was the primary source of this weaving material: "Although martynia grows wild, most of the Indians seed it in their fields, since they find the cultivated plant yields pods with hooks of greater length, finer grain, and a better black. Only a few of the more shiftless Indians gather it wild." Although she knew that when grown the plant had superior qualities, it is uncertain whether she realized that there were two different annual devil's claw: one wild, the other domesticated. Hrdlička (1906:43) called the plant *yi-huk,* or catclaw, and left no doubt about its domesticated status: "The catclaw is cultivated by the Pima in their melon patches, and is occasionally made up into great balls for preservation until needed."

Apparently Castetter and Bell (1942:113) were the first to notice the differences between the white- and black-seeded plants, clearly distinguishing between the domesticate and the wild form: "There were two kinds of devil's claw—a black-seeded variety which grew wild, and a cultivated form with larger pods and white seeds, the longer strips of epidermal tissue being more suitable for making baskets. The white-seeded variety never grew wild but often appeared as a volunteer in cultivated fields." Later (1942: 201–202) they described the planting, cultivation, and harvest of devil's claws, adding that "only the white-seeded form was grown."

In addition to the yellow-flowered wild perennial devil's claw (*P. altheaefolia,* Group G), there are two forms of the annual pink-flowered *P. parviflora*. The one with black seeds grows wild in disturbed but damp places and may be found as a weedy plant in fields. It is the progenitor of the white-seeded variety *hohokamiana* (see Bretting 1981, 1982, 1986). This is the domesticated form that has lost its germination inhibitors and is overall more robust. It would be interesting to follow the occasional feral populations of *P. parviflora* var. *hohokamiana* to see if they can persist indefinitely. The choice of the new varietal name is unfortunate because it means 'pertaining to the Hohokam'; there is no evidence that these earlier people had the white-seeded form. Chronology aside, however, there is a good correspondence between the area occupied by this early culture and the places where later Pimans have been growing their devil's claw crops.

How long has the cultivar been among the Gileños? Castetter and Bell (1942:113) thought it "possible that aboriginally they [the Gila Pima] utilized only wild plants of *Martynia,* and that its cultivation came with the commercial stimulus to increased production of baskets." Gary Nabhan and I (1987) also suggested the likelihood of cultivation by the late 1800s, when the demand for baskets increased and other parts of the Gileño economy were precarious. But Castetter and Bell's Pima consultants may have been correct when they claimed that they had been growing this plant for a very long time. Miksicek (1988) and Gasser (1988) recently discovered seeds of the white-seeded domesticate from excavations at Akĭ Chiñ near Maricopa. Of these, 5 may date to the early 1800s; 14 are from an archaeological feature with charcoal dated to A.D. 1670 (±50 years).

As might be expected of a plant that has been domesticated relatively recently, the folk taxonomy is in a state of transition, with different native speakers using different contrast sets. The usual contrast is between the useless wild form, labeled a Coyote's plant, and the culturally salient form, which is unlabeled. But with the devil's claws, we have a useless small wild species, a larger wild useful species (which is the progenitor of the cultivated), and the still more robust domesticated form. We have proposed three stages in the naming process (Nabhan and Rea 1987).

First, the wild yellow-flowered Coyote's Devil's Claw was called *ban ihug* or *ban*

ihug *Proboscidea* spp.

Top, *P. altheaefolia;* middle, domestic *P. parviflora;* bottom, wild *P. parviflora.*

ihugga, contrasting with the useful Black-seeded Devil's Claw, which was just *ihug.* This simple linguistic contrast set likewise contrasted yellow versus pink flowers, perennial versus annual, short versus long claws. But with the selection and domestication process, a third member was added to the contrast set and became the culturally salient one. We presume that the second step was to include both pink-flowered *(P. parviflora)* forms in the concept of *ihug,* in contrast to yellow-flowered *ban ihugga.* Most Gileños still use this dichotomy. The next step in the linguistic evolution would be to call only the domesticate *ihug,* in contrast to *both* wild forms, which would be *ban ihugga.* No one with whom I worked used this dichotomy.

What is this plant that is so important to Southwestern basket weavers? I followed Ruth Giff one summer morning out into her fields. Next to her long rows of *keli baasho* melons were coarse and clammy plants growing much like datura. The leaves are about as big as my palm, longer than broad. Touching the minute hairs of the heavy stems releases an odd smell. The small pink flowers look much like snapdragons with purple spots on their upper lips and yellow in their throats. The blossoms are hidden modestly among the leaves. Some of the young elongated pods look like a misshapen string bean. Soon they will begin to take on their characteristic shape, still covered with the green husk like a deer antler in velvet. As they mature, this outer covering dries and peels away, leaving the black woody pod with two "claws," the part used by the weavers.

As she pulled a weed here, propped up a melon there, Ruth commented that in the old days, women used to get together and help each other, as when splitting *ihug*. The person they helped would fix lunch. They would sit on the ground and work and talk. Doing the *ihug* was the last thing in basket making that Ruth learned. Tohono O'odham who cannot grow *ihug* buy great wheels of it from her. Things get in and eat Tohono O'odham *ihug*, even when they are grown inside fences, she said. "There's a man in Santan. He's selling them for $50. I don't know what size."

Ruth gave me a wheel of *ihug* she had raised. (I had asked her to make me one from the rejects she grew—ones she would not be using for her own baskets or selling to the Tohono O'odham.) "What do you call these?" I asked.

"*Haḍkiwua*," she answered almost instantaneously. The word *haḍkiwua* she thought could not be broken down further. I suspect it is related to *haḍshshap* 'to adhere, cling, get glued onto' and *haḍshp* 'to adhere, stick, onto a surface' (hence *haḍam* 'sticky'; *haḍshaḍkam* 'Jumping Cholla' *[Opuntia bigelovii]*).

The smaller rings of stripped devil's claw fibers or willow splints Ruth called *hakkoa*, "same as the circle when women used to carry water on their head," she explained. The individual splints of devil's claw she called *taíka*. To scrape the *ihug* strips down to uniform size is *hivshan*. And to split the slender willow twigs in Pima is called *o'otpĭ*.

Selection is an ongoing process. Each year I talked to Ruth, there were certain characteristics of *ihug* that she preferred. She saved the seeds of these for the next year's planting. She discarded those with characteristics she did not like.

One November when I went to her house she was stripping *ihug* claws. "Which ones do you select to plant?" I asked.

"The ones that are wide and not so curled up, like this. It's hard to work with them when you strip them."

"Want them straighter and wider?"

"Uh-huh, and long. Some are the white seeds and thin, like that. Very pointed. Can hardly get a piece off the ends. I got some

from Marvin's [her grandson's] yard, and they're the good kind, *wide*. But I can't even strip them. They're just tight, it won't even peel off, and this lady told me that she had bought some like that. Oh, she said she paid a lot of money for that, but they're no good either. She said it's because they weren't getting enough water. See, this was growing up there, you know, where it's rocky. So that's the problem. They were long enough that I could use. Had to throw them out. I could have saved the seeds because they are *very* wide."

"But they were white-seeded?"

"Uh-huh."

"I wonder how they got here?"

"This year is the first year that I noticed the difference [in them], so I'm doing that now."

"Have you ever planted the black-seeded ones?"

"No, but they come up. They come up by itself, *hejel vuushñim*. They're voluntary and they're *long* enough. But it's just not as good as the white seed."

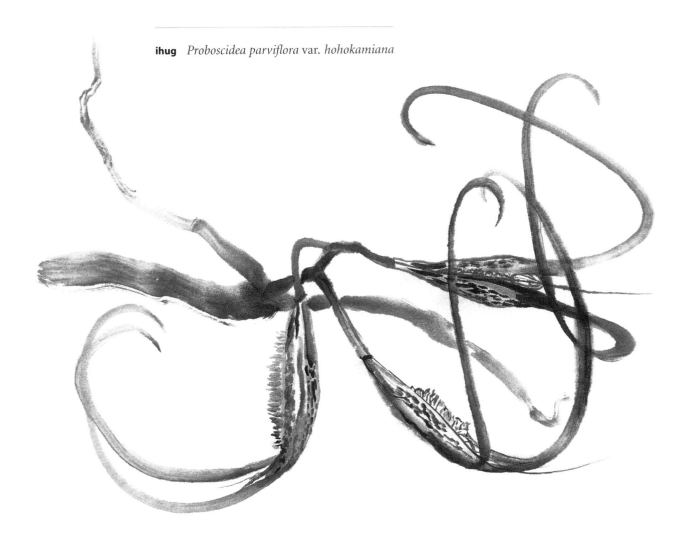

ihug *Proboscidea parviflora* var. *hohokamiana*

"When you first started to make baskets, were they using the white-seeded one?"

"Uh-huh. I'm sure that's what my mother grew. She planted all hers. Every time there's a melon patch, there's rows of that above the borders, and she'll—when she's ready to pick—she'll change her clothes, you know, it's sticky. Go through them. I never did help her, but she used to—she'd feel round the points and if they're hard enough, she knows they're just about ready to pick. She'd pile them and she'd cover them up with grass. And they'll ripen in there. She says that they're softer when they don't ripen on the vine."

"I was going to ask you. I never know when you're supposed to pick them, whether you wait till that green husk starts falling off or before that?"

"I wait till mine are all ripe on the vine."

"Oh, do you?"

"I do, yes. To be sure that they're black and they're not any tougher than the others. But pick them right away, so the sun won't bleach them."

"Just as soon as the green starts peeling?"

"Uh-huh, yeah."

"But your mother used to pick them before that?"

"Yeah." Ruth fished around on a cluttered bureau in her kitchen and handed me an old basket where the black design was uneven. "Because some I notice—on the basket—I notice some of these are kind of faded. Some, like notice, if they're not [mature], I think the black comes off. On the basket, you know. Because they didn't ripen on the bush. That way they won't ever come off. They're *very* soft, that *is* true, they're soft, but the black won't stay on it like the ones that are ripened on the bush. And I heard—this woman asked me if I ever use, oh, *matai* ['ashes'] on the faded *ihug* to make them black, but I think this would be too strong." There are other references to using ashes to blacken the *ihug*. For instance, Castetter and Bell (1942:202) recorded, "The ripe pods were gathered and laid out in piles in the field to dry, and informants said they had seen their mothers spread wood ashes over such piles."

Over the years of watching Ruth split *ihug,* I finally realized what it was she sought in the domesticated ones. Not only should they be long and black and pliant,

but they should be broad all the way to the point so that the two splints she peeled off would be of uniform thickness their entire length.

On my next visit, Ruth was busy preparing the *ihug* splints. She had already stripped two black pieces from each claw. Holding one end of a strip in her teeth, the other in her left hand, she rapidly scraped away any pith from the back side, then pared the strip to a consistent width of a few millimeters. I was amazed that anyone could produce such uniform strips by eye.

If you pry open the two claws of the domesticated *ihug,* you will notice rows of irregular whitish seeds in the capsule. Each looks like a tear-shaped glob of eggs from some insect. These, as with the other kinds of devil's claw seeds, are eaten for their milky oil. Gary Nabhan (personal communication) was told by Tohono O'odham that girls should not eat too many of these seeds or their pubic hair would become stiff like the crests on the seed capsules. George Kyyitan said he had heard that girls were warned about overindulging in eating the seeds, but he never learned what the reason was. No one else I asked among the Gila Pima was familiar with this admonition.

Sylvester Matthias said that a little bit of the point of the claw is put into the skin over an arthritic joint and is then burned. Curtin (1949:108) said the same for rheumatic pains. Hrdlička (1908:246) said instead that the dried petiole or stem of the pod was so used: "Of the *i-huk,* or Martynia, the part used is the dry basal segment of the pod. A bit of this is applied burning to the skin over the sore spot. It burns like a piece of punk, sometimes producing a little blister, but more often leaving only a small eschar or a dark spot." Burning a cottony or punky substance on the skin to relieve pain, a practice called moxa, is a treatment found in many parts of the world.

The immature pods became Pima toys. Sylvester related, "The kids—when they [the pods] are still green—make toys out of them; turn them upside down; the tail is the hook. They would get little sticks, poke it in there, stand [it] up like a horse. And that's their toy! Most everybody plays with it. Make horses out of it. It grows everywhere in the fields. When anyone gets in their way during a rodeo and they can't see, they would say, '*Ihug*

kaviogachda' 'toy horse made out of devil's claw'." Apparently this was a friendly taunt: the person blocking the vision was just a child's toy.

Devil's claw for the black basketry designs has far exceeded any other use historically among the riverine Pima. According to Ruth, descended from several generations of weavers, all three taxa of devil's claw, even the little wild perennial, have been used at some time in the past. But the two readily grown annual forms, the pink-flowered species, were used exclusively in the 19th century. The white-seeded form is classified as *e'es* because it is normally planted. The few remaining basket weavers who still grow their own *ihug* are still selecting seeds from the plants that have the superior qualities preferred for their weaving—proving yet again that domestication is an ongoing process.

Vouchers: *861* (RG, white-seeded cultivar); *969* (RG, black-seeded volunteer).

Ricinus communis L.
Castor-bean
Euphorbiaceae
maamsh

Castor-bean plants are found growing fairly commonly in the yards of Pima houses. Their shiny, dark green, red-tinged leaves are palmate, that is, radiating from the leaf stem like a fan palm. The lobed leaves are deeply incised and may be several times the size of your hand. The flowers are rather inconspicuous, but the spiny coverings of the burs or fruits are red and attractive. Castor-bean plants may develop quite woody stalks that reach up to the eaves of the little adobe houses, but during a severe winter they may freeze back completely.

When I first saw the seeds of Castor-bean, I was at once reminded of blood-engorged ticks, the kind commonly found on dogs. The seeds are attractive, each with a different mottled pattern, but somehow I can never overcome a certain repugnance when handling these "beans." It came as little surprise, then, to learn that the Piman name for 'tick', *maamsh,* is the same word as that for both the

squeeze out oil, put it on sores. Can also use as laxative, but makes you sick. These plants keep the mosquitos away and insects—the Maricopas know this, not the Pimas." (See also *Xanthium strumarium,* Group G.)

Later he told me, "The tribe forbids [the] planting of *maamsh*—because it's poison. And I believe it *is* poison. But on the other hand—the beans is medicine. Children never bother the beans." Nevertheless, he later told me this story that could have ended in tragedy: "There were two little boys. Somebody had Castor-bean in his pocket. And somebody started eating it, after they're in bed. [Here he whispers.] 'Yeah, what is that?' And he gave him a handful and [they] started eating, eating. And the next morning they landed in the hospital—it makes them run. Very serious! Just like diarrhea; one after another go to the toilet. Oh! Makes them sick with headaches. They warn us not to eat Castor-beans. It's not for eating anyway; it's for medicine."

One Pima man had told Curtin (1949:100) that he would eat the seeds "without ill effect," but she was also told, "The Pima poison gophers by placing several beans in the animals' holes." (I have never discovered this practice among contemporaries.)

I consulted my book on poisonous plants (Perkins and Payne 1978). Indeed, the plant is not to be taken lightly: "A few seeds have caused death, yet there have been survivals after ingestion of many chewed seeds. Uncracked seeds may cause only mild symptoms. [Allergic responses] may result from merely handling a seed or leaf. They may be so severe that hospitalization is required." The symptoms, which may appear immediately or be delayed two days, include "burning in the mouth, throat, and abdomen, thirst, nausea, bloody gastroenteritis with profuse bloody vomiting and diarrhea, abdominal pain, headache, dizziness, dull vision, liver and kidney impairment, weakness, convulsions and death" (Perkins and Payne 1978).

On the lower Rio Yaqui in Sonora, the Pima Bajo distinguish two forms of *maamsh:* a green-leaved one, *higuerilla blanca,* and a red- or bronzed-leaved one, *higuerilla roja.* I have found only the normal green form in Gileño country.

Vouchers: *554* (SM); *1089* (Pedro Estrella, Pima Bajo); LSMC *72* (not extant).

maamsh   *Ricinus communis*

Castor-bean plant and its seeds. Presumably, when this tropical plant was introduced into Pimería by Europeans, early Pimans made the same association. This appears to have been an early Hispanic introduction, because the plant's name is the same in lowland Pima Bajo as in Tohono and Akimel O'odham. There is no plural form of the word, according to Sylvester Matthias.

Joseph Giff used this plant to explain to me the concept of *e'es* as opposed to *uus* in Pima. "*Uus,*" he said, "is any kind of sticks or something woody, like a woody plant. It could be a fence post or a house post. But *e'es* means something that someone has planted, as in *ha'ichu e'es* 'something planted', or in other words, 'crops'. Now the Castor-bean, *maamsh,* even though it grows tall like a shrub, is an *e'es,* because they are someone's plants, someone had to plant them, they belong to someone. But over in California, around Banning, I've seen them growing just out in the arroyos, in the washes, just by themselves. Those wouldn't be *e'es,* because they don't belong to anybody. I guess they're just *sha'i* ['bushes'] there. But here they're always *e'es.* That's the normal way they grow."

The Gila Pima grow the shiny green-and-red plants today chiefly as ornamentals, but as with many other Euphorbiaceae, their medicinal values did not go unappreciated. Sylvester Matthias told me, "Roast seeds for oil;

Saccharum officinarum L.
sugarcane
Poaceae (Gramineae)
s-o'oi kaañu

Sugarcane, the crop from southeast Asia that became the economic backbone of a number of tropical countries (particularly in the New World for rum production), has a somewhat nebulous role among the Gila Pima. The plant is called among them *s-o'oi kaañu* (*s-*, intensifier; *o'oi* 'mottled' or 'striped'; *kaañu*, a loan word from Spanish *caño*, sugarcane). The Gileños say their name comes from the red or purple marks or stripes on the green stems.

The Piman classification is complicated. To a botanist, the plant is a grass, but it is not classified with *vashai* 'grasslike plants' by the Pima; rather, it is considered *e'es*, something that is planted, something that must be planted and that belongs to the planter. (A number of other plants also fit this pattern; see *Ricinus communis*.)

We might assume, given its Pima name, that it is a kind of *kaañu* (sweet sorghum, *Sorghum bicolor*).But the literal meaning of a word is no assurance of its folk taxonomic relationship (see *Eschscholzia mexicana*, Group G). While trying to elicit the range of items included in the concept, I asked Sylvester Matthias how many kinds of *kaañu* the Gila Pima had. He gave me the names of the two color varieties (folk species) of sweet sorghum: *s-vepegĭ muḍadkam kaañu* and *s-chuuchk muḍadkam kaañu*. I then asked, "How about [the] *s-o'oi kaañu* you told me about?"

"That's a separate category. It's not included in there with the *kaañu*."

"Why not?" I asked.

"It's separate. It's not a kind of sorghum *kaañu*." The Pima concept of real *kaañu* is a kind of sorghum grain. It makes no difference that *Saccharum officinarum*, through linguistic borrowing, has historic priority. What the People grow is sorghum.

"How did the Pima know sugarcane?"

"They would buy it in Phoenix in the stores—they used to have it. Or those peddlers would come around. Some Mexican used to come around here with fresh produce. Sometimes he would have it."

Ruth Giff also contrasted this with sorgo, saying, "There's another kind that comes from the store, called *s-o'oi kaañu*."

Geoff Levin assures me that occasional references to sugarcane by Anglos traveling through the Pima villages (particularly during the gold rush) would have meant sweet sorghum rather than sugarcane.

But earlier, in the Jesuit period, sugarcane was grown in Pimería Alta. For instance, Nentvig (1980:23) wrote that in 1764 "sugarcane grows in most of the province [of Sonora], more abundantly in the hotter sections. Syrup is extracted from it and *panocha* is made." Pfefferkorn, who worked among the Attí Pima from 1756 to 1763, wrote at some length about sugarcane, but he did not specify how far north the crop was raised: "The beautiful foliage and the delightful greenery with which these canes are decorated are ornaments for the fields and a pleasure to the eye. They grow four to five ells tall [about 8–10 feet] and three to four inches thick" (Pfefferkorn 1949:50). There is no doubt that he is discussing true sugarcane, because he explains the propagation method: "When sugarcane is cut, the stumps which remain attached to their roots are well covered with earth. They again put forth shoots and yield juice in the following year. The plant can be used three times in this way before it deteriorates. Cane will yield sugar the same year it is planted, but the yield is small because the cane is still too delicate and contains little juice. It matures completely only in the third year and for that reason it is more profitable to wait until then; the double yield is the reward of patience" (Pfefferkorn 1949:52). He, too, talks of the production of a syrup and the firmer product, *panocha*. The priest extolled the rich dark brown juice, which he assures us was used daily for dessert. One wonders whether any of this ever made its way by trade north to the Gila. Older Gileños today still know the product *panocha*, or as they say in Pima, *banóóji*.

Once, on a caravan trip down to Pima Bajo country, Ruth Giff, Sally Pablo, Sylvester Matthias, and I stopped in Hermosillo, Sonora, to buy thick canes of *s-o'oi kaañu* from a roadside vendor. The Pima ladies cracked the canes and peeled off the tough outer green covering with their teeth, chewing and sucking the bland sweet juice from the white pith. Sylvester had no teeth, so I stripped some for him. He sucked the pith, remembering the days of youth.

At the Pima Bajo pueblo of Onavas, I looked around for sugarcane but found none growing. But the *panocha* is not forgotten. Men's eyes sparkled when they spoke of it, particularly younger men. They relished it every night, they assured me. And I am sure those who could, enjoyed it for breakfast as well. Curiously, we were unable to find sugarcane and *panocha* at any local village shops.

Solanum tuberosum L.
Irish potato, white potato
Solanaceae
baabas

When the white or Irish potato arrived in Gila Pima country is uncertain, but it may have been relatively recently. In his summary of plants introduced during the Hispanic period, Morrisey (1949) makes no mention of this crop, nor did Bringas (1977) in his 1796–1797 report, nor Grossmann (1873) or Russell (1908) during the early Anglo period. In fact, the plant may not have achieved much importance at all in the northwestern portions of Mexico during early Spanish colonization—quite unlike its rapid dispersal and widespread acceptance about the same time in northern Europe. In 1719 the crop reached the Atlantic seaboard of North America by way of Ireland. Potatoes are native to the high country of the Andes, centering in Peru. Here the Indians maintained dozens of named cultivated varieties of the crop (Brush and associates 1981). In diversity there was defense against the vagaries of weather, pests, and viruses.

There is one record of the potato among the Pima before the Anglo settlement of the Phoenix area: Pancoast (1930:244), a 49er on his way west, mentioned them as one of various crops he saw during a passage through the Pima villages. However, because several other items in this account of his stay at the villages seem to be at such variance with other ethnographic information, I wonder if the account was written from memory somewhat later, with a few embellishments.

The Gila Pima as well as Tohono O'odham name for the potato is *baabas*, a loan word from Spanish *papas*, perhaps suggesting an early origin of the crop among Pimans. The white potato was at least known during the colonial period, because Nentvig (1980:44), writing in 1764, listed *batatas* (potatoes) as well as *camotes* (sweet-potatoes) among the medicinal products of the province of Sonora. Yet Castetter and Bell (1942) make no mention of the crop in their agricultural survey of the Pima and have only a passing reference to "a few potatoes" grown at San Xavier, the Tohono O'odham settlement near Tucson. King and associates (1938:56) wrote from the U.S. Field Station at Sacaton, "Potatoes are not grown on a large scale commercially in the warmer irrigated valleys of Arizona, but experiments over a period of years have shown that they may be grown to advantage on a small scale for local consumption. The Irish Cobbler and the Bliss Triumph are the most reliable producing varieties at the field station."

One January I asked Sylvester Matthias for information about potatoes among the Pima. "Plant *baabas* about this time or in November; harvest in May. They're very important. Stew it or fry it or barbecue." In addition to the regular potato, he mentioned the variety *s-vepegĭ baabas* 'red potato', or the pinkish ones.

Madeleine Mathiot (1973:136) gives four lexically marked preparations in her Tohono O'odham dictionary: *baabas chuhivi* 'mashed potatoes', *baabas hidoḍ* 'potato stew', and *baabas poshol* 'boiled potatoes', in addition to frying in lard. Ruth Giff said she would use only the latter two terms: *baabas hido* when you fry or make soup with it and *baabas poshol* when you boil the potato whole, as when you are going to make potato salad.

At the communal feasts of which the Pima are so fond, huge basins of delicious potato salad are frequently included on the laden tables. The indispensable "chile stew," beef cooked in 10 or 20 gallon containers, often has potatoes as its main vegetable ingredient—even when the red chile is omitted.

Sorghum bicolor (L.) Moench
[= *S. vulgare* Persoon]
sorgo, sweet sorghum, "sugarcane"
Poaceae (Gramineae)
kaañu

Especially around the western villages, I have often found a section of traditional Pima fields planted with a tall sorghum, usually head-height or a bit more. Maybe a quarter- to a half-acre would be planted in rows of this great grass. The plants looked superficially like the congeneric pestiferous Johnson Grass (*S. halepense*) along the ditches and the Sudan Grass (*S. sudanense*) raised for forage. The loose seed heads bore black shiny seeds. But the stalks were thicker, almost resembling those of the Pima flour corn. This esteemed plant, *kaañu*, had long been a favorite among the Pima for making a syrup that must have rivaled the other prime syrup source, Saguaro.

Ruth Giff explained to me about the *kaañu* she and her family had raised for generations: "There are two kinds. The red-seeded *kaañu* is taller, about ten feet [tall], and has bigger seeds. Plant [head] spreads out like a broom [at top]. Not as sweet as the black, but [we] make *sitol* ['syrup'] with it. Black-seeded kind is shorter, has smaller seeds, and the seeds are more together [tighter inflorescence]. A little bit salty sometimes. It's also *kaañu*. Used for *sitol* ['syrup'] also. Boil the juice, then pack it in a *bid ha'a* ['clay olla'], seal it with *bid* over a piece of cloth."

The two cultivated kinds of sweet sorghum were distinguished in Pima as *s-chuk kaañu* 'black cane' and *s-vegĭ kaañu* 'red cane'. Some call the two *s-chuuchk muḍadkam kaañu* 'black [pl.] tassels it has cane' and *s-vepegĭ muḍadkam kaañu* 'red [pl.] tassels it has cane'.

Although Joe and Ruth were still planting the black-seeded sorgo, Ruth said the red-seeded one had been lost for many years. Gary Nabhan and I searched for years among both Akimel and Tohono O'odham agriculturalists without finding the red variety.

One day while I was visiting Pedro Estrella at Onavas, Sonora, he handed me a head of an amber-red cane he called *caña colorada*. In Pima Bajo he called it *siik hun va'ōō*, evidently either 'White-tailed Deer's corn-stirring stick' or 'deer's

corn suckings'. The inflorescence was tighter, heavier than any I had seen in Gila Pima fields, and from each seed projected a whisker or awn. Ruth grew it out for several years in her Komatke field, and I did also in San Diego. But Pedro's *kaañu* was not the lost Pima variety *s-vegĭ kaañu*. It was shorter than the black-seeded one, not taller, and it did not have the right taste.

Pedro's red *kaañu*, now known as Onavas Red, proved perennial in my San Diego garden, getting thicker-stalked each year. "When you used to raise them," I asked Ruth, "would they freeze every winter?"

"Oh, yeah. Because it was cold in those years. Unless it was covered, I think they would. Well, they did that with chile plants. You know, everybody had straw, after their wheat harvest, and they'd just cover them up and it comes up [next season]."

The Piman name *kaañu* is a loan word from Spanish *caña*. Americans began experimenting with this African and Asian crop during the 1850s. Whether the Gileños first became familiar with the true sugarcane, which they now call *s-o'oi kaañu*, or with a sweet sorghum is unknown. This may be an instance of marking reversal (Witkowski and Brown 1983) when sweet sorghum later achieved importance as a local crop. Surely the plant Pfefferkorn (1949) discussed, with stems about three to four inches thick, was *Saccharum officinarum*. In 1857 Emory added "sugarcane" to his 1846 list of Pima crops, but Geoff Levin has assured me that this could well have been sorgo. Russell (1908:74) noted, "*Kanyo*, Sorghum vulgare Pers. Sorghum is cultivated when the water supply permits. It has been obtained recently from the whites, who raise it extensively in the Southwest." Was this sweet sorghum or grain sorghum, now called *huuñ kaañu* or *chuchul huuñ* by Gileños? The Pima name Russell used suggests strongly that this was sweet rather than grain sorghum. Castetter and Bell (1942) apparently overlooked sweet sorghum cultivation among the Pima. But during the previous two decades the U.S. Field Station at Sacaton had been experimenting with sweet sorghum. King and Leding said in 1926:

The interest which was created among the Pima Indians by the field station in the early years of its existence for producing sirup from sorgos (saccharine sorghums) has been maintained. Many of the Indians near Sacaton devote a small portion of their cultivated lands to sorgos for sirup making and for forage.

A test planting was made in 1923 with the best-known sirup-producing varieties. The gross yields (in tons) of the different varieties were as follows: Texas Ribbon, 19.5; Black Amber, 5.8; Red Amber, 9.6; Honey, 11.3; Sumac, 10.3; White African, 13.9; Orange, 8.1.

There is some variation in the form of the Pima name. Among west-end speakers I have heard only *kaañu*. Fr. Antonine (1935) and Russell (1908) both recorded it as *kanyo* from eastern villages, seemingly nearer the original Spanish *caña*. (Neither writer indicated vowel length.) For Tohono O'odham Mathiot (1973:441) gives *kaaña* and *kaaña,* while Saxton, Saxton, and Enos (1983:29) give *kaañia* and *kaañu.*

Mexicans boil down the syrup, or sorghum molasses, to make a solid called *panoche* or *panocha.* The Gileños bought this product, which they called *banóoji,* George Kyyitan explained.

Vouchers: *891, 1010* (RG).

Sorghum bicolor (L.) Moench
    [= *S. vulgare* Persoon]
milo maize, grain sorghum
Poaceae (Gramineae)
huuñ kaañu, chuchul huuñ

Of the various kinds of sorghum cultivated or growing wild in Gila Pima country, grain sorghum is the easiest to recognize. The plants are short, densely leaved, and stocky, with a compact inflorescence of heavy seeds. Grain sorghum is still grown there as a commercial crop. There are two different Pima names for this crop: *huuñ kaañu* 'maize cane' or 'corn cane', and *chuchul huuñ* 'chicken maize'.

It is unclear how early the grain sorghums were introduced into Pima country. The "*kányo*" Russell reported during his 1901–1902 fieldwork as "obtained recently from whites" may have been either this or sweet sorghum

(Russell 1908:74). The two are forms of the same biological species, and he failed to note how the plant he mentioned was used.

At least one variety is recognized linguistically in Pima, the Hegari. It is called *s-toota chuchul huuñ* because of its white heads, said George Kyyitan.

Grain sorghums have ranked as important crops in Pima fields for many decades. "In 1919 the main Pima crops, ranked on the basis of cultivated acreages, were wheat, cotton, alfalfa . . . , barley, milo and beans" (King and Leding 1926:5). In 1923 they ranked sixth as a Pima cash crop. In a 1929 census of Pima farms and crops on the Gila, Castetter and Bell (1942:77) reported, "Grain sorghums, Sudan grass, corn and barley also were grown on a considerable acreage for feed and pasture."

Grain sorghum, or milo maize, was experimented with extensively at the U.S. Field Station near Sacaton, where 14 varieties were planted in the early 1920s, including Standard and Dwarf Hegari (King and Leding 1926:27–28): "It would appear . . . that Standard Yellow milo is the best adapted of all the grain sorghums under the conditions which obtain at Sacaton [46–68 bushels/acre yield]. Although the yields are not materially higher than those of [Dwarf Yellow milo], the advantage which the latter possesses in having a lower water requirement is overbalanced by a distinct disadvantage as an Indian crop in being more susceptible to damage from birds, a factor which is of no little importance with the grain crops grown on the scattered Indian farms of the reservation."

I asked Ruth Giff about traditions of milo maize in her family.

"Joe's family raised it, ours didn't. We call it *chuchul huuñ.* It's grown to sell, I guess—the grain. But then they can turn the horses in to eat the rest. Can feed it to the chickens. Guess that's why we call it that."

She had not heard the name *huuñ kaañu,* but Sylvester Matthias knew both names. He said, "I know an old lady that uses milo maize, *chuchul huuñ,* to roast just like wheat to make pinole out of it. Nobody does but her. I guess nobody likes it." This Pima woman lived long ago in a village that has since disappeared.

Vouchers: *526* (SM).

Triticum aestivum L.
    [= *T. sativum* Lam.]
    [= *T. durum* Desv.]
    [= *T. compactum* Host]
    [= *T. polonicum* L.]
wheat
Poaceae (Gramineae)
pilkañ

Kino's introduction of wheat agriculture at the Piman missions in what is now northern Sonora completely altered the life of the Pimans. Before they had this winter crop, the Tohono O'odham, entirely dependent on the summer rains for agriculture, could raise but a single crop a year—if everything went right. The more fortunate riverine Pimans were double-croppers who planted in the spring when the mesquites leafed out and again in the summer after the Saguaro harvest. Conveniently superimposed on this annual cycle came Old World wheat, which could be planted about November or December after the harvest of aboriginal crops from the second planting. And wheat would be ready for harvest after the spring aboriginal crops were planted. Wheat completed the cycle and meshed all the gears of seasons, crops, river flow, and human work.

Moreover, while corn, the principal aboriginal crop, was a fickle plant that had to be enticed forth with song, ceremony, and votive offerings (yet even then might fail if conditions were not favorable for pollination), wheat was as prosaic and productive as one might wish. It not only filled a seasonal gap but also provided a new backbone to this desert culture's ecosystem.

Presumably wheat and the other European cereals filled an empty niche in indigenous Southwestern agriculture, though there are indications that some native grasses were in stages of incipient domestication aboriginally. Miksicek (1983) found seeds of Wild Barley *(Hordeum leporinum)* that appear abnormally large from Hohokam sites. He suspects selection was taking place. Bohrer (1975, 1991) believes that cool-season grasses gathered from the wild were an important resource throughout the Southwest. Nabhan (1985: 204–205) has traced a domesticated native plant, Sonoran Panicgrass *(Panicum sonorum),* from the

Chemehuevi on the Colorado River to the Guarijío of western Chihuahua, where he found living seed. It has not been recovered, though, from Upper Piman sites, nor is it a cool-season grain. If there were incipient grass domesticates in Pima country, one thing is certain: the introduction of wheat completely eclipsed them. And Eurasian livestock nibbled away at any genetic remnants that may have been left.

The name for wheat in both Akimel and Tohono O'odham is *pilkañ*, or rarely *pilkañi*. Herzog (1941) considered this a loan word from Spanish, but this seemed puzzling. Wheat in Spanish is *trigo* (from Latin *triticum*), which does not seem to be related to *pilkañ*. Among Pimans southward, wheat is *tiligo* (Mountain Pima), *tiricco* (colonial Névome), *tiligi* (Northern Tepehuan), or *tirik* (Tepecano). Names cognate with *trigo* appear among neighboring Uto-Aztecan speakers: in Yaqui it is *tiikom*, and in Tarahumara *tri'ligo*. However, in colonial Eudeve, a Uto-Aztecan group once occupying territory between the Ópata and Pima Bajo, wheat is *perikon* (Lionnet 1986:62), cognate with the northern *pilkañ*. The connection finally becomes apparent from an unexpected source, Guarijío, where wheat is *pirikó*. As preserved in modern languages of Pimería, we can see an evolution from Spanish *trigo* in *tiricco*, *tiligo*, *pirikó*, *perikon*, and finally northern Piman *pilkañ*. Miller (1983:124) observes, "Since the phonetic changes involved are unusual, it is unlikely that each language borrowed directly from Spanish, but rather one borrowed from Spanish, and the word was then passed from one Sonoran language to another. There are a few words for introduced food items that reflect this history."

The Pimas harvested wheat at least by 1702 at Caborca (Bolton 1936:514), but exactly when wheat arrived on the Gila is unknown. Although it is generally assumed that wheat reached the Gila as quickly as Kino did, the first documentary evidence appears in Sedelmayr's *relación* of 1744: "Those [Indians] of Sudacson raise wheat by irrigation" (Sedelmayr 1955:23).

The village Sedelmayr mentioned appears on early Jesuit maps as Sudacson, or, in contemporary Pima, Shuudagĭ Shon 'water beginning' or 'spring'. He seems to have observed the crop only here, and this is the only village where he noted irrigation agriculture. But this is 50 years after Kino's first *entradas* among the Gileños. Were they really this slow in adopting this winter crop? The documents are silent.

But by the Franciscan period (beginning in 1768), everyone was raising wheat and irrigating it as well. In 1775, for example, Garcés (1900:107) wrote of the Gileños: "In all these pueblos they raise large crops of wheat, some of corn, cotton, calabashes, etc., to which end they have constructed good irrigating canals surrounding the fields." The previous year, Díaz, too, had remarked that wheat was the principal crop on the Gila. And Anza wrote, perhaps with some license, that Pima wheat fields were so large that "standing in the middle of them, one cannot see the ends" (Bolton 1930, 1:184). It is clear from all three accounts that wheat had superseded corn as the dominant crop on the Gila.

While it is not certain which variety of wheat first reached the Gileños, there are some clues from Piman missions farther south. Wheat straw is a common binding agent in the manufacture of adobe brick. Dissolving adobes from different construction periods of Hispanic missions, Hendry (1931) found Propo, Little Club, and California Club wheat in Jesuit-period walls in northern Pimería Alta.

Several historical events beginning with the 1849 California gold rush brought great numbers of Anglo-American itinerants through Pima land and provided the Gileños with a ready market for their surplus wheat and other agricultural products for nearly a quarter of a century. Doelle (1975) has compiled statistics on the surplus wheat Anglos recorded buying from the Pima: in 1846, the estimated surplus was 2,500 pounds; by 1858, it had jumped to 100,000 pounds; over the next four years it increased to about 700,000 pounds and continued at this level for the next five years.

In 1862, Ammi M. White, a resident Anglo trader who was temporarily tribal agent (1862–1866), set up a flour mill at Casa Blanca that served through 1868. Whittemore (1893) wrote that Agent Crouse had built a flour mill for the general benefit of the Indians that was capable of producing 25 barrels of flour in 24 hours; this is presumably the same one Russell (1908:90) mentioned at Sacaton. With a demand for flour as well as grain

wheat for feeding horses and other pack animals in these prerailroad days, Pima sales about doubled. Economically, it was a glorious time for the Pima.

But the meteoric rise in the Piman wheat market was followed by an equally rapid collapse due to ecological factors and the underhanded activities of new settlers in the valley (chapter 5). Simultaneous with other destructive processes in the Gila watersheds documented in part 1 (see also Dobyns 1981; Rea 1983) came the major flood of September 1868. It was large enough that the Pima calendar-stick annalist noted it, naming the event *vamaḍ juukĭ* 'snake rain'. Russell (1908:52) added in a footnote that White's "store was more than two miles south of the channel of the river, but it had been built at the foot of a little rise upon which the present village is located and was within reach of the flood." By 1887 the Florence Canal, built by Anglo settlers, had diverted the entire surface flow of the Gila (Spicer 1962:149). The Pima economy collapsed, and many Pimas were reduced to wage labor off-reservation—some even to starvation. Beginning in the 1860s, the Gila basin and its headwaters endured the most serious drought in Arizona's history.

In 1906, the drought years shifted dramatically to a series of wet years. King and Leding (1926) wrote from the U.S. Field Station at Sacaton, "Since wheat can be grown during the winter months when the largest quantity of water can be conserved, this has been the principal crop of the Pimas. In 1919 more than 25 [railroad] carloads of wheat were sold by the Indians in addition to the quantity they consumed. In 1922 the value of the Indian wheat crop was estimated at $84,780."

Wheat harvest demanded a considerable work force. By long-established tradition, the Tohono O'odham would go north to help the Gileños harvest wheat. At least by the 1860s, northern Tohono O'odham were migrating to the Gila and to Sonora to harvest wheat (Annual Report 1864). A reference by Anza (Bolton 1930, 2:17) suggests that this pattern may have obtained as early as 1740, but the Gila as a destination is not specified. Originally, the Tohono O'odham took various trade items from the desert as well as their labor to exchange for native agricultural products. With the great demand for Pima wheat in the mid 1800s, this traditional bond became even more important. Wheat became the chief commodity the Gileños offered in exchange to friends and relatives from the south. Of this exchange, Castetter and Bell (1942:47) wrote, "Thus the Papago almost abandoned the northern part of their desert territory to work for the Pima during wheat harvest on the Gila. This was always a great boon to the Papago for whom this time was the leanest part of the year. Here they remained for about a month, a fourth of which was spent trading and cutting willow for making baskets. For their labor the Papago were given daily food and a share of the crop. Although May and June, the time of wheat harvest, constituted the period of greatest Papago movement to the Gila, some of these Indians returned in late fall to assist in planting, and to pick cotton." This work arrangement continued somewhat into the 20th century. Hoover (1929) wrote, "Until 25 years ago, the Pimas allowed the Papagos to come to their villages and thresh their wheat for a share. The practice is still carried on to a certain extent."

All the people with whom I worked had grown wheat crops in their younger days and had worked in the harvest. Francis Vavages explained, "They usually came without food, the Papago. They know just about time it's ready to be cut. Next year, again. They tell them, there's that much: *tiigi*—maybe one acre. Means allotted so much. [They're] good workers. They bring everything [with them]: cooking pots and everything. Then after it's done, they give them a share of the [main] crop, too. The Papagos came till the [19]20s." This *tiigi* was a parcel of wheat that the Gileños planted specifically for their Piman neighbors to the south, who arrived hungry; the wheat harvest coincided with the height of the spring starvation period. The Tohono O'odham harvested their parcel first, so they would have food while they helped with the larger crop. (Culver Cassa notes the derivation of *tiigi* from Spanish *tequío* 'tax, charge'.)

In addition to assisting with the wheat harvest, the Tohono O'odham would also perform a dance "when hunger forces them out of ruts and huts during the winter or early spring: they come by hundreds, in a long caravan to the Pimas whom they know to be more provident. A small delegation is sent in advance to advertise the Pimas that they are very

hungry, and will soon appear to give them a great dance in exchange for something to eat. . . . They give the Pimas two or three nights' dancing, in return for which each Pima family is expected to give fifty or one hundred pounds of wheat" (Whittemore 1893:82–83).

I asked Sylvester when the Tohono O'odham stopped coming to the Gila to help with the wheat harvest. "In the [19]30s when the mechanical way of doing things like that—thrashing—came in. Back in the [19]20s they were still coming in."

Some Pimas still recall how the brush weirs were constructed on the Gila (see *Atriplex polycarpa*, Group D). One day at Bapchule, I asked George Kyyitan, "Do you remember when they built the dams to make the water come into the irrigation ditches?"

"Oh, it's way back. But I remember them, when they do that. Every time when they had to bring in water, they had to go *way* down there, down here, next to this village, Sweetwater, [upstream] about three miles from there, or two. And then there used to be *lot* of mesquites on the side, where, where there's nothing now. And they used to chop the limbs off from that, and cut, chop some of those poles, so long, and put them in the river."

"How long were they?"

"Long ones. That's why, they pound them there. Way down, so the water won't bring them up."

"Right in the bed of the river?"

"Uh-huh. They put, right across, crisscross that mesquite limbs, the [smaller] limbs of it. Everything. Leaves and everything. Just cut 'em and put 'em right across there. Put 'em crisscross. And the next one *all* the way down. Then, because some of them have a load of straw. At the time they were doing stationary thrashing, by the horses, like that, see. That's why they get all those, they call it *moog*—that's the straw [chaff]. And another way [of saying it] is that *pilkañ va'ug*, the straw, too. These *moog*, they used to make adobes, too. They're good."

"So there would be mesquite branches?"

"Yeah, all the way [across the river]. Then they put all that *moog* right in there with it. So it'll hold the water back and bring it up. And comes evening—late in the night, the water comes in. They already know who's going to get the water there. They're not getting it just for themselves. That's how they get their farm's water."

"You know the name of the dam?" I asked George.

"I don't know. They didn't mention [it], but only say that they will have to do it. They wash away again, then they will put a new one again. In those days *everybody* works—together. That's the way they make it easy. With their wagons, horses, axes."

"What time of year?"

He said it was when the mesquites had leafed out. "Yeah, April. It'll be around there somewhere. That's when they water all their wheat and barley. Because they were using rain in those days. In those days, when they want rain, they have to pray for it. They gather up and meet there, and pray for rain."

The agronomists at the Sacaton Field Station performed experimental wheat plantings under different irrigation regimes; they reported that these attracted considerable attention from the local farmers. A large crop of wheat could be produced with but a single spring irrigation if the field was seeded after a good irrigation. More than two irrigations during growth reduced the yield. According to the agronomists, "Baart and Sonora have yielded consistently well for many years. These two varieties and Club comprise a greater part of the wheat grown by the Indians" (King and Loomis 1932:47).

Pimas to whom I talked usually mentioned four kinds of wheat being grown when they were young, but these were not always the same four. The following varieties were mentioned:

pi ha chiñvo, pi ha chiñvokam 'without beard [attributive]'

olas pilkañ 'round wheat', Sonora wheat

s-chiñvokam 'it has a beard'

s-moik pilkañ, s-moikchu pilkañ 'soft wheat'

s-chuchk pilkañ 'black wheat'

s-kauk pilkañ 'hard wheat'

Rosita Brown at Santa Cruz village preferred *olas pilkañ* for making *haakĭ chu'i*, or pinole. It is round, beardless, and soft. For a while the seed was lost in these villages. When it was discovered again, perhaps in Tohono O'odham country, she and David Brown raised their own crop just for pinole production.

Leonard Soke of Santa Cruz told me that *s-moik pilkañ* is Sonora wheat and *olas pilkañ* is Club wheat. *S-chuchk pilkañ* has dark grains. "*S-chiñvokam* looks like barley," he said. "When it's ready, it bends over, turns down. Good for *poshol* [wheat cooked with teparies]. The dark grains, *s-chuchk pilkañ*, real good, tall, lots of seeds in there. But no more. Some guys used to raise them, but they're all gone."

Sylvester Matthias said, "*Kauk pilkañ* is long, longer than Sonora wheat; good for making *poshol;* brown kernels like ordinary wheat."

Ruth Giff related this on the varieties: "*Olas pilkañ*, Sonora wheat, has no beard, and they're *red*. Boy, does it itch when we work with it! I don't know why, more than the other kinds. We used to go jump in the ditch after working with it." Ruth said of the two beardless varieties that *olas pilkañ* had short heads and round grains while *pi ha-chiñvokam* had longer heads. She thought both were probably from Sonora.

I showed her some plants I found growing in Broad Acres (leased reservation land near Komatke). "This looks like what we call Early Baart—*s-chiñvokam*—it has the beard." The beard is the awn on some varieties of grains. She laughed at the discrepancy between English and Pima, explaining, "But actually means 'mustache', because *eshpo* is 'beard'. And *chiñvo* means 'mustache'." Literally, it means 'mouth hair'.

I asked Francis Vavages what kinds of wheat he remembered. "The first wheat that came, came from Sonora. I guess imported from Spain. *S-moik pilkañ,* softer than *s-chiñvokam pilkañ*. *O'olas pilkañ* is kind of reddish looking; you can call it *s-chiñvokam*. Has *long* [heads]. *S-mohogich pilkañ* 'itchy wheat'. The two kinds without a beard. Others got a beard. *S-chuchk pilkañ* 'black wheat', I think is almost the same as *s-chiñvokam*." Consultants did not always agree on the assignment of English varietal names to wheat, and I did not have examples of the various possible cultivars to serve as a contrast set.

Wheat was preadapted to the conditions along the middle Gila: it does well in the alluvial soils and tolerates some alkalinity. The rising river in winter, diverted onto fields before planting, ensured sufficient soil-stored moisture before the wheat was seeded. In most winters, freezing was not severe enough or of long enough duration to harm winter wheat. In the early years, northern Pimans planted wheat the way they did maize, in hills about six inches apart. "Seed was planted two inches deep,

about four or five grains per hill, and the soil in the hill left loose" (Castetter and Bell 1942:155).

Even though wheat was a secular commodity, outside the "from time immemorial" traditions of some other crops, it still merited some of the growth magic the *mamakai* (medicine men) bestowed on native crops. Russell documented how the *mamakai* went about "securing supernatural aid to insure good crops":

One method of procedure was to gather the people in the large lodge and have some one bring in an olla filled with earth. This the Makai stirred with a willow stick and placed before a clear fire, where it stood all night while rain songs were being sung. At dawn the olla was emptied and was found to contain wheat instead of earth. Four grains were given to each one present, to be buried at the corners of the fields or the four grains together at the center.

For a consideration the Makai would go to a wheat field and perform rites which he assured the owner would result in a heavy yield of wheat. After rolling and smoking a cigarette at each corner of the field, he would go to the center of it and bury a stick (â'mina) [*omina*] 3 or 4 inches long. . . .

Again, the germination and growth of wheat were sometimes imitated by concealing several grains of wheat in the hair and shaking them down upon the soil. Then by a dextrous manipulation of a previously prepared series of young wheat shoots the growth was represented up to the point where a stalk 2 feet in length was slipped from the long coils of hair at the operator's shoulders and shown to the awestricken spectators as a fully developed plant. (Russell 1908:258)

Wheat begins to ripen at the time of the year when the Gileños' resources were in shortest supply. Game was scarce, and stored goods, both cultivated and wild, were coming to an end. Things might be so bad, Joe Giff explained, that people were reduced to eating the very worst of their jerky—dried lungs. At this time, heads of wheat, still in the milk stage, would be cut and roasted directly on the hot coals of the hearth. Ruth Giff later added, "They were starving at that time.

pilkañ   *Triticum aestivum*

They *vohiv,* burn the whiskers off. Just eat it like that or make *poshol* out of it. They didn't wash it in those days—the juice would just be *black!* Or they dry it and cook it like *ga'ivsa* [parched, dried, ground, cooked corn]."

Sylvester also noted three different stages when wheat could be eaten. "Can roast green wheat, run it though the flames, bunch it up, however many you like, kind of burn them up. Get a basket, then just work it with the hands. The grains come out. They call that *dagĭvina* 'work with the hands'. Wheat can be eaten green. When it's green, before it matures, they cut it, dry it in the shade—don't let the sun hit it. And when it's dry, you can thrash it, grains come out *green*. They make that into gravy. They call that *s-cheedagĭ hidoḍ*. It would be green. When it's ready to mature, they thrash it, make that *sitḍoi chu'i*, roasted in a pan. What they do when they roast that *sitḍoi* [is] they have something like a bass drum hammer—a stick with a ball at the end— that's what they use when they put it in there, roast it over the fire. That's *sitḍoi*. It's kind of coarse. But *haakĭ chu'i*, when it's really mature, it's real fine."

Wheat would be dry and ready for harvesting during May and early June. It was a hot, dry season, and the work was heavy. With other crops, men did the cultivating, planting, and irrigating while women generally handled most of the harvesting and storage. But wheat harvesting and threshing were activities for men, assisted by the women.

Francis Vavages explained that the wheat harvest was a time of hard work for everyone. "I hate that. I always get on that thrashing—on the horses. They [his parents] always go [keep on working]. Don't care if the sun goes down. I used to get so tired chasing the horses around— from morning to evening—about four or five horses tied on a pole—chase around. So dry! Then they turn it, the wheat, over. Then we chase around again till the seeds are out. Our folks work just as hard, too. Sometimes a big storm is coming. They stay out till in the morning, working [all night]. In the morning, [there's] a *big* pile of wheat."

I asked Francis for the words used in the harvest. He named these terms with their meanings: *hiik* 'to cut' (*pilkañ hiik,* to cut wheat with a sickle); *voksha,* to pick up sheaves tied with a rope—these were thrown up to the man on the wagon who untied it and threw the rope back down; *kehivia,* to step on it, threshing with horses; *da'ichud,* to throw up for the wind to winnow.

George Kyyitan called this process stationary threshing. In Pima he said threshing with horses' hooves is called *kehivia.* This is his description of the process. "Nothing but their horses. Some, they use the unbroken ones. Sometimes they use five of them. It itches [from the wheat particles]. That's why the kids don't like to ride [the] horses. Yet they like to ride horse, but in that case they don't like [to]. They just have to make a fence around it [the threshing floor] and chase them around, then whenever it's ready, they'd put some more [wheat] in there. Then they run them up again. And when they get through, get little bitty pieces of straw in there that the wind can blow them away. Pile it up. Then they'll throw it up, throw it up. And the wind will blow all the straw away, and the seeds will come right back."

"What do you call that when they throw it up like that?"

"*Ñeenchud*—more than one, throw them up; *da'ichud*—means only one thing [single object]. Whole shovel full, pitchfork full. They use it. Sometimes they were lucky to have wind too much and they can get through earlier, faster. Then in those days they have to sack them [cleaned wheat] up. Before, they use the *vashom* [or] *homda* ['storage bin'] to store away [the wheat]. That's why they never go hungry."

"Which one did they use more for wheat?"

"I think the *vashom*—to kept it long."

The change to a more modern wheat technology George perceived as a serious step backward for the Gila Pima. "But nowadays, with this mechanical way of threshing, we don't see no more wheat here. Never see it again. And whenever they get through with the mechanical way of threshing it, they have to haul it away, right away, right away. And sell it off. Won't left any there. Nothing to save. Then get the money there. In no time, the money will be gone. No more. That's why so many of them go starving. They don't keep no wheat. If they keep wheat, they can go over here to the mill and make *chu'i* ['flour'] out of it." The former subsistence economy had advantages over a cash economy.

The Gileños live in a country where violent windstorms can sweep through a village, blinding people with fine sand, tearing up everything light enough to be blown away. But winds also cooled this hot country and brought the clouds that dropped rain in this thirsty land. Small wonder that Wind, Hevel, was personified by the Pima. Hevel could be violent and destructive, or she could be useful, another example of Piman ambivalence.

Up in the windy crags of the desert sierras dominating the skyline of every Pima village lived Desert Bighorn, called *cheshoñ.* Particular men from each village occasionally hunted Desert Bighorn in these ranges. Both Joe Giff and Sylvester related that the hide as well as the horns and skull must be kept in a safe, respected place; an insult to the animal's remains could result in either a violent wind or a destructive rainstorm.

Winnowing crops such as wheat or teparies required wind in a controlled amount. Joe Giff related the following story: "Old Tashquinth [d. 1954] had a *cheshoñ* hide. Don't think he killed it; he never was a hunter. Someone needed to winnow their threshed wheat but there was no wind at all and he was waiting. Was afraid the rain would come and ruin the grain [exposed on threshing floor]. He went to this Tashquinth, who took out the hide and asked him, 'Which way do you want wind to blow from?' We wouldn't want the chaff to blow on our house. Tashquinth took the hide in that direction and tapped it gently and said, 'This man wants some wind. Will you give him some wind from this direction?' It came. No one has hides anymore."

Wind had her own house, called *hevel kii.* The location varied with different groups in Pimería Alta. Sylvester said that another way of getting wind was by "donating [offering] cornmeal, pinole, or even just wheat flour to a shrine called *hevel kii.* Dangerous! Must avoid abuse." The shrine was a hole in the east base of the Sierra Estrella near Komatke. Joe added that a specific name for Wind's house was *hiha'iñ,* a shrine (a term of respect), a word that can also mean 'burial place', 'cemetery'. Joe said that when he was young, his parents' generation never mentioned where this shrine was because they feared that the new generation (in the early 1900s) might not be careful and might bring some calamity upon the villages. Similar stories of the Wind's house were told by early missionaries in other parts of Pimería Alta: both Velarde and Manje mentioned one near Mission

San Xavier del Bac. In 1716, Velarde (1931:116) called the shrine at Imuris *Uburiqui,* Spanish orthography for *heveḽ kii.*

While unprocessed mesquite pods and unshelled corn were stored outdoors in huge granaries, wheat went into a special basket. Ruth Giff still made the two kinds of storage baskets: the outdoor *homda,* from Arrow-weeds *(Pluchea sericea)* with a willow base; and the indoor *vashom* (pl. *vapshom*), which has round coils of wheat straw bound with strips of Screwbean *(Prosopis pubescens)* bark or, if that is lacking, regular mesquite *(Prosopis velutina)* bark. Screwbean has a paler, grayer bark, the preferred color. Ruth said, "You know, we only use *pilkañ va'ug* ['wheat straw'] for that, never barley or any other kind of straw. It's because they are smooth. They don't have those little knots in them." She made me a small one, with 50 vertical rows of stitches, that holds about 3 gallons of seeds. But most were much larger. For example, Bartlett (1854, 2:236) said they held 10, 12, or even 15 bushels of grain. They were used to store wheat or other seed crops such as teparies. The two types of storage baskets were illustrated by Emory in 1848 (p. 85; see also Bartlett 1854, 2:235–236). Ken Hedges, curator at the Museum of Man, told me that the *vapshom* are unlike any other kind of weaving found in the Southwest and that he doubts they are aboriginal: he has seen almost identical straw baskets in central Europe. It would not surprise me it the early Jesuits, while teaching northern Pimans the technology of wheat production, also taught the local women how to make the *vashom.* These Jesuits, for the most part, were from central Europe. Russell (1908:134) wrote, "Wheat straw is extensively used in the manufacture of the jar-shaped grain baskets. It is of modern introduction, and has not fully supplanted the ancient style of grain bin." The derivation of its name, however, is from the Pima Bajo word for the life form 'grass': *vasho.*

Among each patriarchal cluster of houses, there was, until very recently, a rectangular building called a *kosin,* a storage shed for crops and other stored goods. Often it was made from upright Arrow-weeds, sometimes plastered with mud. In the days of the *o'olas kiikĭ* 'round houses', the *kosin* might be better constructed than the houses people actually slept in. Here stood the *vashom,* which was usually woven within the *kosin* because it was too large to fit through the door. Bartlett (1854:236) sketched the interior of a *kosin,* and Kissell (1916:182) left this description:

> If the year has not been an exceptionally dry one, a Papago or Pima storage shed after the harvest season has past is an enviable sight. On its walls are basket materials, martynia [*Proboscidea* sp.], willow bark, willow splints; standing in a corner are beargrass, cat-tail, wheat straw; from beam to beam hang peppers, red and green; while on the ground are squashes, gourds, and great basket granaries full and running over with grains, beans, and various seeds. All these assure comfort to the family during the season when nothing is supplied by the fields. The preservation of the smaller foods required suitable receptacles, and these coiled granaries have supplied this need and so are greatly prized.

Ruth related this incident of a flood that nearly ruined their stored goods one year: "There was a *kosin* across the Santa Cruz [River] with storage baskets in it and logs piled up in front to close it. It was just below where the [Santa Cruz] community building is now. When I was a little girl there were more floods, when I was about six. The river rose high enough to get the bottoms of the *vapshom* wet, and the stored wheat began to sprout. Someone had to swim across the [Santa Cruz] River to find out what was happening. They removed the wet bottom layers that had begun to sprout, dried them, and ground them to make *matai ch-eḍ chemait* 'bread cooked in ashes'. It was *very* sweet."

Pinole was a primary wheat product among the Gileños. I used to sit enraptured while watching Rosita Brown under her *vatto* (ramada) make *haakĭ chu'i* (pinole). For the process, she needed two items of native manufacture. One was an oval clay parching pan, called a *muta.* This is made by the Maricopa, because the Pima have long since given up on ceramic arts. The other, a shallow, round tray basket, Rosita had made. To begin the process, she sprinkled a pot of whole wheat grains with water, setting it aside to moisten for a while. Into the parching pan she scraped red coals from the mesquite fire. Then she tossed in a few handfuls of the round, soft grains of wheat she preferred for pinole. Holding the *muta* by the earlike handles at each side, she kept everything in motion, the heavier wheat popping and jumping around so nothing burned. As a coal turned to ash, she whiffed it away with her breath. The smell of freshly roasted wheat was soon in the air. There was no stopping, no hesitating, or all would be burnt. An occasional flash showed among the embers as an escaped wheat hull oxidized. This was hot work.

When the wheat was sufficiently toasted, and the popping subsided, she rapidly pushed off the largest, hottest coals from the *muta* onto the ground. Then she flipped out the grains, separate from the lighter coals, into the *hoá,* or flat basket. Then came the cleaning stage, the separation of parched grains from any remaining small coals. Everything still had to be kept in motion or the basket would burn. Her wrists moved just enough to make tiny bits of ash, charcoal, or still-glowing embers jump to the surface, then move to the far edge of the basket. These she brushed away with her hand or blew away with a puff of air. It all looked so simple yet so impossible. For how many thousands of years have Indian women been parching grass grains and other seeds just as Rosita was doing?

The final step was grinding the *haakĭ pilkañ* into *chu'i* 'flour'. Some used their *machchuḍ,* or flat grindstone. Rosita and others I have watched finished the process by grinding the *haakĭ pilkañ* in a store-bought hand miller called a *chu'ikuḍ.*

Ruth explained the difference between two parching methods: *haak* means to roast something with live coals in a *muta* or in a basket, while *sitḍo* means to roast something over a flame with the *muta* or a similarly shaped clay pot (often a piece of a broken olla). Sylvester also discussed these two ways of making pinole from wheat: "*Haakĭ*—roast wheat, put [it on] coals and turn it over [in basket]. *Sitḍoi*—just put [wheat in olla] and stir it around and let it cook. It tastes a little different—*haakĭ* has a little smell, and it's fine; but *sitḍoi* is kind of coarse. I'm not so fond of *sitḍoi,* because when you stir it round [in water], it will just settle; but with the *haakĭ chu'i*—ah, boy!—you can never feel satisfied." He meant that you can never get enough. Others echoed this preference, giving the same reasons.

The prized pinole apparently had a short shelf life. Sylvester explained one day, "They roast it. They use that *muta* and my grandmother does [it]. Just push the coals and they never get burned. And get her *machchuḍ* and grind it. And get a bowl and put the *chu'i* in there. Whenever they want to use it, well. But that don't keep. It won't keep. Because if they keep it too long, it will turn back into *do'i chu'i* '[raw] flour'. See, when the St. Johns band traveled, they took that can of that pinole, they drink it. And they didn't use it up; it turned back into *do'i chu'i* again." Ruth Giff said the same. "But maybe if you keep it in the freezer it will stay," she added.

The last time I visited Rosita, in May 1977, she had just finished making fresh *haakĭ chu'i*. As we talked, she mixed some of the powder-fine wheat into a tall glass tumbler of water and gave it to me to drink. It tasted like a good breakfast cereal, somewhat like puffed wheat but more flavorful.

Rosita is gone, and the art of making *haakĭ chu'i* I think is almost lost among the Gila Pima. If any girls still want to learn, they should apprentice themselves to the few old Akimel or Tohono O'odham women who still make pinole. A little lard bucket or a two-pound coffee can of *haakĭ chu'i* now costs $10. But it's hot, hard work to make pinole.

There is one related art that the Pima are not likely to lose soon, and that is the making of thin wheat flour tortillas. Some call these *chemait*, others *chemat* (in the plural, *chechemait*). No feast is complete without a stack of *chechemait*. Folded into quarters, they appear on every traditional table at every meal. Flour with a high gluten content is the best for making good tortillas, and I have known certain Pima women to drive all the way to "the Gate" on the Mexican border to buy sacks of good-quality flour from Sonora.

At one time, *chemait* dough (*vaaga*) was mixed in special wooden troughs or short-legged dishes carved from mesquite wood called *chu'i vaagakuḍ*. These have now all found their way into a few fortunate museums or the graves of their owners. Today any plastic or aluminum dishpan serves the purpose. White flour, water, shortening, salt, and baking powder are the simple ingredients. Some use lard, while others are aware that it is not good in their diet. Some omit the baking powder and claim equally good results.

After being kneaded, the dough is rolled into fist-sized balls. These are then patted into a pancake shape and finally stretched into the thin tortilla that may reach from a woman's palm up her forearm to her elbow. When sufficiently thin, the *chemait* is thrown onto a hot disc blade that is lying convex side up over a bed of mesquite coals. When properly toasted, the *chemait* is flipped over for a while, then flipped into a half, and finally a quarter, then tucked away under a dish towel.

Bernard Fontana (personal communication) has pointed out that flour tortillas are nearly coextensive with the distribution of Piman languages from the Southwest to Jalisco. Outside this area, corn tortillas are the rule. If you visit a Mountain Pima or Pima Bajo village, you will most likely be served little corn tortillas, but these have long since disappeared from Pimería Alta. I found no one among the Gila Pima who knows how to make them.

The Pima once made several other kinds of *chemait*. For instance, *s-kovk chemait* is a smaller, thick tortilla made from regular or whole wheat flour, more like a thick pancake. One day at the Kyyitans', I was served one of these large biscuits that had been cooked in a frying pan. George said it could also be called *keli chemait* 'old man's tortilla' as well as *s-kovk chemait* 'thick tortilla'. I enquired about the first name, which was new to me. "Old people with no teeth can just break it off and drop it in their beans if they can't chew regular *chemait*," George explained.

Matai ch-eḍ chemait was formerly made by digging a pit in a hot bed of ashes and pouring thin whole wheat dough directly on the coals, then covering it over with more ashes. Russell (1908:68) found this "heavy and indigestible. One loaf was obtained, said to be a comparatively small one, that weighed 14 pounds and yet was only 3 inches thick and 20 inches in diameter." Ruth said, "When it's done, we have to take it out and wash it. No lard, salt, or baking powder!" Several others also told me the ashes were washed from the surface after baking.

Another favorite, found at every Indian fair or rodeo concession stand, is fry bread, or *oámajda* 'something made yellow [or] brown'. This is dough flattened to the size of a large pancake and dropped into boiling grease. White people who think

they are getting something "really Indian" do not realize that there is nothing aboriginal about these scarcely digestible, lard-soaked affairs. As Joe Giff pointed out, wild animals such as deer usually had little fat, and not until Eurasian livestock came among them did Pimans have much lard in their diets. His wife Ruth added that they once used the fat from the horse to cook *oámajda*. Even the name for this type of cooking, *manjugĭ ch-eḍ bahijida*, Joe noted, is a loan word from Spanish (*manjugĭ*, from *manteca*). Real Pima cooking, he told me, is *gai* 'roasting on coals', *chuáma* 'pit roasting', *haak* 'parching with live coals', *sitḍo* 'roasting over a flame in an olla', or *hido* 'boiling [or] stewing' (see chapter 8). Other traditional Pima dishes including wheat are *poshol* (whole wheat soaked, pounded, and cooked with brown teparies) and *atol* (*haashañ kai* 'Saguaro seeds' and whole wheat ground together very fine). These names, derived from *posole* and *atole*, go back to the Contact period, but you are not likely to find either food for sale at an Indian fair these days.

I asked Ruth if there were names for different kinds of wheat flours. She thought for a while, mentioning *haakĭ chu'i*, which she makes and sells. "I guess they say *shondal chu'i* 'white flour' and *hejel chu'i* 'whole wheat flour'." *Shondal* is a loan word from Spanish *soldado* 'soldier'. It was the flour that the Anglo itinerants brought with them on their trek west. And *hejel* signifies here something one makes oneself, one's own ground wheat.

I posed the same question to George Kyyitan, who was from another part of the reservation. He said, "There's *haakĭ chu'i*, toasted, for pinole. *Hejel chu'i* we just grind [ourselves], make it into [whole wheat] flour. And that *milgáán chu'i*, or just say *chu'i*, is white—store flour."

Ruth said when she was young they used to chew kernels of wheat as chewing gum. The elastic gluten remains in the mouth and can be chewed indefinitely.

It is one of the ironies of European colonization that wheat brought by the missionaries helped secularize northern Piman culture. Corn and cornmeal once entered nearly every aspect of O'odham religious ceremonial life. Songs were sung to corn while it grew; masks as well as brown Piman bodies were daubed white to represent its kernels; its meal was spread as a path for the Navĭchu kachina being. And as the winter solstice approached, the community invited the storyteller to come narrate the four-night Creation Epic by sending him a handful of cornmeal. But for the Spaniards, instead, it was wine and bread made from wheat flour that were the two essentials for the holiest of ceremonies, the eucharist: wheat, transubstantiated, became the body of the Savior.

But for the Gileños, wheat required no sacred song: its origins were not encoded in mythic time. In historical times, wheat provided the People with their one great meteoric venture into entrepreneurialism. The enterprise was so successful that it eventually took the stopping of the river by Euro-Americans to humble the Pima.

Vouchers: *157* (GK, *olas pilkañ*); *LSMC 27* (not extant).

Ruth Giff said they used to have an all black *u'us bavĭ*. She and Joe raised it for a number of years; then they lost the seed. It did not have any special marker in Pima folk taxonomy. Ruth told this anecdote of her experiences with black-eyed peas: "When we were kids, we made chewing gum from that *ushabĭ* [resin] from the sunflower *[Helianthus annuus]*, *hivai*, but it tasted awful. Then our mother at lunch made *u'us bavĭ*. And the taste of that *ushabĭ* just came out on our teeth. So I don't like the *u'us bavĭ*, although I would cook it for Joe."

Francis Vavages said, "*U'us bavĭ* is easier to raise than *kalvash* [garbanzos]—big pods. They're easy to grow. When they dry, take them off—they'll tell you. If you

u'us bavĭ   *Vigna unguiculata*

There are apparently no indigenous cultivars of this Old World crop surviving among the Pima, although other tribes to the south maintain stock of various kinds (including an all-black form). The painting is of a robust form of unknown provenience supplied by Pedro Estrella.

Vigna unguiculata (L.) Walp.
  [= *V. sinensis* (L.) Savi ex Hassk.]
black-eyed pea, cowpea
Fabaceae (Leguminosae)
u'us bavĭ

One of Fr. Kino's introductions to Pimería Alta at the end of the 1600s was the black-eyed pea, or cowpea. No one has yet determined when it reached the Gila Pima, but both they and the Tohono O'odham named the new crop *u'us bavĭ* 'stick tepary'. Apparently the Piman name is a reference to the shape of the pods. The great white flowers of *u'us bavĭ* are similar only in shape to the small tepary *(Phaseolus acutifolius)* flower. The pods are hardly wider than a tepary pod but are about three times longer and sticklike. In the evaluation of Castetter and Bell (1942:75), "The [Hispanic] introduced crops which proved to be of greatest significance in early Piman economy were wheat, barley, watermelons and cowpeas, and in that order."

Of *u'us bavĭ*, Sylvester said, "Grow them here; we eat them green. You can buy them in the store, as seeds." He told me of a small variety they used to grow called *yumĭ u'us bavĭga* ('Yuma stick tepary' + *-ga*, ownership marker). This form is supposed to have originated with the Quechan (Yuma) Indians of the lower Colorado River. He said it was about half the size of the regular *u'us bavĭ*.

leave them too long, they're going to lose them [the seeds]. Or [pick them] when green. They call those young ones, when real young, *u'us bavĭ moikdag* ['soft' + attributive]. You cook it with the skin, same as string beans—the shells with it, because they're tender. Can put bacon, cut up—kind of season with it."

"We use *u'us bavĭ* as string beans," said Ruth, "then [later in season] use as regular dry. But cook them! cook them! cook them. Then they'll taste better. I heard the Yuma [Quechan], they like *u'us bavĭ*. They cook them till they turn red."

Castetter and Bell (1942:120) said, "Both the small and large forms of the cowpea were and still are grown, although both tribes [Pima and Tohono O'odham] obtained the small variety first. Solid black seeds of this small form without any spot sometimes occurred, but these merely appeared among the black-eyed ones and rarely were grown as a separate variety. In fact, they were usually discarded entirely even in cooking." These authors noted cowpeas being planted three inches deep in rows 1.5 to 2 feet apart.

Sylvester said that a Tohono O'odham name for *u'us bavĭ* is *huhuḍa s-chuchk* 'black sided [pl.]' (*huḍa* 'side'). "That's a *very* good name for them."

One of the advantages of this crop was differential predation. George Kyyitan was discussing his farming with Leonard Pancott. "Nothing can [will] eat it, too, on that *u'us bavĭ*," said George. "Not a rabbit."

"Jackrabbits used to be the ones that give us trouble on *bavĭ*."

"Yeah," recalled George. "I had some over here on that *kooko'i kii* [Rattlesnakes' House, a shrine near Bapchule], on the next field there. I had some on there. That *u'us bavĭ* and *bavĭ*. I had some. All those rabbits, jackrabbits, came in, never bothered that *u'us bavĭ* there. They go around and eat that *bavĭ*. Boy, they mess [it] up!"

At the Pima Bajo pueblo of Onavas, Sonora, don Pedro Estrella gave me a bag of the small black-eyed peas, which he called *tuk vuupuhikam* 'black eyes it has'. In Spanish he called them *frijol yorimuni*. Here is a curious case where the widespread Uto-Aztecan generic *mun/muñ* 'common bean' has been incorporated into the dominant European language.

Vouchers: *863* (RG).

Zea mays L.
maize, corn
Poaceae (Gramineae)
huuñ, huuñĭ

Pima subsistence, before the advent of whites and wheat, hinged on two crops, one cultivated, the other wild. The first was maize, or corn, called *huuñ* in Pima; the second, mesquite, *kui*. Failure of either one meant hard times on the Gila, while failure of both meant famine and perhaps disaster. So it is no surprise that throughout the Southwest an elaborate ceremonialism evolved around rain and corn. This ceremonialism is best known among the Pueblo tribes to the north, partly because so much of it survives to this day. But it was once found as well among the Gileños and other rancheria Pimans. The puebloid elements, if it is legitimate to call them that, were either better developed or more accurately preserved in Pimería Alta. Winter wheat and a changed economy desacralized the entire subsistence base of the riverine people in the mid 19th century, but not all the parts and pieces were lost.

Corn is one of the four primordial crops in the Creation Story (although in the version narrated by Juan Smith [Hayden 1935], the account that is most detailed in this regard, cotton originated even before these four). Se'ehe designated that his own people, the Pimans, were to be farmers. "That farm that the Pimas made at that time didn't have any ditches on it; they were raising crops from the rains," interjected one of the ancient storytellers (Hayden 1935:10). The People, the O'odham, who occupied the banks of the Gila and lower Salt, would plant during the ebb of the two annual flood cycles along these desert rivers: once in the spring from the melted snows in the eastern mountains, and again during the summer rains. Their original fields were flooded riverbanks and the rain runoff channeled into the mouths of desert washes.

So Se'ehe, who is also called Elder Brother, set about making crops for his People. He rubbed the smooth and oily skin of his chest and brought out two seeds and planted them in the ground: *huuñ* (maize) and its twin, *haal* (pumpkin or squash). Elder Brother sang:

You look in the fields
And you will see corn coming out.
The leaves are swaying back and forth
Made by the wind.

And he sang again:

You look in the fields
And you see that the *haal* is coming out.
The leaves of the *haal* are like the clouds
And the decoration on the pumpkin is like decorated clouds.

And so it was; the slender leaves of the native Pima breeds of corn sway in the hot desert air and help make the wind, or so the singer hopes. And the broad dark green leaves of the native Pima *haal* are splotched with little white replicas of clouds. Even the small *haal maamaḍ* starting to swell are marked with white clouds.

Elder Brother gave the People two more gifts: the little bean called *bavĭ*, the tepary; and another species of bean called *havul*, the lima, or possibly *muuñ*, the common bean. All four were in the *oidag*, the field, singing together. And their song (as well as other crop songs recorded for the Akimel and Tohono O'odham) begins and ends with words that are either so archaic that no modern speakers can interpret them, or else borrowed from some neighboring tribe, perhaps Puebloan. The four crops in the field sang:

He'e yana heo vava heo
They are singing,
The white teparies and the corn are singing,
In the *oidag* they are singing.

Then the beans and the *haal* sang together:

Vava he'o, they are singing,
Haich e ya, ya eena!

But the People do not just plant the crops and leave them to their own devices. The whole process of living on the razor's edge of this desert country means singing along the entire process from planting to harvest. The ground is loosened and the weeds are pulled. But most important, the crops are sung along and the rains are sung in.

Underhill (1938:43–44) wrote of the Tohono O'odham farmers,

Every man has his field at the mouth of a wash where the water comes down after the rains in a swirling red torrent. It floods the land for half a day till the desert sun has sucked it up and the dead dry soil has sucked it down. Then the ground is soft enough for the Papago to thrust into it a sharp-pointed stick which was once his only spade. He stands at the edge of the field, holding his digging stick and a buckskin pouch of corn kernels. Kneeling, he makes his hole and speaks to the seed, in the Papago manner of explaining all acts to Nature lest there be misunderstanding: "Now I place you in the ground. You will grow tall. Then they shall eat, my children and my friends who come from afar."

Carefully he makes the holes, one at each stride, and drops in four corn kernels, and behind him his woman covers the holes with her bare toes. Now the corn will come up "like a feather headdress" and the beans will come "singing together." But not without help. Night after night, the planter walks around his field "singing up the corn." There is a song for corn as high as his knee, for corn waist-high, and for corn with the tassel forming. Sometimes, all the men of a village meet together and sing all night, not only for the corn but also for the beans, the squash, and the wild things.

What Underhill wrote for the Tohono O'odham must have applied to their riverine cousins as well, except that the Gileños and Sobaipuris of the San Pedro and Santa Cruz had the rivers and floodwaters on which to depend, not just the rain.

Even with this, though, life was precarious, and both rains and the corn crop could fail. Hunger was a very real threat for these desert agriculturalists. And so there was the *jujkida*, a formal rain-making ceremony. Even at the turn of the century, the Gila Pima told Russell (1908:347), "When rain is desired one of the leading men who understands the ceremony will notify the medicine-men, the orator or reciter, and the [lead] singer."

Everything is done with the utmost liturgical propriety. Even the wood of the scraping stick *(hivkuḍ)* used to accompany the singers must be from *kauk u'us* (Greasewood, *Sarcobatus vermiculatus*), a tough little bush whose delicate leaves symbolize rain. Surrounding villages are invited.

"When all have assembled in the evening the leader calls out the names one by one of the medicine-men, who take their position behind the fire, facing toward the east. Then the names of those who will sing are called. The leading singer sits behind the medicine-men, and his assistants place themselves on either side of him and around the fire. Then the orator is named and takes his place with the medicine-men" (Russell 1908:347). The role of the *jujkida* narrator was no easy task, for he was expected to relate some ancient speech—some long canticle-like tale of a journey—word for word as he received it. The final part of a journey narration tells what the hero/narrator expects to be accomplished as a result of any particular liturgy. And so the Gila Pima rain-making oration describes this expectation: "Scattering seed, he caused the corn with the large stalk, large leaf, full tassel, good ears to grow and ripen. Then he took it and stored it away. As the sun's rays extend to the plants, so our thoughts reached out to the time when we would enjoy the life-giving corn. With gladness we cooked and ate the corn and, free from hunger and want, were happy. Your worthy sons and daughters, knowing nothing of the starvation periods, have been happy. The old men and the old women will have their lives prolonged yet day after day by the possession of corn" (Russell 1908:352).

Nevertheless, the rains and corn did fail in certain years, if we correctly interpret the story of Corn Man and Tobacco Woman found in the Pima Creation Myth (see also *Nicotiana rustica* and *Cucurbita* spp. accounts, Group I). Corn and Tobacco gambled, then quarreled over the winnings. In anger, Corn told Tobacco that everyone liked him instead. Tobacco too was angry, saying that people liked her because only when the *mamakai* 'shamans' smoked her could rain come to wet the ground for Corn to come out. So Tobacco left toward the west. There was no more rain. Corn expected the People to bring him votive offerings of soft feathers and beads, but they did not. So Corn and his twin, Pumpkin, left the People and went east. Giisobĭ Maakai ('Verdin Shaman') took four votive offerings to entice Tobacco back: soft feathers, eagle tail feathers, the shaman's crystal, and a Hohokam slate palette. She refused to return but sent Verdin back with some seeds. The People were starving, having lost the two essentials for the rain-crop cycle.

Several years passed. A woman out gathering *hannam* (apparently just the pieces of *toota hannam* or Chain-fruit Cholla, *Opuntia fulgida*) was approached by Corn, who shot his arrows into her roasting pit. When she opened it, the pit was filled with cooked corn and pumpkins rather than the starvation food she had put in. Corn sent a young messenger into town to inquire about Tobacco and the prospects of marriage. (In another version, Corn proposed to the beautiful girl out picking *hannam*.) In any event, Corn Man's nuptial union with the woman was to be staged in a special ceremonial house called a *main kii*, made of woven mats, or *petates*. During the night it rained corn and pumpkins into every container set out by the starving People. They were once again replenished, and the child born to Corn and his bride was a special kind of pumpkin-baby. Some children used the pumpkin-baby as a doll, and they accidentally dropped it, breaking its neck. In anger once again, Corn Man left for the east, then came back for a while, and then left again with his pets, the blackbirds, whom he missed when he was away.

So Corn, like Tobacco (see *Nicotiana rustica* account), was a fickle crop that sometimes would not produce. And if the tobacco smoke and wine feast and songs did not produce rains, Corn could not come up either. It was a tenuously balanced cycle for the People to keep in motion.

A spin-off of this cycle is in one version of the Creation Myth. Corn and Tobacco's child is a pumpkin-baby, who, when broken, sinks into the ground and becomes a Saguaro. This is still another rain cycle: Saguaro fruits are essential for the wine feast, a ceremony aimed at perpetuating the summer rains.

One long three-part Gila Pima maize song has been preserved. It is the "Huuñĭ Ñe'i," addressed to Corn Man, who finally returned to the People and made his home in a range called Taatkam, the Picacho

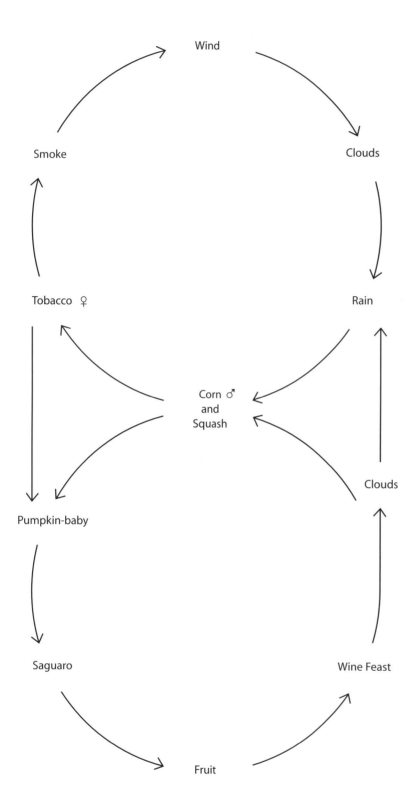

Rain cycle

Mountains. (The name according to Sylvester means 'to have a premonition'.) Russell (1908:333–334), who recorded the song, noted that it was the first ever sung to bring rain. The song was related by Thin Leather, who also was the narrator of the best-preserved Tobacco-Corn segment of the Creation Story.

I

Hi-ilo-o ya-a-a! He who sees everything
  Sees the two stalks of corn standing;
He's my younger brother. Hi-ilo-o
  ya-a-a!
  He who sees everything, sees the
    two squashes;
He's my younger brother. Hi-ilo-o
  ya-a-a!
  On the summit of Taatkam sees
    the corn standing;
He's my younger brother. Hi-ilo-o
  ya-a-a!
  On the summit of Taatkam sees
    the squash standing;
He's my younger brother. Hi-ilo-o
  woiha!

II

Hi-ilo-o ya-a-a! Over Taatkam
  Rise the clouds with their loud
    thundering.
Hi-ilo-o ya-a-a! Over Taatkam
  Rise the clouds with their loud
    raining.
Hi-ilo-o ya-a-a! The Bluebird is holding
  In his talons the clouds that are
    thundering.
Hi-ilo-o ya-a-a! Yellowbird is holding
  In his talons the clouds that are
    raining.

III

Hi-ilo-o ya-a-a! See Elder Brother
  Breathe out the winds that over
    Taatkam
Drive the clouds with their loud
  thundering.
  Hi-ilo-o ya-a-a! See Elder Brother
Breathe out the winds that over
Taatkam
  The welcome storm clouds are
    suspending.
Hi-ilo-o ya-a-a! In the great rain clouds
  Let me sing my song of rejoicing.

Only one other Gila Pima Rain Song was recorded (Russell 1908:331-333), and it, too, associates the twin crops *huuñ* and *haal*.

This one refers to the prayer sticks, *omina,* sometimes planted in the fields, and the primary feathers and down of *ba'ag,* the Golden Eagle, also used in rain supplication.

I

Hi-ihiya naiho-o! Let us begin our
song,
   Let us begin, rejoicing. Hichiya
      yahina-a.
Let us begin our sing, let us begin,
   rejoicing,
      Singing of the large corn. Hichiya
         yahina-a.
Singing of the small corn. Hichiya
      yahina-a.

II

Hi-ihiya naiho-o! The darkness of
   evening
      Falls as we sing before the sacred
         omina.
About us on all sides corn tassels are
   waving.
      Hichiya yahina! The white light
         of day-dawn
Yet finds us singing, while corn tassels
   are waving.
      Hichiya yahina-a! The darkness
         of evening
Falls as we sing before the sacred
   omina,
      About us on all sides corn tassels
         are waving.
Hichiya yahina! The white light of
   day dawn
      Yet finds us singing, while the squash
         leaves are waving.

III

Hi-ihiya naiho-o! The earth is rumbling
   From the beating of our basket
      drums.
The earth is rumbling from the beating
   Of our basket drums, everywhere
      humming.
Earth is rumbling, everywhere raining.

IV

Hi-ihiya naiho-o! Pluck out the feathers
   From the wing of the Eagle and turn
      them
Toward the east where lie the large
   clouds.
      Hichiya yahina-a! Pluck out the soft
         down

From the breast of the Eagle and turn it
   Toward the west where sail the small
      clouds.
Hichiya yahina! Beneath the abode
   Of the rain gods it is thundering;
Large corn is there. Hichiya yahina!
   Beneath the abode of the rain gods
It is raining; small corn is there.

From this rain song, we see that the Gila Pima once shared the same corn-growing cult that Underhill had described for the Tohono O'odham: corn is sung to at dusk, during the night, and at dawn (verse II); the sacred prayer sticks are planted in the field; the overturned tray baskets reverberate from the *hivkuḍ* 'scraping stick', anticipating the rumbling of thunder and the humming of rain falling on powdery earth (verse III); the two long black eagle wing feathers in the hand of the rain shaman beckon the mounting summer cumulus clouds from the east; the fluffy white *viig* 'body feathers' call clouds from the opposite direction (verse IV).

And everything is tied together in this tenuous desert world: the tender grassy leaves of *huuñ;* the white cloud markings of *haal;* the cloud of *viv,* tobacco smoke to bring rain; the *haashañ* or Saguaro fruit to make wine for the rain ceremony; even the little *cheehegam,* the Ladder-backed Woodpecker, pecking for *huuñ vaptopĭdag,* "worms," at the ear tips, "right down into my heart," says Corn in one song. Nothing is extraneous in this precarious unity of plants, rain, and People.

Corn symbolism permeated much of Piman ritual. In various ceremonies the dark brown bodies of participants were daubed with white spots of paint representing kernels of corn (white, not yellow, is the predominant corn color among Southwestern tribes). The various gourd masks worn during the Navĭchu ceremony are often spotted with rows of white paint representing kernels. Cornmeal, the "holy water" of Puebloan ceremonies, appears frequently in the rites of the Pimans. In the Navĭchu ceremony there was a *chu'i wuadam* 'cornmeal spreader' who led the way with a white trail and blessed people, throwing *chu'i* to each side. And in midwinter, when the longest night of the year approached, someone from the rancheria took a handful of cornmeal to the storyteller as a formal request for him to come in four days and narrate the four-night story of how all

things came to be in the Piman world. When staying sickness struck, a handful of cornmeal invited the *makai* diagnostician and then the healing singers.

Ruth Giff looked over the seven types of corn that I brought back from the Mountain Pima of Yepachi, Chihuahua. One had kernels running almost to the very tip. "These are the kind they need when they went to the ocean for salt: with the corn [kernels] clear to the end. They'd grind these and have to sprinkle it all along the trail. Then they put it on the water to make it go down so they could get the salt. Only very fast runners could go, because they had to get out of there fast when the water came up."

Even the salt gathering, according to Underhill (1938:111–133, 1946:211–242), was referred to obliquely as a "corn harvest" so as not to offend the Ocean, who was considered a powerful being. When the Gileños went out locally to get salt from the alkali flats or from a special saline deposit in the river, apparently no particular ceremonies or prohibitions were observed.

If corn was such a touchy crop, how did the People survive? For one thing, all the native Pima maize strains were short-growth varieties that required only 60 days from planting to flowering. If conditions were not favorable for the spring crop, there was always a chance to make up the losses during the summer planting. (In San Diego, with artificial irrigation and no frost, I have raised three crops of Pima corn in a single season, but I do not know whether the People ever attempted this in their own riverine environments.) "If you planted in spring," Sylvester said, "you could harvest in June, early. Also [if you] plant [in the] last part of July and early August, [it] only takes 60 days to have corn to eat—then in mid November would be frost." For another thing, there were two crops of the major wild staple, Velvet Mesquite, to be gathered. If the spring crop was not sufficient, at least the *toomdaj* (second crop) might produce enough. Still, that too could fail.

But the Pima practiced something common to subsistence agriculturalists nearly worldwide, and that is maintaining crop diversity. The Gileños did not raise simply *huuñ,* as a superficial look at stories might suggest; they maintained, at least in the memories of the people with whom I worked, some seven named varieties of maize. (And who knows whether additional varieties had already been lost, after a century of intensive reliance on wheat and a devastating generation of droughts and riparian deterioration?)

I asked Sylvester Matthias what kinds of *huuñ* he remembered the Pima of the western villages raising. He enumerated eight of the native or 60-day varieties, and two 90-day or commercial varieties. (Because there is a nonvoiced *i* at the end of the modifying term, the word *huuñ* is pronounced differently in the combinations *kuupĭ hiuuñ, chechpa'avĭ hiuuñ,* and *chehedagĭ hiuuñ.*) The 60-day varieties of maize included

kuupĭ huuñ 'closed corn': white, shorter than *toota huuñ,* about five-inch cobs

toota huuñ 'white corn': regular Pima white; the only one raised now; longer ears than *kuupĭ huuñ;* good for flour and menudo

s-vopogs huuñ 'spaced corn': gray margins on white; longer than *toota huuñ;* spaced between kernels

chechpa'avĭ huuñ 'hookers' corn': all red; unmarked; long or short ears

chehedagĭ huuñ 'blue corn': blue like Hopi corn (maybe is Hopi corn); for general use; short ears; similar to *toota huuñ*

o'am huuñ 'yellow corn': a native corn; not field (sweet) corn; very rare

hehemdam huuñ 'laughing corn': skinny ears; halfway between yellow and white; good for flour and menudo

taatam huuñ 'tooth corn': a big-eared corn with yellowish kernels that suggest teeth

The 90-day varieties of maize included

ge'egeḑ huuñ 'big corn': June corn; modern, not native here; came in the 1920s

kavk huuñ 'hard corn': big; kernel color between white and yellowish; big leaves; long ears like June corn of the commercial varieties; good for *poshol* or menudo

Some of these varieties must be very old, for not only were most of them not known to others I asked, but even the meanings of some of the words were obscure. I asked Sylvester what *kuupĭ* means. He said, "It's something [the word], but it's not used anymore." He knew this variety from references in legends from his grandmother. Mathiot (n.d.:17) gives *kuup[i]* as 'to be closed, locked in one place, shut, stuffed up'. Apparently, as applied to corn, the kernels were expanded on the cob so that the spaces between them were closed. This description suggests Onaveño.

Kuupĭ huuñ and *s-vopog huuñ* appear to stand in opposition to each other. Sylvester glossed *s-vopog* as 'deep in the groove', saying, "It's full ears, but spaced between the kernels. A young man or boy isn't supposed to eat this kind or the Apaches are going to get him." This must have been an important taboo. He mentioned it every time we discussed corn varieties. He identified specimens of Tuxpeño as *s-vopog huuñ.*

"*Ge'egeḑ huuñ* is June corn. Means 'big corn,'" he said. "That's modern. But I don't know where that June corn comes from. It's not a native here—not an Indian [Pima] corn. *All* the rest are 60-day and you can have fresh corn, but June corn, oh boy! It takes ages!"

While looking at a specimen of Ocho de Carrero or Harinoso de Ocho, Sylvester recalled another type of Pima maize he called *taatam huuñ.* "They're large [kernels]; they're like teeth and they call it [that]. It's similar to this; they're kind of yellowish, white-yellowish. Big . . . a little larger than this [one]." The ear he was holding was 12 inches long.

Sylvester knew a Pima man from Komatke who grew blue teparies but would not sell them. He also maintained the Pima *o'am huuñ.* "He kept it and used it himself, the short one. It's a rare corn, too. Nobody raises it [now]."

Sylvester was amused by an apparent euphemism. Of *chechpa'avĭ huuñ* he said, "That's 'prostitute's [corn]' or 'whore's corn'. But I read it in one of the books; they put it this way: 'crazy corn'. They didn't want to use 'prostitute'!" However, Mathiot (1973:405) gives *totonto huuñ* 'crazy corn', as a coequal of *chechpa'avĭ huuñ,* at least as used by the Tohono O'odham.

There is some difference of opinion as to what exactly 60- or 90-day signifies. Castetter and Bell (1942:81–82) decided this must refer to the time from planting to flowering. "Although the native Pima and Papago corns produce silks and tassels within fifty to sixty days after planting, our field studies consistently have failed to reveal any variety of corn among Pima, Papago, Yuma, Mohave and Maricopa which matured in this length of time."

I asked George Kyyitan what 60-day meant. "When it starts to put on seeds, on the ears."

"You can eat it in 60 days?"

"Naw. Two, three more weeks. Same as milo maize *[Sorghum bicolor]*—60 days is early [variety]; others take longer."

There is a qualitative difference between the ears grown during the early and late summer plantings. I have noticed this in the Pima *toota huuñ* that I have grown over the years. The earlier crop has larger kernels on more robust ears. Pimas picked out from my samples the second crop, which has smaller ears, often with irregular rows.

One day George explained to me how the pollen from the corn's tassels fell or was carried in the wind to the corn's silk to make the seeds inside the ears. "See, that's [the] way it is. That's why we plant the [different kinds of] pumpkins way over there and the [different kinds of] corn way over here. So you'll have good seeds, if you'll plant it separately, apart." He had just been discussing intercropping.

"How many kinds of corn did they keep apart?"

"I think those white one, and yellow."

"*S-toota huuñ, s-o'am huuñ?*"

"Yeah. In the olden days they say the yellow ones are quicker, faster than the white ones. Be ready first, before the white ones."

"What other kinds did they have?"

"They got all those—*chechpa'avĭ huuñ.*"

"What does it look like?"

"Those that [have] mixed seeds—purple, white, and all that, like the Hopi grow, [and] red ones. And I think those popcorn—they call them *s-vapshuḑ huuñ?*"

"What does that mean?"

"Ah, . . . kind of blistered. Kind of shiny on top." Mathiot (n.d.:239) gives *vapshshuḑag* as 'blister(s)'.

"Did the Pima raise that?"

"I think they do because that's why they know it's *s-vapshuḑ huuñ.* But they never use it [as popped corn]—just eat it. Roast 'em." Gary Nabhan tells me this is Reventador.

"How about *s-chehedagĭ huuñ?* Blue?" I asked.

"I don't know it. There's some called *s-i'ibhaimagĭ huuñ*—that's 'purple'. *S-chehedagĭ* is 'blue' and 'green'—they're all together like 'blue' and 'green'. And *s-vegĭ* is red."

"Did they have a *s-vepegĭ huuñ?*"

"Uh-huh. Like those *chechpa'avĭ huuñ.* They got red [kernels]."

"What happened? Today the only Pima corn left is the *s-toota huuñ.*"

"Oh, yeah. Because that's what *they* use, mostly. That's the main crop." George provided the name of still another variety: "*S-i'ovĭ huuñ* is 'sweet corn'. Nobody likes it because it's not good for *ga'ivsa* [porridge]—it tastes some way. I don't think anyone raised it. White man brought it out. Well, it's okay, just to roast, but not for *ga'ivsa.*" Others echoed his evaluation. (For the preparation of *ga'ivsa*, a staple Pima food, see beyond.) He also mentioned *al ha'as huuñ* 'smallish corn', which was a 60-day white or yellow flour corn.

The soft flour corns were most prized by the Pima. These were usually *toota huuñ*, or rarely *vepegĭ huuñ.* (There is some doubt whether this last, red, was a variety deliberately maintained by the Pima or just an occasional variant.) These are all called together *s-moik huuñ* 'soft corn' or *a'al huuñ* 'small corn'. Sylvester used the cover term *kiikam huuñ* for all native or 60-day maize. He was almost the only Gileño consultant who knew varieties other than the several flour corns and *ge'egeḍ huuñ.*

The importance of these soft flour corns corroborates the findings of Castetter and Bell (1942:79–89) half a century earlier. Apparently the other named maize varieties *(s-vopogs huuñ, taatam huuñ,* and *kuupĭ huuñ)* dropped from Pima agriculture during the 100 years that wheat raising was prominent or during the 40 years when they suffered water loss.

In addition to the Pima *toota huuñ*, Ruth also raised a Hopi white corn, brought down from the high country by relatives who had married into that tribe. Superficially it looks like the usual white Pima soft corn, but when germinating, it sends out a powerful taproot (to penetrate deep into the sand of Hopi fields). "Pima *huuñ* is a 60-day corn," she said, "but Hopi corn is maybe 80 or 90 days—like a June [commercial variety of] corn."

The history of popcorn among the Gila Pima is equivocal. Ruth said she does not remember ever having real popcorn. "But when my kids were small, we grew some popcorn, but we didn't know it had to be kept away from the others, so it didn't come out right. Then we learned and planted it far away from the other ones. We didn't have it before, so we didn't have a name for it." Sylvester agreed.

But it may have come and gone once much earlier. Hrdlička (1908:261) said, "The Pima plant pop corn, having learned to do so, they say, long ago from the Mexican Indians. They roast the pop corn in a pot and add salt. Pinole *[ga'ivsa]* is prepared in a simple manner by roasting and grinding. Nothing is mixed with it until the pinole is to be eaten, when some add salt and others sugar." Nearly 40 years later Castetter and Bell (1942:84–85) found no popcorn among the Gileños and little among the Tohono O'odham, who "said the tribe has had pop corn for only about forty years and that it was originally obtained from the Mexicans." Both Akimel and Tohono O'odham told the two agronomists that the closely related flint corn "was not Indian corn, but of Mexican origin, and had come into their area comparatively recently. Both tribes grow it to some extent today and it does well." They add, "Also the Pima grow a hard, flinty, yellow variety now known as Santan Yellow, accepted in some quarters as an aboriginal variety." Unfortunately its Pima name was not recorded.

Pima *o'am huuñ* may well be what King and Loomis (1932:50) called Santan Yellow, a local flint corn. Twice these agronomists contrasted this native variety with Sacaton June, a non-Pima variety developed at the station. Although they were enthusiastic in promotion of their hybrid strain, they admitted that it "requires a longer growing season, however, and therefore requires more irrigation water for production. Now that there is an ample water supply for a greater part of the Indian lands on the Gila River Reservation, the Indians should find it more profitable to grow Sacaton June than the native varieties." Earlier, King and Leding (1926:28) had noted, "The superiority of the Sacaton June variety can hardly be questioned in the light of these results [roughly double the yield per acre], but for the Indians who are dependent on the irregular flow of the Gila River some of the quicker maturing varieties fill an important need. Many of the Indians are able to produce substantial yields on the Pima soft and the San Tan Yellow flint varieties with only one irrigation, because of the short period required for development and their low water requirements." Perhaps the Pima in their centuries of farming in desert environments of the Greater Southwest had mastered some lessons the newcomers have yet to learn.

As might be expected of a crop that was one of the mainstays of a culture, there were many ways of preparing *huuñ.* Castetter and Bell's (1942:188) conclusion, "The standard Pima meal consisted largely of cornmeal gruel," sounds like a pretty dull bill of fare! My own impression after eating with Pimas for over a quarter-century is that they had an interesting, varied, and nutritious diet that changed throughout the year with the availability of various wild and cultivated crops.

The standard travel food, when going to war or on hunting expeditions, for example, has always been pinole or *chu'i.* Nearly every myth and narrative involving travel mentions pinole. Although replaced later by *haakĭ chu'i (pilkañ chu'i* 'wheat pinole'), the original *chu'i* was from *huuñ.* Ruth Giff told how they used to make it: "The O'odham used to parch the corn; it's not really like popcorn. They say— you know, way back—that you give a child a handful, and that would be a meal for him.

"*Huuñ haakĭ*, put it in the coals; and they do grind the *haakĭ.* It's funny, they don't drink it like the *pilkañ chu'i*; they just put it in their mouth dry and drink water after it. I don't know why they do that—the old men. Toast [parch it] in a basket with coals; then grind it on the *machchuḍ* ['grindstone']; we don't use the *chepa* [wooden mortar] for that purpose, like the black people. I see in *Roots*, that's how they grind their corn; [we use the *chepa*] just for *vihog* [mesquite pods], to make flour out of it."

Francis Vavages noted, "To make *huuñ haakĭ*, have to do it with soft corn, not hard."

Sylvester and others called the process of eating corn pinole *chiñhum*, from *chiñ* 'mouth'. He too commented on the difference between eating this and wheat pinole. "Just put *huuñ haakĭ* in mouth and drink water *after.* I would say, mix it in their mouth." He mimed how one would take a bit of powdered maize with the fingertips of one hand, then chase it with a sip of water from a container in the opposite hand. "The Yaquis mix it like *haakĭ chu'i,* but the Pimas don't."

Among the Mountain Pima and the lowland Pima Bajo, the staple of every meal is the small round corn tortilla, usually served hot. Dried corn is boiled with slaked lime and is next washed, drained, and ground on a metate. The resulting *masa (nixtamal)* is then ready to be patted by hand into the flat round tortillas that are cooked on a hot *komal* 'something flat'. I have never known an Upper Pima to make a corn tortilla. All the *chechemait* 'tortillas' I have eaten among Akimel and Tohono O'odham have been the large thin ones made from wheat flour.

Perhaps I had just overlooked them; I visited Ruth Giff to inquire. She had made a number of visits with me to the Pima Bajo village of Onavas and had watched María Estrella, don Pedro's daughter, spend many hours a day preparing corn *masa* and making tortillas. "No, I never heard of anyone doing that here, not for tortillas, just the little patties, *huuñ chechemait.* My mother used to make them from the kernels, when they're just starting to dry but still have the milk. Scrape them off or just *kel,* that means 'take the kernels off'; or it's like that word we were talking about, when *shuug* [mockingbird] told the women, *Keliv! Keliv!* 'Pick! Pick!', but it's just *kel.* But I suppose it means about the same thing [see *Lycium* spp. account, Group D]. And she'd grind that on the *hodai* [stone metate], mix it with some shortening, wrap them in *huuñ eldag* 'corn husks', then steam it. Like little tiny tamales, called *huuñ chechemait.*"

Still, it was hard for me to imagine that Gileños did not once make corn tortillas. At the turn of the century, corn still held some of its earlier importance among the Pima. Russell (1908:86) mentioned, "In olden time maize was ground upon the flat metates and formed into loaves." But in his day, the Pima still had "numerous varieties [which] are all prepared in about the same manner. As the husked corn is brought in by the women, it is piled on a thin layer of brush and roasted by burning the latter, after which it is cut from the cob, dried, and stored away for future use. The shelled corn is ground on the metate and baked in large cakes in the ashes. Corn is also boiled with ashes, dried, and the hulls washed off, then thoroughly

dried and parched with coals or over the fire. It is then made into a gruel, but it is not so highly regarded as the wheat pinole" (Russell 1908:72–73). But did the Pima once make corn tortillas, the mainstay at every meal prepared by desert Pima Bajo and Mountain Pima today as well as by their neighboring tribes? It would be surprising if they did not, though many years of using wheat flour may have obliterated the memory of it. Possibly they did, because Russell (1908:68) wrote, "No knowledge of the pueblo wafer [piki] breads exist among the Pimas, who confine their treatment of mesquite, corn, wheat, and other flour to baking as tortillas or as loaves in the ashes, frying in suet, or boiling, either in water to form a gruel or mush, or with other foods in the shape of dumplings." His statement is still ambiguous: were all these flours prepared in all these ways? No contemporary Pima has ever heard of a mesquite tortilla, for instance.

Fortunately, one observer left information that resolves the ambiguity. In 1903 J. William Lloyd (1911:60) spent a little over two months living with a Pima family at Lower Santan. He wrote: "The *nah-dah-kote [naadakuḍ],* or fire-place, was an affair of stones and adobe mud to support the earthen pots for cooking or to support the earthen plates on which the thin cakes of corn or wheat meal were baked. These were what the Mexicans call tortillas."

The *naadakuḍ* is a thing of the past among the Gila Pima. And the clay cooking plate or *komal* has now been supplanted by disk blades or occasionally the cut-out top of a 50-gallon metal drum, but Russell (1908:pl. 19b) illustrated one, and I have found them still in use in some Pima Bajo homes. Lloyd added: "*Taw-mahls,* or corn-cakes of ground green corn, wrapped in husks and roasted in the ashes, or boiled, were also favorites." These are what Ruth called *huuñ chechemait.*

But in the 20th century, most corn is made into *ga'ivsa.* "Store corn is not very good for *ga'ivsa,* not sweet enough," Ruth said. So she grows her own *toota huuñ* each summer. "When field corn is picked, it is now steamed. Before they would barbecue [pit roast] it. Burn wood for five or four hours. Cut some cornstalks. When the coals have burned down, put over the coals first a layer of tin, then put a layer of these cornstalks in the pit [over the tin].

Then throw in all the ears, left in their husks. You cover over again with a piece of tin, then seal [it] with dirt and leave till next morning. Then they are steamed. When it cools off, take them out, peel them of their husks." The process is called *chuáma.* "This is a better way," Ruth added, "than they used to—just throw pieces of wood, limbs, leftover tips of mesquite [branches], anything, together and set it on fire, and throw the corn ears on top of that to roast; but there's lots of waste—some burned black and others were not cooked at all. Then we had to stand around a fire and roast by hand the ones that weren't done."

At this stage the steamed ears could be eaten, but most were for use during the next year or so. I had seen Ruth dry great quantities of both her own *toota huuñ* (which had long been in her family) and a similar white Hopi corn. "Then you spread them out to dry, then pack away on the ear. Store like *vihog* [mesquite pods] in the [big baskets]. Shell them as needed. Seeds are kept *on* the cob because they last longer. If you shell them, they'll get wormy within a year, I think."

A woman could go any time to these great storage baskets to get the partially prepared corn. It is next ground on a metate or with a metal hand miller; then the coarse meal is cooked with water. This finished *ga'ivsa* looks something like cornmeal mush but has quite a different flavor and texture than the commercially available yellow field corn.

One summer day I drove into George Kyyitan's yard and found him roasting ears of green corn, stripped of their husks, directly on a bed of mesquite coals in the outdoor barbecue. With a shovel, George fished out several very blackened ears, and several of his boys scraped off the charcoal and began eating them with relish. The other ears George put into a screen-bottomed box hanging on the clothesline with some he had roasted the previous day. "You roast the corn still on the ear," he explained, "then dry it. Store it still on the cob. But it has to be *dry,*" he emphasized, "or they'll rot. Then use a grinder." In the 115° heat of the afternoon, I suspected that his *ga'ivsa* would not take long to dry thoroughly.

Still other dishes could be made with the corn. "Just whole corn and *oám bavĭ,* those brown beans [teparies], are cooked," said Ruth. "The skins [of the corn] will come off. That's called *poshol*." I asked whether the Pima used ashes or lye to make the hominy-type corn as many other tribes do, but she said no. Just the cooking produces the *menudo*-style kernels. However, at Bapchule, George remembered chemical processing of maize. "They put the ashes in there and wash it with it in there. And after that come off the peelings of the corn. Some of them, they use *lime*. I think after they cook it. I don't know how much. Those that they know, they used that. They're used to making [it] the hard way, to make tamales at that time. After that have to use the grinder, too. For menudo, take peelings off, then wash it again—*poshol*." This alkali processing allows the amino acid tryptophan in corn to be converted into niacin, essential to good health (see Nabhan 1993:98-109). Pima *poshol* is the widespread *posole*, still served throughout much of Mexico as well as in the more traditional restaurants in the Southwest. *Posole,* at least in the Southwest, is basically white corn with pork, while *menudo* is the same fluffy white kernels cooked with beef tripe.

Tamales are another food from corn that appear frequently in Pima food sales and on feast days. These are made of beef cooked with red chile, surrounded by some *masa* (dough), then folded into a packet of corn husks, plastered with more *masa,* covered with another corn husk, and tied with a little ribbon of husk. Tamale preparation is time-consuming and so is usually a social activity. When all are ready, they are boiled several hours in a large pot. Tamales are of Mexican Indian origin, so they presumably should have been a local Piman dish. I asked Ruth about this.

"Nobody was making *tatamal* when I was a *little* girl. The first woman to make tamales lived over here, Mrs. John Michael. She used [plain] cornmeal. She taught us in the [19]30s; when I got married [1931] was when we started making them. I don't know where she learned it— from the Jujkam ['Mexicans'], I guess. We made them especially for New Year's. Later, women got together and started buying *masa* [commercially prepared corn dough]. Even buy the *huuñ eldag,* because they're larger. Ours [were] too small. We raised only the small Indian corn."

Traditional growers prefer their own corn varieties to the commercially available crops, which have a high forage yield. Ruth calls the traditional white Pima corn "June corn" in English. It is the only one she plants, because she does not like the taste and texture of the commercial varieties. July 18 is the deadline for planting it, she said, but she planted some about a week later than this one year. "But you know, now the frosts come later, in November. So maybe there's enough time." I asked about how many seeds she plants together. Just one, she replied, but she used to plant three together (of anything as a rule: corn, watermelon, squash, beans). I was really surprised it was not four, traditionally the Piman number for completion (Castetter and Bell 1942:152; Underhill 1951:61). But when she married and moved over to Joe's place, they were planting just single kernels of corn in each hole, the way the mechanical planter did, so she started doing this too. It drops just one seed at a time, and she thinks that is why Joe's family started planting single kernels, even by hand, because that was the way the machine did it.

---

huuñ, huuñĭ   *Zea mays*

Although the long growing season meant that theoretically a number of corn crops could be raised each year, the Pima farmer realized that not all of the frost-free season was equally suitable for successful pollination. Sylvester explained, "Two seasons for planting corn—put in at wrong time and they'll grow but won't have kernels on the cobs [ready by] June. Second planting, *shopol esiabig,* July through August for fall, even late on tail end of August, after Feast of the Assumption [15 August], and the frost may not get it. Get ripe in November."

Both Joe and Francis emphasized that crops should be in by the Feast of the Assumption to expect maturation before the killing frosts set in. The People in all villages watched the position of the sun setting in the Estrellas after it began moving back southward (following the summer solstice). When it reached a certain peak, indicating mid August, seeds had to be in the ground.

Still, freezing weather might catch the second corn planting before the Pima harvested. "When corn not quite matured, when frost comes, in that fix, they stack them up, whole plants," said Sylvester. "Later in winter, boil them like fresh [green] corn. Called *chutka*."

Corn served the Gileños for purposes other than just food. Cornstalks were frequently mentioned as thatching material for the beehive-shaped *olas kii* 'round house'. Other materials included Arrow-weed *(Pluchea sericea)*, cattail reeds, and wheat straw (Russell 1908:154). In post-European times, corn husks were used as fodder, but Pima corn was never selected for large plants with big leaves: instead, high ear productivity, a short growing season, and drought resistance were the criteria. During the many years I have grown Pima corn and seen it in the Pimas' fields, I have constantly been amazed at how short and grassy the plants usually are.

Maize husks, called *huuñ eldag,* were useful things to be stored away until needed. Joe Giff explained one use. "They use the thin corn husk, you know, next to the kernels of the corn, the real thin ones, like paper, and they use that to roll tobacco in, which is good paper. You have to wet it, [or] it'll crack and break."

Corncobs are called *huuñ kuumkuḍ* 'corn something that has been chewed on'. George Kyyitan said that they were stored for fuel to be used in *haak* 'parching' because the resulting coals were not too hot. Sylvester recalled the same practice from the western villages. "Use *huuñ kuumkuḍ* to roast the corn with, either green or dry." Sylvester also mentioned another use: "Take a corncob—clean it up, run a stick through it, and use it as a back scratcher, *hukshanakuḍ*." No one recalled corncobs ever being used in lieu of toilet paper, as among some other societies.

In addition to the *eskuḍ,* the stick for planting corn and other crops, there was another instrument used throughout Pima-Tepehuan country for ripping open the husks surrounding the ear. I have found them among the lowland Pima Bajo and Mountain Pima, and Pennington (1969:75) shows one from the Northern Tepehuan. Each group made them somewhat differently. I asked Sylvester whether the Gila Pima had such a tool. "Ah, the *huuñ elpigakuḍ*." He knew it well and immediately began making the motions of slitting the husks of an imaginary ear, pulling them aside, then twisting the ear off and tossing it in a pile. "Can be made of bone or wood. If wood, they use *kavk u'us*" (Greasewood, *Sarcobatus vermiculatus;* see Group D). I had him make some. At the time he had a little patch of corn to the side of his mud house. When he finished the instrument, he showed me how it works. The corn husker is strapped to the palm so the harvester still has both hands free with which to work. George Kyyitan of Bapchule knew it, too, but used a different name, *huuñ hihipgakuḍ.* "Means to rip out like when ripping open an animal to take out the intestines." He likewise said it was made from *kauk u'us.* Ruth Giff, who still raises corn, confessed apologetically, "We used to use that, but now I just use a butcher knife in my hand."

Vouchers: *890* (SM, RG).

Group J, Ha'ichu Iibdag
planted fruit trees

While segregating plants into categories, Sylvester Matthias pulled out a set of trees that are raised for their fruits. These were separate from native trees such as cottonwoods and paloverdes, which he considered to be *u'us chuuchim* (see Group E). And at another level, he distinguished them from *e'es* (see Group I) as crops, such as wheat and beans, which are for the most part annuals that need to be seeded each year. Tree crops were clearly a conceptual category, and he called these all *ha'ichu iibdag* 'some kind of fruit'. However, it was also clear that *e'es* is used polysemously at two hierarchical levels: inclusively for everything that is somebody's plant, either seed crops or orchard trees, and more specifically for crops as distinct from fruit trees (as well as any other kind of plant). And while it is clear from when Pimas speak English that peach and orange trees are considered to be trees, the more important thing in Piman thinking is the special category of "plantedness" and ownership rather than treeness.

All the trees in this category save one are Old World introductions adapted relatively recently by Gileños. A few, such as the peach and apricot, may have been taken over during Spanish colonial times (and many of the names date from that period), but most cultivation of fruit trees dates from the Anglo period, so the need for the conceptual category *ha'ichu iibdag* is probably recent. Unlike Group I, in which Old World species were merely added to an existing repertory of crops, fruit trees introduced something entirely new: no tree had been cultivated aboriginally for food in Gila Pima country.

Ha'ichu iibdag includes one plant, the grape, that to Western thinking is a vine, and another, the date, that botanically is a palm. There are no palm or vine categories in Gila Pima. The only native palm in Arizona, *Washingtonia filifera*, does not occur naturally in Gila Pima country, although it is now widely cultivated both in major urban areas and on the reservation. It is categorized as an *uus keekam* 'tree' rather than as someone's plant.

Carya illinoinensis (Wangenh.) C.
    Koch
pecan
Juglans regia Linn.
English walnut
Juglandaceae
ki'ichdam, e-kuumdam

When the Anglos founded the little
town of Phoenix, a new oasis arose amid
the rubble and dry canals left by the
Hohokam centuries earlier. At least by
1895, if not earlier, both pecans and
English walnuts were producing crops in
the valley (Hughes 1895:100). Eventually,
great pecan orchards were to be found in
the Salt River Valley, including the vicinity
of Cashion, Buckeye, and Avondale where
the Pima used to conduct communal
hunts and fire drives.

One time, driving through this area
with Sylvester Matthias, I picked a pecan
branch and showed it to him. "This is
e-kuumdam 'cracking', name for pecan
[and] English walnut but not for peanut,
[called] *kakawáádi; e-kuumdam je'ech* is
the tree, 'the pecan's mother'. Pima didn't
raise them—not native around here."
When I asked him again some years
later he just said, "I think *e-kuumdam*—
something you have to chew." *Kuum* is a
specialized verb that means "to eat, chew
on something that comes in little pieces
such as corn, popcorn, pieces of candy"
(Mathiot n.d.:15), hence *huuñ kuumkuḍ*
'corncob'.

No one else at the west end seemed
to know pecan in Pima. I asked George
Kyyitan of Bapchule if he ever heard
a name for pecan. "Yeah, *ki'ichdam*
'something you crack'. That's walnut,
pecan, anything like that. You crack them
with your teeth, *ki'ich*. But not peanuts,
because you can just crack them with
your fingers." George did not recognize
the name *e-kuumdam,* and Sylvester had
never heard of *ki'ichdam.*

About four months later I handed
George a pecan, asking if he knew it.
"*Ki'ichdam.* That's pecan, English walnut,
anything cracked like that, not with your
fingers like peanuts."

"Would it include the pecan tree, too?"

"Umhum. That's the same. Also, what
do they call that? Pistachio? Over here
[toward Casa Blanca] they grow them one
time around here."

"How about *uupio* [Arizona Black
Walnut, *Juglans major*]?"

"Naw. Is there something in there, like
that?" Curiously, he knew this nut, but not
that there was something edible inside it.
He also called almonds and hazelnuts
ki'ichdam.

A few days later I stopped to see Francis
Vavages, to leave off some rawhide for
sandals. Although it was still early, the
whole house was astir. Francis was leaving
for an arts and crafts show where he
would be demonstrating flute making at
the tribal museum. I took the opportunity
to try a pecan on him. So many people
did not know a name for it that I was
beginning to wonder whether what I had
so far was idiosyncratic.

"Do you know what this is called in
O'odham?" I asked, handing him the nut.

"Yeah. Ah . . . let me think. Ki. . . . Oh,
I'm forgetting my language. Golly. It's
ki'ichdam. Because you bite on it," he said.

"Is that the name for the pecan tree,
too?"

"Yeah. And for the walnut."

Frank Jim, a Tohono O'odham living at
Sacaton, gave me several other names for
nuts. *Naakwadag* is the English walnut;
the name refers to the inside of the 'ear',
naak, and *vadag* 'juice, moistness'. And
the Brazil nut is *s-chukchu huch,* literally
'Negro's nail', in reference to a thickened
toenail. But no one I asked among the
Gileños recognized these words as plant
names.

After Christmas one year there was a
great bowl of mixed unshelled nuts in
George's living room. Pecans, almonds,
and even hazelnuts he called *ki'ichdam.*
I pulled out some dark Brazil nuts, not
exactly what you would want to bite
with your teeth. George said, "Those
are *s-chukchu taḍshaḍgig.*" The name
means 'Negro's toe'. Whether this is an
independently invented name or a
back-translation of an English name I
do not know.

I have found no pecan or walnut trees
on the Gila Reservation, but plantings
were begun at the Sacaton Field Station in
1908, and by the early 1920s there were 18
varieties being grown (King and Leding
1926:40). Field Station agronomists noted,
"The great number of young trees that
have perished before getting started and
the general behavior of those that have
survived give no basis for the assumption
that pecans will soon become one of the

principal orchard commodities of this
region" (King and Leding 1926:40).
Culture techniques improved. Those
authors would be surprised at the many
thousands of acres of pecan groves now
found in the vicinity of Maricopa and the
Eloy-Picacho region just south of Pima
country. Like cotton, pecans are great
water consumers in the hot season, but
that never stops modern agribusiness.

Vouchers: *1658* (SM), *1954* (GK, FV).

Citrus limon (L.) Burm. f.
lemon
Rutaceae
s-he'ekchu

The lemon is called simply *s-he'ekchu*
'something sour' (from *s-he'ek* 'sour',
'tart', 'acid'). Sylvester Matthias explained,
"*S-he'ekchu* means 'vinegar' in Papago but
means 'lemon' in our language." Jesuits
introduced both lemons and limes to their
missions in Pimería Alta. But Gila Pima
contact with this citrus seems to stem
from the late Anglo contact period rather
than from Hispanic mission times.
Lemons were among the citrus planted at
the Sacaton Field Station starting in 1919
(King and Leding 1926). Eureka and Meyer
were the varieties grown here. How many
plants were ultimately distributed to the
Indians, the stated goal of the station, is
not recorded. Although used commonly
by Gileños today, the fruits are purchased
almost exclusively from outside sources.
"They buy it by the crate or in bags if it's
for home use," Sylvester said. "They make
lemonade. Always at feasts, especially." But
now syrupy red tropical punch seems to
be the drink of choice at these community
events.

To the south, in Mexico, the Spanish
word *limón* refers not to the English
speaker's lemon but rather to what we call
in English limes *(C. aurantifolia)*. At the
Pima Bajo pueblo of Onavas, Sonora, little
green limes are grown in nearly everyone's
yard and are frequently used at meals. The
Pima Bajo call their limes variously *liim,*
limoon, or *limo'on,* all loan words from
Spanish. But neither the word nor the
plant seems ever to have made it as far
north as Gila country.

s-he'ekchu   *Citrus limon*

Citrus x *paradisi* Macfady
grapefruit
Rutaceae
**s-oám elidkam, (pl.) s-o'am
   e'elidkam**

The grapefruit appears not to have been among the several citrus species introduced into Pimería Alta early in the Jesuit mission period, beginning late in the 1600s. Gila Pima contact with the plant probably came during the Anglo period of early Phoenix.

The Gileño name for grapefruit, *s-oám elidkam* (pl. *s-o'am e'elidkam*), means 'yellow skin it has'. *Eldag* means something's skin or bark or, in this case, rind.

The word can be used for corn husks or an animal's hide as well.

Starting in 1919, Marsh and several other varieties of grapefruit were planted at the seed farm of the Sacaton Field Station (King and associates 1938). While I was writing this book at Sacaton, one ancient tree in the neighbor's yard, reported to be more than 60 years old, kept us supplied daily with fresh juice for breakfast from January through June. It is surprising that more are not grown here. The Pima say they purchase their grapefruits from stores or directly from the groves surrounding the Phoenix area.

The Gila Pima use grapefruit not only as a food. Sylvester Matthias told me, "Grapefruit is good for a cough. It was a remedy from the Mexicans. I don't know

who found out, because it was the Papagos who first started doing that for cough. They get one teaspoonful of honey, one grapefruit in a good-sized glass. Squeeze it. Then make a hot drink: *pi i'ihogig kulañ*, cough medicine. Your cough stops. It really helps."

Citrus sinensis (L.) Osbeck.
orange
Rutaceae
nalash, olinj

Orange trees are found sparingly across the reservation although they are raised abundantly in the Phoenix region. There are various trees around the agency buildings at Sacaton, but these seem mostly to be inedible ornamental oranges. Oranges are frequently eaten, but they are obtained from off-reservation sources.

Knowledge of this fruit must date from early in the Hispanic period. Oranges, limes, and lemons are listed among the fruit trees the early Jesuits took to their Pimería Alta missions (Morrisey 1949). As Ezell (1961:138) wrote, though, "Of the long list of crops which were available to them, the Gila Pimas chose wheat, a few of the vegetables, and melons. They failed to take over many of the other cultivated plants, their most notable rejections being fruit trees and grapes." Perhaps the Gileños saw no need to pamper this tropical exotic tree when a local bush, *kuávul* (wolfberry, *Lycium* spp.), each April produced more berries, also high in vitamin C, than the women could ever hope to pick and preserve at that season.

But the Pima did take the name from the Spaniards. "*Nalash*," explained Sylvester Matthias, "comes from the Spanish *naranja* 'orange'. Were raised— Casa Blanca—in rocky places." The word is quite well known and is the same with the Tohono O'odham. Often it sounds to my Portuguese ear like the People are saying *naraz*. The Pima Bajo of Onavas pueblo in southern Sonora say *nara'az*. Fr. Antonine (1935) recorded the Pima name as *olenj*, a loan from English, one of the few English words to enter the Gileño ethnobotany.

Francis Vavages explained this bit of historico-linguistics: "First oranges were called *nalash,* then *s-oámchu* ['something orange or yellow'], then later they were called *olinj.*" In this sequence, the initial version is a Spanish-derived word, the next a Pima descriptive name, and the last an English loan word.

"How about grapefruit?" I asked Francis.

"That wasn't known at that time. Later on. I don't know what state that came from. Maybe California or Florida or somewhere. That was brought in later, the grapefruit. But they [were] first raised right by the Camelback Mountain [north Phoenix], right down below there, and oranges first grown. And people from here go over there. They go on a wagon and stop at Salt River south of Mesa and next day they go over there and get so many. Especially on Christmas or Thanksgiving Day. Then they come back and spend the night again [at Salt River]. Then they start off next day, they start off [to Sacaton]. They'd be a lot of them [at home] asking for them. 'You get me some. . . . ' You know. They give them all the money. Sometimes somebody would offer his wagon, 'You take my wagon. And you take my horse. And you take the other horse. Bring back some fruit.' Big day. They used to all help together."

I asked Sylvester how they used to use oranges.

"They just peel it and eat it, or else make juice out of it."

"Was there any other use? Any medicine?"

"It's from the Mexicans. A long time [ago] there's a Mexican sister there at St. John's, Sister Euphrasia. About middle teens to 1939. She taught the Pimas that the orange peeling, if you lay it on the hot stove, it makes something like a toast. And you eat [it for colds]. And it's good for cough, too. The Mexican style. Now they don't know. Anybody got a bad cold, they just run to the clinic. There's remedy around."

The planting of oranges and other citrus trees at the Sacaton Field Station began in 1919, with unusually cold winters taking their toll (King and Leding 1926). By 1935 some 110 citrus trees had been propagated and distributed on the two Pima reservations (King and associates 1938:10).

Ficus carica L.
domesticated fig
Moraceae
suuna

Although at least three wild fig species grow in the Pima Bajo country of southern Sonora, no species makes it northward into Pimería Alta. At the end of the 17th century, with the establishment of Jesuit missions in this region, domesticated figs from the Mediterranean were one of the introduced fruit trees. Figs grew fast and were readily reproduced by cuttings. They probably spread rapidly, even though there were no missions as far north as the Gileño country. This was a black variety that later became known as the Mission fig, so named because of its association with the Franciscan missions of southern California. Other varieties were imported by various Phoenix settlers (for instance, the Smyrna in 1880), and 11 varieties were tried at the Sacaton Field Station, 5 with good results.

About the first of March, on the reservation, figs burst their buds, and in a matter of days there are little green leaves as well as fig buds on the naked branches. The large, rough, deeply cleft leaves attached to thick, pale gray branches should be unmistakable. The "fruit" or the fig that you eat is really the flower that does not open. There is a little opening on the end for special wasps to enter and fertilize it.

Throughout Piman country the Old World fig is called *suuna.* (There are different, unrelated names for the various wild figs that grow in lowland Pima Bajo country, but the domesticated one there, too, is *suuna.*) Supposedly *suuna* is borrowed from Yaqui *chuuna* (Johnson 1962; Saxton, Saxton, and Enos 1983:55). I suspect that the word must be borrowed from the large cultivated prickly-pear (*Opuntia ficus-indica),* called in Spanish *tuna* throughout its vast range. (The Spanish name for fig is *higo,* so the Piman names are not from that source.)

When the old Komatke village was established (or reestablished) in 1874 on the west side of the Gila along the base of the Sierra Estrella, there were fig, peach, and apricot trees, recalled Sylvester Matthias. But the floods of 1906 washed out the orchards and narrowed the flood-plains. When the village was gradually

reestablished on the east side of the river, where it is now, peaches, apricots, pomegranates, and grapes were the principal fruits planted.

I have found only a few fig trees being raised by modern Pimas, mostly in the Sacaton region. Ruth Giff planted two small trees from nursery stock, but gophers damaged the roots of one in 1987 when the tree was just beginning to bear.

I asked Sylvester Matthias what he remembered of fig trees in the early days at the west end. "There were about three at Komatke who had *suuna.* They were *big* ones. And along Central Avenue [in Phoenix]—that was just a dirt road then—they had *suuna,* lined up. They were free to pick—people go there and pick *suuna.*"

"Did you use them fresh or dried?"

"Fresh. I did see some old people, they do that, they dry them. And they eat it and drink it with pinole instead of *panocha.* And they used that dried *suuna* for *panocha.* People used to have a few figs in their yard [in Phoenix]."

"The ones you knew were black?" I asked.

"They were black. I thought it's always black—that's the only kind of figs we saw—and in the [Salt River] Valley, too. They were black."

Ruth Giff said, "Drying figs is very difficult. They get wormy very easy."

Malus pumila Mill.
apple
Rosaceae
aabals

Sylvester Matthias classified apples as *ha'ichu iibdag* 'planted fruit trees' even though they are not raised on the reservation. The Gila Pima name is *aabals.* "It's from English," he said, "Papago would say *mansáána;* it comes from Spanish." This is one of the few loan words in Gileño that I know of that came from English rather than Spanish.

Ruth Giff told me, "Apples are *aabals.* No one raised them. Also called *mansáána.*" People at the east end told me the same. Fr. Antonine (1935) recorded both forms. Saxton, Saxton, and Enos (1983) give *ablis* as the Pima form of the word, but none of the Gila Pima with whom I worked recognized it this way.

Sylvester said, "One time a school-teacher tried an apple orchard, but no success. Nothing doing. Not the climate here." Nor have I found any on the reservation. Apples are grown in Arizona, but at much higher elevations. George Kyyitan also reiterated that people bought apples in town rather than raising them. He mentioned Aabals Atkam, someone's nickname.

Morus alba L.
mulberry tree
Moraceae
siipuḍ s-wulvañ, gōhi

One of the common shade trees grown around Pima houses having irrigation is the mulberry tree. It is easily started from cuttings, has large leaves to provide shade in summer, and is deciduous to let in plenty of sun in the winter. As the native cottonwoods have declined as shade trees, the Athel Tamarisk *(Tamarix aphylla)* and secondarily the mulberry and Chinaberry *(Melia azedarach)* have increased.

I was visiting Carmelita Raphael, a woman of Santa Cruz village whose place still resembles the more traditional ranchos that I used to see more frequently on the reservation a quarter of a century ago, with their corrals, adobe outbuildings, permanent pastures, and gardens. On one side of her house was a series of mulberry trees. I asked if the tree had a name in Pima, and she started laughing. "They say they can make bows out of them," she said, but she would not tell me the name, saying Sylvester Matthias would tell me.

He did: *siipuḍ s-wulvañ* 'wrinkled anus'. "I wonder, how did it *ever* get that name," he added. Curiously, the introduced shade tree, *Morus alba,* has an entirely new name rather than some modification of the native Piman name *gōhi,* for the congeneric Desert Mulberry, *Morus microphylla,* that is found in the southeast Arizona mountains. (Francis Vavages of Sacaton, however, used the name *gōhi* for both the wild and domesticated mulberry species.)

I wondered if *siipuḍ s-wulvañ* was just a local name, but others from villages across the reservation also gave me the same name. Ruth Giff told me, "There was one big tree at Santa Cruz when we were kids—with great big leaves." The berries were eaten.

I fingered the leaves of the tree behind George Kyyitan's house.

He asked, "Do you know what the name of that is?"

"*Siipuḍ s-wulvañ,* but I don't know why it's called that. Is it because of the rough leaves?"

"No, I think it's from the berries. They look like strawberries." That is the closest I have come to learning why the Gileños call this tree 'wrinkled anus', a name some Gileños apply to the strawberry (see *Fragaria* sp., Group I).

Presumably the growing of mulberries dates from the settling of the lower Salt River Valley by whites, although the tree is listed among the new crops that Fr. Kino (1919:457) brought to Pimería Alta.

Musa x *paradisiaca* L.
banana
Musaceae
i'ipig chu'igam, haivañ a'ag, hovij chu'igam

When Anglos were first settling Phoenix, the Pima thronged into markets looking for exciting new goods. One that they took back to their villages was bananas. I asked Sylvester Matthias about the name of bananas. He said, "The decent way they call it is *haivañ a'ag* ['cow's horn']. Other name came in the early teens and early twenties, *i'ipig chu'igam* 'cocks looks like'. There's another name . . . ," but he could not remember it. Something seemed amiss with Sylvester's gloss: 'penis' is *viha,* not *i'ipig,* a word that appeared in none of my Pima/Papago dictionaries. Everyone seemed to recognize the word, but no one offered any help. Herzog (1941:67) recorded a name for banana that tran-slated as "foreskin pulled back (the vulgar expression)," but he gave no Pima words in this work, just English glosses.

I asked George Kyyitan if he knew *i'ipig chu'igam.* The corners of his eyes wrinkled

in laughter, and he asked, "Who told you that?"

"Oh, someone at the west end."

"Iiiih!"

Still puzzled five years later, I happened to be thumbing through Pennington's (1979) vocabulary of colonial Névome compiled apparently in the 1660s. My eye ran across *ipicuna,* which the Jesuit glossed, "Hacer acciones carnales por la parte del cuerpo conocida para esta efecto" (Pennington 1979:110). ("To have carnal actions for the part of the body known for that effect.") Coming from the pen of a vowed celibate, that was sufficiently circumlocutious. But the rest was more blunt: "*Ipica* [es] estraer la cabeza del miembro viril" ("*Ipica* [is] to extract the head of the male member"). Herzog's gloss was quite accurate. Over time, table talk must have required a compromise, for what I hear most of the time now is *haivañ a'ag* 'cow's horns'.

There is still another variant, *hovij chu'igam,* a name that incorporates another 'it looks like' comparison, this time with the Banana Yucca *(Yucca baccata)* found in Tohono O'odham country. Between the two languages we have a delightful language twist whereby banana becomes 'Banana Yucca it looks like'.

But naming can take still another step—marking reversal—and at least some of the Tohono O'odham seem to have taken it with the banana. Frank Jim, a Tohono O'odham, told me that the banana today may be shortened to simply *howij,* and the original desert plant designated *O'odham howij* 'the People's banana'. Even though the Akimel O'odham do not use this phrase to name bananas, some are aware of it. When discussing the various Pima names, Sylvester added, "Papagos just call it *hovij,* same as the yucca."

To biologists, the banana is not truly a tree because its apparent trunk is just leaf stems (petioles) clasping each other; it is not woody. One day while discussing why different plants were classified into various groups, Sylvester told this story of why he considered this plant a *ha'ichu iibdag:* "On 15th Avenue [of Phoenix] toward the fairgrounds, there is a *big*-sized banana tree. Then [they] asked the lady that owns the property. She says, 'That's a banana tree.' But they won't make no bananas, because they only make bananas in the tropical country—it takes eight to ten months to make a banana, and here the winter comes in and makes—just got 'em [they freeze]. But in those countries, well, it's just warm, warm, warm, and you have the bananas. But here we have the winter!"

Phoenix dactylifera L.
date palm
Arecaceae (Palmae)
chukuḍ shōsha

Date palms are planted rather commonly across the reservation, usually around peoples' homes. They are now uncared for (they have to be cross-pollinated to yield satisfactory crops), and I doubt many dates are produced. Those that are, are enjoyed mostly by the *hikvig* (Gila Woodpeckers) and *kuḍat* (Red-shafted and Gilded Flickers). The Pima as well as the Tohono O'odham name for date is *chukuḍ shōsha* 'Great Horned Owl's snot'. The name applies to both the tree and the fruit. (It would be interesting to see if any other Tepiman peoples use this name.) The word *chukuḍ* has two meanings. As most frequently used, it signifies Great Horned Owl. The other meaning is 'deceased person' or 'ghost'. The two meanings are related, and a Pima hearing a *chukuḍ* at night may be listening to the bird or may be being called to the other world by some close relative who has died.

Old World date palms were introduced probably early in the Jesuit period, almost three centuries ago. This would explain the uniformity of the name in Pimería Alta (at least as far as is known).

The Sacaton Field Station was active in searching for dates that would produce most satisfactorily under local conditions. They maintained 24 varieties. Kustawi, Halawy, Khadrawy, and Hayany varieties proved the most suitable. From 1931 to 1935, more than a thousand date trees were distributed on the reservation (King and Leding 1926; King and associates 1938).

Sometimes a good dose of humility is needed when working with an unfamiliar language. Years ago, in the course of my investigations of the birds of the Gila Pima country, Sylvester Matthias mentioned, in the context of the Great Horned Owl, the Pima name for dates. I knew the first word but not *shōsha,* and I inquired about its meaning. Sylvester, apparently not knowing a proper word for it in English, put his crooked index finger up to the side of his nose with an outward curving motion. "Ah, the owl's *beak,*" I said. And so I wrote it, Sylvester not protesting. Later in the day when Sally Pablo was going over my list of new bird words, she came to that and burst out in laughter.

"What's wrong?"

"Who told you this?"

"Sylvester."

"*Shōsha* doesn't mean 'beak'; it means 'snot'!" Indeed, Saxton and Saxton (1969) glossed it 'owl nasal discharge'.

But *chukuḍ shōsha* has a precursor, and this name is applied only in an allegorical manner to the date palm. Both Sylvester and Ruth Giff explained that 'owl's snot' was something brown, about as thick as your finger or thumb joint, found glued to thin branches, particularly on mesquite trees. "I don't know how they get there, but something puts them there," explained Sylvester. I thought for a moment, then realized what the real *chukuḍ shōsha* was: the oblong egg cases of the praying mantis (Mantidae). The Pimans first encountering the Old World date palm had made quite appropriate comparison between the mantid egg cases with which they were already familiar and the oblong palm fruits.

Vouchers: *995* (sm).

chukuḍ shōsha   *Phoenix dactylifera*

Prunus armeniaca L.
apricot
Rosaceae
vilgóódi

Late in the 17th century, when the Jesuits established the first missions in Pimería Alta almost to within 100 miles of Gileño country, the apricot was one of the fruit trees introduced. The Pima name for the fruit and the tree itself, *vilgóódi,* apparently stems from this early contact period. The Spanish name is *albaricoque.* That must have been a mouthful for the Akimel and Tohono O'odham, who settled on their own transformation of the last syllables. The Tohono O'odham, with closer Hispanic contacts, said *wiilagóóki.* Fr. Antonine (1935) recorded it as *vilkodi* among Gileños. The plural is *vipilgógodi,* an unusual form. Native linguistic consultant Culver Cassa notes that this structure has parallels among other loan words such as *pistóóli,* pl. *pipistótoli* 'pistol'.

These and peaches apparently were much easier to raise than the various kinds of citrus trees. The searingly hot summers, with scarcely ever a day below 100°F, and the frosty winters were rough on many other fruit trees but seemed to suit the apricot.

The apricot tree is usually short and spreading, with spade-shaped leaves that look like those of poplar or aspen. I know of only two apricot trees on the reservation in the mid 1980s, both in Sacaton yards, but the tree was once more widespread in the various rancherias.

Ruth Giff said, "Oh, there was a place there where this man, he had a lot of trees: peaches and apricots, and I think *kalnááyu* [pomegranate]. He lived there in Santa Cruz—where Lupe Avery lives. There was a *big* apricot tree—almost like a cottonwood. And this girl that lives there was my playmate. And we used to go there and eat all the fruit. And what my mother did, when the seeds fall, there would be small ones; she got some and transplanted them, when we were just babies then, so when we grew up they had fruit on it. She had about six apricot and peaches. And then they just dried up. I guess they didn't take care of it. And she dried the fruit, both the peaches and apricots. Sometimes horses would get in there and eat the fruits [still on the tree]. They prune the trees once in

vilgóódi   *Prunus armeniaca*

a while and the fruit come[s] out *big.* And there was a place just below the [Gila Crossing] day school with lots of trees. I think that was Leonard Soke's grandfather. And there was another orchard down in Coop. *Kalnááyu*—there used to be lots of it. Peaches were small, little tiny ones. Only one man had the big ones."

Although women preferred the wild Arizona Black Walnut *(Juglans major)* for their *kams* to keep in their mouth, the apricot pit could be used in its place. Apparently there was some competition for these pits, as Sylvester Matthias related: "We didn't eat prune seeds—but did apricot—*vilgóódi.* Eaten like nuts. We tried a peach, but can't get much out of it. When we were kids, we learned it from the older boys: every time we find an apricot seed, we saved it. Women wanted them to keep their mouth wet, for their *kams.*"

The Pima writer George Webb (1959:58), who was once Ruth Giff's neighbor at Santa Cruz, recorded similar recollections from his own childhood:

To the north side of the house was a small orchard of about a dozen trees, mostly apricots. I remember the first time these apricots ripened. Kelihi had given us strict orders not to go into the orchard to eat apricots until they were good and ripe. But we, like any other children, would go into the orchard when he was away and eat green apricots. By the time he got back, we would be all doubled up with a tummy-ache.

He would laugh at us and say: "You will know better next time!"

And we did know better, too. We would never go into the orchard unless we were told to, to pick fruit for my mother, Rainbow's Ends.

Rainbow's Ends would dry the apricots on racks in the sun. No one told us not to pass by the racks, so we would pass by them very often until our stomachs told us that we had better think of something else to do.

Castetter and Bell (1942) make no mention of apricots among either the Akimel or Tohono O'odham. But this was one of the fruit trees being experimented with at the seed farm at the Sacaton Field Station: "The varieties of apricots which have fruited are the Newcastle, Blenheim, Tilton, Moorpark, and Royal. The Newcastle is the earliest to ripen, followed by Blenheim and others from two to three weeks later. All bore light crops in 1922 and 1923. In 1924 the Newcastle variety produced an average of 80 pounds per tree and the Blenheim and Royal trees about 50 pounds each. The other two varieties yielded somewhat less" (King and Leding 1926:33).

Prunus domestica L.
plum, prune
Rosaceae
gogoks viipiḍ

The Pima sometimes extended their nick-naming proclivities to plants they encountered. The plum and prune, for instance, found among the Anglos on the lower Salt River Valley, they appropriately named *gogoks viipiḍ* 'dogs' testicles'. The tree's name is always in the plural; putting it in the singular, I am told, would sound bad. Immediately to the south, the Tohono O'odham say *gogs wiipiḍŏ*.

Starting in 1917, the seed farm at the Sacaton Field Station began experimentation with seven varieties of plums and five of prunes (King and Leding 1926). Prunes were not very successful, but among the plum varieties, Santa Rosa appeared to be the best suited to local conditions.

I have never found anyone raising trees of *gogoks viipiḍ* on the reservation. But the fruit is still known. It is occasionally purchased in stores, dried or canned, and it is available through government surplus commodity programs.

Sylvester Matthias once told me, "That soup or juice or what's left after *gogoks viipiḍ* is cooked, you can use that for a laxative."

"When did you learn about this?"

"Not very long ago. Because hardly anybody use[d] that *gogoks viipiḍ*. But the schools . . . St. John's. . . . That was

every morning—had *gogoks viipiḍ* and dried peaches besides our oatmeal. That's the time I learned it's *to m-vakuichud* [laxative]."

Prunus persica (L.) Batsch.
peach
Rosaceae
ñulash

The slender leaves of the peach tree are unmistakable: they look like those of a large willow that have been folded down the midrib. These are quite unlike apricot leaves, which instead resemble cottonwood leaves. In late winter or early spring, peach trees are covered with pink blossoms. Fruiting time varies according to variety.

Upper Pima contact with the peach came early. Peaches were among the fruit

gogoks viipiḍ   *Prunus domestica*

trees Fr. Kino introduced at the Pimería Alta missions. Peaches themselves, as well as the name *ñulash*, were probably adopted by Gileños about this time. *Ñulash* is a loan word from American Spanish *durazno*. Grossmann (n.d.) recorded *ñulashan* on the Gila about 1870. The Tohono O'odham call peaches either *ñulash* or *julashan*, both likewise derived from *durazno*. At least by the 1890s, Pimans living at Ak Chiñ just south of the Gila (near modern Maricopa) had peaches, leaving the pits in debris archaeobotanists have analyzed (Robert Gasser, personal communication).

According to the Pimas from the western villages with whom I worked, peaches were raised fairly commonly in their orchards at the beginning of this century, though to a lesser extent than apricots. For instance, Ruth Giff said, "*Ñulash* is peach. They were little tiny ones. Some man at Santa Cruz had large ones." Her description of his orchard is given in the apricot account.

Sylvester Matthias told me, "*Ñulash* is the Pima peach. They're little—smaller than a lemon—about the size of an ordinary apricot. [It's] the right climate for them here." Apparently these were all of the original Hispanic stock, grown from seed and perhaps selected over several centuries for the particular growing conditions in these desert villages. But nothing survives of this early peach except for a few memories.

"They could use a peach seed as *kams*," Sylvester said. "When they go to work on the basket, they had to use that *kams* to keep their mouth wet. And that walnut [Arizona Black Walnut, *Juglans major*], *uupio*, and the peach seed, *ñulash kai*." He translated *kams*, the object Pima women kept in their mouths, as 'gargle'. The peach pit has a rough surface, so it was preferred over the apricot pit for this purpose.

Starting in 1917, the seed farm at the Sacaton Field Station experimented with different types of peaches and by 1924 had 24 varieties under cultivation (King and Leding 1926:32–33). Crawford, Elberta, Cling, and Lukens Honey produced well. The number of cuttings ultimately distributed to the Pima is not given. But Castetter and Bell (1942:78) make no

mention of peaches among the fruit trees
the Pima were raising during their studies,
nor have I found a single surviving tree
in 30 years on the reservation.

Still, peaches as well as fruit cocktail
are the standard desserts at village feasts.
These come in big one-pound cans, packed
in heavy syrup, of course.

Punica granatum L.
pomegranate
Punicaceae
kalnááyu, kalnááyō

In addition to supplying a delicious fruit,
pomegranate shrubs bear quite attractive
flowers in late spring. The entire bud or
calyx is scarlet. This bursts open at the tip,
producing a Star of David that appears as
if sharply cut from thick red leather.
Indeed, these are the buds that inspire
the so-called squash blossoms of
Southwestern Indian silversmiths, a
pattern that came by way of Hispanic
Moors. Tightly folded within each bud like
crushed silk are six delicate petals, also
red. One petal is attached to each basal
corner of the star. After these unfold, a
brush of scarlet stamens with yellow
anthers is revealed. The leathery sepals
and ephemeral silky petals contrast at
one level, but the entire bright scarlet
flower is set off by a bush of small, bright
green, shiny leaves, providing another
level of contrast. During the summer the
red leathery fruit swells to grapefruit size,
and by fall it is full of juicy seeds.

The plant to the Gila Pima, as well as
to their desert cousins, is *kalnááyu,* a loan
word from Spanish *granada.* Tohono
O'odham *galnááyo* is a bit closer to the
source word (Saxton, Saxton, and Enos
1983:97). It was an introduction from the
Kino mission era, and I suspect that it
has been with the Pima ever since. These
are tough heat-adapted desert plants from
Old World oases. Given groundwater and
a minimum amount of attention, they
are highly productive here.

Kalnááyu has long been a very popular
fruit among the Gileños. According
to everyone I asked, pomegranates were
eaten fresh rather than prepared in

ñulash   *Prunus persica*

any way. No Pima with whom I worked
had any recollection of the juice being
fermented.

This was one of the fruit trees tried
out at the Sacaton Field Station. King
and Leding (1926:34) reported success:
"At the field station, the Wonderful is the
most popular variety with the Indians,
with the Greenrind, Sweet *(Sweet Fruited),*
and Hermosillo also in great demand.

Many cuttings and rooted plants of these
varieties are distributed among the
Indians each year by the field station."
Between 1931 and 1935 more than two
thousand rooted cuttings were distributed
(King and associates 1938).

And when I went to the reservation
more than a quarter-century ago, they
still were everywhere, the dense, some-

kalnááyu, kalnááyō *Punica granatum*

what spiny shrubs or trees doubling as ornamental plantings and as living fences or hedges in yards, at the edges of fields, and on practically every little rancho. Today they are certainly far fewer, but the attractive red-and-green bushes still grace many Pima yards.

Pyrus communis L.
pear
Rosaceae
vakoa chu'igam, (pl.) vapkoa chu'igam, chuchul maamaḍ

Pears as well as apples seem not well suited to the desert climate, although both are grown in Arizona at higher elevations; nevertheless, the Gila Pima have a name for them. To these name-inventive people, the pear fruits looked like little bottle gourds hanging in a tree, so they called them *vakoa chu'igam* 'gourd it looks like' or, in the plural, *vapkoa chu'igam.* One of the types of gourds, a canteen gourd, is indeed pear shaped.

Sylvester Matthias told me the name *vapkoa chu'igam,* but few other Gila Pima seem to remember it. However, in his 1935 vocabulary, Fr. Antonine recorded exactly this form.

The seed farm at the Sacaton Field Station began experimenting with pears as well as other potential fruit crops shortly after its founding in 1917. By 1926 King and Leding reported seven varieties being grown. "Pears give indications of being poorly suited to the soil at the seed farm. However, the Bartlett and Winter Nelis varieties have produced heavily in some years at the field station" (King and Leding 1926:33). Apparently none of these have survived among the Pima, because I have found no pear trees on the reservation nor could others tell me of any still being grown. Most contact with pears today comes from off-reservation produce stores or as canned goods.

Another name for pears, Sylvester said, was *chuchul maamaḍ* 'baby chicks' (*chuchul* is the domestic chicken). This is the name George Kyyitan knew at Bapchule. The Pima, seeing yellow pears hanging in a tree, likened them to so many chicks gone to roost. This allegory has nothing to do with the medieval English minstrel song celebrating the twelve days from the Nativity to Epiphany.

Vitis vinifera L.
wine grape, Old World grape
Vitaceae
huuḍvis, uuḍvis

Among the many plants the Spanish missionaries introduced into New Spain was the common wine grape, which thrived in the Mediterranean countries from where many of these people came. Grapes and wheat were necessary to produce the essentials for the eucharistic celebration: bread and wine. Wheat was readily transported by seed, and grapes in winter could be propagated from cuttings, those long straight, jointed canes that were cut back on the old vines each fall. One disgruntled missionary sent to Brazil wrote complainingly back to Europe saying that the Lord surely never intended them to be there among those heathens (Crosby 1972). And the surest sign of this was that grapevines would not grow in that tropical climate and the essential for the consecration of the sacred species was unavailable.

Whether the Gileños ever acquired grape cuttings from any of the Jesuit and later Franciscan missions to the south is unknown, but at least they borrowed the word, huuḍvis, occasionally heard as uuḍvis, from the Spanish name uva 'grape' and perhaps vid 'grapevine'. Ezell (1961:139), writing of the Hispanic influence on the Gileños, said that grapes were one of the new crops they rejected: "The successful propagation of grape vines and fruit trees demanded knowledge and skills not possessed by the Indians." Perhaps the Gileños were more skilled than that assessment would indicate.

"Did anyone here have grapevines?" I inquired of Sylvester Matthias.

"There used to be one down there," he said, pointing off across the Komatke flats. "They [a family] used to live across the river before the flood [of 1906]—name is Sweet. They moved across [here] after the flood—not exactly across from Helen Allison's. There was a big vine at their old place."

"Did they know how to prune and everything?"

"That's why they're nice growing every year. That's the only grapevine I saw here. Outside of that, they're just peaches, pomegranates, apricots."

Another time Sylvester said, "Huuḍvis, [from] uvas, [comes] from the Mexicans. Not common. Some people—whites— may have a little vineyard at their houses, and they sell it out on the road on [the] way to Phoenix, on Central Avenue—[it] was a narrow [road] at that time! There's none [here at Komatke] till after the flood—1906—[they] start their orchards again. Those people, who are now Co-op [Colony people, called] chichino."

By 1920 there were some 20 varieties of grapes being grown at the Sacaton Field Station and another 15 varieties in the seed farm vineyard. These proved productive, and by 1935 more than a thousand rooted cuttings had been distributed to the Salt and Gila River Reservation Indians (King and Leding 1926; King and associates 1938).

The traditional midsummer production of navait 'wine' was solely from Saguaro fruit. There was some limited production of o'okĭ navait 'women's wine', from mesquite pods (see Prosopis velutina, Group E). Sylvester Matthias said wine was made just from raisins and other dried fruits. However, Palmer (ca. 1885) mentioned that the Gila Pima fermented "pitahays" [pitahaya, Saguaro], agave, corn, prickly pear, and grapes. Whether the grapes were of the Pimas' own raising is not mentioned.

I questioned Sylvester about the possible use of agave in fermenting beverages during his lifetime, but he was unaware of such a practice. "They bake a'uḍ [Agave deserti hearts] in [the] ground to eat. Don't know for alcohol. Pima used only haashañ [Saguaro] or grape or raisin. But they say, they'll make you run!" He broke out in laughter at the home brew of the Pima giving people diarrhea. "You drink too much of it. . . . Even if you pass out there, you'll still let it out!" Apparently the Pima had not mastered the art of wine making, or else they drank it too soon while the yeasts were still fermenting. But vomiting and diarrhea were both symbolic of what the People wanted the clouds to do: shed their moisture.

Grape wine making seems never to have caught on, and one sees grapevines on the reservation today only occasionally, these planted probably as much for shade as for their fruit.

Group K, Category Unknown

miscellaneous unaffiliated plants

This is a category of things that are considered to be plants but are not assigned to any other group; in folk taxonomic terms, these are unaffiliated generic taxa. They are things that 'grow up', *ha'ichu vuushdag* or *ha'ichu vuushñim* (Mathiot n.d.:323 lists *vuushñim* as 'coming out of the ground, emerging'). As a group, they might be considered fungi or fungi-like, because none contains chlorophyll. Sand-food is a root parasite whose flowering cap appears on the surface of sand dunes in spring. It is doubtful whether Gila Pima had much personal experience with this plant, except perhaps dried as a trade item. But they referred to the westernmost band of Papago as Hiá Tatk O'odham 'Sand-food People' and knew this plant was important to their survival. The remainder are all fungi: three of these 'grow up' and hence are *ha'ichu vuushdag*. The fourth, Corn Smut, does not, but its remaining characteristics apparently are sufficient to merit its perceptual inclusion with puffballs and mushrooms. Or its inclusion with *ha'ichu vuushdag* might stem from its essential association with corn plants, parallel to the placement of mistletoe in the life form 'tree'.

mushroom
Fungus
kumul, naakado

Once, when the Gila River ran some 65 miles across the Pima lands and a gallery forest of willows and giant cottonwoods edged much of its course, with a thick understory of mesquites, Arrow-weed (*Pluchea sericea*), and Seep-willow (*Baccharis salicifolia*), you likely would have found a white mushroom growing in moist, shady microhabitats. But don't look for them now: they're all gone.

Some old Pimas remember their edible mushrooms, called simply *kumul*. According to Sylvester Matthias, "*Kumul* [were] used for food; they were white— grew around willow trees; [a] *lot* of them under the willow trees. My grandmother used to get a lot, grind them, smash them, with ground whole wheat; [it was] white and black, black when gravy was done. I heard some were dangerous, but I think the ones that grew around here were not poisonous or anything."

Ruth Giff told me, "We used to eat it. I've never eaten it. They knew the kind that you cook. I'm sure they put flour in it. Don't know what time of year they come. I used to see them on willow trunks, because in those days there's a lot of willows. Ones in the store are fat, but the *kumul* are thin, like a little umbrella; maybe they picked them when they were tiny, before they open out."

Five years later, Sylvester Matthias told me again about their mushroom: "Other *kumul* grows on the wet ground. Find them after a flood—any time of the year. At a fallen willow log. Or a big willow tree. It's a mushroom. White and black. Take it and smash it up and mix it with a few wheaten flour in the *machchuḍ* ['grind-stone'] and make a sort of a gravy-like— and appear to be halfway dark and halfway brown. It tastes *good*." By draw-ings, he indicated that the cap was quite rounded or domed and that the edges of the cap were smooth, not ragged as in some species. His *kumul* was definitely not like some Oyster Mushrooms (*Pleurotus ostreatus*) I showed him.

"Did anyone ever eat (or hear of) poisonous ones?" I asked him again.

"No, there are some *kumul* that are poisonous. They know. The old people know which are which."

I showed him some bracken fungi from the cottonwood-willow groves along the lower Salt River, but he did not recognize them.

On another occasion when I asked Sylvester about the kinds of *kumul*, he said, "Just one. Find [them] in damp ground, at base of willow trees. They're good-sized ones. Top white, underneath black."

"How big around are they?"

"The largest you could find are about five to six inches [in] diameter."

"And how tall?"

"About that tall [5–6 inches]. Of course, they can be smaller. There used to be other [kinds], but that one is very popu-lar—that is found at the base of a willow tree."

But I could not find even one specimen for people to verify. Pressed for more identifying characters, Sylvester said, "Top rough, all white, may reach ground. Thick in middle; stem short. Bottom is like umbrella—black. Along willow trees—trunks and fallen limbs." All this points pretty clearly to the Parasol Mush-room (*Lepiota rachotes*), according to Dr. Robert L. Gilbertson, mycologist at the University of Arizona.

Nearly a century ago, Russell (1908:75) recorded the following unidentified food plant: "The heads are gathered and washed, sometimes twice, then boiled in an olla with a little water. Wheat flour and a seasoning of salt are added and the whole is stirred until the heads fall to pieces." Russell had no idea what the plant was. He wrote the name for this plant as *kómult*. But his *ko-*'s in this list are all *ku-* words, and he writes *haal* 'squash' as *halt*; so this is the word for mushroom in a different orthography.

In 1992 I took Sylvester a box of dried Parasol Mushrooms, *Lepiota rachotes*. We had been working with fungi for at least a decade. He picked out the larger specimens and surprised me, calling them *naakado*. He had never used that term before. "*Naakado* [are] damp place mushrooms . . . willow logs. Not on cottonwoods. They're white; white on top, black underneath. Cooked with ground wheat; can salt it." He took another out of the box and pointed to the cap. "*Naakado* [have] large flat top, almost touch the ground. A kind of *kumul*."

The name Sylvester was using is cog-nate with the base word for 'mushroom'

in Southern Tepehuan, *nakai* (Gonzalez Elizondo 1991). In their folk taxonomy these upland Pimans from the south dis-tinguish 15 species of edible mushrooms, 6 with *nakai* as the modified base term. I suggested the relationship to the word *naak* 'ear', or *naank* 'ears', but Sylvester did not think there was a connection.

For years everyone with whom I talked had used the word *kumul* for mushroom, and that for a specific edible species. Sylvester was using the lexeme *naakado* for a particular fungus, calling it a kind of *kumul*. Then he continued, "*Okstakuḍ* is another kind of *kumul*, has like ochre. There is *kumul* on corn, certain time when they plant it, kind of a black color, but inside it's clear—*huuñ kumuldag*—and also on the wheat. *S-kumuldag*—gets on the tassel, but inside it's wheat, in the kernel. The Pima don't think it's fit to eat. In some cases the Pimas are superstitious about eating certain kinds of grain, berries." In this context, Sylvester used the term *kumul* in the more distributive sense of fungi, including Stalked Puffball (*Podaxis* or *Battarrea* spp.), Corn Smut (*Ustilago maydis*), and the Black Stem Rust (*Puccinia graminis*) of wheat, as well as the edible mushroom.

Vouchers: *1978* (SM, *Agaricus* sp.), *1979* (SM, *Schizophyllum commune* Fr., *che'ul kumulga* 'willow its fungus'), *1982* (SM, *Lepiota rachodes*). Taxa excluded from *kumul* by SM: *1977a* (*Inonotus munzii* (Lloyd) Gilb.), *1977b* (*Ganoderma lucidum* (Curtis: Fr.) P. Karst.), *1980* (*Coriolopsis gallica* (Fr.) Ryvarden), *1981* (*Tremella* sp.).

Pholisma sonorae (A. Gray) Yatskievych
[= *Ammobroma sonorae* Torr.]
Sand-root, Sand-food
Lennoaceae
hiá tatk

In the sand dunes along both sides of the lower Colorado River lives a plant nearly buried in the sand. Only a little mushroomlike cap (which bears, in season, circles of purplish flowers) shows on the surface. Even this is dusted with sand grains. But below the surface is a long, thick watery stem, like an elongated daikon (Japanese radish) that has gone

hiá tatk   *Pholisma sonorae*

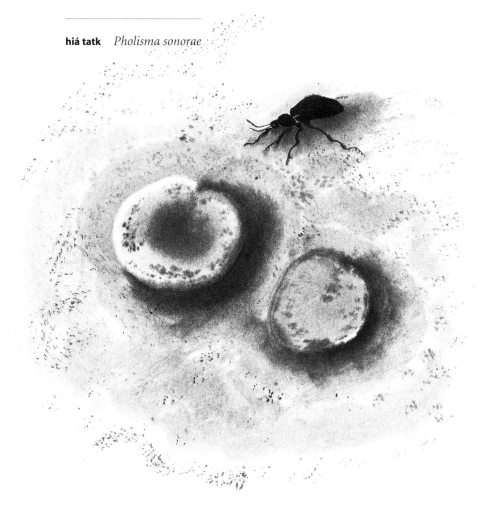

dunes. They have a slight tang of parsnip. Or they can be roasted over coals for about 20 minutes. (I have not had them roasted, but they are reported to taste then like sweet-potato.) The stalk can also be dried and stored. If Sand-food reached the Gileños, as I suspect it did by trade, it would have been in this dried form.

Sylvester Matthias said of *hiá tatk,* "It's the root that they eat."

"Did the Sand People trade it to the Pima?"

"No. It's similar to that *shaaḍ* [Saya, *Amoreuxia palmatifida,* which was traded northward]. See, we call those people from down that way S-oobmakam ['not friendly, bellicose']. They are Papagos. [Also called] Otomkal Kuá'adam—they live in the sand, 'lizard eaters.'" Sylvester's reference here is to the Desert Iguana, a vegetarian reptile eaten by these Piman bands.

About five years later, Sylvester commented once again on these people and their food: "There were some people who came to my grandmother's place at Sacate—S-oobmakam, Sand Papagos. They speak two languages, Pima and their own language [a Piman dialect]. Right along the gulf [of California], where there is sand. Must be around Rocky Point or somewhere in that area. They live on those roots and those lizards. Not *shaaḍ.* Something different. It grew in the sand. I guess it's close to the Cúcapa [Cocopah], because the Cúcapa comes to Sacate, too." Until the late 19th century, Sacate was the westernmost settlement of riverine Pimans, except at first contact, when villages with Pima names extended down the Gila to the Mohawk Mountains, Yuma County. The Sacate Pima apparently maintained contact longest with these groups from the lower Colorado River valley. In both conversations, Sylvester contrasts Sand-food, which he does not know, with Saya, a plant he knows personally. The other tribe to which Sylvester refers, known in English as the Cocopah, is a group of Yuman speakers from the lower Colorado River, who were enemies of the Quechan (also known as the Yuma). The Quechan battled with the Maricopa and middle Gila Pima (see Kroeber and Fontana 1986).

If you do search the hot sands for *hiá tatk,* you will be in a long line of desert mystics who have sought the mysterious plant, including the earlier Sand People

wild. To find Sand-food you must come on foot, then sit and meditate. Maybe in an hour, or even two, the little purple crown might appear to you out of the sand. With your bare hands, you may dig in the sand and, like one of the nomadic Pimans who once thrived on Sand-food, learn the secrets of this shy plant. Probably few Gila Pimas ever saw *hiá tatk.* But they knew of it.

Joseph Giff sat telling me about these desert Pimans: "Hiá Tatk Kuá'adam, the 'Sand-root Eaters'. They lived on some kind of root that grew in the dunes, the *hiá tatk* plant. They live back there where they [the O'odham] went for the salt. When the People going after salt would camp, they [Sand-root Eaters] would come and sit around; they wouldn't ask them, but they wanted something to eat. And they [O'odham] used to take them something besides what they needed for their [own] journey, and they give them some and they'd eat it, then go away. Every night they come. They're wild; they wouldn't come around in the daytime, just

at night. This Papago man told me that they got some of the children, the boys— they wanted to give them the girls to raise, but they didn't want them—they just brought some boys to take to school. But they got sick, and they had to take them back. They are wild by nature, and couldn't get along in school, and they had to take them back.

"They could talk with the Tohono O'odham. I guess they're still down there. They're still poor."

The People to whom Joe was referring are also called the Hiá ch-eḍ O'odham, Sand People or Sand Papago; they were the westernmost group of desert Pimans who lived a nomadic life between the head of the Gulf of California and the lower Gila in some of the most forbidding desert of North America. The turgid, juicy *hiá tatk* stalks were available from March to the beginning of summer, depending on the rains. Unlike the bitter *mo'otaḍk,* or Desert Broom-rape *(Orobanche cooperi),* the watery and subtly sweet *hiá tatk* can be eaten raw as they come from the sand who, in their "poverty," had divested themselves of every shred of material nonessentials. (Their nudity shocked the first Spaniards to encounter them!) Thackery and Gilman (1931) sought its mysteries, in particular wondering "just how this parasitic plant makes its contact with the host root, usually from 2 to 5 feet below the surface." (This is the same Gilman who taught for seven years in Gila Pima country, discovering some plants there that no other botanists have been able to find on the reservation.) My friends and latter-day mystics Wayne Armstrong (1980a, 1980b) and Gary Nabhan (1980, 1985) have also pondered these and other questions of Sand-food.

Podaxis sp.
? *Battarrea* sp.
Stalked Puffball
Fungus
okstakuḍ, kelitakuḍ

Deserts are not the places where you would expect to find fungi growing, and indeed few kinds grow there. When rains thoroughly moisten the desert, particularly during late winter and again during the summer, white puffballs, ghostlike, may appear commonly along the gravel edges of roads and on bare alluvial flats. At either season they seem about as incongruous as they are spontaneous and widespread. These are often as tall as your hand, but smaller ones may be only as tall as your palm. They have a distinct pencil-thin stalk and an elongated white cap, the puffball, as thick as a man's thumb. The inside is filled with powderlike spores, so fine when dry as to be smokelike.

Sylvester Matthias showed me one with fine blackish rufous spores that he called *okstakuḍ* 'old woman maker'. "It has powder inside; comes out something like ochre. They say it's very good for open sores. You can just pour it on the sore and bandage it up. They said it is very good—it can heal up in no time. [I] never tried it, but I *heard*." Later I collected more, including the little Tulostoma Puffball (see next species account), all of which he considered to be *okstakuḍ*. "They can be used to control bleeding from a cut, and for open

sores." Once when I asked him about arrow wounds, he said, "Can use *okstakuḍ* to *stop* the bleeding." The use of the spores from puffballs as a blood-clotting agent and as an antiseptic was widespread among North American Indians and even among people of other continents (Burk 1983).

I have never been able to find out for certain the connection between the plant's name, 'old woman maker', and any actual belief. Irene Hendricks of Casa Blanca laughingly said, "Oh! I don't want to look at that. I don't want to become an old lady." This seemed no more than a joke.

"Well, then, if you call it *kelitakuḍ*, I don't want to look at it either," I responded.

Still it seemed there was some vague revulsion there. Culver Cassa of Sacaton said his understanding was that if you were a woman, touching the *okstakuḍ* would make you into an old lady, as the name literally says, and that if you were a man, it signifies that you will marry an old lady.

For a long time I thought that the alternate name, *kelitakuḍ* 'old man maker', was used only by the Tohono O'odham. But Myrtel Harvier of Sacaton used this form exclusively, saying that it did not include the Tulostoma Puffball. I asked George Kyyitan how he used the words. He thought that *okstakuḍ* was the one with the yellow-ochre spores and *kelitakuḍ* the one with the rufous-black spores. Francis Vavages used *okstakuḍ*, which he said was the same as *kelitakuḍ*. "It smells like sweat," he added. For the Tohono O'odham, Saxton and Saxton (1969:24) give *kelitakuḍ* and *okstakuḍ* as the same thing, but both they and Mathiot (n.d.: 463) misidentify the plant as a mushroom or toadstool, quite a different form of fungus from the Stalked Puffball.

Vouchers: *1983* (SM, RG, *Podaxis pistillaris* L. ex Persoon), *1985* (SM, cf. *Podaxis pistillaris*).

Tulostoma sp.
Tulostoma Puffball
Fungus
jeweḍ mōōtō [?]

Out in the flats where the water from rains collects, a very short stalked puffball may appear, one much less conspicuous than the large whitish *okstakuḍ* (or *kelitakuḍ*).

This little fungus is about the size and shape of a leather coat-button, with a small opening at the top like a navel. This lowly plant found an important use among the Gileños.

In 1987, while discussing *kelitakuḍ*, Myrtel Harvier told me this story of an incident from her childhood: "There were some little ones, they stand about that high [indicating about two inches]. Something happened to me one time and I never forgot it. And I always think about those little ones. They grow up about that high—you know, something like that *mo'otaḍk* [Desert Broom-rape, *Orobanche cooperi*]. And they got little bitty, round tops—like—the ones you eat, *kumul*, but they're not the eating kind.

"One time, we used to live under the Estrella Mountains there, a long time—that's where my grandmother used to live. And we used to cross the river there—the Gila River—then we'd go on, then we'd cross the Santa Cruz River, then go up to the foot of the mountains. Grandmother wanted to see an aunt. And we rode in wagons then. There's a big old wagon. And they put me in the box. They told me to hang on to that side of that—and watch the seat, you know, that I didn't get under it. So I sat back there. And I got excited about things, you know, and I was moving around. And I guess I got my hand too close to that seat—you know they got little springs on them—and we went through the riverbed and the rocks, you know, and that thing bounced in there and caught my little finger—this one," she held up her left index, "it's got a funny nail—it caught and it mashed this finger, and pulled out my nail. Oh! I was screaming. I remember that. I never did forget that ache.

"And we went across. Oh, poor Grandma. She was so excited that she went and tore a piece of her dress and she wrapped it up. We went and—I don't know who was driving us—but anyway, we got across there and she said to look for some of those little things like *kelitakuḍ* along the way. We stopped somewhere, and she looked around, and she said, 'Look, I think I see some of the ones I want,' she said. She got down, and she looked around and she got one of those *little* tiny ones, with the little round tops. It's not like those big ones, *kelitakuḍ*, it's a different thing. She got it and she ripped off the top. She poured all that little brown—it's got little brown powder in it—she poured it *all* over my little sore finger, and she wrapped it up, and she took some of them, took some of those little things. And we went on up to where the people lived. And we stayed there. And she used to dress that with that stuff on it. And this thing got well. And my finger began to grow out nice—and a little nail started to come on."

Myrtel laughed. "And now I got my nail back. And now can point!" She laughed again. "I always think about those little things, that they are medicines. And I think about the Creator—and he created things that are supposed to be medicines for people, those 'weeds'. Maybe they called it *kumul*. Those are round tops. I think she said *kumul*."

I was out with Sylvester Matthias on the flats between the Lone Butte dunes and found the little round puffball with the navel on top, as well as the tall stalked one. He said, "Both kinds are called *okstakuḍ* and both are used to control bleeding and for open sores." But according to Myrtel, only the smaller one was used for that.

One time when Ruth Giff was discussing Stalked Puffball, *okstakuḍ*, she said, "There's something else in this group," but she could not remember what its name was.

Apparently the taxonomy is collapsing, but through the details left by Hrdlička (1908:245), perhaps written at the very time that Myrtel was being treated, we

can rescue this folk taxon. Hrdlička (1908:245) gave a different name: "*Che-wa-te mo-to-a-te* ('earth carries on head': Tulostoma) is a little fungus which grows somewhat like a puffball, but has a well-defined stem. The dark yellow pollen of this fungus is applied by the Pima about the cord of the newborn infant, both as a preventative of inflammation and as a remedy when inflammation or suppuration has developed." Sylvester had never heard this name for the plant but understood the words Hrdlička had written, reinterpreting them as *jeweḍ mōōtō* 'earth on head carries'. To carry anything on the head is *mōōtō*. The image of carrying earth on the head may seem strange to us

jeweḍ mōōtō (?)   *Tulostoma* sp.

today, but to early irrigation agriculturalists with a much simpler technology, it was a common sight. When the great ditches were being dug throughout Pima country, the earth was loosened with digging sticks and was carried away in large tray baskets balanced on the head. Similarly, this little fungus carries a load of earth-colored spores balanced atop its stem.

Vouchers: *1984* (*Tulostoma* sp., not recognized by SM, GK, RG, FV).

Ustilago maydis DC.
Corn Smut
Fungus
huuñ kumuldag

A blackish fungus sometimes infests the ears or other parts of the corn plant. In this dry climate, this is not a common problem, but when the rains come at just the right time in late summer and the humidity rises, the corn may be attacked. I have been given several names for Corn Smut. Ruth Giff calls it just *kumul*, the same word she uses for the edible mushroom. In July she told me, "I haven't seen any this year yet. They just knock it off and don't do anything with it. Never heard of cooking and eating it." But by that October there were a few growths on the lower parts of her cornstalks. Then the *tasi'ikol* (Javelinas) got in the garden and took care of the smut as well as the rest of her crop.

Sylvester Matthias once called it *huuñ kumul-takuḍ*, which is an odd construction signifying 'corn fungus its instrument' as in *hivkuḍ*, the scraping stick. I think he slipped, because later I heard only *huuñ kumuldag*, meaning 'corn fungus its object'. I asked him about it. "Not anything. Not bad," he answered, "Not poison. Get a knife and whack it off— takes strength of the corn. Don't eat it [as far as he had ever heard]."

Castetter and Bell (1942:188) recorded, "One informant [whether Akimel or Tohono O'odham is not specified] said that as a boy, he used to see people putting ears heavily infested with corn smut on coals to roast, then taking off the smut

and eating it without other preparation." Whiting (1939:100) recorded a similar use among the Hopi, but their humorous suggestion of causality apparently has no counterpart among the Gileños.

Don Pedro Estrella, the last Pima Bajo speaker at the pueblo of Onavas, southern Sonora, gave me only *matori*, as Pima for *hongo de maíz*, this being the same word given to Pennington (1980: 184). There the white form or stage (but not the black, which is reputed to be poisonous) is fried and eaten. Among the Mountain Pima, I was told that the fungus is called simply *huunga* 'corn's thing' (*huun* 'corn' + *-ga*, possession marker). It is fried in grease and eaten.

I asked among the Gila Pima, but no one had heard of it being eaten by them. Ruth Giff said, "I've heard someone did, but not the Pima."

---

huuñ kumuldag   *Ustilago maydis*

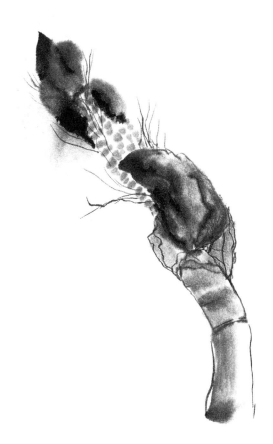

Group L, Unassigned Organisms

not *ha'ichu vuushdag* 'things that grow up'

This is a group of unassigned organisms that are not considered *ha'ichu vuushdag* 'things that grow up', or plants, but are considered living things. They are included in this work for the sake of completeness because Western botanists consider them to be plants in the broad or traditional sense. One group, the *hodai eldag,* are the things that grow on rocks, such as lichens and mosses. Unlike Corn Smut, which is included in Group K with other fungi, the mildew or molds are considered here as unassigned. *Jeweḍ eldag,* the terrestrial mosses and liverworts, are green and would seem to bear a more obvious relationship to conventional plants.

One organism, a still unidentified lichen, is of considerable cultural salience and perhaps falls into the category of sacred things that are dangerous in that they may produce something other than the intended effect. Only a few plants such as the daturas are so characterized, but many animals are.

hodai eldag   *Bryophyta* spp. (lichens)

Bryophyta
rock lichens and mosses
Bryophyta
hodai eldag

Sylvester Matthias was the first to provide me with the name *hodai eldag* one day when we drove up to the top of the Sierra Estrella bajada looking for plants along the edge of an *akĭ* 'arroyo'. It was a February day, and mosses and other mesophytic plants were thriving in the moist shade of large rocks and Desert Ironwood *(Olneya tesota)* trees. The fine, short mosses we found growing on the rocks he called *hodai eldag,* which he glossed as 'rock skin'. Sylvester favored the concept of this folk genus as a moss—any moss that grows on rocks. This he contrasted with *jeweḍ hiósig* 'earth flowers', a mystical and elusive lichen (see lichen species account).

When George Kyyitan was telling me about *jeweḍ hiósig,* I asked, "Do you know *hodai eldag*—the little plant that grows on rocks?"

"Yeah. [It's] different colors. It's different [from *jeweḍ hiósig*]. Not in one [color]; it's just white and yellowish and kind of pink or reddish."

So one person considered the name to apply primarily to mosses; the other restricted it to lichens. All seemed to contrast it with *jeweḍ hiósig.* The name may apply to any plant that grows on the rock as its "skin," so that status or location is more critical than strict morphology.

One day in the field, George called three different kinds of lichens growing on a rock (but not the mosses nearby) *hodai hiósig.* He was not sure of the relationship between this category and *jeweḍ hiósig.*

Bryophyta
terrestrial mosses and liverworts
Bryophyta
jeweḍ eldag

On a midwinter outing along a wash streaking down the side of the Sierra Estrella, Sylvester Matthias looked over plants growing on the shady banks. He pointed out some thick green mosses, saying they were *jeweḍ eldag* 'earth's skin'. It is surprising how much moss and how many kinds of mosses live in these extreme xeric situations, particularly where soils are undisturbed for long periods. They

may occur even in the shade of sand dunes. I never inquired of anyone about soil algae found in similar situations.

I found some flat green bodies of liverworts and showed them to him. He knew them and figured that they probably were not distinguished from the green mosses. He thought both would be called *jeweḍ eldag.* (This does not mean, of course, that they never were distinguished in the Pima lexicon of plants—only that they are not now distinguished.) The largest and most conspicuous moss growing in these desert sierras is Arizona Spike-moss, *Selaginella arizonica* Maxon.

I asked Ruth Giff if she knew of *jeweḍ eldag.* "Something like velvet," she commented. Then she said she did not know that was a formal name, but she knew the words and knew what it meant. What more can we ask of a name?

One of the most popular and frequently recounted episodes from the Creation Story is the tale of a gambler who is turned into Eagle Man (see Lloyd 1911:72–88; Russell 1908:218–221). The eagle becomes a predator of humans, dragging his prey to an inaccessible cliff. After Se'ehe (or I'itoi) at length conquers and kills the monster, he goes about the business of resurrecting the pile of corpses in various stages of decomposition. According to one version Sylvester Matthias told, "Some were just bones at the bottom, all white. And he stoops down and grabbed the bark of the earth, the bark or skin, *jeweḍ eldag,* and a pointed stick. And he wrote something. He wrote an inscription on the *jeweḍ*

jeweḍ eldag   *Selaginella arizonica*

mamtoḍgi mold (on *Cucurbita moschata*)

eldag. And he said, 'You go east and you talk together with this [writing].' Then he grabbed [them] and throw them across the ocean. But then they come right back." And this is how the white people came to be, from the corpses that were so rotten that they could not remember anything without writing it down on something. In another version, Sylvester said, Se'ehe made marks on a piece of leather. Then he added, "It's just a joke!"

Although of another language family, the Seri of coastal Sonora group soil algae, lichens, and mosses in a similar manner, even calling one terrestrial algae 'soil its-skin' (Felger and Moser 1985:217).

Fungus
mildew, mold
Fungus
mamtoḍgi

Finding the name for a folk taxon may take years of patient waiting. I had asked four of the people with whom I regularly worked about a name for mold such as grows on bread or fruit. They all assured me, "There *is* a name in Pima, but I can't think of it right now." Ruth Giff told me the name for mold was the same as for mildew on laundry.

One day Sylvester Matthias and I were discussing the process of decaying. He gave me the word *jevalig* for 'spoiled' or 'decayed'. "You say *pan jevalig* for 'decayed bread'. Or *miiliñ jevalgaj* for 'decayed/moldy/spoiled watermelon'." And then it finally came. This context jarred his memory.

"And *mamtoḍgi* can go with that 'decayed'. It would be similar to that *pan jevalig,* decayed bread. . . . " He was stuck.

"What would you say in English?" I pressed.

"Let's see, what would you call it? When you leave out wet clothes too long and whether you soak it too long—or the water dried out. What would you call that when. . . . " He paused.

"Mildew?"

"Ah! Mildew! *Mamtoḍgi* is mildew!" He was as excited at discovering the English name as I was at finding the Pima word at last.

"Would you say that's the stuff that's *on* the bread or the food when it spoils?"

"Yeah. That would be that *mamtoḍgi* on bread. Even on rotten melons, could be *mamtoḍgi,* too."

"Would you call that a kind of a plant? Or a living thing, or what?"

"No, it wouldn't be classified as a plant."

"Do you think it is a living thing?"

"Yeah, living thing."

"Well, how should we classify it? Where shall we put it?" I asked.

"With *huuñ kumuldag* [Corn Smut]— they go together." He paused, then added, "And I think *kumul* [mushroom] goes in there too." This spontaneous association of three unlike fungi surprised me. Also surprising was the discrepancy in not calling mildew a plant, when he considered mushrooms and Corn Smut to be plants.

Although classified quite differently in the Piman hierarchy, *mamtoḍgi* appears to be etymologically related to *mamtoḍ,* the common spongy algae. From an etymological perspective Culver Cassa notes this progressive derivation: *mamtoḍ* 'algae' (noun) to *mamtoḍag* 'being covered with *mamtoḍ*' (stative verb) to *mamtoḍgi* 'to become like *mamtoḍ*' (intransitive verb).

Lichen
unidentified lichen
Lichen
jeweḍ hiósig

Up in the desert sierras, those improbable rocky upthrusts from the smooth alluvial plains, grows some still unidentified species of lichen that was highly significant to both the Akimel and Tohono O'odham. This is called *jeweḍ hiósig* or *jeveḍ hiósig* 'earth flower(s)'. A considerable cult surrounds this lowly plant.

Some Piman concepts are difficult to understand, particularly when they stand in direct contradiction to something that European cultures uphold as a value: something as simple as quantity. "More is better" is so pervasive an axiom in Western culture that it dominates nearly every waking hour. If a word of prayer is good, then why not flood heaven with a torrent? If several plowed acres of culti-vated desert ground prove productive, why not denude the entire desert and pump every last bit of "fossil" water? If a particular desert grassland can support 5 head of cattle, why not graze 25? Contrast this with a concept that holds that a little of something is good but more is dangerous. For Pimans, more—rather than being a "good"—is an ambi-valence that can produce just the opposite, an "evil."

The *maakai* 'shaman', for instance, achieved power in the community because of his ability to cure certain kinds of illnesses. He was treated with respect and was paid well but was also feared. He could bring good, or he could bring evil. He was a dangerous being. The Creation Story and the Pima calendar-stick histories are filled with occasions where the community slaughtered its *mamakai* 'shamans'. Likewise, bravery was esteemed; the Gileños were ever vigilant for the dreaded Apache. But killing one made the victor "dangerous," so that he was required to stay outside the community during 16 days of rigorous purifications. Further, ownership of things—accumulation of material wealth—was looked upon as a vice rather than a virtue. Russell (1908) wrote, "Stinginess could not be more abhorred. The chiefs, especially, were expected to bestow liberally all gifts within their control. The present [1901] chief has had a canny sense of thrift and possesses a large bank account, which renders him much less popular than he might be otherwise."

This Piman concept is necessary for an understanding of *jeweḍ hiósig*. Everyone who knew about this plant considered it dangerous and mentioned the potential for insanity, always with a sort of approach-avoidance awe. One day Joseph Giff told me, "I had a little pouch, smaller than your little finger, someone had given me, with *jeweḍ hiósig* sewn in it. [It] can drive you crazy—go out of your head. It was used also in curing the sick. [It came] from the Sierra Estrella—between the two peaks. Ambrose Juan [Joe's wife's brother] and I went up on the mountains. When we got back, people smelled it on us." The little pouch that Joe mentioned was worn in the belt or carried in a pocket to attract women, particularly at dances.

But young man seeking amorous affairs ran a risk. David Brown explained about the plant: "There are three kinds of *jeweḍ hiósig* 'ground flower'—one good to use; other two, they're no good. Mix a little *jeweḍ hiósig* with Bull Durham and the smoke spreads out and the women will come around. The other will make a man or boy out a 'queer', *uikwuad* ['transves-tite']. But the good one, if you have it in your pocket, can bring the ladies. A mesquite or any kind of tree that looks good on [the] outside but is some way on [the] inside is called *uikwuad*. If you cook, wash dishes, and make tortillas—they'll call you *uikwuad*. [You can keep] *jeweḍ hiósig* in [your] pocket—use like body lotion, but you can't smell it yourself, but others in the crowd can and it smells good." But the ambivalent lichen may smell good to other men as well as to women. Note the element of danger here; the plant can turn to the opposite of the intended effect. Even the words between the intended and the potential effects of the plant are related: *uvimkam* is a 'wom-anizer', and *uikwuad* is a 'transvestite', one who assumes the role of a woman.

George Kyyitan and I were sitting out-side one spring evening discussing datura, marijuana, and paint- and gas-sniffing. "They all make you crazy," he said. I asked whether he considered *jeweḍ hiósig* to be in this same general class of things. He certainly did not.

"*Jeweḍ hiósig* has a sweet [smell] . . . smells good. You could smell it over in Prescott. There's a lot of it there. When the wind blows from there you could smell—it's from the limbs of that pines there. From that—that's where it gets on there. See that time when I was working over here on tribal farms—and they bought cedar posts and they [the posts] come in with [it]—on their peelings or whatever they call it. . . . "

"The *eldag* ['bark']?"

"Yeah, they were on that. And some of those workers there, they scrape them off and put them in [a] Bull Durham sack." Prescott is the mile-high town some 99 miles northwest of Phoenix. The Pima call Prescott S-hukak 'lots of pines' from *huk* 'pine'. This is the only reference I ever heard of this magical lichen growing on trees.

"What color is *jeweḍ hiósig*?" I asked George.

"It's kind of just white. White. You can smoke it. Just mix it with Bull Durham and smoke it. And it smells *good*. That's the only way they can [use it]. But I heard this way, that some people say that some people, those, they were medicine man, they use that one. But it's just the crazy thoughts that way that they could use it for, to get somebody's wife—or something like that, to make her crazy like that."

"What about the *jeweḍ hiósig* they would scrape off the rocks? Do you know what that one is?"

"It's same, over here on that hill—there's a butte over here. But I never go over there, because I never use it. I just smell it, though, and I seen it. They use it. There's a guy who used to live right there. He used to go over there always to get it."

"Which butte?"

"Not the first one, Fox Butte, they call it, but further on—way over there on that mountain there."

"In the Sacaton Mountains?"

"*Hau'u* [yes]."

"Does anybody use that *jeweḍ hiósig* anymore?"

"I don't think so. I don't think there's anybody. But there might be. But if they know. But I don't think they know. But I know—I could smell it over there, when we go to Prescott, there. The wind kind of blows—you could smell it. Especially when it's kind of damp—or rains."

"So they say that at the dances, if they carry that around, it would bring in the ladies? Make them *chechpa'aĭ* [promiscu-ous women]?" I asked.

"Uh-huh, yeah." He laughed. "Not just the dances, but anytime. 'Cause it just smells *good*. But if you had too much in there, it'll smell awful. But just a little bit in there and it smells alright." Here again, in George's account, is the ambivalence of this plant to which nearly everyone referred.

Several years later I mentioned *jeweḍ hiósig* to George. "It's no good, if you use it every day; makes your lips sore. You have to mix it with Bull Durham. If you put too much, [it's] no good. Just use it once in a while. It sure smells good, but it's no good. Makes ladies crazy. It's just like a perfume; have a little piece of it, put it in your pocket."

"Have you ever heard the story that if you get the wrong kind of *jeweḍ hiósig* that it will make you, turn you into a *uikwuad*?"

"Hah!" He laughed. "I never heard that."

One day when Francis Vavages and I were discussing a pink-flowered tobacco that he recalled from his youth, he mentioned something else.

"They call it *jeweḍ hiósig*—you hear [of] that?"

"Yeah."

"There's lots of that up there in the Estrella Mountains—Komaḍk."

"What color is the *jeweḍ hiósig?*"

"It's sort of a green, greenish looking when it was forming, I guess, but it turns kind of a white—light looking after and that's what they scrape off. And *that's* what they use. It's got a funny smell. And they said when you wanted to get acquainted with a girl, you stand where the wind is blowing so you can smoke that. You've heard that?"

"Yeah," I said.

Francis laughed at my recognition, then continued. "It's got little—when it really gets done or ripe or whatever—you can scrape it off—there are little black things in there, then you crush it," he rubbed his palms together, "then put it in your tobacco and smoke it. It's got sort of a pretty smell. It's the smell what gets them, not what you smoke. It's the smell."

Sylvester Matthias had firsthand knowledge of this mysterious plant. One day I quizzed him on its properties.

"There's different kinds, different colors, but the smell is the same." One could find it in the Sierra Estrella, but not the lower, dryer South Mountains. He used to have some when he was young.

"Did you try it out?"

"Yeah."

"Did it work?"

"No. Well, . . . it *did* work, in a way."

"You mean the women didn't come chasing?" At this, the old Pima elaborated no further.

I asked, "Would the people who went up to get it sell it?"

"Yeah, they can be sold or given away. The Mexicans do believe it that way, too. And they use it for medicine. I don't know what for cure, but they use it for medicine. But I know another medicine man, a Pima one, and he said it can cure sores. Like, I don't know what kind of a sore you had— maybe you had a sore for years and years and *jeweḍ hiósig* would cure it."

"How would you use it on a sore?"

"Just grind it up and make a powder and spread it over [the sore]. I don't think the young generation don't know about *jeweḍ hiósig*. They wouldn't go after it." Even topically applied, it must be used with caution (Curtin 1949:78).

But the lichen's better-known use was in a social context. I asked Sylvester, "Sometimes they carried it on them and sometimes they smoked it with tobacco?"

"Yeah, that's what they do. Sometimes they'd mix it with Bull Durham and smoke it."

"How did you use yours when you had it?"

"Just carried it."

"Could you smell it? Or could somebody else smell it?"

"Could smell it and somebody else could smell it also. They said it smells good, but it don't smell good. But they had a smell. Not like perfume or anything else that smells good. It's just kind of a peculiar odor. During the dances, the social dance, I could smell it—somebody's using it."

Castetter and Bell (1942:112) said, "The lichen used was described by our informants as having a strong sweet smell, being somewhat narcotic and of such magical properties that smoking a little of it mixed with tobacco would attract game while hunting." Sylvester had no knowledge of the use of this lichen by hunters.

With the expansion of the agricultural enterprise among the Gileños, particularly by gold rush times, deer hunting that would take men long away from their fields declined. However, their desert cousins, the Tohono O'odham, have continued to hunt on a more pervasive scale. Underhill was still able to record extensive information on deer-hunting ritual and technique among the Tohono O'odham in the early 1930s. Some hunters had hunting fetishes that some animal in a dream told them about.

Those who had had no vision sometimes owned a less reputable form of magic, though they would not confess it. This came from the lichen called 'earth flowers,' which grew on rocks in inaccessible places. Men climbed to get it, and, after asking its permission, rubbed off a little with a rock and wore it in the folds of the breech clout.

It attracted both deer and women, and its odor on a man's sweaty hands in the drinking dance would bring any woman to him, they said. But it was dangerous, and a man who used it too long would ultimately sicken and die. No man ever admitted possessing it, and the few whom the writer knew to be suspected of owning it were regarded with fear. (Underhill 1946:88)

I asked Frank Jim, a Tohono O'odham friend, what he knew about *jeweḍ hiósig*.

"It's like pasted on rock. Greenish. Not exactly green. Scrape off with knife. Another one grows inside cave. Same color. More like powder, on rock. Scrape it off, too."

"How deep a cave? In the back or close to the front?"

"Just inside." What he described seemed more like a rock cavity or recess. "You can smell it—like perfume—not really like a perfume, but it smells good. Entrance to cave. You can just go in and sit and you'll smell it. Deer like that stuff; they lick it. They like it. Smell it and come. Hunters spread it on body—on guns. If you see cave, you'll see the deer tracks where they come, licking it. [It] blows from the hunters to the deer and they'll come around that way."

The origin of the knowledge of *jeweḍ hiósig* is given in the Creation Story as recorded by Russell (1908:206–237) but inexplicably is missing from that by Lloyd (1911), although both versions were told by Thin Leather. (Joe Giff narrated this segment to me several times.) After Ho'ok, the part-human witch, was killed and cremated in her cave, the People had a feast but did not invite an old woman and her two grandsons. She told the boys to go look for Ho'ok's blood, which they should put in a dish. This turned into two parrots or macaws. She told them that the people would come and try to destroy her and the boys to get their beautiful birds, but that they should escape with the birds to the east; when the boys return, they should bury her in the sand and watch her grave for four days. Pursued by the people, the boys at last let loose the birds, which flew up into the mountains. The boys followed, hiding in the forest, eluding the people. They returned to find their grandmother dead, then carried out her instructions. In four days a tobacco plant appeared on her grave in the arroyo. Se'ehe (Elder Brother) inquired after the boys and went to visit them. They did not know yet what to do with the tobacco into which their grandmother had transformed. In Russell's version (1908:224), Elder Brother instructed the boys: "'These are to be rolled in corn husks and smoked,' he said; 'I will give you, also, earth flowers to mix with the tobacco when you smoke if you desire to gain the favor of the women.' He also showed them how to collect the bark of the tree which induces sleep *[kosdakuḏ]*. 'Make this into a powder,' he said, 'and when you wish to overpower anyone just shake this before them.' Then Elder Brother left the youths, who followed his instructions and found the love philter and the sleeping powder to be irresistible. But the people were incensed at their use of the charms and finally killed them." Even at its initial use, the lichen had an ambivalent effect, first as an aphrodisiac, but subsequently bringing about the boys' own destruction.

Jeweḏ hiósig has eluded identification. Russell (1908:224n.) said it is "a whitish lichen gathered by the Pimas and kept in little bags or in hollow reeds 3 or 4 inches long." Castetter and Underhill (1935:27) said that the lichen used by the Tohono O'odham was "a yellow growth found on the north side of rocks" on the Baboquivaris and had to be scraped off with a knife.

Lewis Manuel of Blackwater told Curtin (n.d.) that it is the color of gray ashes, has a strong smell, and grows on rocks and dead wood in only certain spots on the mountains. Isaac Howard of Sacate was 50 when Curtin interviewed him in 1941. He told her, "*Jeweḏ hiósig* is reddish and white and different colors and it smells like violets" (Curtin 1949:78). And 47 years later, George Kyyitan, of Bapchule, just a few miles upstream from Sacate, told me it was "kind of just white." Another time Sylvester Matthias told me, "*Jeweḏ hiósig* smells similar to incense; have to use chisel or knife to get them off rocks. Some purple, some gray; [grows] over a cliff or cave or rocks; not everywhere, just on rocky mountains." Still another time Sylvester told me, "It's not in the South Mountains. It's found in three places in the Estrellas— under part of rock, in the shale *[sic]*. It's reddish . . . or gray speckled."

I went out into the low Sacaton Mountains one day with Irene Hendricks, who pointed out rocks with what she considered to be *jeweḏ hiósig*. This was a common bright greenish yellow lichen. I gathered some of the lichen-covered rocks and took them back to my little room in Sacaton, leaving them just outside the door, which I left open at night. Nothing happened for a few days.

Then early one morning, before the sun cracked over the horizon, I was showering in the outside washroom when I heard an insistent and powerful knocking on the door. It sounded urgent. I cut the water and stuck my head out.

There stood a short dumpy woman with straight black hair cut short, wearing a worn plaid work shirt, a baggy pair of pants, and men's big black work shoes. "Where's Father Richard?" she demanded in a husky voice. "I've got to talk to him. I'm in a bad way." The light morning breeze wafted in to me the distinctive evidence of an all-nighter.

"He's still asleep. He went down to Tohono country for a funeral and didn't get back till three this morning. He won't be up for a long time."

"But I've gotta talk to somebody," she said, eyeing me once from top to bottom. "I guess you'll do. Get your *clothes* on and get *out* here!" From that day we became fast friends, but I knew right then that the greenish yellow lichen we had found in the low desert mountains was one of the wrong kinds of *jeweḏ hiósig*.

Appendices

Appendix A

The Gila Pima Ethnobotanical Lexicon

Appendix Table A.1.   Plant Products in the Lexicon

Product	Scientific Name	Pima Name	Source[1]
chewing gum	various trees	*ki'ivi* or *viibam* (see *Sarco-stemma* account)	Pima: 'chewed' or 'milkweed vine'
cocoa, chocolate	*Theobroma cacao*	—	
coffee	*Coffea arabica*	*kavhíí*	< Spanish *café*
lumber, boards	various trees	*huk komalkadam*	Pima: 'pine something flattened'
	various trees	*uus komalk*	flat wood [or] 'wood flat'
paper	various trees	*tapial*	< Spanish *papel* (?)
pepper	*Piper nigrum*	*vavish ha chu'i*	Pima: '*Anemopsis californica* their powder'
rice	*Oryza sativa*	*totoň maamaḍ, totoň nonha*	Pima: 'ant babies', 'ant eggs'
rubber	*Hevea brasiliensis*	*uuli* (RG)	< Spanish *hule* from Nahuatl *ulli*[2]
sugar	*Saccharum officinarum*	*asúúga* (see Group I) (some: *s-i'ovchu*)	< Spanish *azucar* (Pima: 'sweetness')
tea	*Camellia sinensis*	*tii*	< English tea
vanilla	*Vanilla plantifolia*	—	

[1] < indicates "derived from" or "loan."
[2] Pima/Papago possibly cognate with Nahuatl.

Appendix Table A.2.　Tohono O'odham Folk Taxa Not Recognized by Gila Pimas

Folk Taxon	Scientific Name	Source[1]
aḍ[2]	Apodanthera undulata	GPN 398
babad iiwagĭ	Heliotropium curassavicum[3]	GPN
ban chiñshañ	Sisymbrium irio; Lyrocarpa coulteri	GPN 332; GPN-2; SS&E:6
ban iibda	(unid. flower)	S&S:175
ban shuushk	Proboscidea altheaefolia[3]	GPN
chaango mo'o	Cocos nucifera	(commonly known)
cheḍkodam	Verbena sp.?	SS&E:109
chekapul tak	(root Navĭchu smoked)	MM-1:189
chuchk muḍagkam	Bouteloua barbata	S&S:172
chuchul i'ispul	Delphinium sp.[3]	MM-1:214
chuuwi wuipo, chuuvĭ wuipo[2]	Calliandra sp.?	S&S:172
chuuvi vuupui[2]	(Cucurbit, wild?)	MM-1:227
chuuwĭ wuupui	Solanum nodiflorum[3]	GPN 159
chuuwi bahi sha'i	Erioneuron pluchellum	AMR 1721
daḍpk washai	Bouteloua curtipendula	S&S:172
gepĭ	Citrullus lanatus[3]	S&S:100; SS&E:110; MM-1:281
gevul	Cirsium spp.	MM-1:286
gi'in	(unidentified grass [sic])	S&S:172
giḍag[2]	Acacia constricta	SS&E:16
giiko	Baileya multiradiata	MM-1:292
giikoda	Baileya spp.	S&S:172, 173
giikota	Sphaeralcea laxa[3]	SS&E:16
gisokĭ	Platyopuntia (sp.?)	SS&E:16
haaḍ kos	Calochortus kennedyi	SS&E:18
hauk u'us	Bebbia juncea	GPN 296; GPN-2
havañ taataḍ	Erodium cicutarium (sic)	SS&E:19
hejel e'eshaḍam	Salsola australis[3]	SS&E:20
hejel e'eshadam	Salsola australis[3]	GPN; GPN-2
heko taḍk	?	S&S:(ex GPN)
hewel e'es	Penstemon spp.	S&S:21, 172; MM; GPN; AMR 1629
hiikimul (iibda)	(unid. flower)	S&S:175
hina hitam[2]	Sapium biloculare	AMR 194; GPN-2
hiwichuis	Chenopodium murale (sic, misid.?)	C&U
hohowai, hihovai[2]	Simmondsia californica	C&U; MM-1:376; SS&E:23
hoi	Yucca baccata[3]	C&U
hovij je'e	Yucca baccata[3] (plant itself)	SS&E:24, 112
huuñ va'ug	Platyopuntia (sp.?)	SS&E:24
ina hitá	Sapium biloculare	GPN 136; GPN-2
iol[2]	Arctostaphylos spp.	MM-2:443
ioligam[2]	Arctostaphylos spp.	SS&E:26
jeweḍ ho'iḍag	Tribulus terrestris[3]	SS&E:27
ka'al[2]	Quercus sp.? (white oak)[3]	SS&E:29
kaakowanĭ	Lepidium lasiocarpum[3]	S&S:; GPN 452
kalistp	(unid. sp. paloverde [sic])	S&S:170(unid.); SS&E:30
kalisp	Cercidium floridum[3]	GPN-2
kaska-loon (ex Sp. cascarón)	Abronia villosa[3]	SS&E:31
kawk koawul[2]	Condalia globosa	GPN 269
kawpdam	Chenopodium murale	GPN(notes)
ko'okmagĭ shaaḍ[2]	Jatropha macrorhiza	S&S:172(unid.); GPN(notes)
ko'ovĭ taatami	Cassia (Senna)	SS&E:34
komagĭ waas[2]	Jatropha cinerea	GPN; GPN-2
kookod hoho'ibad	Erodium cicutarium[3]	SS&E:33
kookod oipij	Erodium cicutarium[3]	SS&E:33
ku'upak	Asclepias subulata	AMR 198 (misid. for Ephedra?)

(Appendix Table A.2 continued on page 381)

Appendix Table A.2. (cont.). Tohono O'odham Folk Taxa Not Recognized by Gila Pima

Folk Taxon	Scientific Name	Source[1]
kuksho vuupülim	*Delphinium* sp.[3]	MM-1:8, 214, 320; SS&E:36
kukuvid ha-haad[2]	*Rudbeckia hirta (sic)*	SS&E:36
kuukkpaḑag	*Muhlenbergia porteri*	SS&E:35
kuukp	(a grass, unid.)	MM-2:14
kuukpaḑag	*Muhlenbergia porteri*	MM-2:14
maahagam	*Washingtonia californica*[3]	MM-2:32
mansáana	*Malus pumila*[3]	MM-2:49
mansaniiya (ex Sp. manzanilla)	*Matricaria matricariodies*[3]	C&U; GPN
mimblii'a	*Cydonia oblonga*	AMR
mo'owhani	*Echinocereus* sp.[3]	SS&E:43
moostas [=mostois?]	*Brassica tournefortii*	GPN-2
naank je'ejegashakuḑ	*Bouteloua eriopoda*	MM-2:92
ohki	*Sporobolus airoides/S. wrightii*[3]	C&U
onk kui	*Tamarix ramosissima*[3]	SS&E:106
onk washai	*Distichlis spicata*[3]	*GPN 144*
oobgam	*Parkinsonia aculeata*	S&S:170; SS&E:47; MM-2:466; GPN; *AMR 1126, 1170,* Pima Bajo
piilos (ex Sp. peras)	*Pyrus communis*[3]	SS&E:50; MM-2:148
puuhl[2]	*Trifolium repens*	*GPN 233*
s-i'ovi vihol	*Lathyrus odoratus?*	MM-2:170
s-o'oi muuñ	*Phaseolus vulgaris* (pintos)[3]	MM-2:170
shapijk	*Cucurbita* sp.? (cult.)	S&S:178
shuuch	(a cactus)	MM-2:355
siwi taḑshagi[2]	*Gutierrezia sarothrae*	GPN
taḑshshagĭ	*Ambrosia deltoidea*[3]	MM-2:181; S&S 170(?); *AMR 1966*
taiaroti	*Solanum* sp.?	C&U
tatk	*Cassia* sp.?	MM-2:186
tatshshagi	("bitterweed")	S&S:170
toa/doa	*Quercus emoryi*	C&U
toki, tookĭ [=tohi?]	*Phoradendron californicum*[3] *[=Tillandsia?]*	GPN; GPN-2
toki sha'i (vashai)	*Erioneuron pluchellum*	*AMR 1721*
utko, utko jeej	*Yucca elata?*[3]	S&S:45
utko jeej[2]	*Yucca schottii*	SS&E:60; GPN
vupiostakuḑ	*Penstemon* sp.	MM-2:316; SS&E:67
vupuistakuḑ	*Penstemon* sp.	MM-2:316; *AMR 1629*
wepegĭ washai	*Cuscuta* spp.[3]	GPN; GPN-2; MM-2:259
wiopulĭ	*Nicotiana trigonophylla*[3]	C&U; Lumholtz 1912

[1] Sources:
S&S   Saxton and Saxton 1973
SS&E   Saxton, Saxton, and Enos 1983
MM-1   Mathiot, vol. 1, 1973
MM-2   Mathiot, vol. 2, n.d.
C&U   Castetter and Underhill 1935
GPN   Gary Paul Nabhan 1983b (specimen if *GPN*)
GPN-2   Gary Paul Nabhan (in Felger and associates 1992)
AMR   A. M. Rea specimen
[2] Extralimital to Gila Pima country
[3] Alternative name exists in Gila River Pima

Appendix Table A.3.   Plants Apparently Not Named in the Lexicon

English Name	Scientific Name
Introduced Old World Crops or Weeds	
almond	*Prunus dulcis*
aloe	*Aloe* spp.
anise	*Foeniculum vulgare* var. *azoricum*
artichoke	*Cynara scolymus*
Athel Tamarisk	*Tamarix aphylla*
Australian Saltbush	*Atriplex semibaccata*
canna	*Canna* x *generalis*
cherry	*Prunus cerasus*
cilantro	*Coriandrum sativum*
coconut	*Cocos nucifera*
cucumber	*Cucumis sativus*
flax	*Linum usitatissimum*
Foxtail Brome	*Bromus rubens*
hollyhock	*Alcea rosea*
honeysuckle	*Lonicera japonica*
horehound	*Marrubium vulgare*
iris	*Iris* spp.
lily	*Lilium* spp.
lime	*Citrus aurantifolia*
Malta Star Thistle	*Centaurea melitensis*
Milk Thistle	*Silybum marianum*
mint	*Mentha* spp.
mustards	*Brassica* spp.
okra	*Abelmoschus esculentus*
oleander	*Nerium oleander*
olive	*Olea europaea*
oregano	*Origanum vulgare*
Prostrate Knotweed	*Polygonum aviculare*
quince	*Cydonia oblonga*
roses	*Rosa* spp.
Spicebush, Butterflybush	*Vitex agnus-castus*
Other Plants	
Arizona Sycamore	*Platanus wrightii*
Jojoba	*Simmondsia chinensis*
juniper	*Juniperus* spp.

Appendix Table A.4. Plants That Should Have Been in the Lexicon

English Name	Scientific Name	Reason[1]
Mexican Oregano, Wright's Lippia	*Aloysia (Lippia) wrightii*	2
White Bursage	*Ambrosia dumosa*	1
Indian Root	*Aristolochia porphyrophylla (watsoni)*	4
White-stemmed/Desert Milkweed	*Asclepias albicans/A. subulata*	2,5(?)
Narrow-leafed Saltbush	*Atriplex linearis*	1
Salton Saltbush	*Atriplex fasciculata*	1
Desert/Woolly Marigold	*Baileya multiradiata/B. pleniradiata*	1,3
Chuckwalla's Delight	*Bebbia juncca*	1,3
spiderlings	*Boerhavia* spp. (except *erecta/intermedia*)	1
Bromegrasses	*Bromus* spp.	1
Indian Paintbrushes	*Castilleja* spp.	2
Episote, Mexican-tea	*Chenopodium ambrosioides*	1,4
New Mexican Thistle	*Cirsium neomexicanum*	1,3
Virgin's Bower	*Clematis drummondii*	1,
Plicate Coldenia	*Coldenia (Tequilia) plicata*	1
Spike Rush	*Eleocharis* spp.	1
Buckwheats	*Eriogonum* spp.	2
Fluff Grass	*Erioneuron pulchellum*	1,3
Fagonia	*Fagonia* spp.	2
Desert Bedstraw	*Galium stellatum*	2
Water Star-grass	*Heteranthera dubia*	1
Camphorweed	*Heterotheca subaxillaris*	1
Big Galetta	*Hilaria rigida*	1
Water Pennywort	*Hydrocotyle verticillata*	1,5
Janusia	*Janusia gracilis*	2
Bush Penstemon	*Keckiella (Penstemon) antirrhinoides*	2
Yellow Water Weed	*Ludwigia peploides*	1
Menadora	*Menodora scabra*	2
Yellow Monkey Flower	*Mimulus* spp.	1
Olive	*Olea europea*	1,3
Mexican Palo-verde	*Parkinsonia aculeata*	1,3
Marsh Fleabane	*Pluchea odorata*	1,5
Willow Smartweed	*Polygonum lapathifolium*	1
Odora	*Porophyllum gracile*	2,?3
Athel Tamarisk	*Tamarix aphylla*	1
Woolly Tidestromia	*Tidestromia lanuginosa*	1
Arizona Rosewood	*Vauquelinia californica*	2

[1]Reasons

1. Occurs conspicuously where Pima lived (floodplains & riparian).
2. Occurs conspicuously where Pima hunted (local sierras and bajadas).
3. Is named in Tohono O'odham.
4. Has widespread use in folk medicine, etc.
5. Is believed to have had a name, but now forgotten.

Appendix B

Unidentified Gila Pima Folk Plant Taxa

akimel baasho vashai

In August 1987 I was pulling Crab Grass *(Digitaria sanguinalis)* from my yard in San Diego when Sylvester Matthias asked if that grass grew in the desert. A grass similar to it (or the same perhaps) grew on the edge of the creeks but not on the river, he said. "In spring, summer, and late fall but not in winter. Name means 'creek side grass'. Same tassels [as *Digitaria*]. We used to cut the tassels off [for play]. Here and there along creeks with [*Pluchea odorata* (L.) Cass, whose name in Pima he could not remember but which he pointed out along the lower Colorado], some ladies might be washing clothes, others bathing where this grew. The 1927 floods covered up everything, then the creeks went dry." Apparently this was the end of *akimel baasho vashai*, too. This is almost certainly *Paspalum paspalodes* (Michx.) Scribn., a native perennial sometimes found in marshes of Pima country.

auppa nanhagio

The Pima name means 'cottonwood's earrings'. In September 1990 Sylvester commented on something blackish hanging on the small limbs of Frémont's Cottonwoods *(Populus fremontii)* but not on the main trunks. He mentioned this in the context of fungi but did not know the texture of the plants, because they were way up in the tree. His grandmother pointed them out to him when he was small. He thought this might possibly have been a foliose lichen, or perhaps even a chrysalis.

awkivik

This name might be *okvik* in modern orthography. Russell (1908:149) mentioned this plant in an episode from the Creation Story: "An unidentified species of grass [*sic*], called *awkivik* [short final *i*] by the Pimas, is said to have been spun into thread in ancient times. According to the myth, it supplied the son of Corn Woman with material for his bow-string. When there is sufficient rain, this grass grows on the Mo'hatuk hills, north of Gila Crossing [Muhaḍag, South Mountains]." The plant very well may not be a true grass; anything in the Pima category *vashai* would have been glossed 'grass' in English.

baaban vippi

In June 1991 Sylvester Matthias began enumerating "Coyote's plants." About the fourth one he mentioned was *baaban vippi* 'Coyotes' breasts [or] nipples'. "Sort of a bush; it's leafy; small—four inches *at the most.* It's not common. Sort of gray and green—a mix up between gray and green."

"Does it have flowers?"

"Maybe *muḍadag* ['tassels']. I never paid much attention because I wasn't interested in plants then."

"Does it grow in fields or wet places?"

"Out in the desert. Not near water. I think the Mexicans use it for medicine, but [the] Pima don't."

Six weeks later, the subject came up again when Sylvester and I visited George Kyyitan at Bapchule. Sylvester said it was a plant that we had found once at 115th Avenue (at the Salt-Gila confluence). George did not know it; nor did Ruth Giff and Francis Vavages. Frank Jim (a Tohono O'odham) was curious about the name and asked other Tohono O'odham he knew, none of whom had ever heard of it. The name suggests a plant with a milky sap, like a *Euphorbia.*

ban aḍavĭ

One day while I was discussing Finger-leaved Gourds, *aḍavĭ (Cucurbita digitata)* with Francis Vavages of Sacaton, he asked me if I knew of *ban aḍavĭ.* (I did not.) "Well, it's a little smaller than [regular] *aḍavĭ.* About dollar-sized fruits, hard and striped. It spreads out—on the desert under a tree—in the flats. It's hard to find now." Later he added that regular *aḍavĭ* has greener leaves and is found in wetter places such as in fields and along the river, while *ban aḍavĭ* has more whitish leaves and grows out in the desert. We never identified this plant in the field, and the name was unknown to George Kyyitan, Herbert Narcia, and Sylvester Matthias.

cheenashaḍ

This name might be *jeeñshaḍ* in modern orthography. This apparently hallucinogenic or narcotic plant has never been identified, although it occurs in both Akimel and Tohono O'odham stories and song. When Elder Brother's invading army was destroying pueblos in the country the Pima now inhabit, the army moved down the Gila as far as Gila Crossing, where they sang this song:

I am the magician who with the sacred pipe

Of cheenashaḍ increase my magic power.

I am the magician of the downy feather

With the soothing sacred pipe

I bring sleep upon my enemy. (Russell 1908:229)

The defeated chief of the Gila Crossing pueblo is called Cheenashaḍ Siivañĭ. This person appears again at the end of Thin Leather's Creation Story (Lloyd 1911: 235–239), in which S-cheedag Siivañ ('Blue or Green Chief') invites Tcheunassat (Cheenashaḍ) Siivañ to gamble his wives. Lloyd made no comment on the derivation of this chief's name. (For details of this episode, see *Acacia greggii* account, Group E.)

When Elder Brother, Se'ehe, is called in to destroy the witch Ho'ok, who is eating babies and children, he instructs the people to sing a song with him. He makes cane cigarettes or pipes of sleep-inducing *cheenashaḍ* (Russell 1908:279). Lloyd's (1911:114) translation of the Creation Story does not mention the plant by name: "And Hawak [Ho'ok] was a great smoker, and Ee-ee-toy [I'itoi] made cigarettes for her that had something in them that would make folks sleep. And he smoked these himself, a little, to assure her, but cautiously and moderately, not inhaling the smoke, but she inhaled the smoke, and before the four nights [of the dance] were up she was so sleepy that the people were dragging her around as they danced, and then she got so fast asleep that Ee-ee-toy carried her on his shoulder."

The story occurs also in Juan Smith's version of the Creation Story (Hayden 1935:26):

So this Siuuhu [Se'ehe, Elder Brother] went to work and made four cigarettes. And one of those cigarettes were called chuna-suk [*cheenashaḍ*] (one kind of tobacco in this tube, sweet tobacco . . . chuna-suk mixed with regular tobacco).

And the other one is called heu-ko-takt—some kind of roots to make her weak. And the third was some kind of tobacco that is dope for sleeping. And the fourth was a weed that grows in damp places—roots of Jimsonweed [*Datura* sp.].

Se'ehe's song mentioning the four cigarettes is "untranslatable . . . Old Pima Words."

Russell (1908:278–279) recorded the song in the original song language given by Thin Leather. *Cheenashat* is mentioned but is noted as "the name of a former Pueblo chief *[siivañ]* who lived near Móhatûk [Muhaḍag, now South] Mountain." Actually, the *siivañ* was probably named after the plant, and it is the plant name that is intended in the song. Another cigarette is called *kâsinakon,* most likely *kosidakuḍ* in normal speech, 'something to make sleep with'. This kind of tobacco may come from a plant with the same name. I doubt that the other two repeated words in this stanza have meaning, at least in Pima.

Underhill (1946:297) also makes reference to the plant: "Two other plants [in addition to *pihuri,* or Peyote *(Lophophora williamsii)]* take human form and cause hallucinations. One is *Datura.* . . . The other plant is the unidentified *tcĭnacat* [= *cheenashaḍ* in a different orthography], which has here been called lizard weed. For this no song was obtained."

Bahr and Haefer (1978) recorded a set of five *Datura*-sickness-curing songs. Structurally matched with Jimsonweed in the last song is *jenshat.* They note, "The word translated as 'lizard' is actually the name of a particular species of lizard. Bahr is informed that it is a real creature which lives in the mountains and hence is little seen by people. He has not seen one and has no zoological identification for it. He cannot explain why it of all animals was chosen to 'quiet' the semantic disturbance set in the first song." And later, "Bahr's sources hold that it is a lizard and not a plant. They apparently view it as an accessory in the jimsonweed complex."

However, in light of the well-established role of *cheenashaḍ* (or *jeeñshaḍ*) as a narcotic tobacco throughout so much of Piman literature, I believe that this is indeed a plant. Its semantic pairing with the narcotic *Datura* in the song set Bahr recorded would thus make more sense. What is unexplained, though, is why Underhill chose, ostensibly arbitrarily, the English name 'lizard weed' and why Bahr's consultants insisted it was a lizard. Perhaps there is some allegorical relationship between a narcotic plant and an actual reptile. The Pima word for 'to smoke' is *jeeñ*, and 'to take a smoke break', *jeñshaḏ* (Mathiot 1973:417). This also lends credence to the botanical rather than the zoological interpretation.

chekapul tak

There are two references to this unidentified plant. Mathiot (1973:89) defined it as "a root said to have been used formerly by the navijhu [*sic*] for medicinal purposes." It appears also in Thin Leather's account of the origin of the Navĭchu (Lloyd 1911:211): "And on the mountain [Twisted Neck Mountain] he felt rather faint, and he put his hand in his pouch and found a root and chewed it, the root *Cheek-kuh-pool-tak*, and breathed it out and it stopped raining." The name was not recognized by Sylvester Matthias.

chivichuch e'es

The Pima name means 'Killdeer's plant'. I heard this name once at the west end of the reservation. Sylvester Matthias asked one day when we were discussing water plants: "*Chivichuch e'es!* We haven't got yet *chivichuch e'es?*"

"'Killdeer's plant'? No, I never heard of that. Where does it grow?"

"In the water . . . or *under* the water. . . ."

Is this really a plant's name, or just a play on words with *sivijuls* (Canaigre, *Rumex hymenosepalus*)? In the Sacate region, where Sylvester's mother originated, the Killdeer is called *sivichuch*. I inquired repeatedly over several years, but Sylvester never recalled this name again, and no one else had heard of it.

chuchk muḏadkam

The Pima name means 'black tassels it has'. Sylvester Matthias said, "Heard of it from the Papago but don't know what it looks like." Perhaps it is not actually a Gila Pima folk taxon. Mathiot (n.d.:79) spells 'one with tassels' *muḏadkam*. Saxton and Saxton (1969:179) identify this as Six-week Grama (*Bouteloua barbata*, which I have found only in a single colony at Blackwater). But this attribution needs verification.

daḏhakam vashai

Along Boundary Creek north of Jacumba and Interstate 8 in California, Sylvester Matthias pointed out a specimen of Sacaton Grass (*Sporobolus wrightii*, *noḏ* in Pima) that was smaller and somewhat differently shaped; he said it was *daḏhakam vashai* 'sitting down grass'. "*Noḏ* grows *up*, but this one droops *down*."

"Where would you find it?"

"Anywhere on the desert—like back of the Estrellas. A desert grass." He then told of some old missionary bringing his team up from Tohono O'odham country to the Gila River Pima Reservation, and the only grass available to feed his animals was *daḏhakam vashai*. He also said it was the same as *biibhinol vashai* (which I doubt; he had already identified this in the field as *Heteropogon contortus*, an erect-growing plant).

The next time we talked about this plant he said it was found anywhere, back of the Sierra Estrella, in the flat country now called Rainbow Valley. Almost certainly this is Alkali Sacaton (*Sporobolus airoides*), which now occurs uncommonly in Pima country at several seeps on the steep slopes of the Estrellas.

dapk vashai

The Pima name literally means 'naked grass' or 'smooth/slippery grass'. This grass is distinguished by the Tohono O'odham, but I do not know what is its biological referent. (Saxton and Saxton 1969:172 called it Side Oats Grama, which is *Bouteloua curtipendula*.) George Kyyitan had never heard of it. Sylvester Matthias had, but he did not know what it looked like. Francis Vavages said it is a Pima grass, a small grass, probably a bunch grass, with whitish tassels. He was not sure where it grew but thought it was seasonal, probably growing during the summer rains. Another time Francis described *dapk vashai* as a short grass, about 8–10 inches tall, that grows close to the mountains after rains. "Sort of white tassels—cattle feed on it." At the Desert Botanical Garden in Phoenix, Sylvester Matthias called a bristlegrass (*Setaria macrostachya* H.B.K.) *dapk*. The word *dapk* when used alone names an annual sage (*Salvia columbariae*) with small gray edible seeds and purple flowers, called in Spanish and English Chia.

Puzzling specimens of a robust-seeded bristlegrass have been found in the vicinity of Blackwater. These have been identified by James Rominger as *Setaria leucopila* (Scribn. & Merr.) K. Schum., closely related to *S. macrostachya*. Both are perennial. One wonders, in light of work by archaeologists Vorsila Bohrer (1975, 1991) and Charles Miksicek (1983), if these might be relics of some ancient incipient cultivar.

goks haal

This is presumably a folk species or cultivar of the genus *haal*. In Thin Leather's version of the Creation Story occurs an account of Tobacco Woman and Corn Man's pumpkin baby, glossed 'dog-pumpkin' (Lloyd 1911:225; for details see *Cucurbita* account, Group I). 'Dog pumpkin' would be *goks haal* in Pima. Apparently this kind of pumpkin is now completely forgotten; it was unknown to George Kyyitan, Sylvester Matthias, and Ruth Giff. Culver Cassa suggests that during the translating, Lloyd might have heard *daapk haal*, a type of *Cucurbita argyrosperma*, and mistook it for "dog" *haal* or 'pumpkin'. According to the version of this story known to Sylvester Matthias, Corn Man's baby was a *sha'ashkadk haal* (*Cucurbita moschata*).

heuko tatk

Recorded by Hayden (1935:26) as a root smoked in legend to make Ho'ok weak. The second word means 'root'. The first word is unidentified. A remote possibility is *utko* (*Yucca* spp.), a term still used in some Tohono O'odham dialects.

hikimulĭ

This plant is mentioned by Russell (1908:308). In a song, harlots are described wearing these flowers in their hair. Culver Cassa and others recognized the literal meaning of the words as 'something cut into irregular pieces', apparently in reference to the shape of the flower or leaves. Herbert Narcia said, "That's Papago; I've heard that in Papago." Such a plant was not recognized by Sylvester Matthias, Francis Vavages, or George Kyyitan.

kapichk

The Pima name means 'something narrow'. According to Sylvester Matthias, this is a plant that grows up in the Sierra Estrella. He thinks it may be a fern. The largest ferns there *(Notholaena sinuata)* indeed have very long and narrow fronds, up to two feet long. I have not been able to learn any northern Piman name for fern.

katikum

This Pima name was given by Lloyd (1911:221) as *kah-tee-kum* in the story of Corn and Tobacco. It might be *katikam* or even *katkam*. At the evening assembly, the father of Tobacco said that he smelled something new. "And one said, 'Perhaps it is some greens that I ate today that you smell,' and he breathed toward him. But the mahkai said, 'That is not it.'" The following evening we hear again from "the man who had eaten the greens *kah-tee-kum* the day before." This name was unknown to George Kyyitan and Sylvester Matthias.

kauk vaapk

The Pima name means 'hard reed'. Possibly this was a name given to Giant Reed *(Arundo donax)* when it first appeared, to distinguish it from the native Common Reed or Carrizo, *Phragmites australis*. The Pima call both these plants "bamboo" in English. *Arundo* is larger and tougher and grows in dryer places than *Phragmites*.

ko'oi tatk

This plant is known to both Akimel and Tohono O'odham. In April 1985, Tohono O'odham Frank Jim said: "*Ko'oi tatk* [is a] white root found on the *akĭ* ['arroyo'] banks, used to repel snakes; has no taste; found way up on the Ajo Mountains and other places. Soft root." A few years later he said, "It's a root, a white root, very long," and he indicated from half a meter to a meter. "Only as big around as your thumb. Can be used for protection against snakes or as a medicine. Take it for headache. It's very good. Take a piece of that and boil it and drink it. It's very light, porous, hardly weighs anything." He

reiterated that it is found way up on Ajo Mountain, but he does not know what the plant looks like or the color of its flowers.

Two years later Frank noted: "*Ko'oi tatk,* more like tree. Ajo Mountains, shady where it stays moist. Roots come out on [surface]—white. Gray bark, but scrape off—real white. Let it dry. Put piece in pocket. If [a] horse [is] bitten, pound it and spit on it [to make paste] and put it on. Or headache. Chew some. It's way up there, like that *ko'okol* [Chiltepín, *Capsicum annuum*]. I used to carry some in my chaps."

Several months later, the name of this plant came up again with Frank: "It will keep away [rattle]snakes; if bitten, boil it or grind it, put in around bite and wrap it up. It's light [in weight]—white root, no smell."

I asked Sylvester Matthias about this plant in April 1987. He knew it right away. "It's something like a root—*ko'oi tatk*—you chew on the root if you are going to travel in snake country—spit it on the legs; if riding a horse, use it on the legs, up to knees of the horse. In the family of *sha'i* [small brushy plants]; about the size of a carrot or a little bit larger, but that shape. I've heard of it."

Later he told me: "*Ko'oi tatk* is protection from rattlesnakes. Spray [it] on horse's legs. It's a root. The one I saw was something like—goes straight down, about the color of a sweet-potato."

Russell (1908:354) recorded what is apparently the same plant in I'itoi's resurrection speech, but he translates his *kokoi tak* as 'ghost root'. (*Ko'oi* is 'rattlesnake' and *ko'i* is 'ghost'.) Here Jeweḍ Maakai, Earth Doctor, blows smoke made from the roots of this plant and *bawui* (Coral-bean, *Erythrina flabelliformis*) on Elder Brother's breast.

Mathiot (1973:504) glossed *ko'ovĭ-taatk* or *ko'ovĭ-taatam* 'Desert Senna, Rattlesnakeweed' *(Senna (Cassia) covesii* A. Gray). She listed *ko'ovĭ* (dialectical variant, *ko'ol* [= *ko'oi?*]), pl. *kooko'ovĭ,* as 'rattlesnake, venomous snake'. This must be the same folk taxon. Both Sylvester and Frank associated the plant with rattlesnake. But Mathiot's identification needs verification.

Five years later Sylvester added: "There is also a plant there—that's to keep the rattlesnakes [away]—so if anybody want[s] to travel at night or on foot to avoid the rattlesnake, they can chew on

that root and spit it on, or paint it on the hooves of the horses when they go out and the rattlesnakes would give a warning, so they can keep the rattlesnakes away. That's a root, a plant. If he's walking, the rattlesnakes would give a warning."

"What's it called?"

"*Ko'oi [tatk].*"

"Where would they get the *ko'oi tatk?*"

"They grow wild in the washes."

"*Hebai?* [Where?]"

"Anywhere on the desert."

"In Akimel country or Tohono?"

"In Tohono [O'odham country]. They don't grow anything like that here. If you want a *ko'oi tatk* you have to buy it from the Papagos."

This folk lore seems to have entered the Hispanic culture of Papaguería. Dan Woods (21–22 March 90, *Tri-Valley Dispatch*, p. 2) spoke of a Mexican cowboy from the Sasabe area, Sonora, using Hierba-de-víbora; if you carried this root in your pocket, a snake would not harm you. "To me it looked like the root of a milkweed plant," Woods commented. Oswaldo Méndez Soto, a rancher from near Caborca, Sonora, identified Hierba-de-la-víbora as *Asclepias subulata* (AMR *1667*). Neither this nor Desert Senna has the root characters Pimans ascribe to their folk taxon.

komevat

Given by Russell (1908:73) as *kâ'meûvat.* "After the August rains this seed is gathered, parched over coals in the parching pan, ground on the metate, and eaten as pinole." This name was not recognized by my Pima consultants.

kukuvid hannamga

The Pima name means 'Pronghorns' cholla' (*kukuvid* 'Pronghorn' + *hannam* 'Buckhorn Cholla' + -*ga,* possessed marker). Sylvester Matthias gave it as "a species of cholla." He said it grows "just in one place, on the foothills of the Estrellas. It's taller. I don't see it, but someone saw it." He also noted that it had big fruit.

Another time I asked him, "What is different about it? What were the fruits like?"

"They were longer, and tall, longer and tall. The buds are kind of tall too. [Heard] from my father. Same place as the *giishul* [unidentified colonial cactus]. Right across, the foothills of the Santa Cruz River."

Carmelita Raphael had heard of it but did not know the plant. Juanita Norris of the Akĭ Chiñ Reservation said this was a type of cholla, but she could not remember how it differed from *hannam* (Buckhorn Cholla, *Opuntia acanthocarpa?*).

Owl's-feathers

The Pima name may possibly be *chukuḍ aa'an*. In Thin Leather's version of the Creation Epic (Lloyd 1911:62) occurs a story of a contrary orphan named Braided-Feather, who takes commands literally. For instance, "One day his grandmother sent him to get some of the vegetable called 'owl's-feathers,' which the O'odham cook by making it into a sort of tortilla, baked on the hot ground where a fire has just been. And he went and found an owl and pulled its feathers out and brought them to the old woman, and she said, 'This is not what I want! It is a vegetable that I mean!'" Unfortunately, the Pima name for the owl is not given here, and there are at least five folk taxa of owls. The largest and perhaps the best known is the *chukuḍ* or Great Horned Owl. (The word *chukuḍ* also means the spirit of a deceased person.) No one with whom I worked recognized the name of this plant. Culver Cassa suggests that 'owl's-feathers' is homologous to 17th-century Névome *tucuru vopo* 'owl's body feathers' (Pennington 1979:61), a folk taxon I never encountered while working with the last speakers of this language. At Onavas, Sonora, Pedro Estrella identified a specimen of *Physalis acutifolia* (Miers) Sandwith. as *tukuḍ wuupuhi* 'Great Horned Owl's eyes' *(AMR 489)*.

shashañ wuupui (?)

This plant is mentioned only in English translation in Thin Leather's account of an orphan who lived between the Santan and Superstition Mountains. He was mischievous, taking orders from his grandmother or others literally, contrary to intention. For instance, the old woman "sent him for the vegetables named 'crow's feet' and 'blackbird's eyes', saying to him that they were very good cooked together. And the mischievous orphan went and got the feet of some real crows and the eyes of real blackbirds and brought them to her" (Lloyd 1911:63). The first of these is *Phacelia* spp., a plant still widely known (see Group F), but no one recognizes the 'blackbird's eyes'. Presumably this would be *shashañ wuupui* in Pima, *shashañ* meaning Red-winged Blackbird specifically but any species of blackbird distributively.

s-hekuvchu aḍavĭ

Fr. Antonine (1935) recorded a species of wild gourd named *sekufchu adav*. Although no one with whom I worked knew a cucurbit by this name, there were several ideas as to what it might be. Ruth Giff suggested *s-he'ekchu aḍavĭ*, "those big ones, up higher and in California." She was referring to the larger triangle-leaved Buffalo Gourd *(Cucurbita foetidissima)*, a foul-smelling plant. Ruth's derivation is through *s-he'ek* 'sour'. Culver Cassa suggested a derivation through *s-hekuvchu* 'smelly or odoriferous underarm', also in reference to the species from higher elevations in Arizona. I showed Sylvester this species in the field on several occasions, but he did not know any special name for it. However, he realized it was some kind of *aḍavĭ*.

sheshelik

Russell (1908:76) lists this among the plant foods, saying only, "This is used as greens with similar plants." The word is in plural form, indistinguishable from the Piman name for Round-tailed Ground Squirrel *(Spermophilus tereticaudus)*, which Russell mentions beyond in the singular as an animal food. No one with whom I worked recognized this as a plant name, so this *iivagĭ* must be forgotten.

s-oám hiósigam

The Pima name means 'yellow flowers it has' (*s-*, intensifier + *oám* 'yellow' + *hiósig* 'flowers' + *-gam*, attributive marker). I heard reference to this plant many times over the years, particularly by Sylvester Matthias, but I was never able to establish its botanical identity conclusively. The name was given by Russell (1908:80) as "*soam hiâseikkam* 'yellow flower'. An infusion made from the flowers of this plant is used as a remedy for sore eyes." (For possible identification, see *Pectis papposa* and *Verbesina encelioides* accounts, Group G.) Francis Vavages (January 1989) said, "*S-oám hiósigam* kind of smells. Comes out in summer. Some are flat and some are like a ball. Kind of smells like something like medicine. Grows right around where it's wet. Flowers like a little ball—like out of a shell. Lots of flowers. Only about a foot and a half tall." The plant is almost certain to prove a member of the Asteraceae (Compositae). In southern Sonora, Pima Bajo Pedro Estrella identified a tall roadside plant, *Lagascea decipiens*, as *s-oam heosig* (AMR 1070).

tooḍk

Sylvester Matthias told me, "*Tooḍk* used to grow on a sandy soil. Not on the creek. But along where it's close to the water. And I heard of it and I never know what it look like. It used to grow some right below where we used to live in our field— and they called it *tooḍk*, just like 'horsefly' we call it *tooḍk*. And I *heard* people talking about *tooḍk*. There's a patch of them right there." The plant name is identical with *tooḍk*, the large biting horsefly, *Tabanus* sp.

toota muḍadkam

This is an unidentified grass with white tassels. Sylvester Matthias mentioned this name various times in discussions of the country about Chandler. He said, "Like a hygeria [hegari, a grain sorghum?]; something like a *vashai*. Likes moisture, lots of moisture. Where Chandler is now—Queen Creek floods all that area. I think the stock like it." This is the white-tasseled grass that Russell (1908:91, 359) associated with cultivated oats. No contemporary Pima considered *toota muḍadkam* a folk synonym for oats (see Group C). The species might be Cottontop, *Digitaria californica*, or Cane Beardgrass, *Bothriochloa (Andropogon) barbinodis*, both perennials. When shown a specimen of Feather Fingergrass, *Choris crinita (AMR 1574)*, taken near Blackwater, Sylvester said, "Maybe this is *toota muḍadkam*." This perennial grass is found "usually in heavy alluvial soils of bottomlands" (Gould and Moran 1981). The Pima name, he explained, applies both to a kind of grass and to the particular area about Chandler where a grassland occurred before settlement.

vuplo sha'i, vupilo sha'i

This name, meaning 'burros' bush', was mentioned by George Kyyitan. "I heard that one, but I never pay no attention to how does it look," he said.

"[Did it grow] around here?"

"I think they do. That's why they knew it."

Tohono O'odham would seem to be more likely to compound a plant's name from the Spanish loan word for burro, but I have been unable to find this name in any Tohono O'odham dictionary. It is not likely that this is a back-translation from English Burro-brush (*Hymenoclea* spp.) or Burro-weed (*Isocoma* spp.) or Burro-bush *(Ambrosia dumosa)*. This name was unknown to Sylvester Matthias.

Scientific Names of Animals Mentioned in Text

Mammals

Harris Antelope Squirrel, *Ammospermophilus harrisii*

Pronghorn, *Antilocapra americana*

Coyote, *Canis latrans*

North American Beaver, *Castor canadensis*

Black-tailed Prairie-dog, *Cynomys ludovicianus*

Javelina, Collared Peccary, *Dicotyles tajacu*

American Porcupine, *Erethizon dorsatum*

Long-nosed Bat, *Leptonycteris sanborni*

Antelope Jackrabbit, *Lepus alleni*

Black-tailed Jackrabbit, *Lepus californicus*

Striped Skunk, *Mephitis mephitis*

packrat, woodrat, *Neotoma* spp.

Mule Deer, *Odocoileus hemionus*

White-tailed Deer, *Odocoileus virginianus*

Desert Bighorn, *Ovis canadensis*

pocket mouse, *Perognathus* spp.

Raccoon, *Procyon lotor*

Hispid Cotton Rat, *Sigmodon hispidus*

Round-tailed Ground Squirrel, *Spermophilus tereticaudus*

Desert Cottontail, *Sylvilagus auduboni*

North American Badger, *Taxidea taxus*

Pocket Gopher, *Thomomys bottae*

Black (Brown) Bear, *Ursus americanus*

Reptiles

Banded Gecko, *Cleonyx variegatus*

Desert Iguana, *Dipsosaurus dorsalis*

Gila Monster, *Heloderma suspectum*

Red Racer, *Masticophis flagellum*

Desert Spiny Lizard, *Sceloporus magister*

Birds

Red-winged Blackbird, *Agelaius phoeniceus*

Golden Eagle, *Aquila chrysaetos*

Scarlet Macaw, *Ara macao*

Costa's Hummingbird, *Archilochus costae*

Green Heron, *Ardeola virescens*

Verdin, *Auriparus flaviceps*

Great Horned Owl, *Bubo virginianus*

Red-tailed Hawk, *Buteo jamaicensis*

Lark Bunting, *Calamospiza melanocorys*

Gambel's Quail, *Callipepla gambelii*

Scaled Quail, *Callipepla squamata*

Cactus Wren, *Campylorhynchos brunneicapillus*

Pyrrhuloxia, *Cardinalis sinuatus*

Turkey Vulture ("Buzzard"), *Cathartes aura*

Common Nighthawk, *Chordeiles acutipennis*

Killdeer, *Chradrius vociferus*

Marsh Wren, *Cistothorus palustris*

Yellow-billed Cuckoo, *Coccyzus americanus*

Common Flicker, *Colaptes auratus*

Gilded Flicker, *Colaptes auratus mearnsi*

Red-shafted Flicker, *Colaptes auratus* ssp.

Common Crow, *Corvus brachyrhynchos*

Common Raven, *Corvus corax*

White-necked Raven, *Corvus cryptoleucus*

Ladder-backed Woodpecker, *Dendrocopos scalaris*

Peregrine Falcon, *Falco peregrinus*

American Coot, *Fulica americana*

Common Gallinule, *Gallinula chloropus*

Greater Roadrunner, *Geococcyx californianus*

Common Yellowthroat, *Geothlypis trichas*

Yellow-breasted Chat, *Icteria virens*

Hooded Oriole, *Icterus cucullatus*

Bullock's Oriole, *Icterus galbula*

Gila Woodpecker, *Melanerpes uropygialis*

Common Turkey, *Meleagris gallopavo*

Northern Mockingbird, *Mimus polyglottos*

Western Screech-owl, *Otus kennikotii (asio)*

English (House) Sparrow, *Passer domesticus*

Song Sparrow, *Passerella melodia*

Blue Grosbeak, *Passerina caerulea*

Common Peafowl, peacock, *Pavo cristatus*

Phainopepla, *Phainopepla nitens*

Abert's Towhee, *Pipilo aberti*

Purple Martin, *Progne subis*

Virginia Rail, *Rallus limicola*

Thick-billed Parrot, *Rhynchopsitta pachyrhyncha*

Rock Wren, *Salpinctes obsoletus*

Black Phoebe, *Sayornis nigricans*

Rufous Hummingbird, *Selasphorus rufus*

Red-naped Sapsucker, *Sphyrapicus ruber*

Common Starling, *Sturnus vulgarus*

Bendire's Thrasher, *Toxostoma bendirei*

LeConte's Thrasher, *Toxostoma lecontei*

Western Kingbird, *Tyrannus verticalis*

Bell's Vireo, *Vireo bellii*

White-winged Dove, *Zenaida asiatica*

Mourning Dove, *Zenaida macroura*

White-crowned Sparrow, *Zonotrichia leucophrys*

Fish

Colorado Squawfish, *Ptychocheilus lucius*

Razorback Sucker, *Xyrauchen texanus*

Appendix D

Folk Taxonomic Adjustments in a Changing Biotic Environment

From a folk taxonomist's point of view, there are two ways of studying the historical processes by which organisms are named in folk taxonomies. One is by looking at the vocabulary of a people who invade a new area, encountering and naming new species of the biota. An example would be the development of plant and animal vernacular names as English-speaking immigrants colonized North America. Another way is by looking at how a speech community in situ linguistically accommodates new species entering the local biota as the result of contact. The Gila River folk taxonomy affords an opportunity to analyze this second situation because many named Pima plants are not native.

The linguistic processes by which a culture accommodates changes and innovations from another culture—linguistic acculturation—have been studied for a number of groups (for instance, Johnson

1943 and Spicer 1943 for the Yaqui; Lee 1943 for the Wintu; Casagrande 1954a, 1954b, 1955 for the Comanche; and, most important in the context of this study, Herzog 1941 for the Gila Pima). Salzmann (1954) has addressed the general problems of lexical acculturation.

In postconquest times, the Piman language continuum of riverine, semiriverine, desert, and oasis peoples (see chapter 1) eventually became fragmented into separated entities. Warfare, mestizoization, land usurpation, and, presumably most important at first contact, population losses due to European diseases promoted this fragmentation. Displacement of other tribes into the Piman corridor during the early contact period probably assisted segmentation (Spicer 1962:86–92). What was perhaps once a continuous ribbon of mutually intelligible Piman languages in northwestern Mexico stretching from the modern state of Jalisco, Mexico, to southern Arizona, United States (see figure 1.1), became ever-diminishing clusters of speech communities, each with its own linguistic evolution during the past 100–200 or more years. Most of the tribes immediately along the Piman corridor were also Uto-Aztecan speakers (Cora, Huichol, Yaqui, Mayo, Tarahumara, Guarijío, Ópata, Eudeve). From a linguistic perspective, these probably facilitated the early diffusion of plant names throughout northwestern Mexico.

Except in a very few cases where they illustrate important processes of naming, crop varieties are omitted from this analysis. However, even though they are the same biological species, I am taking the liberty of including here as separate forms several plants with very different cultivars. The identity of the supposedly "wild" onion (*Allium cepa, aggregatum* group) was learned only in 1988, when several plants bloomed. It is conspecific with the common onion from Europe. Sugar sorgos *(Sorghum bicolor)*, called "sugar cane" by the Pima, arrived during the Hispanic period, but the conspecific grain sorghums became an important Pima crop apparently only well into the Anglo period. Native cotton, called Sacaton Aboriginal, was long considered a separate species, *Gossypium hopi*, but is now relegated to a line of *G. hirsutum.* The Pima contrast their aboriginal crop with the various short cultivars now grown abundantly in their region. By an ironic twist, what is called today "Pima cotton" is an Anglo cultivar that was bred at an agricultural research station on the reservation. Pima varieties of maize *(Zea mays)* are all 60-day forms. Those obtained from Anglo or Mexican sources are 90-day varieties. For statistical purposes, the *Allium, Sorghum, Gossypium,* and *Zea* sets have been treated here as if they were different species, lexically marked novelties.

As a result of European contact, at least 82 biological species of plants (see below) have entered the formal Gila Pima lexicon—some 36 during the Hispanic period (1694–1854) and 39 during the Anglo period (1854–present). We are still searching for clues as to when a few others arrived (5 species). Plants that entered the lexicon during the Hispanic period:

Allium cepa, aggregatum group, "wild" onion

Allium cepa, cepa group, onion

Allium sativum, garlic

Arundo donax (?), Giant Reed

Avena spp., Wild Oat, oat

Cicer arietinum, garbanzo, chick-pea

Citrullus lanatus, watermelon

Citrus limon, lemon

Citrus sinensis, orange

Cucumis melo, muskmelon

Erodium cicutarium, Filaree, Alfilaria

Ficus carica, domesticated fig

Hordeum leporinum, Common Foxtail, Wild Barley

Hordeum vulgare, barley

Lens culinaris, lentil

Lycopersicon esculentum, tomato

Malva parviflora, Cheese-weed

Medicago polymorpha, Bur Clover

Melilotus indica, Sweet Clover, Sour Clover

Nicotiana glauca, Tree Tobacco

Nicotiana tabacum, cultivated tobacco

Phoenix dactylifera, date palm

Pisum sativum, garden pea

Prunus armeniaca, apricot

Prunus domestica, plum, prune

Prunus persica, peach

Punica granatum, pomegranate

Pyrus communis, pear

Ricinus communis, Castor-bean

Saccharum officinarum, sugarcane

Sisymbrium irio, London Rocket

Sonchus asper, Spiny Sowthistle

Sorghum bicolor, sorgo, sweet sorghum

Triticum aestivum, wheat

Vigna unguiculata, black-eyed pea, cowpea

Vitis vinifera, domesticated grape

Plants that entered the lexicon during the Anglo period:

Ananas comosus, pineapple

Asparagus officinalis, asparagus

Beta vulgaris, beet

Brassica oleracea var. *capitata,* cabbage

Caesalpinia gilliesii, Yellow Bird-of-paradise

Cannabis sativa, marijuana, hemp

Capsicum annuum (some cultivars), chile

Citrus x *paradisi,* grapefruit

Convolvulus arvensis, Field Bindweed

Conyza canadensis, Horseweed

Cynodon dactylon, Bermuda Grass

Cyperus odoratus, C. rotundus, flat-sedges, Chufa

Daucus carota, carrot

Eucalyptus spp., eucalyptus, gum

Gossypium hirsutum (nonaboriginal cultivars), cotton

Helianthus annuus var. *macrocarpus,* Russian sunflower

Ipomoea carnea ssp. *fistulosa,* bush morning-glory

Lactuca sativa, lettuce

Lathyrus odoratus, sweet-pea

Ligusticum porteri, Oshá, Chuchupate

Maclura pomifera, Osage-orange

Malus pumila, apple

Medicago sativa, alfalfa

Melia azedarach, Chinaberry

Morus alba, mulberry

Musa x *paradisiaca,* banana

Opuntia ficus-indica, tuna, domesticated prickly-pear

Oryza sativa, rice

Phalaris minor, Littleseed Canary Grass

Polypogon monspeliensis, Rabbitfoot Grass

Populus nigra, Lombardy Poplar

Salsola australis, Russian-thistle, Tumbleweed

Solanum tuberosum, Irish potato, white potato (possibly earlier)

Sorghum bicolor, grain sorghum, milo maize

Sorghum halepense, Johnson Grass

Sorghum sudanense, Sudan Grass

Tamarix ramosissima, Salt-cedar

Tribulus terrestris, Puncture-vine, Bullhead

Zea mays (some cultivars, 90-day varieties), corn, maize

Plants for which the arrival time is unknown or uncertain:

Arachis hypogaea, peanut

Echinochloa colonum, Junglerice, Watergrass

Ipomoea batatas, sweet-potato, "yam"

Rumex crispus, etc., dock, sorrel

Sonchus oleraceus, Annual Sowthistle

Determination of arrival time was based on various sources of information, including analysis of plant remains from dated adobe bricks in Arizona, California, and Sonora (Hendry 1931; Hendry and Kelly 1925; O'Rourke 1983); documentary relations, particularly from the Jesuit period (Morrisey 1949); early herbarium specimens, which sometimes demonstrate clearly the progress of invasion by weedy species (Frenkel 1970); and for later introductions, direct oral history from native consultants included in the species accounts. For instance, the people with whom I worked still remembered when two of the now most abundant weedy species invaded their country: *Salsola australis* and *Tamarix ramosissima.* Some plants long believed to have been introduced Old World species have been demonstrated by archaeobotanists actually to be native (for example, *Portulaca oleracea*). Some native U.S. species seem to have invaded the Southwest (or Pima country in particular) only during historic times (for example, *Conyza canadensis*). These are considered here. Some of the crop species were available to the Gila Pima from the early mission period but appear to have been adopted only within the past century (for example, cabbage and lettuce). Their Pima names date from the Anglo period.

New crops sometimes arrived in advance of the Europeans themselves, and it is possible that names for them were passed along with the seed. Southern Pimans had at least a century of contact with established missions before the Jesuits reached Pimería Alta. When Kino and Manje reached southern Arizona, for example, they found crops such as melons already well established.

New species of plants were encoded into the Pima ethnobotanical inventory by several linguistic and perceptual methods. These processes are

1. Inclusion of low-salience referent without marking (extension)
2. Direct borrowing from Spanish and English (loan word)
3. Simple descriptive name (allegory)
4. Marking of familiar referent: old referent + 'similar to'; wild or aboriginal taxon + 'large'; anomalous taxon marked 'wild'; other types of comparisons

5. Transferal of name from old to new referent (replacement)
6. Marking reversal with referents in transition

As already noted, the Gila Pima show a high degree of conservatism and specificity with regard to the domains of their folk taxa. Characteristically, a name applies to one and only one biological species. Rarely does one encounter a folk taxon that maps to two or more biological species of plants, even among relatively difficult to distinguish groups such as grasses, *Boerhavia*, and *Atriplex*. Some exceptions are the spurges (*Euphorbia* or *Chamaesyce*) and *Cyperus.* In these cases, species identification usually requires a binocular dissection scope.

Not all plants new to Pima country (and apparently even some conspicuous native ones) have been named. Gileños will usually respond, when asked about an unnamed plant, that it is merely *ha'ichu hiósig* 'some flower' or *ha'ichu sha'i* 'some bush' but will hasten to add that that is not the plant's name, but it is "just a bush." With linguistic erosion, some former distinctions are now being lost. For instance, many speakers call all *Lycium* spp. by a single name, the generic, whereas four folk species were formerly distinguished. However, language loss is proceeding so rapidly among the younger generation that complete loss of Pima is now eclipsing erosional loss of specifics.

Languages differ in their relative amounts of linguistic acculturation by borrowing, extending meanings, marking old referents, and coining new lexemes. The late Joseph Giff, whose first language was Gila Pima, explained an important linguistic difference he perceived between the Akimel O'odham and Tohono O'odham. Where the Tohono O'odham were more likely to use a loan word from Spanish for a postconquest cultural item, the Gileños were more prone to invent a new word within the framework of Pima rules for word compounding. He provided numerous examples, indicating that he had broad knowledge of both Spanish and Tohono O'odham in addition to English and Pima. An example that occurs in a plant's name is illustrative. Tohono O'odham call the monkey *chaango*, from Spanish, whereas Gileños say *goks o'odham* 'dog person'. Thus, an ornamental bush morning-glory, with seeds resembling

monkey heads, is called *goks o'odham wuhiosha* 'monkey face' in Gila Pima.

The Hopi, a Puebloan group of northern Arizona, were missionized by Franciscans in 1629, more than 60 years before the first Jesuit entradas into Gileño country. Hopi was a possible source for new names as well as new plants coming into Pima country. However, Whiting's (1939) comprehensive ethnobotany suggests no evidence of plant names reaching the Pima via the Hopi. There are several similar loan words (*melóni, mansana, sió'uyi, tomáti*), but there is no reason to believe these were not independently derived. True neologisms differ completely between these two groups.

However, some of the newly coined names (for example, *chukuḍ shōsha*, dates; *gogoks viipiḍ*, prunes; *maamsh*, Castorbean) are shared between the Akimel and Tohono O'odham, indicating a common origin for these early folk taxa.

Process 1: Inclusion of Low-Salience Referent without Marking

In Pima, extensions of the domain of a word to include a new species are rare, perhaps because of the importance of such marking mechanisms as *chu'igam* ('it looks like'; see process 4). Also, the Pima are characteristically highly restrictive with domains, preferring a narrow domain even though that might exclude new but similar species. Occasionally a new plant is included within an existing lexeme without marking of any sort; the new referent never achieved sufficient cultural salience to merit marking or even a complete referent transferal. In each case, the new item is perceived as somehow similar to the original referent. Examples (given by Pima name, original referent, and extended domain) include

a'uḍ: Agave deserti, Ananas comosus

iikovĭ: Hoffmanseggia glauca, Ipomoea batatas

iivagĭ: wild greens eaten, Lactuca sativa

vihol: Pisum sativum, Lathyrus odoratus

These differ from simple inclusion of new species into a multispecies domain in that the compared plants are readily recognized as different. In the agave-pineapple case, the harvested heart of the wild native plant ready for pit roasting is similar in appearance to the pineapple fruit. Also, the baked agave and the pineapple are both very sweet, with a similar flavor. *Hoffmanseggia glauca* (Camote-de-raton) produces a small edible tuber; *Ipomoea batatas* (sweet-potato) is its larger analog. *Iivagĭ* is used polysemously for an important wild category as well as for lettuce, which is of minor importance to the Pima. Only the pea example comes close to an actual perceptual (as well as lexemic) domain extension—and both the crop *(Pisum sativum)* and the ornamental *(Lathyrus odoratus)* are introduced species.

Some new species were so similar to the already familiar ones that they were included within an existing folk taxon, apparently without comment or discrimination. These are not included in the calculations, because they involve no conscious attempt at linguistic acculturation. They are botanical subtleties rather than examples of intentional broadening of referents (given by Pima name, followed by the old referent and then the new referent):

baabkam: Phalaris angusta, P. minor

hōho'ipaḍ: Erodium texanum, E. cicutarium

vaapk: Phragmites australis, Arundo donax

vakoandam: Rumex violascens, etc., *R. crispus,* etc.

vashai s-uuv: Cyperus esculentus, C. rotundus, C. odoratus

Even some of these cases may be simplifications. Perhaps at the initial introduction the new species was linguistically marked. There is some evidence of this with the reeds: *Arundo* may have been marked *kauk vaapk* when it first appeared, but about this time the rivers were drying and *Phragmites,* a riparian species, almost totally disappeared. This may be a case of simple substitution of one referent for another. With the others, also, there may have been a restriction to the original referent that has broken down as people have become less familiar with their natural environment. (The *Sisymbrium* case discussed in connection with process 5 is an example of this.)

Process 2: Direct Borrowing from Spanish and English

The simplest and most direct way for any new organism to enter a folk taxonomy formally is through direct lexical borrowing, or as a loan word. The loan words in the Gila Pima plant lexicon derived from Spanish are

sivol: cebolla

aahō: ajo

kakawáádi: cacahuate

chuális: chuale

kalvash: garbanzo

miiliñ, miiloñ: melon (not *sandía*)

nalash: naranja

suuna: tuna

sivááyu: cebada

kamóódi: camote

laanji, laanjikĭ: lenteja

tomáádi: tomate

sanwán: palo San Juan

vilgóódi: albaricoque

ñulash: durazno

kalnááyu: granada

kaañu: caña

pilkañ: trigo

[h]uuḍvis: uva

jujubáádi: chuchupate

baabas: papas

The loan words in the Gila Pima plant lexicon derived from English are

aabals: apples

kabij: cabbage

maliwáána: marijuana (Spanish via English?)

olinj: orange

With the Pima, an interesting pattern emerges. Although the number of newly named plants is almost the same between those arriving in the Hispanic and Anglo periods (36 vs. 39), there is a disproportionate number (20 vs. 3 or possibly 4) borrowed from Spanish rather than English. The Tohono O'odham use even more loan words from Spanish. This is because they were geographically more proximate to Hispanic settlements than

were the Gila Pima and because they had longer temporal contact with that language (until well into the 20th century). Even today, an older Tohono O'odham is occasionally encountered whose second language is Spanish rather than English.

Most of the Spanish loans appear to have been coined broadly throughout at least the northern half of Piman country, indicating rapid acceptance of these plants during the Hispanic period. Even groups, such as the Gila Pima, that were not directly involved in the mission system sent representatives considerable distances to Piman missions. These representatives brought back many crops. Such loan words, with their dialectical differences, are found among Pima Bajo, Tohono O'odham, and Gila Pima, the three speech communities that have survived into modern times. Ethnobotanical studies among the Mountain Pima, Northern Tepehuan, and Southern Tepehuan may ultimately show that many of the Hispanic loans are pan-Piman.

Some of the Spanish-derived names may have undergone several phonological mutations as they were passed northward from one group to another: *vilgóódi* from *albaricoque* (see *Prunus armeniaca,* Group J) and *kalnááyu* from *granada* (see *Punica granatum,* Group J) are examples. In spite of phonological modifications, a number of the Spanish loans still violate the Piman pattern of accenting the first syllable (see preceding list).

Without doubt, the most important crop introduced during the Hispanic era was wheat. Its Pima name, *pilkañ,* is of Spanish origin (Herzog 1941). Starting from Spanish *trigo* and moving through the Uto-Aztecan languages along the Tepiman corridor we have reconstructed the cognate sequence *tiricco, tiligo, pirikó, perikon,* and *pilkañ* (described more fully in the species account; *Triticum aestivum,* Group I). Similarly, although to a lesser extent than wheat, the domesticated (Old World) fig tree (in Spanish, *higo*) spread widely with the conquest. In Yaqui it is *chuuna,* Tarahumara *chuná,* Ópata and Eudeve *chúna;* but among colonial Pimería Alta speakers, then lacking the *ch,* fig became *suuna. Tuna,* a word widespread in New World Spanish that identifies the edible fruit of prickly-pear (particularly *Opuntia ficus-indica*), is supposedly of Caribbean derivation. Ethnobotanist Robert A. Bye suspects that as Spaniards dispersed northward with both the fig and the cultivated cactus, native peoples applied the name from one to the other, the communality being fleshy fruits with "seedy" interiors. Similarly, Spaniards at contact called the fruit of the nopal *higo de tuna* 'fig of the *tuna* [or] prickly-pear' or *higo índico* (Hernández 1959, 2:311).

Several plant names appear to have originated in Nahuatl, a dominant Uto-Aztecan language of Mexico, became incorporated into Mexican Spanish, then were reintroduced into Piman. Piman *jujubáádi*, from Spanish *chuchupate,* is apparently from Nahuatl *chuchupatle*. Piman *chuális,* from Spanish *chuale,* may be derived from Nahuatl *tzoalli*. Piman *wahai* or *huáhi* may be cognate with another Nahuatl quelite, *uauhtli*. Piman *kakawáádi*, by way of Spanish *cacahuate,* comes from Nahuatl *tlalcacahuatl*. Piman *kamóódi* likewise comes from Nahuatl *camohtli* by way of Spanish *camote*.

The loan word *miiliñ* (among the western villages) or *miiloñ* (eastern villages) is more complicated than appears at first sight. In Spanish, *sandía* labels only watermelons (*Citrullus lanatus),* and *melón* labels the various muskmelons (*Cucumis melo)*. In Gila Pima, *do'ig hugĭ* as well as *miiliñ/miiloñ* are the generic for watermelon, of which there were at least four named cultivars, two including the lexeme *miiloñ*. And *s-uuv miiloñ/miiliñ* 'smelly melon' is the generic for muskmelon, of which there were at least six named cultivars, some including the word *miiloñ*. Thus the loan word is used polysemously in both crop species, even though the Pima unequivocally segregate the two crops conceptually. This is a good example of where it is critical to discover the emic domains of words and not rely exclusively on purely lexical data; the latter do not necessarily reflect the speakers' concepts.

Few loan words have entered Pima from English in spite of long contact with a nearby major Anglo settlement (Phoenix) and several smaller ones. Among plant names, *aabals* is the one in most frequent use. Older Pimas recognize the name *mansáána*, from Spanish, as used by the Tohono O'odham, but they are quick to point out, "That's a Papago word, not Pima." Most Pimas use the English loan for cabbage, as noted by Herzog (1941), while others use a descriptive term *chichino iivagga,* 'Chinese [pl.] greens' (see process 3).

Gila Pima *maliwáána* is ultimately derived from Spanish but entered the Pima lexicon quite late in the Anglo period, probably as much from English speakers and the media as from Mexican field workers. The plant has had little cultural significance.

The orange is labeled with loan words from both Spanish and English *(nalash* and *olinj).* I think *nalash* predominates and is the form offered when the Pima name for this citrus is asked.

Process 3: Simple Descriptive Name

With Gila Pima, there is a tendency to use descriptive names rather than direct loan words. Some 40 descriptive folk taxa map to 32 biological species. (Folk synonymies account for the numerical discrepancies.) Several of these descriptive plant names contain Spanish loan words: *koli* (from *corral*), *tool* (*toro;* bull), *chichino* (*Chinos;* Chinese). Spicer (1943:412) noted a similar root compounding in Yaqui. Examples of plant names with their descriptive glosses include

akshpĭ muḍadkam 'backwards tassels': *Avena barbatus*

aatoks muḍadkam 'hanging down tassels': *Avena* spp.

koksham 'it has a coat/jacket': *Avena* spp.

s-vepeg daadaidkam 'red roots it has': *Beta vulgaris*

chichino iivagga 'Chinese [pl.] their greens': *Brassica oleracea* var. *capitata*

aadŏ o'ohoḍag 'macaw/peacock's color pattern': *Caesalpinia gilliesii*

s-oám eḍakam 'yellow insides': *Citrullus lanatus*

s-oám eldakam 'yellow skins it has': *Citrus* x *paradisi*

s-he'ekchu 'sourness': *Citrus limon*

vopōksha 'quiver' or 'stepchildren': *Conyza canadensis*

keli baasho 'old man's chest': *Cucumis melo*

oámakam 'yellow it has': *Cucumis melo*

a'ai hiimdam vashai 'it spreads out in all directions grass': *Cynodon dactylon*

kii wecho vashai 'beneath house grass': *Cynodon dactylon*

komal himdam vashai 'it spreads out in a direction grass': *Cynodon dactylon*

s-oám daadaidkam 'yellow/orange roots it has': *Daucus carota*

s-o'oi vashai 'striped/spotted grass': *Echinochloa colonum*

chev uus 'tall tree': *Eucalyptus* spp.

s-vepegĭ iibadkam 'red fruit it has': *Fragaria* sp.

koson bahi 'packrat's tail': *Hordeum leporinum*

sheshelik baabhai 'ground squirrels' tails': *Hordeum leporinum*

goks o'odham wuhiosha 'monkey's face': *Ipomoea carnea* var. *fistulosa*

s-ho'idkam koli 'thorny it has corral': *Maclura pomifera*

ge'egeḍ haahagam 'big leaves it has': *Malva parviflora*

s-taḍañ haahag '[it has] broad leaves': *Malva parviflora*

s-eehekak 'very shady': *Melia azedarach*

sipuḍ s-wulvañ 'anus wrinkled': *Morus alba* (some speakers, *Fragaria* sp.)

haivañ a'ag 'cow's horns': *Musa* x *paradisi-aca*

totoñ maamaḍ 'ants babies [pupae]': *Oryza sativa*

akimel baasho vashai 'riverbank grass': *Paspalum* sp.?

chukuḍ shosha 'Great Horned Owl's snot': *Phoenix dactylifera*

ban bahi 'Coyote's tail': *Polypogon monspeliensis*

shelik bahi (pl. *sheshelik baabhai*) 'ground squirrel's tail': *Polypogon monspeliensis*

gogoks viipiḍ 'dogs' testicles': *Prunus domestica*

chuchul maamaḍ 'chicken babies': *Pyrus communis*

maamsh 'ticks': *Ricinus communis*

vopodam sha'i 'rolling brush': *Salsola australis*

s-ho'idkam iivagĭ 'thorny greens': *Sonchus asper*

huai hehevo 'Mule Deer's eyelashes': *Sonchus oleraceus*

vepegĭ u'us 'red sticks/branches': *Tamarix ramosissima*

tool a'ag 'bull horns': *Tribulus terrestris*

Process 4: Marking of Familiar Referent

A common method incorporating a new item into the lexicon is by linguistically marking a familiar referent. The marked or modified referent implies some sort of relationship to the unmarked original form. Marking can involve any hierarchical level from life form down. Gila Pima 'tree' (*uus keekam*) is an example of marking at the life-form level. Here *uus* means 'stick' or 'branch', *keek* means 'standing', and *-kam* is an attributive, so that the phrase *uus keekam* means 'wood whose property it is to be standing upright' (see introduction to Group E). Varietals are typically marked, but these are rare in folk taxonomy. Marking as a process in biological folk nomenclature has been discussed by Brown (1984, see especially pp. 83–97). Marking may not be the final linguistic solution to naming a new biological species (see process 6). The Pima employ several different classes of linguistic marking to name newly acquired plants.

Old Referent + 'Similar To'

One formula uses a simple comparison: some old referent + *chu'igam* 'like' or 'similar to'. The comparison may range from quite realistic (a new *Malva*, *ñiádam chu'igam*, looks like a native mallow) to something fanciful (the name for pear, *vakoa chu'igam*, means 'bottle-gourd it looks like'). Although the folk etymology is given as 'it looks like', Culver Cassa notes that the Pima term implies qualitative similarities rather than physical similarities; hence 'is like' is a more precise rendering of the meaning. There are at least ten such folk taxa recorded:

a'ud chu'igam 'agave [heart] it looks like': *Ananas comosus*

mo'otadk chu'igam 'Orobanche it looks like': *Asparagus officinalis*

ñiádam chu'igam 'Eremalche it looks like': *Malva parviflora*

s-puluvĭ chu'igam vashai 'Sweet Clover it looks like': *Medicago sativa*

hovij chu'igam 'Banana Yucca it looks like': *Musa* x *paradisiaca*

i'ipig chu'igam 'erection it looks like': *Musa* x *paradisiaca*

vakoa chu'igam 'bottle-gourd it looks like': *Pyrus communis*

kaañu chu'igam 'sweet sorghum it looks like': *Sorghum halepense*

vaapk chu'igam [vashai] 'Carrizo it looks like [grass]': *Sorghum halepense*

kaañu chu'igam vashai 'cane it looks like grass': *Sorghum sudanense*

Wild or Aboriginal Taxon + 'Large'

There is another formula where the robust domesticated form is marked *ge'eged* 'large' + wild plant name:

ge'eged hivai: Russian sunflower (*Helianthus annuus* var. *macrocarpus*)

ge'eged iibhai: domesticated prickly-pear (*Opuntia ficus-indica*)

ge'eged ko'okol: bell pepper (*Capsicum annuum*)

ge'eged huuñ: 90-day corn (commercial corn) (*Zea mays*)

Ge'eged ko'okol is but one of seven cultivated chiles acquired in postcontact times. The bell pepper is the largest of these. The *ko'okol* itself, the tiny wild Chiltepín, is now often secondarily marked *olas ko'okol* or *al ko'okol*, 'round' or 'small' chile. *Ge'eged huuñ* and *kauk huuñ*, both 90-day corn, contrast with seven named varieties of native maize, which are all 60-day forms. Unmarked, *hivai* names the Wild Sunflower and *iibhai* indicates the local wild prickly-pears. The cultivated forms are classified differently from their wild relatives.

Anomalous Taxon Marked 'Wild'

Another way of coining a name is with the 'anomalous wild' term *mischiñ* + a comparative term. *Mischiñ* signifies something that is expected to be domesticated but for some reason is "wild," somewhat in the English sense of feral or unbroken (in the case of horses). My Dutch Sonoran friend and scholar Dr. Eric Mellink puzzled for some time over the word, then sent me this analysis: "*Mistien [mischiñ]* seems to be coming from *mesteño*, which comes from *mesta*, and means without a known owner; it is said of feral animals. *Mesta* was the Spanish association of cattlemen that took care of the growing of domestic stock, large and small, and nomadic grazing. The word derives from *mixta* and *micēre* (Lat., 'to mix'). *Mestizo* derives from the same word. My idea of mustang deriving from *mesteño* is also held by the dictionary. A Spanish synonym of *mesteño* is *mostrenco*, currently in use in Sonora." This Piman word has quite a pedigree.

The Pima plant names that include *mischiñ* are *mischiñ siivol*, "wild" onion (*Allium cepa*, *aggregatum* group), and *mischiñ uudvis*, Arizona Grape (*Vitis arizonica*). Curiously, these cases involve a Spanish loan word as the compared species base: *uudvis* from Spanish *uva* 'grape' and perhaps *vid* 'grapevine', and *siivol* from Spanish *cebolla*. There may once have been original names for both the native wild grape and the true native onion, both species found upriver at higher elevations from the Pima. The important point is that the organism is perceived as being somehow anomalous: the native Arizona or Canyon Grape, *Vitis arizonica*, now encountered during Pima travels outside the desert, is different from the more familiar cultivated grape, *Vitis vinifera*, known since early contact times.

Other Types of Comparisons

Some marking involves a comparison with another crop. While the new referent is similar to the familiar species, it is not a "kind of" the compared base. Some examples of Old World crops named in this manner are

chuchul huuñ 'chicken maize': *Sorghum bicolor* (milo)

huuñ kaañu 'maize cane': *Sorghum bicolor* (milo)

s-o'oi kaañu 'spotted/striped cane': *Saccharum officinarum*

u'us bavĭ 'stick tepary': *Vigna unguiculata*

Several varieties of a native bean named *baví* (tepary, *Phaseolus acutifolius*) were grown in Piman country. When black-eyed pea *(Vigna unguiculata)* was introduced, it was compared to a tepary with sticklike pods. When the kinds of tepary are elicited, however, *u'us baví* is excluded: *u'us baví* is not a kind of tepary for the Gila Pima any more than, in English, milo maize is a kind of corn or prickly-pear is a kind of pear.

In a very few cases, though, the marked form apparently does imply a kind of the unmarked base. Two examples (both involving the same biological species, *Populus nigra*) are *kavichk auppa* 'narrow cottonwood' and *moomli auppa* 'Mormon cottonwood'. The compared base or unmarked form is *auppa*, Frémont's Cottonwood *(Populus fremontii)*, a common native species.

Process 5: Transferal of Name from Old to New Referent

This section is closely related to process 1 (extension) and process 6 (marking reversal) but differs in that no marking is involved and that the two referents are not simultaneously included. In the following list, the Pima name is given first, followed by the old referent and then the new referent.

shuu'uvaḍ: Descurainia pinnata, Sisymbrium irio

(?) *naví:* any prickly-pear (*Opuntia* spp.), *Opuntia leptocaulis*

(?) *biibhi'ag:* some native vine (*Ipomoea?*), *Convolvulus arvensis*

The first case is the clearest. Introduced *Sisymbrium*, a weedy mustard, is now much more abundant and conspicuous than *Descurainia*, a native mustard. Most families of Pima, Tohono O'odham, and Pima Bajo now refer to *Sisymbrium* as *shuu'uvaḍ*. However, some Gila Pimas will say, "Well, *he* may call that plant *shuu'uvaḍ*, but the *real shuu'uvaḍ* is this one [*Descurainia*]." This appears to be a transferal in progress, with some still remembering the original referent and denying the new one. I have never found anyone who calls both the native and the introduced species *shuu'uvaḍ*.

The *Opuntia* situation is less clear. Among southern Pimans, *nav* or *naví* is the base for naming any flat-stemmed *Opuntia*, or prickly-pear. Yet some Pimas use *a'aji naví* as the name for *Opuntia leptocaulis*, a small cholla. This apparent anomaly is discussed in detail in the *O. engelmannii* account, Group H.

The lexeme *biibhi'ag* occurs among all the Piman groups I have studied and is applied to some specific plant that is characteristically viny. Etymologically the word might be glossed 'entwineness'. The Gila Pima now use the name for a European species that is an important weedy pest in their fields. I suspect that it was simply transferred from some native vine, perhaps *Convolvulus equitans* or an *Ipomoea*, that is now locally extirpated due to changes in land use. There are probably more examples of this naming process, but our time depth on folk taxonomies is extremely shallow.

Process 6: Marking Reversal with Referents in Transition

Marking reversal denotes an evolutionary process in linguistic acculturation described by Witkowski and Brown (1983). The name for a new cultural item is initially the marked form of some similar preexisting item (see example in Group E, *Maclura pomifera* account). Any of the plant names involving marking (process 4, 5, or 6) might become a candidate for marking reversal. Surprisingly few Pima plant names involve this process, and these appear to be in transition, that is, the marking is not yet driven to completion:

ihug, ban ihugga (3 biological taxa)

viv, ban vivga (4 biological spp.)

s-puluví, spuluví chu'igam vashai

toki, che'echev toki, maskal toki

The first of these, dealing with the fiber plant *Proboscidea*, devil's claw, has been discussed at length by Nabhan and Rea (1987). This marking reversal involves a root perennial wild species with inferior fiber, an annual black-seeded wild species of intermediate quality, and a domesticated white-seeded variety of the annual species (see *Proboscidea* accounts, Groups G and I). Domestication appears to be so recent that the marking reversal is in transition (Nabhan and Rea 1987).

Some people call both wild species *ban ihugga*, reserving *ihug* for the domesticate. Most now call only the yellow-flowered perennial *ban ihugga*, while including both varieties of the pink-flowered species (black-seeded and white-seeded) in the unmarked lexeme *ihug*. We presume that the original contrast was between wild yellow *ban ihugga* with small claws and wild pink *ihug* with longer claws.

The tobacco situation is even more complicated because four biological species are involved, early investigators failed to prepare voucher specimens for the folk taxa or failed to elicit complete contrast sets, and most Pimas who knew the distinctions are now dead. At the simplest level, *viv* and *ban vivga* contrast a cultivated and a wild species of *Nicotiana*. But eventually there were four species available to the Pima: *N. trigonophylla*, which is wild; *N. attenuata*, supposedly wild but perhaps semicultivated or even cultivated; *N. tabacum*, a cultivated form; and *N. rustica*, another cultivated form (see *Nicotiana* accounts, Groups G and I). Some contemporaries distinguish *ban vivga* as the inferior cultivar and *viv* as the superior cultivar, usually the commercial product. There are additional labeled forms, but no one seems to remember a complete biological contrast set.

The marking *ban* 'Coyote' is used to designate forms of a plant, usually wild, that are perceived to be related to a domesticated form, as for *ban ihugga* and *ban vivga*. These forms are also perceived as inferior or perhaps completely useless. This appears to be a pan-Piman convention (Nabhan 1981). In some cases, the wild species is the progenitor of the domesticate and back-crossing still occurs between the wild and adjacent cultivated forms.

Some Pima distinguish *s-puluví* (Sweet Clover, *Melilotus indica*), introduced early in the Hispanic period, from *s-puluví chu'igam vashai* 's-puluví it looks like grass' (alfalfa, *Medicago sativa*). Some call alfalfa *kiikam s-puluvam* 'about the house s-puluv', indicating its cultivated status in contrast to *Melilotus*, which grows wild. Others now apply *s-puluv* directly to alfalfa, a crop introduced later in the Anglo period. And still others apply *vashai* polysemously to alfalfa and the life form "grassy plants."

The cotton example well illustrates the intermediate stage of a marking reversal. *Toki* is the folk generic for any cotton (*Gossypium* spp.). A very fine quality of aboriginal cotton was an important Gileño crop. It dropped out of cultivation during the early quarter of the 20th century. Some decades later, Anglo-Americans developed a short cultivar that could be harvested mechanically. This the Pima called *maskal toki* 'silky cotton'. They then called their own cultivar *che'echev toki* 'tall cotton'. Because only short cotton is grown today, the unmarked generic is now used in most contexts for all commercial cotton.

Among the Tohono O'odham, the wild or native form of a plant that has undergone a marking reversal is sometimes called I'itoi's plant. I'itoi is the major culture hero (usually called Se'ehe among the Gila Pima). The sense is that his plant is the aboriginal form in contrast to the unmarked introduced form. I have found no examples of such labeling among the Akimel O'odham.

One might add one more method of accommodating new biological species, a nonlinguistic one, and that is simply ignoring the species in the native lexicon. There are quite a number of highly visible, sometimes even culturally salient plants that fall into this category now on the Gila River Reservation, including Athel Tamarisk *(Tamarix aphylla)* and olive *(Olea europea)* (see appendix table A.3). Many of these are late 19th or early 20th century arrivals, so it is doubtful a name has been coined and already lost. Olives were introduced to Pimans to the south during the Jesuit period but seem not to have been accepted at the northern periphery of Pimería Alta until recently. They require a processing that the Gileños apparently never undertook. But the other plants exhibiting this type of non-naming are common throughout the rancherias. Reference to them is made only in English, with no attempt at forcing the English name into Piman phonetic structures.

The Gila Pima have encoded into their folk taxonomy probably all Hispanic-introduced crops that were accepted and most significant weeds. Some weedy species arriving during the subsequent Anglo period have been named, while others have not. In this regard it must be pointed out that not every native plant in the Pima's environment is named. Some common plants, such as *Tidestromia lanuginosa* and native species of *Bromus* are unnamed (see appendix table A.4).

Variation within the Gila Pima Speech Communities: Folk Synonymies

Synonymies (simultaneous use of two or more different names for the same biological species) for native plants are almost unknown among Gileño Pima speech communities. There are variants of the same or a closely related lexeme, such as *vakoa hai, vakoa hahaisig,* and *vakoa hahaiñig* for White Horse-nettle *(Solanum elaeagnifolium).* I have heard three variants of the name for dodder (*Cuscuta* spp.): *vamaḍ giikoa* 'snake's crown', *vamaḍ givuḍ* 'snake's belt', and *vamaḍ vijina* 'snake's thread [or] web'. And there are labeled varietal names (folk specifics) for variable cultivars such as for teparies *(Phaseolus acutifolius)* and maize. But I know of no true folk synonym for an indigenous plant that is not clearly the result of a collapsing taxonomy (and even those cases are very rare and idiosyncratic). The degree of concordance in naming between Pima and Tohono O'odham cannot be worked out because no one has done a comprehensive ethnobotany of the Tohono O'odham groups yet. Some interdialectical folk synonymies exist (see Appendix Table A.2).

In contrast, synonymies are relatively common among introduced species. Excluding variations in the names for cultivars, 8.3% of the Hispanic-introduced flora and 17.9% of the Anglo-introduced flora have synonymies. Some biological species may be named by double synonyms, others by triple. Only two species of plants that arrived during the Hispanic period have three currently used synonyms, while three plants from the Anglo period have three current names; no plant has more than three. Considering the total number of synonyms, 18.6% of the names for plants presumed to have originated during the Hispanic period are reduplicative (name the same biological species), and 34.7% of the names for plants presumed to have originated during the Anglo period are reduplicative. In other words, there are fewer Pima names in use for plants acquired during the early contact period than there are for those acquired more recently in history. Note that these are alternative names used in the speech community that in no case overclassify a biological species.

Of some 97 Pima names that have been applied to 81 plant species or groups new to Pima country, 4 involve a conscious extension of a domain without marking, 4 are probably unintentional domain extensions involving cryptic species, 24 are loans (20 Spanish, 4 English), 40 are neologisms or new descriptive names, 20 involve marking of some other plant referent, 4 involve marking reversals in transition, and at least 1 is a name transferal not yet gone to completion. The idea that the Gila Pima depend strongly on inventing new descriptive names is borne out: these are twice as common as marking (41 vs. 20, or 24 including the reversals) and almost twice as common as loans (41 vs. 24). There are five times more loan words from Spanish than from English.

New plant names were being incorporated into the Pima lexicon as late as the 1920s and 1930s, but no new plants seem to have been named since then, even though some of these species are now both conspicuous and abundant and sometimes troublesome. Substitution of pure English, unmodified, would be used for these by Pima speakers.

Bibliography

Alamán, Lucas. 1825. Memoria sobre los Pimas Gileños y Cocomaricopas. Unpub. ms. dated 17 July 1825, Archivo Militar, Mexico City.

Alcock, John. 1987. The lively death of a Saguaro cactus. Pacific Discovery 40:28–33.

Aldrich, Lorenzo D. 1950. A Journal of the Overland Route to California and the Gold Mines. Dawson's Book Shop, Los Angeles.

Allyn, Joseph P. 1974. The Arizona of Joseph Pratt Allyn: Letters from a Pioneer Judge. John Nicolson (editor). University of Arizona Press, Tucson.

Annual Report. 1863. Annual Report to the Commissioner of Indian Affairs for the Year 1863. Washington, D.C.

Antisell, Thomas. 1856. Geological report. Pp. 1–204 in vol. 7 of Explorations and Surveys for a Railroad Route from the Mississippi River to the Pacific Ocean. 33d Cong., 2d sess., House Exec. Doc. 91.

Antonine, Fr. See Willenbrink, Antonine.

Arlegui, José de. 1737 (n.v.). Crónica de la provincia de N.S.P.S. Francisco de Zacatecas. Reprint, Mexico City, 1851.

Armstrong, Wayne P. 1980a. Sand food: A strange plant of the Algodones Dunes. Fremontia 7:3–9.

_____. 1980b. More about sand food. Fremontia 8:30–31.

Attí vocabulary. ca. 1774. Unpub. partial ms., M-M475 Bancroft Library.

Audubon, J. W. 1906. Audubon's Western Journal, 1849–1850. Maria R. Audubon (editor). Arthur H. Clark, Cleveland.

Bahr, Donald M. 1983a. Pima and Papago social organization. Pp. 178–192 in Handbook of North American Indians, vol. 10, Southwest. Alfonso Ortiz (editor). Smithsonian Institution, Washington, D.C.

_____. 1983b. Pima and Papago medicine and philosophy. Pp. 193–200 in Handbook of North American Indians, vol. 10, Southwest. Alfonso Ortiz (editor). Smithsonian Institution, Washington, D.C.

Bahr, Donald M., Joseph Giff, and Manuel Havier. 1979. Piman songs on hunting. Ethnomusicology 23:245–296.

Bahr, Donald M., Juan Gregorio, David Lopez, and Albert Alvarez. 1974. Piman Shamanism and Staying Sickness (Ka:cim Mumkidag). University of Arizona Press, Tucson.

Bahr, Donald M., and J. Richard Haefer. 1978. Song in Piman curing. Ethnomusicology 22:89–122.

Bahre, Conrad Joseph. 1991. A Legacy of Change: Historic Human Impact on Vegetation of the Arizona Borderlands. University of Arizona Press, Tucson.

Bancroft, Hubert H. 1886. The Native Races [of the Pacific States]. The Works of Hubert Howe Bancroft, vol. 3. The History Co., San Francisco.

_____. 1889. History of Arizona and New Mexico, 1530–1888. The Works of Hubert Howe Bancroft, vol. 17. The History Co., San Francisco.

Bandelier, Adolph F. A. 1884. Reports by A. F. Bandelier on his investigations in New Mexico during the years 1883–84. Annual Report of the Executive Committee of the Archaeological Institute of America 5:55–98.

_____. 1890–1892. Final Report of Investigations among the Indians of the Southwestern United States, Carried on Mainly in the Years from 1880–1885. 2 vols. Papers, Archaeological Institute of America, American Series 3 and 4. Cambridge, Mass.

Bartlett, John Russell. 1854. Personal Narrative of Explorations and Incidents in Texas, New Mexico, California, Sonora, and Chihuahua. 2 vols. D. Appleton, New York.

Basso, Keith H. (editor). 1971. Western Apache Raiding and Warfare from the Notes of Greenville Goodwin. University of Arizona Press, Tucson.

_____. 1983. Western Apache. Pp. 462–488 in Handbook of North American Indians, vol. 10, Southwest. Alfonso Ortiz (editor). Smithsonian Institution, Washington, D.C.

Bates, John M. 1992. Frugivory on Bursera microphylla (Burseraceae) by wintering Gray Vireos (Vireo vicinior, Vireonidae) in the coastal deserts of Sonora, Mexico. Southwestern Naturalist 37:252–258.

Bauer, R. W. 1971. The Papago cattle economy: Implications for economic and community development in arid lands. Pp. 79–102 in Food, Fiber, and the Arid Lands. W. G. McGinnies, B. J. Goldman, and Patricia Paylore (editors). University of Arizona Press, Tucson.

Bauml, James A., Gilbert Voss, and Peter Collings. 1990. 'uxa identified. Journal of Ethnobiology 10:99–101.

Beals, Ralph L. 1933. The Acaxee: A mountain tribe of Durango and Sinaloa. Ibero-Americana 6:1–36.

Bean, Lowell John, and Katherine Siva Saubel. 1972. Temalpakh: Cahuilla Indian Knowledge and Usage of Plants. Malki Museum Press, Morongo, Calif.

Beauchamp, R. Mitchel. 1986. A Flora of San Diego County, California. Sweetwater River Press, National City, Calif.

Behnke, Robert J., and Robert F. Raleigh. 1978. Grazing and the riparian zone: Impact and management perspectives. Pp. 263–267 in USDA General Technical Report WO-12. R. Roy Johnson and J. Frank McCormick (editors). U.S. Forest Service, Washington, D.C.

Bell, William A. 1869. New Tracks in North America: A Journal of Travel and Adventure Whilst Engaged in the Survey for a Southern Railroad to the Pacific Ocean during 1867–8. Chapman and Hall, London.

Bell, Willis H., and Edward F. Castetter. 1937. Ethnobiological studies in the American Southwest V: The utilization of mesquite and screwbean by the aborigines in the American Southwest. University of New Mexico Bulletin 314. University of New Mexico Biological Series 5:1–55.

Bennett, Wendell C., and Robert M. Zingg. 1935. The Tarahumara: An Indian Tribe of Northern Mexico. University of Chicago Press, Chicago.

Benson, Lyman. 1969. The Cacti of Arizona. 3d edition. University of Arizona Press, Tucson.

Benson, Lyman, and Robert A. Darrow. 1981. Trees and Shrubs of the Southwestern Deserts. 3d edition. University of Arizona Press, Tucson.

Berlin, Brent. 1972. Speculations on the growth of ethnobotanical nomenclature. Language and Society 1:51–86.

_____. 1992. Ethnobiological Classification: Principles of Categorization of Plants and Animals in Traditional Societies. Princeton University Press, Princeton, N.J.

Berlin, Brent, Dennis E. Breedlove, and Peter H. Raven. 1973. General principles of classification and nomenclature in folk biology. American Anthropologist 75:214–242.

Berry, Claudia F., and William S. Marmaduke. 1982. The Middle Gila Basin: An Archaeological and Historical Overview. U.S. Interior Department, Bureau of Reclamation, Central Arizona Project, Indian Distribution Division, Phoenix.

Berry, Wendell. 1981. The Gift of Good Land: Further Essays Cultural and Agricultural. North Point Press, San Francisco.

Bigler, Henry William. 1962. Bigler's Chronicle of the West: The Conquest of California, Discovery of Gold, and Mormon Settlement as Reflected in Henry William Bigler's Diaries. Erwin G. Gudde (editor). University of California Press, Berkeley.

Blake, Leonard W. 1981. Early acceptance of watermelons by Indians of the United States. Journal of Ethnobiology 1:193–199.

Bohrer, Vorsila L. 1970a. Ethnobotanical aspects of Snaketown, a Hohokam village in southern Arizona. American Antiquity 35:413–431.

_____. 1970b. Paleoecology of Snaketown. Kiva 36:11–15.

_____. 1975. The prehistoric and historic role of the cool-season grasses in the Southwest. Economic Botany 29:199–207.

_____. 1991. Recently recognized cultivated and encouraged plants among the Hohokam. Kiva 56:227–235.

Bolton, Herbert Eugene. 1930. Anza's California Expeditions. 5 vols. University of California Press, Berkeley.

_____. 1936. Rim of Christendom: A Biography of Eusebio Francisco Kino, Pacific Coast Pioneer. Macmillan, New York. Reprinted 1984, University of Arizona Press, Tucson.

Bourke, John G. 1891. On the Border with Crook. Charles Scribner's Sons, New York. Reprinted 1962, Rio Grande Press, Chicago.

Bowden, Charles. 1977. Killing the Hidden Waters. University of Texas Press, Austin.

Brand, Donald D. 1971. The sweet potato: An exercise in methodology. Pp. 343–365 in Man across the Sea: Problems of Pre-Columbian Contacts. Carroll L. Riley, J. Charles Kelly, Campbell W. Pennington, and Robert L. Rands (editors). University of Texas Press, Austin.

Breninger, George F. 1901. A list of birds observed on the Pima Indian Reservation, Arizona. Condor 4[=3]:44–46.

Bretting, Peter K. 1981. A Systematic and Ethnobotanical Study of Proboscidea and Allied Genera of the Martyniaceae. Ph.D. diss., Botany Department, Indiana University, Bloomington.

_____. 1982. Morphological differentiation of Proboscidea parviflora ssp. parviflora (Martyniaceae) under domestication. American Journal of Botany 69:1531–1537.

_____. 1986. Changes in fruit shape in Proboscidea parviflora ssp. parviflora (Martyniaceae) with domestication. Economic Botany 40:170–176.

Bringas de Manzaneda y Encinas, Diego Miguel. 1977. Friar Bringas Reports to the King: Methods of Indoctrination on the Frontier of New Spain, 1796–97. Daniel S. Matson and Bernard L. Fontana (editors and translators). University of Arizona Press, Tucson.

Brown, Cecil H. 1984. Language and Living Things: Uniformities in Folk Classification and Naming. Rutgers University Press, New Brunswick, N.J.

Brown, David E. 1978. The vegetation and occurrence of chaparral and woodland flora on isolated mountains within the Sonoran and Mohave deserts in Arizona. Journal of the Arizona-Nevada Academy of Science 13:7–12.

Brown, Herbert. 1900. The conditions governing bird life in Arizona. Auk 17:31–34.

_____. 1906. A Pima-Maricopa ceremony. American Anthropologist 8:688–690.

Browne, J. Ross. 1864. A tour through Arizona. Harper's New Monthly Magazine 29 (173): 553–574.

_____. 1869. Adventures in the Apache Country: A Tour through Arizona and Sonora, with Notes on the Silver Regions of Nevada. Harper & Brothers, New York. Reprinted, in part, 1974, University of Arizona Press, Tucson.

Brugge, David M. 1961. History, huki, and warfare: Some random data on the Lower Pima. Kiva 26:6–16.

Brush, Stephen B. 1986. Genetic diversity and conservation in traditional farming systems. Journal of Ethnobiology 6:151–167.

Brush, Stephen B., Heath J. Carney, and Zósimo Huamán. 1981. Dynamics of Andean potato agriculure. Economic Botany 35:70–85.

Bryan, Kirk. 1922. Erosion and sedimentation in the Papago country, Arizona, with a sketch of the geology. Pp. 19–90 *in* U.S. Geological Service Bulletin 730-B, Washington, D.C.

Buelna, Eustaquio (editor). 1890. Arte de la lengua cahita por un padre de la Compañía de Jesús. Reprinted 1989, Siglo Veintiuno editores, Mexico City.

Buhrow, R. 1983. The wild beans of southwestern North America. Desert Plants 5:67–71, 82–88.

Burk, William R. 1983. Puffball usages among North American Indians. Journal of Ethnobiology 3:55–62.

Casagrande, Joseph B. 1954a. Comanche linguistic acculturation I. International Journal of American Linguistics 20:140–151.

_____. 1954b. Comanche linguistic acculturation II. International Journal of American Linguistics 20:217–237.

_____. 1955. Comanche linguistic acculturation III. International Journal of American Linguistics 21:8–25.

Castetter, Edward F., and Willis H. Bell. 1942. Pima and Papago Indian Agriculture. Inter-Americana Studies 1. University of New Mexico Press, Albuquerque. Reprinted 1980, AMS Press, New York.

Castetter, Edward F., Willis H. Bell, and A. R. Grove. 1938. The Early Utilization and Distribution of *Agave* in the American Southwest. University of New Mexico Bulletin, Biological Series, vol. 5, no. 4. Albuquerque.

Castetter, Edward F., and Ruth M. Underhill. 1935. The Ethnobiology of the Papago Indians. University of New Mexico Bulletin 275, Ethnobiological Studies in the American Southwest 2, pp. 3–84. Reprinted 1978, AMS Press, New York.

Chapman, Stephen R., and Lark P. Carter. 1976. Crop Production: Principles and Practices. W. H. Freeman & Co., San Francisco.

Colton, William F. 1869. The valley of the Rio Gila, and country lying between the Rio Colorado of the West and the Pacific Ocean. Pp. 78–121 *in* New Tracks in North America, vol. 2. William A. Bell (editor). Chapman and Hall, London.

Conklin, E. 1878. Picturesque Arizona. Continental Stereoscopic Company, New York.

Conner, Daniel Ellis. 1956. Joseph Reddeford Walker and the Arizona Adventure. Donald J. Berthrong and Odessa Davenport (editors). University of Oklahoma Press, Norman.

Cook, Minnie A. 1976. Apostle to the Pimas: The Story of Charles H. Cook, the First Missionary to the Pimas. Omega Books, Tiburon, Calif.

Cooke, Philip St. George. 1848. Report of Lieut. Col. P. St. George Cooke of His March from Santa Fe, New Mexico, to San Diego, Upper California. 30th Cong., 1st sess., House Exec. Doc. 41, pp. 549–563.

_____. 1878. The Conquest of New Mexico and California: An Historical and Personal Narrative. Putnam's Sons, New York. Reprinted 1964, Horn and Wallace, Albuquerque, N.Mex., and Rio Grande Press, Chicago.

_____. 1938. Cooke's journal of the march of the Mormon Battalion, 1846–1847. Pp. 65–240 *in* Exploring Southwestern Trails, 1846–1854. Ralph P. Bieber and A. B. Bender (editors). Southwestern Historical Series, vol. 7. Arthur H. Clark Co., Glendale, Calif.

Cozzens, Samuel Woodworth. 1874. The Marvellous Country; Or, Three Years in Arizona and New Mexico, the Apache's Home, Ancient Cibola. Lee & Shepard, Boston.

Cremony, John C. 1868. Life among the Apaches. A. Roman & Co., San Francisco. Reprinted 1951, Arizona Silhouettes, Tucson, and 1969, Rio Grande Press, Glorieta, N.Mex.

Crosby, Alfred W. 1972. The Columbian exchange: Biological and cultural consequences of 1492. Greenwood, Westport, Conn.

Crosswhite, Frank S. 1980. The annual Saguaro harvest and crop cycle of the Papago, with reference to ecology and symbolism. Desert Plants 2:3–61.

_____. 1981. Desert plants, habitat, and agriculture in relation to the major pattern of cultural differentiation in the O'odham people of the Sonoran Desert. Desert Plants 3:47–76.

Crumrine, N. Ross. 1983. Mayo. Pp. 264–275 *in* Handbook of North American Indians, vol. 10, Southwest. Alfonso Ortiz (editor). Smithsonian Institution, Washington, D.C.

Cruse, Thomas. 1941. Apache Days and After. Eugene Cunningham (editor). Caxton Printers, Caldwell, Idaho.

Curtin, Leonora S. M. 1949. By the Prophet of the Earth: Ethnobotany of the Pima. San Vincente Foundation, Santa Fe, N.Mex. Reprinted 1984, University of Arizona Press, Tucson.

_____. n.d. Notes for specimens used in By the Prophet of the Earth. Unpub. ms., University of Arizona Library, Special Collections, Tucson.

Curtis, Edward S. 1908. The North American Indian, vol. 2. University Press, Cambridge, Mass.

Cushing, Frank H. 1896. Outlines of Zuni Creation Myths. 13th Annual Report of the Bureau of American Ethnology, 1891–92, pp. 321–447. Washington, D.C.

_____. 1988. The Mythic World of the Zuni. Barton Wright (editor and illustrator). University of New Mexico Press, Albuquerque.

Dahl, Kevin. 1990. Corn Soot Woman's timeless lesson: Eat your smut. Permaculture Drylands 11:10–11, 14.

Daniel, Thomas F., and Mary L. Butterwick. 1992. Flora of the South Mountains of South-central Arizona. Desert Plants 10:99–119.

Davy, J. Burtt. 1898. Bermuda Grass in Arizona. Erythea 6:24–25.

Dean, Jeffrey S. 1991. Thoughts on Hohokam chronology. Pp. 61–149 *in* Exploring the Hohokam. George J. Gumerman (editor). Amerind Foundation Publication. University of New Mexico Press, Albuquerque.

Densmore, Frances. 1929. Papago Music. Bulletin of the Bureau of American Ethnography 90. Smithsonian Institution, Washington, D.C.

DiPeso, Charles C. 1956. The Upper Pima of San Cayetano del Tumacacori: An Archaeohistorical Reconstruction of the Ootam of Pimería Alta. Amerind Foundation, Dragoon, Ariz.

_____. 1979. Prehistory: O'otam. Pp. 91–99 *in* Handbook of North American Indians, vol. 9, Southwest. Alfonso Ortiz (editor). Smithsonian Institution, Washington, D.C.

Dobyns, Henry F. 1963. Indian extinction in the middle Santa Cruz River Valley, Arizona. New Mexico Historical Review 38:163–181.

_____. 1972. The Papago People. Indian Tribal Series, Phoenix.

_____. 1974. The Kohatk: Oasis and akchin horticulturalists. Ethnohistory 21:317–327.

_____. 1978. Who killed the Gila? Journal of Arizona History 19:17–30.

_____. 1981. From Fire to Flood: Historic Human Destruction of Sonoran Desert Riverine Oases. Anthropological Papers 20. Ballena Press, Socorro, N.Mex.

_____. 1988. Piman Indian historic agave cultivation. Desert Plants 9:49–53.

Dobyns, Henry F., and Robert C. Euler. 1980. Indians of the Southwest: A Critical Bibliography. Newberry Library Publications. Indiana University Press, Bloomington.

Doelle, William. 1975. The adaptation of wheat by the Gila Pima: A study in agricultural change. Unpub. ms., Arizona State Museum Library, University of Arizona, Tucson.

Dole, William P. 1865. Annual Report of the Commissioner of Indian Affairs to the Secretary of the Interior, 1864. Washington, D.C.

Doyel, David E. 1976a. Salado cultural developments in the Tonto Basin and Globe-Miami areas, central Arizona. Kiva 42:5–16.

_____. 1976b. Classic period Hohokam in the Gila River Basin, Arizona. Kiva 42:27–37.

_____. 1979. The prehistoric Hohokam of the Arizona desert. American Scientist 67:544–554.

_____. 1991. Hohokam cultural evolution in the Phoenix Basin. Pp. 231–278 in Exploring the Hohokam. George J. Gumerman (editor). Amerind Foundation Publication. University of New Mexico Press, Albuquerque.

Dunne, Peter Masten. 1944. Pioneer Jesuits in Northern Mexico. University of California Press, Berkeley.

Dunnigan, Timothy. 1981. Adaptive strategies of peasant Indians in a biethnic Mexican community: A study in Mountain Pima assimilation. Pp. 36–49 in Themes of Indigenous Acculturation in Northwest Mexico. Thomas B. Hinton and Phil C. Weigand (editors). Anthropological Papers of the University of Arizona 38. University of Arizona Press, Tucson.

_____. 1983. Lower Pima. Pp. 217–229 in Handbook of North American Indians, vol. 10, Southwest. Alfonso Ortiz (editor). Smithsonian Institution, Washington, D.C.

Durivage, John E. 1937. Letters and journal of John E. Durivage. Pp. 159–255 in Southern Trails to California in 1849. R. P. Bieber (editor). Arthur H. Clark, Glendale, Calif.

Ebeling, Walter. 1986. Handbook of Indian Foods and Fibers of Arid America. University of California Press, Berkeley.

Eccleston, Robert. 1950. Overland to California on the Southwestern Trail, 1849: Diary of Robert Eccleston. George P. Hammond and Edward H. Howes (editors). University of California Press, Berkeley.

Emory, William H. 1848. Notes of a Military Reconnaissance, from Fort Leavenworth, in Missouri, to San Diego, in California, including Parts of Arkansas, del Norte, and Gila Rivers; Made in 1846–1847. 30th Cong., 1st sess., Senate Exec. Doc. 7 (serial no. 505).

_____. 1857. Report on the United States and Mexican Boundary Survey, vol. 1, pt. 1. 34th Cong., 1st sess., House Exec. Doc. 135.

Euler, R. C., and V. H. Jones. 1956. Hermetic sealing as a technique for food preservation among the Indians of the American Southwest. Proceedings of the American Philosophical Society 100:87–99.

Evans, G.W.B. 1945. Mexican Gold Trail: The Journal of a Forty-niner. G. S. Dumke (editor). Huntington Library, San Marino, Calif.

Evers, Larry (editor). 1980. The South Corner of Time: Hopi, Navajo, Papago, Yaqui Tribal Literature. University of Arizona Press, Tucson.

Ezell, Paul H. 1958. An early geographer of the Southwest: Father Diego Bringas. El Museo (n.s.) 2:18–30.

_____. 1961. The Hispanic Acculturation of the Gila River Pima. Memoirs of the American Anthropological Association 90. Menasha, Wisc.

_____. 1963. Is there a Hohokam-Pima culture continuum? American Antiquity 29:61–66.

_____. 1983. History of the Pima Indians. Pp. 149–160 in Handbook of North American Indians, vol. 10, Southwest. Alfonso Ortiz (editor). Smithsonian Institution, Washington, D.C.

Feinman, Gary M. 1991. Hohokam archaeology in the eighties: An outside view. Pp. 461–483 in Exploring the Hohokam. George J. Gumerman (editor). Amerind Foundation Publication. University of New Mexico Press, Albuquerque.

Felger, Richard S. 1977. Mesquite in Indian cultures of southwestern North America. Pp. 150–176 in Mesquite: Its Biology in Two Desert Scrub Ecosystems. B. B. Simpson (editor). Dowden, Hutchinson, and Ross, Stroudsburg, Penn.

Felger, Richard S., and Mary Beck Moser. 1985. People of the Desert and Sea: Ethnobotany of the Seri Indians. University of Arizona Press, Tucson.

Felger, Richard S., Peter L. Warren, L. Susan Anderson, and Gary P. Nabhan. 1992. Vascular plants of a desert oasis: Flora and ethnobotany of Quitobaquito, Organ Pipe Cactus National Monument, Arizona. Proceedings of the San Diego Society of Natural History 8:1–39.

Ferg, Alan (editor). 1987. Western Apache Material Culture: The Goodwin and Guenther Collections. University of Arizona Press, Tucson.

Fish, Suzanne K. 1993. Pollen. Pp. 241–246 in The Maricopa Road Site: A Pre-Classic Hohokam Village. John Ravesloot and Annick Lascaux (editors). Arizona State University Anthropological Field Studies 28. Tempe.

Fish, Suzanne K., Paul R. Fish, Charles Miksicek, and John Madsen. 1985. Prehistoric agave cultivation in southern Arizona. Desert Plants 7:100, 107–112.

Font, Pedro. 1931. Font's Complete Diary: A Chronicle of the Founding of San Francisco; Diary kept by the Father Preacher Fray Pedro Font . . . During the Journey Which He Made to Monterey, vol. 4 of Anza's California Expeditions. Herbert E. Bolton (editor and translator). University of California Press, Berkeley.

_____. 1975. Letters of Friar Pedro Font, 1776–1777. Matson, Dan S. (editor). Ethnohistory 22:263–293.

Fontana, Bernard. 1976. Desertification of Papagueria: Cattle and the Papago. Pp. 59–69 in Desertification: Process, Problems, Perspectives. P. Paylore and R. A. Haney, Jr. (editors). University of Arizona Office of Arid Land Studies, Tucson.

_____. 1979. The Material World of the Tarahumara. Arizona State Museum, University of Arizona, Tucson.

_____. 1983a. Pima and Papago: Introduction. Pp. 125–136 in Handbook of North American Indians, vol. 10, Southwest. Alfonso Ortiz (editor). Smithsonian Institution, Washington, D.C.

_____. 1983b. History of the Papago. Pp. 137–148 in Handbook of North American Indians, vol. 10, Southwest. Alfonso Ortiz (editor). Smithsonian Institution, Washington, D.C.

Fontana, Bernard L., William J. Robinson, Charles W. Cormack, and Ernest E. Leavitt, Jr. 1962. Papago Indian Pottery. University of Washington Press, Seattle.

Fowler, Catherine S. 1983. Some lexical clues to Uto-Aztecan prehistory. International Journal of American Linguistics 49:224–257.

_____. 1994. Corn, beans, and squash: Some linguistic perspectives from Uto-Aztecan. Pp. 445–467 in Corn and Culture in the Prehistoric New World. Sissel Johannessen and Christine A. Hastorf (editors). Westview Press, Boulder, Colo.

Franklin, Hayward H., and W. Bruce Masse. 1976. The San Pedro Salado: A case of prehistoric migration. Kiva 42:47–55.

Freeman, George F. 1912. Southwestern beans and teparies. University of Arizona Agricultural Experiment Station Bulletin 68, pp. 1–55. Tucson.

_____. 1918. Southwestern beans and teparies (revised). University of Arizona Agricultural Experiment Station Bulletin 68, pp. 573–619. Tucson.

Frenkel, Robert E. 1970. Ruderal vegetation along some California roadsides. University of California Publications in Geography 20:1–163. Reprinted 1977, California Library Reprint Series.

Fritts, H. C. 1965. Tree-ring evidence for climatic changes in western North America. Monthly Weather Review 7:421–443.

Fritz, Gordon L. 1990. The ecological significance of early Piman immigration to southern Arizona. The Artifact 27:51–108.

Fuller, Wallace H. 1975. Soils of the Desert Southwest. University of Arizona Press, Tucson.

Garcés, Francisco. 1900. On the Trail of a Spanish Pioneer: The Diary and Itinerary of Francisco Garcés in His Travels through Sonora, Arizona, and California, 1775–1776. 2 vols. Elliott Coues (editor and translator). Harper, New York.

_____. 1965. A Record of Travel in Arizona and California, 1775–1776: A New Translation. John Galvin (editor and translator). John Howell, San Francisco.

Gasser, Robert E. 1988. Farming, gathering, and hunting at Ak-Chin: Evidence for change, seasonality, and comparison. Ms. on file, Soil Systems, Inc., Phoenix.

Gasser, Robert E., and Scott M. Kwiatkowski. 1991. Food for thought: Recognizing patterns in Hohokam subsistence. Pp. 417–459 *in* Exploring the Hohokam. George J. Gumerman (editor). Amerind Foundation Publication. University of New Mexico Press, Albuquerque.

Gentry, Howard Scott. 1942. Rio Mayo Plants. Carnegie Institution of Washington Publications 527. Washington, D.C.

_____. 1963. The Warihio Indians of Sonora-Chihuahua: An Ethnographic Survey. Bureau of American Ethnology Bulletin 186, Anthropological Papers no. 65. Smithsonian Institution, Washington, D.C.

_____. 1982. Agaves of Continental North America. University of Arizona Press, Tucson.

Giff, Joseph. 1980. Pima Blue Swallow Songs of Gratitude. Pp. 127–139 *in* Speaking, Singing, and Teaching: A Multidisciplinary Approach to Language Variation. Florence Barkin and Elizabeth Brandt (editors). Proceedings of the 8th Annual Southwestern Areal Language and Linguistics Workshop. Swallow 8. Arizona State University Anthropological Research Papers 20. Tempe.

Gilman, M. French. 1909. Among the thrashers in Arizona. Condor 11:49–54.

_____. 1915. Woodpeckers of the Arizona lowlands. Condor 17:151-163.

Gladwin, Harold S., Emil W. Haury, E. B. Sayles, and N. Gladwin. 1937. Excavations at Snaketown, vol. 1, Material Culture. Gila Pueblo Medallion Paper 25. Globe, Ariz. Reprinted 1965, Arizona State Museum, University of Arizona, Tucson.

Golder, Frank Alfred. ca. 1928. The March of the Mormon Battalion from Council Bluffs to California; Taken from the Journal of Henry Standage, in Collaboration with Thomas A. Bailey and J. Lyman Smith. Century Co., New York.

Gonzáles Ramos, Gilardo. 1972. Los Coras. Instituto Nacional Indigenista, Mexico City.

González, Luis. 1977. Etnología y misión en la Pimería Alta, 1715–1740. Instituto de Investigaciones Históricas, Universidad Nacionál Autónoma de México, Mexico City.

Goodwin, Grenville. 1971. Western Apache Raiding and Warfare: From the Notes of Grenville Goodwin. Keith H. Basso (editor). University of Arizona Press, Tucson.

Gosper, John J. 1881. Report of the Acting Governor of Arizona [Territory]. Pp. 915–937 *in* Report to the Secretary of the Interior. Washington, D.C.

Gould, Frank W., and Reid Moran. 1981. The Grasses of Baja California, Mexico. San Diego Society of Natural History Memoir 12.

Gracy, David B., III, and Helen J. H. Rugeley. 1965. From the Mississippi to the Pacific: An Englishman in the Mormon Battalion. Arizona and the West 7:127–160.

Graham, James D. 1852. Report of Lieutenant Colonel Graham on the Subject of the Boundary Line between the United States and Mexico. 32d Cong., 1st sess., Senate Exec. Doc. 121.

Grant, Verne, and Karen A. Grant. 1971. Dynamics of clonal microspecies in cholla cactus. Evolution 25:144–155.

Gray, Asa. 1855. *Planta Novae Thurberianae:* The characters of some new genera and species of plants in a collection made by George Thurber, Esq., of the late Mexican Boundary Commission, chiefly in New Mexico and Sonora. Memoirs of the American Academy of Arts and Sciences (n.s.) 5:297–328.

Gray, A. B. 1855. Report and Map Relative to the Mexican Boundary. 33d Cong., 2d sess., Senate Exec. Doc. 55, pp. 1–35.

_____. 1856. Southern Pacific Railroad: Survey of a Route for the Southern Pacific R.R. on the Thirty-Second Parallel by A. B. Gray, for the Texas Western R.R. Company. Cincinnati. Reprinted 1963, Westernlore Press, Los Angeles.

Grebinger, Paul. 1976. Salado: Perspectives from the Middle Santa Cruz Valley. Kiva 42:39–46.

Gregonis, Linda M., and Karl J. Reinhard. 1979. Hohokam Indians of the Tucson Basin. University of Arizona Press, Tucson.

Griffen, William B. 1983. Southern Periphery: East. Pp. 329–342 in Handbook of North American Indians, vol. 10, Southwest. Alfonso Ortiz (editor). Smithsonian Institution, Washington, D.C.

Griffin, John S. 1943. A Doctor Comes to California. California Historical Society, San Francisco.

Grossmann, Fredrick E. 1872. Report to E. S. Parker, Commissioner of Indian Affairs, Washington, D.C., 16 August 1871. Pp. 358–363 *in* Annual Report of the Commissioner of Indian Affairs the Year 1871. GPO, Washington, D.C.

_____. 1873. The Pima Indians of Arizona. Pp. 407–419 *in* Annual Rept. of the Board of Regents of the Smithsonian Institution . . . for . . . 1871. GPO, Washington, D.C.

_____. n.d. [Pima dictionary]. Document no. 30464, Bureau of American Ethnology, Smithsonian Institution, Washington, D.C.

Gumerman, George J. 1991. Understanding the Hohokam. Pp. 1–27 *in* Exploring the Hohokam. George J. Gumerman (editor). Amerind Foundation Publication. University of New Mexico Press, Albuquerque.

Gumerman, George J., and Emil W. Haury. 1979. Prehistory: Hohokam. Pp. 75–90 *in* Handbook of North American Indians, vol. 9, Southwest. Alfonso Ortiz (editor). Smithsonian Institution, Washington, D.C.

Gumerman, George J., and Carol S. Weed. 1976. The question of Salado in the Agua Fria and New River drainages of central Arizona. Kiva 42:105–112.

Hackenberg, Robert A. 1983. Pima and Papago ecological adaptations. Pp. 161–177 *in* Handbook of North American Indians, vol. 10, Southwest. Alfonso Ortiz (editor). Smithsonian Institution, Washington, D.C.

Hackenberg, Robert A., and Bernard L. Fontana. 1974. Aboriginal Land Use and Occupancy of the Pima-Maricopa Indians. 2 vols. American Indian Ethnohistory: Indians of the Southwest. Garland Press, New York.

Hale, Kenneth. n.d. Breve vocabulario del idioma Pima de Ónavas. Unpub. ms., Arizona State Museum Library, University of Arizona, Tucson.

Hale, Kenneth, and David Harris. 1979. Historical linguistics and archaeology. Pp. 170–177 in Handbook of North American Indians, vol. 9, Southwest. Alfonso Ortiz (editor). Smithsonian Institution, Washington, D.C.

Hall, Harvey M., and Frederic E. Clements. 1923. The phylogenetic method in taxonomy: Genus *Atriplex*. Carnegie Institution of Washington Publications 326, pp. 235–346.

Hall, Sharlot M. 1907. The story of a Pima record rod. Out West 26:413–423.

Hammack, Laurence C. 1969. A preliminary report of the excavations at Las Colinas. Kiva 35:11–28.

Harris, Benjamin B. 1960. The Gila Trail: The Texas Argonauts and the Gold Rush. Richard H. Dillon (editor). University of Oklahoma Press, Norman.

Harris, D. R. 1966. Recent plant invasions in the arid and semi-arid Southwest of the United States. Annals of the Association of American Geographers 56:408–423.

Hart, Elisabeth. 1934. Pima Cookery. Mimeograph of unpub. ms. Reprinted 1949, Pueblo Grande Museum, Phoenix.

Harvard, Valery. 1896. Drink plants of the North American Indians. Bulletin of the Torrey Botanical Club 23:33–46.

Hastings, J. R., and R. M. Turner. 1965. The Changing Mile: An Ecological Study of Vegetation Change with Time in the Lower Mile of an Arid and Semiarid Region. University of Arizona Press, Tucson.

Hastings, J. R., Raymond M. Turner, and Douglas K. Warren. 1972. An Atlas of Some Plant Distributions in the Sonoran Desert. Technical Reports on the Meteorology and Climatology of Arid Regions 21. University of Arizona Institute of Atmospheric Physics, Tucson.

Haury, Emil W. 1945. Excavations of Los Muertos and Neighboring Ruins in the Salt River Valley, Southern Arizona: Based on the Work of the Hemenway Southwestern Archaeological Expedition of 1887–1888. Papers of the Peabody Museum of American Archaeology and Ethnology 24. Harvard University.

_____. 1950. The Stratigraphy and Archaeology of Ventana Cave, Arizona. University of New Mexico Press, Albuquerque. Reprinted 1975, University of Arizona Press, Tucson.

_____. 1976. The Hohokam: Desert Farmers and Craftsmen; Excavations at Snaketown, 1964–1965. University of Arizona Press, Tucson.

Hayden, C. 1965. A History of the Pima Indians and the San Carlos Irrigation Project. 89th Cong., 1st sess., Senate Doc. 11. Compiled in 1924.

Hayden, Julian D. 1935. Pima Creation Myth, as told by Juan Smith, Snaketown, Arizona. Unpub. ms., University of Arizona Library, Tucson.

_____. 1969. Gyratory crushers of the Sierra Pinacate, Sonora. American Antiquity 34:154–161.

_____. 1970. Of Hohokam origins and other matters. American Antiquity 35:87–93.

Heiser, Charles B., Jr. 1976. The Sunflower. University of Oklahoma Press, Norman.

_____. 1979. The Gourd Book. University of Oklahoma Press, Norman.

_____. 1985. Of Plants and People. University of Oklahoma Press, Norman.

Hendrickson, Dean A., and W. L. Minckley. 1984 [1985]. Ciénegas: Vanishing climax communities of the American Southwest. Desert Plants 6:131–175.

Hendry, George W. 1931. The adobe brick as a historical source: Reporting further studies in adobe brick analysis. Agricultural History 5:110–127.

Hendry, George W., and Margaret P. Kelly. 1925. The plant content of adobe brick. California Historical Society Quarterly 4:361–345.

Hernández, Francisco. 1959. Historia de las plantas de Nueva España, vols. 1 and 2 of Historia natural de Nueva España. Obras Completas, vols. 2 and 3. Universidad Nacional Autónoma de México, Mexico City.

Herzog, George. 1941. Culture changes and language: Shifts in the Pima vocabulary. Pp. 66–74 in Language, Culture, and Personality: Essays in Memory of Edward Sapir. Leslie Sapir, A. Irving Hallowell, and Stanley S. Newman (editors). Sapir Memorial Publication Fund, Menasha, Wisc. Reprinted 1960.

Hickman, James C. (editor). 1993. The Jepson Manual: Higher Plants of California. University of California Press, Berkeley.

Hine, Robert V. 1968. Bartlett's West: Drawing the Mexican Boundary. Yale University Press, New Haven.

Hinton, Thomas B. 1964. The Cora Village: A civil religious hierarchy in Northern Mexico. Pp. 44–62 in Culture Change and Stability: Essays in Memory of Olive Ruth Barker and George C. Barker, Jr. Ralph C. Beals (editor). University of California Department of Anthropology, Los Angeles.

_____. 1971. An analysis of religious syncretism among the Cora of Nayarit. Pp. 275–279 in Proceedings of the 38th International Congress of Americanists, vol. 3. Munich.

_____. 1981. Cultural Visibility and the Cora. Pp. 1–3 in Themes of Indigenous Acculturation in Northwest Mexico. T. B. Hinton and Phil C. Weigand (editors). Anthropological Papers of the University of Arizona 38. University of Arizona Press, Tucson.

_____. 1983. Southern Periphery: West. Pp. 315–328 in Handbook of North American Indians, vol. 10, Southwest. Alfonso Ortiz (editor). Smithsonian Institution, Washington, D.C.

Hodge, Hiram C. 1877. Arizona as It Is; or, The Coming Country. Hurd and Houghton, N.Y. Reprinted 1965, Rio Grande Press, Chicago.

Hoover, J. W. 1929. The Indian country of southern Arizona. Geographical Review 19:38–60.

Hrdlička, Aleš. 1903. The region of the ancient Chichimecos, with notes on the Tepecanos and the ruins of La Quemada, Mexico. American Anthropologist 5:384–440.

_____. 1906. Notes on the Pima of Arizona. American Anthropologist (n.s.) 8:39–46.

_____. 1908. Physiological and medical observations among the Indians of the southwestern United States and northern Mexico. Bureau of American Ethnology Bulletin 34, pp. 1–266. Smithsonian Institution, Washington, D.C.

_____. 1974. Peyote Hunt: The Sacred Journey of the Huichol Indians. Cornell University Press, Ithaca, N.Y.

Hughes, Louis C. 1895. Report of the Governor of Arizona [Territory] to the Secretary of the Interior. Pp. 1–112 in Report of the Secretary of the Interior. Washington, D.C.

Johnson, Jean B. 1943. A clear case of linguistic acculturation. American Anthropologist (n.s.) 45:427–434.

_____. 1950. The Opata: An Inland Tribe of Sonora. University of New Mexico Publications in Anthropology 6. Albuquerque.

_____. 1962. El idioma Yaqui. Instituto Nacional de Antropología e Historia, Mexico City.

Johnston, Abraham R. 1848. Journal of Abraham R. Johnston, First Dragoons. 30th Cong., 1st sess., House Exec. Doc. 41, pp. 567–614.

Jones, Nathaniel V. 1931. The journal of Nathaniel V. Jones with the Mormon Battalion. Utah Historical Quarterly 4(1):3–23.

Jones, Volney H. 1936. A summary of data on aboriginal cotton of the Southwest: Symposium on prehistoric agriculture. Bulletin of the University of New Mexico, Anthropological Series vol. 1, no. 5, pp. 51–64. Albuquerque.

Joseph, Alice, Rosamund B. Spicer, and Jane Chesky. 1949. The Desert People: A Study of the Papago Indians. University of Chicago Press, Chicago.

Judd, I. B. 1971. The lethal decline of mesquite on the Casa Grande Ruins National Monument. Southwestern Naturalist 31:153–159.

Kearney, Thomas H., Robert H. Peebles, and collaborators. 1960. Arizona Flora. 2d edition. University of California Press, Berkeley and Los Angeles.

King, C. V., R. E. Beckett, and Orlan Parker. 1938. Agricultural Investigations at the United States Field Station, Sacaton, Ariz., 1931–35. USDA Circular 479.

King, C. V., and A. R. Leding. 1926. Agricultural Investigations at the United States Field Station, Ariz., 1922, 1923, and 1924. USDA Circular 372.

King, C. V., and H. F. Loomis. 1932. Agricultural Investigations at the United States Field Station, Sacaton, Ariz., 1925–1930. USDA Circular 206.

Kino, Eusebio Francisco. 1919. Kino's Historical Memoir of Pimería Alta, vol. 1. H. E. Bolton (editor and translator). A. H. Clark, Cleveland.

_____. 1985. Crónica de la Pimería Alta: Favores celestiales. Gobierno del Estado de Sonora, Hermosillo.

_____. 1989. Las misiones de Sonora y Arizona: Comprendiendo "Favores celestiales" y "Relación diaria de la entrada al noroeste" . . . Editorial Porrúa, Mexico City.

Kirchhoff, Paul. 1944. Los recolectores: Cazadores del norte de Mexico. Pp. 133–144 in El norte de México y el sur de Estados Unidos: Tercera reunión de mesa redonda sobre problemas antropológicos de México y Centro América. Sociedad Mexicana de Antropología, Mexico City.

Kissell, Mary Lois. 1916. Basketry of the Papago and Pima. Anthropological Papers of the American Museum of Natural History 17:115–264. Reprinted 1972, Rio Grande Press, Glorieta, N.Mex.

Kroeber, A. L. 1934. Uto-Aztecan languages of Mexico. Ibero-Americana 8:1–28.

_____. 1939. Cultural and Natural Areas of Native North America. University of California Publications in American Archaeology and Ethnology 38. Berkeley.

Kroeber, Clifton B., and Bernard L. Fontana. 1986. Massacre on the Gila: An Account of the Last Major Battle between American Indians, with Reflections on the Origin of War. University of Arizona Press, Tucson.

Laferrière, Joseph E. 1991. Optimal use of ethnobotanical resources by the Mountain Pima of Chihuahua, Mexico. Ph.D. diss. Department of Ecology and Evolutionary Biology, University of Arizona, Tucson.

Laferrière, Joseph E., Charles W. Weber, and Edwin A. Kohlhepp. 1991. Use and nutritional composition of some traditional Mountain Pima plant foods. Journal of Ethnobiology 11:93–114.

Lanner, Ronald M. 1984. Trees of the Great Basin. University of Nevada Press, Reno.

LeBlanc, Steven, and Ben Nelson. 1976. The Salado in southwestern New Mexico. Kiva 42:71–79.

Lee, D. D. 1943. The linguistic aspect of Wintu acculturation. American Anthropologist (n.s.) 45:435–440.

Lee, Richard B., and Irven DeVore (editors). 1968. Man the Hunter. Aldine Publishing Co., New York.

Lehr, J. Harry. 1978. A Catalogue of the Flora of Arizona. Desert Botanical Garden, Phoenix.

Lewton, Frederick L. 1912. The cotton of the Hopi Indians: A new species of Gossypium. Smithsonian Miscellaneous Collections 60(6):1–10.

Linares, Olga F. 1976. "Garden hunting" in the American tropics. Human Ecology 46:331–349.

Lionnet, Andrés. 1986. Un idioma extinto de Sonora: El Eudeve. Instituto de Investigaciones Anthropológicas, Lingüística, Serie Anthropológica 60. Universidad Nacional Autónoma de México, Mexico City.

Lloyd, J. William. 1911. Aw-aw-tam Indian Nights. The Lloyd Group, Westfield, N.J.

Lumholtz, Carl. 1898. The Huichol Indians of Mexico. Bulletin of the American Museum of Natural History 10:1–14.

_____. 1902. Unknown Mexico, vol. 1. Scribner's, New York.

_____. 1912. New Trails in Mexico: An Account of One Year's Exploration in North-Western Sonora, Mexico, and South-Western Arizona, 1909–1910. Scribner's, New York. Reprinted 1990, University of Arizona Press, Tucson.

Manje, Juan Mateo. 1926. Luz de tierra incógnita: Version paleográfica por Francisco Fernández del Castillo. Publicaciones del Archivo General de la Nación 10. Mexico City.

_____. 1954. Unknown Arizona and Sonora, 1693–1721, from the Francisco Fernández del Castillo. Harry J. Karns and associates (translators). Arizona Silhouettes, Tucson.

Martínez, Maximino. 1979. Catálogo de nombres vulgares y científicos de plantas mexicanos. 3d edition. Fundo de Cultura Económica, Mexico City.

Mason, J. Alden. 1912. The Tepehuán Indians of Azqueltán. Pp. 344–351 in Proceedings of the 18th International Congress of Americanists.

_____. 1917. Tepecano: A Piman language of western Mexico. Annals of the New York Academy of Sciences 25:309–416.

_____. 1920. The Papago harvest festival. American Anthropologist 22:13–25.

_____. 1921. The Papago migration legend. Journal of American Folk-Lore 34:254–268.

_____. 1948. The Tepehuán and the other aborigines of the Mexican Sierra Madre Occidental. América Indígena 8:288–300.

_____. 1952. Notes and observations on the Tepehuán. América Indígena 12:33–53.

_____. 1981. The ceremonialism of the Tepecan Indians of Azqueltán, Jalisco. Phil C. Weigand (editor). Pp. 62–76 in Themes of Indigenous Acculturation in Northwest Mexico. Thomas B. Hinton and Phil C. Weigand (editors). Anthropological Papers of the University of Arizona 38. University of Arizona Press, Tucson.

Mason, J. Alden, and George Agogino. 1972. The ceremonialism of the Tepehuán. Eastern New Mexico University Contributions in Anthropology 4:1–44.

Masse, W. Bruce. 1981. A reappraisal of the protohistoric Sobaipuri Indians of Southeastern Arizona. Pp. 28–56 in The Protohistoric Period in the North American Southwest. David R. Wilcox and W. Bruce Masse (editors). Arizona State University Anthropological Research Papers 24. Tempe.

Mathiot, Madeleine. 1962. Noun classes and folk taxonomy in Papago. American Anthropology 64:340–350. Reprinted 1964 as pp. 154–163 in Language in Culture and Society. Dell Hymes (editor). Harper and Row, New York.

_____. 1973. A Dictionary of Papago Usage, vol. 1: B–K. Language Science Monographs, vol. 8. Indiana University Publications. Bloomington.

_____. n.d. A Dictionary of Papago Usage, vol. 2: Ku–'u. Language Science Monographs, vol. 8. Indiana University Publications. Bloomington.

McClintock, James H. 1916. Arizona: Prehistoric, Aboriginal, Pioneer, Modern. Vol. 1. S. J. Clarke Publishing, Chicago.

McCormick, Richard C. 1865. Arizona: Its Resources and Prospects. New York Tribune, letter to the editor, 26 June 1865, p. 15.

McGuire, Randall H. 1991. On the outside looking in: The concept of periphery in Hohokam archaeology. Pp. 347–382 in Exploring the Hohokam. George J. Gumerman (editor). Amerind Foundation Publication. University of New Mexico Press, Albuquerque.

McVaugh, Rogers. 1956. Edward Palmer: Plant explorer of the American West. University of Oklahoma Press, Norman.

Merrick, Laura C. 1990. Crop genetic diversity and its conservation in traditional agroecosystems. Pp. 3–11 in Agroecology and Small Farm Development. M. A. Altiere and S. B. Hecht (editors). CRC Press, Boca Raton, Fla.

_____. 1991. Systematics, Evolution, and Ethnobotany of a Domesticated Squash, Cucurbita argyrosperma. Ph.D. diss., Cornell University.

Michler, Nathaniel. 1857. From the 111th meridian of longitude to the Pacific Ocean. Pp. 101–125 in Report on the United States and Mexican Boundary Survey, vol. 1, pt. 1. William H. Emory (editor). 34th Cong., 1st sess., House Exec. Doc. 135.

Miksicek, Charles. 1983. Archaeobotanical aspects of Las Fosas: A statistical approach to prehistoric plant remains. Pp. 671–700 in Habitat Sites on the Gila River, vol. 6, Hohokam Archaeology along the Salt-Gila Aqueduct, Central Arizona Project. Lynn S. Teague and Patricia L. Crown (editors). Arizona State Museum Archaeological Series 150. University of Arizona, Tucson.

_____. 1987. Late Sedentary–Early Classic period Hohokam agriculture: Plant remains from the Marana Community Complex. Pp. 197–216 in Studies in the Hohokam Community of Marana. Glen E. Rice (editor). Arizona State University Anthropological Field Studies 15. Tempe.

_____. 1988. Pioneer to present: Archaeobotanical remains from Ak-Chin, Arizona. Ms. on file, Soil Systems, Inc., Phoenix.

Miller, Wick R. 1983. Uto-Aztecan Languages. Pp. 113–124 in Handbook of North American Indians, vol. 10, Southwest. Alfonso Ortiz (editor). Smithsonian Institution, Washington, D.C.

Mindeleff, Victor. 1891. A Study of Pueblo Architecture: Tusayan and Cibola. 8th Annual Report of the Bureau of American Ethnology, 1886–87, pp. 3–228. Washington, D.C.

Minnis, Paul E. 1991. Famine foods of the North American desert borderlands in historic context. Journal of Ethnobiology 11:231–257.

Mobley, George F. 1980. Las Sierras, los Volcanes. Pp. 134–157 in America's Magnificent Mountains. Robert Breeden (editor). National Geographic Society, Special Publications Division, Washington, D.C.

Molina Molina, Flavio. 1989. Diccionario de flora y fauna indígena de Sonora. Gobierno del Estado de Sonora and the Instituto Sonorense de Cultura, Hermosillo.

Morrisey, Richard J. 1949. Early agriculture in Pimería Alta. Mid-America 31:101–108.

Mowry, Sylvester. 1858a. Letter to J. W. Denver, Commissioner of Indian Affairs, Washington, D.C., 10 November 1857. Pp. 296–305 in Annual Report of the Commissioner of Indian Affairs the Year 1857. 35th Cong., 1st sess., House Exec. Doc. 2.

_____. 1858b. Report to James W. Denver, Commissioner of Indian Affairs, 10 November 1857. Pp. 584–593 in Annual Report of the Commissioner of Indian Affairs the Year 1857. 35th Cong., 1st sess., House Exec. Doc. 2.

_____. 1859. Letter to Alfred B. Greenwood, Commissioner of Indian Affairs, Washington, D.C., 21 November 1859. Pp. 721–730 in Report of the Commissioner of Indian Affairs. 36th Cong., 1st sess., Senate Exec. Doc. 2, vol. 1.

_____. 1864. Arizona and Sonora. Harper & Bros., New York.

Myerhoff, Barbara G. 1970. The deer-maize-peyote symbol complex among the Huichol Indians of Mexico. Anthropological Quarterly 43:64–78.

_____. 1974. Peyote Hunt: The Sacred Journey of the Huichol Indians. Cornell University Press, Ithaca, N.Y.

Nabhan, Gary Paul. 1979a. Ecology of floodwater farming in southwestern North America. Agro-Ecosystems 5:245–255.

_____. 1979b. Tepary beans—the effects of domestication on adaptations to arid environments. Arid Lands Newsletter 10:11–16.

_____. 1980. Ammobroma sonorae: An endangered parasitic plant in extremely arid North America. Desert Plants 2:188–196.

_____. 1982. The Desert Smells Like Rain: A Naturalist in Papago Indian Country. North Point Press, San Francisco.

_____. 1983a. Guest editorial: A special issue on the tepary bean. Desert Plants 5:2.

_____. 1983b. Papago Indian fields: Arid lands ethnobotany and agricultural ecology. Ph.D. diss., Department of Arid Lands Resources, University of Arizona, Tucson.

_____. 1985. Gathering the Desert. University of Arizona Press, Tucson.

_____. 1993. Songbirds, Truffles, and Wolves: An American Naturalist in Italy. Pantheon Books, New York.

Nabhan, G. P., J. W. Berry, C. Anson, and C. W. Weber. 1980. Papago Indian floodwater fields and tepary bean yields. Ecology of Food and Nutrition 10:71.

Nabhan, Gary P., and Richard S. Felger. 1978. Teparies in southwestern North America: A biographical and ethnohistorical study. Economic Botany 32:2–19.

Nabhan, Gary Paul, and Amadeo Rea. 1987. Plant domestication and folk-biological change: The Upper Piman/devil's claw example. American Anthropologist 89:57–73.

Nabhan, Gary P., Amadeo M. Rea, Karen L. Reichhardt, Eric Mellink, and Charles F. Hutchinson. 1982. Papago influences on habitat and biotic diversity: Quitovac Oasis ethnoecology. Journal of Ethnobiology 2:124–143.

Nabhan, Gary P., and Thomas E. Sheridan. 1977. Living fencerows of the Rio San Miguel, Sonora, Mexico: Traditional technology for floodplain management. Human Ecology 5:97–111.

Nabhan, Gary, Alfred Whiting, Henry Dobyns, Richard Hevly, and Robert Euler. 1981. Devil's claw domestication: Evidence from southwestern Indian fields. Journal of Ethnobiology 1:135–164.

Neff, J. A. 1940. Notes on nesting and other habits of the White-Winged Dove in Arizona. Journal of Wildlife Management 4:279–290.

Nentvig, Juan. 1971. Descripción geográfica . . . de Sonora. Germán Viveros (editor). Publicaciones del Archivo General de la Nación, 2d ser., no. 1. Mexico City.

_____. 1980. Rudo Ensayo: A Description of Sonora and Arizona in 1764. Albert Francisco Pradeau and Robert R. Rasmussen (editor and translators). University of Arizona Press, Tucson.

Nesom, Guy L. 1991. Taxonomy of Isocoma (Compositae: Astereae). Phytologia 70:69–114.

Niethammer, Carolyn. 1983. Tepary cuisine. Desert Plants 5:8–10.

Olmstead, Frank H. 1919. Gila River Flood Control. 65th Cong., 3d sess., Senate Doc. 436.

O'Neale, Lila M., and Juan Dolores. 1943. Notes on Papago color designations. American Anthropologist (n.s.) 45:387–397.

Ormsby, Waterman L. 1942. The Butterfield Overland Mail. Lyle H. Wright and Josephine M. Bynum (editors). Huntington Library, San Marino, Calif.

O'Rourke, Mary Kay. 1983. Pollen from adobe brick. Journal of Ethnobiology 3:39–48.

Palmer, Edward. 1871. Food products of the North American Indians. Pp. 404–428 in Report of the Commissioner of Agriculture, 1870. GPO, Washington, D.C.

_____. 1878a. Fish-hooks of the Mohave Indians. American Naturalist 12:403.

_____. 1878b. Plants used by the Indians of the United States. American Naturalist 12:593–606, 646–655.

_____. 1881. Chia. Zoe 2:140–142.

_____. ca. 1885. [Miscellaneous notes on the Pima]. University of Arizona Library Archives, Special Collections. Tucson.

Pancoast, Charles E. 1930. A Quaker Forty-Niner: The Adventures of Charles Edward Pancoast on the American Frontier. Anna Paschall Hannum (editor). University of Pennsylvania Press, Philadelphia.

Parke, John G. 1857. Report of Explorations for Railroad Routes . . . the Pimas Villages on the Gila to the Rio Grande. . . . Pt. 1 [dated 1854–1855]. Pp. 1–42 in vol. 7 of Explorations and Surveys for a Railroad Route from the Mississippi River to the Pacific Ocean. 33d Cong., 2d sess., House Exec. Doc. 91.

Parker, Kittie. 1972. An Illustrated Guide to Arizona Weeds. University of Arizona Press, Tucson.

Parry, C. C. 1857. General Geological Features of the Country. Pp. 1-23 in Report on the United States and Mexican Boundary Survey, vol. 1, pt. 1, ch. 1. William H. Emory (editor). 34th Cong., 1st sess., House Exec. Doc. 135.

Parsons, Mary Elizabeth. 1907. The Wild Flowers of California: Their Names, Haunts, and Habits. Cunningham, Curtis & Welch, San Francisco. Reprinted 1966, Dover, New York.

Pattie, James O. 1930. The Personal Narrative of James O. Pattie of Kentucky. Timothy Flint (editor); Milo M. Quaife (annotator). Lakeside Press, Chicago. Originally published 1833, E. H. Flint, Cincinnati.

Pennington, Campbell W. 1963a. Medicinal plants utilized by the Tepehuán of southern Chihuahua. América Indígena 23:31–47.

_____. 1963b. The Tarahumara of Mexico: Their Environment and Material Culture. University of Utah Press, Salt Lake City.

_____. 1969. The Tepehuán of Chihuahua: Their Material Culture. University of Utah Press, Salt Lake City.

_____. 1979. Vocabulario en la lengua Névome. Vol. 2 of The Pima Bajo of Central Sonora, Mexico. University of Utah Press, Salt Lake City.

_____. 1980. The Material Culture. Vol. 1 of The Pima Bajo of Central Sonora, Mexico. University of Utah Press, Salt Lake City.

_____. (editor). 1981. Arte y vocabulario de la lengua dohema, heve o eudeve: Manuscript of an Anonymous Eighteenth Century Author. Universidad Autónoma de México, Instituto de Investigaciones Filológicas, Mexico City.

_____. 1983a. Tarahumara. Pp. 276–289 in Handbook of North American Indians, vol. 10, Southwest. Alfonso Ortiz (editor). Smithsonian Institution, Washington, D.C.

_____. 1983b. Northern Tepehuan. Pp. 306–314 in Handbook of North American Indians, vol. 10, Southwest. Alfonso Ortiz (editor). Smithsonian Institution, Washington, D.C.

_____. n.d. A vocabulary made at Yepáchic, Chihuahua, among the Pima Bajo (1968–1970) [and] A vocabulary made at Maicoba, Sonora, among the Pima Bajo (1968, 1970). Unpub. mss.

Pérez de Ribas, Andrés. 1645. Historia de los triumphos de nuestra Santa Fe e entre gentes las más barbaras y fieras del Nuevo Orbe. A. de Parades, Madrid. Reprinted 1944, Editorial Layac, Mexico City.

Perkins, Kent D., and Willard W. Payne. 1978. Guide to the poisonous and irritant plants of Florida. Florida Cooperative Extension Service, Circular 441. University of Florida, Gainesville.

Pfefferkorn, Ignaz. 1949. Sonora: A Description of the Province. Theodore E. Treutlein (translator). Coronado Cuarto Centennial Publications, 1540–1940, vol. 12. University of New Mexico Press, Albuquerque. Reprinted 1989, University of Arizona Press, Tucson.

Pilcher, William W. 1967. Some comments on the folk taxonomy of the Papago. American Anthropologist 69:204–208.

Poston, Charles. 1865. Speech of Honorable Charles D. Poston, of Arizona, on Indian Affairs. Delivered in the House of Representatives, Thursday, 2 March 1865. Edmund Jones and Company, New York. (Congressional Globe 36, pt. 2:1319–1320).

Powledge, Fred. 1982. Water: The Nature, Uses, and Future of Our Most Precious and Abused Resource. Farrar Straus Giroux, New York.

Pratt, Richard C., and Gary Paul Nabhan. 1988. Evolution and diversity of Phaseolus acutifolius genetic resources. Pp. 409–440 in Genetic Resources of Phaseolus Beans: Their Maintenance, Domestication, Evolution, and Utilization. Paul Gepts (editor). Kluwer Academic Publishers, Dordrecht, Holland.

Preuss, Konrad T. 1912. Die Nayarit: Expedition, vol. 1, Die Religion des Cora: Indianer. Treubner, Leipzig.

Pritchard, Hayden N., and Patricia T. Bradt. 1984. Biology of Nonvascular Plants. Times Mirror/Mosby, College Publications, St. Louis.

Ravesloot, John C., and Stephanie M. Whittlesey. 1987. Inferring the protohistoric period in southern Arizona. Pp. 81–98 in The archaeology of the San Xavier Bridge Site (AZ BB:13:14), Tucson Basin, Southern Arizona. John C. Ravesloot (editor). Arizona State Museum Archaeological Series 171. University of Arizona, Tucson.

Ravussin, Eric, and William C. Knowler. 1993. Obesity in Pima Indians. Paper presented at the conference "Type 2 Diabetes among Desert-Dwelling Australian Aborigines and Native Americans: Parallels in Genetics, Diet, and Culture," 10–12 May, Toowoon Bay and Sydney, Australia. Ms. available at Native Seed/SEARCH, Tucson, Ariz.

Rea, Amadeo M. 1978. The ecology of Pima fields. Environment Southwest 484:8–13.

———. 1979. Hunting lexemic categories of the Pima Indians. Kiva 44:113–119.

———. 1983. Once a River: Bird Life and Habitat Changes on the Middle Gila. University of Arizona Press, Tucson.

———. 1988. Habitat restoration and avian recolonization from wastewater on the middle Gila River, Arizona. Pp. 1395–1405 in Arid Lands: Today and Tomorrow. E. E. Whitehead, C. F. Hutchinson, B. N. Timmermann, and R. G. Varady (editors). Westbutte Press, Boulder, Colo.

———. 1990. Naming as process: Taxonomic adjustments in a changing biotic environment. Pp. 61–79 in Proceedings of the First International Congress of Ethnobiology, Bélem, Brazil, vol. 1. Darrell A. Posey and associates (editors). Museu Paraense Emílio Goeldi.

———. 1991. Gila River Pima Dietary Reconstruction. Arid Lands Newsletter (fall/winter) 31:3–10.

Reff, Daniel T. 1991. Disease, Depopulation, and Culture Change in Northwestern New Spain, 1518–1764. University of Utah Press, Salt Lake City.

Reichhardt, Karen. 1992. Natural Vegetation of Casa Grande Ruins National Monument. Technical Report NPS/WRUA/NRTR-92/45. National Park Service, Cooperative National Park Resources Studies Unit, School of Renewable Resources, University of Arizona, Tucson.

Reid, J. C. 1858. Reid's Tramp; or, A Journal of the Incidents of Ten Months Travel Through Texas, New Mexico, Arizona, Sonora, and California. John Hardy, Selma, Ala. Reprinted 1935, Steck, Austin.

Riley, Carroll L. 1969. The Southern Tepehuan and Tepecano. Pp. 814–821 in Handbook of Middle American Indians, vol. 8, pt. 2. E. Vogt (editor). University of Texas Press, Austin.

Robbins, W. W. 1940. Alien Plants Growing without Cultivation in California. University of California, Berkeley, Agricultural Experiment Station Bulletin 637.

Roberts, Leslie. 1989. Disease and death in the New World. Science 246:1245–1247.

Ross, Clyde P. 1923. The Lower Gila Region, Ariz. U.S. Geological Survey Water Supply Paper no. 498. Washington, D.C.

Rusling, James F. 1874. Across America; or, The Great West and the Pacific Coast. Sheldon and Co., New York.

Russell, Frank. 1908. The Pima Indians. Annual Report of the Bureau of American Ethnology 26:3–389. Reprinted 1975, University of Arizona Press, Tucson.

Salzmann, Zdenêk. 1954. The problem of lexical acculturation. International Journal of American Linguistics 20:137–139.

Sauer, Carl O. 1934. The distribution of aboriginal tribes and languages in northwestern Mexico. Ibero-Americana 5:1–94.

Sauer, Jonathan D. 1967. The grain amaranths and their relatives: A revised taxonomic and geographic survey. Annals of the Missouri Botanical Garden 54:103–137.

Saxton, Dean F., and Lucille Saxton. 1969. Dictionary: Papago and Pima to English, English to Papago and Pima; O'odham–Mil-gahn, Mil-gahn–O'odham. University of Arizona Press, Tucson.

———. 1973. O'othham Hoho'ok A'agitha: Legends and Lore of the Papago and Pima Indians. University of Arizona Press, Tucson.

Saxton, Dean, Lucille Saxton, and Susie Enos. 1983. Dictionary: Papago/Pima–English, O'othham–Mil-gahn; English–Papago/Pima, Mil-gahn–O'othham. 2d edition. R. L. Cherry (editor). University of Arizona Press, Tucson.

Scheerens, J. C., A. M. Tinsley, I. R. Abbas, C. W. Webber, and J. W. Berry. 1983. The nutritional significance of tepary bean consumption. Desert Plants 5:11–14, 51–56.

Schroeder, Albert H. 1954. Comment on Reed's "Transition to History in The Pueblo Southwest." American Anthropologist 56:597–599.

Schultes, Richard Evans. 1984. Amazonian cultigens and their northward and westward migration in pre-Columbian times. Pp. 19–37 in Pre-Columbian Plant Migration. Doris Stone (editor). Papers of the Peabody Museum of Archaeology and Ethnology 76. Harvard University.

Sedelmayr, Jacobo. 1939. Sedelmayr's relación of 1746. Ronald L. Ives (translator). Bulletin of the Bureau of American Ethnology 123, Anthropology Papers 9, pp. 101–117. Smithsonian Institution, Washington, D.C.

———. 1955. Jacobo Sedelmayr, 1744–1751: Missionary, Frontiersman, Explorer in Arizona and Sonora; Four Original Manuscript Narratives. P. M. Donne (editor and translator). Arizona Pioneers' Historical Society, Tucson. (Includes 1744 draft of the relación.)

Segesser, Philipp. 1945. Document: The relation of Philipp Segesser. Theodore E. Treutlein (translator). Mid-America 27:139–187, 257–260.

Setchell, William Alpert. 1921. Aboriginal tobaccos. American Anthropologist 23:397–414.

Shantz, H. L., and R. L. Piemeisel. 1924. Indicator significance of the natural vegetation of the southwestern desert region. Journal of Agricultural Research 28(8):721–779.

Shapiro, Eric-Anders. 1989. Cotton in Arizona: A historical geography. M.S. thesis, Department of Geography and Regional Development, University of Arizona.

Shaul, David L. 1986. Topics in Nevome Syntax. University of California Publications in Linguistics. University of California Press, Berkeley.

Shaw, Anna Moore. 1968. Pima Indian Legends. University of Arizona Press, Tucson.

Sheridan, David. 1981. Desertification of the United States. Council of Environmental Quality, Washington, D.C.

Shreve, Forrest, and Ira L. Wiggins. 1964. Vegetation and Flora of the Sonoran Desert. 2 vols. Stanford University Press, Stanford, Calif.

Simpson, Ruth D. 1946. Those who have gone still live: The Hohokam since 1400 A.D. Masterkey 20:73–80.

Smith, Buckingham (translator). 1863. Rudo Ensayo. Records of the American Catholic Society, Philadelphia, vol. 5. Reprinted 1951, Arizona Silhouettes, Tucson.

Smith, Cynthia J., Elaine M. Manahan, and Sally G. Pablo. n.d. Food habits and cultural change among the Pima Indians. Ms. available at Native Seed/SEARCH, Tucson, Ariz.

Smith, Cynthia, and Sally Pablo. 1993. Food habits and culture change among the Pima Indians. Paper presented at the conference "Type 2 Diabetes among Desert-Dwelling Australian Aborigines and Native Americans: Parallels in Genetics, Diet, and Culture," 10–12 May, Toowoon Bay and Sydney, Australia. Ms. available at Native Seed/SEARCH, Tucson, Ariz.

Smith, Fay Jackson, John L. Kessell, and Francis J. Fox. 1966. Father Kino in Arizona. Arizona Historical Foundation, Phoenix.

Sobarzo, Horacio. 1966. Vocabulario Sonorense. Editorial Porruz. Mexico City.

Southworth, C. H. 1914–1915. Maps produced for the U.S. Indian Irrigation Service. On file (#3819) at the Arizona Historical Society, Tucson.

_____. 1919. The history of irrigation along the Gila River. Pp. 103–223 in Indians of the United States: Hearings before the Committee on Indian Affairs, vol. 2, app. A. House Doc. 66th Cong., 1st sess., 2 vols.

_____. 1931. A Pima calendar stick. Arizona Historical Review 4:45–52.

Spicer, Edward H. 1943. Linguistic aspects of Yaqui acculturation. American Anthropologist (n.s.) 45:410–426.

_____. 1962. Cycles of Conquest: The Impact of Spain, Mexico, and the United States on the Indians of the Southwest, 1533–1960. University of Arizona Press, Tucson.

_____. 1980. The Yaquis: A Cultural History. University of Arizona Press, Tucson.

_____. 1983. Yaqui. Pp. 250–263 in Handbook of North American Indians, vol. 10, Southwest. Alfonso Ortiz (editor). Smithsonian Institution, Washington, D.C.

Spier, Leslie. 1933. Yuman Tribes of the Gila River. University of Chicago Press, Chicago. Reprinted 1978, Dover, New York.

Stabler, D. Frederic. 1985. Increasing summer flow in small streams through management of riparian areas and adjacent vegetation: A synthesis. Pp. 206–210 in Riparian Ecosystems and Their Management: Reconciling Conflicting Uses. R. Roy Johnson, Charles D. Ziebel, David R. Patton, Peter F. Ffolliott, and R. H. Hamre (editors). USDA Forest Service General Technical Report RM-120. Rocky Mountain Forest and Range Experimental Station, Fort Collins, Colo.

Standley, Paul C., and Louis O. Williams. 1970. Convolvulaceae: Flora of Guatemala. Fieldiana: Botany 24, pt. 9:4–85.

Stone, Eric. 1932. Medicine among the American Indians. Clio Medica, vol. 7. Hoeber, New York. Reprint 1962, Hafner, New York.

Stout, J. H. 1873. Report of the Commissioner of Indian Affairs, Accompanying the Annual Report of the Secretary of the Interior, for the year 1873, pp. 281–283. GPO, Washington, D.C.

Sundell, Eric G. 1974. Vegetation and Flora of the Sierra Estrella Regional Park, Maricopa County, Arizona. M.S. thesis, Botany Department, Arizona State University, Tempe.

Thackery, Franklin A., and M. French Gilman. 1931. A rare parasitic food plant of the Southwest. Annual Report of the Smithsonian Institution, 1930, no. 3094, pp. 409–416.

Timmermann, B. N. 1977. Practical uses of Larrea. Pp. 252–259 in Creosote Bush: Biology and chemistry of Larrea in New World deserts. J. T. Mabry, J. H. Hunziker, and D. R. DiFeo, Jr. (editors). US/IBP Synthesis Series 6. Dowden, Hutchinson & Ross, Stroudsburg, Penn.

Trager, George L. 1939. 'Cottonwood' = 'tree': A southwestern linguistic trait. International Journal of American Linguistics 9:117–118.

Turner, Christy G., II. 1993. Southwest Indian teeth. National Geographic Research & Exploration 9:32–53.

Turner, Christy G., II, and Joel D. Irish. 1989. Further assessment of Hohokam affinity: The Classic period populations of the Grand Canal and Casa Buena Sites, Phoenix, Arizona. Pp. 775–792 in Archaeological investigations at the Grand Canal ruins: A Classic period site in Phoenix, Arizona, vol. 2. Douglas R. Mitchell (editor). Soil Systems Publications in Archaeology no. 12. Phoenix.

Turner, Henry Smith. 1966. Original journals of Henry Smith Turner, with Stephen Watts Kearny to New Mexico and California. D. L. Clarke (editor). University of Oklahoma Press, Norman.

Tyler, Daniel. 1881. A Concise History of the Mormon Battalion in the Mexican War, 1846–1847. Privately printed, Salt Lake City, Utah.

Underhill, Ruth Murray. 1938a. A Papago calendar record. University of New Mexico Bulletin 322, Anthropology Series vol. 2, no 5. Albuquerque.

_____. 1938b. Singing for Power: The Song Magic of the Papago Indians of Southern Arizona. Reprinted 1993, University of Arizona Press, Tucson.

_____. 1939. Social Organization of the Papago Indians. Columbia University Contributions in Anthropology 30. New York.

_____. 1946. Papago Indian Religion. Columbia University Press, New York.

Underhill, Ruth M., Donald M. Bahr, Baptisto Lopez, Jose Pancho, David Lopez. 1979. Rainhouse and Ocean: Speeches for the Papago Year. American Tribal Religions, vol. 4. Museum of Northern Arizona Press, Flagstaff.

Van Devender, Thomas R., Robert S. Thompson, and Julio L. Betancourt. 1987. Vegetation history of the deserts of south-western North America: The nature and timing of the Late Wisconsin–Holocene transition. Pp. 323–352 in North America and Adjacent Oceans during the Last Deglaciation. W. F. Ruddiman and H. E. Wright, Jr. (editors). The Geology of North America, vol. K-3. Geological Society of America, Boulder, Colo.

Velarde, Luis. 1931. Padre Luis Velarde's relación of Pimería Alta. R. K. Wyllys (editor). New Mexico Historical Review 6:111–157.

Wagner, H. R. 1966. Spanish Voyages to the Northwest Coast of America in the Sixteenth Century. N. Israel, Amsterdam.

Wagoner, Jay J. 1952. History of the Cattle Industry in Southern Arizona, 1540–1940. University of Arizona Social Science Bulletin 20, vol. 23, no. 2. Tucson.

Weaver, Donald E., Jr. 1976. Salado influences in the lower Salt River Valley. Kiva 42:17–26.

_____. 1977. Investigations concerning the Hohokam Classic period in the Lower Salt River Valley, Arizona. Ariz. Arch. 9.

Webb, George. 1959. A Pima Remembers. Reprinted 1994, University of Arizona Press, Tucson.

Weber, David J. 1971. The Taos Trappers: The Fur Trade in the Far Southwest, 1540–1846. University of Oklahoma Press, Norman.

Weil, Andrew. 1980. The Marriage of the Sun and the Moon. Houghton Mifflin, Boston.

Whipple, Amiel W. 1856. Extracts from the [preliminary] report. Pp. 1–32 in vol. 3 of Reports of Explorations and Surveys for a Railroad Route from the Mississippi River to the Pacific Ocean. 33d Cong., 2d sess., House Exec. Doc. 91.

Whitaker, Thomas W. 1971. Endemism and Pre-Columbian migration of the bottle-gourd, Lagenaria siceraria (Mol.) Standl. Pp. 320–327 in Men across the Sea: Problems of Pre-Columbian Contacts. Carroll L. Riley, J. Charles Kelley, Campbell W. Pennington, and Robert L. Rands (editors). University of Texas Press, Austin.

Whitaker, Thomas W., and Glen N. Davis. 1962. Cucurbits: Botany, Cultivation, and Utilization. World Crops Books. Interscience Publishers, New York.

Whiting, Alfred F. 1939. Ethnobotany of the Hopi. Museum of Northern Arizona, Flagstaff. Reprinted 1966.

_____. n.d. Field notes [unpublished notes on the Water or Seed Clan of the Hopi]. A. F. Whiting Indian Archives, Northern Arizona University, Flagstaff.

Whittemore, Isaac T. 1893. The Pima Indians: Their Manners and Customs. Pp. 51–96 *in* Among the Pimas; or, The Mission to the Pima and Maricopa Indians. Charles H. Cook (editor). Ladies Union Mission School Association, Albany, N.Y.

Wilcox, David R. 1981. The entry of Athapaskans into the American Southwest: The problem today. Pp. 213–256 *in* The Protohistoric Period in the North American Southwest. David R. Wilcox and W. Bruce Masse (editors). Arizona State University Anthropological Research Papers 24. Tempe.

Wilhelm, Stephen. 1974. The garden strawberry: A study of its origin. American Scientist 62:264–271.

Willenbrink, Antonine. 1935. Notes on the Pima Indian Language. The Franciscan Fathers of California. Unpub. ms. in Santa Barbara Mission archives.

Wilson, Hugh D. 1980. Artificial hybridization among species of *Chenopodium* section *Chenopodium*. Systematic Botany 5:253–263.

_____. 1981. Domesticated *Chenopodium* of the Ozark bluff dwellers. Economic Botany 35:233–239.

Wilson, Hugh D., and C. B. Heiser. 1979. The origin of evolutionary relationships of "huauzontle" (*Chenopodium nuttalliae* Safford), domesticated chenopod of Mexico. American Journal of Botany 66:198–206.

Winter, Joseph C. 1973. Cultural modifications of the Gila Pima, A.D. 1697–A.D. 1846. Ethnohistory 20:67–77.

Witkowski, Stanley R., and Cecil H. Brown. 1983. Marking reversals and cultural importance. Language 59:569–581.

Witkowski, Stanley R., Cecil H. Brown, and Paul K. Chase. 1981. Where do tree terms come from? Man (n.s.) 16:1–14.

Woods, Dan. 1990. Tri-Valley Dispatch, March 21-22, p. 2. March 21$\frac{1}{N}$ 22, p. 2. Casa Grande, Ariz.

Wright, Harold B. 1929. Long Ago Told: Legends of the Papago Indians. D. Appleton & Co., New York.

Yen, Douglas E. 1971. Construction of the hypothesis for distribution of the sweet potato. Pp. 328–342 *in* Men across the Sea: Problems of Pre-Columbian Contacts. Carroll L. Riley, J. Charles Kelly, Campbell W. Pennington, and Robert L. Rands (editors). University of Texas Press, Austin.

Zepeda, Ofelia. 1983. A Papago Grammar. University of Arizona Press, Tucson.

Zingg, Robert M. 1982. Los Huicholes: Una tribu de artistas. Serie Clásicos de la Anthropología 12. vol. 1. Instituto Nacional Indigenista, Mexico City.

Index

About the Author

Amadeo M. Rea is an ornithologist and ethnobiologist. From 1977 through 1991 he was curator of birds and mammals at the San Diego Natural History Museum. He has since become a freelance consultant. He holds a Ph.D. in zoology and anthropology from the University of Arizona. For more than three decades he has worked on the Gila River Indian Reservation to enable the Pima to pass along their knowledge and understanding of the plants and animals of their desert environment. Rea is the author of *Once a River: Bird Life and Habitat Changes on the Middle Gila* (University of Arizona Press, 1983).